Essentials of
Oceanography

TWELFTH EDITION

Alan P. Trujillo

DISTINGUISHED TEACHING PROFESSOR

PALOMAR COLLEGE

Harold V. Thurman

FORMER PROFESSOR EMERITUS

MT. SAN ANTONIO COLLEGE

Senior Acquisitions Editor: Andrew Dunaway
Executive Product Marketing Manager: Neena Bali
Executive Marketing Manager: Mary Salzman
Project Manager: Connie Long
Program Manager: Sarah Shefveland
Development Editor: Barbara Price
Art Development Editor: Jay McElroy
Editorial Assistant: Michelle Koski
Development Manager: Jennifer Hart
Program Management Team Lead: Kristen Flathman
Project Management Team Lead: David Zielonka
Production Management: Lumina Datamatics

Compositor: Lumina Datamatics
Design Manager: Derek Bacchus
Interior Designer: Gary Hespenheide
Cover Designer: Gary Hespenheide
Illustrators: International Mapping Associates, Peter Bull Art
 Studio, Justin Hofman
Rights & Permissions Management: Rachel Youdelman
Photo Researcher: Kristin Piljay
Manufacturing Buyer: Maura Zaldivar-Garcia

Cover Photo Credit: Justin Hofman

Library of Congress Cataloging-in-Publication Data

Names: Trujillo, Alan P. | Thurman, Harold V.
Title: Essentials of oceanography / Alan P. Trujillo, Distinguished Teaching
 Professor, Palomar College, Harold V. Thurman, Professor Emeritus,
 Mt. San Antonio College.
Description: Twelth edition. | Boston : Pearson, [2018] | Includes
 bibliographical references and index.
Identifiers: LCCN 2015035841| ISBN 9780134073545 (alk. paper) | ISBN
 0134073541 (alk. paper)
Subjects: LCSH: Oceanography—Textbooks.
Classification: LCC GC11.2 .T49 2018 | DDC 551.46—dc23
LC record available at http://lccn.loc.gov/2015035841

1 2 3 4 5 6 7 8 9 10—V011—18 17 16

PEARSON

www.pearsonhighered.com

ISBN 10: **0-134-07354-1**; ISBN 13: **978-0-134-07354-5** (Student edition)
ISBN 10: **0-134-25395-7**; ISBN 13: **978-0-134-25395-4** (Instructor's Review Copy)
ISBN 10: **0-134-26742-7**; ISBN 13: **978-0-134-26742-5** (NASTA)

**Dedicated to my father Dr. Anthony P. Trujillo,
mentor, role model, teacher, and friend**

—Al Trujillo

In memory of Hal Thurman (1934–2012)

ABOUT THE AUTHORS

ALAN P. TRUJILLO Al Trujillo is a Distinguished Teaching Professor in the Earth, Space, and Aviation Sciences Department at Palomar College in San Marcos, California. He received his bachelor's degree in geology from the University of California at Davis and his master's degree in geology from Northern Arizona University, afterward working for several years in industry. Al began teaching at Palomar in 1990. In 1997, he was awarded Palomar's Distinguished Faculty Award for Excellence in Teaching, and in 2005 he received Palomar's Faculty Research Award. He has coauthored *Introductory Oceanography* with Hal Thurman and is a contributing author for the textbooks *Earth* and *Earth Science*. In addition to writing and teaching, Al works as a naturalist and lecturer aboard natural history expedition vessels for Lindblad Expeditions/National Geographic in Alaska, Iceland, and the Sea of Cortez/Baja California. His research interests include beach processes, sea cliff erosion, and active teaching techniques. He enjoys photography, and he collects sand as a hobby. Al and his wife, Sandy, have two children, Karl and Eva.

HAROLD V. THURMAN Hal Thurman's interest in geology led to a bachelor's degree from Oklahoma A&M University, followed by seven years working as a petroleum geologist, mainly in the Gulf of Mexico, where his interest in the oceans developed. He earned a master's degree from California State University at Los Angeles. Hal began teaching at Mt. San Antonio College in Walnut, California, in 1968 as a temporary teacher and taught Physics 1 (a surveying class) and three Physical Geology labs. In 1970, he taught his first class of General Oceanography. It was from this experience that he decided to write a textbook on oceanography and received a contract with Charles E. Merrill Publishing Company in 1973. The first edition of his book *Introductory Oceanography* was released in 1975. Harold authored or coauthored over 20 editions of textbooks that include *Introductory Oceanography*, *Essentials of Oceanography*, *Physical Geology*, *Marine Biology*, and *Oceanography Laboratory Manual*, many of which are still being used today throughout the world. In addition, he contributed to the *World Book Encyclopedia* on the topics of "Arctic Ocean," "Atlantic Ocean," "Indian Ocean," and "Pacific Ocean." Hal Thurman retired in May 1994, after 24 years of teaching, and moved to be closer to family in Oklahoma, then to Florida. Hal passed away at the age of 78 on December 29, 2012. His writing expertise, knowledge about the ocean, and easy-going demeanor will be dearly missed.

BRIEF CONTENTS

CONTENTS

"The sea, once it casts its spell, holds one in its net of wonder forever."

—Jacques-Yves Cousteau, oceanographer, underwater videographer, and explorer (circa 1963)

To the Student

Welcome! You're about to embark on a journey that is far from ordinary. Over the course of this term, you will discover the central role the oceans play in the vast global system of which you are a part.

This book's content was carefully developed to provide a foundation in science by examining the vast body of oceanic knowledge. This knowledge includes information from a variety of scientific disciplines—geology, chemistry, physics, and biology—as they relate to the oceans. However, no formal background in any of these disciplines is required to successfully master the subject matter contained within this book. Our desire is to have you take away from your oceanography course much more than just a collection of facts. Instead, we want you to develop a fundamental understanding of how the oceans work and why the oceans behave the way that they do.

This book is intended to help you in your quest to know more about the oceans. Taken as a whole, the components of the ocean—its sea floor, chemical constituents, physical components, and life-forms—comprise one of Earth's largest interacting, interrelated, and interdependent systems. Because human activities impact Earth systems, it is important to understand not only how the oceans operate but also how the oceans interact with Earth's other systems (such as its atmosphere, biosphere, and hydrosphere) as part of a larger picture. Thus, this book uses a systems approach to highlight the interdisciplinary relationships among oceanographic phenomena and how those phenomena affect other Earth systems.

DIVING DEEPER PREFACE.1

A USER'S GUIDE FOR STUDENTS: HOW TO READ A SCIENCE TEXTBOOK

Have you known someone who could scan a reading assignment or sleep with it under their pillow and somehow absorb all the information? Studies have shown that those people haven't really committed anything to long-term memory. For most of us, it takes a focused, concentrated effort to gain knowledge through reading. Interestingly, if you have the proper motivation and reading techniques, you can develop excellent reading comprehension. What is the best way to read a science textbook such as this one that contains many new and unfamiliar terms?

One common mistake is to approach reading a science textbook as one would read a newspaper, magazine, or novel. Instead, many reading instructors suggest using the SQ4R reading technique, which is based on research about how the brain learns. The SQ4R technique includes these steps:

1. **Survey:** Read the title, introduction, major headings, first sentences, concept statements, review questions, summary, and study aids to become familiar with the content in advance.

2. **Question:** Have questions in mind when you read. If you can't think of any good questions, use the chapter questions as a guide.

3. **Read:** Read flexibly through the chapter, using short time periods to accomplish the task one section at a time (not all in one sitting).

4. **Recite:** Answer the chapter questions. Take notes after each section and review your notes before you move on.

5. **(w)Rite:** Write summaries and/or reflections on what you've read. Write answers to the questions in Step 2.

6. **Review:** Review the text using the strategy in the survey step. Take the time to review your end-of-section notes as well as your summaries.

To help you study most effectively, this textbook includes many study aids that are designed to be used with the SQ4R technique. For example, each chapter includes a word cloud of key terms, a list of learning objectives that are tied to the Essential Concepts throughout the chapter, review Concept Check questions embedded at the end of each section, and an Essential Concepts Review that includes a chapter summary, study resources, and critical thinking questions.

Here are some additional reading tips that may seem like common sense but are often overlooked:

- Don't attempt to do your reading when you are tired, distracted, or agitated.

- Break up your reading into manageable sections. Don't save it all until the last minute.

- Take a short break if your concentration begins to fade. Listen to music, call a friend, have a snack, or drink some water. Then return to your reading.

Remember that every person is different, so experiment with new study techniques to discover what works best for you. In addition, being a successful student is hard work; it is not something one does in his/her spare time. With a little effort in applying the SQ4R reading technique, you will begin to see a difference in what you remember from your reading.

To that end—and to help you make the most of your study time—we focused the presentation in this book by organizing the material around three essential components:

1. **CONCEPTS:** General ideas derived or inferred from specific instances or occurrences (for instance, the concept of density can be used to explain why the oceans are layered)

2. **PROCESSES:** Actions or occurrences that bring about a result (for instance, the process of waves breaking at an angle to the shore results in the movement of sediment along the shoreline)

3. **PRINCIPLES:** Rules or laws concerning the functioning of natural phenomena or mechanical processes (for instance, the principle of sea floor spreading suggests that the geographic positions of the continents have changed through time)

Interwoven within these concepts, processes, and principles are hundreds of photographs, illustrations, real-world examples, and applications that make the material relevant and accessible (and maybe sometimes even *entertaining*) by bringing science to life.

Ultimately, it is our hope that by understanding how the oceans work, you will develop a new awareness and appreciation of all aspects of the marine environment and its role in Earth systems. To this end, the book has been written for you, a student of the oceans. So enjoy and immerse yourself! You're in for an exciting ride.

Al Trujillo

To the Instructor

This twelfth edition of *Essentials of Oceanography* is designed to accompany an introductory college-level course in oceanography taught to students who have no formal background in mathematics or science. As in previous editions, the goal of this edition of the textbook is to clearly present the relationships of scientific principles to ocean phenomena in an engaging and meaningful way.

This edition has greatly benefited from being thoroughly reviewed by hundreds of students who made numerous suggestions for improvement. Comments by former students about the book include, "*I have really enjoyed the oceanography book we've used this semester. It had just the right mix of graphics, text, and user-friendliness that really held my interest,*" and "*What I really liked about the book is that it's a welcoming textbook—open and airy. You could almost read it at bedtime like a story because of all the interesting pictures.*"

This edition has been reviewed in detail by a host of instructors from leading institutions across the country. Reviewers of the eleventh edition described the text as follows: "*Clean, sleek, easy-to-read text with engaging photos, figures, text features, and animations/videos that will "hook" students in and get them excited about the material,*" and "*I think the text is very well put together. It does a nice job presenting the material and supporting it with many pictures, illustrations, and graphs. The text is well-organized and laid out in an easy-to-use fashion. I would recommend this text to a colleague for teaching Introductory Oceanography.*"

In 2012, the tenth edition of *Essentials of Oceanography* received a Textbook Excellence Award, called a "Texty," from the Text and Academic Authors Association (TAA). The Texty award recognizes written works for their excellence in the areas of content, presentation, appeal, and teachability. The publisher, Pearson Education, nominated the book for the award, and the textbook was critically reviewed by a panel of expert judges.

The 16-chapter format of this textbook is designed for easy coverage of the material in a 15- or 16-week semester. For courses taught on a 10-week quarter system, instructors may need to select those chapters that cover the topics and concepts of primary relevance to their course. Chapters are self-contained and can thus be covered in any order. Following the introductory chapter (Chapter 1, which covers the general geography of the oceans; a historical perspective of oceanography; the reasoning behind the scientific method; and a discussion of the origin of Earth, the atmosphere, the oceans, and life itself), the

DIVING DEEPER PREFACE.2

OCEAN LITERACY: WHAT SHOULD PEOPLE KNOW ABOUT THE OCEAN?

The ocean is the defining feature of our planet. Accordingly, there is great interest in developing *ocean literacy*, which means understanding the ocean's influence on humans as well as humans' influence on the ocean. For example, scientists and educators agree that an ocean-literate person:

- Understands the essential principles and fundamental concepts about the functioning of the ocean.
- Can communicate about the ocean in a meaningful way.
- Is able to make informed and responsible decisions regarding the ocean and its resources.

To achieve this goal, ocean educators and experts have developed the **Seven Principles of Ocean Literacy**. The following ideas are what everyone—especially those who successfully pass a college course in oceanography or marine science—should understand about the ocean:

1. Earth has one big ocean with many features.

2. The ocean and life in the ocean shape the features of Earth.

3. The ocean is a major influence on weather and climate.

4. The ocean makes Earth habitable.

5. The ocean supports a great diversity of life and ecosystems.

6. The ocean and humans are inextricably interconnected.

7. The ocean is largely unexplored.

This book is intended to help all people achieve ocean literacy. For more information about the Seven Principles of Ocean Literacy, see **http://oceanliteracy.wp2.coexploration.org/**

https://goo.gl/eMPIXD

four major academic disciplines of oceanography are represented in the following chapters:

- Geological oceanography (Chapters 2–4 and Chapter 10)
- Chemical oceanography (Chapter 5 and Chapter 11)
- Physical oceanography (Chapters 6–9)
- Biological oceanography (Chapters 12–15)
- Interdisciplinary oceanography: Climate change (Chapter 16)

We strongly believe that oceanography is at its best when it links together several scientific disciplines and shows how they are interrelated in the oceans. Therefore, this interdisciplinary approach is a key element of every chapter, particularly Chapter 16, "The Oceans and Climate Change."

Reorganization of What's New in This Edition Section

Changes in this edition are designed to increase the readability, relevance, and appeal of this book. Major changes include the following:

1. NEW! Hybridization of the Textbook

- Inclusion of more than 70 Web Animations from Pearson's Geoscience Animations Library, which include state-of-the-art computer animations that have been created by Al Trujillo and a panel of geoscience educators
- Addition of seven new Geoscience Animations that have been specifically designed for this edition to help students visualize some of the most challenging oceanographic concepts:
 - Formation of Earth's Oceans (Chapter 1)
 - How Salt Dissolves in Water (Chapter 5)
 - Three Types of Breakers (Chapter 8)
 - Effects of Elliptical Orbits (Chapter 9)
 - Osmosis (Chapter 12)
 - Feeding in Baleen Whales (Chapter 14)
 - Latitude and Longitude on Earth (Appendix III)
- Inclusion of links to more than 50 hand-picked Web videos that show important oceanographic processes in action
- Addition of QR codes embedded in the text that allow students to use their mobile devices to link directly to MasteringOceanography Animations, SmartFigures, SmartTables, and Web Videos
- Select Diving Deeper features have been migrated online to MasteringOceanography as Bonus Web Content in an effort to reduce the length of the text
- The addition in each chapter of a series of new SmartFigures and SmartTables which provide a video explanation of difficult-to-understand oceanographic concepts and numerical data by an oceanography teaching expert

2. NEW! Content/Art Revisions

- A thoroughly updated Chapter 16 "The Oceans and Climate Change," including new information about ocean acidification, the most recent findings of the IPCC, five new "Students Sometimes Ask … " features that address student misconceptions and concerns regarding climate change, a new Diving Deeper Box about the father-son team of Charles David and Ralph Keeling who created and maintain the Keeling curve of atmospheric carbon dioxide, an expanded discussion on the role of orbital parameters in creating natural cycles of climate change, and a new section on the effect of aerosols on global warming

- A redesigned and updated Chapter 13, "Biological Productivity and Energy Transfer," including contributions by Dr. Angel Rodriguez of Broward College in Florida highlighting issues of fisheries sustainability

- A new name and focus for Chapter 11 "Marine Pollution," which reframes the chapter discussion along environmental themes

- Reorganization of Chapter 10, adding content about the properties of the coastal ocean from Chapter 11 and renaming the chapter "Beaches, Shoreline Processes, and the Coastal Ocean"

- Greater emphasis on the ocean's role in Earth systems

- Addition of a revised word cloud at the beginning of each chapter that uses different font sizes to show the most important vocabulary terms within the chapter and directs students to the glossary at the end of the book to discover the meaning of any terms they don't already know

- A detailed list of specific chapter-by-chapter changes is available at **www2.palomar.edu/users/atrujillo**.

3. NEW! Pedagogical Enhancements

- A stronger learning path that directly links the learning objectives listed at the beginning of each chapter to the end-of-section "Concept Checks," which allow and encourage students to pause and test their knowledge as they proceed through the chapter

- Addition of a new "Recap" feature that summarizes key points throughout the text that making studying easier

- A new active learning pedagogy that divides chapter material into easily digestible chunks, which makes studying easier and assists student learning (cognitive science research shows that the ability to "chunk" information is essential to enhancing learning and memory)

- The addition of one or more "Give It Some Thought" assessment questions to each "Diving Deeper" boxed feature

- The addition of a new "Climate Connection" icon that alerts students to topics that are related to the overarching theme of the ocean's importance to global climate change

- A new multidisciplinary icon that flags content related to two or more of the sub-disciplines in oceanography: geological oceanography, chemical oceanography, physical oceanography, and biological oceanography

- In all Essential Concept Review (end-of-chapter) materials, the revision of existing "Critical Thinking Questions" and the addition of new "Active Learning Exercise" questions that can be used for in-class group activities

- Updating of information throughout the text to include some of the most recent and critical developments in oceanography

- Addition of an array of new "Students Sometimes Ask ... " questions throughout the book
- Diving Deeper features appearing in the book are organized around the following four themes:
 - **HISTORICAL FEATURES**, which focus on historical developments in oceanography that tie into chapter topics
 - **RESEARCH METHODS IN OCEANOGRAPHY**, which highlight how oceanographic knowledge is obtained
 - **OCEANS AND PEOPLE**, which illustrate the interaction of humans and the ocean environment
 - **FOCUS ON THE ENVIRONMENT**, which emphasizes environmental issues that are an increasingly important component of ocean studies
- All text in the chapters has been thoroughly reviewed and edited by students and oceanography instructors in a continued effort to refine the style and clarity of the writing
- In addition, this edition continues to offer some of the previous edition's most popular features, including the following:
- Scientifically accurate and thorough coverage of oceanography topics
- "Students Sometimes Ask ... " questions, which present actual student questions along with the authors' answers
- Use of the international metric system (Système International [SI] units), with comparable English system units in parentheses
- Explanation of word etymons (*etumon* = sense of a word) as new terms are introduced, in an effort to demystify scientific terms by showing what the terms actually mean
- Use of **bold print** on key terms, which are defined when they are introduced and are described in the glossary
- A reorganized "Essential Concepts Review" summary at the end of each chapter
- Mastering**Oceanography**, which features chapter-specific Essential Concepts, eText, Bonus Web Content, Geoscience Animations, Web Videos, Web Destinations, and two Test Yourself quiz modules

4. NEW! Squidtoons

- A new, comic-styled *Squidtoons* cartoon in each chapter; created by a team of graduate students at Scripps Institution of Oceanography in California, each *Squidtoons* highlights an important marine organism relevant to the chapter's content; the cartoon links to a poster-like presentation that uses engaging graphics and humor to discuss interesting aspects of each creature

For the Student

- Mastering**Oceanography** delivers engaging, dynamic learning opportunities—focused on course objectives and responsive to each student's progress—that are proven to help students absorb course material and understand difficult concepts. MasteringOceanography and MyLab & Mastering are customized learning resources that include:
 - **Student Study Area**, which is designed to be a one-stop resource for students to acquire study help and serve as a launching pad

for further exploration. Content for the site was written by author Al Trujillo and is tied, chapter-by-chapter, to the text. The Student Study Area is organized around a four-step learning pathway:

1. *Review*, which contains **Essential Concepts** as learning objectives
2. *Read*, which contains the **eText** and **Bonus Web Content**
3. *Visualize*, which contains **Geoscience Animations**, **Web Videos**, and **Web Destinations**. **Geoscience Animations** were created by a team of geoscience educators and include an array of more than 70 visualizations that help students understand complex oceanographic concepts and processes by allowing the user to control the action. For example, students can fully examine how an animation develops by replaying it, controlling its pace, and stopping and starting the animation anywhere in its sequence. In order to facilitate effective study, Al Trujillo has written an accompanying narration and assessment quiz questions including hints and specific wrong-answer feedback for each animation. **Web Videos** include more than 50 hand-selected short video clips of oceanographic processes in action. **Web Destinations** include links to some of the best oceanography sites on the Web.
4. *Test Yourself*, which contains three **Test Yourself** modules, including multiple-choice and true/false, multiple-answer, and image-labeling exercises. Answers, once submitted, are automatically graded for instant feedback.

- **RSS Feeds**, which allow students to subscribe and stay up-to-date on oceanographic discoveries
- **Study Tools** such as flashcards and a searchable online glossary to help make the most of students' study time

- **THE PEARSON eTEXT** gives students complete access to a digital version of the text whenever and wherever you have access to the Internet. eText pages look exactly like the printed text, offering powerful new portability and functionality.

For the Instructor

- Mastering**Oceanography**: **CONTINUOUS LEARNING BEFORE, DURING, AND AFTER CLASS** MasteringOceanography is an online homework, tutorials, and assessments program designed to improve results by helping students quickly master oceanography concepts. Students will benefit from self-paced tutorials that feature immediate wrong-answer feedback and hints that emulate the office-hour experience to help keep them on track. With a wide range of interactive, engaging, and assignable activities, students will be encouraged to actively learn and retain tough course concepts:
 - **SmartFigures** and **SmartTables**, which are short instructional videos that examine and explain the most important concepts illustrated by the figure or data table. With nearly 100 of these SmartFigures/SmartTables inside the text, students can stop, pause, and replay the videos multiple times to help them learn about important concepts and real oceanographic data.

- **Mobile Interactive Geoscience Animations**, which include more than 70 Geoscience Animations of difficult-to-understand concepts that are embedded throughout the text using mobile-friendly QR codes
- **GeoTutors**, which coach students through difficult concepts
- **Encounter Oceans Activities**, which provide interactive explorations of oceanography concepts using Google Earth™. Students work through the activities in Google Earth and then test their knowledge by answering the assessment questions, which include hints and specific wrong-answer feedback.
- **Geoscience Animations**, which illuminate the most difficult-to-understand topics in oceanography and were created by an expert team of geoscience educators. The animation activities include audio narration, a text transcript, and assignable multiple-choice questions with specific wrong-answer feedback.
- **Dynamic Study Modules**, which help students study effectively on their own by continuously assessing their activity and performance in real time. Here's how it works: Students complete a set of questions with a unique answer format that also asks them to indicate their confidence level. Questions repeat until the student can answer them all correctly and confidently. Once completed, Dynamic Study Modules explain the concept using materials from the text. These are available as graded assignments prior to class, and accessible on smartphones, tablets, and computers.
- **Learning Catalytics™**, which are an interactive student response tool that uses students' smartphones, tablets, or laptops to engage them in more sophisticated tasks and thinking. Now included with MyLab & Mastering and eText, Learning Catalytics™ enables you to generate classroom discussion, guide your lecture, and promote peer-to-peer learning with real-time analytics.

- **STUDENT PERFORMANCE ANALYTICS** MasteringOceanography allows an instructor to gain easy access to information about student performance and their ability to meet student learning outcomes. Instructors can quickly add their own learning outcomes, or use publisher-provided ones, to track student performance.
- **INSTRUCTOR MANUAL (DOWNLOAD ONLY)** This resource contains learning objectives, chapter outlines, answers to embedded end-of-section questions, and suggested teaching tips to spice up your lectures.
- **TESTGEN® COMPUTERIZED TEST BANK (DOWNLOAD ONLY)** This resource is a computerized test generator that lets instructors view and edit *Test Bank* questions, transfer questions to tests, and print the test in a variety of customized formats. The *Test Bank* includes over 1200 multiple-choice, matching, and short-answer/essay questions. All questions are tied to the chapter's learning outcomes, include a rating based on Bloom's taxonomy of learning domains (Bloom's 1–6) and contain the section number in which each question's answer can be found.
- **INSTRUCTOR POWERPOINT® PRESENTATIONS (DOWNLOAD ONLY)** Instructor Resource Materials include the following three PowerPoint® files for each chapter so that you can cut down on your preparation time, no matter what your lecture needs:

1. **EXCLUSIVELY ART:** This file provides all the photos, art, and tables from the text, in order, loaded into PowerPoint® slides.
2. **LECTURE OUTLINE:** This file averages 50 PowerPoint® slides per chapter and includes customizable lecture outlines with supporting art.
3. **CLASSROOM RESPONSE SYSTEM (CRS) QUESTIONS:** Authored for use in conjunction with classroom response systems, this PowerPoint® file allows you to electronically poll your class for responses to questions, pop quizzes, attendance, and more.

For more information about these instructor resources, contact your Pearson textbook representative.

Acknowledgments

I am indebted to many individuals for their helpful comments and suggestions during the revision of this book. I am particularly indebted to Development Editor Dr. Barbara Price of Pearson Education for her encouragement, ideas, and tireless advocacy that she provided to improve the book. It was a pleasure working with you, Barbara! Laura Faye Tenenbaum did an outstanding job of creating and narrating the SmartFigure and SmartTable videos that are found as QR code links throughout this book. Jenny Duncan did an excellent job of updating all the study area questions in MasteringOceanography. Thanks also go to Dr. Angel Rodriquez of Broward College in Florida for his ideas and contributions to Chapter 13. In addition, Garfield Kwan did a superb job of bringing scientific information to life through his *Squidtoons* infographic comics, some of which are included as links in each chapter of this book.

Many people were instrumental in helping the text evolve from its manuscript stage. My chief liaison at Pearson Education, Senior Geoscience Editor Andrew Dunaway, suggested many of the new ideas in the book to make it more student-friendly and expertly guided the project. The copy editors at Lumina Datamatics did a superb job of editing the manuscript, catching many English and other grammar errors, including obscure errors that had persisted throughout several previous editions. Program Manager Sarah Shefveland worked behind the scenes to manage the quality of this text in terms of accuracy, budget, and the achievement of program goals. Project Manager Connie Long kept the book on track by making sure deadlines were met along the way and facilitated the distribution of various versions of the manuscripts. Media Producer Mia Sullivan helped create the electronic supplements that accompany this book, including MasteringOceanography and all of its outstanding features. The animations studio Thought Café - http://thoughtcafe.ca/#sthash.TQCfCo9V.dpbs crafted the new animations and added additional ideas, which led to great improvements. International Mapping Associates and Peter Bull Art Studio did a beautiful job of modernizing and updating all maps and most of the figures to add annotations that help tell the story of the content through the art. Art Development Editor Jay McElroy reviewed every single piece of art throughout the text and suggested many improvements to make the figures more clear. Marine biologist and talented Digital Graphic Artist Justin Hofman supplied a host of new figures featuring realistic marine organisms that greatly enhanced the art program. The artful design elements of

the text, including its color scheme, text wrapping, and end-of-chapter features, was developed by Layout Designer Gary Hespenheide in conjunction with Pearson's Design Manager Derek Bacchus. New photos were researched and secured by Photo Researcher Kristin Piljay. Last but not least, Senior Production Manager Lindsay Bethoney of Lumina Datamatics deserves special recognition for her persistence and encouragement during the many long hours of turning the manuscript into the book you see today.

I thank my students, whose questions provided the material for the "Students Sometimes Ask … " sections and whose continued input has proved invaluable for improving the text. Because scientists (and all good teachers) are always experimenting, thanks also for allowing yourselves to be a captive audience with which to conduct my experiments.

I also thank my patient and understanding family for putting up with my absence during the long hours of preparing "The Book." Finally, appreciation is extended to the chocolate manufacturers Hershey, See's, and Ghirardelli, for providing inspiration. A heartfelt thanks to all of you!

Many other individuals (including dozens of anonymous reviewers) have provided valuable technical reviews for this and previous works. The following reviewers are gratefully acknowledged:

Patty Anderson, *Scripps Institution of Oceanography*
Shirley Baker, *University of Florida*
William Balsam, *University of Texas at Arlington*
Tsing Bardin, *City College of San Francisco*
Tony Barros, *Miami-Dade Community College*
Steven Benham, *Pacific Lutheran University*
Lori Bettison-Varga, *College of Wooster*
Thomas Bianchi, *Tulane University*
David Black, *University of Akron*
Mark Boryta, *Consumnes River College*
Laurie Brown, *University of Massachusetts*
Kathleen Browne, *Rider University*
Aurora Burd, *Green River Community College*
Nancy Bushell, *Kauai Community College*
Chatham Callan, *Hawaii Pacific University*
Mark Chiappone, *Miami-Dade College–Homestead Campus*
Chris Cirmo, *State University of New York, Cortland*
G. Kent Colbath, *Cerritos Community College*
Thomas Cramer, *Brookdale Community College*
Richard Crooker, *Kutztown University*
Cynthia Cudaback, *North Carolina State University*
Warren Currie, *Ohio University*
Hans Dam, *University of Connecticut*
Dan Deocampo, *California State University, Sacramento*
Richard Dixon, *Texas State University*
Holly Dodson, *Sierra College*
Joachim Dorsch, *St. Louis Community College*
Wallace Drexler, *Shippensburg University*
Walter Dudley, *University of Hawaii*
Iver Duedall, *Florida Institute of Technology*
Jessica Dutton, *Adelphi University*
Charles Ebert, *State University of New York, Buffalo*
Ted Eckmann, *University of Portland*
Charles Epifanio, *University of Delaware*
Jiasong Fang, *Hawaii Pacific University*
Diego Figueroa, *Florida State University*
Kenneth Finger, *Irvine Valley College*

Catrina Frey, *Broward College*
Jessica Garza, *MiraCosta College*
Benjamin Giese, *Texas A&M University*
Cari Gomes, *MiraCosta College*
Dave Gosse, *University of Virginia*
Carla Grandy, *City College of San Francisco*
John Griffin, *University of Nebraska, Lincoln*
Elizabeth Griffith, *University of Texas at Arlington*
Gary Griggs, *University of California, Santa Cruz*
Joseph Holliday, *El Camino Community College*
Mary Anne Holmes, *University of Nebraska, Lincoln*
Timothy Horner, *California State University, Sacramento*
Alan Jacobs, *Youngstown State University*
Ron Johnson, *Old Dominion University*
Uwe Richard Kackstaetter, *Metropolitan State University of Denver*
Charlotte Kelchner, *Oakton Community College*
Matthew Kleban, *New York University*
Eryn Klosko, *State University of New York, Westchester Community College*
M. John Kocurko, *Midwestern State University*
Lawrence Krissek, *Ohio State University*
Jason Krumholz, *NOAA/University of Rhode Island*
Paul LaRock, *Louisiana State University*
Gary Lash, *State University of New York, Fredonia*
Richard Laws, *University of North Carolina*
Richard Little, *Greenfield Community College*
Stephen Macko, *University of Virginia, Charlottesville*
Chris Marone, *Pennsylvania State University*
Jonathan McKenzie, *Edison State College–Lee Campus*
Matthew McMackin, *San Jose State University*
James McWhorter, *Miami-Dade Community College*
Gregory Mead, *University of Florida*
Keith Meldahl, *MiraCosta College*
Nancy Mesner, *Utah State University*
Chris Metzler, *MiraCosta College*
Johnnie Moore, *University of Montana*
P. Graham Mortyn, *California State University, Fresno*
Andrew Muller, *Millersville University*
Andrew Muller, *Utah State University*
Daniel Murphy, *Eastfield College*
Jay Muza, *Florida Atlantic University*
Jennifer Nelson, *Indiana University–Purdue University at Indianapolis*
Jim Noyes, *El Camino Community College*
Sarah O'Malley, *Maine Maritime Academy*
B. L. Oostdam, *Millersville University*
William Orr, *University of Oregon*
Joseph Osborn, *Century College*
Donald Palmer, *Kent State University*
Nancy Penncavage, *Suffolk County Community College*
Curt Peterson, *Portland State University*
Adam Petrusek, *Charles University, Prague, Czech Republic*
Edward Ponto, *Onondaga Community College*
Donald Reed, *San Jose State University*
Randal Reed, *Shasta College*
M. Hassan Rezaie Boroon, *California State University, Los Angeles*
Cathryn Rhodes, *University of California, Davis*
James Rine, *University of South Carolina*
Felix Rizk, *Manatee Community College*
Angel Rodriguez, *Broward College*
Diane Shepherd, *Shepherd Veterinary Clinic, Hawaii*
Beth Simmons, *Metropolitan State College of Denver*
Jill Singer, *State University of New York, Buffalo*
Arthur Snoke, *Virginia Polytechnic Institute*

Pamela Stephens, *Midwestern State University*
Dean Stockwell, *University of Alaska, Fairbanks*
Scott Stone, *Fairfax High School, Virginia*
Lenore Tedesco, *Indiana University–Purdue University at Indianapolis*
Shelly Thompson, *West High School*
Craig Tobias, *University of North Carolina, Wilmington*
M. Craig VanBoskirk, *Florida Community College at Jacksonville*
Paul Vincent, *Oregon State University*
George Voulgaris, *University of South Carolina*
Bess Ward, *Princeton University*
Jackie Watkins, *Midwestern State University*
Jamieson Webb, *Gulf Coast State College*
Arthur Wegweiser, *Edinboro University of Pennsylvania*
Diana Wenzel, *Seminole State College of Florida*
John White, *Louisiana State University*
Katryn Wiese, *City College of San Francisco*
Raymond Wiggers, *College of Lake County*
John Wormuth, *Texas A&M University*
Memorie Yasuda, *Scripps Institution of Oceanography*

Although this book has benefited from careful review by many individuals, the accuracy of the information rests with the authors. If you find errors or have comments about the text, please contact me.

Al Trujillo
Department of Earth, Space, and Aviation Sciences
Palomar College
1140 W. Mission Rd.
San Marcos, CA 92069
atrujillo@palomar.edu
www2.palomar.edu/users/atrujillo

"If there is magic on this planet, it is contained in water."
—Loren Eiseley, American educator and
natural science writer (1907–1977)

SmartFigures

SmartTables

New Geoscience Animations Specifically Designed for this Edition

OCEANOGRAPHY JUST GOT
REAL!

WITH AL TRUJILLO,
ESSENTIALS OF OCEANOGRAPHY
TWELFTH EDITION

DYNAMIC VISUALS AND INTEGRATED MEDIA
BRING OCEANOGRAPHY TO LIFE

Highly visual and interactive tools make oceanography approachable, enabling students to see oceanographic processes in action.

NEW! **SmartFigures** and **SmartTables** are 3- to 4-minute mini video lessons containing explanations of difficult-to-understand oceanographic concepts and numerical data directed by an oceanography teaching expert and NASA Science Communicator. By scanning the accompanying QR code, or typing in the short URL, students now have a multitude of ways to learn from art and data tables, all designed to teach.

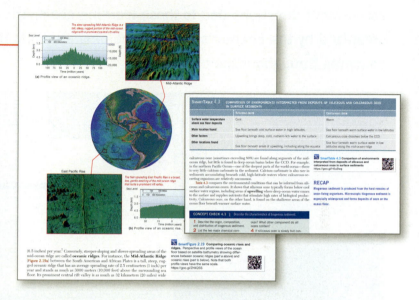

NEW! **Squidtoons**, a comic-styled call-out created by a team of graduate students at Scripps Institution of Oceanography in California, are featured in each chapter. These infographics highlight an important marine organism related to each chapter's content using graphical representation to display recent discoveries by researchers in an interesting and captivating manner. By scanning the associated QR code or typing in the short URL in the text, students will be taken to the digital space to view the full cartoon.

NEW! **Enhanced illustration program**, with new art incorporating the research-proven technique of strategically placing annotations and labels within the key figures, allows students to focus on the most relevant visual information and helps them interpret complex art. Overall, nearly 90% of the entire book's artwork has been updated or is new, including new figures that provide visual summaries of essential processes and concepts.

ESSENTIAL ELEMENTS FORM A PATH TO SUCCESSFUL LEARNING

Each chapter is organized into easily digestible chunks, making studying easier and assisting student learning. Chapter material begins with learning goals and ends with assessment questions tied to those learning goals. The end-of-chapter material is also organized by the chapter's sections, helping students remain focused on the essential concepts throughout the chapter.

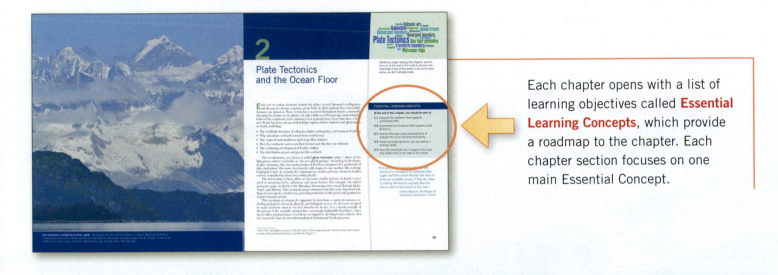

Each chapter opens with a list of learning objectives called **Essential Learning Concepts**, which provide a roadmap to the chapter. Each chapter section focuses on one main Essential Concept.

Concept Check questions at the end of each section are designed to let students check their understanding of the Essential Learning Concept. By stopping and answering questions, students ensure that they have a thorough understanding of key points before moving on to the next section.

CONCEPT CHECK 2.1 | Evaluate the evidence that supports continental drift.

1 When did the supercontinent of Pangaea exist? What was the ocean that surrounded the supercontinent called?

2 Regarding glacial ages, why is it unlikely that the entire world was covered by ice 300 million years ago?

3 Cite the lines of evidence Alfred Wegener used to support his idea of continental drift. Why did scientists of the time doubt that continents had drifted?

RECAP

The Coriolis effect causes moving objects to curve to the right in the Northern Hemisphere and to the left in the Southern Hemisphere. It is at its maximum at the poles and is nonexistent at the equator.

NEW! A **Recap** feature now appears throughout each chapter, summarizing essential concepts. This is a great tool for directing students' study and review.

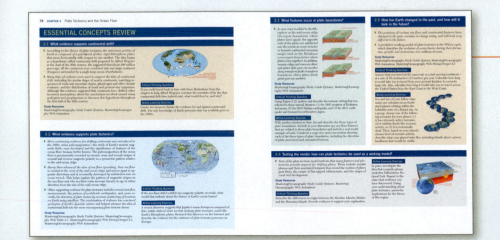

Each chapter ends with the **Essential Concepts Review**, which simplifies the study process. Also organized by section, this review highlights a key figure from the chapter and provides a summary of the chapter's key concepts. It also includes study resources, Critical Thinking Questions, and **NEW!** Active Learning Exercises.

TURNING INTEREST **INTO ENGAGEMENT**

Everyday topics in a real world context help students relate oceanography to their lives while engaging them in how oceanography is studied.

NEW! **Climate Connection:** This icon shows how various sections of the text relate to the overarching theme of the importance of Earth's oceans to global climate change.

NEW! **Interdisciplinary Relationship:** This icon shows how various sections of the text relate to two or more sub-disciplines in oceanography: geological oceanography, biological oceanography, physical oceanography, and chemical oceanography.

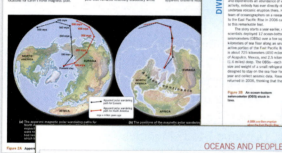

The new edition includes a variety of **Diving Deeper** features, including Historical Features, Research Methods in Oceanography, Oceans and People, and Focus on the Environment. These features foster multi-dimensional understanding with captivating examples and stories. Each Diving Deeper feature now includes one or more "Give It Some Thought" assessment questions.

The popular **Students Sometimes Ask** features answer often entertaining questions posed by real students.

STUDENTS SOMETIMES ASK . . .

What happened to the recent Malaysian Airlines flight that vanished after takeoff?

It's still a mystery. Malaysian Airlines flight MH370 went missing on March 8, 2014, while in route from Kuala Lumpur, Malaysia to Beijing, China. Satellite communications suggest that the flight veered south and ended up running out of fuel and crash-landing in the Indian Ocean west of Australia. Unfortunately, the suspected area of the crash is large, remote, and deep, and the region's rugged sea floor is very poorly explored, all of which has hampered recovery efforts. In the days following

CONTINUOUS LEARNING
BEFORE, DURING, AND AFTER CLASS
WITH MasteringOceanography™

MasteringOceanography delivers engaging, dynamic learning opportunities—focusing on course objectives and responsive to each student's progress—that are proven to help students absorb oceanography course materials and understand challenging physical processes and oceanography concepts.

BEFORE CLASS — DYNAMIC STUDY MODULES AND ETEXT 2.0 PROVIDE STUDENTS WITH A PREVIEW OF WHAT'S TO COME.

NEW! **Dynamic Study Modules** enable students to study effectively on their own in an adaptive format. Students receive an initial set of questions with a unique answer format asking them to indicate their confidence. Once completed, Dynamic Study Modules include explanations using material taken directly from the text.

NEW! **Interactive eText 2.0** comes complete with embedded media and is both mobile friendly and ADA accessible.

- Now available on smartphones and tablets.
- Seamlessly integrated videos and other rich media.
- Fully accessible (screen-reader ready).
- Configurable reading settings, including resizable type and night reading mode.
- Facilitates instructor and student note-taking, highlighting, bookmarking, and search.

DURING CLASS ENGAGE STUDENTS WITH LearningCatalytics

NEW! **LearningCatalytics,** a "bring your own device" student engagement, assessment, and classroom intelligence system (PRS), allows students to use their smartphone, tablet, or laptop to respond to questions in class without the need for a "clicker."

AFTER CLASS HELPING STUDENTS VISUALIZE OCEANOGRAPHY CONCEPTS THAT CAN BE EASILY ASSIGNABLE.

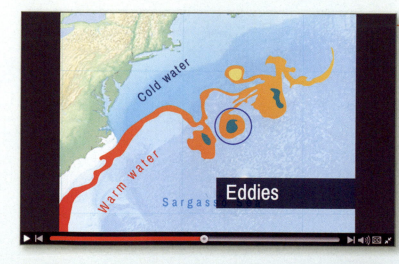

NEW! The following 7 **geoscience animations** have been specifically designed for this edition:

- Formation of Earth's Oceans
- How Salt Dissolves in Water
- Three Types of Breakers
- Effects of Elliptical Orbits
- Osmosis
- Feeding in Baleen Whales
- Latitude and Longitude on Earth

More than 70 geoscience animations are associated with the text, and all include audio narration, a text transcript, and assignable multiple-choice questions with specific wrong-answer feedback in Mastering.

Select key animations have been refreshed and made compatible for Mastering and mobile devices.

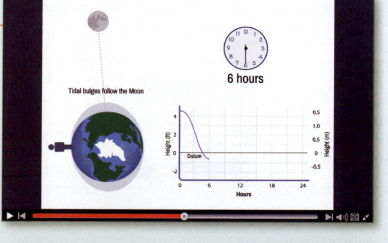

MasteringOceanography™ HELPS STUDENTS LEARN...

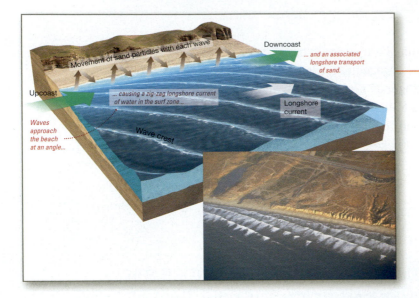

NEW! **SmartFigures** bring key chapter illustrations to life! These videos are accessible on mobile devices via scannable Quick Response (QR) codes printed in the text and through the Study Area in MasteringOceanography. Paired with other assessments in Mastering, these videos become assignable and assessable learning objects that can either prepare students for lecture or assess what they have learned.

NEW! **SmartTables** are engaging tutorial videos that explain the relevance of the real data found in tables within the textbook. Paired with other assessments in Mastering, these become assignable and assessable learning objects that allow students to interpret real data sets.

TABLE 2.1	CHARACTERISTICS, TECTONIC PROCESSES, FEATURES, AND EXAMPLES OF PLATE BOUNDARIES					
Plate boundary	Plate movement	Crust types	Sea floor created or destroyed?	Tectonic process	Sea floor feature(s)	Geographic examples
Divergent plate boundaries	Apart	Oceanic–oceanic	New sea floor is created	Sea floor spreading	Mid-ocean ridge; volcanoes; young lava flows	Mid-Atlantic Ridge, East Pacific Rise
		Continental–continental	As a continent splits apart, new sea floor is created	Continental rifting	Rift valley; volcanoes; young lava flows	East Africa Rift Valleys, Red Sea, Gulf of California
Convergent plate boundaries	Together	Oceanic–continental	Old sea floor is destroyed	Subduction	Trench; volcanic arc on land	Peru–Chile Trench, Andes Mountains
		Oceanic–oceanic	Old sea floor is destroyed	Subduction	Trench; volcanic arc as islands	Mariana Trench, Aleutian Islands
		Continental–continental	N/A	Collision	Tall mountains	Himalaya Mountains, Alps
Transform plate boundaries	Past each other	Oceanic	N/A	Transform faulting	Fault	Mendocino Fault, Eltanin Fault (between mid-ocean ridges)
		Continental	N/A	Transform faulting	Fault	San Andreas Fault, Alpine Fault (New Zealand)

An excellent study guide!

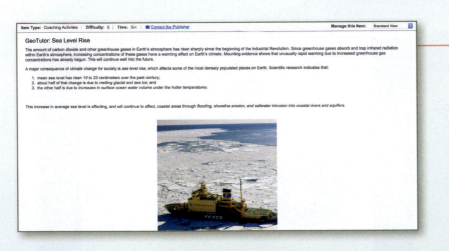

GeoTutor Coaching Activities are based on data collected from educators across the country and address the topics most frequently ranked as tough for students to understand. These activities guide students towards mastery of these topics, using highly visual, kinesthetic, and interactive activities.

...AND MASTER THE SCIENCE OF OCEANOGRAPHY

Encounter Activities provide rich, interactive explorations of Oceanography concepts using the dynamic features of Google Earth™ to visualize and explore Earth's Oceans. Dynamic assessment includes multiple-choice and short-answer questions related to core geology concepts. All explorations include corresponding media files, and questions include hints and specific wrong-answer feedback to help coach students towards mastery of the concepts.

Student Study Area Resources in MasteringOceanography include:

- Practice quizzes
- Interactive Animations
- Oceanography Videos—*A series of studio demo and field segment videos created by author Al Trujillo; most of the studio demos were created as 2-part interactive videos and the field segments show real oceanographic processes in action.*
- Web Video links
- RSS Feeds from ScienceDaily and Scientific American

Learning Outcomes: All of the MasteringOceanography assignable content is tagged to book content. Instructors also have the ability to add their own learning outcomes to assessments and keep track of student performance relative to those learning outcomes. Mastering offers a data-supported measure to quantify students' learning gains and to share those results quickly and easily with colleagues and administrators.

The blue marble, next generation. This composite image of satellite data shows Earth's interrelated atmosphere, oceans, and land—including human presence. Its various layers include the land surface, sea ice, ocean, cloud cover, city lights, and the hazy edge of Earth's atmosphere.

1

Introduction To Planet "Earth"

Before you begin reading this chapter, use the glossary at the end of this book to discover the meanings of any of the words in the word cloud above you don't already know.[1]

The **oceans**[2] are the largest and most prominent feature on Earth. In fact, they are the single most defining feature of our planet. As viewed from space, our planet is a beautiful blue, white, and brown globe (see this chapter's opening photo). The abundance of liquid water on Earth's surface is a distinguishing characteristic of our home planet.

Yet it seems perplexing that our planet is called "Earth" when 70.8% of its surface is covered by oceans. Many early human cultures that lived near the Mediterranean (*medi* = middle, *terra* = land) Sea envisioned the world as being composed of large landmasses surrounded by marginal bodies of water. From their viewpoint, landmasses—not oceans—dominated the surface of Earth. How surprised they must have been when they ventured into the larger oceans of the world. Our planet is misnamed "Earth" because we live on the land portion of the planet. If we were marine animals, our planet would probably be called "Ocean," "Water," "Hydro," "Aqua," or even "Oceanus" to indicate the prominence of Earth's oceans. Let's begin our study of the oceans by examining some of the unique geographic characteristics of our watery world.

1.1 How Are Earth's Oceans Unique?

In all of the planets and moons in our solar system, Earth is the only one that has oceans of liquid water on its surface. No other body in the solar system has a confirmed ocean, but recent satellite missions to other planets have revealed some tantalizing possibilities. For example, the spidery network of fluid-filled cracks on Jupiter's moon Europa (**Figure 1.1**) almost certainly betrays the presence of an ocean of liquid water beneath its icy surface. In fact, a recent analysis of the icy blocks that cover Europa's surface indicates that the blocks are actively being reshaped in a process analogous to plate tectonics on Earth. Two other moons of Jupiter, Ganymede and Callisto, may also have liquid oceans of water beneath their cold, icy crust. Yet another possibility for a nearby world with an ocean beneath its icy surface is Saturn's tiny moon Enceladus, which displays geysers of water vapor and ice that have

[1]The most commonly used words in this chapter are shown by larger font sizes in this *word cloud,* which is a visual aid for identifying important terms. Look for word clouds of important vocabulary terms on the opening page of each chapter throughout this book.

[2]Note that all **bolded** words are key vocabulary terms that are defined in the glossary at the end of this book.

ESSENTIAL LEARNING CONCEPTS

At the end of this chapter, you should be able to:

1.1 Compare the characteristics of Earth's oceans.

1.2 Discuss how early exploration of the oceans was achieved.

1.3 Explain why oceanography is considered an interdisciplinary science.

1.4 Describe the nature of scientific inquiry.

1.5 Explain how Earth and the solar system formed.

1.6 Explain how Earth's atmosphere and oceans formed.

1.7 Discuss why life is thought to have originated in the oceans.

1.8 Demonstrate an understanding of how old Earth is.

"When you're circling the Earth every 90 minutes, what becomes clearest is that it's mostly water; the continents look like they're floating objects."
—Loren Shriver, NASA astronaut (2008)

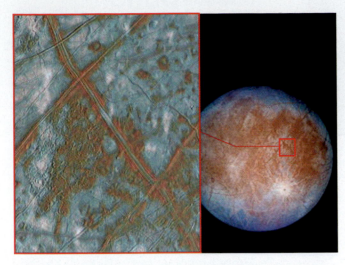

Figure 1.1 **Jupiter's moon Europa.** Europa's network of dark fluid-filled cracks suggests the presence of an ocean beneath its icy surface.

1.1 Squidtoons

https://goo.gl/xwUwNX

recently been analyzed and, remarkably, contain salt. Recent analysis of the gravity field of Enceladus suggests the presence of a 10-kilometer (6.2-mile) deep saltwater ocean beneath a thick layer of surface ice. Also contained in the geysers' icy spray are tiny mineral grains, and in 2015 analysis of these particles indicated that the dust-sized grains likely form when hot, mineral-laden water from the moon's rocky interior travels upward, coming into contact with cooler water. This evidence of subsurface hydrothermal activity is reminiscent of underwater hot springs in the deep oceans on Earth, a place that may have been key to the development of life on Earth. And evidence continues to mount that Saturn's giant moon Titan has small seas of liquid hydrocarbons, suggesting that Titan may be the only other body in the solar system besides Earth known to have stable liquid at its surface. All these moons are enticing targets for space missions to search for signs of extraterrestrial life. Still, the fact that our planet has so much water, *and in the liquid form,* is unique in the solar system.

Earth's Amazing Oceans

Earth's oceans have had a profound effect on our planet and continue to shape our planet in critical ways. The oceans are essential to all life-forms and are in large part responsible for the development of life on Earth, providing a stable environment in which life could evolve over billions of years. Today, the oceans contain the greatest number of living things on the planet, from microscopic bacteria and algae to the largest life-form alive today (the blue whale). Interestingly, water is the major component of nearly every life-form on Earth, and our own body fluid chemistry is remarkably similar to the chemistry of seawater.

The oceans influence climate and weather all over the globe—even in continental areas far from any ocean—through an intricate pattern of currents and heating/cooling mechanisms, some of which scientists are only now beginning to understand. The oceans are also the "lungs" of the planet, taking carbon dioxide gas out of the atmosphere and replacing it with oxygen gas. Scientists have estimated that the oceans supply as much as 70% of the oxygen that humans breathe.

The oceans determine where our continents end and have thus shaped political boundaries and human history. The oceans conceal many features; in fact, the majority of Earth's geographic features are on the ocean floor. Remarkably, there was once more known about the surface of the Moon than about the floor of the oceans! Fortunately, our knowledge of both has increased dramatically over the past several decades.

The oceans also hold many secrets waiting to be discovered, and new scientific discoveries about the oceans are made nearly every day. The oceans are a source of food, minerals, and energy that remains largely untapped. More than half of the world population lives in coastal areas near the oceans, taking advantage of the mild climate, an inexpensive form of transportation, proximity to food resources, and vast recreational opportunities. Unfortunately, the oceans are also the dumping ground for many of society's wastes. In fact, the oceans are currently showing alarming changes caused by pollution, overfishing, invasive species, and climate change, among other things. All of these and many other topics are contained within this book.

How Many Oceans Exist on Earth?

The oceans are a common metaphor for vastness. When one examines a world map (**Figure 1.2**), it's easy to appreciate the impressive extent of Earth's oceans. Notice that *the oceans dominate the surface area of the globe.* For those people who have traveled by boat across an ocean (or even flown across one in an airplane), the one thing that immediately strikes them is that the oceans are enormous. Notice, also, that *the oceans are interconnected* and form a single continuous body of seawater, which is why the oceans are commonly referred to as a "world ocean" (singular, not plural).

For instance, a vessel at sea can travel from one ocean to another, whereas it is impossible to travel on land from one continent to most others without crossing an ocean. In addition, *the volume of the oceans is immense.* For example, the oceans comprise the planet's largest habitat and contain 97.2% of all the water on or near Earth's surface (**Figure 1.3**).

The Arctic Ocean is the smallest and shallowest ocean.

The Pacific Ocean is the world's largest and deepest ocean.

The Atlantic Ocean is the second-largest ocean.

The Indian Ocean exists mostly in the Southern Hemisphere.

The Southern Ocean surrounds Antarctica; its boundary is defined by the Antarctic Convergence.

The Four Principal Oceans, Plus One

Our world ocean can be divided into four principal oceans plus an additional ocean, based on the shape of the ocean basins and the positions of the continents (Figure 1.2).

SmartFigure 1.2 **Earth's oceans.** Map showing the four principal oceans plus the Southern Ocean, or Antarctic Ocean. https://goo.gl/BJXqyt

PACIFIC OCEAN The **Pacific Ocean** is the world's largest ocean, covering more than half of the ocean surface area on Earth (**Figure 1.4b**). The Pacific Ocean is the single largest geographic feature on the planet, spanning more than one-third of Earth's entire surface. The Pacific Ocean is so large that *all* of the continents could fit into the space occupied by it—with room left over! Although the Pacific Ocean is also the deepest ocean in the world (**Figure 1.4c**), it contains many small tropical islands. It was named in 1520 by explorer Ferdinand Magellan's party in honor of the fine weather they encountered while crossing into the Pacific (*paci* = peace) Ocean.

ATLANTIC OCEAN The **Atlantic Ocean** is about half the size of the Pacific Ocean and is not quite as deep (Figure 1.4c). It separates the Old World (Europe, Asia, and Africa) from the New World (North and South America). The Atlantic Ocean was named after Atlas, who was one of the Titans in Greek mythology.

INDIAN OCEAN The **Indian Ocean** is slightly smaller than the Atlantic Ocean and has about the same average depth (Figure 1.4c). It is mostly in the Southern Hemisphere (south of the equator, or below 0 degrees latitude in Figure 1.2). The Indian Ocean was named for its proximity to the subcontinent of India.

ARCTIC OCEAN The **Arctic Ocean** is about 7% the size of the Pacific Ocean and is only a little more than one-quarter as deep as the rest of the oceans (Figure 1.4c). Although it has a permanent layer of sea ice at the surface, the ice is only a few meters thick. The Arctic Ocean was named after its location in the Arctic region, which exists beneath the northern constellation Ursa Major, otherwise known as the Big Dipper, or the Bear (*arktos* = bear).

SOUTHERN OCEAN, OR ANTARCTIC OCEAN Oceanographers recognize an additional ocean near the continent of Antarctica in the Southern Hemisphere (Figure 1.2). Defined by the meeting of currents near Antarctica called the Antarctic Convergence, the **Southern Ocean**, or **Antarctic Ocean**, is really the portions of the Pacific, Atlantic, and Indian Oceans south of about 50 degrees south latitude. This ocean was named for its location in the Southern Hemisphere.

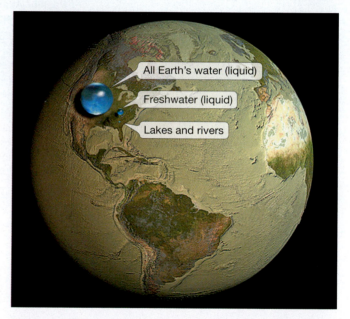

All Earth's water (liquid)

Freshwater (liquid)

Lakes and rivers

Figure 1.3 Relative sizes of the spheres of water on Earth. This image shows all of Earth's liquid water using three blue spheres of proportional sizes. The big sphere is all liquid water in the world, 97% of which is seawater. The next smallest sphere represents a subset of the larger sphere, showing freshwater in the ground, lakes, swamps, and rivers. The tiny speck below it represents an even smaller subset of all the water—just the freshwater in lakes and rivers.

RECAP

The four principal oceans are the Pacific, Atlantic, Indian, and Arctic Oceans. An additional ocean, the Southern Ocean, or Antarctic Ocean, is also recognized.

Figure 1.4 Ocean size and depth. (a) Relative proportions of land and ocean on Earth's surface. **(b)** Relative size of the four principal oceans. **(c)** Average ocean depth. **(d)** Comparing average and maximum depth of the oceans to average and maximum height of land.

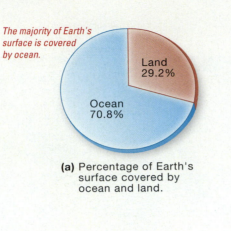

The majority of Earth's surface is covered by ocean.

Land 29.2%

Ocean 70.8%

(a) Percentage of Earth's surface covered by ocean and land.

Arctic 3.4%

Indian 20.5%

Atlantic 26.0%

Pacific 50.1%

The Pacific Ocean comprises about half of all oceans.

(b) Comparing the relative size of each ocean.

The Arctic Ocean isn't very deep.

Pacific Atlantic Indian Arctic

Pacific: 3940 meters (12,927 feet)
Atlantic: 3844 meters (12,612 feet)
Indian: 3840 meters (12,598 feet)
Arctic: 1117 meters (3665 feet)

The Pacific Ocean is the deepest ocean.

(c) Comparing the average depth of each ocean.

Web Animation
Earth's Water and the Hydrologic Cycle
http://goo.gl/kAo8FC

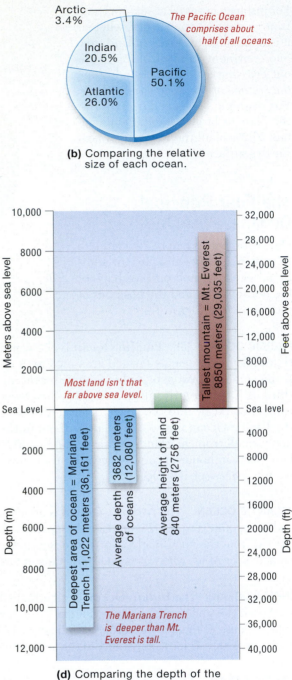

Tallest mountain = Mt. Everest 8850 meters (29,035 feet)

Most land isn't that far above sea level.

Deepest area of ocean = Mariana Trench 11,022 meters (36,161 feet)

Average depth of oceans 3682 meters (12,080 feet)

Average height of land 840 meters (2756 feet)

The Mariana Trench is deeper than Mt. Everest is tall.

(d) Comparing the depth of the oceans to the height of land.

Oceans versus Seas

What is the difference between an ocean and a sea? In common use, the terms *sea* and *ocean* are often used interchangeably. For instance, a *sea* star lives in the *ocean*, the *ocean* is full of *sea* water, *sea* ice forms in the *ocean*, and one might stroll the *sea* shore while living on *ocean*-front property. Technically, however, a *sea* is defined as follows:

- Smaller and shallower than an ocean (this is why the Arctic Ocean might be more appropriately considered a sea)
- Composed of salt water (although some inland "seas," such as the Caspian Sea in Asia, are actually large lakes with relatively high salinity)

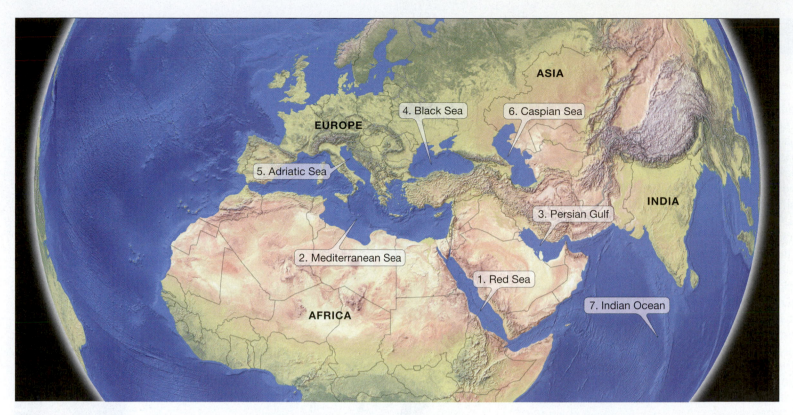

Figure 1.5 Map of the ancient seven seas. This map represents the extent of the known world to Europeans before the 15th century.

- Somewhat enclosed by land (although some seas, such as the Sargasso Sea in the Atlantic Ocean, are defined by strong ocean currents rather than by land)
- Directly connected to the world ocean

COMPARING THE OCEANS TO THE CONTINENTS **Figure 1.4d** shows that the average depth of the world's oceans is 3682 meters[3] (12,080 feet). This means that there must be some extremely deep areas in the ocean to offset the shallow areas close to shore. Figure 1.4d also shows that the deepest depth in the oceans (the Challenger Deep region of the Mariana Trench, which is near Guam) is a staggering 11,022 meters (36,161 feet) below sea level.

How do the continents compare to the oceans? Figure 1.4d shows that the average height of the continents is only 840 meters (2756 feet), illustrating that the average height of the land is not very far above sea level. The highest mountain in the world (the mountain with the greatest height above sea level) is Mount Everest in the Himalaya Mountains of Asia, at 8850 meters (29,035 feet). Even so, Mount Everest is a full 2172 meters (7126 feet) shorter than the Mariana Trench is deep. The mountain with the *greatest total height* from base to top is Mauna Kea on the island of Hawaii in the United States. It measures 4206 meters (13,800 feet) above sea level and 5426 meters (17,800 feet) from sea level down to its base, for a total height of 9632 meters (31,601 feet). The total height of Mauna Kea is 782 meters (2566 feet) higher than Mount Everest, but it is still 1390 meters (4560 feet) shorter than the Mariana Trench is deep. Therefore, no mountain on Earth is taller than the Mariana Trench is deep.

STUDENTS SOMETIMES ASK . . .

Where are the seven seas?

❝Sailing the seven seas" is a familiar phrase in literature and song, but the origin of the saying is shaded in antiquity. To the ancients, the term "seven" often meant "many," and before the 15th century, Europeans considered these the main seas of the world (**Figure 1.5**):

1. The Red Sea
2. The Mediterranean Sea
3. The Persian Gulf
4. The Black Sea
5. The Adriatic Sea
6. The Caspian Sea
7. The Indian Ocean (notice how "ocean" and "sea" are used interchangeably)

Today, however, more than 100 seas, bays, and gulfs are recognized worldwide, nearly all of them smaller portions of the huge interconnected world ocean.

[3]Throughout this book, metric measurements are used (and the corresponding English measurements follow in parentheses). See Appendix I, "Metric and English Units Compared," for conversion factors between the two systems of units.

STUDENTS SOMETIMES ASK ...

Have humans ever explored the deepest ocean trenches? Could anything live there?

Humans have indeed visited the deepest part of the oceans—where there is crushing high pressure, complete darkness, and near-freezing water temperatures—and they first did so over half a century ago! In January 1960, U.S. Navy Lt. Don Walsh and explorer Jacques Piccard descended to the bottom of the Challenger Deep region of the Mariana Trench in the *Trieste*, a deep-diving bathyscaphe (*bathos* = depth, *scaphe* = a small ship) (**Figure 1.6**). At 9906 meters (32,500 feet), the men heard a loud cracking sound that shook the cabin. They were unable to see that a 7.6-centimeter (3-inch) Plexiglas viewing port had cracked (miraculously, it held for the rest of the dive). More than five hours after leaving the surface, they reached the bottom, at 10,912 meters (35,800 feet)—a record depth for human descent. They did observe some small organisms that are adapted to life in the deep: a flatfish, a shrimp, and some jellies.

In 2012, film icon James Cameron made a historic solo dive to the Mariana Trench in his submersible *DEEPSEA CHALLENGER* (**Figure 1.7**). On the seven-hour round-trip voyage, Cameron spent about three hours at the deepest spot on the planet to take photographs and collect samples for scientific research. Other notable voyages to the deep ocean in submersibles are discussed in MasteringOceanography **Web Diving Deeper 1.3**.

RECAP

The deepest part of the ocean is the Mariana Trench in the Pacific Ocean. It is 11,022 meters (36,161 feet) deep and has been visited only twice by humans: once in 1960 and more recently in 2012.

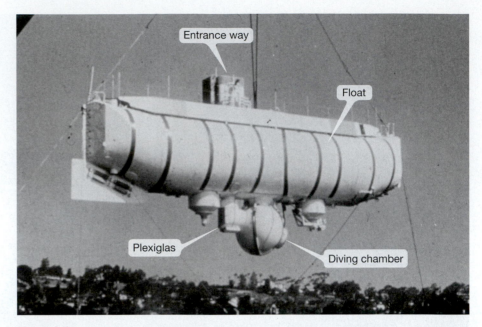

Figure 1.6 **The U.S. Navy's bathyscaphe Trieste.** The *Trieste* suspended on a crane before its record-setting deep dive in 1960. The 1.8-meter (6-foot) diameter diving chamber (the round ball below the float) accommodated two people and had steel walls 7.6 centimeters (3 inches) thick.

Figure 1.7 **James Cameron emerges from the submersible *DEEPSEA CHALLENGER* after his solo dive to the Mariana Trench.** In 2012, famous moviemaker James Cameron completed a record-breaking solo dive to the bottom of the Mariana Trench, becoming only the third human to visit the deepest spot on Earth.

CONCEPT CHECK 1.1 | Compare the characteristics of Earth's oceans.

1 How did the view of the ocean by early Mediterranean cultures influence the naming of planet Earth?

2 Although the terms *ocean* and *sea* are sometimes used interchangeably, what is the technical difference between an ocean and a sea?

3 Where is the deepest part of the ocean? How deep is it, and how does it compare to the height of the tallest mountain on Earth?

1.2 How Was Early Exploration of the Oceans Achieved?

The ocean's huge extent over the surface area of Earth has not prevented humans from exploring its furthest reaches. Since early times, humans have developed technology that has allowed civilizations to travel across large stretches of open ocean. Today, we can cross even the Pacific Ocean in less than a day by airplane. Even so, much of the deep ocean remains out of reach and woefully unexplored. In fact, the surface of the Moon has been mapped more accurately than most parts of the sea floor. Yet satellites at great distances above Earth are being used to gain knowledge about our watery home.

Early History

Humankind probably first viewed the oceans as a source of food. Archeological evidence suggests that when boat technology was developed about 40,000 years ago, people probably traveled the oceans. Most likely, their vessels were built to move upon the ocean's surface and transport oceangoing people to new fishing grounds. The oceans also provided an inexpensive and efficient way to move large and heavy objects, facilitating trade and interaction between cultures.

PACIFIC NAVIGATORS The peopling of the Pacific Islands (Oceania) is somewhat perplexing because there is no evidence that people actually evolved on these islands. Their presence required travel over hundreds or even thousands of kilometers of open ocean from the continents (probably in small vessels of that time—double canoes, outrigger canoes, or balsa rafts) as well as remarkable navigation skills (**Diving Deeper 1.1**). The islands in the Pacific Ocean are widely scattered, so it is likely that only a fortunate few of the voyagers made landfall and that many others perished during voyages. **Figure 1.8** shows the three major inhabited island regions in the Pacific Ocean: Micronesia (*micro* = small, *nesia* = islands), Melanesia (*mela* = black, *nesia* = islands), and Polynesia (*poly* = many, *nesia* = islands), which covers the largest area.

No written records of Pacific human history have been found prior to the arrival of Europeans in the 16th century. Nevertheless, the movement of Asian peoples into Micronesia and Melanesia is easy to imagine because distances between islands are relatively short. In Polynesia, however, large distances separate island groups, which must have presented great challenges to ocean voyagers. Easter Island, for example, at the southeastern corner of the triangular-shaped Polynesian Islands region, is more than 1600 kilometers (1000 miles) from Pitcairn Island, the next nearest island. Clearly, a voyage to the Hawaiian Islands must have been one of the most difficult because Hawaii is more than 3000 kilometers (2000 miles) from the nearest inhabited islands, the Marquesas Islands (Figure 1.8).

Archeological evidence suggests that humans from New Guinea may have occupied New Ireland as early as 4000 or 5000 B.C. However, there is little evidence of human travel farther into the Pacific Ocean before 1100 B.C. By then, the *Lapita people*,[4] a group of early settlers who produced a distinctive type of

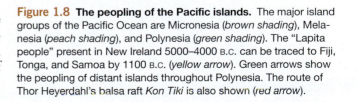

Figure 1.8 The peopling of the Pacific islands. The major island groups of the Pacific Ocean are Micronesia (*brown shading*), Melanesia (*peach shading*), and Polynesia (*green shading*). The "Lapita people" present in New Ireland 5000–4000 B.C. can be traced to Fiji, Tonga, and Samoa by 1100 B.C. (*yellow arrow*). Green arrows show the peopling of distant islands throughout Polynesia. The route of Thor Heyerdahl's balsa raft *Kon Tiki* is also shown (*red arrow*).

[4]In recent years, a combination of genetic, linguistic, and archaeological evidence has suggested that the forebears of the Lapita people—and thus Polynesians—originated in Taiwan, just off the coast of China.

HOW DO SAILORS KNOW WHERE THEY ARE AT SEA? FROM STICK CHARTS TO SATELLITES

How do you know where you are in the ocean, without roads, signposts, or any land in sight? How do you determine the distance to a destination? How do you find your way back to a good fishing spot or where you have discovered sunken treasure? Sailors have relied on a variety of navigation tools to help answer questions such as these by being able to locate where they are at sea.

Some of the first navigators were the Polynesians. Remarkably, the Polynesians were able to successfully navigate to small islands located at great distances across the Pacific Ocean. These early navigators must have been very aware of the marine environment and been able to read subtle differences in the ocean and sky. The tools they used to help them navigate between islands included the Sun and Moon, the nighttime stars, the behavior of marine organisms, various ocean properties, and an ingenious device called a *stick chart* (**Figure 1A**). Stick charts are like a map that depicts the dominant pattern of ocean waves. By orienting their vessels relative to these regular ocean wave directions, sailors could successfully navigate at sea. The bent wave directions let them know when they were getting close to an island—even one that was located beyond the horizon.

Figure 1A Navigational stick chart. This bamboo stick chart of Micronesia's Marshall Islands shows islands (represented by shells at the junctions of the sticks), regular ocean wave direction (represented by the straight strips), and waves that bend around islands (represented by the curved strips). Similar stick charts were used by early Polynesian navigators.

The importance of knowing where you are at sea is illustrated by a tragic incident in 1707, when a British battle fleet was more than 160 kilometers (100 miles) off course and ran aground in the Isles of Sicily near England, with the loss of four ships and nearly 2000 men. **Latitude** (location north or south) was relatively easy to determine at sea by measuring the position of the Sun and stars using a device called a *sextant* (*sextant* = sixth, in reference to the instrument's arc, which is one-sixth of a circle) (**Figure 1B**).

The accident occurred because the ship's crew had no way of keeping track of their **longitude** (location east or west; see Appendix III, "Latitude and Longitude on Earth"). To determine longitude, which is a function of time, it was necessary to know the time difference between a reference meridian and when the Sun was directly overhead of a ship at sea (noon local time). The pendulum-driven clocks in use in the early 1700s, however, would not work for long on a rocking ship at sea. In 1714, the British Parliament offered a £20,000 prize (about $20 million today) for developing a device that would work well enough at sea to determine longitude within half a degree, or 30 nautical miles (34.5 statute miles), after a voyage to the West Indies.

A cabinetmaker in Lincolnshire, England, named **John Harrison** began working in 1728 on such a timepiece, which was dubbed the *chronometer* (*chrono* = time, *meter* = measure). Harrison's first chronometer, H-1, was successfully tested in 1736, but he received only £500 of the prize because the device was deemed too complex, costly, and fragile. Eventually, his more compact fourth version, H-4—which resembles an oversized

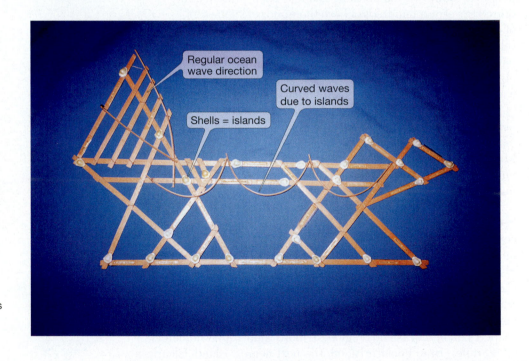

Regular ocean wave direction

Curved waves due to islands

Shells = islands

pottery, had traveled on to Fiji, Tonga, and Samoa (Figure 1.8, *yellow arrow*). From there, Polynesians sailed on to the Marquesas (about 30 B.C.), which appear to have been the starting point for voyages to other islands in the far reaches of the Pacific (Figure 1.8, *green arrows*), including the Hawaiian Islands (about 300 A.D.) and New Zealand (about 800 A.D.). Surprisingly, new genetic research suggests that Polynesians populated Easter Island relatively recently, about 1200 A.D.

Despite the obvious Polynesian backgrounds of the Hawaiians, the Maori of New Zealand, and the Easter Islanders, an adventurous biologist/anthropologist

pocket watch (**Figure 1C**)—was tested during a trans-Atlantic voyage in 1761. Upon reaching Jamaica, it was so accurate that it had lost only *five seconds* of time, a longitude error of only 0.02 degree, or 1.2 nautical miles (1.4 statute miles)! Although Harrison's chronometer greatly exceeded the requirements of the government, the committee in charge of the prize withheld payment, mostly because the astronomers on the committee wanted the solution to come from measurement of the stars. Because the committee refused to award him the prize without further proof, a second sea trial was conducted in 1764, which confirmed his success. Harrison was reluctantly granted £10,000. Only when King George III intervened in 1773 did Harrison finally receive the remaining prize money and recognition for his life work—at age 80.

Today, navigating at sea relies on the *Global Positioning System (GPS)*, which was initiated in the 1970s by the U.S. Department of Defense. Initially designed for military purposes but now available for a variety of civilian uses, GPS relies on a system of 24 satellites that send continuous radio signals to the surface. Position is determined by very accurate measurement of the time of travel of radio signals from at least four of the satellites to receivers on board a ship (or on land). Thus, a vessel can determine its exact latitude and longitude to within a few meters—a small fraction of the length of most ships. Navigators from days gone by would be amazed at how quickly and accurately a vessel's location can be determined, but they

might say that it has taken all the adventure out of navigating at sea.

GIVE IT SOME THOUGHT

1. Why was longitude difficult to determine at sea?
2. Even though his invention solved the problem of determining longitude at sea, why did John Harrison receive only a portion of the British Parliament longitude prize?

Figure 1B **Using a handheld sextant.** This sextant is similar to the ones used by early navigators to determine latitude.

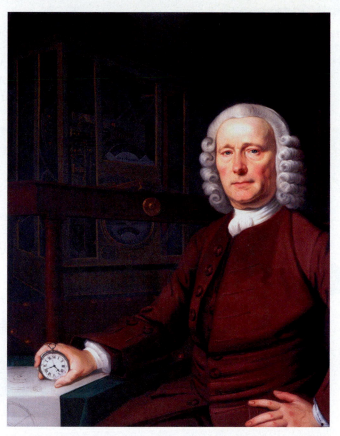

Figure 1C **John Harrison and his chronometer H-4.** Painting (circa 1735) of John Harrison holding his chronometer H-4, which was his life's work. The timepiece H-4 proved to be a vital technological breakthrough that allowed the determination of longitude at sea and won Harrison the prize for solving the longitude problem.

named **Thor Heyerdahl** proposed that voyagers from South America may have reached islands of the South Pacific before the coming of the Polynesians. To prove his point, in 1947 he sailed the ***Kon Tiki***—a balsa raft designed like those that were used by South American navigators at the time of European discovery (**Figure 1.9**)—from South America to the Tuamotu Islands, a journey of more than 11,300 kilometers (7000 miles) (Figure 1.8, *red arrow*). Although the remarkable voyage of the *Kon Tiki* demonstrates that early South Americans could have traveled to Polynesia just as easily as early Asian cultures, anthropologists can find no

Figure 1.9 **The balsa raft *Kon Tiki*.** In 1947, Thor Heyerdahl sailed this authentic wooden balsa raft named *Kon Tiki* from South America to Polynesia to show that ancient South American cultures may have completed similar voyages.

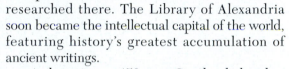

Figure 1.10 **Ptolomy's map of the world.** In about 150 A.D., an Egyptian-Greek geographer named Claudius Ptolomy produced this map of the world that showed the extent of Roman geographic knowledge. Note the use of a coordinate system on land, similar to latitude and longitude used today.

evidence of such a migration. Further, comparative DNA studies show a strong genetic relationship between the peoples of Easter Island and Polynesia but none between these groups and natives in coastal North or South America.

EUROPEAN NAVIGATORS The first Mediterranean people known to have developed the art of navigation were the **Phoenicians**, who lived at the eastern end of the Mediterranean Sea, in the present-day area of Egypt, Syria, Lebanon, and Israel. As early as 2000 B.C., they investigated the Mediterranean Sea, the Red Sea, and the Indian Ocean. The first recorded circumnavigation of Africa, in 590 B.C., was made by the Phoenicians, who had also sailed as far north as the British Isles.

The Greek astronomer-geographer **Pytheas** sailed northward in 325 B.C. using a simple yet elegant method for determining latitude (one's position north or south) in the Northern Hemisphere. His method involved measuring the angle between an observer's line of sight to the North Star and line of sight to the northern horizon.[5] Despite Pytheas's method for determining latitude, it was still impossible to accurately determine longitude (one's position east or west).

One of the key repositories of scientific knowledge at the time was the **Library of Alexandria** in Alexandria, Egypt, which was founded in the 3rd century B.C. by *Alexander the Great*. It housed an impressive collection of written knowledge that attracted scientists, poets, philosophers, artists, and writers who studied and researched there. The Library of Alexandria soon became the intellectual capital of the world, featuring history's greatest accumulation of ancient writings.

As long ago as 450 B.C., Greek scholars became convinced that Earth was round, using lines of evidence such as the way ships disappeared beyond the horizon and the shadows of Earth that appeared during eclipses of the Moon. This inspired the Greek **Eratosthenes** (pronounced "AIR-uh-TOS-thuh-neez") (276–192 B.C.), the second librarian at the Library of Alexandria, to cleverly use the shadow of a stick in a hole in the ground and elementary geometry to determine Earth's circumference. His value of 40,000 kilometers (24,840 miles) compares remarkably well with the true value of 40,032 kilometers (24,875 miles) known today.

An Egyptian-Greek geographer named **Claudius Ptolemy** (c. 85 a.d.–c. 165 a.d.) produced a map of the world in about 150 a.d. that represented the extent of Roman knowledge at that time (**Figure 1.10**). The map not only included the continents of Europe, Asia, and Africa, as did earlier Greek maps, but it also included vertical lines of longitude and horizontal lines of latitude, which had been developed by Alexandrian scholars. Moreover, Ptolemy showed the known seas to be surrounded by land, much of which was as yet unknown and proved to be a great enticement to explorers.

Ptolemy also introduced an (erroneous) update to Eratosthenes's surprisingly accurate estimate of Earth's circumference. Ptolemy wrongly depended on flawed calculations and an overestimation of the size of Asia, and as a result, he determined Earth's circumference to be 29,000 kilometers (18,000 miles), which is about 28% too small. Nearly 1500 years later, Ptolemy's error caused explorer Christopher Columbus to believe he had encountered parts of Asia rather than a new world.

[5]Pytheas's method of determining latitude is featured in Appendix III, "Latitude and Longitude on Earth."

The Middle Ages

After the destruction of the Library of Alexandria in 415 A.D. (in which all of its contents were burned) and the fall of the Roman Empire in 476 A.D., the achievements of the Phoenicians, Greeks, and Romans were mostly lost. Some of the knowledge, however, was retained by the *Arabs*, who controlled northern Africa and Spain. The Arabs used this knowledge to become the dominant navigators in the Mediterranean Sea area and to trade extensively with East Africa, India, and Southeast Asia. The Arabs were able to trade across the Indian Ocean because they had learned how to take advantage of the seasonal patterns of monsoon winds. During the summer, when monsoon winds blow from the southwest, ships laden with goods would leave the Arabian ports and sail eastward across the Indian Ocean. During the winter, when the trade winds blow from the northeast, ships would return west.[6]

Meanwhile, in the rest of southern and eastern Europe, Christianity was on the rise. Scientific inquiry counter to religious teachings was actively suppressed, and the knowledge gained by previous civilizations was either lost or ignored. As a result, the Western concept of world geography degenerated considerably during these so-called *Dark Ages*. For example, one notion envisioned the world as a disk with Jerusalem at the center.

In northern Europe, the **Vikings** of Scandinavia, who had excellent ships and good navigation skills, actively explored the Atlantic Ocean (**Figure 1.11**). Late in the 10th century, aided by a period of worldwide climatic warming, the Vikings colonized Iceland. In about 981 A.D., **Erik "the Red" Thorvaldson** sailed westward from Iceland and discovered Greenland. He may also have traveled further westward to Baffin Island. He returned to Iceland and led the first wave of Viking colonists to Greenland in 985 A.D. **Bjarni Herjólfsson** sailed from Iceland to join the colonists, but he sailed too far southwest and is thought to be the first Viking to have seen what is now called Newfoundland. Bjarni did not land but instead returned to the new colony at Greenland. **Leif Eriksson**, son of Erik the Red, became intrigued by Bjarni's stories about the new land Bjarni had seen. In 995 A.D., Leif bought Bjarni's ship and set out from Greenland for the land that Bjarni had seen to the southwest. Leif spent the winter in that portion of North America and named the land *Vinland* (now Newfoundland, Canada) after the grapes that were found there. Climatic cooling and inappropriate farming practices for the region caused these Viking colonies in Greenland and Vinland to struggle and die out by about 1450.

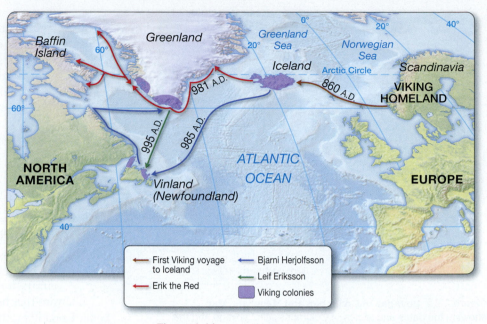

Figure 1.11 **Viking colonies in the North Atlantic.** Map showing the routes and dates of Viking explorations and the locations of the colonies that were established in Iceland, Greenland, and parts of North America.

The Age of Discovery in Europe

The 30-year period from 1492 to 1522 is known as Europe's **Age of Discovery**. During this time, Europeans explored the continents of North and South America, and the globe was circumnavigated for the first time. As a result, Europeans learned the true extent of the world's oceans and that human populations existed elsewhere on newly "discovered" continents and islands with cultures vastly different from those familiar to European voyagers.

Why was there such an increase in ocean exploration during Europe's Age of Discovery? One reason was that Sultan Mohammed II had captured Constantinople (the capital of eastern Christendom) in 1453, a conquest that isolated Mediterranean port cities from the riches of India, Asia, and the East Indies (modern-day Indonesia). As a result, the Western world had to search for new eastern trade routes by sea.

[6]More details about Indian Ocean monsoons can be found in Chapter 7, "Ocean Circulation."

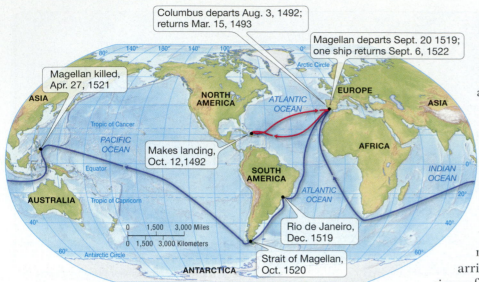

Columbus departs Aug. 3, 1492; returns Mar. 15, 1493

Magellan departs Sept. 20 1519; one ship returns Sept. 6, 1522

Magellan killed, Apr. 27, 1521

Makes landing, Oct. 12,1492

Rio de Janeiro, Dec. 1519

Strait of Magellan, Oct. 1520

Figure 1.12 Voyages of Columbus and Magellan. Map showing the dates and routes of Columbus's first voyage and the first circumnavigation of the globe by Magellan's party.

The Portuguese, under the leadership of **Prince Henry the Navigator** (1392–1460), led a renewed effort to explore outside Europe. The prince established a marine institution at Sagres to improve Portuguese sailing skills. The treacherous journey around the tip of Africa was a great obstacle to an alternative trade route. Cape Agulhas (at the southern tip of Africa) was first rounded by **Bartholomeu Diaz** in 1486. He was followed in 1498 by **Vasco da Gama**, who continued around the tip of Africa to India, thus establishing a new eastern trade route to Asia.

Meanwhile, the Italian navigator and explorer **Christopher Columbus** was financed by Spanish monarchs to find a new route to the East Indies across the Atlantic Ocean. During Columbus's first voyage in 1492, he sailed west from Spain and made landfall after a two-month journey (**Figure 1.12**). Columbus believed that he had arrived in the East Indies somewhere near India, but Earth's circumference had been substantially underestimated, so he was unaware that he had actually arrived in uncharted territory in the Caribbean. Upon his return to Spain and the announcement of his discovery, additional voyages were planned. During the next 10 years, Columbus made three more trips across the Atlantic.

Even though Christopher Columbus is widely credited with discovering North America, he never actually set foot on the continent.[7] Still, his journeys inspired other navigators to explore the "New World." For example, in 1497, only five years after Columbus's first voyage, the Italian navigator and explorer **Giovanni Caboto**, who was also known as **John Cabot**, landed somewhere on the northeastern coast of North America. Later, Europeans first saw the Pacific Ocean in 1513, when **Vasco Núñez de Balboa** attempted a land crossing of the Isthmus of Panama and sighted a large ocean to the west from atop a mountain.

The culmination of the Age of Discovery was a remarkable circumnavigation of the globe initiated by **Ferdinand Magellan** (Figure 1.12). Magellan left Spain in September 1519, with five ships and 280 sailors. He crossed the Atlantic Ocean, sailed down the eastern coast of South America, and traveled through a passage to the Pacific Ocean at 52 degrees south latitude, now named the Strait of Magellan in his honor. After landing in the Philippines in March, 1521, Magellan was killed about a month later in a fight with the inhabitants of these islands. **Juan Sebastian del Caño** completed the circumnavigation by taking the last of the ships, the *Victoria*, across the Indian Ocean, around Africa, and back to Spain in 1522. After three years, just one ship and 18 men completed the voyage.

Following these expeditions, the Spanish initiated many other voyages to take gold from the Aztec and Inca cultures in Mexico and South America. The English and Dutch, meanwhile, used smaller, more maneuverable ships to rob the gold from bulky Spanish galleons, which resulted in many confrontations at sea. The maritime dominance of Spain ended when the English defeated the Spanish Armada in 1588. With control of the seas, the English thus became the dominant world power—a status they retained until early in the 20th century.

The Beginning of Voyaging for Science

The English realized that increasing their scientific knowledge of the oceans would help maintain their maritime superiority. For this reason, Captain **James Cook** (1728–1779), an English navigator and prolific explorer (**Figure 1.13**), undertook three voyages of scientific discovery with the ships *Endeavour*,

[7]For more information about the voyages of Columbus, see Diving Deeper 6.1 in Chapter 6, "Air–Sea Interaction."

Resolution, and *Adventure* between 1768 and 1779. He searched for the continent Terra Australis ("Southern Land," or Antarctica) and concluded that it lay beneath or beyond the extensive ice fields of the southern oceans, if it existed at all. Cook also mapped many islands previously unknown to Europeans, including the South Georgia, South Sandwich, and Hawaiian Islands. During his last voyage, Cook searched for the fabled "northwest passage" from the Pacific Ocean to the Atlantic Ocean and stopped in Hawaii, where he was killed in a skirmish with native Hawaiians.

Cook's expeditions added greatly to the scientific knowledge of the oceans. He determined the outline of the Pacific Ocean and was the first person known to cross the Antarctic Circle in his search for Antarctica. Cook initiated systematic sampling of subsurface water temperatures, measuring winds and currents, taking *soundings* (which are depth measurements that, at the time, were taken by lowering a long rope with a weight on the end to the sea floor), and collecting data on coral reefs. Cook also discovered that a shipboard diet containing the German staple sauerkraut prevented his crew from contracting scurvy, a disease that incapacitated sailors. Scurvy is caused by a vitamin C deficiency, and the cabbage used to make sauerkraut contains large quantities of vitamin C. Prior to Cook's discovery about preventing scurvy, the malady claimed more lives than all other types of deaths at sea, including contagious disease, gunfire, and shipwreck. In addition, by proving the value of John Harrison's chronometer as a means of determining longitude (see Diving Deeper 1.1), Cook made possible the first accurate maps of Earth's surface, some of which are still in use today.

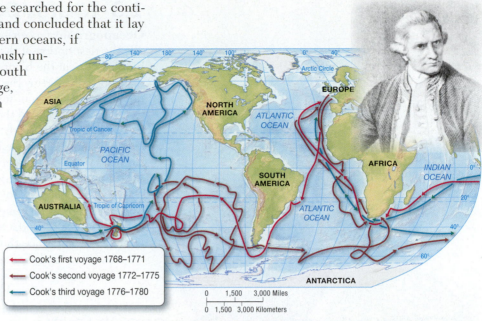

Figure 1.13 **Captain James Cook (1728–1779) and his voyages of exploration.** Routes taken by Captain James Cook (*inset*) on his three scientific voyages, which initiated scientific exploration of the oceans. Cook was killed in 1779 in Hawaii during his third voyage.

Legend:
← Cook's first voyage 1768–1771
← Cook's second voyage 1772–1775
← Cook's third voyage 1776–1780

History of Oceanography . . . To Be Continued

Much has changed since the early days of studying the oceans, when scientists used buckets, nets, and lines deployed from ships. And yet, some things remain the same. For example, going to sea aboard ships continues to be a mainstay of ocean science. Also, even though efforts to monitor the ocean are getting bigger and more sophisticated, vast swaths of the marine world remain unknown.

Today, oceanographers employ many high-technology tools, such as state-of-the-art research vessels that routinely use sonar to map the sea floor, remotely operated data collection devices, drifting buoys, robotics, sea floor observation networks, sophisticated computer models, and Earth-orbiting satellites. Many of these tools are featured throughout this book. Further, additional events in the history of oceanography can be found as Diving Deeper features in subsequent chapters. These boxed features are identified by the "Historical Feature" theme, and each introduces an important historical event that is related to the subject of that particular chapter.

STUDENTS SOMETIMES ASK ...

What is NOAA? What is its role in oceanographic research?

NOAA (pronounced "NO-ah") stands for National Oceanic and Atmospheric Administration and is the branch of the U.S. Department of Commerce that oversees oceanographic research. Scientists at NOAA work to ensure wise use of ocean resources through the National Ocean Service, the National Oceanographic Data Center, the National Marine Fisheries Service, and the National Sea Grant Office. Other U.S. government agencies that work with oceanographic data include the U.S. Naval Oceanographic Office, the Office of Naval Research, the U.S. Coast Guard, and the U.S. Geological Survey (coastal processes and marine geology). The NOAA Website is at **www.noaa.gov**. In 2013, federal officials developed the National Ocean Policy Implementation Plan, which proposes moving NOAA to the Department of the Interior so that agencies dealing with natural resources would all be grouped within the same department.

CONCEPT CHECK 1.2 | Discuss how early exploration of the oceans was achieved.

1 While the Arabs dominated the Mediterranean region during the Middle Ages, what were the most significant ocean-related events taking place in northern Europe?

2 Describe the important events in oceanography that occurred during the Age of Discovery in Europe.

3 List some of the major achievements of Captain James Cook.

RECAP

The ocean's large size did not prohibit early explorers from venturing into all parts of the ocean for discovery, trade, or conquest. Voyaging for science began relatively recently, and many parts of the ocean remain unknown.

1.3 What Is Oceanography?

Oceanography (*ocean* = the marine environment, *graphy* = description of) is literally the description of the marine environment. Although the term was first coined in the 1870s, at the beginning of scientific exploration of the oceans, this definition does not fully portray the extent of what oceanography encompasses: Oceanography does much more than just *describe* marine phenomena. Oceanography could be more accurately called the scientific study of all aspects of the marine environment. Hence, the field of study called oceanography could (and maybe *should*) be called oceanology (*ocean* = the marine environment, *ology* = the study of). However, the science of studying the oceans has traditionally been called *oceanography*. It is also called *marine science* and includes the study of the water of the ocean, the life within it, and the (not so) solid Earth beneath it.

Since prehistoric time, people have used the oceans as a means of transportation and as a source of food. Ocean processes, on the other hand, have been studied using technology only since the 1930s, beginning with the search for offshore petroleum and then expanding greatly during World War II with the emphasis on ocean warfare. The recognition of the importance of marine problems by governments, their readiness to make money available for research, the growth in the number of ocean scientists at work, and the increasing sophistication of scientific equipment have all made it feasible to study the ocean on a scale and to a degree of complexity never before attempted nor even possible.

Consider, for example, the logical assumption that those who make their living fishing in the ocean will go where the physical processes of the oceans offer good fishing. How ocean geology, chemistry, and physics work together with biology to create good fishing grounds has been more or less a mystery until only recently, when scientists from those disciplines began to investigate the oceans with new technology. One insight from these studies was the realization of how much of an impact humans are beginning to have on the ocean. As a result, much recent research has been concerned with documenting human impacts on the ocean.

Oceanography is traditionally divided into different academic disciplines (or subfields) of study. The four main disciplines of oceanography that are covered in this book are as follows:

- *Geological oceanography,* which is the study of the structure of the sea floor and how the sea floor has changed through time; the creation of sea floor features; and the history of sediments deposited on it
- *Chemical oceanography,* which is the study of the chemical composition and properties of seawater, how to extract certain chemicals from seawater, and the effects of pollutants
- *Physical oceanography,* which is the study of waves, tides, and currents; the ocean–atmosphere relationship that influences weather and climate; and the transmission of light and sound in the oceans
- *Biological oceanography,* which is the study of the various oceanic life-forms and their relationships to one another, their adaptations to the marine environment, and developing sustainable methods of harvesting seafood

Other disciplines include ocean engineering, marine archaeology, and marine policy. Because the study of oceanography often examines in detail all the different disciplines of oceanography, it is frequently described as being an *interdisciplinary* science, or one covering all the disciplines of science as they apply to the oceans (**Figure 1.14**). In essence, this is a book about *all* aspects of the oceans.

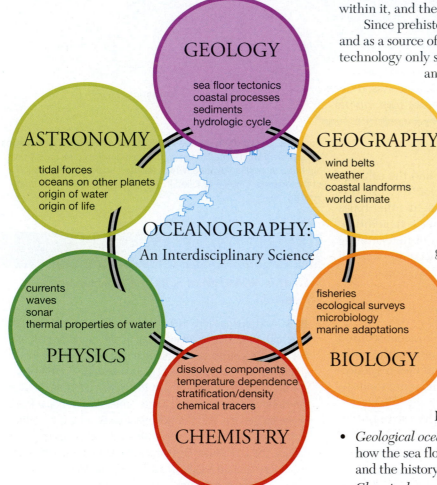

Figure 1.14 **A Venn diagram showing the interdisciplinary nature of oceanography.** Oceanography is an interdisciplinary science that overlaps into many scientific disciplines.

GEOLOGY
sea floor tectonics
coastal processes
sediments
hydrologic cycle

ASTRONOMY
tidal forces
oceans on other planets
origin of water
origin of life

GEOGRAPHY
wind belts
weather
coastal landforms
world climate

OCEANOGRAPHY:
An Interdisciplinary Science

PHYSICS
currents
waves
sonar
thermal properties of water

BIOLOGY
fisheries
ecological surveys
microbiology
marine adaptations

CHEMISTRY
dissolved components
temperature dependence
stratification/density
chemical tracers

Throughout this book you will see a multicolored interdisciplinary icon showing sections where interdisciplinary science is a featured topic. Two or more parts of the icon—geology, chemistry, physics, and biology—will be highlighted to show which disciplines in particular lend insights to the discussion.

Interdisciplinary

Relationship

RECAP

A broad range of interdisciplinary science topics from the diverse fields of geology, chemistry, physics, and biology are included in the study of oceanography.

CONCEPT CHECK 1.3	Explain why oceanography is considered an interdisciplinary science.

1 What was the impetus for studying ocean processes that led to the great expansion of the science of oceanography?

2 What are the four main disciplines or subfields of study in oceanography?

What other marine-related disciplines exist?

3 What does it mean when oceanography is called an interdisciplinary science?

1.4 What Is the Nature of Scientific Inquiry?

In modern society, scientific studies are increasingly used to substantiate the need for action. However, there is often little understanding of how science operates. For instance, how certain are we about a particular scientific theory? How are facts different from theories?

The overall goal of science is to discover underlying patterns in the natural world and then to use this knowledge to make predictions about what should or should not be expected to happen given a certain set of circumstances. Scientists develop explanations about the causes and effects of various natural phenomena (such as why Earth has seasons or what the structure of matter is). This work is based on an assumption that all natural phenomena are controlled by understandable physical processes and the same physical processes operating today have been operating throughout time. Consequently, science has demonstrated remarkable power in allowing scientists to describe the natural world accurately, to identify the underlying causes of natural phenomena, and to better predict future events that rely on natural processes.

Science supports the explanation of the natural world that best explains all available observations. Scientific inquiry is formalized into what is -observations called the **scientific method** (**Figure 1.15**), which is used to formulate scientific theories and separate science from pseudoscience, fact from fiction.

Observations

The scientific method begins with *observations*, which are occurrences we can measure with our senses. They are things we can manipulate, see, touch, hear, taste, or smell, often by experimenting with them directly or by using sophisticated tools (such as a microscope or telescope) to sense them. If an observation is repeatedly confirmed—that is, made so many times that it is assumed to be completely valid—then it can be called a *scientific fact*.

Hypothesis

As observations are being made, the human mind attempts to sort out the observations in a way that reveals some underlying order or pattern in the observations or phenomena. This sorting process—which involves a lot of trial and error—seems to be driven by a fundamental human urge to make sense of our world. This is how **hypotheses** (*hypo* = under, *thesis* = an arranging) are made.

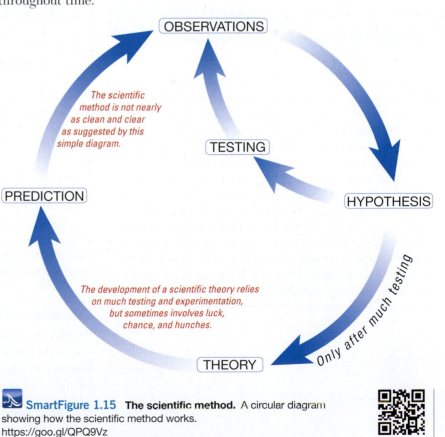

OBSERVATIONS

TESTING

PREDICTION

HYPOTHESIS

THEORY

The scientific method is not nearly as clean and clear as suggested by this simple diagram.

The development of a scientific theory relies on much testing and experimentation, but sometimes involves luck, chance, and hunches.

Only after much testing

SmartFigure 1.15 **The scientific method.** A circular diagram showing how the scientific method works. https://goo.gl/QPQ9Vz

Figure 1.16 A breaching humpback whale (*Megaptera novaeangliae*).

A hypothesis is sometimes labeled as an informed or educated guess, but it is more than that. A hypothesis is a tentative, testable statement about the general nature of reconsidered and modified observed. In other words, a hypothesis is an initial idea of how or why things happen in nature.

Suppose we want to understand why whales *breach* (that is, why whales sometimes leap entirely out of water; **Figure 1.16**). After scientists observe breaching many times, they can organize their observations into a hypothesis. For instance, one hypothesis is that a breaching whale is trying to dislodge parasites from its body. Scientists often have multiple working hypotheses (for example, whales may use breaching to communicate with other whales). If a hypothesis cannot be tested, it is not scientifically useful, no matter how interesting it might seem.

Testing

Hypotheses are used to understand certain occurrences that lead to further research and the refinement of those hypotheses. For instance, the hypothesis that a breaching whale is trying to dislodge its parasites suggests that breaching whales have more parasites than whales that don't breach. Analyzing the number of parasites on breaching versus nonbreaching whales would either support that hypothesis or cause it to be reconsidered and modified. If observations clearly suggest that the hypothesis is incorrect (that is, the hypothesis is *falsified*), then it must be dropped, and other alternative explanations of the facts must be considered.

In science, the validity of any explanation is determined by its coherence with observations in the natural world and its ability to integrate further observations. Only after much testing and experimentation—usually done by many experimenters using a wide variety of repeatable tests—does a hypothesis gain validity where it can be advanced to the next step.

Theory

If a hypothesis has been strengthened by additional observations and if it is successful in explaining additional phenomena, then it can be advanced to what is called a **theory** (*theoria* = a looking at). A theory is a well-substantiated explanation of some aspect of the natural world that can incorporate facts, laws (descriptive generalizations about the behavior of an aspect of the natural world), logical inferences, and tested hypotheses. A theory is not a guess or a hunch. Rather, it is an understanding that develops from extensive observation, experimentation, and creative reflection.

In science, theories are formalized only after many years of testing and verification. Thus, scientific theories have been rigorously scrutinized to the point where most scientists agree that they are the best explanation of certain observable facts. Examples of prominent, well-accepted theories that are held with a very high degree of confidence include biology's theory of evolution (which is discussed later in this chapter) and geology's theory of plate tectonics (which is covered in the next chapter).

Theories also have predictive value, that is, they are useful in predicting what should happen given a certain set of circumstances. If a theory makes no predictions at all, then it has little scientific value. But as is often the case, predictions lead to new observations and a continuation of the cycle that is the process of science.

Theories and the Truth

We've seen how the scientific method is used to develop theories, but does science ever arrive at the undisputed "truth"? Science never reaches an absolute truth because we can never be certain that we have all the observations, especially considering that new technology will be available in the future to examine phenomena in different ways. Notice that there is no end point to the process depicted here. New observations are always possible, so the nature of scientific truth is subject to change. Therefore, it is more accurate to say that science arrives at that which is *probably* true, based on the available observations.

It is not a downfall or weakness of science that scientific ideas are modified as more observations are collected. In fact, the opposite is true. Science is a process that depends on reexamining ideas as new observations are made. Thus, science progresses when new observations yield new hypotheses and modification of theories. As a result, science is littered with hypotheses that have been abandoned in favor of later explanations that fit new observations. One of the best known is the idea that Earth was at the center of the universe, a proposal that was supported by the apparent daily motion of the Sun, Moon, and stars around Earth.

The statements of science should never be accepted as the "final truth." Over time, however, they generally form a sequence of increasingly more accurate statements. Theories are the endpoints in science and do not turn into facts through accumulation of evidence. Nevertheless, the data can become so convincing that the accuracy of a theory is no longer questioned. For instance, the *heliocentric* (*helios* = sun, *centric* = center) *theory* of our solar system states that Earth revolves around the Sun rather than vice versa. Such concepts are supported by such abundant observational and experimental evidence that they are no longer questioned in science.

Is there really such a formal method to science as the scientific method suggests? Actually, the work of scientists is much less formal and is not always done in a clearly logical and systematic manner. Like detectives analyzing a crime scene, scientists use ingenuity and serendipity, visualize models, and sometimes follow hunches in order to unravel the mysteries of nature.

Finally, a key component of verifying scientific ideas is through the peer review process. Once scientists make a discovery, their aim is to get the word out to the scientific community about their results. This is typically done via a published paper, but a draft of the manuscript is first checked by other experts to see if the work has been conducted according to scientific standards and the conclusions are valid. Normally, corrections are suggested and the paper is revised before it is published. This process is a strength of the scientific community and helps weed out inaccurate or poorly formed ideas.

STUDENTS SOMETIMES ASK …

If a theory is proven again and again, does it become a law?

No, that's a common misconception. In science, we collect facts, or observations, we use natural laws to describe them (often using mathematics), and we use a theory to explain them. Natural laws are typically conclusions based on repeated scientific experiments and observations over many years and which have become accepted universally within the scientific community. For example, the *law of gravity* is a description of the force; then there is the *theory of gravitational attraction,* which explains why the force occurs. Theories don't get "promoted" to a law by an abundance of proof, and so a theory never becomes a law. They're really two separate things.

RECAP

Science supports the explanation of the natural world that best explains all available observations. Because new observations can modify existing theories, science is always developing.

CONCEPT CHECK 1.4 | Describe the nature of scientific inquiry.

1 Describe the steps involved in the scientific method.

2 What is the difference between a hypothesis and a theory?

3 Briefly comment on the phrase "scientific certainty." Is it an oxymoron (a combination of contradictory words),

or are scientific theories considered to be the absolute truth?

4 Can a theory ever be so well established that it becomes a fact? Explain. or slight changes in the emitted light of distant stars, such as the decrease in brightness as planets pass in front of them.

1.5 How Were Earth and the Solar System Formed?

Earth is the third of eight major planets[8] in our **solar system** that revolve around the Sun (**Figure 1.17**). Evidence suggests that the Sun and the rest of the solar system formed about 5 billion years ago from a huge cloud of gas and space dust called a **nebula** (*nebula* = a cloud). Astronomers base this hypothesis on the orderly nature of our solar system and the consistent age of meteorites (pieces of the early solar system). Using sophisticated telescopes, astronomers have also been able to observe distant nebula and planetary systems in various stages of formation elsewhere in our galaxy (**Figure 1.18**). In addition, more than 2000 planets have been discovered outside our solar system—including several that are about the size of Earth—by detecting the telltale wobble of distant stars or slight changes in the emitted light of remote stars, such as the decrease in brightness as planets pass in front of them.

[8]Pluto, which used to be considered the ninth planet in our solar system, was reclassified by the International Astronomical Union as a "dwarf planet" in 2006, along with other similar bodies.

The Nebular Hypothesis

According to the **nebular hypothesis** (**Figure 1.19**), all bodies in the solar system formed from an enormous cloud composed mostly of hydrogen and helium, with only a small percentage of heavip elements. As this huge accumulation of gas and dust revolved around its center, it began to contract under its own gravity, becoming hotter and denser, eventually forming the Sun.

As the nebular matter that formed the Sun contracted, small amounts of it were left behind in swirling eddies, which are similar to small whirlpools in a stream. The material in these eddies was the beginning of the **protoplanets** (*proto* = original, *planetes* = wanderer) and their orbiting satellites, which later consolidated into the present planets and their moons.

Proto-Earth

Proto-Earth looked very different from Earth today. Its size was larger than today's Earth, and there were neither oceans nor any life on the planet. In addition, the structure of the deep proto-Earth is thought to have been *homogenous* (*homo* = alike, *genous* = producing), which means that it had a uniform composition throughout. The structure of proto-Earth changed, however, as its heavier constituents sank toward the center to form a heavy core.

During this early stage of formation, many meteorites and comets from space bombarded proto-Earth (**Figure 1.20**). In fact, a leading theory states that the Moon was born in the aftermath of a titanic collision between a Mars-size planet named *Theia* and proto-Earth. While most of Theia was swallowed up and incorporated into the magma ocean it created on impact, the collision also flung a small world's worth of vaporized and molten rock into orbit. Over time, this debris coalesced into a sphere and created Earth's orbiting companion, the Moon.

During this early formation of the protoplanets and their satellites, the Sun condensed into a body so massive and hot that pressure within its core initiated the process of **thermonuclear fusion** (*thermo* = hot, *nucleos* = a little nut; *fusus* = melted). Thermonuclear fusion occurs when temperatures reach tens of millions of degrees and hydrogen **atoms** (*a* = not, *tomos* = cut) combine to form helium atoms, releasing enormous amounts of energy.[9] Not only does the Sun emit light, it also emits *ionized* (electrically charged) particles that make up the *solar wind*. During the early stages of

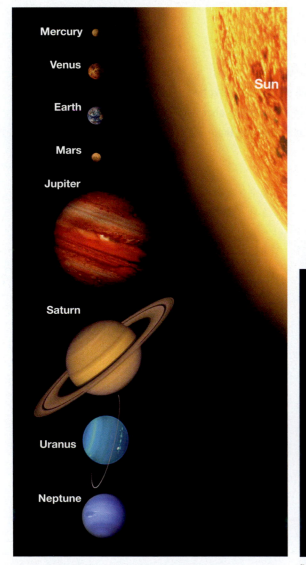

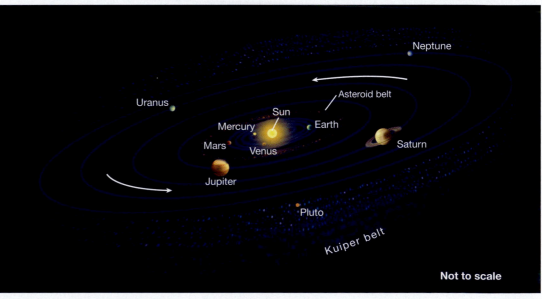

(a) Features and relative sizes of the Sun and the eight major planets of the solar system.

(b) Orbits and relative positions of various features of the solar system.

Figure 1.17 **The solar system.** Schematic views of the solar system, which includes the Sun and eight major planets.

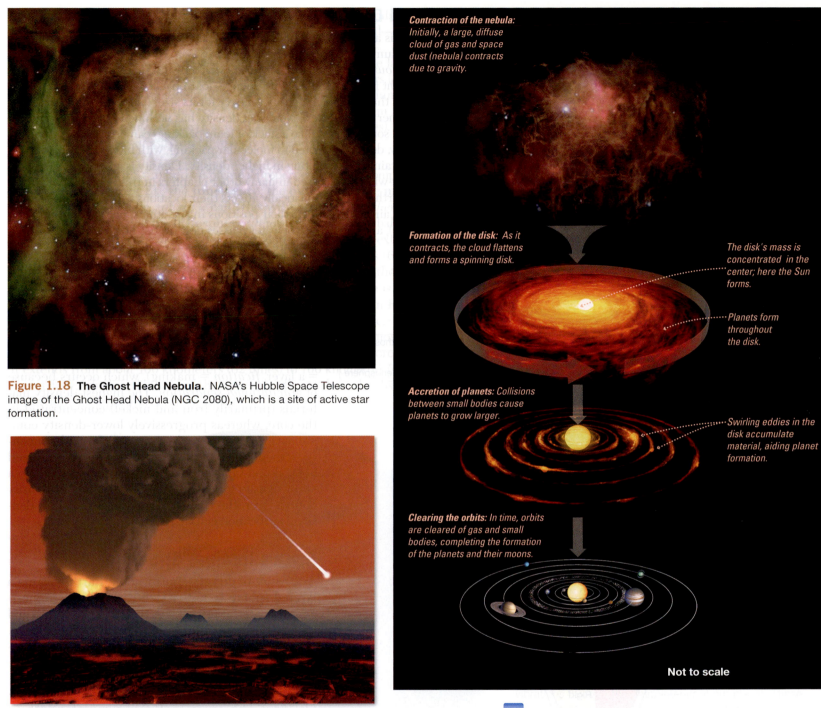

Figure 1.18 The Ghost Head Nebula. NASA's Hubble Space Telescope image of the Ghost Head Nebula (NGC 2080), which is a site of active star formation.

Contraction of the nebula: Initially, a large, diffuse cloud of gas and space dust (nebula) contracts due to gravity.

Formation of the disk: As it contracts, the cloud flattens and forms a spinning disk.

The disk's mass is concentrated in the center; here the Sun forms.

Planets form throughout the disk.

Accretion of planets: Collisions between small bodies cause planets to grow larger.

Swirling eddies in the disk accumulate material, aiding planet formation.

Clearing the orbits: In time, orbits are cleared of gas and small bodies, completing the formation of the planets and their moons.

Not to scale

Figure 1.20 Proto-Earth. An artist's conception of what Earth may have looked like early in its development.

 SmartFigure 1.19 The nebular hypothesis of solar system formation. According to the nebular hypothesis, our solar system formed from the gravitational contraction of an interstellar cloud of gas and space dust called a *nebula*.
https://goo.gl/FoY7Yt

Web Animation

The Nebular Hypothesis of Solar System Formation
http://goo.gl/KObsRK

formation of the solar system, this solar wind blew away the nebular gas that remained from the formation of the planets and their satellites.

The protoplanets closest to the Sun (including Earth) also lost their initial atmospheres (mostly hydrogen and helium), blown away by the bombardment by ionized solar radiation. At the same time, these rocky protoplanets were gradually cooling, causing them to contract and drastically shrink in size. As the protoplanets continued to contract, another source of heat was produced deep within their cores from the spontaneous disintegration of atoms, called *radioactivity* (*radio* = ray, *acti* = to cause).

[9]Thermonuclear fusion in stars also creates larger and more complex elements, such as carbon. It is interesting to note that as a result, all matter—even the matter that comprises our bodies—originated as stardust long ago.

Figure 1.26 **Creation of organic molecules.**

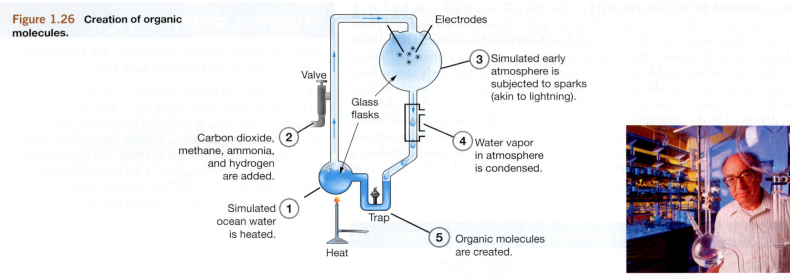

② Carbon dioxide, methane, ammonia, and hydrogen are added.

Valve

Glass flasks

Electrodes

③ Simulated early atmosphere is subjected to sparks (akin to lightning).

④ Water vapor in atmosphere is condensed.

① Simulated ocean water is heated.

Trap

Heat

⑤ Organic molecules are created.

(a) Laboratory apparatus used by Stanley Miller to simulate the conditions of the early atmosphere and the oceans. The experiment produced various organic molecules and suggests that the basic components of life were created in a "prebiotic soup" in the oceans.

(b) Stanley Miller in 1999, with his famous apparatus in the foreground.

surface of Earth from most of the Sun's harmful ultraviolet radiation (which is why the atmospheric ozone hole over Antarctica has generated such concern).

Evidence suggests that Earth's early atmosphere (the product of outgassing) was different from Earth's initial hydrogen–helium atmosphere and different from the mostly nitrogen–oxygen atmosphere of today. The early atmosphere probably contained large percentages of water vapor and carbon dioxide and smaller percentages of hydrogen, methane, and ammonia but very little free oxygen (oxygen that is not chemically bound to other atoms). Why was there so little free oxygen in the early atmosphere? Oxygen may well have been outgassed, but oxygen and iron have a strong affinity for each other.[15] As a result, iron in Earth's early crust would have reacted with the outgassed oxygen immediately, removing it from the atmosphere.

Without oxygen in Earth's early atmosphere, moreover, there would have been no ozone layer to block most of the Sun's ultraviolet radiation. The lack of a protective ozone layer may, in fact, have played a crucial role in several of life's most important developmental milestones.

Stanley Miller's Experiment

In 1952, **Stanley Miller** (**Figure 1.26b**)—then a 22-year-old graduate student of chemist Harold Urey at the University of Chicago—conducted a laboratory experiment that had profound implications about the development of life on Earth. In Miller's experiment, he exposed a mixture of carbon dioxide, methane, ammonia, hydrogen, and water (the components of the early atmosphere and ocean) to ultraviolet light (from the Sun) and an electrical spark (to imitate lightning) (**Figure 1.26a**). By the end of the first day, the mixture turned pink, and after a week it was a deep, muddy brown, indicating the formation of a large assortment of organic molecules, including amino acids—which are the basic components of life—and other biologically significant compounds.

Interdisciplinary

Relationship

Miller's now-famous laboratory experiment of a simulated primitive Earth in a bottle—which has been duplicated and confirmed numerous times

[15]As an example of the strong affinity of iron and oxygen, consider how common rust—a compound of iron and oxygen—is on Earth's surface.

since—demonstrated that vast amounts of organic molecules could have been produced in Earth's early oceans, often called a "prebiotic soup." This prebiotic soup, perhaps spiced by extraterrestrial molecules aboard comets, meteorites, or interplanetary dust, was fueled by raw materials from volcanoes, certain minerals in sea floor rocks, and undersea hydrothermal vents. On early Earth, the mixture was energized by lightning, cosmic rays, and the planet's own internal heat, and it is thought to have created life's precursor molecules about 4 billion years ago.

Exactly how these simple organic compounds in the prebiotic soup assembled themselves into more complex molecules—such as proteins and DNA—and then into the first living entities remains one of the most tantalizing questions in science. Research suggests that with the vast array of organic compounds available in the prebiotic soup, several kinds of chemical reactions led to increasingly elaborate molecular structures. In fact, research suggests that small, simple molecules could have acted as templates, or "molecular midwives," in helping the building blocks of life's genetic material form long chains and thus may have assisted in the formation of longer, more elaborate molecular complexes. Among these complexes, some began to carry out functions associated with the basic molecules of life. As the products of one generation became the building blocks for another, even more complex molecules, or polymers, emerged over many generations that could store and transfer information. Such genetic polymers ultimately became encapsulated within cell-like membranes that were also present in Earth's primitive broth. The resulting cell-like complexes thereby housed self-replicating molecules capable of multiplying—and hence evolving—genetic information. Many specialists consider this emergence of genetic replication to be the true origin of life.

Evolution and Natural Selection

Every living organism that inhabits Earth today is the result of **evolution** by the process of **natural selection** that has been occurring since life first existed on Earth. The theory of evolution states that groups of organisms adapt and change with the passage of time, causing descendants to differ morphologically and physiologically from their ancestors (**Diving Deeper 1.2**). Certain advantageous traits are naturally selected and passed from one generation to the next. Evolution is the process by which various **species** (*species* = a kind) have been able to inhabit increasingly numerous environments on Earth.

As we shall see, when species adapt to Earth's various environments, they can also modify the environments in which they live. This modification can be localized or nearly global in scale. For example, when plants emerged from the oceans and inhabited the land, they changed Earth from a harsh and bleak landscape as barren as that of the Moon to one that is green and lush.

Plants and Animals Evolve

The very earliest forms of life were probably **heterotrophs** (*hetero* = different, *tropho* = nourishment). Heterotrophs require an external food supply, which was abundantly available in the form of nonliving organic matter in the ocean around them. **Autotrophs** (*auto* = self, *tropho* = nourishment), which can manufacture their own food supply, evolved later. The first autotrophs were probably similar to present-day **anaerobic** (*an* = without, *aero* = air) bacteria, which live without atmospheric oxygen. They may have been able to derive energy from inorganic compounds at deep-water hydrothermal vents using a process called **chemosynthesis** (*chemo* = chemistry, *syn* = with, *thesis* = an arranging).[16] In fact, the detection of microbes deep within the ocean crust as well as the discovery of 3.2-billion-year-old microfossils of bacteria from deep-water marine rocks support the idea of life's origin on the deep-ocean floor in the absence of light.

Interdisciplinary

Relationship

RECAP

Organic molecules were produced in a simulation of Earth's early atmosphere and ocean, suggesting that life most likely originated in the oceans.

[16]More details about chemosynthesis are discussed in Chapter 15, "Animals of the Benthic Environment."

1.3 What is oceanography?

▶ *Oceanography, or marine science, is the scientific study of all aspects of the marine environment.* During World War II, *a tactical advantage was gained by studying ocean processes,* leading to great advances in technology and the ability to observe and study the oceans in more detail. *Today, much study is focused on human impacts on the ocean.*

▶ *Oceanography is traditionally divided into four academic disciplines (or subfields) of study.* These four disciplines are: (1) *geological oceanography,* (2) *chemical oceanography,* (3) *physical oceanography,* and (4) *biological oceanography. Oceanography is frequently described as being an interdisciplinary science* because it encompasses all the different disciplines of science as they apply to the oceans.

Study Resources
MasteringOceanography Study Area Quizzes

Critical Thinking Question
Describe one of today's ocean problems that encompasses at least two of the different disciplines in the multidisciplinary science that is oceanography.

Active Learning Exercise
With another student in class, make a list of all the types of careers you would be qualified for with a degree in oceanography or marine science. As an example of someone who works in oceanography or marine science, consider your instructor.

1.4 What is the nature of scientific inquiry?

▶ The *scientific method* is used to understand the occurrence of physical events or phenomena and can be stated as *science supports the explanation of the natural world that best explains all available observations.* Steps in the scientific method include making *observations* and establishing *scientific facts;* forming one or more *hypotheses* (a tentative, testable statement about the general nature of the phenomena observed); extensive *testing* and *modification of hypotheses;* and, finally, developing a *theory* (a well-substantiated explanation of some aspect of the natural world that can incorporate facts, laws, logical inferences, and tested hypotheses). Science never arrives at the absolute "truth"; rather, *science arrives at what is probably true* based on the available observations and can *continually change because of new observations.*

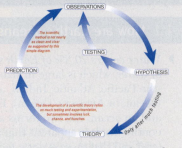

Study Resources
MasteringOceanography Study Area Quizzes

Critical Thinking Question
What is the difference between a fact and a theory? Can either (or both) be revised?

Active Learning Exercise
With another student in class, discuss if you believe nature is simple enough for humans to truly understand. Give reasons why or why not. If not, do you think it is still reasonable for scientists to make this assumption in applying the scientific method in their work?

1.5 How were Earth and the solar system formed?

▶ Our *solar system,* consisting of *the Sun and eight major planets,* probably formed from a *huge cloud of gas and space dust* called a *nebula.* According to the *nebular hypothesis,* the nebular matter contracted to form the Sun, and the planets were formed from eddies of material that remained. The Sun, composed of hydrogen and helium, was massive enough and concentrated enough to emit *large amounts of energy* from fusion. The Sun also emitted *ionized particles that swept away any nebular gas* that remained from the formation of the planets and their satellites.

▶ *Proto-Earth,* more massive and larger than Earth today, *was molten and homogenous.* The *initial atmosphere,* composed mostly of hydrogen and helium, *was later driven off into space* by intense solar radiation. Proto-Earth began a period of rearrangement called *density stratification* and formed a *layered internal structure based on density,* resulting in the development of the *crust, mantle,* and *core.* Studies of Earth's internal structure indicate that brittle plates of the *lithosphere* are riding on a plastic, high-viscosity *asthenosphere.* Near the surface, the lithosphere is composed of *continental* and *oceanic crust.* Continental crust consists mostly of granite and oceanic crust consists mostly of basalt. *Continental crust is lower in density, lighter in color, and thicker than*

oceanic crust. Both types of crust float isostatically on the denser mantle below.

Study Resources
MasteringOceanography Study Area Quizzes, MasteringOceanography Web Animations, MasteringOceanograhy Web Videos

Critical Thinking Question
Describe how the chemical composition of Earth's interior differs from its physical properties. Include specific examples.

Active Learning Exercise
The nebular hypothesis of solar system formation is a scientific hypothesis. Based on your understanding of the scientific method, describe to another student in class how sure of this hypothesis you think scientists really are. Why would scientists have this level of certainty?

1.6 How were Earth's atmosphere and oceans formed?

▶ *Outgassing produced an early atmosphere* rich in water vapor and carbon dioxide. Once Earth's surface cooled sufficiently, the *water vapor condensed and accumulated to give Earth its oceans. Rainfall on the surface dissolved compounds that,* when carried to the ocean, *made it salty.*

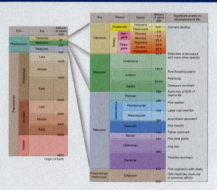

Water vapor and other gases

Early in Earth's history, volcanic activity released large amounts of water vapor into the atmosphere.

Water vapor and other gases

Water vapor condensed into clouds.

Liquid water fell to Earth's surface, where it accumulated in low areas and over time formed the oceans.

Study Resources
MasteringOceanography Study Area Quizzes

Critical Thinking Question
Compare the two ways in which Earth was supplied with enough water to have an ocean. Which is likely to have contributed most of the water on Earth?

Active Learning Exercise
With another student in class, describe in your own words how Earth's oceans became salty.

1.8 How old is Earth?

▶ *Radiometric age dating* is used to determine the age of most rocks. Information from extinctions of organisms and from age dating rocks comprises the *geologic time scale,* which indicates that Earth has experienced a long history of changes since *its origin 4.6 billion years ago.*

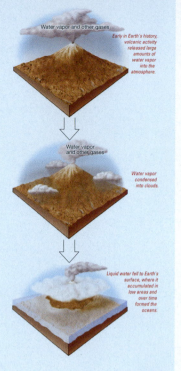

Study Resources
MasteringOceanography Study Area Quizzes, MasteringOceanography Web Diving Deeper 1.2, MasteringOceanography Web Animation

Critical Thinking Question
Explain how radiometric age dating works. Why does the parent material never totally disappear completely, even after many half-lives?

Active Learning Exercise
Working as a team, construct a representation of the geologic time scale, using an appropriate quantity of any substance (other than dollar bills or toilet paper, which are used as examples in MasteringOceanography Web Diving Deeper 1.2). Be sure to indicate some of the major changes that have occurred on Earth since its origin, such as "Origin of Earth," "Origin of oceans," "Earliest known life-forms," "Oxygen-rich atmosphere first occurs," "First organisms with shells," "Dinosaurs die out," and "Age of humans."

1.7 Did life begin in the oceans?

▶ *Life is thought to have begun in the oceans.* Stanley Miller's experiment showed that ultraviolet radiation from the Sun and hydrogen, carbon dioxide, methane, ammonia, and inorganic molecules from the oceans may have combined to produce *organic molecules such as amino acids.* Certain combinations of these molecules eventually produced *heterotrophic organisms* (which cannot make their own food) that were probably similar to present-day anaerobic bacteria. Eventually, *autotrophs evolved* that had the ability to make their own food through *chemosynthesis.* Later, some cells developed *chlorophyll,* which made *photosynthesis* possible and led to the *development of plants.*

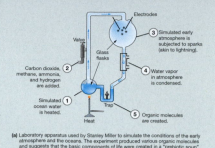

Electrodes

③ Simulated early atmosphere is subjected to sparks (akin to lightning).

Valve

Glass flasks

② Carbon dioxide, methane, ammonia, and hydrogen are added.

④ Water vapor in atmosphere is condensed.

① Simulated ocean water is heated.

Heat Trap

⑤ Organic molecules are created.

(a) Laboratory apparatus used by Stanley Miller to simulate the conditions of the early atmosphere and the oceans. The experiment produced various organic molecules and suggests that the basic components of life were created in a "prebiotic soup" in the oceans.

▶ *Photosynthetic organisms altered the environment* by extracting carbon dioxide from the atmosphere and also by releasing free oxygen, thereby creating today's *oxygen-rich atmosphere.* Eventually, both *plants and animals evolved* into forms that could survive on land.

Study Resources
MasteringOceanography Study Area Quizzes

Critical Thinking Question
How would you answer the accusation, made by some religious groups, that scientific theories such as Stanley Miller's theory on the origin of life on Earth are inherently weak because it is a historic event that no one actually observed? Please explain your answer in detail.

Active Learning Exercise
With another student in class, discuss which of these two statements has more validity: (1) the greatest environmental crisis of all time was the build-up of toxic oxygen in Earth's atmosphere 2 billion years ago or (2) humans are causing the greatest environmental crisis of all time.

MasteringOceanography™
www.masteringoceanography.com

Looking for additional review and test prep materials? With individualized coaching on the toughest topics of the course, MasteringOceanography offers a wide variety of ways for you to move beyond memorization and deeply grasp the underlying processes of how the oceans work. Visit the Study Area in www.masteringoceanography.com to find practice quizzes, study tools, and multimedia that will improve your understanding of this chapter's content. Sign in today to enjoy the following features: Self Study Quizzes, SmartFigures, SmartTables, Oceanography Videos, Squidtoons, Geoscience Animation Library, RSS Feeds, Digital Study Modules, and an optional Pearson eText.

Tall mountains created by tectonic uplift. Tall coastal mountains such as these in Glacier Bay National Park in southeastern Alaska have been uplifted by plate tectonic processes, creating a large amount of relief. Some of the uplifted rocks here have come from distant areas and include parts of the sea floor.

Convection **Volcanic arc** Pangaea
Continental drift **Subduction** Transform fault **Ocean trench**
Convergent boundary Asthenosphere **Divergent boundary**
Hotspot **Alfred Wegener**
Plate Tectonics **Sea floor spreading**
Nematath **Transform boundary** Panthalassa
Lithosphere
Atoll **Mid-ocean ridge**

Plate Tectonics and the Ocean Floor

Before you begin reading this chapter, use the glossary at the end of this book to discover the meanings of any of the words in the word cloud above you don't already know.

Each year at various locations around the globe, several thousand earthquakes and dozens of volcanic eruptions occur, both of which indicate how remarkably dynamic our planet is. These events have occurred throughout history, constantly changing the surface of our planet, yet only a little over 50 years ago, most scientists believed the continents were stationary over geologic time. Since that time, a bold new theory has been advanced that helps explain surface features and phenomena on Earth, including:

- The worldwide locations of volcanoes, faults, earthquakes, and mountain building
- Why mountains on Earth haven't been eroded away
- The origin of most landforms and ocean floor features
- How the continents and ocean floor formed and why they are different
- The continuing development of Earth's surface
- The distribution of past and present life on Earth

This revolutionary new theory is called **plate tectonics** (*plate* = plates of the lithosphere; *tekton* = to build), or "the new global geology." According to the theory of plate tectonics, the outermost portion of Earth is composed of a patchwork of thin, rigid plates[1] that move horizontally with respect to one another, like icebergs floating on water. As a result, the continents are mobile and move about on Earth's surface, controlled by forces deep within Earth.

The interaction of these plates as they move builds features of Earth's crust (such as mountain belts, volcanoes, and ocean basins). For example, the tallest mountain range on Earth is the Himalaya Mountains that extend through India, Nepal, and Bhutan. This mountain range contains rocks that were deposited millions of years ago in a shallow sea, providing testimony of the power and persistence of plate tectonic activity.

Plate tectonics is extensively supported by data from a variety of sciences, including geological, chemical, physical, and biological sources. Yet it wasn't accepted by many scientists when it was first introduced. In fact, it is a classic example of the process of the scientific method: how a seemingly implausible hypothesis, when faced with a preponderance of evidence to support it, developed into a theory that now forms the basis of our understanding of fundamental Earth processes.

ESSENTIAL LEARNING CONCEPTS

At the end of this chapter, you should be able to:

2.1 Evaluate the evidence that supports continental drift.

2.2 Summarize the evidence that supports plate tectonics.

2.3 Discuss the origin and characteristics of features that occur at plate boundaries.

2.4 Show how plate tectonics can be used as a working model.

2.5 Describe how Earth has changed in the past and predict how it will look in the future.

"It is just as if we were to refit the torn pieces of a newspaper by matching their edges and then check whether the lines of print run smoothly across. If they do, there is nothing left but to conclude that the pieces were in fact joined in this way."

—*Alfred Wegener,* The Origins of Continents and Oceans *(1915)*

[1]These thin, rigid plates are pieces of the *lithosphere* that comprise Earth's outermost layer and contain oceanic and/or continental crust, as described in Chapter 1.

2.1 What Evidence Supports Continental Drift?

Alfred Wegener (**Figure 2.1**), a German meteorologist and geophysicist, was the first to advance the idea of mobile continents in 1912. He envisioned that the continents were slowly drifting across the globe and called his idea **continental drift**. Let's examine the evidence that Wegener compiled that led him to formulate the idea of drifting continents.

Fit of the Continents

The idea that continents—particularly South America and Africa—fit together like pieces of a jigsaw puzzle originated with the development of reasonably accurate world maps. As far back as 1620, Sir Francis Bacon wrote about how the continents appeared to fit together. However, little significance was given to this idea until 1912, when Wegener used the shapes of matching shorelines on different continents as a supporting piece of evidence for continental drift.

Wegener suggested that during the geologic past, the continents collided to form a large landmass, which he named **Pangaea** (*pan* = all, *gaea* = Earth) (**Figure 2.2**). Further, a huge ocean, called **Panthalassa** (*pan* = all, *thalassa* = sea), surrounded Pangaea. Panthalassa included several smaller seas, including the **Tethys Sea** (*Tethys* = a Greek sea goddess). Wegener's evidence indicated that as Pangaea began to split apart, the various continental masses started to drift toward their present geographic positions.

Wegener's attempt at matching shorelines revealed considerable areas of crustal overlap and large gaps. Some of the differences could be explained by material deposited by rivers or eroded from coastlines. What Wegener didn't know at the time was that the shallow parts of the ocean floor close to shore are underlain by materials similar to those beneath continents. In the early 1960s, Sir Edward Bullard and two associates used a computer program to fit the continents together (**Figure 2.3**). Instead of using the shorelines of the continents as Wegener had done, Bullard achieved the best fit (for example, with minimal overlaps or gaps) by using a depth of 2000 meters (6560 feet) below sea level. This depth corresponds to halfway between the shoreline and the deep-ocean basins; as such, it represents the true edge of the continents. By using this depth, the continents fit together remarkably well.

Matching Sequences of Rocks and Mountain Chains

If the continents were once together, as Wegener had hypothesized, then evidence should appear in rock sequences that were originally continuous but are now separated by large distances. To test the idea of drifting continents, geologists began comparing the rocks along the edges of continents with rocks found in adjacent positions on matching continents. They wanted to see if the rocks had similar types, ages, and structural styles (the type and degree of deformation). In some areas, younger rocks had been deposited during the millions of years since the continents separated, covering the rocks that held the key to the past history of the continents. In other areas, the rocks had been eroded away. Nevertheless, in many other areas, the key rocks were present.

Figure 2.1 **Alfred Wegener, circa 1912–1913.** Alfred Wegener (1880–1930), shown here in his research station in Greenland, developed the idea of continental drift. He was one of the first scientists to use multiple lines of evidence to suggest that continents are mobile.

Moreover, these studies showed that many rock sequences from one continent were identical to rock sequences on an adjacent continent—although the two were separated by an ocean. In addition, mountain ranges that terminated abruptly at

the edge of a continent continued on another continent across an ocean basin, with identical rock sequences, ages, and structural styles. **Figure 2.4** shows, for example, how similar rocks from the Appalachian Mountains in North America match up with identical rocks from the British Isles and the Caledonian Mountains in Europe.

Wegener noted the similarities in rock sequences on both sides of the Atlantic and used the information as a supporting piece of evidence for continental drift. He suggested that mountains such as those seen on opposite sides of the Atlantic formed during the collision when Pangaea was formed. Later, when the continents split apart, once-continuous mountain ranges were separated. Confirmation of this idea exists in a similar match with mountains extending from South America through Antarctica and across Australia.

Glacial Ages and Other Climate Evidence

Wegener also noticed the occurrence of past glacial activity in areas that are now tropical and suggested that it, too, provided supporting evidence for drifting continents. Currently, the only places in the world where thick continental *ice sheets* occur are in the polar regions of Greenland and Antarctica. However, evidence of ancient glaciation is found in the lower-latitude regions of South America, Africa, India, and Australia.

These deposits, which have been dated at 300 million years old, indicate one of two possibilities: (1) There was a worldwide **ice age** at that time, and even tropical areas were covered by thick ice, or (2) some continents that are now in tropical areas were once located much closer to one of the poles. It is unlikely that the entire world was covered by ice 300 million years ago because coal deposits from the same geologic age now present in North America and Europe originated as vast semitropical swamps. Thus, a reasonable conclusion is that some of the continents must have been closer to the poles than they are today.

Another type of glacial evidence indicates that certain continents have moved from more polar regions during the past 300 million years. When glaciers flow, they move and abrade the underlying rocks, leaving grooves that indicate the direction of flow. The arrows in **Figure 2.5a** show how the glaciers would have flowed away from the South Pole on Pangaea 300 million years ago. The direction of flow is consistent with the grooves found on many continents today (**Figure 2.5b**), providing additional evidence for drifting continents.

Many examples of plant and animal fossils indicate very different climates than today. Two such examples are fossil palm trees in Arctic Spitsbergen and coal deposits in Antarctica. Earth's past environments can be interpreted from these rocks because plants and animals need specific environmental conditions in which to live. Corals, for example, generally need seawater above 18 degrees centigrade (°C) or 64 degrees Fahrenheit (°F) in order to survive. When fossil corals are found in areas that are cold today, two explanations seem most plausible: (1) Worldwide climate has changed dramatically or (2) the rocks have moved from their original location.

As explained in Chapter 16, "The Oceans and Climate Change," natural processes have caused Earth's climate to change in the geologic past. Although dramatic shifts in Earth's climate might help explain climate evidence such as fossils that seem out of place today, the distribution of these fossils could also be explained by drifting continents. Unaware of the changes in Earth's climate that are known by Earth scientists today, Wegener suggested that the out-of-place fossils as well as other climate evidence provided support for the slow movement of the continents and added another item to a growing list of evidence.

Distribution of Organisms

To add credibility to his argument for the existence of the supercontinent of Pangaea, Wegener cited documented cases of several fossil organisms found on different landmasses that could not have crossed the vast oceans presently separating the

(a) The positions of the continents today.

(b) The positions of the continents about 200 million years ago, showing the supercontinent of Pangaea and the single large ocean, Panthalassa.

Figure 2.2 Reconstruction of Pangaea.

Climate

Connection

Figure 2.3 An early computer fit of the continents. Map showing the 1960s fit of the continents using a depth of 2000 meters (6560 feet) (*black lines*), which is the true edge of the ocean basin. The results indicate a remarkable match, with few overlaps and minimal gaps. Note that the present-day shorelines of the continents are shown with blue lines.

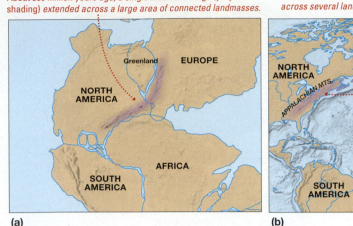

About 300 million years ago, a single mountain range (purple shading) extended across a large area of connected landmasses.

Today, this once-continuous mountain range is scattered across several landmasses and is separated by an ocean.

(a) (b)

Figure 2.4 Matching mountain ranges across the North Atlantic Ocean.

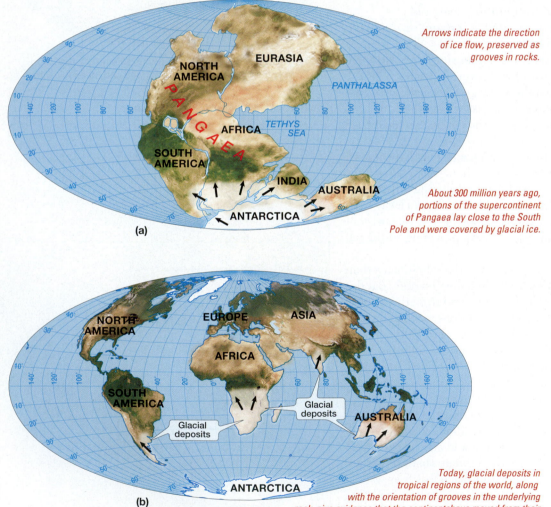

Arrows indicate the direction of ice flow, preserved as grooves in rocks.

(a)

About 300 million years ago, portions of the supercontinent of Pangaea lay close to the South Pole and were covered by glacial ice.

Today, glacial deposits in tropical regions of the world, along with the orientation of grooves in the underlying rock, give evidence that the continents have moved from their former positions.

(b)

Figure 2.5 Ice age on Pangaea.

Web Animation
Breakup of Pangaea
http://goo.gl/egACqz

continents. For example, the fossil remains of **Mesosaurus** (*meso* = middle, *saurus* = lizard), an extinct, presumably aquatic reptile that lived about 250 million years ago, are located only in eastern South America and western Africa (**Figure 2.6**). If *Mesosaurus* had been strong enough to swim across an ocean, why aren't its remains more widely distributed?

Wegener's idea of continental drift provided an elegant solution to this problem. He suggested that the continents were closer together in the geologic past, so *Mesosaurus* didn't have to be a good swimmer to leave remains on two different continents. Later, after *Mesosaurus* became extinct, the continents moved to their present-day positions, and a large ocean now separates the once-connected landmasses. Other examples of similar fossils on different continents include those of plants, which would have had a difficult time traversing a large ocean.

Before continental drift, several ideas were proposed to help explain the curious pattern of these fossils, such as the existence of island stepping stones or a land bridge. It was even suggested that at least one pair of land-dwelling *Mesosaurus* survived the arduous journey across several thousand kilometers of open ocean by rafting on floating logs. However, there is no evidence to support the idea of island stepping stones or a land bridge, and the idea of *Mesosaurus* rafting across an ocean seems implausible.

Wegener also cited the distribution of present-day organisms as evidence to support the concept of drifting continents. For example, modern organisms with similar ancestries clearly had to evolve in isolation during the past few million years. Most obvious of these are the Australian marsupials (such as kangaroos, koalas, and wombats), which have a distinct similarity to the marsupial opossums found in the Americas.

Objections to the Continental Drift Model

Wegener first published his ideas in *The Origins of Continents and Oceans* in 1915, but the book did not attract much attention until it was translated into English, French, Spanish, and Russian in 1924. From that point until his death in 1930,[2] Wegener's drift hypothesis

[2] Wegener perished in 1930 while trying to establish a year-round meteorological station atop the Greenland ice sheet.

received much hostile criticism—and sometimes open ridicule—from the scientific community because of the mechanism he proposed for the movement of the continents. Wegener suggested that the continents plowed through the ocean basins to reach their present-day positions and that the leading edges of the continents deformed into mountain ridges because of the drag imposed by ocean rocks. Further, the driving mechanism he proposed was a combination of the gravitational attraction of Earth's equatorial bulge and tidal forces from the Sun and Moon.

Scientists rejected the idea as too fantastic and contrary to the laws of physics. Debate over the mechanism of drift concentrated on the long-term behavior of the substrate and the forces that could move continents laterally. Material strength calculations, for example, showed that ocean rock was too strong for continental rock to plow through it. Further, analysis of gravitational and tidal forces indicated that they were too small to move the great continental landmasses. Even without an acceptable mechanism, many geologists who studied rocks in South America and Africa accepted continental drift because it was consistent with the rock record. North American geologists—most of whom were unfamiliar with these Southern Hemisphere rock sequences—remained highly skeptical.

As compelling as his evidence may seem today, Wegener was unable to convince the scientific community as a whole of the validity of his ideas. Although his hypothesis was correct in principle, it contained several incorrect details, such as the driving mechanism for continental motion and how continents move across ocean basins. In order for any scientific viewpoint to gain wide acceptance, it must explain all available observations and have supporting evidence from a wide variety of scientific fields. This supporting evidence would not come until more details of the nature of the ocean floor were revealed, which, along with new technology that enabled scientists to determine the original positions of rocks on Earth, provided additional observations in support of drifting continents.

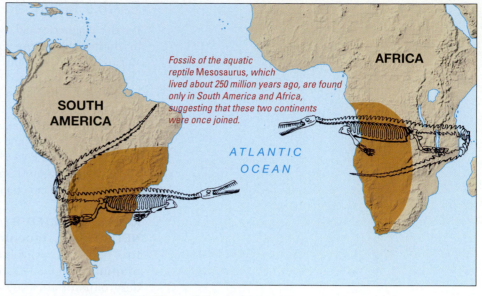

Fossils of the aquatic reptile Mesosaurus, which lived about 250 million years ago, are found only in South America and Africa, suggesting that these two continents were once joined.

Figure 2.6 **Fossils of Mesosaurus.**

RECAP

Alfred Wegener used a variety of interdisciplinary Earth science information to support continental drift. However, he did not have a suitable mechanism or any information about the sea floor, and his idea was widely criticized.

CONCEPT CHECK 2.1 | Evaluate the evidence that supports continental drift.

1 When did the supercontinent of Pangaea exist? What was the ocean that surrounded the supercontinent called?

2 Regarding glacial ages, why is it unlikely that the entire world was covered by ice 300 million years ago?

3 Cite the lines of evidence Alfred Wegener used to support his idea of continental drift. Why did scientists of the time doubt that continents had drifted?

2.2 What Evidence Supports Plate Tectonics?

Very little new information about Wegener's continental drift hypothesis was introduced between the time of Wegener's death in 1930 and the early 1950s. However, studies of the sea floor using sonar that were initiated during World War II and continued after the war provided critical evidence in support of drifting continents. In addition, technology unavailable in Wegener's time enabled scientists to analyze the way rocks retained the signature of Earth's **magnetic field**. These developments caused scientists to reexamine continental drift and advance it into the more encompassing theory of plate tectonics.

STUDENTS SOMETIMES ASK . . .

What causes Earth's magnetic field?

Studies of Earth's magnetic field and research in the field of *magnetodynamics* suggest that convective movement of fluids in Earth's liquid iron–nickel outer core is the cause of Earth's magnetic field. The most widely accepted view is that Earth's magnetic field is created by strong electrical currents generated by a dynamo process resulting from the convective flow of molten iron in Earth's outer core. Earth's magnetic field is so complex that it has only recently been successfully modeled using some of the world's most powerful computers. In our solar system, the Sun and most other planets (and even some planets' moons) also exhibit magnetic fields. Interestingly, recent research based on ancient rocks in South Africa reveal that Earth's magnetic field must have been present by 3.45 billion years ago.

Earth's Magnetic Field and Paleomagnetism

Earth's magnetic field, which is shown in **Figure 2.7**, plays a crucial role in guiding navigators and also protects Earth's life-forms from solar storms. The invisible lines of magnetic force that originate within Earth and travel out into space resemble the magnetic field produced by a large bar magnet.[3] Similar to Earth's magnetic field, the ends of a bar magnet have opposite polarities (labeled either + and − or N for north and S for south) that cause magnetic objects to align parallel to its magnetic field. In addition, notice in Figures 2.7b and 2.7c that Earth's geographic North Pole (the rotational axis) and Earth's magnetic north pole (magnetic north) do not coincide.

Interdisciplinary Relationship

ROCKS AFFECTED BY EARTH'S MAGNETIC FIELD **Igneous rocks** (*igne* = fire, *ous* = full of) solidify from molten **magma** (*magma* = a mass) either underground or after volcanic eruptions at the surface that produce **lava** (*lavare* = to wash). Nearly all igneous rocks contain **magnetite**, a naturally magnetic iron mineral. Particles of magnetite in magma align themselves with Earth's magnetic field because magma and lava are fluid. Once molten material is cooled to a certain temperature,[4] however,

Web Animation
Flipping of Earth's Magnetic Field
http://goo.gl/2SpTZ1

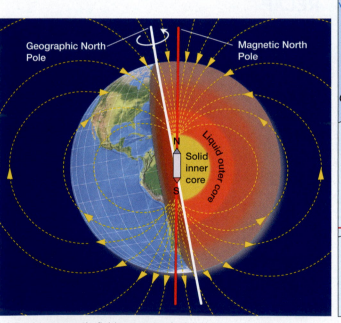

(a) Earth's magnetic field generates invisible lines of magnetic force similar to a large bar magnet. Note that the Geographic North Pole and the Magnetic North Pole are not in exactly the same location.

Figure 2.7 **Earth's magnetic field.**

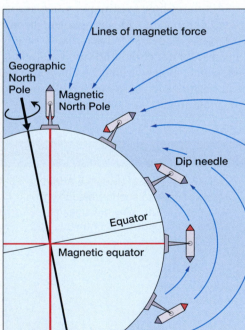

(b) Earth's magnetic field causes a dip needle to align parallel to the lines of magnetic force and change orientation with increasing latitude. Consequently, an approximation of latitude can be determined based on the dip angle.

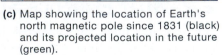

(c) Map showing the location of Earth's north magnetic pole since 1831 (black) and its projected location in the future (green).

[3]The properties of a magnetic field can be explored easily enough with a bar magnet and some iron particles. Place the iron particles on a table and place a bar magnet nearby. Depending on the strength of the magnet, you should get a pattern resembling that in Figure 2.7a.

[4]This temperature is called the *Curie point* and is named after French physicist Pierre Curie. For typical rocky Earth materials, it is about 550°C or 1022°F.

internal magnetite particles are frozen into position, thereby recording the angle of Earth's magnetic field at that place and time. In essence, grains of magnetite serve as tiny compass needles that record the strength and orientation of Earth's magnetic field. Unless the rock is heated to the temperature where magnetite grains are again mobile, these magnetite grains contain information about the magnetic field where the rock originated, regardless of where the rock subsequently moves.

Magnetite is also deposited in sediments. As long as the sediment is surrounded by water, the magnetite particles can align themselves with Earth's magnetic field. After sediment is buried and solidifies into **sedimentary rock** (*sedimentum* = settling), the particles are no longer able to realign themselves if they are subsequently moved. Thus, magnetite grains in sedimentary rocks also contain information about the magnetic field where the rock originated. Although other rock types have been used successfully to reveal information about Earth's ancient magnetic field, the most reliable ones are igneous rocks that have high concentrations of magnetite such as **basalt**, which is the rock type that comprises oceanic crust.

PALEOMAGNETISM The study of Earth's ancient magnetic field is called **paleomagnetism** (*paleo* = ancient). Scientists who study paleomagnetism analyze magnetite particles in rocks to determine not only their north–south direction but also their angle relative to Earth's surface. The degree to which a magnetite particle points into Earth is called its **magnetic dip**, or *magnetic inclination*.

Magnetic dip is directly related to latitude. Figure 2.7b shows that a dip needle does not dip at all at Earth's magnetic equator. Instead, the needle lies horizontal to Earth's surface. At Earth's magnetic north pole, however, a dip needle points straight into the ground. A dip needle at Earth's magnetic south pole is also vertical to the surface, but it points out instead of in. Thus, magnetic dip increases with increasing latitude, from 0 degrees at the magnetic equator to 90 degrees at the magnetic poles. Because magnetic dip is retained in magnetically oriented rocks, measuring the dip angle reveals the latitude at which the rock initially formed. Done with care, paleomagnetism is an extremely powerful tool for interpreting where rocks first formed. Based on paleomagnetic studies, convincing arguments could finally be made that the continents had drifted relative to one another **(Diving Deeper 2.1)**.

MAGNETIC POLARITY REVERSALS Magnetic compasses on Earth today follow lines of magnetic force and point toward magnetic north. It turns out, however, that the **polarity** (the north-south orientation of the magnetic field) has reversed itself periodically throughout geologic time. In essence, the north and south magnetic poles *reverse* or *switch* so that magnetic north becomes magnetic south and vice versa. **Figure 2.8** shows how ancient rocks have recorded the switching of Earth's magnetic polarity through time.

Why does Earth's magnetic field switch polarity? Geophysicists who study Earth's magnetic field do not yet fully understand the process of magnetic polarity reversals, but they are in agreement that Earth's rotation causes the electrically conducting liquid iron outer core to generate a self-sustaining magnetic field. Every so often, the flow of liquid iron is disturbed locally and twists part of the magnetic field in the opposite direction, weakening it. What triggers these disturbances is unknown; it may be because of turbulent flow conditions, or it may be just an inevitable consequence of a

2.1 Squidtoons

How does the green sea turtle navigate home?

https://goo.gl/utFcXO

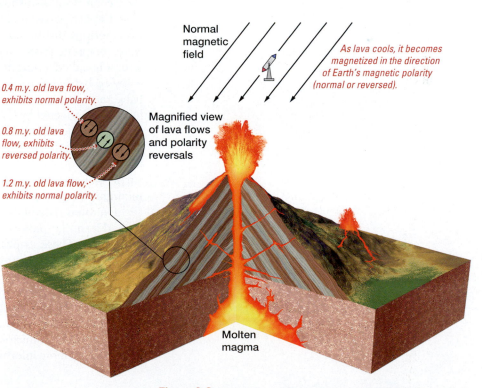

Figure 2.8 Paleomagnetism preserved in rocks. The switching of Earth's magnetic polarity through time is preserved in a sequence of rocks such as these lava flows, which are produced successively from the volcano. Note that m.y. = million years.

Normal magnetic field

As lava cools, it becomes magnetized in the direction of Earth's magnetic polarity (normal or reversed).

Magnified view of lava flows and polarity reversals

0.4 m.y. old lava flow, exhibits normal polarity.

0.8 m.y. old lava flow, exhibits reversed polarity.

1.2 m.y. old lava flow, exhibits normal polarity.

Molten magma

STUDENTS SOMETIMES ASK . . .

What changes to Earth's environment would occur when the magnetic poles reverse?

During a reversal, compasses would likely show incorrect directions, and people could have difficulty navigating. The same goes for some fish, birds, and mammals that sense the magnetic field during migrations (see MasteringOceanography **Web Diving Deeper 2.1**). The decrease in strength of the magnetic field also reduces the protection that the field provides for life-forms against cosmic rays and particles coming from the Sun, and this could disrupt low-Earth-orbiting satellites as well as some communication and power grid systems. Also, Earth's *aurora borealis* (the Northern Lights) and its counterpart *aurora australis* (the Southern Lights), which are natural light displays in the sky, might be visible at much lower latitudes. On the bright side, we know that life on Earth has successfully survived previous magnetic reversals, so reversals might not be as dangerous as they are sometimes portrayed (such as in the 2003 science fiction film *The Core*, which is full of scientific inaccuracies).

naturally chaotic system. Interestingly, computer simulations of Earth's core reveal frequent flipping of Earth's magnetic field.

Paleomagnetic studies reveal that 184 major reversals have occurred in the past 83 million years. The pattern of switching of Earth's magnetic field is highly irregular and ranges from 25,000 years to more than 30 million years. Even though the pattern has been described as random, on average a reversal occurs about every 450,000 years or so. The flipping of Earth's magnetic field takes an average of about 5000 years; it can happen as quickly as 1000 years or as slowly as 20,000 years. Changes in Earth's magnetic polarity are identified in rock sequences by a gradual decrease in the intensity of the magnetic field of one polarity, followed by a gradual increase in the intensity of the magnetic field of opposite polarity. Interestingly, there have been several documented instances of false starts where the weakening of the magnetic field does not lead to a full flip.

Earth's magnetic north pole—which does not coincide with the geographic North Pole—was first located near Boothia Peninsula in the Canadian Arctic in 1831; since that time, it's been migrating northwest by about 50 kilometers (30 miles) each year (Figure 2.7c). If this rate continues, Earth's magnetic pole will pass within 400 kilometers (250 miles) of the geographic North Pole in 2018 and will be in Siberia by 2050. In addition, geologic evidence indicates that Earth's magnetic field has also been weakening during the past 2000 years. New satellite analysis reveals that Earth's magnetic field is losing strength at a rate of about 5% per decade, which is more rapidly than previously thought. Geophysicists think that the diminishing strength of Earth's magnetic field may be an indication that Earth's current "normal" polarity may reverse itself. In fact, the last major reversal of Earth's magnetic poles occurred 780,000 years ago, which suggests that the next one is overdue.

PALEOMAGNETISM AND THE OCEAN FLOOR Paleomagnetism had certainly proved its usefulness on land, but, up until the mid-1950s, paleomagnetic studies had only been conducted on continental rocks. Would the ocean floor also show variations in magnetic polarity? To test this idea, the U.S. Coast and Geodetic Survey, in conjunction with scientists from Scripps Institution of Oceanography, undertook an extensive deepwater mapping program off Oregon and Washington in 1955. Using a sensitive instrument called a **magnetometer** (*magneto* = magnetism, *meter* = measure), which is towed behind a research vessel, the scientists spent several weeks at sea, moving back and forth in a regularly spaced pattern, measuring Earth's magnetic field and how it was affected by the magnetic properties of rocks on the ocean floor.

Interdisciplinary

Relationship

When the scientists analyzed their data, they found that the entire surveyed area had a pattern of north–south stripes in a surprisingly regular and alternating pattern of above-average and below-average magnetism. What was even more surprising was that the pattern appeared to be symmetrical with respect to a long mountain range that was fortuitously in the middle of their survey area.

Detailed paleomagnetic studies of this and other areas of the sea floor confirmed a similar pattern of alternating stripes of above-average and below-average magnetism. These stripes are called **magnetic anomalies** (*a* = without, *nomo* = law; an anomaly is a departure from normal conditions). The ocean floor had embedded in it a regular pattern of alternating magnetic stripes unlike anywhere on land.

Researchers had a difficult time explaining why the ocean floor had such a regular pattern of magnetic anomalies. Nor could they explain how the sequence on one side of the underwater mountain range matched the sequence on the opposite side—in essence, they were a mirror image of each other. To understand how this pattern could have formed, more information was needed about ocean floor features and their origin.

Sea Floor Spreading and Features of the Ocean Basins

Geologist **Harry Hess** (1906–1969), when he was a U.S. Navy captain in World War II, developed the habit of leaving his depth recorder on at all times while his ship was traveling at sea. After the war, compilation of these and many other depth

USING MOVING CONTINENTS TO RESOLVE AN APPARENT DILEMMA: DID EARTH EVER HAVE TWO WANDERING NORTH MAGNETIC POLES?

Interdisciplinary Relationship

The theory of plate tectonics has proved to be helpful in resolving some apparent dilemmas about Earth history. A classic example of this occurred when magnetic dip data for rocks on various continents were used to determine the ancient position of the magnetic north pole on Earth. Scientists who analyzed the data concluded that Earth's north magnetic pole must be wandering, or moving, through time. Further, the data suggested that rocks on different continents pointed to two different locations for Earth's north magnetic pole.

Figure 2A (*part a, left*) shows the magnetic **polar wandering paths**—sometimes called *polar wandering curves*—for North America and Eurasia. Notice how both paths have a similar shape but, for all rocks older than about 70 million years, the pole determined from North American rocks lies to the west of that determined from Eurasian rocks. From this data, it appeared that Earth had two separate magnetic poles in the geologic past, which would be remarkably different than today, where Earth has a single north magnetic pole. In fact, geophysical data indicates that only one north magnetic pole can exist at any given time and that it is unlikely that its position has changed very much through time because it must remain very closely aligned with Earth's rotational axis. Earth scientists were initially puzzled by these findings until they realized that the discrepancy could be resolved by having a single magnetic pole that remains relatively stationary while

North America and Eurasia moved relative to the pole and relative to each other. So it wasn't Earth's magnetic field that was moving; instead, it was the continents themselves that were moving. That's why the magnetic polarity paths are called *apparent* polar wandering paths.

Figure 2A (*part b, right*) shows that when the continents are moved into the positions they occupied when they were part of Pangaea, the two wandering paths match up, providing strong evidence that there were never two magnetic north poles on Earth. A more reasonable conclusion in light of plate tectonic is that the *continents* had moved relative to each other throughout geologic time.

GIVE IT SOME THOUGHT

1. What puzzled scientists about Earth's ancient magnetic field? How was the apparent dilemma resolved?

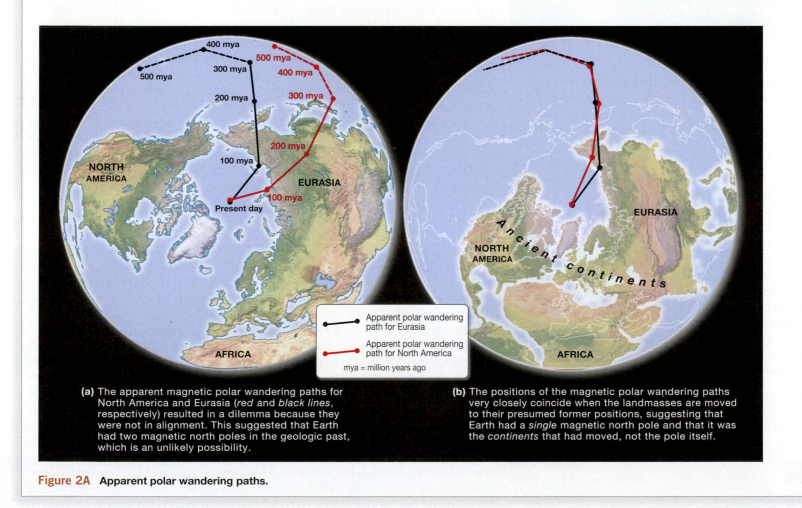

(a) The apparent magnetic polar wandering paths for North America and Eurasia (*red* and *black lines*, respectively) resulted in a dilemma because they were not in alignment. This suggested that Earth had two magnetic north poles in the geologic past, which is an unlikely possibility.

(b) The positions of the magnetic polar wandering paths very closely coincide when the landmasses are moved to their presumed former positions, suggesting that Earth had a *single* magnetic north pole and that it was the *continents* that had moved, not the pole itself.

Figure 2A Apparent polar wandering paths.

Figure 2.9 Processes and resulting features of plate tectonics.

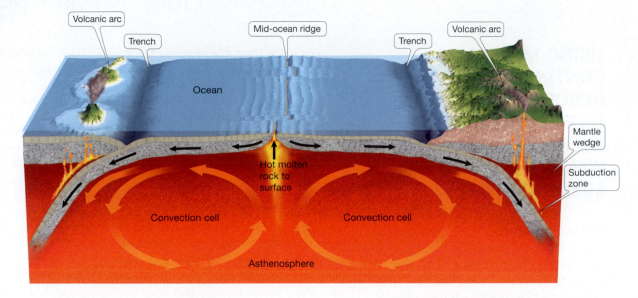

Web Animation
Sea Floor Spreading and Plate Boundaries
http://goo.gl/9iEcQD

STUDENTS SOMETIMES ASK . . .

Figure 2.9 shows that the mantle is moving in large circles. Is the mantle molten?

No. Because the mantle is often depicted as flowing in convective motion, a common misconception is that the mantle is molten. Seismic studies reveal that the mantle is unambiguously greater than 99% solid, although it does have the ability to flow S-L-O-W-L-Y over time (hence the arrows in the figure). The only places where the mantle is partially molten are (1) underneath the mid-ocean ridge, where release of pressure causes molten material to form; (2) in the mantle wedge above a subducting plate, where water released from the downgoing oceanic plate causes melting; and (3) in isolated mantle plumes, which are discussed later in this chapter. Make no mistake about it: The vast majority of the mantle is composed of hot, solid rock. But even that rock can flow if enough pressure is applied to it. Think of how a blacksmith can deform and shape a red-hot piece of solid iron by using the pressure of repeated hammerings. Imagine how much greater the pressure is inside Earth to cause hot, solid rock to deform and flow!

Web Animation
Convection in a Lava Lamp
p://goo.gl/dacqQL

records showed extensive mountain ridges near the centers of ocean basins and extremely deep, narrow trenches at the edges of ocean basins. In 1962, Hess published *History of Ocean Basins*, which contained the idea of **sea floor spreading** and the associated circular movement of rock material in the mantle—**convection cells** (*con* = with, *vect* = carried)—as the driving mechanism (**Figure 2.9**). He suggested that new ocean crust was created at the ridges, split apart, moved away from the ridges, and later disappeared back into the deep Earth at trenches. Mindful of the resistance of North American scientists to the idea of continental drift, Hess referred to his own work as "geopoetry."

As it turns out, Hess's initial ideas about sea floor spreading have been confirmed. The **mid-ocean ridge** (Figure 2.9) is a continuous underwater mountain range that winds through every ocean basin in the world and resembles the seam on a baseball. It is entirely volcanic in origin, wraps one-and-a-half times around the globe, and rises more than 2.5 kilometers (1.5 miles) above the surrounding deep-ocean floor. It even rises above sea level in places such as Iceland. New ocean floor forms at the crest, or axis, of the mid-ocean ridge. By the process of sea floor spreading, new ocean floor is split in two and carried away from the axis, replaced by the upwelling of volcanic material that fills the void with new strips of sea floor. Sea floor spreading occurs along the axis of the mid-ocean ridge, which is referred to as a **spreading center**. One way to think of the mid-ocean ridge is as a zipper that is being pulled apart. Thus, Earth's zipper (the mid-ocean ridge) is becoming unzipped!

At the same time, ocean floor is being destroyed at deep **ocean trenches**. Trenches are the deepest parts of the ocean floor and, on a map of the sea floor, resemble a narrow crease or trough (Figure 2.9). The largest earthquakes in the world occur near these trenches; they are caused by a plate bending downward and slowly plunging back into Earth's interior. This process is called **subduction** (*sub* = under, *duct* = lead), and the sloping area from the trench along the downward-moving plate is called a **subduction zone**.

In 1963, geologists **Frederick Vine** and **Drummond Matthews** of Cambridge University combined the seemingly unrelated pattern of magnetic sea floor stripes with the process of sea floor spreading to explain the perplexing pattern of alternating and symmetric magnetic stripes on the sea floor (**Figure 2.10**). Vine and Matthews interpreted the pattern of above-average and below-average magnetic polarity episodes embedded in sea floor rocks to be caused by Earth's magnetic field alternating between "normal" polarity (similar to today's magnetic pole position in the north) and "reversed" polarity (with the magnetic pole to the south). They proposed that the pattern could be created when newly formed rocks at the mid-ocean ridge are magnetized with whichever polarity exists on Earth during their formation. As those rocks are slowly moved away from the crest of the mid-ocean ridge, they maintain their original polarity, and subsequent rocks record the

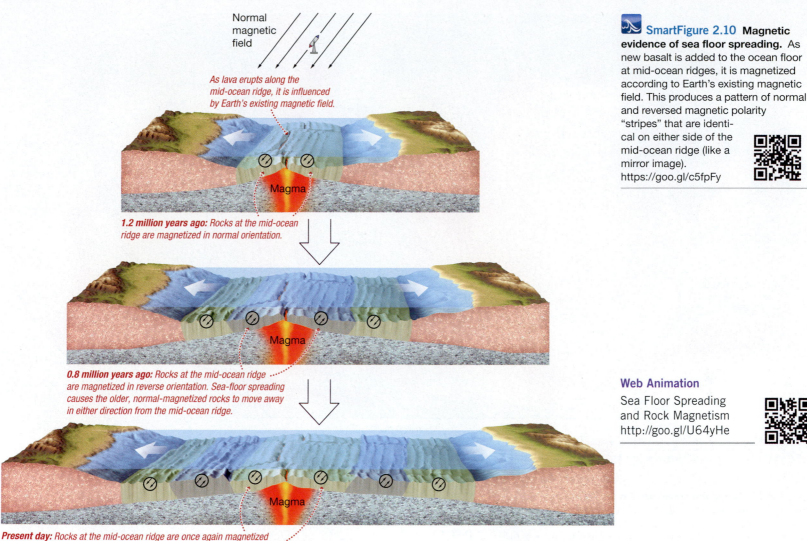

Normal magnetic field

As lava erupts along the mid-ocean ridge, it is influenced by Earth's existing magnetic field.

Magma

1.2 million years ago: *Rocks at the mid-ocean ridge are magnetized in normal orientation.*

Magma

0.8 million years ago: *Rocks at the mid-ocean ridge are magnetized in reverse orientation. Sea-floor spreading causes the older, normal-magnetized rocks to move away in either direction from the mid-ocean ridge.*

Magma

Present day: *Rocks at the mid-ocean ridge are once again magnetized in normal orientation, continuing the symmetric pattern of normal and reversed magnetic polarity "stripes" on either side of the mid-ocean ridge.*

SmartFigure 2.10 Magnetic evidence of sea floor spreading. As new basalt is added to the ocean floor at mid-ocean ridges, it is magnetized according to Earth's existing magnetic field. This produces a pattern of normal and reversed magnetic polarity "stripes" that are identical on either side of the mid-ocean ridge (like a mirror image).
https://goo.gl/c5fpFy

Web Animation
Sea Floor Spreading and Rock Magnetism
http://goo.gl/U64yHe

periodic switches of Earth's magnetic polarity. The result is an alternating pattern of magnetic polarity stripes that are symmetric with respect to the mid-ocean ridge.

The pattern of alternating reversals of Earth's magnetic field as recorded in the sea floor was the most convincing piece of evidence set forth to support the concept of sea floor spreading—and, as a result, continental drift. However, the continents weren't plowing through the ocean basins as Wegener had envisioned. Instead, the ocean floor was a conveyer belt that was being continuously formed at the mid-ocean ridge and destroyed at the trenches, with the continents just passively riding along on the conveyer. By the late 1960s, most geologists had changed their stand on continental drift in light of this new evidence, which is a prime example of how the scientific method works.

Other Evidence from the Ocean Basins

Even though the tide of scientific opinion had indeed switched to favor a mobile Earth, additional evidence from the ocean floor would further support the ideas of continental drift and sea floor spreading.

AGE OF THE OCEAN FLOOR In the late 1960s, an ambitious deep-sea drilling program was initiated to test the existence of sea floor spreading. One of the program's primary missions was to drill into and collect ocean floor rocks for radiometric age dating. If sea floor spreading does indeed occur, then the youngest sea floor rocks would be atop the mid-ocean

RECAP

The plate tectonic model states that new sea floor is created at the mid-ocean ridge, where it moves outward by the process of sea floor spreading and is destroyed by subduction into ocean trenches.

Interdisciplinary

Relationship

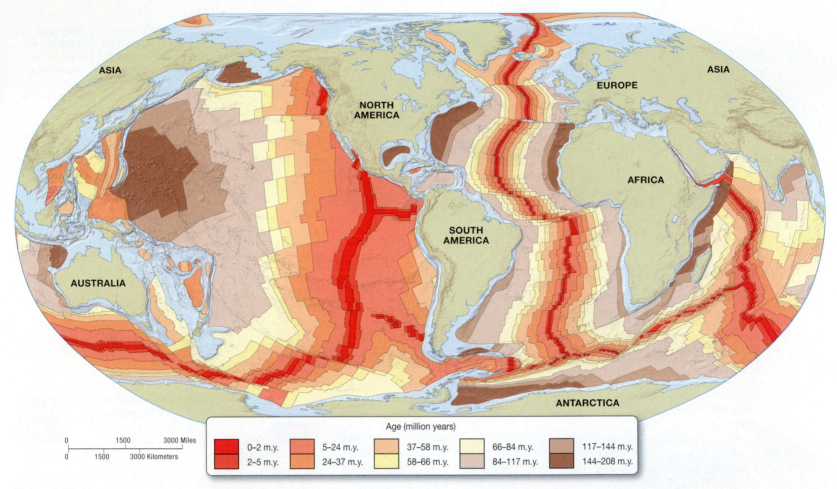

Age (million years)

0–2 m.y.	5–24 m.y.	37–58 m.y.	66–84 m.y.	117–144 m.y.
2–5 m.y.	24–37 m.y.	58–66 m.y.	84–117 m.y.	144–208 m.y.

Figure 2.11 **Age of the ocean crust beneath deep-sea deposits.** The youngest rocks (*bright red areas*) are found along the mid-ocean ridge. Farther away from the mid-ocean ridge, the rocks increase linearly in age in either direction. Ages shown are in millions of years before present.

ridge, and the ages of rocks would increase on either side of the ridge in a symmetric pattern.

The map in **Figure 2.11**, showing the age of the ocean floor beneath deep-sea deposits, is based on the pattern of magnetic stripes verified with thousands of radiometrically age-dated samples. It shows that the ocean floor is youngest along the mid-ocean ridge, where new ocean floor is created, and the age of rocks increases with increasing distance in either direction away from the axis of the ridge. The symmetric pattern of ocean floor ages confirms that the process of sea floor spreading must indeed be occurring.

The Atlantic Ocean has the simplest and most symmetric pattern of age distribution in Figure 2.11. The pattern results from the newly formed Mid-Atlantic Ridge that rifted Pangaea apart. The Pacific Ocean has the least symmetric pattern because many subduction zones surround it. For example, ocean floor east of the East Pacific Rise that is older than 40 million years has already been subducted. The ocean floor in the northwestern Pacific, about 180 million years old, has not yet been subducted. A portion of the East Pacific Rise has even disappeared under North America. The age bands in the Pacific Ocean are wider than those in the Atlantic and Indian Oceans, which suggests that the rate of sea floor spreading is greatest in the Pacific Ocean.

Recall from Chapter 1 that the ocean is at least 4 billion years old. However, the oldest ocean floor is only 180 million years old (or 0.18 billion years old), and the majority of the ocean floor is not even half that old (see Figure 2.11). How could the ocean floor be so incredibly young, while the oceans themselves are so phenomenally old? According to plate tectonic theory, new ocean floor is created at the mid-ocean ridge by sea floor spreading and moves off the ridge to eventually be subducted and remelted in the mantle. In this way, the ocean floor keeps regenerating itself. The floor beneath the oceans today is not the same one that existed beneath the oceans 4 billion years ago.

If the rocks that comprise the ocean floor are so young, why are continental rocks so old? Using radiometric age dating, scientists have determined that the oldest rocks on land are about 4 billion years old. Many other continental rocks approach this age, implying that the same processes that constantly renew the sea floor do not operate on land. Rather, evidence suggests that continental rocks, because of their low density, do not get recycled by the process of sea floor spreading, and thus they remain at Earth's surface for long periods of time.

HEAT FLOW The heat from Earth's interior is released to the surface as **heat flow**. Current models indicate that this heat moves to the surface with magma in convective motion. Most of the heat is carried to regions of the mid-ocean ridge spreading centers (see Figure 2.9). Cooler portions of the mantle descend along subduction zones to complete each circular-moving convection cell. Interdisciplinary / Relationship

Heat flow measurements show that the amount of heat flowing to the surface along the mid-ocean ridge can be up to eight times greater than the average amount flowing to other parts of Earth's crust. Additionally, heat flow at deep-sea trenches, where ocean floor is subducted, can be as little as one-tenth the average. Increased heat flow at the mid-ocean ridge and decreased heat flow at subduction zones is what would be expected based on thin crust at the mid-ocean ridge and a double thickness of crust at the trenches (see Figure 2.9).

WORLDWIDE EARTHQUAKES *Earthquakes* are sudden releases of energy caused by fault movement or volcanic eruptions. The map in **Figure 2.12a** shows that most large earthquakes occur along ocean trenches, reflecting the energy released during subduction. Other earthquakes occur along the mid-ocean ridge, reflecting the energy released during sea floor spreading. Still others occur along major faults in the sea floor and on land, reflecting the energy released when moving plates contact other plates along their edges. When you examine the two maps in Figure 2.12, notice how closely the pattern of major earthquakes matches the locations of plate boundaries. This is because most earthquakes worldwide are created by plates interacting with each other at their margins.

STUDENTS SOMETIMES ASK . . .

How fast do plates move, and have they always moved at the same rate?

Currently, plates move an average of 2 to 12 centimeters (1 to 5 inches) per year, which is about as fast as a person's fingernails grow. A person's fingernail growth is dependent on many factors, including heredity, gender, diet, and amount of exercise, but averages about 8 centimeters (3 inches) per year. This may not sound very fast, but the plates have been moving for millions of years. Over a very long time, even an object moving slowly will eventually travel a great distance. For instance, fingernails growing at a rate of 8 centimeters (3 inches) per year for 1 million years would be 80 kilometers (50 miles) long!

Evidence shows that the plates were moving faster millions of years ago than they are moving today. Geologists can determine the rate of plate motion in the past by analyzing the width of new oceanic crust produced by sea floor spreading, since fast spreading produces more sea floor rock. (By using this relationship and examining Figure 2.11, you should be able to determine whether the Pacific Ocean or the Atlantic Ocean had a faster spreading rate.) Recent studies using this same technique indicate that about 50 million years ago, India attained a speed of 19 centimeters (7.5 inches) per year. Other research indicates that about 530 million years ago, plate motions may have been as high as 30 centimeters (1 foot) per year! What caused these rapid bursts of plate motion? Geologists are not sure why plates moved more rapidly in the past, but increased heat release from Earth's interior is a likely mechanism.

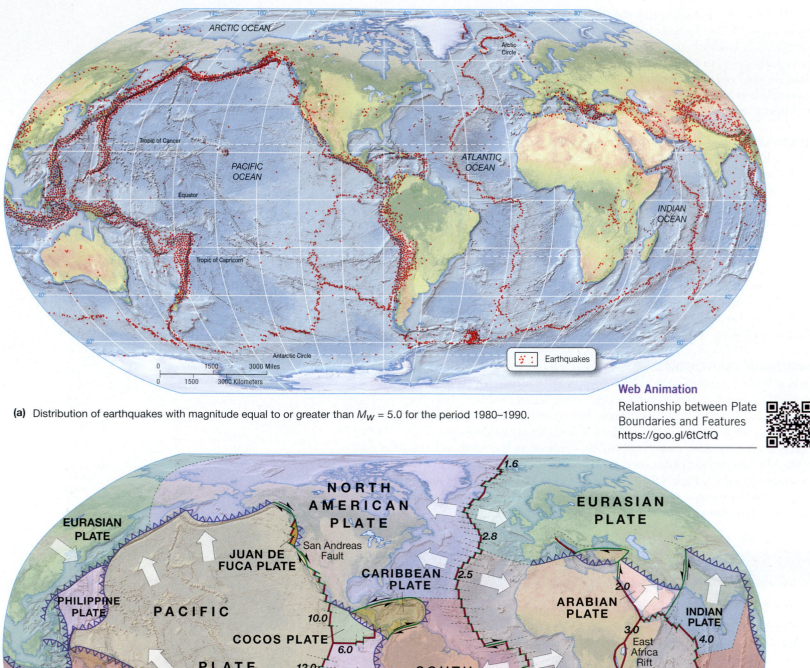

(a) Distribution of earthquakes with magnitude equal to or greater than M_W = 5.0 for the period 1980–1990.

Web Animation

Relationship between Plate
Boundaries and Features
https://goo.gl/6tCtfQ

(b) Plate boundaries define the major tectonic plates (shaded), with arrows indicating the direction of motion
and numbers representing the rate of motion in centimeters per year.

◁◁◁ Convergent boundaries		⬅ Direction of plate movement
— Divergent boundaries		*0.5* Spreading rate (cm/yr)
— Transform fault boundaries		---- Diffuse plate boundary

🌊 **SmartFigure 2.12** **Earthquakes and tectonic plate boundaries.** World maps showing **(a)** earthquakes and **(b)** tectonic plates.
Comparison of the two maps shows that most earthquakes occur along plate boundaries.
https://goo.gl/aYGmK9

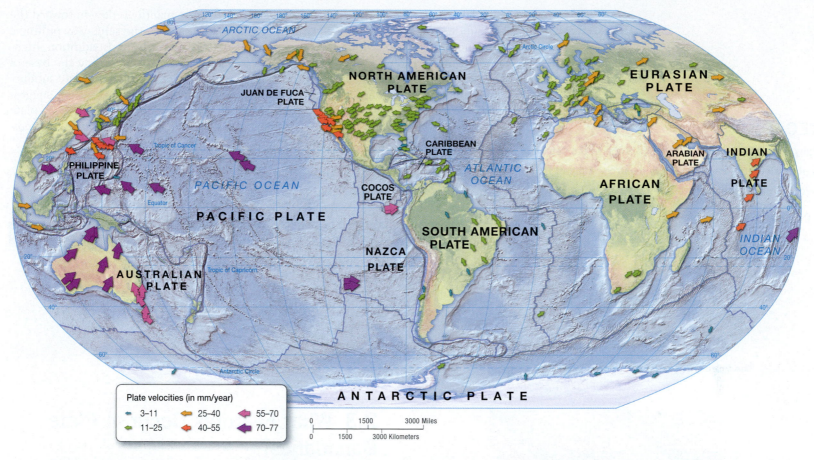

Plate velocities (in mm/year)

3–11	25–40	55–70
11–25	40–55	70–77

Figure 2.13 **Satellite positioning of locations on Earth.** Arrows show the direction of motion based on repeated satellite measurement of positions on Earth. The rate of plate motion in millimeters per year is indicated with different-colored arrows (*see legend*). Plate boundaries are shown with blue lines and are dashed where uncertain or diffuse.

Detecting Plate Motion with Satellites

Since the late 1970s, orbiting satellites have allowed the accurate positioning of locations on Earth. (This technique is also used for navigation by ships at sea; see Diving Deeper 1.2.) If the plates are moving, satellite positioning should show this movement over time. The map in **Figure 2.13** shows numerous locations that have been measured in this manner and confirms that regions on Earth are indeed moving in good agreement with the direction and rate of motion predicted by plate tectonics. The successful prediction of how locations on Earth are moving with respect to one another very strongly supports plate tectonic theory.

The Acceptance of a Theory

The accumulation of lines of evidence such as those mentioned in this section, along with many other lines of evidence in support of moving continents, has convinced scientists of the validity of continental drift. Since the late 1960s, the concepts of continental drift and sea floor spreading have been united into a much more encompassing theory known as plate tectonics, which describes the movement of the outermost portion of Earth and the resulting creation of continental and sea floor features. These tectonic plates are pieces of the **lithosphere** (*lithos* = rock, *sphere* = ball) that float on the more fluid **asthenosphere** (*asthenos* = weak, *sphere* = ball) below.[5]

What forces drive plate motion? Although several mechanisms have been proposed for the force (or forces) responsible for driving this motion, none of them are able to explain all aspects of plate movement. However, scientific studies based on a simple model of lithosphere and mantle interactions suggest that two major tectonic forces may act in unison on subducting plates (slabs): (1) *slab pull,* which is generated by the pull of the weight of a plate as it sinks underneath an overlying plate, pulling the rest of the plate behind it in a similar fashion to how a heavy comforter often slides off a bed onto the floor, and (2) *slab suction,* which is created as a subducting

[5]See Chapter 1 for a discussion of properties of the lithosphere and asthenosphere.

plate drags against the viscous mantle and causes the mantle to flow in toward the subduction zone, thereby sucking in nearby plates much in the same way pulling a plug from a full bathtub draws floating objects toward its drain. In addition, high-resolution seismic studies have located a weak, partially molten layer at the base of the lithosphere that aids sliding and may reduce the force required for plate subduction. Other modeling studies that include upper mantle viscosity variations suggest that mantle flow differences contribute to either reinforcing or counteracting plate motions. Although researchers continue to model the forces that drive plate motions, these studies are hampered by the inaccessibility and complexity of Earth's mantle.

Since the acceptance of the theory of plate tectonics, much research has focused on understanding various features associated with plate boundaries, both on the sea floor and on land.

RECAP

Many independent lines of evidence, such as the detection of plate motion by satellites, provide strong support for the theory of plate tectonics.

Web Animation
Motion at Plate Boundaries
http://goo.gl/LNnG80

CONCEPT CHECK 2.2 I Summarize the evidence that supports plate tectonics.

1 Describe what Earth's magnetic field looks like and how it has changed through time.

2 Describe sea floor spreading and why it was an important piece of evidence in support of plate tectonics.

3 Why does a map of worldwide earthquakes closely match the locations of worldwide plate boundaries?

Figure 2.14 **The three types of lithospheric plate boundaries.**

The three main types of plate boundaries are...

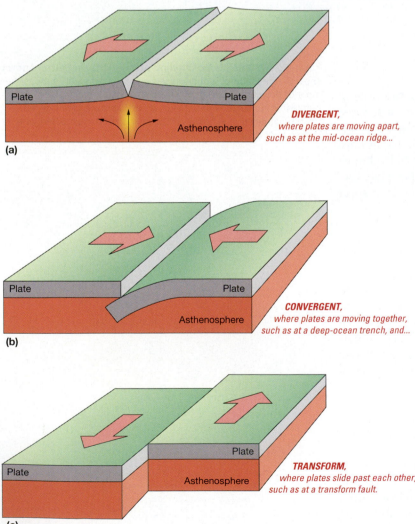

(a) *DIVERGENT, where plates are moving apart, such as at the mid-ocean ridge...*

(b) *CONVERGENT, where plates are moving together, such as at a deep-ocean trench, and...*

(c) *TRANSFORM, where plates slide past each other, such as at a transform fault.*

2.3 What Features Occur at Plate Boundaries?

Plate boundaries—where plates interact with each other—are associated with a great deal of tectonic activity, such as mountain building, volcanic activity, and earthquakes. In fact, the first clues to the locations of plate boundaries were the dramatic tectonic events that occur there. For example, Figure 2.12 shows the close correspondence between worldwide earthquakes and plate boundaries. Further, **Figure 2.12b** shows that Earth's surface is composed of seven major plates along with many smaller ones. Close examination of Figure 2.12b shows that the boundaries of plates do not always follow coastlines and, as a consequence, nearly all plates contain both oceanic and continental crust.[6] Notice also that about 90% of plate boundaries occur on the sea floor.

There are three types of plate boundaries, as shown in **Figure 2.14**. **Divergent boundaries** (*di* = apart, *vergere* = to incline) are found along oceanic ridges where new lithosphere is being added. **Convergent boundaries** (*con* = together, *vergere* = to incline) are found where plates are moving together and one plate subducts beneath the other. **Transform boundaries** (*trans* = across, *form* = shape) are found where lithospheric plates slowly grind past one another. **Table 2.1** summarizes characteristics, tectonic processes, features, and examples of these plate boundaries.

[6]For a review of the differences between (basaltic) oceanic and (granitic) continental crust, see Chapter 1.

SMARTTABLE 2.1 CHARACTERISTICS, TECTONIC PROCESSES, FEATURES, AND EXAMPLES OF PLATE BOUNDARIES

Plate boundary	Plate movement	Crust types	Sea floor created or destroyed?	Tectonic process	Sea floor feature(s)	Geographic examples
Divergent plate boundaries	Apart	Oceanic–oceanic	New sea floor is created	Sea floor spreading	Mid-ocean ridge; volcanoes; young lava flows	Mid-Atlantic Ridge, East Pacific Rise
		Continental–continental	As a continent splits apart, new sea floor is created	Continental rifting	Rift valley; volcanoes; young lava flows	East Africa Rift Valleys, Red Sea, Gulf of California
Convergent plate boundaries	Together	Oceanic–continental	Old sea floor is destroyed	Subduction	Trench; volcanic arc on land	Peru–Chile Trench, Andes Mountains
		Oceanic–oceanic	Old sea floor is destroyed	Subduction	Trench; volcanic arc as islands	Mariana Trench, Aleutian Islands
		Continental–continental	N/A	Collision	Tall mountains	Himalaya Mountains, Alps
Transform plate boundaries	Past each other	Oceanic	N/A	Transform faulting	Fault	Mendocino Fault, Eltanin Fault (between mid-ocean ridges)
		Continental	N/A	Transform faulting	Fault	San Andreas Fault, Alpine Fault (New Zealand)

SmartTable 2.1 Characteristics, tectonic process, features, and examples of plate boundaries
https://goo.gl/Lj7TyM

Divergent Boundary Features

Divergent plate boundaries occur where two plates move apart, such as along the crest of the mid-ocean ridge, where sea floor spreading creates new oceanic lithosphere (Figure 2.15). A common feature along the crest of the mid-ocean ridge is a **rift valley**, which is a central downdropped linear depression (Figure 2.16). Pull-apart faults located along the central rift valley show that the plates are *continuously being pulled apart* rather than being pushed apart by the upwelling of material beneath the mid-ocean ridge. Upwelling of magma beneath the mid-ocean ridge is simply filling in the void left by the separating plates of lithosphere. In the process, sea floor spreading produces about 20 cubic kilometers (4.8 cubic miles) of new ocean crust worldwide each year.

Figure 2.17 shows how the development of a mid-ocean ridge creates an ocean basin. Initially, molten material rises to the surface, causing upwarping and thinning of the crust. Volcanic activity produces vast quantities of high-density basaltic rock. As the plates begin to move apart, a linear rift valley is formed, and volcanism continues. Further splitting apart of the land—a process called **rifting**—and more spreading cause the area to drop below sea level. When this occurs, the rift valley eventually floods with seawater, and a young linear sea is formed. After millions of years of sea floor spreading, a full-fledged ocean basin is created, with a mid-ocean ridge in the middle of the two landmasses.

Figure 2.24 Aerial view of the San Andreas Fault in California. The San Andreas Fault cuts through coastal southern and central California and produces many earthquakes. This aerial view of the Carrizo Plain in central California shows the San Andreas Fault as a long linear scar; arrows show relative fault motion.

RECAP

The three main types of plate boundaries are divergent (plates moving apart, such as at the mid-ocean ridge), convergent (plates moving together, such as at an ocean trench), and transform (plates sliding past each other, such as at a transform fault).

shallow but often strong earthquakes in the lithosphere. Magnitudes of $M_w = 7.0$ have been recorded along some oceanic transform faults. One of the best-studied faults in the world is California's **San Andreas Fault** (Figure 2.24), a continental transform fault that runs from the Gulf of California through coastal southern and central California past San Francisco and continues offshore parallel to the coast in northern California. Because the San Andreas Fault cuts through continental crust, which is much thicker than oceanic crust, earthquakes are considerably larger than those produced by oceanic transform faults, sometimes up to $M_w = 8.5$.

CONCEPT CHECK 2.3 | Discuss the origin and characteristics of features that occur at plate boundaries.

1 Most lithospheric plates contain both oceanic- and continental-type crust. Use plate boundaries to explain why this is true.

2 Describe the differences between oceanic ridges and oceanic rises. Include in your answer why these differences exist.

3 Using the profile view of the Mid-Atlantic Ridge in Figure 2.19a, calculate its total spreading rate over the past 50 million years (divide total distance by time). Then do a similar calculation for the East Pacific Rise (Figure 2.19b) and compare the two.

4 Convergent boundaries can be divided into three types, based on the type of crust contained on the two colliding plates. Compare and contrast the different types of convergent boundaries that result from these collisions.

5 Describe the differences in earthquake magnitudes that occur between the three types of plate boundaries and explain why these differences occur.

2.4 Testing the Model: How Can Plate Tectonics Be Used as a Working Model?

One of the strengths of plate tectonic theory is how it unifies so many seemingly separate processes and features into a single consistent model. Let's look at a few examples that illustrate how plate tectonic processes can be used to explain the origin of features that, up until the acceptance of plate tectonics, were difficult to explain.

Hotspots and Mantle Plumes

Although the theory of plate tectonics helped explain the origin of many features near plate boundaries, it did not seem to explain the origin of **intraplate features** (*intra* = within, *plate* = plate of the lithosphere) that are far from any plate boundary. For instance, how can plate tectonics explain volcanic islands near the middle of a plate? Areas of intense volcanic activity that remain in more or less the same location over long periods of geologic time and are unrelated to plate boundaries are called **hotspots**.[9] For example, the continuing volcanism in Yellowstone National Park and Hawaii is caused by hotspots.

Interdisciplinary

Relationship

Why is there so much volcanic activity at hotspots? The plate tectonic model infers that hotspot volcanism is caused by the presence of **mantle plumes** (*pluma* = a soft feather), which are vertical tube-shaped areas of hot molten rock that arise from deep within the mantle (**Figure 2.25**). Mantle plumes can be identified by researchers who measure how fast seismic waves from earthquakes travel below ground; the underlying principle is that seismic waves move more slowly through hot rock than cold. Seismic studies suggest that several types of mantle plumes exist: Some come from the core–mantle boundary, while others have a shallower source. Geophysical research reveals that the core–mantle boundary is not a simple, smooth dividing zone but has many regional variations, which has implications for the development of mantle plumes. In addition, new research suggests that the asthenosphere may actually play a more significant role than the core–mantle boundary in the development of hotspots. Because

This sharp bend [in the] Hawaiian-Empe[ror] chain was creat[ed] by a combinatio[n] of the changing motion of the Pacific Plate and the slow movement of the Hawaiian hotspot itself.

To hel[p] chain. Eve[n] cano Kilau[ea] chain. The[n] (Figure 2.[) (65 millio[n] Trench.

These[e] westward, [...] The result[...] plate mov[e...] forming, y[...] progressiv[e...] = thread, [...] suggests t[...] to a north[...] (large elb[ow] from the [...] tracks thr[...] time, but [...]

Recen[t] do not re[...] hotspots [...] may have[...]

SmartFigure 2.25 Origin and development of mantle plumes and hotspots. Schematic cross-sectional views of Earth showing the development of a mantle plume and hotspot according to the plume hypothesis.
https://goo.gl/SWPXyr

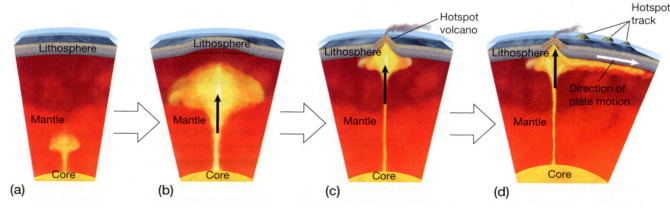

A plume of hot buoyant material detaches from the deep mantle or the core-mantle boundary.

The plume rises more rapidly in its conduit than the plume head can push through the viscous mantle, which inflates the head and elevates Earth's surface.

Decompression near the surface partially melts the plume head, which comes to the surface and creates a hotspot volcano.

The volcano is carried away by plate motion as the plume continues to feed subsequent volcanoes, creating a hotspot track (nematath).

(a) Lithosphere / Mantle / Core

(b) Lithosphere / Mantle / Core

(c) Hotspot volcano / Lithosphere / Mantle / Core

(d) Hotspot track / Lithosphere / Direction of plate motion / Mantle / Core

[9]Note that a hotspot is different from either a volcanic arc or a mid-ocean ridge (both of which are related to plate boundaries), even though all are marked by a high degree of volcanic activity.

Figure 2.29 Stages of development in coral reefs. Cross-sectional view (*above*) and map view/aerial photographs (*below*) of (**a**) a fringing reef, (**b**) a barrier reef, and (**c**) an atoll. With the right conditions for coral growth and enough time, a coral reef progresses from fringing reef to barrier reef to atoll.

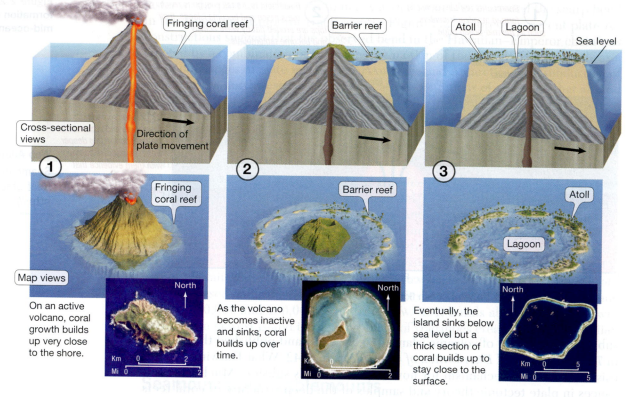

Cross-sectional views | Direction of plate movement

Fringing coral reef | Barrier reef | Atoll | Lagoon | Sea level

Map views

① Fringing coral reef — On an active volcano, coral growth builds up very close to the shore.

North — Km 0 — 2 | Mi 0 — 2

② Barrier reef — As the volcano becomes inactive and sinks, coral builds up over time.

North — Km 0 — 2 | Mi 0 — 2

③ Atoll — Lagoon — Eventually, the island sinks below sea level but a thick section of coral builds up to stay close to the surface.

North — Km 0 — 5 | Mi 0 — 5

The largest reef system in the world is Australia's **Great Barrier Reef**, a series of more than 3000 individual reefs collectively in the barrier reef stage of development, home to hundreds of coral species and thousands of other reef-dwelling organisms. The Great Barrier Reef lies 40 kilometers (25 miles) or more offshore, averages 150 kilometers (90 miles) in width, and extends for more than 2000 kilometers (1200 miles) along Australia's shallow northeastern coast. The effects of the Indian–Australian Plate moving north toward the equator from colder Antarctic waters are clearly visible in the age and structure of the Great Barrier Reef (**Figure 2.30**). It is oldest (around 25 million years old) and thickest at its northern end because the northern part of Australia reached water warm enough to grow coral before the southern parts did. In other areas of the Pacific, Indian, and Atlantic Oceans, smaller barrier reefs are found around the tall volcanic peaks that form tropical islands.

The **atoll** (*atar* = crowded together) stage (**Figure 2.29c**) comes after the barrier reef stage. As a barrier reef around a volcano continues to subside, coral builds up toward the surface. After millions of years, the volcano becomes completely submerged, but the coral reef continues to grow. If the rate of subsidence is slow enough for the coral to keep up, a circular reef called an atoll is formed. The atoll encloses a lagoon usually not more than 30 to 50 meters (100 to 165 feet) deep. The reef generally has many channels that allow circulation between the lagoon and the open ocean. Buildups of crushed-coral debris often form narrow islands that encircle the central lagoon and are large enough to allow human habitation.

Alternatively, a new theory has been put forward to explain the origin of coral atolls. The theory suggests that glacial cycles cause sea level to fluctuate,

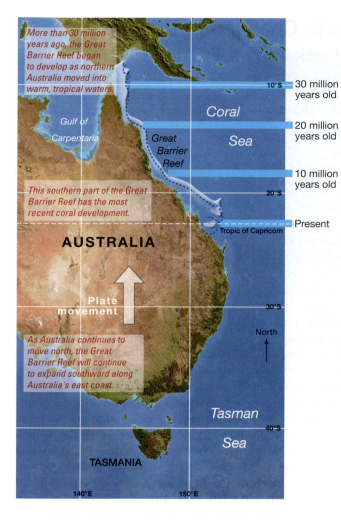

Figure 2.30 **Australia's Great Barrier Reef records plate movement.**

More than 30 million years ago, the Great Barrier Reef began to develop as northern Australia moved into warm, tropical waters.

This southern part of the Great Barrier Reef has the most recent coral development.

As Australia continues to move north, the Great Barrier Reef will continue to expand southward along Australia's east coast.

leading to episodes of reef exposure and dissolution when global sea level is lower during ice ages, alternating with coral reef submergence and deposition when sea level is higher during interglacial stages. Instead of the slow growth of ring-shaped coral above a sinking volcanic island, this alternating cycle may be responsible for the formation of coral atolls. Sea level change is discussed further in Chapter 10, "Beaches, Shoreline Processes, and the Coastal Ocean," and Chapter 16, "The Oceans and Climate Change."

RECAP

Mantle plumes create hotspots at Earth's surface, which produce volcanic chains called nemataths that record the motions of plates.

CONCEPT CHECK 2.4 | Show how plate tectonics can be used as a working model.

1 How is the age distribution pattern of the Hawaiian Islands–Emperor Seamount chain explained by the position of the Hawaiian hotspot? What could have caused the curious bend in the chain?

2 What are the differences between a mid-ocean ridge and a hotspot?

3 How can plate tectonics be used to help explain the difference between a seamount and a tablemount?

4 Draw and describe each of the three stages of coral reef development. How does this sequence tie into the plate tectonic model?

Soundings

The first recorded attempt to measure the ocean's depth was conducted in the Mediterranean Sea in about 85 B.C. by a Greek explorer named Posidonius. His mission was to answer an age-old question: How deep is the ocean? Posidonius's crew made a **sounding**[1] by letting out nearly 2 kilometers (1.2 miles) of line before the heavy weight on the end of the line touched bottom. For the next 2000 years, voyagers used sounding lines to probe the ocean's depths. The standard unit of ocean depth is the **fathom** (*fathme* = outstretched arms[2]), which is equal to 1.8 meters (6 feet).

The first systematic bathymetric measurements of the oceans were made in 1872 aboard the HMS *Challenger*, during its historic three-and-a-half-year voyage.[3] Every so often, *Challenger*'s crew stopped and measured the depth, along with many other ocean properties. These measurements indicated that the deep-ocean floor was not flat but had significant *relief* (variations in elevation), just as dry land does. However, determining bathymetry by making occasional soundings rarely gives a complete picture of the ocean floor. For instance, imagine trying to determine what the surface features on land look like while flying in a blimp at an altitude of several kilometers on a foggy night, using only a long weighted rope to determine your height above the surface. This is similar to how bathymetric measurements were collected from ships using sounding lines.

Web Animation
Sonar and Echolocation
http://goo.gl/sGFcIJ

Echo Soundings

The presence of mid-ocean undersea mountains had long been known, but recognition of their full extent into a connected worldwide system had to await the invention and use of the **echo sounder**, or *fathometer*, in the early 1900s. An echo sounder sends a sound signal (called a **ping**) from the ship downward into the ocean, where it produces echoes when it bounces off any density difference, such as marine organisms or the ocean floor (**Figure 3.1**). Water is a good transmitter of sound, so the time it takes for the echoes to return[4] is used to determine the depth and as a result, the corresponding shape of the ocean floor. In 1925, for example, the German vessel *Meteor* used echo sounding to identify the underwater mountain range running through the center of the South Atlantic Ocean.

Echo sounding, however, lacks detail and often gives an inaccurate view of the relief of the sea floor. For instance, the sound beam emitted from a ship 4000 meters (13,100 feet) above the ocean floor widens to a diameter of about 4600 meters (15,000 feet) at the bottom. Consequently, the first echoes to return from the bottom are usually from the closest (highest) peak within this broad area. Nonetheless, most of our knowledge of ocean bathymetry has been provided by the echo sounder.

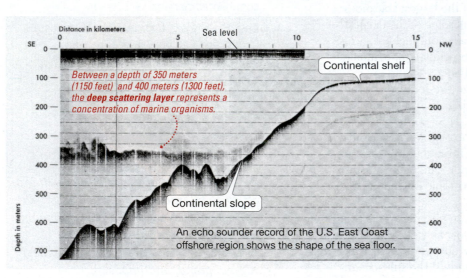

Figure 3.1 **An echo sounder record.** Vertical exaggeration (the amount of expansion of the vertical scale) is 12 times.

[1]A *sounding* refers to a probe of the environment for scientific observation and was borrowed from atmospheric scientists, who released probes called soundings into the atmosphere. Ironically, the term does not actually refer to sound; the use of sound to measure ocean depths came later.

[2]This term is derived from the method used to bring depth sounding lines back on board a vessel by hand. While hauling in the line, workers counted the number of arm-lengths collected. By measuring the length of the person's outstretched arms, the amount of line taken in could be calculated. Much later, the distance of 1 fathom was standardized to equal exactly 6 feet.

[3]For more information about the accomplishments of the *Challenger* expedition, see MasteringOceanography **Web Diving Deeper 5.2**.

[4]This technique uses the speed of sound in seawater, which varies with salinity, pressure, and temperature but averages about 1507 meters (4945 feet) per second.

Sounds produced by echo sounders bounce off any density difference in the ocean. Because of this property, it was soon discovered that echo sounders could detect and track submarines. During World War II, antisubmarine warfare inspired many improvements in the technology of "seeing" into the ocean using sound.

During and after World War II, there was great improvement in sonar technology. For example, the **precision depth recorder (PDR)**, which was developed in the 1950s, uses a focused high-frequency sound beam to measure depths to a resolution of about 1 meter (3.3 feet). Throughout the 1960s, PDRs were used extensively and provided a reasonably good representation of the ocean floor. From thousands of research vessel tracks, the first reliable global maps of sea floor bathymetry were produced. These maps helped confirm the ideas of sea floor spreading and plate tectonics.

Modern *acoustic* (*akouein* = to hear) instruments that use sound to map the sea floor include *multibeam echo sounders* (which use multiple frequencies of sound simultaneously) and side-scan **sonar** (an acronym for *sound navigation and ranging*). **Seabeam**—the first multibeam echo sounder—made it possible for a survey ship to map the features of the ocean floor along a strip up to 60 kilometers (37 miles) wide. Multibeam systems use sound emitters directed away from both sides of a survey ship, with receivers permanently mounted on the ship's hull. Multibeam instruments emit multiple beams of sound waves, which are reflected off the ocean floor. As the sound waves bounce back with different strengths and timing, computers analyze these differences to determine the depth and shape of the sea floor and whether the bottom is rock, sand, or mud (**Figure 3.2**). In this way, multibeam surveying provides incredibly detailed imagery of the seabed. Because its beams of sound spread out with depth, multibeam systems have resolution limitations in deep water.

In deep water or where a detailed survey is required, side-scan sonar can provide enhanced views of the sea floor. A side-scan sonar instrument is towed behind a survey ship and can be lowered to just above the ocean floor to produce a detailed strip map of ocean floor bathymetry (**Figure 3.3**). To maximize its resolution, the side-scan instrument can be lowered on its cable so that it "flies" just

SmartFigure 3.2 Multibeam sonar. An artist's depiction of how a survey vessel uses multibeam sonar to map the ocean floor. Colors on the sea floor represent different elevations. https://goo.gl/2oMV73

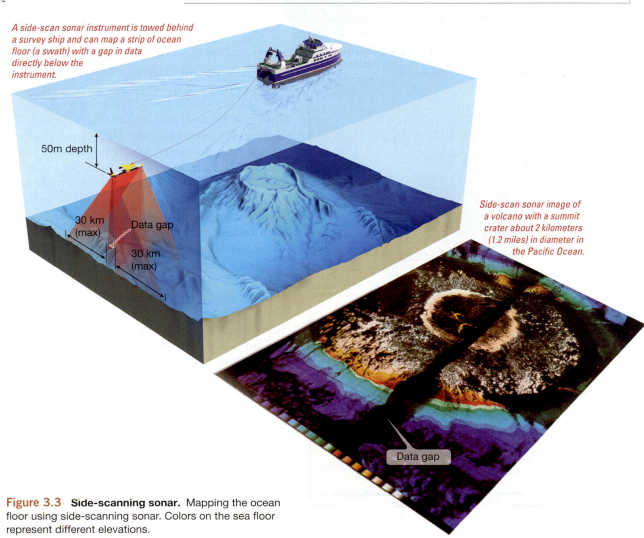

Figure 3.3 Side-scanning sonar. Mapping the ocean floor using side-scanning sonar. Colors on the sea floor represent different elevations.

RECAP

Plate tectonic processes are responsible for producing most ocean floor features.

ESSENTIAL CONCEPTS REVIEW

3.1 What techniques are used to determine ocean bathymetry?

▸ *Bathymetry is the measurement of ocean depths and the charting of ocean floor topography.* The varied bathymetry of the ocean floor was first determined using *soundings* to measure water depth. Later, the development of the *echo sounder* gave ocean scientists a more detailed representation of the sea floor.

▸ Today, much of our knowledge of the ocean floor has been obtained using various *multibeam echo sounders* or *side-scan sonar instruments* (to make detailed bathymetric maps of a small area of the ocean floor), *satellite measurement* of the ocean surface (to produce maps of the world ocean floor), and *seismic reflection profiles* (to examine Earth structure beneath the sea floor).

Study Resources

MasteringOceanography Study Area Quizzes, MasteringOceanography Web Table 3.1, MasteringOceanography Web Animation, Web Video

Critical Thinking Question

Describe how satellite measurements of the ocean *surface* allow oceanographers to create a map of the sea *floor*.

Active Learning Exercise

Use the Internet to research how a "fish finder" works on modern sport-fishing boats. How do these techniques compare to the sonar techniques described in this chapter?

3.2 What features exist on continental margins?

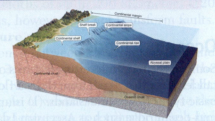

▸ *Passive continental margins are not associated with any plate boundaries.* Features of passive margins include, from the shore outward, *the continental shelf, the continental slope*, and *the continental rise*. The continental shelf is generally shallow, low relief, and gently sloping; it can also contain various features such as coastal islands, reefs, and banks. The boundary between the continental slope and the continental shelf is marked by an increase in slope that occurs at the *shelf break*.

▸ Cutting deep into the continental slopes are *submarine canyons*, which resemble canyons on land but are created by erosive turbidity currents. *Turbidity currents* deposit their sediment load at the base of the continental slope, creating deep-sea fans that merge to produce a gently sloping continental rise. The deposits from turbidity currents (called *turbidite deposits*) have characteristic sequences of graded bedding. Active margins have similar features although they are modified by their associated plate boundary.

▸ *Active continental margins have some features in common with passive margins*, although they are modified by their associated plate boundary (either convergent or transform). Active continental margins are associated with a *high degree of tectonic activity* such as *earthquakes, volcanoes, tall mountains*, and in some cases *deep trenches* located close to shore.

Study Resources

MasteringOceanography Study Area Quizzes, MasteringOceanography Web Diving Deeper 3.3, MasteringOceanography Web Animation, Web Videos

Critical Thinking Question

To help reinforce your knowledge of continental margins, draw and describe differences between passive and active continental margins from memory. Be sure to include a real-world example of each type, associated features, and how these features relate to plate tectonics.

Active Learning Exercise

Archeologists have discovered evidence of a submerged prehistoric human settlement including fire rings and pottery shards on a continental shelf that is now 50 kilometers (30 miles) offshore and in water that is 60 meters (200 feet) deep. With another student in class, discuss how this would be possible.

3.3 What features exist in the deep-ocean basins?

▶ The *continental rises* gradually become flat, extensive, deep-ocean *abyssal plains*, which form by *suspension settling* of fine sediment. Poking through the sediment cover of the abyssal plains are numerous *volcanic peaks*, including volcanic islands, seamounts, tablemounts, and abyssal hills. In the Pacific Ocean, where sedimentation rates are low, abyssal plains are not extensively developed, and abyssal hill provinces cover broad expanses of ocean floor.

▶ *Along the margins of many continents*—especially those around the *Pacific Ring of Fire*—are deep linear scars called *ocean trenches* that are associated with convergent plate boundaries and volcanic arcs.

Study Resources
MasteringOceanography Study Area Quizzes, MasteringOceanography Web Diving Deeper 3.2

Critical Thinking Question
In which ocean basin are most ocean trenches found? Use plate tectonic processes to help explain why.

Active Learning Exercise
About 170 kilometers (105 miles) offshore of Southern California is a famous big wave surf spot known as Cortez Bank. Working with another student in class, find Cortez Bank on Figure 3.11. Discuss whether or not Cortez Bank is part of the Southern California continental shelf.

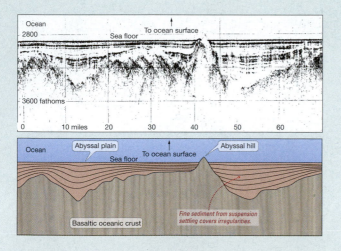

3.4 What features exist along the mid-ocean ridge?

▶ The *mid-ocean ridge is a continuous mountain range* that winds through all ocean basins and is entirely *volcanic in origin*. Common features associated with the mid-ocean ridge include a *central rift valley, faults and fissures, seamounts, pillow basalts, hydrothermal vents, deposits of metal sulfides,* and *unusual life forms*. Segments of the mid-ocean ridge are either *oceanic ridges* if steep with rugged slopes (indicative of slow sea floor spreading) or *oceanic rises* if sloped gently and less rugged (indicative of fast spreading).

▶ *Long linear zones of weakness—fracture zones and transform faults*—cut across vast distances of ocean floor and *offset the axes of the mid-ocean ridge*. Fracture zones and transform faults are differentiated from one another based on the direction of movement across the feature. *Fracture zones* (an intraplate feature) *have movement in the same direction*, while *transform faults* (a transform plate boundary) *have movement in opposite directions*.

Study Resources
MasteringOceanography Study Area Quizzes, MasteringOceanography Web Animation, Web Videos

Critical Thinking Question
Using a labeled diagram, describe the differences between fracture zones and transform faults. Along which feature are earthquakes more likely to occur?

Active Learning Exercise
Working with another student, discuss how volcanoes on land are different from volcanoes at the mid-ocean ridge.

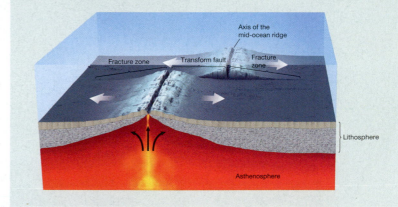

MasteringOceanography™
www.masteringoceanography.com

Looking for additional review and test prep materials? With individualized coaching on the toughest topics of the course, MasteringOceanography offers a wide variety of ways for you to move beyond memorization and deeply grasp the underlying processes of how the oceans work. Visit the Study Area in **www.masteringoceanography.com** to find practice quizzes, study tools, and multimedia that will improve your understanding of this chapter's content. Sign in today to enjoy the following features: Self Study Quizzes, SmartFigures, SmartTables, Oceanography Videos, Squidtoons, Geoscience Animation Library, RSS Feeds, Digital Study Modules, and an optional Pearson eText.

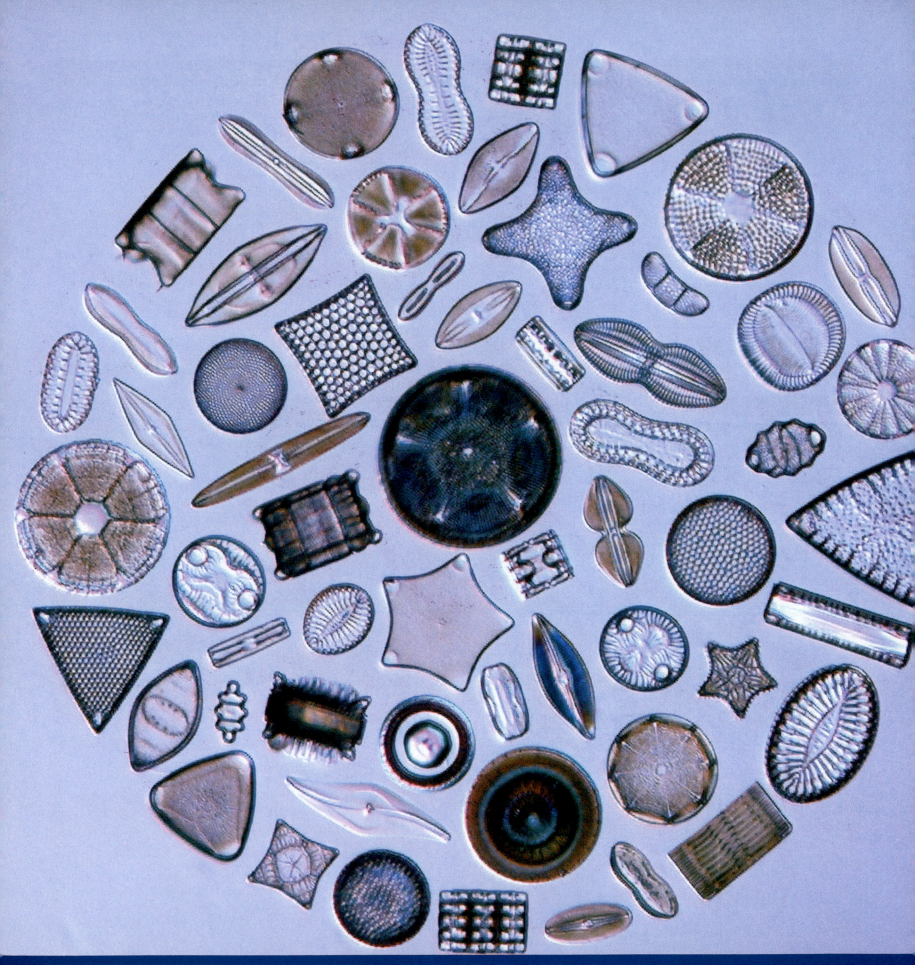

Microscopic view of arranged diatoms.The objects in this photomicrograph are diatoms, a type of microscopic marine algae that exists in incredible abundance in the ocean. This image shows various species of diatoms magnified several hundred times and was made by carefully arranging them under a microscope.

4

Marine Sediments

Coccolithophore **Calcareous ooze** · Radiolarian
foraminifer
Test Biogenous sediment
Calcite compensation depth (CCD) Metal sulfide Diatomaceous earth
Lithogenous sediment Siliceous ooze Rotary drilling
Evaporite Methane hydrate **Cosmogenous sediment Diatom**
Abyssal clay
Hydrogenous sediment Manganese nodule

Before you begin reading this chapter, use the glossary at the end of this book to discover the meanings of any of the words in the word cloud above you don't already know.

Why are **sediments** (*sedimentum* = settling) interesting to oceanographers? Although ocean sediments appear to be little more than eroded particles and fragments of dirt, dust, and other debris that have slowly settled out of the water by the process of **suspension settling** and accumulated on the ocean floor (**Figure 4.1**), they reveal much about Earth's history. For example, over millions of years, the thick deposit of sediment that accumulates on the ocean floor often contains microscopic fossils that provide clues to the past geographic distributions of marine organisms. Marine sediments are also useful for determining the pattern of ancient ocean circulation, the movement of the sea floor, and even the timing and severity of global extinction events. Further, marine sediments reveal a detailed history of Earth's past climate, thus providing insight into today's climate changes. Remarkably, sediments that accumulate over time on the sea floor comprise a nearly continuous, undisturbed record of Earth history unlike anything on land. In essence, marine sediments represent Earth's largest museum with displays of Earth history dating back millions of years.

Over time, sediments can become *lithified* (*lithos* = stone, *fic* = making)—turned to rock—and form *sedimentary rock*. More than half of the rocks exposed on the continents are sedimentary rocks deposited in ancient ocean environments and uplifted onto land by plate tectonic processes. Perhaps surprisingly, even the tallest mountains on the continents—far from any ocean—contain telltale marine fossils, which indicate that these rocks originated on the ocean floor in the geologic past. For example, the summit of the world's tallest mountain (Mount Everest in the Himalaya Mountains) consists of limestone, which is a type of rock that originated as sea floor deposits.

Climate Connection

Particles of marine sediment come from worn pieces of rocks, as well as living organisms, minerals dissolved in water, and even from outer space. Clues to sediment origin are found in its mineral composition and its **texture** (the size and shape of its particles).

This chapter begins with a brief discussion about how marine sediments are collected and the important information they reveal about Earth history. Then the four main types of sediment are examined, with regard to their characteristics, origin, and distribution (**Table 4.1**). Note that Table 4.1 summarizes much of the content within this chapter and so it can be used as a road map of topics to help you organize information as you learn about marine sediments. Mixtures of marine sediment and sediment distribution are also considered. Finally, the chapter concludes with a discussion of the resources that marine sediments provide.

For each of the four main types of sediment (*first column*), the table shows important aspects of its composition (*second column*), sources/origin (*third column*), and distribution/main locations found (*fourth column*).

ESSENTIAL LEARNING CONCEPTS

At the end of this chapter, you should be able to:

4.1 Demonstrate an understanding of how marine sediments are collected and what historical events they reveal.

4.2 Describe the characteristics of lithogenous sediment.

4.3 Describe the characteristics of biogenous sediment.

4.4 Describe the characteristics of hydrogenous sediment.

4.5 Describe the characteristics of cosmogenous sediment.

4.6 Specify how the distribution of pelagic and neritic deposits is determined by proximity to sediment sources and mechanisms of transport.

4.7 Identify the various resources that marine sediments provide.

"From the sediments the history of the ocean emerged with all its wonders . . . "

—*Wolf H. Berger,* Oceans: Reflections on a Century of Exploration (*2009*)

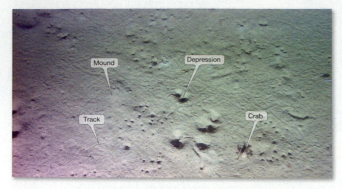

Figure 4.1　Oceanic sediment. View of typical deep-ocean floor, which is covered with a thick blanket of fine particles that have slowly settled onto the sea floor by the process of suspension settling. The depressions, mounds, and tracks are made by bottom-dwelling organisms. The crab in the lower right is about 4 inches (10 cm) across.

4.1　How Are Marine Sediments Collected, and What Historical Events Do They Reveal?

One of the difficulties of studying marine sediments is collecting adequate samples from the deep-ocean floor. Until relatively recently, the inaccessibility of the deep ocean has hindered the collection of marine sediments, especially those beneath the surface of the sea floor.

Collecting Marine Sediments

Collecting sediments suitable for analysis from the deep ocean is an arduous process. During early exploration of the oceans, a bucket-like device called a *dredge* was used to scoop up sediment from the deep-ocean floor for analysis. This technique, however, was tedious and had many limitations. For example, it often didn't work right and the dredge came up empty. It also disturbed the sediment and could only gather samples from the surface of the ocean floor. Later, the *gravity corer*—a hollow steel tube with a heavy weight on top—was thrust into the sea floor to collect

TABLE 4.1　CLASSIFICATION OF MARINE SEDIMENTS

Type		Composition	Sources/Origin		Distribution/Main locations where sediment currently forms
Lithogenous	Continental margin	Rock fragments Quartz sand Quartz silt Clay	Rivers; coastal erosion; landslides		Continental shelf
			Glaciers		Continental shelf in high latitudes
			Turbidity currents		Continental slope and rise; ocean basin margins
	Oceanic	Quartz silt Clay	Wind-blown dust; rivers		Abyssal plains and other regions of the deep-ocean basins
		Volcanic ash	Volcanic eruptions		
Biogenous	Calcium carbonate/ calcite ($CaCO_3$)	Calcareous ooze (microscopic)	Warm surface waters	Coccolithophores (algae) Foraminifers (protozoans)	Low-latitude regions; sea floor above CCD; along mid-ocean ridges and the tops of submarine volcanic peaks
		Shells and coral fragments (macroscopic)		Macroscopic shell-producing organisms	Continental shelf; beaches
				Coral reefs	Shallow low-latitude regions
	Silica ($SiO_2 \cdot nH_2O$)	Siliceous ooze	Cold surface waters	Diatoms (algae) Radiolarians (protozoans)	High-latitude regions; sea floor below CCD; upwelling areas where cold, deep water rises to the surface, especially that caused by surface current divergence near the equator
Hydrogenous		Manganese nodules (manganese, iron, copper, nickel, cobalt)	Precipitation of dissolved materials directly from seawater due to chemical reactions		Abyssal plain
		Phosphorite (phosphorous)			Continental shelf
		Oolites ($CaCO_3$)			Shallow shelf in low-latitude regions
		Metal sulfides (iron, nickel, copper, zinc, silver)			Hydrothermal vents at mid-ocean ridges
		Evaporites (gypsum, halite, other salts)			Shallow restricted basins where evaporation is high in low-latitude regions
Cosmogenous		Iron–nickel spherules Tektites (silica glass)	Space dust		In very small proportions mixed with all types of sediment and in all marine environments
		Iron–nickel meteorites	Meteors		Localized near meteor impact structures

the first **cores** (cylinders of sediment and rock). Although the gravity corer could sample below the surface, its depth of penetration was limited. Today, specially designed ships perform **rotary drilling** to collect cores from the deep ocean.

In 1963, the U.S. National Science Foundation began funding a program that borrowed drilling technology from the offshore oil industry to obtain long sections of core from deep below the surface of the ocean floor. The program united four leading oceanographic institutions (Scripps Institution of Oceanography in California; Rosenstiel School of Atmospheric and Oceanic Studies at the University of Miami, Florida; Lamont-Doherty Earth Observatory of Columbia University in New York; and the Woods Hole Oceanographic Institution in Massachusetts) to form the *Joint Oceanographic Institutions for Deep Earth Sampling (JOIDES)*. The oceanography departments of several other leading universities later joined JOIDES.

The first phase of the **Deep Sea Drilling Project (DSDP)** was initiated in 1966, when the specially designed drill ship *Glomar Challenger* was launched. It had a tall drilling rig resembling a steel tower. Cores could be collected by drilling into the ocean floor in water up to 6000 meters (3.7 miles) deep. From the initial cores collected, scientists confirmed the existence of sea floor spreading by documenting that (1) the age of the ocean floor increased progressively with distance from the mid-ocean ridge (see Figure 2.11), (2) sediment thickness increased progressively with distance from the mid-ocean ridge (see Figure 4.24), and (3) Earth's magnetic field polarity reversals were recorded in ocean floor rocks (see Figure 2.10).

Although the oceanographic research program was initially financed by the U.S. government, it became international in 1975, when West Germany, France, Japan, the United Kingdom, and the Soviet Union also provided financial and scientific support. In 1983, the Deep Sea Drilling Project became the **Ocean Drilling Program (ODP)**, with 20 participating countries under the supervision of Texas A&M University and a broader objective of drilling the thick sediment layers near the continental margins.

In 1985, the *Glomar Challenger* was decommissioned and replaced by the drill ship *JOIDES Resolution* (**Figure 4.2**). The new ship also has a tall metal drilling rig to conduct *rotary drilling*. The drill pipe is in individual sections of 9.5 meters (31 feet), and sections can be screwed together to make a single string of pipe up to 8200 meters (27,000 feet) long (Figure 4.2). The drill bit, located at the end of the pipe string, rotates as it is pressed against the ocean bottom and can drill up to 2100 meters (6900 feet) into the sea floor. Like twirling a soda straw into a layer cake, the drilling operation crushes the rock around the outside and retains a cylinder of rock (a *core sample*) on the inside of the hollow pipe. A core can then be raised to the surface from inside the pipe, cut in half, and analyzed using state-of-the-art laboratory facilities on board the *Resolution*. Worldwide, more than 2000 holes have been drilled into the sea floor using this method, allowing the collection of cores (**Figure 4.3**) that provide scientists with valuable information about Earth history, as recorded in sea floor sediments.

In 2003, the ODP was replaced by the **Integrated Ocean Drilling Program (IODP)**, and, in 2013, its name was updated to *International Ocean Discovery Program (IODP): Exploring the Earth under the Sea*. This new international effort continues over five decades of scientific collaboration that seeks to recover geological data and samples from beneath the ocean floor to study the history and dynamics of Planet Earth. In addition, the program does not rely on just one drill ship but uses multiple vessels for exploration. For example, one

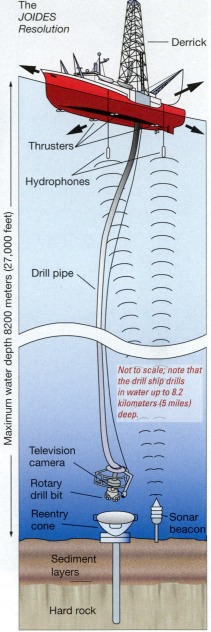

Figure 4.2 **Rotary drilling from the *JOIDES Resolution*.** Using its array of thrusters, the JOIDES Resolution (*photo*) can remain in one place at the surface while performing rotary drilling, which is shown diagrammatically (*right*).

Each sediment layer represents a unique event in Earth's history.

Figure 4.3 An ocean sediment core. Cylinders of sediment and rock called *cores* are retrieved from the ocean floor and then cut in half for examination. Oldest layers are at the bottom of the core and youngest are at the top.

RECAP

Marine sediments accumulate on the ocean floor and contain a record of Earth history, including past environmental conditions.

of the new vessels that began operations in 2007 is a state-of-the-art drill ship named *Chikyu* (which means "Planet Earth" in Japanese) that can drill up to 7000 meters (23,000 feet) into the sea floor. Plans to upgrade the vessel with new drilling technology will allow it to drill even deeper, perhaps as deep as through Earth's crust into the mantle. The program's primary mission is to collect cores that will allow scientists to better understand the properties of the deep crust, the microbiology of the deep-ocean floor, Earth's climate change patterns, and earthquake mechanisms. For example, shortly after the devastating 2011 Tohoku-Oki Earthquake and resulting tsunami, the *Chikyu* drillship began an expedition to the Japan Trench to drill into the fault zone near the site of the earthquake to study earthquake-generated heat from friction by taking detailed temperature measurements beneath the sea floor.

Environmental Conditions Revealed by Marine Sediments

Marine sediments provide a wealth of information about past conditions on Earth. As sediment accumulates on the ocean floor, it preserves the materials—and the conditions of the environment—that existed in the overlying water column. By carefully analyzing cylindrical cores of sediment collected from the sea floor and interpreting them (**Figure 4.4**), Earth scientists can infer past environmental conditions such as sea surface temperature, nutrient supply, abundance of marine life, atmospheric winds, ocean current patterns, volcanic eruptions, major extinction events, changes in Earth's climate, and the movement of tectonic plates. In fact, most of what is known of Earth's past geology, climate, and biology has been learned through studying ancient marine sediments.

Climate

Connection

Interdisciplinary

Relationship

Paleoceanography

The study of how the ocean, atmosphere, and land have interacted in the past to produce changes in ocean chemistry, circulation, biology, and climate is called **paleoceanography** (*paleo* = ancient, *ocean* = the marine environment, *graphy* = description of), a branch of oceanography that relies on sea floor sediments to gain insight into these past changes. Recent paleoceanographic studies, for example, have linked changes in deep-ocean circulation with rapid climate change. In the North Atlantic Ocean, cold, relatively salty water sinks and forms a body of water called *North Atlantic Deep Water*. Water in this deep current circulates through the global ocean, driving deep-ocean circulation and global heat transport, which, in turn, impacts global climate. This is widely viewed as one of the most climatically sensitive regions on Earth, and North Atlantic sea floor sediments from the past several million years have revealed that the region has experienced abrupt changes to its ocean–atmosphere system, triggered by fluctuations of freshwater from melting glaciers. Understanding the timing, mechanisms, and causes of this abrupt climate change is one of the major challenges facing paleoceanography today.

Climate

Connection

CONCEPT CHECK 4.1 | Demonstrate an understanding of how marine sediments are collected and what historical events they reveal.

1 Using Table 4.1, list and describe the characteristics of the four main types of marine sediment.

2 Describe the process of how a drill ship like the *JOIDES Resolution*

obtains core samples from the deep-ocean floor.

3 What types of past environmental conditions can be inferred by studying cores of sediment?

4.2 What Are the Characteristics of Lithogenous Sediment?

Lithogenous sediment (*lithos* = stone, *generare* = to produce) is derived from preexisting rock material that originates on the continents or islands from erosion, volcanic eruptions, or blown dust. Note that lithogenous sediment is sometimes referred to as **terrigenous sediment** (*terra* = land, *generare* = to produce).

Origin of Lithogenous Sediment

Lithogenous sediment begins as rocks on continents or islands. Over time, **weathering** agents such as water, temperature extremes, and chemical effects break rocks into smaller pieces, as shown in **Figure 4.5**. When rocks are in smaller pieces, they can be more easily **eroded** (picked up) and transported. This eroded material is the basic component of which all lithogenous sediment is composed.

Eroded material from the continents is carried to the oceans by streams, wind, glaciers, and gravity (**Figure 4.6**). Each year, stream flow alone carries about 20 billion metric tons (44 trillion pounds) of sediment to Earth's continental margins; almost 40% is provided by runoff from Asia.

Transported sediment can be deposited in many environments, including bays or lagoons near the ocean, as deltas at the mouths of rivers, along beaches at the shoreline, or further offshore across the continental margin. It can also be carried beyond the continental margin to the deep-ocean basin by turbidity currents, as discussed in Chapter 3.

The greatest quantity of lithogenous material is found around the margins of the continents, where it is constantly moved by high-energy currents along the shoreline and in deeper turbidity currents. Lower-energy currents distribute finer components that settle out onto the deep-ocean basins. Microscopic particles from wind-blown dust or volcanic eruptions can even be carried far out over the open ocean by prevailing winds. These particles are deposited into the ocean either as the wind speed decreases or when they serve as nuclei around which raindrops and snowflakes form, and ultimately settle onto the sea floor as fine layers of sediment.

Figure 4.4 **Examining deep-ocean sediment cores.** Sediment cores reveal interesting aspects of Earth history such as the past geographic distributions of marine organisms, ocean circulation changes, major extinctions, and Earth's past climate.

Composition of Lithogenous Sediment

The composition of lithogenous sediment reflects the material from which it was derived. All rocks are composed of discrete crystals of naturally occurring compounds called *minerals*. One of the most abundant, chemically stable, and durable minerals in Earth's crust is **quartz**, composed of silicon and oxygen in the form of SiO_2—the same composition as ordinary glass. Quartz is a major component of most rocks. Because quartz is resistant to abrasion, it can be transported long distances and deposited far from its source area. The majority of lithogenous deposits—such as beach sands—are composed primarily of quartz (**Figure 4.7**).

A large percentage of lithogenous particles that find their way into deep-ocean sediments far from continents are transported by prevailing winds that remove small particles from the continents' subtropical desert regions. The map in **Figure 4.8** shows a close relationship between

Figure 4.5 **Weathering of a rock outcrop.**

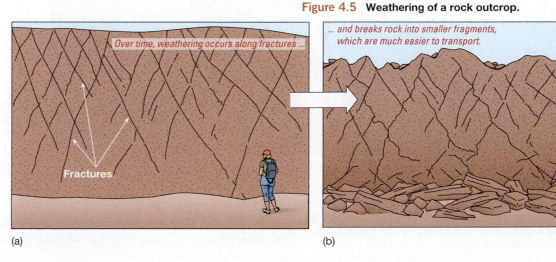

(a) *Over time, weathering occurs along fractures ...*

Fractures

(b) *... and breaks rock into smaller fragments, which are much easier to transport.*

Light colors are suspended sediment in the water.

(a) Stream: Po River, Italy, which displays a prominent delta and a visible sediment plume in the water.

(b) Wind: Dust storm approaching a military base, Australia.

Dark stripe is crushed rock debris.

(c) Glacier: Riggs Glacier, Glacier Bay National Park, Alaska, which displays a dark stripe of sediment along its length called a *medial moraine*.

(d) Gravity, which creates landslides: Del Mar, California.

Figure 4.6 **Sediment transport mechanisms.** Photos showing various ways sediment can be transported, including **(a)** streams, **(b)** wind, **(c)** glaciers, and **(d)** gravity.

the location of microscopic fragments of lithogenous quartz in the surface sediments of the ocean floor and the strong prevailing winds in the desert regions of Africa, Asia, and Australia. Satellite observations of dust storms (Figure 4.8, *inset*) confirm this relationship. Sediment is not the only item transported by wind. In fact, scientists have documented the transportation of a variety of airborne substances—including viruses, pollutants, and even living insects—from Africa all the way across the Atlantic Ocean to North America.

Sediment Texture

One of the most important properties of lithogenous sediment is its texture, including its **grain size**.[1] The **Wentworth scale of grain size** (Table 4.2) indicates that particles can be classified as boulders (largest), cobbles, pebbles, granules, sand, silt, or clay (smallest). Sediment size is proportional to the energy needed to lay down a deposit. Deposits laid down where wave action is strong (areas of high energy) may be composed primarily of larger particles—cobbles and boulders. Fine-grained particles, on the other hand, are deposited where the energy level is low and the current speed is minimal. When clay-sized particles—many of which are flat—are deposited, they tend to stick together by cohesive forces. Consequently, higher-energy conditions than what would be expected based on grain size alone are required to erode and transport clays. In general, however, lithogenous sediment tends to become finer with increasing distance from shore. This relationship is mostly because high-energy transporting mechanisms predominate close to shore and lower-energy conditions exist in the deep-ocean basins.

The texture of lithogenous sediment also depends on its **sorting**. Sorting is a measure of the uniformity of grain sizes and indicates the

[1]Sediment grains are also known as particles, fragments, or clasts.

selectivity of the transportation process. For example, sediments composed of particles that are primarily the same size are well sorted—such as in coastal sand dunes, where winds can only pick up a certain size particle. Poorly sorted deposits, on the other hand, contain a variety of different sized particles and indicate a transportation process capable of picking up clay- to boulder-sized particles. An example of poorly sorted sediment is that which is carried by a glacier and left behind when the glacier melts.

Distribution of Lithogenous Sediment

Marine sedimentary deposits can be categorized as either neritic or pelagic. **Neritic deposits** (*neritos* = of the coast) are found on continental shelves and in shallow water near islands; these deposits are generally coarse grained. Alternatively, **pelagic deposits** (*pelagios* = of the sea) are found in the deep-ocean basins and are typically fine grained. Moreover, lithogenous sediment in the ocean is ubiquitous: At least a small percentage of lithogenous sediment is found nearly everywhere on the ocean floor.

NERITIC DEPOSITS Lithogenous sediment dominates most neritic deposits. Lithogenous sediment is derived from rocks on nearby landmasses, consists of coarse-grained deposits, and accumulates rapidly on the continental shelf, slope, and rise. Examples of lithogenous neritic deposits include beach deposits, continental shelf deposits, turbidite deposits, and glacial deposits.

Beach Deposits Beaches are made of whatever materials are locally available. Beach materials are composed mostly of quartz-rich sand that is washed down to the coast by rivers but can also be composed of a wide variety of sizes and compositions. This material is transported by waves that crash against the shoreline, especially during storms.

Continental Shelf Deposits At the end of the last ice age (about 10,000 years ago), glaciers melted and sea level rose. As a result, many rivers of the world today drop their sediment in drowned river mouths rather than carry it onto the continental shelf as they did during the geologic past. This explains why, in many areas, the sediments that cover the continental shelf—called *relict* (*relict* = left behind) *sediments*—were deposited from 3000 to 7000 years ago and are not covered by sediments discharged by rivers today. These relict sediments presently cover about 70% of the world's continental shelves. In other areas, deposits of sand ridges on the continental shelves appear to have been formed more recently than the most recent ice age and at present water depths.

Climate
Connection

Turbidite Deposits As discussed in Chapter 3, **turbidity currents** are underwater avalanches that periodically move down the continental slopes and carve submarine canyons. Turbidity currents also carry vast amounts of neritic material. This material spreads out as deep-sea fans, comprises the continental rise, and gradually thins toward the abyssal plains. These deposits are called **turbidite deposits** and are composed of characteristic layering called *graded bedding* (see Figure 3.12).

Glacial Deposits Poorly sorted deposits containing particles ranging from boulders to clays may be found in the high-latitude[2] portions of the continental shelf. These **glacial deposits** were laid down during the most recent ice age by glaciers that covered the continental shelf and eventually melted. Glacial deposits are currently

[2]High-latitude regions are those far from the equator (either north or south); low latitudes are areas close to the equator.

Figure 4.7 Lithogenous beach sand. Photomicrograph of well-sorted lithogenous beach sand, which is composed mostly of particles of white quartz plus small amounts of other minerals. This sand, from North Beach, Hampton, New Hampshire, is magnified approximately 23 times.

STUDENTS SOMETIMES ASK . . .

How effective is wind as a transporting agent?

Any material that gets into the atmosphere—including dust from dust storms, soot from forest fires, specks of pollution, and ash from volcanic eruptions—is transported by wind and can be found as deposits on the ocean floor. Every year, wind storms lift an estimated 3 billion metric tons (6.6 trillion pounds) of this material into the atmosphere, where it gets transported around the globe. As much as three-quarters of these particles—mostly dust—come from Africa's Sahara Desert; once airborne, they are carried out across the Atlantic Ocean (see Figure 4.8). Much of this dust falls in the Atlantic, and that's why ships traveling downwind from the Sahara Desert often arrive at their destinations quite dusty. Some of it falls in the Caribbean (where the pathogens it contains have been linked to stress and disease among coral reefs), in Bermuda (where past accumulations have produced the island's red soils), the Amazon (where its iron and phosphorus fertilize nutrient-poor soil), and across the southern United States as far west as New Mexico. The dust also contains bacteria and pesticides—even African desert locusts have been transported alive across the Atlantic Ocean during strong wind storms!

Figure 4.8 Lithogenous quartz in surface sediments of the world's oceans and transport by wind.

Web Video
Saharan Dust Cloud Travels
Across the Atlantic
https://goo.gl/1BXnDP

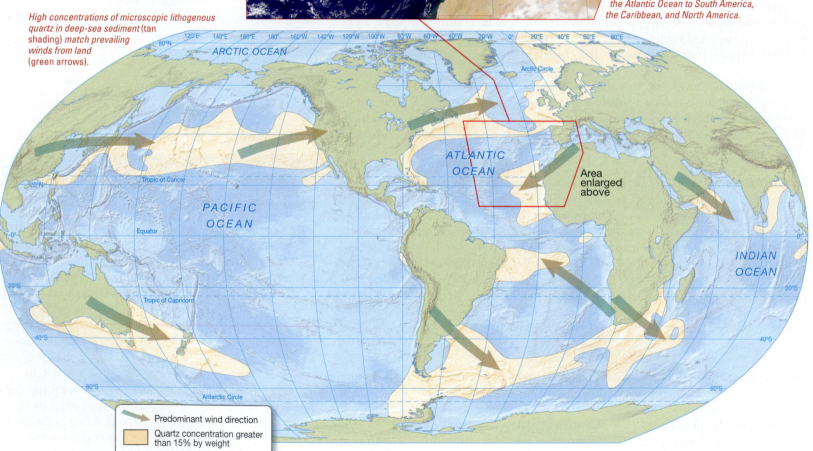

SeaStar SeaWiFS satellite photo from February 26, 2000, shows a wind storm that has blown dust from the Sahara Desert off the northwest coast of Africa. Some of this dust is transported across the Atlantic Ocean to South America, the Caribbean, and North America.

High concentrations of microscopic lithogenous quartz in deep-sea sediment (tan shading) match prevailing winds from land (green arrows).

Area enlarged above

→ Predominant wind direction
▨ Quartz concentration greater than 15% by weight

forming around the continent of Antarctica and around Greenland by **ice rafting**. In this process, rock particles trapped in glacial ice are carried out to sea by icebergs that break away from coastal glaciers. As the icebergs melt, lithogenous particles of many sizes are released and settle onto the ocean floor.

PELAGIC DEPOSITS Turbidite deposits of neritic sediment on the continental rise can spill over into the deep-ocean basin. However, most pelagic deposits are composed of fine-grained material that accumulates slowly on the deep-ocean floor.

TABLE 4.2 WENTWORTH SCALE OF GRAIN SIZE FOR SEDIMENTS				
Size range (millimeters)	Particle name	Grain size	Example	Energy of the depositional environment
Above 256	Boulder	Coarse-grained	Coarse material found in streambeds near the source areas of rivers and along some beaches	High energy
64 to 256	Cobble			
4 to 64	Pebble			
2 to 4	Granule			
$1/16$ to 2	Sand		Beach sand	
$1/256$ to $1/16$	Silt		Feels gritty in teeth	
$1/4096$ to $1/256$	Clay	Fine-grained	Microscopic; feels sticky	Low energy

Scale in millimeters: 0 10 20 30 40 50 60

Pelagic lithogenous sediment includes particles that have come from volcanic eruptions, windblown dust, and fine material that is carried by deep-ocean currents.

Abyssal Clay **Abyssal clay** is composed of at least 70% (by weight) fine, clay-sized particles from the continents. Even though they are far from land, deep abyssal plains contain thick sequences of abyssal clay deposits composed of particles transported great distances by winds or ocean currents and deposited on the deep-ocean floor. Because abyssal clays contain oxidized iron, they are commonly red-brown or buff in color and are sometimes referred to as **red clays**. The predominance of abyssal clay on abyssal plains is caused not by an abundance of clay settling on the ocean floor but by the absence of other material that would otherwise dilute it.

RECAP

Lithogenous sediment is produced from preexisting rock material, is found on most parts of the ocean floor, and can occur as thick deposits close to land.

CONCEPT CHECK 4.2 | Describe the characteristics of lithogenous sediment.

1 Describe the origin, composition, texture, and distribution of lithogenous sediment.

2 Why is most lithogenous sediment composed of quartz grains? What is the chemical composition of quartz?

3 What is the difference between neritic and pelagic deposits? Give examples of lithogenous sediment found in each.

4.3 What Are the Characteristics of Biogenous Sediment?

Biogenous sediment (*bio* = life, *generare* = to produce) (also called *biogenic sediment*) is derived from the remains of hard parts of once-living organisms.

Origin of Biogenous Sediment

Biogenous sediment begins as the hard parts (shells, bones, and teeth) of living organisms ranging from minute algae and protozoans to fish and whales. When organisms that produce hard parts die, their remains settle onto the ocean floor and can accumulate as biogenous sediment.

Interdisciplinary
Relationship

Biogenous sediment can be classified as either macroscopic or microscopic. **Macroscopic biogenous sediment** is large enough to be seen without the aid of a microscope and includes shells, bones, and teeth of large organisms.

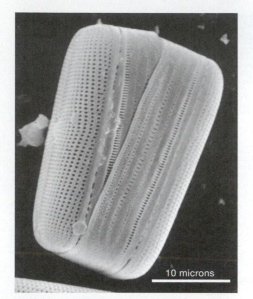

(a) Diatom, showing how the two parts of the diatom's test fit together.

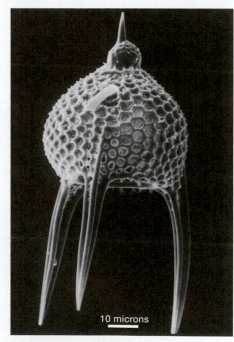

(b) Radiolarian.

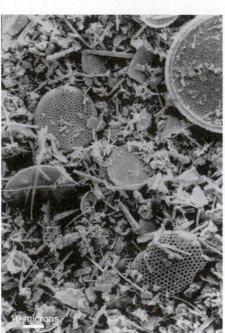

(c) Siliceous ooze, showing mostly fragments of diatom tests.

Figure 4.9 **Microscopic siliceous tests.** Scanning electron micrographs of various siliceous tests.

Except in certain tropical beach localities where shells and coral fragments are numerous, this type of sediment is relatively rare in the marine environment, especially in deep water where fewer organisms live. Much more abundant is **microscopic biogenous sediment**, which contains particles so small they can be seen well only through a microscope. Microscopic organisms produce tiny shells called **tests** (*testa* = shell) that begin to sink after the organisms die and continually rain down in great numbers onto the ocean floor. These microscopic tests can accumulate on the deep-ocean floor and form deposits called **ooze** (*wose* = juice). As its name implies, ooze resembles very fine-grained, mushy material.[3] Technically, biogenous ooze must contain at least 30% biogenous test material by weight. What comprises the other part—up to 70%—of an ooze? Commonly, it is fine-grained lithogenous clay that is deposited along with biogenous tests in the deep ocean. By volume, much more microscopic ooze than macroscopic biogenous sediment exists on the ocean floor.

The organisms that contribute to biogenous sediment are chiefly **algae** (*alga* = seaweed) and **protozoans** (*proto* = first, *zoa* = animal). Algae are primarily aquatic, eukaryotic,[4] photosynthetic organisms, ranging in size from microscopic single cells to large organisms like giant kelp. Protozoans are any of a large group of single-celled, eukaryotic, usually microscopic organisms that are generally not photosynthetic.

Composition of Biogenous Sediment

The two most common chemical compounds in biogenous sediment are **calcium carbonate** ($CaCO_3$, which forms the mineral **calcite**) and **silica** (SiO_2). Often, the silica is chemically combined with water to produce $SiO_2 \cdot nH_2O$, the hydrated form of silica, which is called *opal*.

Interdisciplinary

Relationship

SILICA Most of the silica in biogenous ooze comes from microscopic algae called **diatoms** (*diatoma* = cut in half) and protozoans called **radiolarians** (*radio* = a spoke or ray).

Because diatoms photosynthesize, they need strong sunlight and are found only within the upper, sunlit surface waters of the ocean. Most diatoms are free-floating, or **planktonic** (*planktos* = wandering). The living organism builds a glass greenhouse out of silica as a protective cov-ering and lives inside. Most species have two parts to their test that fit together like a petri dish or pillbox (**Figure 4.9a**). The tiny tests are perforated with small holes in intricate patterns to allow nutrients to pass in and waste products to pass out. Where diatoms are abundant at the ocean surface, thick deposits of diatom-rich ooze can accumulate below on the ocean floor. When this ooze lithifies, it becomes **diatomaceous earth**,[5] a lightweight white rock composed of diatom tests and clay (**Diving Deeper 4.1**).

Radiolarians are microscopic single-celled protozoans, most of which are also planktonic. As their name implies, they often have long spikes or rays of silica protruding from their siliceous shell (**Figure 4.9b**). They do not photosynthesize but rely on external food sources such as bacteria and other plankton. Radiolarians typically

[3]Ooze has the consistency of toothpaste mixed about half and half with water. As a way to remember this term, imagine walking barefoot across the deep-ocean floor and having the fine sediment there *ooze* between your toes.

[4]Eukaryotic (*eu* = good, *karyo* = the nucleus) cells contain a distinct membrane-bound nucleus.

[5]Diatomaceous earth is also called diatomite, tripolite, or kieselguhr.

DIATOMS: THE MOST IMPORTANT THINGS YOU HAVE (PROBABLY) NEVER HEARD OF

"Few objects are more beautiful than the minute siliceous cases of the diatomaceae: were these created that they might be examined and admired under the higher powers of the microscope?"

—*Charles Darwin (1872)*

Diatoms are microscopic single-celled photosynthetic organisms. Each one lives inside a protective silica test, most of which contain two halves that fit together like a shoebox and its lid. First described with the aid of a microscope in 1702, their tests are exquisitely ornamented with holes, ribs, and radiating spines unique to individual species. The fossil record indicates that diatoms have been on Earth since the Jurassic Period (180 million years ago), and more than 70,000 species of diatoms have been identified.

Diatoms live for a few days to as much as a week, can reproduce sexually or asexually, and occur individually or linked together into long communities. They are found in great abundance floating in the ocean and in certain freshwater lakes but can also be found in many diverse environments, such as on the undersides of polar ice, on the skins of whales, in soil, in thermal springs, and even on brick walls.

When marine diatoms die, their tests rain down and accumulate on the sea floor as siliceous ooze. Hardened deposits of siliceous ooze, called *diatomaceous earth*, can be as much as 900 meters (3000 feet) thick. Diatomaceous earth consists of billions of minute silica tests and has many unusual properties: It is lightweight, has an inert chemical composition, is resistant to high temperatures, and has excellent filtering properties. Diatomaceous earth is used to produce a variety of common products (**Figure 4A**). The main uses of diatomaceous earth include:

- Filters (for refining sugar, separating impurities from wine, straining yeast from beer, and filtering swimming pool water)
- Mild abrasives (in toothpaste, facial scrubs, matches, and household cleaning and polishing compounds)
- Absorbents (for chemical spills, in cat litter, and as a soil conditioner)
- Chemical carriers (in pharmaceuticals, paint, and even dynamite)

Other products from diatomaceous earth include optical-quality glass (because of the pure silica content of diatoms) and space shuttle tiles (because they are lightweight and provide good insulation). Diatomaceous earth is also used as an additive in concrete, a filler in tires, an anticaking agent, a natural pesticide, and even a building stone in the construction of houses.

Further, the vast majority of oxygen that all animals breathe is a by-product of photosynthesis by diatoms. In addition, each living diatom contains a tiny droplet of oil. When diatoms die, their tests containing droplets of oil accumulate on the sea floor and are the beginnings of petroleum deposits, such as those found offshore of California.

Given their many practical applications, it is difficult to imagine how different our lives would be without diatoms!

GIVE IT SOME THOUGHT

1. What are several reasons diatoms are so remarkable? List products that contain or are produced using diatomaceous earth.

Figure 4A **Products containing or produced using diatomaceous earth** (diatom *Thalassiosira eccentrica*, inset).

20 microns

display well-developed symmetry, which is why they have been described as the "living snowflakes of the sea."

The accumulation of siliceous tests of diatoms, radiolarians, and other silica-secreting organisms produces **siliceous ooze** (**Figure 4.9c**).

CALCIUM CARBONATE Two significant sources of calcium carbonate biogenous ooze are the **foraminifers** (*foramen* = an opening)—close relatives of radiolarians—and microscopic algae called **coccolithophores** (*coccus* = berry, *lithos* = stone, *phorid* = carrying).

Coccolithophores are single-celled algae, most of which are planktonic. Coccolithophores produce thin plates or shields made of calcium carbonate, 20 or 30 of which overlap to produce a spherical test (**Figure 4.10a**). Like diatoms, coccolithophores photosynthesize, so they need sunlight to live. Coccolithophores are really, *really* small. In fact, coccolithophores are about 10 to 100 times smaller than most diatoms (**Figure 4.10b**), which is why coccolithophores are often called **nannoplankton** (*nanno* = dwarf, *planktos* = wandering).

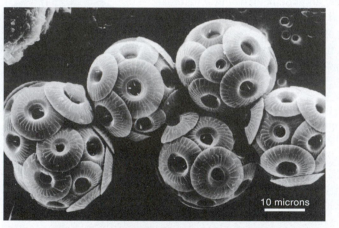

(a) Coccolithophores, which resemble tiny spheres.

(b) Diatom (siliceous) surrounded by coccoliths (calcareous).

(c) Foraminifers, which resemble tiny shells found at a beach.

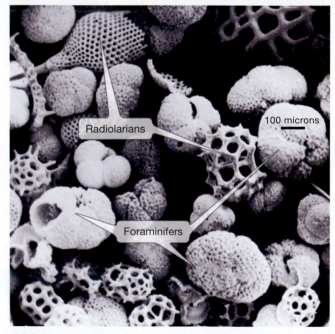

(d) Calcareous ooze, which also includes some siliceous radiolarian tests.

Figure 4.10 Microscopic calcareous tests. Scanning electron micrographs (*above*) and photomicrographs (*below*) of various calcareous tests.

When the organism dies, the individual plates (called **coccoliths**) disaggregate and can accumulate on the ocean floor as coccolith-rich ooze. When this ooze lithifies over time, it forms a white deposit called **chalk**, which is used for a variety of purposes (including writing on chalkboards). The White Cliffs of southern England are composed of hardened, coccolith-rich calcium carbonate ooze, which was deposited on the ocean floor and has been uplifted onto land (**Figure 4.11**). Deposits of chalk the same age as the White Cliffs are so common throughout Europe, North America, Australia, and the Middle East that the geologic period in which these deposits formed is named the Cretaceous (*creta* = chalk) Period.

Foraminifers are single-celled protozoans, many of which are planktonic, ranging in size from microscopic to macroscopic. They do not photosynthesize, so they must ingest other organisms for food. Foraminifers produce a hard calcium carbonate test in which the organism lives (**Figure 4.10c**). Most foraminifers produce a segmented or chambered test, and all tests have a prominent opening in one end. Although very small in size, the tests of foraminifers resemble the large shells that one might find at a beach.

Deposits comprised primarily of tests of foraminifers, coccoliths, and other calcareous-secreting organisms are called **calcareous ooze** (**Figure 4.10d**).

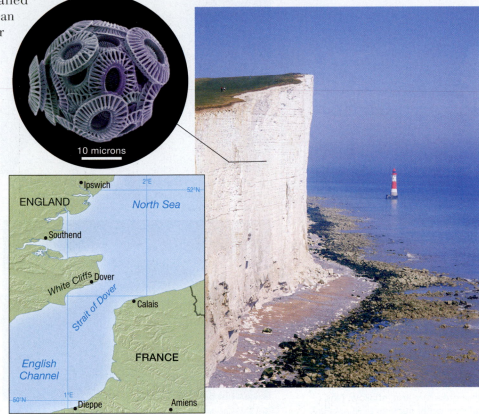

Figure 4.11 **The White Cliffs of southern England.** The White Cliffs near Dover in southern England are composed of chalk, which is hardened coccolith-rich calcareous ooze. Inset shows a colored image of the coccolithophore *Emiliana huxleyi*.

Distribution of Biogenous Sediment

Biogenous sediment is one of the most common types of pelagic deposits. The distribution of biogenous sediment on the ocean floor depends on three fundamental processes: (1) productivity, (2) destruction, and (3) dilution.

Productivity is the number of organisms present in the surface water above the ocean floor. Surface waters with high biologic productivity contain many living and reproducing organisms—conditions that are likely to produce biogenous sediments. Conversely, surface waters with low biologic productivity contain too few organisms to produce biogenous oozes on the ocean floor.

Destruction occurs when skeletal remains (tests) dissolve in seawater at depth. In some cases, biogenous sediment dissolves before ever reaching the sea floor; in other cases, it is dissolved before it has a chance to accumulate into deposits on the sea floor.

Dilution occurs when the deposition of other sediments decreases the percentage of the biogenous sediment found in marine deposits. For example, other types of sediments can dilute biogenous test material below the 30% necessary to classify it as ooze. Dilution occurs most often because of the abundance of coarse-grained lithogenous material in neritic environments, so biogenous oozes are uncommon along continental margins.

NERITIC DEPOSITS Although neritic deposits are dominated by lithogenous sediment, both microscopic and macroscopic biogenous material may be incorporated into lithogenous sediment in neritic deposits. In addition, biogenous carbonate deposits are common in some areas.

Carbonate Deposits **Carbonate** minerals are those that contain CO_3 in their chemical formula—such as calcium carbonate, $CaCO_3$. Rocks from the marine environment composed primarily of calcium carbonate are called **limestones**. Most limestones contain fossil marine shells, suggesting a biogenous origin, while other carbonate-containing rocks

Interdisciplinary

Relationship

(a) Location map of Shark Bay, Australia.

Figure 4.12 Stromatolites.
Stromatolites are bulbous algal mats that grow in warm, shallow, high-salinity water such as in Shark Bay, Australia.

(b) Shark Bay stromatolites, which form in high-salinity tidal pools and reach a maximum height of about 1 meter (3.3 feet).

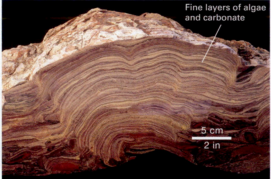

Fine layers of algae and carbonate

5 cm
2 in

(c) Profile view through a stromatolite, showing its internal fine layering.

Web Animation
The Accumulation of Siliceous Ooze
https://goo.gl/2XEQww

appear to have formed directly from seawater without the help of any marine organism. Modern environments where calcium carbonate is currently being deposited (such as in the Bahama Banks, Australia's Great Barrier Reef, and the Persian Gulf) suggest that carbonate deposits are formed in shallow, warm-water shelves and around tropical islands as coral reefs and beaches.

Ancient marine carbonate deposits constitute 2% of Earth's crust and 25% of all sedimentary rocks on Earth. In fact, marine limestones form the underlying bedrock of Florida and many Midwestern states, from Kentucky to Michigan and from Pennsylvania to Colorado. Percolation of groundwater through these deposits has dissolved the limestone to produce sinkholes and, in some cases, spectacular caverns.

Stromatolites **Stromatolites** are lobate structures consisting of fine layers of carbonate that form in specific warm, shallow-water environments such as the high salinity tidal pools in Shark Bay, Western Australia (**Figure 4.12**). Cyanobacteria[6] produce these deposits by trapping fine sediment in mucous mats. Other types of algae produce long filaments that bind carbonate particles together. Like tree rings being added as a tree grows, layer upon layer of these algae colonize the surface, forming a bulbous structure. In the geologic past—particularly from about 1 to 3 billion years ago—conditions were ideal for the development of stromatolites, so stromatolite structures hundreds of meters high can be found in rocks from these ages.

PELAGIC DEPOSITS Microscopic biogenous sediment (ooze) is common on the deep-ocean floor because there is so little lithogenous sediment deposited at great distances from the continents that could dilute the biogenous material.

Siliceous Ooze Siliceous ooze contains at least 30% of the hard remains of silica-secreting organisms. When the siliceous ooze consists mostly of diatoms, it is called *diatomaceous ooze*. When it consists mostly of radiolarians, it is called *radiolarian ooze*. When it consists mostly of single-celled silicoflagellates—another type of protozoan—it is called *silicoflagellate ooze*.

The ocean is undersaturated with silica at all depths, which means that any solid particle made up of silica will tend to dissolve in seawater. In fact, if living diatoms, radiolarians, and silicoflagellates were not hard at work creating their silica-containing tests, they would dissolve, too! As a consequence, the destruction of siliceous biogenous particles (that is, the tests of dead organisms drifting to the sea floor), by dissolving in seawater, occurs continuously and slowly at all depths. How can siliceous ooze accumulate on the ocean floor if it is being dissolved? One way is to accumulate the siliceous tests faster than seawater can dissolve them. For instance, many tests sinking at the same time will create a deposit of siliceous ooze on the sea floor below (**Figure 4.13**).[7] Once buried beneath other siliceous tests, they are no

[6]Cyanobacteria (*kuanos* = dark blue) are simple, ancient creatures whose ancestry can be traced back to some of the first photosynthetic organisms on Earth.

[7]An analogy to this is trying to get a layer of sugar to form on the bottom of a cup of hot coffee. If a few grains of sugar are slowly dropped into the cup, a layer of sugar won't accumulate. If a whole bowl full of sugar is dumped into the coffee, however, a thick layer of sugar will form on the bottom of the cup.

longer exposed to the dissolving effects of seawater. Thus, siliceous ooze is commonly found in areas below surface waters with high biologic productivity of silica-secreting organisms.

Calcareous Ooze and the CCD Calcareous ooze contains at least 30% of the hard remains of calcareous-secreting organisms. When it consists mostly of coccolithophores, it is called *coccolith ooze*. When it consists mostly of foraminifers, it is called *foraminifer ooze*. One of the most common types of foraminifer ooze is *Globigerina ooze*, named for a foraminifer that is especially widespread in the Atlantic and South Pacific oceans. Other calcareous oozes include *pteropod oozes* and *ostracod oozes*.

The destruction of calcium carbonate (calcite) varies with depth. At the warmer surface and in the shallow parts of the ocean, seawater is generally saturated with calcium carbonate, so calcite does not dissolve. In the deep ocean, however, the colder water contains greater amounts of carbon dioxide, which forms carbonic acid and causes calcareous material to dissolve. The higher pressure at depth also helps speed the dissolution of calcium carbonate.

The depth in the ocean at which the pressure is high enough, and the amount of carbon dioxide in deep-ocean waters is great enough, to begin dissolving calcium carbonate is called the **lysocline** (*lusis* = a loosening, *cline* = slope). Below the lysocline, calcium carbonate dissolves at an increasing rate with increasing depth until the **calcite compensation depth (CCD)**[8] is reached (**Figure 4.14**). At the CCD and greater depths, sediment does not usually contain much calcite because it readily dissolves; even the thick tests of foraminifers dissolve within a day or two. In essence, calcite accumulates only near the tops of the tall peaks that rise off the sea floor and extend above the CCD but dissolves at deeper depths associated with the base of the peaks. This situation creates the marine equivalent of a mountain's "snow line," but with deposits of light-colored calcite on the mountaintop instead of frozen water.

The CCD, on average, is 4500 meters (15,000 feet) below sea level, but depending on the chemistry of the deep ocean, it may be as deep as 6000 meters (20,000 feet) in portions of the Atlantic Ocean or as shallow as 3500 meters (11,500 feet) in the Pacific Ocean. The depth of the lysocline also varies from ocean to ocean but averages about 4000 meters (13,100 feet).

In the geologic past, higher concentrations of carbon dioxide in the atmosphere have led to increased amounts of dissolved carbon dioxide in the ocean, thereby making the ocean more acidic and causing the CCD to rise. Currently, scientists have documented an increase in ocean acidity due to higher levels of atmospheric carbon dioxide caused by human-caused emissions. Increased ocean acidity and its effect on marine life are discussed in Chapter 16, "The Oceans and Climate Change."

Because of the CCD, modern carbonate oozes are generally rare below 5000 meters (16,400 feet). Still, buried deposits of ancient calcareous ooze are found beneath the CCD. How can calcareous ooze exist below the CCD? The necessary conditions are shown in **Figure 4.15**. The mid-ocean ridge is a topographically high

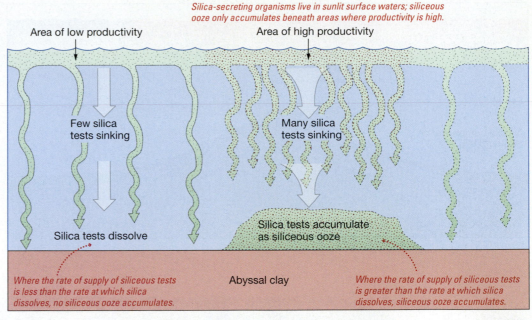

Silica-secreting organisms live in sunlit surface waters; siliceous ooze only accumulates beneath areas where productivity is high.

Area of low productivity — Area of high productivity

Few silica tests sinking — Many silica tests sinking

Silica tests dissolve — Silica tests accumulate as siliceous ooze

Abyssal clay

Where the rate of supply of siliceous tests is less than the rate at which silica dissolves, no siliceous ooze accumulates.

Where the rate of supply of siliceous tests is greater than the rate at which silica dissolves, siliceous ooze accumulates.

SmartFigure 4.13 Accumulation of siliceous ooze.
https://goo.gl/4iUEf1

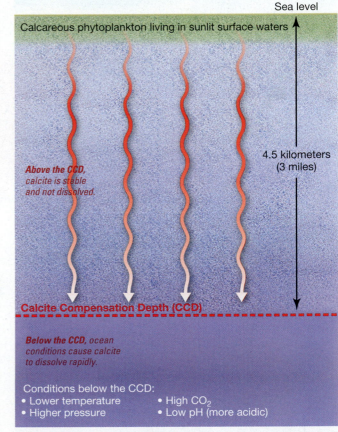

Ocean pressure increases and the properties of seawater change below the CCD, affecting where calcite dissolves and where it is deposited.

Sea level

Calcareous phytoplankton living in sunlit surface waters

4.5 kilometers (3 miles)

Above the CCD, calcite is stable and not dissolved.

Calcite Compensation Depth (CCD)

Below the CCD, ocean conditions cause calcite to dissolve rapidly.

Conditions below the CCD:
• Lower temperature • High CO_2
• Higher pressure • Low pH (more acidic)

Figure 4.14 Characteristics of water above and below the calcite compensation depth (CCD).

Interdisciplinary / Relationship

Climate Connection

[8]Because the mineral calcite is composed of calcium carbonate, the *calcite compensation depth* is also known as the *calcium carbonate compensation depth* or the *carbonate compensation depth*. All go by the handy abbreviation CCD.

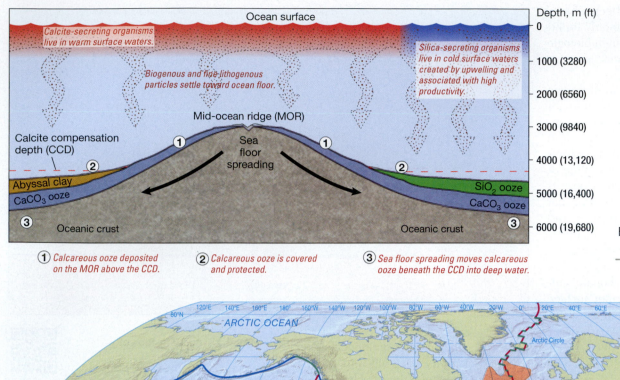

SmartFigure 4.15 **Sea floor spreading and sediment accumulation.** Relationships among carbonate compensation depth (CCD), the mid-ocean ridge, sea floor spreading, productivity, and destruction that allow calcareous ooze to be preserved below the CCD. https://goo.gl/s9vlsw

Web Animation

How Calcareous Ooze Can Be Found Beneath the CCD
https://goo.gl/h3rDxA

① *Calcareous ooze deposited on the MOR above the CCD.* ② *Calcareous ooze is covered and protected.* ③ *Sea floor spreading moves calcareous ooze beneath the CCD into deep water.*

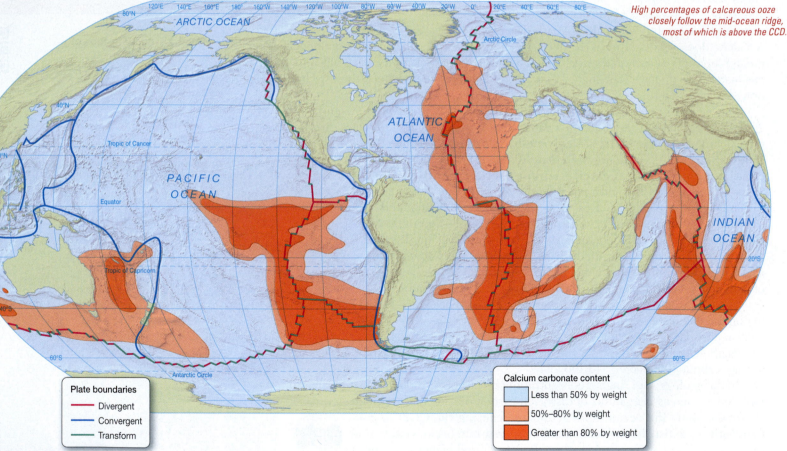

High percentages of calcareous ooze closely follow the mid-ocean ridge, most of which is above the CCD.

Plate boundaries
— Divergent
— Convergent
— Transform

Calcium carbonate content
Less than 50% by weight
50%–80% by weight
Greater than 80% by weight

Figure 4.16 **Distribution of calcium carbonate in modern surface sediments.**

feature that rises above the sea floor. It often pokes up above the CCD, even though the surrounding deep-ocean floor is below the CCD. Thus, calcareous ooze deposited on top of the mid-ocean ridge does not dissolve. However, sea floor spreading causes the newly created sea floor and the calcareous sediment on top of it to move into deeper water away from the ridge, eventually being transported below the CCD. This calcareous sediment will dissolve below the CCD unless it is covered by some deposit that is unaffected by the CCD (such as siliceous ooze or abyssal clay).

The map in **Figure 4.16** shows the percentage (by weight) of calcium carbonate in the modern surface sediments of the ocean basins. High concentrations of

SMARTTABLE 4.3	COMPARISON OF ENVIRONMENTS INTERPRETED FROM DEPOSITS OF SILICEOUS AND CALCAREOUS OOZE IN SURFACE SEDIMENTS	
	Siliceous ooze	**Calcareous ooze**
Surface water temperature above sea floor deposits	Cool	Warm
Main location found	Sea floor beneath cool surface water in high latitudes	Sea floor beneath warm surface water in low latitudes
Other factors	Upwelling brings deep, cold, nutrient-rich water to the surface	Calcareous ooze dissolves below the CCD
Other locations found	Sea floor beneath areas of upwelling, including along the equator	Sea floor beneath warm surface water in low latitudes along the mid-ocean ridge

calcareous ooze (sometimes exceeding 80%) are found along segments of the mid-ocean ridge, but little is found in deep-ocean basins below the CCD. For example, in the northern Pacific Ocean—one of the deepest parts of the world ocean—there is very little calcium carbonate in the sediment. Calcium carbonate is also rare in sediments accumulating beneath cold, high-latitude waters where calcareous-secreting organisms are relatively uncommon.

Table 4.3 compares the environmental conditions that can be inferred from siliceous and calcareous oozes. It shows that siliceous ooze typically forms below cool surface water regions, including areas of **upwelling** where deep-ocean water comes to the surface and supplies nutrients that stimulate high rates of biological productivity. Calcareous ooze, on the other hand, is found on the shallower areas of the ocean floor beneath warmer surface water.

 SmartTable 4.3 **Comparison of environments interpreted from deposits of siliceous and calcareous ooze in surface sediments.**
https://goo.gl/H5uBwg

RECAP

Biogenous sediment is produced from the hard remains of once-living organisms. Microscopic biogenous sediment is especially widespread and forms deposits of ooze on the ocean floor.

CONCEPT CHECK 4.3 | Describe the characteristics of biogenous sediment.

1 Describe the origin, composition, and distribution of biogenous sediment.

2 List the two major chemical compounds of which most biogenous sediment is composed and two examples of the microscopic organisms that produce them. Sketch and label these organisms.

3 Technically, what must a biogenous ooze contain to be classified as ooze? What other component do all oozes contain?

4 If siliceous ooze is slowly but constantly dissolving in seawater, how can deposits of siliceous ooze accumulate on the ocean floor?

5 Explain the stages of progression that result in calcareous ooze existing below the CCD.

4.4 What Are the Characteristics of Hydrogenous Sediment?

Hydrogenous sediment (*hydro* = water, *generare* = to produce) is derived from the dissolved material in water.

Origin of Hydrogenous Sediment

Seawater contains many dissolved materials. Chemical reactions within seawater cause certain minerals to come out of solution, or **precipitate** (change from the dissolved to the solid state). Precipitation usually occurs when there is a *change in conditions*, such as a change in temperature or pressure or the addition of chemically active fluids. To make rock candy, for instance, a pan of water is heated and sugar is added. When the water is hot and the sugar dissolved, the pan is removed from the heat, and the sugar water is allowed to cool. The *change in temperature* causes the sugar to become

Interdisciplinary

Relationship

(a) Manganese nodules, including some that are cut in half.

(b) Close-up of a baseball-sized manganese nodule cut in half, revealing its central nucleation object and layered internal structure.

(c) An abundance of manganese nodules on a portion of the deep South Pacific Ocean floor about 4 meters (13 feet) across.

Figure 4.17 Manganese nodules.

oversaturated, which causes it to precipitate. As the water cools, the sugar precipitates on anything that is put in the pan, such as pieces of string or kitchen utensils.

Composition and Distribution of Hydrogenous Sediment

Although hydrogenous sediments represent a relatively small portion of the overall sediment in the ocean, they have many different compositions and are distributed in diverse environments of deposition.

MANGANESE NODULES Manganese nodules are rounded, hard lumps of manganese, iron, and other metals typically 5 centimeters (2 inches) in diameter up to a maximum of about 20 centimeters (8 inches). When cut in half, they often reveal a layered structure formed by precipitation around a central nucleation object (**Figures 4.17a** and **4.17b**). The nucleation object may be a piece of lithogenous sediment, coral, volcanic rock, a fish bone, or a shark's tooth. Manganese nodules lie on sediment in vast expanses of abyssal plains, potentially covering some 60% of the ocean basin at a typical water depth of about 5 kilometers (3.1 miles). Manganese nodules can sometimes occur in concentrations of about 100 nodules per square meter (square yard); in rare cases, they exist in even greater abundance (**Figure 4.17c**), resembling a scattered field of golf ball- to baseball-sized nodules. The formation of manganese nodules requires extremely low rates of lithogenous or biogenous input so that the nodules are not buried.

The major components of these nodules are manganese hydroxide (around 30% by weight) and iron hydroxide (around 20%). The element manganese is important for making high-strength steel alloys. Other accessory metals present in manganese nodules include copper (used in electrical wiring, in pipes, and to make brass and bronze), nickel (used to make stainless steel), and cobalt (used as an alloy with iron to make strong magnets and steel tools). Although the concentration of these accessory metals is usually less than 1%, they can exceed 2% by weight, which may make them attractive sources for these metals in the future.

The origin of manganese nodules has puzzled oceanographers since manganese nodules were first discovered in 1872 during the voyage of HMS *Challenger*.[9] If manganese nodules are truly hydrogenous and precipitate from seawater, then how can they have such high concentrations of manganese (which occurs in seawater at concentrations often too small to measure accurately)? Furthermore, why are the nodules on *top* of ocean floor sediment and not buried by the constant rain of sedimentary particles?

Unfortunately, nobody has definitive answers to these questions. Perhaps manganese nodules are created by one of the slowest chemical reactions known—on average, they grow at a rate of about 5 millimeters (0.2 inch) per *million years*. Scientific studies suggest that the formation of manganese nodules may be aided by bacteria and an as-yet-unidentified marine organism that intermittently lifts and

[9]For more information about the accomplishments of the *Challenger* expedition, see Mastering-Oceanography Web Diving Deeper 5.2.

rotates them. Other studies reveal that the nodules don't form continuously over time but in spurts that are related to specific conditions such as a low sedimentation rate of lithogenous clay and strong deep-water currents. Remarkably, the larger the nodules are, the faster they grow. The origin of manganese nodules is widely considered the most interesting unresolved problem in marine chemistry.

PHOSPHATES Phosphorus-bearing compounds (**phosphates**) occur abundantly as coatings on rocks and as nodules on the continental shelf and on banks at depths shallower than 1000 meters (3300 feet). Concentrations of phosphates in such deposits commonly reach 30% by weight and indicate abundant biological activity in surface water above where they accumulate. Because phosphates are valuable as fertilizers, ancient marine phosphate deposits that have been uplifted onto land are extensively mined to supply agricultural needs.

CARBONATES The two most important carbonate minerals in marine sediment are calcite and **aragonite**. Both are composed of calcium carbonate ($CaCO_3$), but aragonite has a different crystalline structure that is less stable and transforms into calcite over time. Carbonates are widely used in the construction industry, in the production of cement, and they are commonly used medicinally as calcium supplements or antacids.

As previously discussed, most carbonate deposits are biogenous in origin. However, hydrogenous carbonate deposits can precipitate directly from seawater in tropical climates to form aragonite crystals less than 2 millimeters (0.08 inch) long. In addition, **oolites** (*oo* = egg, *lithos* = rock) are small calcite spheres 2 millimeters (0.08 inch) or less in diameter that have layers like an onion and form in some shallow tropical waters where concentrations of $CaCO_3$ are high. Oolites are thought to precipitate around a nucleus and grow larger as they roll back and forth on beaches by wave action, but some evidence suggests that a type of algae may aid their formation.

METAL SULFIDES Deposits of **metal sulfides** are associated with hydrothermal vents and black smokers along the mid-ocean ridge. These deposits contain iron, nickel, copper, zinc, silver, and other metals in varying proportions. Transported away from the mid-ocean ridge by sea floor spreading, these deposits can be found throughout the ocean floor and can even be uplifted onto continents(see MasteringOceanography **Web Diving Deeper 2.3**).

EVAPORITES Generally, **evaporite minerals** form wherever there are high evaporation rates (dry climates) accompanied by restricted open ocean circulation. One such example is the Mediterranean Sea, which contains thick deposits of evaporites on its floor that suggest that sometime in the geologic past, the sea completely dried up (see MasteringOceanography **Web Diving Deeper 4.1**). As water evaporates in these dry areas, the remaining seawater becomes saturated with dissolved minerals, which then begin to precipitate (form a solid). Because they are heavier than seawater, the minerals sink to the bottom or form a white crust of evaporite minerals around the edges of these areas (**Figure 4.18**). Collectively termed "salts," some evaporite minerals, such as *halite* (common table salt, NaCl), taste salty, and some, such as the calcium sulfate minerals *anhydrite* ($CaSO_4$) and *gypsum* ($CaSO_4 \cdot H_2O$), do not.

Figure 4.18 Evaporative salts cover the floor of a seasonally flooded basin. After seasonal rains at Death Valley, California, the high evaporation rate causes salts (white material) to precipitate out, resulting in this extensive salt flat.

RECAP

Hydrogenous sediment is produced when dissolved materials precipitate out of solution, producing a variety of materials, and are found in localized concentrations on the ocean floor.

CONCEPT CHECK 4.4 | Describe the characteristics of hydrogenous sediment.

1 Describe the origin, composition, and distribution of hydrogenous sediment.

2 Describe manganese nodules, including what is currently known about how they form.

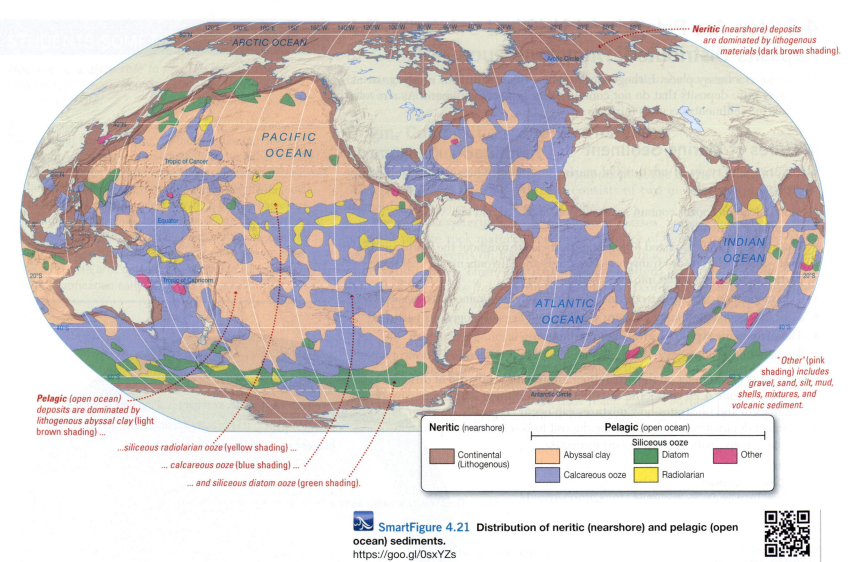

Neritic (nearshore) deposits are dominated by lithogenous materials (dark brown shading).

Pelagic (open ocean) deposits are dominated by lithogenous abyssal clay (light brown shading) ...

...siliceous radiolarian ooze (yellow shading) ...

... calcareous ooze (blue shading) ...

... and siliceous diatom ooze (green shading).

"Other" (pink shading) includes gravel, sand, silt, mud, shells, mixtures, and volcanic sediment.

Neritic (nearshore)	Pelagic (open ocean)		
		Siliceous ooze	
Continental (Lithogenous)	Abyssal clay	Diatom	Other
	Calcareous ooze	Radiolarian	

SmartFigure 4.21 **Distribution of neritic (nearshore) and pelagic (open ocean) sediments.**
https://goo.gl/0sxYZs

Neritic Deposits

Neritic (nearshore) deposits cover about one-quarter of the ocean floor, and pelagic (deep-ocean basin) deposits cover the other three-quarters. The map in **Figure 4.21** shows the distribution of neritic and pelagic deposits in the world's oceans. Coarse-grained lithogenous neritic deposits dominate continental margin areas (*dark brown shading*), which is not surprising because lithogenous sediment is derived from nearby continents. Although neritic deposits usually contain biogenous, hydrogenous, and cosmogenous particles, these constitute only a minor percentage of the total sediment mass.

Pelagic Deposits

Figure 4.21 shows that pelagic deposits are dominated by biogenous calcareous oozes (*blue shading*), which are found on the relatively shallow deep-ocean areas along the mid-ocean ridge. Biogenous siliceous oozes are found beneath areas of unusually high biological productivity such as the northernmost North Pacific Ocean, surrounding Antarctica (*green shading,* where diatomaceous ooze occurs), and the equatorial Pacific (*yellow shading,* where radiolarian ooze occurs). Fine lithogenous pelagic deposits of abyssal clays (*light brown shading*) are common in deeper areas of the ocean basins, such as in the North Pacific. Hydrogenous and cosmogenous sediment comprise only a small proportion of pelagic deposits in the ocean.

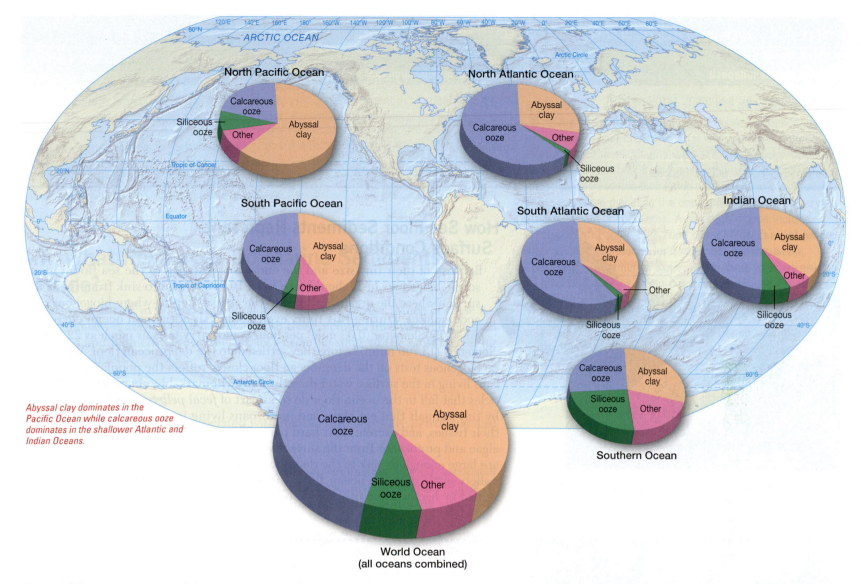

Abyssal clay dominates in the Pacific Ocean while calcareous ooze dominates in the shallower Atlantic and Indian Oceans.

Figure 4.22 Pelagic sediment types within each major ocean. World map and accompanying pie charts showing the relative amount of deep-ocean floor covered by each of the three main pelagic sediment types: abyssal clay, siliceous ooze, and calcareous ooze. Pie chart for the world ocean (*below left*) displays data from all oceans combined.

Figure 4.22 shows the proportion of each ocean floor that is covered by the pelagic deposits abyssal clay, calcareous ooze, and siliceous ooze. The world ocean (combined) pie chart shows that calcareous ooze is the most dominant sediment worldwide, covering about 45% of the deep-ocean floor. The world ocean pie chart also shows that abyssal clay covers about 38% and siliceous ooze about 8% of the world ocean floor area. If you examine the individual ocean pie charts, they show that the amount of ocean basin floor covered by calcareous ooze decreases in deeper ocean basins because they generally lie beneath the CCD. The dominant oceanic sediment in the deepest basin—the North Pacific—is abyssal clay (see also Figure 4.21). Conversely, calcareous ooze is the most widely deposited sediment in the shallower Atlantic and Indian Oceans. Note that siliceous oozes cover a smaller percentage of the ocean floor because regions of high productivity of organisms that produce silica tests are generally restricted to the equatorial region (for radiolarians) and the high latitudes such as near Antarctica and the far northern Pacific (for diatoms). **Table 4.4** shows the average rates of deposition of selected marine sediments in neritic and pelagic deposits.

RECAP

Neritic deposits occur close to shore and are dominated by coarse lithogenous material. Pelagic deposits occur in the deep ocean and are dominated by biogenous oozes and fine lithogenous clay.

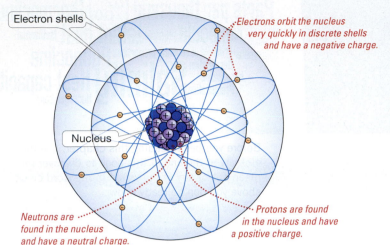

Figure 5.1 Simplified model of an atom. An atom consists of a central nucleus composed of protons and neutrons that is encircled by high-speed electrons.

form of matter. Additional study has revealed that atoms are composed of even smaller particles, called subatomic particles.[1] As shown in Figure 5.1, the **nucleus** (*nucleos* = a little nut) of an atom is composed of **protons** (*protos* = first) and **neutrons** (*neutr* = neutral) that are bound together by strong forces. Protons have a positive electrical charge, whereas neutrons have no electrical charge. Both protons and neutrons have about the same mass, which is extremely small. Surrounding the nucleus are particles called **electrons** (*electro* = electricity), which have about $1/2000$ the mass of either protons or neutrons. Electrical attraction between positively charged protons and negatively charged electrons holds electrons in layers, or shells, around the nucleus.

The overall electrical charge of individual atoms is balanced because each atom contains an equal number of protons and electrons. An oxygen atom, for example, has eight protons and eight electrons. Most oxygen atoms also have eight neutrons, which do not affect the overall electrical charge because neutrons are electrically neutral. The number of protons is what distinguishes atoms of the 118 known chemical elements from one another. For example, an oxygen atom (and only an oxygen atom) has eight protons. Similarly, a hydrogen atom (and only a hydrogen atom) has one proton, a helium atom has two protons, and so on (for more details, see Appendix IV, "A Chemical Background: Why Water Has 2 Hs and 1 O"). In some cases, an atom will lose or gain one or more electrons and thus have an overall electrical charge, in which case it is called an **ion** (*ienai* = to go).

The Water Molecule

A **molecule** (*molecula* = a mass) is a group of two or more atoms held together by mutually shared electrons. It is the smallest piece of a substance that can exist yet still retain the original properties of that substance. When atoms combine with other atoms to form molecules, they share or trade electrons and establish chemical bonds. For instance, the chemical formula for water—H_2O—indicates that a water molecule is composed of two hydrogen atoms chemically bonded to one oxygen atom.

GEOMETRY Atoms can be represented as spheres of various sizes, and a general rule of thumb is that the more electrons an atom contains, the larger its sphere. It turns out that an oxygen atom (with eight electrons) is about twice the size of a hydrogen atom (with one electron). A water molecule consists of a central oxygen atom covalently bonded to the two hydrogen atoms, which are separated by an angle of about 105 degrees (**Figure 5.2a**). The **covalent bonds** (*co* = with, *valere* = to be strong) in a water molecule are due to the sharing of electrons between oxygen and each hydrogen atom. They are relatively strong chemical bonds, so a lot of energy is needed to break them.

Figure 5.2b shows a water molecule in a more compact representation, and in **Figure 5.2c** letter symbols are used to represent the atoms in water (*O* for oxygen, *H* for hydrogen). Instead of water's atoms being in a straight line like most other molecules, *both hydrogen atoms are on the same side of the oxygen atom.* This unusual bend in the geometry of the water molecule is the underlying cause of most of the unique properties of water.

POLARITY The bent geometry of the water molecule gives a slight overall negative charge to the end that contains the oxygen atom and a slight overall positive charge to the other side that contains the hydrogen atoms (Figure 5.2a). This slight separation of charges gives the entire molecule an electrical **polarity** (*polus* = pole, *ity* = having the

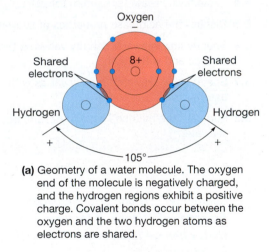

(a) Geometry of a water molecule. The oxygen end of the molecule is negatively charged, and the hydrogen regions exhibit a positive charge. Covalent bonds occur between the oxygen and the two hydrogen atoms as electrons are shared.

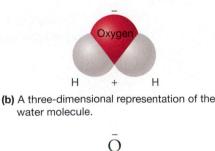

(b) A three-dimensional representation of the water molecule.

$$\overset{-}{O}$$
$$H \quad_{+} \quad H$$

(c) The water molecule represented by letters (*H* = hydrogen, *O* = oxygen).

Figure 5.2 Representations of the water molecule.

[1]It has been discovered that subatomic particles themselves are composed of a variety of even smaller particles, such as *quarks*, *leptons*, and *bosons*.

quality of), so water molecules are **dipolar** (*di* = two, *polus* = pole). Other common dipolar objects are flashlight batteries, car batteries, and bar magnets. In fact, a good way to visualize the polarity of water molecules is to view them as if they contain a tiny, weak bar magnet.

INTERCONNECTIONS OF MOLECULES If you've ever experimented with bar magnets, you know they have polarity and orient themselves relative to one another such that the positive end of one bar magnet is attracted to the negative end of another. Water molecules have polarity, too, and as a result they orient themselves relative to one another. In water, the positively charged hydrogen area of one water molecule interacts with the negatively charged oxygen end of an adjacent water molecule, forming a **hydrogen bond** (**Figure 5.3**). The hydrogen bonds between water molecules are much weaker than the covalent bonds that hold the hydrogen and oxygen atoms of water molecules together. In essence, weaker hydrogen bonds form *between* adjacent water molecules, and stronger covalent bonds occur *within* water molecules.

Even though hydrogen bonds are weaker than covalent bonds, they are strong enough to cause water molecules to stick to one another and exhibit **cohesion** (*cohaesus* = a clinging together). The cohesive properties of water cause it to "bead up" on a waxed surface, such as a freshly waxed car. They also give water its **surface tension.** Water's surface has a thin "skin" that allows a glass to be filled just above the brim without spilling any of the water. Surface tension results from the formation of hydrogen bonds between the outermost layer of water molecules and the underlying molecules. Water's ability to form hydrogen bonds causes it to have the highest surface tension of any liquid except the element mercury.[2]

WATER: THE UNIVERSAL SOLVENT Water molecules stick not only to other water molecules but also to other polar chemical compounds. In doing so, water molecules can reduce the attraction between ions of opposite charges by as much as 80 times. For instance, ordinary table salt—sodium[3] chloride, NaCl—consists of an alternating array of positively charged sodium ions and negatively charged chloride ions (**Figure 5.4a**). The **electrostatic attraction** (*electro* = electricity, *stasis* = standing) between oppositely charged ions produces an **ionic bond** (*ienai* = to go). When solid NaCl is placed in water, the electrostatic attraction (ionic bonding) between the sodium and chloride ions is reduced by 80 times. This, in turn, makes it much easier for the sodium ions and chloride ions to separate. When the ions separate, the positively charged sodium ions become attracted to the negative ends of the water molecules, the negatively charged chloride ions become attracted to the positive ends of the water molecules (**Figure 5.4b**), and the salt is dissolved in water. The process by which water molecules completely surround ions is called *hydration* (*hydra* = water, *ation* = action or process).

Because water molecules interact with other water molecules and other polar molecules, water is able to dissolve nearly everything.[4] Given enough time, water can dissolve more substances and in greater quantity than any other known substance. This is why water is called the universal solvent. It is also why the ocean contains so much dissolved material—an estimated 50 quadrillion metric tons (110 quintillion pounds) of salt—which makes seawater taste salty.

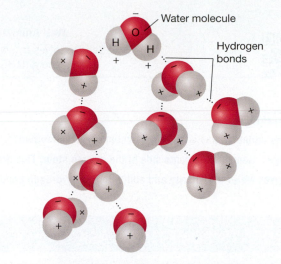

Figure 5.3 **Hydrogen bonding in water.** Dashed lines indicate locations of hydrogen bonds, which occur between water molecules.

STUDENTS SOMETIMES ASK ...

Why does a water molecule have the unusual shape that it does?

Based on simple symmetry considerations and charge separations, a water molecule should have its two hydrogen atoms on opposite sides of the oxygen atom, thus producing a linear shape, like the shape of many other molecules. But water's odd shape where both hydrogen atoms are on the same side of the oxygen atom stems from the fact that oxygen has four bonding sites, which are evenly spaced around the oxygen atom. No matter which two bonding sites are occupied by hydrogen atoms, there is a curious bend in each water molecule.

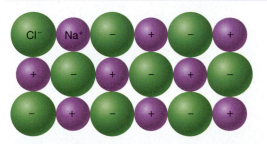

(a) Molecular structure of table salt, which is composed of sodium chloride (Na^+ = sodium ion, Cl^- = chlorine ion).

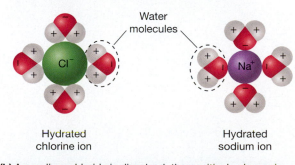

(b) As sodium chloride is dissolved, the positively charged ends of water molecules are attracted to the negatively charged Cl^- ion, while the negatively charged ends are attracted to the positively charged Na^+ ion.

Figure 5.4 **Water as a solvent.**

[2]Mercury is the only metal that is a liquid at normal surface temperatures, which is why it was commonly used in older thermometers. Today, mercury thermometers have been replaced by digital thermometers, which do not contain toxic mercury.

[3]Sodium is represented by the letters *Na* because the Latin term for sodium is *natrium*.

[4]If water is such a good solvent, why doesn't oil dissolve in water? As you might have guessed, the chemical structure of oil is remarkably nonpolar. With no positive or negative ends to attract the polar water molecule, oil will not dissolve in water.

Web Animation
How Salt Dissolves in Water
https://goo.gl/IT8Sd3

RECAP

A water molecule has a bend in its geometry, with the two hydrogen atoms on the same side of the oxygen atom. This property gives water its polarity and ability to form hydrogen bonds.

STUDENTS SOMETIMES ASK ...

How can it be that water—a liquid at room temperature—can be created by combining hydrogen and oxygen—two gases at room temperature?

It is true that combining two parts hydrogen gas with one part oxygen gas produces liquid water. This can be accomplished as a chemistry experiment, although care should be taken because much energy is released during the reaction (don't try this at home!). Oftentimes, when combining two elements, the product has very different properties than the pure substances. For instance, combining elemental sodium (Na), a highly reactive metal, with pure chlorine (Cl_2), a toxic nerve gas, produces cubes of harmless table salt (NaCl). This is what most people find amazing about chemistry.

CONCEPT CHECK 5.1 | Specify water's unique chemical properties.

1 Sketch a model of an atom, showing the positions of the subatomic particles protons, neutrons, and electrons.

2 Describe what condition exists in water molecules to make them dipolar.

3 Sketch several water molecules, showing all covalent and hydrogen bonds. Be sure to indicate the polarity of each water molecule.

4 How does hydrogen bonding produce the surface tension phenomenon of water?

5 Discuss how the dipolar nature of a water molecule makes it such an effective solvent of ionic compounds.

5.2 What Important Physical Properties Does Water Possess?

Water's important physical properties include its thermal properties (such as water's freezing and boiling points, heat capacity, and latent heats) and how water's thermal contraction affects its density.

Water's Thermal Properties

Water exists on Earth as a solid, a liquid, and a gas and has the ability to store and release great amounts of heat. Water's thermal properties influence the world's heat budget and are in part responsible for the development of tropical cyclones, worldwide wind belts, and ocean surface currents.

HEAT, TEMPERATURE, AND CHANGES OF STATE Matter around us is usually in one of the three common states: solid, liquid, or gas.[5] What must happen to change the state of a compound? The attractive forces between molecules or ions in a substance must be overcome if the state of the substance is to be changed from solid to liquid or from liquid to gas. These attractive forces include hydrogen bonds and van der Waals forces. The **van der Waals forces**—named for Dutch physicist Johannes Diderik van der Waals (1837–1923)—are relatively weak interactions that become significant only when molecules are very close together, as in the solid and liquid states (but not the gaseous state). Energy must be added to the molecules or ions so they can move fast enough to overcome these attractions.

What form of energy changes the state of matter? Very simply, adding or removing heat causes a substance to change its state of matter. For instance, adding heat to ice cubes causes them to melt, and removing heat from water causes ice to form. Before proceeding, let's clarify the difference between heat and temperature:

- **Heat** is defined as the *amount of energy transferred from one body to another due to a difference in temperature*. Heat is proportional to the average **kinetic energy** (*kinetos* = moving) of the molecules in a body. For example, water can exist as a solid, liquid, or gas, depending on the amount of heat added. Heat may be generated by combustion (a chemical reaction commonly called burning), through other chemical reactions, by friction, or from radioactivity; it can be transferred by conduction, by convection, or by radiation. A **calorie** (*calor* = heat) is the amount of heat required to raise the temperature of 1 gram of water[6] by 1 degree centigrade. The familiar "calorie" used to measure the energy content of foods is actually a *kilocalorie*, or 1000 calories. Although the metric

[5]*Plasma* is widely recognized as a fourth state of matter distinct from solids, liquids, and normal gases. Plasma is a gaseous substance in which atoms have been ionized—that is to say, stripped of electrons. Plasma television screens take advantage of the fact that plasmas are strongly influenced by electric currents.

[6]1 gram (0.035 ounce) of water is equal to about 10 drops.

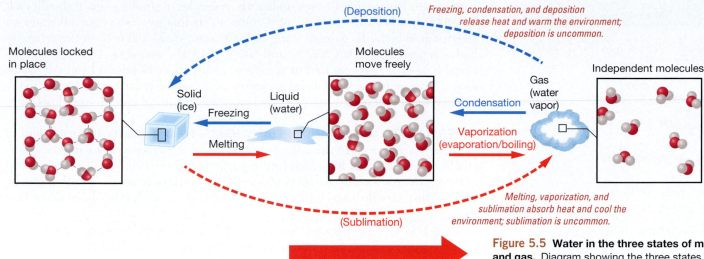

(Deposition)

Freezing, condensation, and deposition release heat and warm the environment; deposition is uncommon.

Molecules locked in place

Molecules move freely

Independent molecules

Solid (ice)

Liquid (water)

Gas (water vapor)

Condensation

Freezing

Melting

Vaporization (evaporation/boiling)

Melting, vaporization, and sublimation absorb heat and cool the environment; sublimation is uncommon.

(Sublimation)

Increasing molecular energy

Figure 5.5 Water in the three states of matter: solid, liquid, and gas. Diagram showing the three states of matter in which water is found on Earth and the processes associated with changes from one state to another.

unit for thermal energy is the *joule*, calories are directly tied to some of water's thermal properties, as will be discussed in the next section.

- **Temperature** is the *direct measure of the average kinetic energy of the molecules that make up a substance.* The greater the temperature, the greater the kinetic energy of the substance. Temperature changes when heat energy is added to or removed from a substance. Temperature is usually measured in degrees centigrade (°C) or degrees Fahrenheit (°F).

Web Animation

Phase Changes of Water
http://goo.gl/gRT6NV

Figure 5.5 shows water molecules in the solid, liquid, and gaseous states. In the *solid state* (ice), water has a rigid structure and does not normally flow over short time scales. Intermolecular bonds are constantly being broken and reformed, but the molecules remain firmly attached. That is, the molecules vibrate with energy but remain in relatively fixed positions. As a result, solids do not conform to the shape of their container.

In the *liquid state* (water), water molecules still interact with each other, but they have enough kinetic energy to flow past each other and take the shape of their container. Intermolecular bonds are being formed and broken at a much greater rate than in the solid state.

In the *gaseous state* (water **vapor**), water molecules no longer interact with one another except during random collisions. Water vapor molecules flow very freely, filling the volume of whatever container they are placed in.

WATER'S FREEZING AND BOILING POINTS If enough heat energy is added to a solid, it melts to a liquid. The temperature at which melting occurs is the substance's **melting point.** If enough heat energy is removed from a liquid, it freezes to a solid. The temperature at which freezing occurs is the substance's **freezing point,** which is the same temperature as the melting point (Figure 5.5). For pure water, melting and freezing occur at 0°C (32°F).[7]

If enough heat energy is added to a liquid, it converts to a gas. The temperature at which boiling occurs is the substance's **boiling point.** If enough heat energy is removed from a gas, it **condenses** to a liquid. The highest temperature at which condensation occurs is the substance's **condensation point,** which is the same temperature as the boiling point (Figure 5.5). For pure water, boiling and condensation occur at 100°C (212°F).

Both the freezing and boiling points of water are unusually high compared to those of similar chemical substances. As shown in **Figure 5.6**, if water followed the

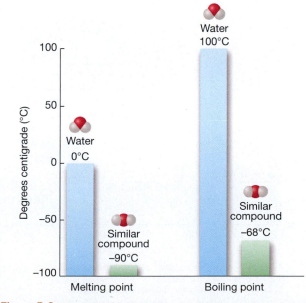

Figure 5.6 Comparison of melting and boiling points of water with similar chemical compounds. Bar graph showing the melting and boiling points of water compared to the melting and boiling points of similar chemical compounds. Note that water would have properties like those of similar chemical compounds if water molecules did not have their unique geometry and resulting polarity.

[7]All melting/freezing/boiling points discussed in this chapter assume a standard sea level pressure of 1 atmosphere (14.7 pounds per square inch).

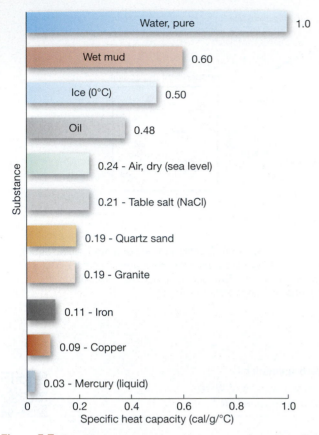

Figure 5.7 **Specific heat capacity of common substances.** Bar graph showing the specific heat capacity of common substances at 20°C (68°F). Note that water has a very high specific heat capacity, which means it takes a lot of energy to increase water temperature.

*As water boils, it reaches a plateau where all energy added is used to break intermolecular bonds in water, not increase its temperature. This is called the **latent heat of vaporization**.*

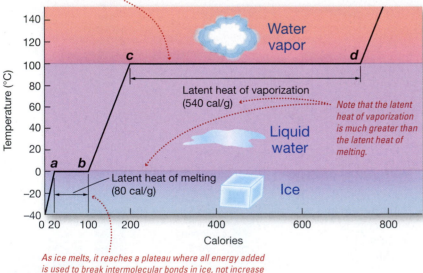

*As ice melts, it reaches a plateau where all energy added is used to break intermolecular bonds in ice, not increase its temperature. This is called the **latent heat of melting**.*

 SmartFigure 5.8 **Latent heats and changes of state of water.** The latent heat of melting (80 calories per gram) is much less than the latent heat of vaporization (540 calories per gram). See text for description of points *a, b, c,* and *d.*
https://goo.gl/18osWM

pattern of other chemical compounds with molecules of similar mass, it should melt at –90°C (–130°F) and boil at –68°C (–90°F). If that were the case, all water on Earth would be in the gaseous state. Instead, water melts and boils at the relatively high temperatures of 0°C (32°F) and 100°C (212°F),[8] respectively, because additional heat energy is required to overcome its hydrogen bonds and van der Waals forces. Thus, if not for the unusual geometry and resulting polarity of the water molecule, all water on Earth would be boiled away, and life as we know it would not exist.

WATER'S HEAT CAPACITY AND SPECIFIC HEAT **Heat capacity** is *the amount of heat energy required to raise the temperature of a substance by 1 degree centigrade.* Substances that have high heat capacity can absorb (or lose) large quantities of heat with only a small change in temperature. Conversely, substances that change temperature rapidly when heat is applied—such as oil or metals—have lower heat capacity.

The heat capacity per unit mass of a body, called *specific heat capacity* or, more simply, **specific heat,** is used to more directly compare the heat capacity of substances. For example, as shown in **Figure 5.7**, pure water has a high specific heat capacity that is exactly 1 calorie per gram,[9] whereas other common substances have much lower specific heats. Note that metals such as iron and copper—which heat up rapidly when heat is applied—have capacity values that are about 10 times lower than that of water.

Why does water have such high heat capacity? The reason is because it takes more energy to increase the kinetic energy of hydrogen-bonded water molecules than it does for substances in which the dominant intermolecular interaction is the much weaker van der Waals force. As a result, water gains or loses much more heat than other common substances while undergoing an equal temperature change. In addition, water resists any change in temperature, as you may have observed when heating a large pot of water. When heat is applied to the pot, which is made of metal that has a low heat capacity, the pot heats up quickly. The water *inside* the pot, however, takes a long time to heat up (hence, the saying that a watched pot never boils but an unwatched pot boils over!). Making the water boil takes even more heat because all the hydrogen bonds must be broken. The exceptional capacity of water to absorb large quantities of heat helps explain why water is used in home heating, industrial and automobile cooling systems, and home cooking applications.

WATER'S LATENT HEATS When water undergoes a change of state—that is, when ice melts or water freezes, or when water boils or water vapor condenses—a large amount of heat is absorbed or released. The amount of heat absorbed or released is due to water's high latent (*latent* = hidden) heats and is closely related to water's unusually high heat capacity. As water evaporates from your skin, it cools your body by absorbing heat (this is why sweating cools your body). Conversely, if you have ever been scalded by water vapor—steam—you know that steam releases an enormous amount of latent heat when it condenses to a liquid.

LATENT HEAT OF MELTING The graph in **Figure 5.8** shows how latent heat affects the amount of energy needed to increase water temperature and change the state of water. Beginning with 1 gram of ice (*lower left*), the addition of 20 calories of

[8]Note that the temperature scale *centigrade* (*centi* = a hundred, *grad* = step) is based on 100 even divisions between the melting and boiling points of pure water. It is also called the Celsius scale, after its founder (see Appendix I, "Metric and English Units Compared").

[9]Note that the specific heat capacity of water is used as the unit of heat quantity, the calorie. Thus, water is the standard against which the specific heats of other substances are compared.

heat raises the temperature of the ice by 40 degrees, from −40°C to 0°C (point *a* on the graph). The temperature remains at 0°C (32°F) even though more heat is being added, as shown by the plateau on the graph between points *a* and *b*. The temperature of the water does not change until 80 more calories of heat energy have been added. The **latent heat of melting** is the energy needed to break the intermolecular bonds that hold water molecules rigidly in place in ice crystals. The temperature remains unchanged until most of the bonds are broken and the mixture of ice and water has changed completely to 1 gram of water.

After the change from ice to liquid water has occurred at 0°C (32°F), additional heat raises the water temperature between points *b* and *c* in Figure 5.8. As it does, it takes 1 calorie of heat to raise the temperature of the gram of water 1°C (or 1.8°F). Therefore, another 100 calories must be added before the gram of water reaches the boiling point of 100°C (212°F). So far, a total of 200 calories has been added to reach point *c*.

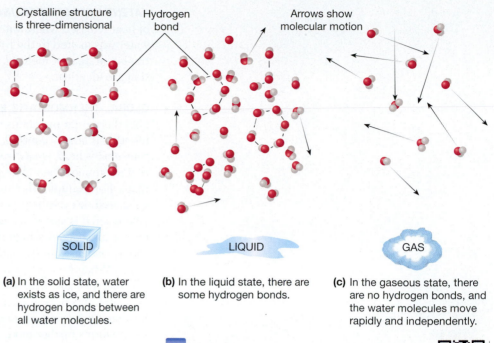

(a) In the solid state, water exists as ice, and there are hydrogen bonds between all water molecules.

(b) In the liquid state, there are some hydrogen bonds.

(c) In the gaseous state, there are no hydrogen bonds, and the water molecules move rapidly and independently.

 SmartFigure 5.9 **Hydrogen bonds in H$_2$O and the three states of matter.**
https://goo.gl/Gp5JK5

LATENT HEAT OF VAPORIZATION The graph in Figure 5.8 flattens out again at 100°C (212°F), between points *c* and *d*. This plateau represents the **latent heat of vaporization,** which is 540 calories per gram for water. This is the amount of heat that must be added to 1 gram of a substance at its boiling point to break the intermolecular bonds and complete the change of state from liquid to vapor (gas).

The drawings in **Figure 5.9**, which show the structure of water molecules in the solid, liquid, and gaseous states, help explain why the latent heat of vaporization is so much greater than the latent heat of melting. To go from a solid to a liquid, just enough hydrogen bonds must be broken to allow water molecules to slide past one another (see **Figure 5.9b**). To go from a liquid to a gas, however, all of the hydrogen bonds must be completely broken so that individual water molecules can move about freely (see **Figure 5.9c**).

LATENT HEAT OF EVAPORATION Sea surface temperatures average 20°C (68°F) or less. How, then, does liquid water convert to vapor at the surface of the ocean? The conversion of a liquid to a gas below the boiling point is called **evaporation.** At ocean surface temperatures, individual molecules converted from the liquid to the gaseous state have less energy than do water molecules at 100°C (212°F). To gain the additional energy necessary to break free of the surrounding ocean water molecules, an individual molecule must capture heat energy from its neighbors. In other words, the molecules left behind have lost heat energy to those that evaporate, which explains the cooling effect of evaporation.

It takes more than 540 calories of heat to produce 1 gram of water vapor from the ocean surface at temperatures less than 100°C (212°F). At 20°C (68°F), for instance, the **latent heat of evaporation** is 585 calories per gram. More heat is required because more hydrogen bonds must be broken. At higher temperatures, liquid water has fewer hydrogen bonds because the molecules are vibrating and jostling about more.

LATENT HEAT OF CONDENSATION When water vapor is cooled sufficiently, it condenses to a liquid and releases its **latent heat of condensation** into the surrounding air. On a small scale, the heat released is enough to cook food; this is how a steamer works. On a large scale, the heat released is sufficient to power large thunderstorms and even hurricanes (see Chapter 6, "Air–Sea Interaction").

LATENT HEAT OF FREEZING Heat is also released when water freezes. The amount of heat released when water freezes is the same amount that was absorbed when the water was melted in the first place. Thus, the **latent heat of freezing** is identical to the latent heat of melting. Similarly, the latent heats of vaporization and condensation are identical.

GLOBAL THERMOSTATIC EFFECTS Most people are familiar with the way a household thermostat maintains temperature inside a house. Earth has a natural thermostat, too, that is largely controlled by the properties of water. These **thermostatic effects** (*thermos* = heat, *stasis* = standing) of water include the unique properties of water that act to moderate changes in global temperature, which in turn affect Earth's climate. For example, the huge amount of heat energy exchanged in the evaporation–condensation cycle helps make life possible on Earth. The Sun radiates energy to Earth, where some is stored in the oceans. Evaporation removes this heat energy from the oceans and carries it high into the atmosphere. In the cooler upper atmosphere, water vapor condenses into clouds, which are the source of **precipitation** (mostly rain and snow). When precipitation occurs, it also releases water's latent heat of condensation. The map in **Figure 5.10** shows how this cycle of evaporation and condensation removes huge amounts of heat energy from the low-latitude oceans and adds huge amounts of heat energy to the heat-deficient higher latitudes. In addition, the heat released when sea ice forms further moderates Earth's high-latitude regions near the poles.

The exchange of latent heat between ocean and atmosphere is very efficient. For every gram of water that condenses in cooler latitudes, the amount of heat released to warm these regions equals the amount of heat removed from the tropical ocean when that gram of water was evaporated initially. The end result is that the thermal

Figure 5.10 **Atmospheric transport of surplus heat from low latitudes into heat-deficient high latitudes.**

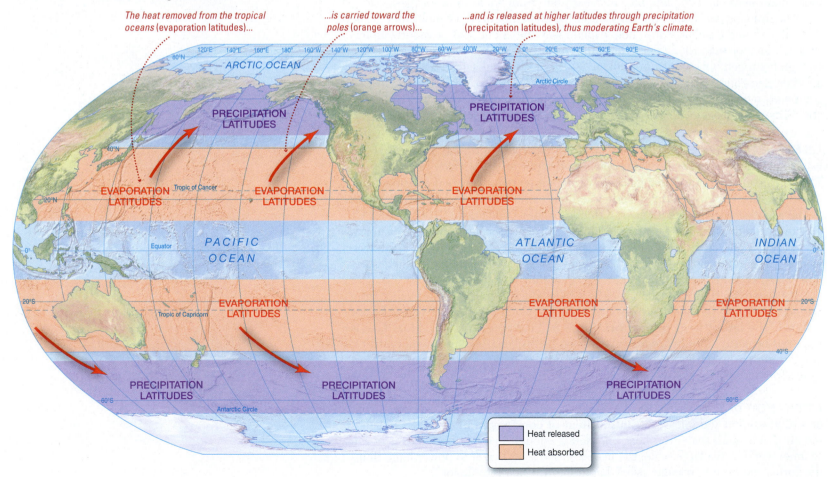

For a
erties of s

5.3 H

What is th
ter? One
ter contain
salty tast
sodium ch
salts, met
contain e
layer mor
height of
content of
irrigating
to many n

Salinity

Salinity (
cluding di
atures) bu
particles b
because t
solved sub

The s
than fresh
96.5% pu
ter, its ph
variations

Figur
the sulfat
dissolved
tified in s
naturally
solved con

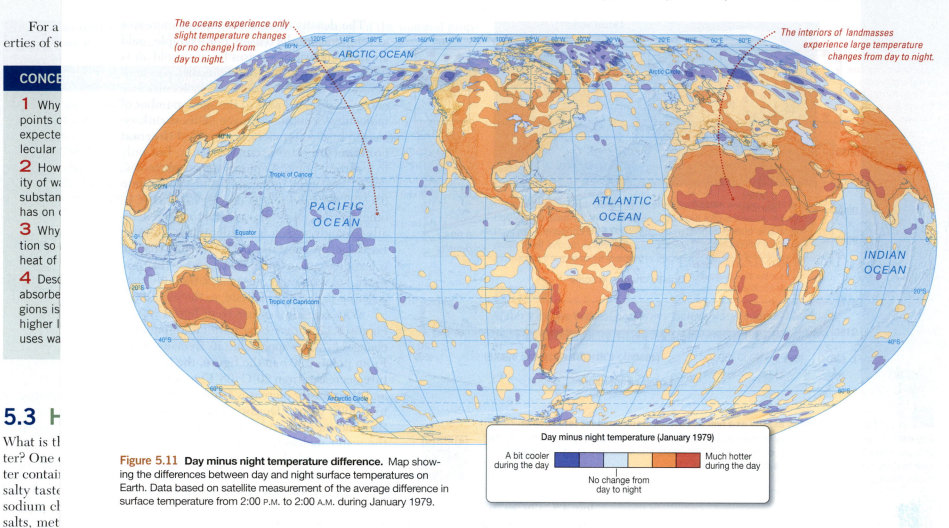

The oceans experience only slight temperature changes (or no change) from day to night.

The interiors of landmasses experience large temperature changes from day to night.

Figure 5.11 **Day minus night temperature difference.** Map showing the differences between day and night surface temperatures on Earth. Data based on satellite measurement of the average difference in surface temperature from 2:00 P.M. to 2:00 A.M. during January 1979.

Day minus night temperature (January 1979)

A bit cooler during the day | No change from day to night | Much hotter during the day

properties of water have prevented wide variations in Earth's temperature, thus moderating Earth's climate. Because rapid environmental changes can often result in the death of many life-forms, our planet's moderated climate is one of the main reasons life exists on Earth.

Climate

Connection

Another thermostatic effect of the ocean can be seen in **Figure 5.11**, which shows the temperature difference between day and night. The map shows that in the ocean, there is only a small difference in temperature between day and night, while the land experiences a much greater variation. This difference between ocean and land is due to the higher heat capacity of water, which gives it the ability to absorb the daily gains and minimize the daily losses of heat energy much more easily than the dirt and rock of landmasses. The ability of the oceans to moderate temperatures along coastlines and islands is referred to as a **marine effect**. Alternatively, areas less affected by the sea and therefore having a greater range of temperature differences—both daily and yearly—are said to experience a **continental effect.**

Water Density as a Result of Thermal Contraction

Recall from Chapter 1 that density is mass per unit volume and can be thought of as *how heavy something is for its size*. Ultimately, density is related to how tightly the molecules or ions of a substance are packed together. Typical units of density are grams per cubic centimeter (g/cm^3). Pure water, for example, has a density of 1.0 g/cm^3. Note that temperature, salinity, and pressure all affect water density.

TABLE 5.1 SELECTED DISSOLVED MATERIALS IN 35‰ SEAWATER

1. Major constituents (in parts per thousand by weight, ‰)

Constituent	Concentration (‰)	Ratio of constituent/total salts (%)
Chloride (Cl^-)	19.2	55.04
Sodium (Na^+)	10.6	30.61
Sulfate (SO_4^{2-})	2.7	7.68
Magnesium (Mg^{2+})	1.3	3.69
Calcium (Ca^{2+})	0.40	1.16
Potassium (K^+)	0.38	1.10
Total	**34.58‰**	**99.28%**

2. Minor constituents (in parts per million by weight, ppm[a])

Gases		Nutrients		Others	
Constituent	Concentration (ppm)	Constituent	Concentration (ppm)	Constituent	Concentration (ppm)
Carbon dioxide (CO_2)	90	Silicon (Si)	3.0	Bromide (Br^-)	65.0
Nitrogen (N_2)	14	Nitrogen (N)	0.5	Carbon (C)	28.0
Oxygen (O_2)	6	Phosphorus (P)	0.07	Strontium (Sr)	8.0
		Iron (Fe)	0.002	Boron (B)	4.6

3. Trace constituents (in parts per billion by weight, ppb[b])

Constituent	Concentration (ppb)	Constituent	Concentration (ppb)	Constituent	Concentration (ppb)
Lithium (Li)	185	Zinc (Zn)	10	Lead (Pb)	0.03
Rubidium (Rb)	120	Aluminum (Al)	2	Mercury (Hg)	0.03
Iodine (I)	60	Manganese (Mn)	2	Gold (Au)	0.005

[a]Note that 1000 ppm = 1‰.

[b]Note that 1000 ppb = 1 ppm.

Salinity is often expressed in **parts per thousand (‰)**. For example, as 1% is 1 part in 100, 1‰ is 1 part in 1000. When converting from percent to parts per thousand, the decimal is simply moved one place to the right. For instance, typical seawater salinity of 3.5% is the same as 35‰. Advantages of expressing salinity in parts per thousand are that decimals are often avoided and values convert directly to grams of salt per kilogram of seawater. For example, 35‰ seawater has 35 grams of salt in every 1000 grams of seawater.[10]

Determining Salinity

Early methods of determining seawater salinity involved evaporating a carefully weighed amount of seawater and weighing the salts that precipitated from it. However, the accuracy of this time-consuming method is limited because some water can remain bonded to salts that precipitate and some substances can evaporate along with the water.

Another way to measure salinity is to use the *principle of constant proportions*, which was firmly established by chemist William Dittmar (1859–1951) when he analyzed the water samples collected during the *Challenger* Expedition (see MasteringOceanography **Web Diving Deeper 5.2**). The **principle of constant**

5.1 Squidtoons

My big tooth
is a
salinometer!

https://goo.gl/kbHv7K

[10]Note that the units "parts per thousand" are effectively parts per thousand by weight. Salinity values, however, lack units because the salinity of a water sample is determined as the *ratio* of the electrical conductivity of the sample to the electrical conductivity of a standard. Thus, salinity values are sometimes reported in *p.s.u.*, or *practical salinity units*, which are equivalent to parts per thousand.

Figure 5.13
of actual snow
flakes mirror th
held together b

HOW TO AVOID GOITERS

Interdisciplinary

Relationship

The nutritional label on containers of salt usually proclaims "this product contains iodine, a necessary nutrient." Why is iodine necessary in our diets? It turns out that if a person's diet contains an insufficient amount of iodine, a potentially life-threatening affliction called **goiters** (*guttur* = throat) may result (**Figure 5A**).

Iodine is used by the thyroid gland, which is a butterfly-shaped organ located in the neck in front of and on either side of the trachea (windpipe). The thyroid gland manufactures hormones that regulate cellular metabolism essential for mental development and physical growth. If people lack iodine in their diet, their thyroid glands cannot function properly. Often, this results in the enlargement or swelling of the thyroid gland. Severe symptoms include dry skin, loss of hair, puffy face, weakness of muscles, weight increase, diminished vigor, mental sluggishness, and a large nodular growth on the neck called a

goiter. If proper steps are not taken to correct this disease, it can lead to cancer. Iodine ingested regularly often begins to reverse the effects. In advanced stages, surgery to remove

Figure 5A A woman with goiters.

the goiter or exposure to radioactivity is the only course of action.

How can you avoid goiters? Fortunately, goiters can be prevented with a diet that contains just *trace amounts* of iodine. Where can you get iodine in your diet? All products from the sea contain trace amounts of iodine because iodine is one of the many elements dissolved in seawater. Sea salt, seafood, seaweed, and other sea products contain plenty of iodine to help prevent goiters. Although goiters are rarely a problem in developed nations like the United States, goiters pose a serious health hazard in many underdeveloped nations, especially those far from the sea. In the United States, however, many people get too much iodine in their diet, leading to the overproduction of hormones by the thyroid gland. That's why most stores that sell iodized salt also carry noniodized salt for those people who have a *hyperthyroid* (*hyper* = excessive, *thyroid* = the thyroid gland) condition and must restrict their intake of iodine.

GIVE IT SOME THOUGHT

1. What are goiters? How can they be avoided?

proportions states that the major dissolved constituents responsible for the salinity of seawater occur nearly everywhere in the ocean in exactly the same proportions, independent of salinity. The ocean, therefore, is well mixed. When salinity changes, moreover, the salts don't leave (or enter) the ocean, but water molecules do. Seawater has *constancy of composition*, so the concentration of a single major constituent can be measured to determine the total salinity of a given water sample. The constituent that occurs in the greatest abundance and is the easiest to measure accurately is the chloride ion, Cl^-. The weight of this ion in a water sample is its **chlorinity**.

In any sample of ocean water worldwide, the chloride ion accounts for 55.04% of the total proportion of dissolved solids (Figure 5.15 and Table 5.1). Therefore, by measuring only the chloride ion concentration, the total salinity of a seawater sample can be determined using the following relationship:

$$\text{Salinity (‰)} = 1.80655 \times \text{chlorinity (‰)}° \qquad (5.1)$$

For example, the average chlorinity of the ocean is 19.2‰, so the average salinity is 1.80655×19.2‰, which rounds to 34.7‰. In other words, on average there are 34.7 parts of dissolved material in every 1000 parts of seawater.

Standard seawater consists of ocean water analyzed for chloride ion content to the nearest ten-thousandth of a part per thousand by the Institute of Oceanographic Services in Wormley, England. It is then sealed in small glass vials called *ampules*

°The number 1.80655 comes from dividing 1 by 0.5504 (the chloride ion's proportion in seawater of 55.04%). However, if you actually divide this, you will get 1.81686, which is different from the original value by 0.57%. Empirically, oceanographers have found that seawater's constancy of composition is an approximation and have agreed to use 1.80655 because it more accurately represents the total salinity of seawater.

STUDENTS SOMETIMES ASK . . .

What is the strategy behind adding salt to a pot of water when making pasta? Does it make the water boil faster?

Adding salt to water will not make the water boil faster. It will, however, make the water boil at a slightly higher temperature because dissolved substances raise its boiling point (and, in fact, also *lower* its freezing point; see Table 5.2). Thus, the pasta will cook in slightly less time. In addition, the salt adds flavoring, so the pasta may taste better, too. Be sure to add the salt after the water has come to a boil, though, or it will take longer to reach a boil. This is a wonderful use of chemical principles—helping you to cook better!

Electrode

Freshwater is not electrically conductive, so the bulb does not light up.

Adding dissolved salts to water increases its conductivity, so the bulb lights up.

Freshwater The more dissolved salts, the brighter the bulb shines. Saltwater

Figure 5.16 **Salinity affects water conductivity.**

A similar egg sinks in freshwater because of fresh-water's lower density.

An egg floats in saltwater because of saltwater's high density.

Saltwater Freshwater

Figure 5.17 **An egg floats in saltwater but sinks in freshwater.**

SmartTable 5.2 **Comparison of selected properties of pure water and seawater.**
https://goo.gl/qecylF

and sent to laboratories throughout the world for use as a reference standard in calibrating analytical equipment.

Seawater salinity can be measured very accurately with modern oceanographic instruments such as a **salinometer** (*salinus* = salt, *meter* = measure). Most salinometers measure seawater's *electrical conductivity* (the ability of a substance to transmit electric current), which increases as more substances are dissolved in water (**Figure 5.16**). Salinometers can determine salinity to resolutions of better than 0.003‰.

Comparing Pure Water and Seawater

Table 5.2 compares various properties of pure water and seawater. Because seawater is 96.5% water, most of its physical properties are very similar to those of pure water. For instance, adding a small amount of dissolved salt to water does not change its transparency, and so the color of pure water and seawater is identical.

The dissolved substances in seawater, however, give it slightly different yet important physical properties, as compared to pure water. For example, recall that dissolved substances interfere with pure water changing state. The freezing points and boiling points in Table 5.2 show that dissolved substances decrease the freezing point and increase the boiling point of water. Thus, seawater freezes at a temperature of –1.9°C (28.6°F), which is lower than the freezing point of pure water (0°C [32°F]). Similarly, seawater boils at a temperature of 100.6°C (213.1°F), which is higher than the boiling point of pure water (100°C [212°F]). In effect, the salts in seawater extend the range of temperatures in which water is a liquid. This same principle applies to antifreeze used in automobile radiators. Antifreeze lowers the freezing point of the water in a radiator and increases the boiling point, thus extending the range over which the water remains in the liquid state. Antifreeze, therefore, protects your radiator from freezing in the winter *and* from boiling over in the summer.

Density is another property that exhibits small but remarkable differences between pure water and seawater. Recall that density is defined as mass per unit volume. When substances are added to water and dissolved, the water's density increases because more mass has been added per unit volume. Although the difference in density between pure water and seawater seems negligible (Table 5.2 shows an increase of only 0.028 g/cm³), a simple experiment with an egg in two different glasses of water shows how dramatically small differences in density can affect floating objects (**Figure 5.17**).

Other important properties of seawater (such as its pH and how seawater density varies with depth) are discussed later in this chapter.

SMARTTABLE 5.2	COMPARISON OF SELECTED PROPERTIES OF PURE WATER AND SEAWATER		
Property		**Pure water**	**35‰ seawater**
Color (light transmission)	Small quantities of water	Clear (high transparency)	Same as for pure water
	Large quantities of water	Blue-green because water molecules scatter blue and green wavelengths best	Same as for pure water
Odor		Odorless	Distinctly marine
Taste		Tasteless	Distinctly salty
pH		7.0 (neutral)	Surface waters, range = 8.0–8.3; average = 8.1 (slightly alkaline)
Freezing point		0°C (32°F)	–1.9°C (28.6°F)
Boiling point		100°C (212°F)	100.6°C (213.1°F)
Density at 4°C (39°F)		1.000 g/cm³	1.028 g/cm³

CONCEPT CHECK 5.3 | Demonstrate an understanding of what salinity is and how salinity is measured.

1 What is the average salinity of seawater? What units are normally used, and why are those units useful?

2 What condition of salinity makes it possible to determine the total salinity of ocean water by measuring the concentration of only one constituent, the chloride ion?

3 In what ways are seawater and pure water similar? How are the two different?

5.4 Why Does Seawater Salinity Vary?

Using salinometers and other techniques, oceanographers have determined that salinity varies from place to place in the oceans. What are the patterns of seawater salinity, and what causes them?

Salinity Variations

In the open ocean far from land, salinity varies between about 33 and 38‰. In coastal areas, salinity variations can be extreme. In the Baltic Sea, for example, salinity averages only 10‰ because physical conditions create **brackish** (*brak* = salt, *ish* = somewhat) water. Brackish water is produced in areas where freshwater (from rivers and high rainfall) and seawater mix. In the Red Sea, on the other hand, salinity averages 42‰ because physical conditions produce **hypersaline** (*hyper* = excessive, *salinus* = salt) water. Hypersaline water is typical of seas and inland bodies of water that experience high evaporation rates and limited open-ocean circulation.

Some of the most hypersaline water in the world is found in inland lakes, which are often called seas because they are so salty. The Great Salt Lake in Utah, for example, has a salinity of 280‰, and the Dead Sea on the border of Israel and Jordan has a salinity of 330‰. The water in the Dead Sea, therefore, contains 33% dissolved solids and is almost *10 times saltier than seawater*. As a result, hypersaline waters are so dense and buoyant that one can easily float (**Figure 5.18**), even with arms and legs sticking up above water level! Hypersaline waters also taste much saltier than seawater.

Salinity of seawater in coastal areas also varies seasonally. For example, the salinity of seawater off Miami Beach, Florida, varies from about 34.8‰ in October to 36.4‰ in May and June, when evaporation is high. Offshore of Astoria, Oregon, seawater salinity is always extremely low because of the vast freshwater input from the Columbia River. Here, surface water salinity can be as low as 0.3‰ in April and May (when the Columbia River is at its maximum flow rate) and 2.6‰ in October (the dry season, when freshwater input is reduced).

Other types of water have much lower salinity. Tap water, for instance, has salinity somewhere below 0.8‰, and good-tasting tap water is usually below 0.6‰. Salinity of premium bottled water is on the order of 0.3‰, with the salinity often displayed

RECAP

Seawater salinity can be measured using a salinometer and averages 35‰. The dissolved components in seawater give it different yet important physical properties as compared to pure water.

STUDENTS SOMETIMES ASK ...

I've seen the labels on electric cords warning against using electrical appliances close to water. Are these warnings because water's polarity allows electricity to be transmitted through it?

Yes and no. Water molecules are polar, so you might assume that water is a good conductor of electricity. Pure water is a very poor conductor, however, because water molecules are neutral overall and will not move toward the negatively or positively charged pole in an electrical system. If an electrical appliance is dropped into a tub of absolutely pure water, the water molecules will transmit no electricity. Instead, the water molecules will simply orient their positively charged hydrogen ends toward the negative pole of the appliance and their negatively charged oxygen ends toward its positive pole, which tends to neutralize the electric field. Interestingly, it is the dissolved substances that transmit electrical current through water (see Figure 5.16). Even slight amounts, such as those in tap water, allow electricity to be transmitted. That's why there are warning labels on the electric cords of household appliances that are commonly used in the bathroom, such as blow dryers, electric razors, and heaters. That's also why it is recommended to stay out of any water—including a bathtub or shower—during a lightning storm!

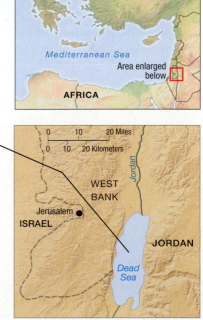

Figure 5.18 High-salinity water of the Dead Sea allows swimmers to easily float. The Dead Sea, which has 330‰ salinity (almost 10 times the salinity of seawater), has high density. As a result, it also has high buoyancy that allows swimmers to float easily.

What would happen to a person if he or she drank seawater?

It depends on the quantity. The salinity of seawater is about four times greater than that of your body fluids. In your body, seawater causes your internal membranes to lose water through *osmosis* (*osmos* = to push), which transports water molecules from higher concentrations (the normal body chemistry of your internal fluids) to areas of lower concentrations (your digestive tract containing seawater). Thus, your natural body fluids would move into your digestive tract and eventually be expelled, causing dehydration.

Don't worry too much if you've inadvertently swallowed some seawater. As a nutritional drink, seawater provides seven important nutrients and contains no fat, cholesterol, or calories. Some people even claim that drinking a small amount of seawater daily gives them good health! However, beware of microbial contaminants in seawater, such as viruses and bacteria that can often exist in great quantities.

SmartTable 5.3 **Processes that affect seawater salinity.**
https://goo.gl/fZpUXm

prominently on its label, usually as total dissolved solids (TDS) in units of parts per million (ppm), where 1000 ppm equals 1‰.

Processes Affecting Seawater Salinity

Processes affecting seawater salinity change either the amount of water (H_2O molecules) or the amount of dissolved substances in the water. Adding more water, for instance, dilutes the dissolved component and lowers the salinity of the sample. Conversely, removing water increases salinity. Changing the salinity in these ways does not affect the *amount* or the *composition* of the dissolved components, which remain in constant proportions. Let's first examine processes that affect the amount of water in seawater before turning our attention to processes that influence dissolved components.

PROCESSES THAT DECREASE SEAWATER SALINITY Table 5.3 summarizes the processes that affect seawater salinity. Precipitation, **runoff** (stream discharge), melting icebergs, and melting sea ice *decrease* seawater salinity by adding more freshwater to the ocean. Precipitation is the way atmospheric water returns to Earth as rain, snow, sleet, and hail. Worldwide, about three-quarters of all precipitation falls directly back into the ocean and one-quarter falls onto land. Precipitation falling directly into the oceans adds freshwater, reducing seawater salinity.

Most of the precipitation that falls on land returns to the oceans indirectly as stream runoff. Even though this water dissolves minerals on land, the runoff is relatively pure water, as shown in **Table 5.4.** Runoff, therefore, adds mostly water to the ocean, causing seawater salinity to decrease.

Icebergs are chunks of ice that have broken free (*calved*) from a glacier when it flows into an ocean or marginal sea and begins to melt. Glacial ice originates as snowfall in high mountain areas, so icebergs are composed of freshwater. When icebergs melt in the ocean, they add freshwater, which is another way in which seawater salinity is reduced.

Sea ice forms when ocean water freezes in high-latitude regions and is composed primarily of freshwater. When warmer temperatures return to high-latitude

SmartTable 5.3	PROCESSES THAT AFFECT SEAWATER SALINITY					
Process	How accomplished	Adds or removes	Effect on salt in seawater	Effect on H_2O in seawater	Salinity increase or decrease?	Source of freshwater from the sea?
Precipitation	Rain, sleet, hail, or snow falls directly on the ocean	Adds very fresh water	None	More H_2O	Decrease	N/A
Runoff	Streams carry water to the ocean	Adds mostly fresh water	Negligible addition of salt	More H_2O	Decrease	N/A
Icebergs melting	Glacial ice calves into the ocean and melts	Adds very fresh water	None	More H_2O	Decrease	Yes, icebergs from the Antarctic have been towed to South America
Sea ice melting	Sea ice melts in the ocean	Adds mostly fresh water and some salt	Adds a small amount of salt	More H_2O	Decrease	Yes, sea ice can be melted and is better than drinking seawater
Sea ice forming	Seawater freezes in cold ocean areas	Removes mostly freshwater	30% of salts in seawater are retained in ice	Less H_2O	Increase	Yes, through multiple freezings, called *freeze separation*
Evaporation	Seawater evaporates in hot climates	Removes very pure water	None (essentially all salts are left behind)	Less H_2O	Increase	Yes, through evaporation of seawater and condensation of water vapor, called *distillation*

TABLE 5.4	COMPARISON OF MAJOR DISSOLVED COMPONENTS IN STREAMS WITH THOSE IN SEAWATER	
Constituent	**Concentration in streams (parts per million by weight)**	**Concentration in seawater (parts per million by weight)**
Bicarbonate ion (HCO_3^-)	58.4	trace
Calcium ion (Ca^{2+})	15.0	400
Silicate (SiO_2)	13.1	3
Sulfate ion (SO_4^{2-})	11.2	2700
Chloride ion (Cl^-)	7.8	19,200
Sodium ion (Na^+)	6.3	10,600
Magnesium ion (Mg^{2+})	4.1	1300
Potassium ion (K^+)	2.3	380
Total (parts per million)	119.2 ppm	34,793 ppm
Total (‰)	0.1192‰	34.8‰

regions in the summer, sea ice melts in the ocean, adding mostly freshwater with a small amount of salt to the ocean. Seawater salinity, therefore, is decreased.

PROCESSES THAT INCREASE SEAWATER SALINITY The formation of sea ice and evaporation *increase* seawater salinity by removing water from the ocean (Table 5.3). Sea ice forms when seawater freezes. Depending on the salinity of seawater and the rate of ice formation, about 30% of the dissolved components in seawater are retained in sea ice. This means that 35‰ seawater creates sea ice with about 10‰ salinity (30% of 35‰ is 10‰). Consequently, the formation of sea ice removes mostly freshwater from seawater, increasing the salinity of the remaining unfrozen water. High-salinity water also has a high density, so it sinks below the surface.

Recall that evaporation is the conversion of water molecules from the liquid state to the vapor state at temperatures below the boiling point. Evaporation removes water from the ocean, leaving its dissolved substances behind. Evaporation, therefore, increases seawater salinity. Worldwide, about 86% of all evaporation occurs in the oceans.

THE HYDROLOGIC CYCLE Figure 5.19 shows the **hydrologic cycle,** (*hydro* = water, *logos* = study of) which describes the continual movement of water on, above, and below the surface of Earth. The movement of water through various components of the hydrologic cycle involves processes that recycle water among the ocean, the atmosphere, and the continents, illustrating that water is in constant motion between the different components (or *reservoirs*) of the hydrologic cycle. Note that many of the processes of the hydrologic cycle affect seawater salinity. For example, river runoff into the ocean changes seawater salinity in that region. The figure also shows that of Earth's reservoirs, the vast majority of water at or near Earth's surface is contained in the ocean.

Interdisciplinary
Relationship

In addition, Figure 5.19 shows the average yearly amounts of transfer, or *flux*, of water between various reservoirs.

Dissolved Components Added to and Removed from Seawater

Seawater salinity is a function of the amount of dissolved components in seawater. Interestingly, dissolved substances do not remain in the ocean forever. Instead, they are cycled into and out of seawater by the processes shown in Figure 5.20. These processes include stream runoff, in which streams dissolve ions from continental rocks and carry

RECAP

Various surface processes either decrease seawater salinity (precipitation, runoff, icebergs melting, or sea ice melting) or increase seawater salinity (sea ice forming and evaporation).

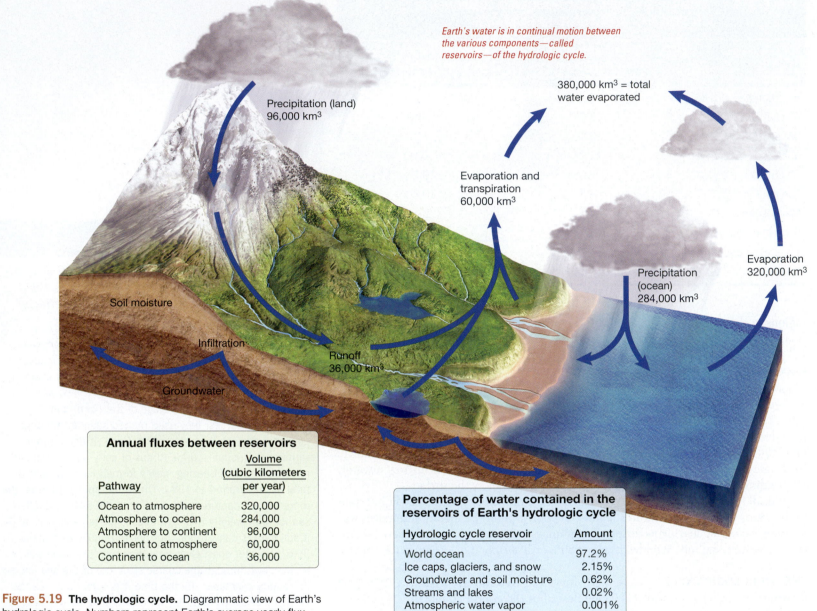

Earth's water is in continual motion between the various components—called reservoirs—of the hydrologic cycle.

Precipitation (land)
96,000 km³

Evaporation and transpiration
60,000 km³

380,000 km³ = total water evaporated

Evaporation
320,000 km³

Precipitation (ocean)
284,000 km³

Soil moisture

Infiltration

Runoff
36,000 km³

Groundwater

Annual fluxes between reservoirs

Pathway	Volume (cubic kilometers per year)
Ocean to atmosphere	320,000
Atmosphere to ocean	284,000
Atmosphere to continent	96,000
Continent to atmosphere	60,000
Continent to ocean	36,000

Percentage of water contained in the reservoirs of Earth's hydrologic cycle

Hydrologic cycle reservoir	Amount
World ocean	97.2%
Ice caps, glaciers, and snow	2.15%
Groundwater and soil moisture	0.62%
Streams and lakes	0.02%
Atmospheric water vapor	0.001%

Figure 5.19 The hydrologic cycle. Diagrammatic view of Earth's hydrologic cycle. Numbers represent Earth's average yearly flux (volume of water moved between reservoirs) in cubic kilometers. Left table shows average yearly flux between reservoirs; right table shows the percentage of Earth's water in each reservoir.

Web Animation
Earth's Water and the Hydrologic Cycle
http://goo.gl/3Ra4l2

them to the sea, and volcanic eruptions, both on the land and on the sea floor. Other sources include the atmosphere (which contributes gases) and biological interactions.

Stream runoff is the primary method by which dissolved substances are added to the oceans. Table 5.4 compares the major components dissolved in stream water with those in seawater. It shows that streams have far lower salinity and a vastly different composition of dissolved substances than seawater. For example, bicarbonate ion (HCO_3^-) is the most abundant dissolved constituent in stream water yet is found in only trace amounts in seawater. Conversely, the most abundant dissolved component in seawater is the chloride ion (Cl^-), which exists in very small concentrations in streams.

If stream water is the main source of dissolved substances in seawater, why do the components of the two not match each other more closely? One of the reasons is that some dissolved substances stay in the ocean and accumulate over time. **Residence time** is the average length of time that a substance resides in the ocean. Long residence times lead to higher concentrations of the dissolved substance. The sodium ion (Na^+), for instance, has a residence time of 260 million years and, as a result, has a high concentration in the ocean. Other elements such as aluminum have a residence time of only 100 years and occur in seawater in much lower concentrations.

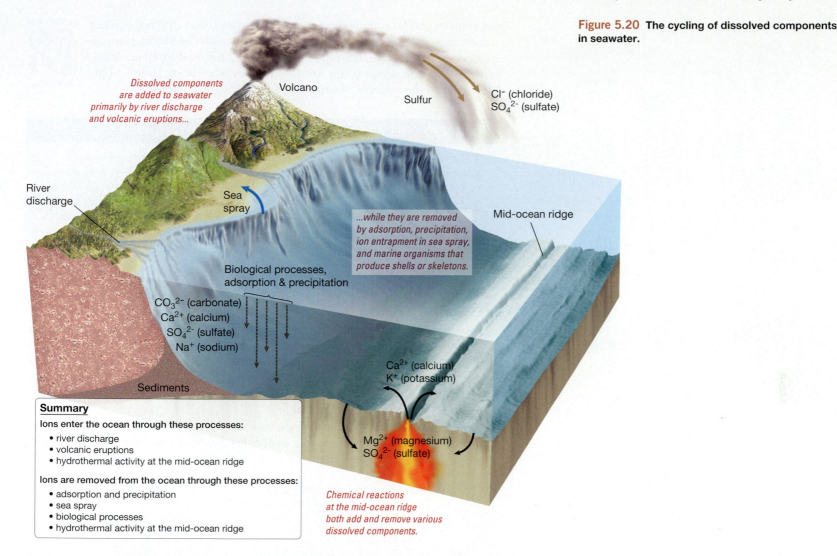

Figure 5.20 **The cycling of dissolved components in seawater.**

Dissolved components are added to seawater primarily by river discharge and volcanic eruptions...

Volcano

Sulfur

Cl⁻ (chloride)
SO₄²⁻ (sulfate)

River discharge

Sea spray

...while they are removed by adsorption, precipitation, ion entrapment in sea spray, and marine organisms that produce shells or skeletons.

Mid-ocean ridge

Biological processes, adsorption & precipitation

CO₃²⁻ (carbonate)
Ca²⁺ (calcium)
SO₄²⁻ (sulfate)
Na⁺ (sodium)

Ca²⁺ (calcium)
K⁺ (potassium)

Sediments

Mg²⁺ (magnesium)
SO₄²⁻ (sulfate)

Chemical reactions at the mid-ocean ridge both add and remove various dissolved components.

Summary

Ions enter the ocean through these processes:

- river discharge
- volcanic eruptions
- hydrothermal activity at the mid-ocean ridge

Ions are removed from the ocean through these processes:

- adsorption and precipitation
- sea spray
- biological processes
- hydrothermal activity at the mid-ocean ridge

Are the oceans becoming saltier through time? This might seem logical since new dissolved components are constantly being added to the oceans and because most salts have long residence times. However, analysis of ancient marine organisms and sea floor sediments suggests that the oceans have not increased in salinity over time. This must be because the rate at which an element is added to the ocean equals the rate at which it is removed, so the average *amounts* of various elements remain constant (this is called a *steady-state* condition).

Materials added to the oceans are counteracted by several processes that cycle dissolved substances out of seawater. When waves break at sea, for example, sea spray releases tiny salt particles into the atmosphere, where they may be blown over land before being washed back to Earth. The amount of material leaving the ocean in this way is enormous: According to a recent study, as much as 3.3 billion metric tons (7.3 trillion pounds) of salt as sea spray enter the atmosphere each year. Another example is the infiltration of seawater along mid-ocean ridges near hydrothermal vents (see Figure 5.20), which incorporates magnesium and sulfate ions into sea floor mineral deposits. In fact, chemical studies of seawater indicate that the *entire volume of ocean water* is recycled through this hydrothermal circulation system at the mid-ocean ridge approximately every 3 million years. As a result, the chemical exchange between ocean water and the basaltic crust has a major influence on the composition of ocean water.

Interdisciplinary

Relationship

Dissolved substances are also removed from seawater in other ways. Calcium, carbonate, sulfate, sodium, and silicon are deposited in ocean sediments within the shells of dead microscopic organisms and animal feces. Vast amounts of dissolved

substances can be removed when inland arms of seas dry up, leaving salt deposits called *evaporites* (such as those beneath the Mediterranean Sea; see MasteringOceanography Web Diving Deeper 4.1). In addition, ions dissolved in ocean water are removed by adsorption (physical attachment) to the surfaces of sinking clay and biological particles.

Interdisciplinary Relationship

CONCEPT CHECK 5.4 I Explain why seawater salinity varies.

1 What physical conditions create brackish water in the Baltic Sea and hypersaline water in the Red Sea?

2 Describe the ways in which dissolved components are added and removed from seawater.

3 List the components (reservoirs) of the hydrologic cycle that hold water on Earth and the percentage of Earth's water in each one. Describe the processes by which water moves among these reservoirs.

5.5 Is Seawater Acidic or Basic?

An **acid** is a compound that releases hydrogen ions (H^+) when dissolved in water. The resulting solution is said to be *acidic*. A strong acid readily and completely releases hydrogen ions when dissolved in water. An **alkaline**, or a **base**, is a compound that releases hydroxide ions (OH^-) when dissolved in water. The resulting solution is said to be *alkaline*, or *basic*. A strong base readily and completely releases hydroxide ions when dissolved in water.

Both hydrogen ions and hydroxide ions are present in extremely small amounts at all times in water because water molecules dissociate and reform. Chemically, this is represented by the equation:

$$H_2O \underset{\text{reform}}{\overset{\text{dissociate}}{\rightleftharpoons}} H^+ + OH^- \tag{5.2}$$

Note that if the hydrogen ions and hydroxide ions in a solution are due only to the dissociation of water molecules, they are always found in equal concentrations, and the solution is consequently neutral.

When substances dissociate in water, they can make the solution acidic or basic. For example, if hydrochloric acid (HCl) is added to water, the resulting solution will be acidic because there will be a large excess of hydrogen ions from the dissociation of the HCl molecules. Conversely, if a base such as baking soda (sodium bicarbonate, $NaHCO_3$) is added to water, the resulting solution will be basic because there will be an excess of hydroxide ions (OH^-) from the dissociation of $NaHCO_3$ molecules.

The pH Scale

Figure 5.21 shows the **pH** (power of hydrogen) **scale**, which is a measure of the hydrogen ion concentration of a solution. Values for pH range from 0 (strongly acidic) to 14 (strongly alkaline or basic), and the pH of a **neutral** solution such as pure water is 7.0. The pH scale is not linear: A decrease of 1.0 pH unit corresponds to a 10-fold increase in the concentration of hydrogen ions, making the water more acidic, whereas a change of 1.0 unit upward corresponds to a 10-fold decrease, making the water more alkaline.

Ocean surface waters have a pH that averages about 8.1 and ranges from about 8.0 to 8.3, so seawater is slightly alkaline. At depth, seawater pH is generally lower than surface waters (**Figure 5.22**) Water in the ocean combines with carbon dioxide to form a weak acid, called carbonic acid (H_2CO_3), which dissociates and releases hydrogen ions (H^+):

$$H_2O + CO_2 \rightarrow H_2CO_3 \rightarrow H^+ + HCO_3^- \tag{5.3}$$

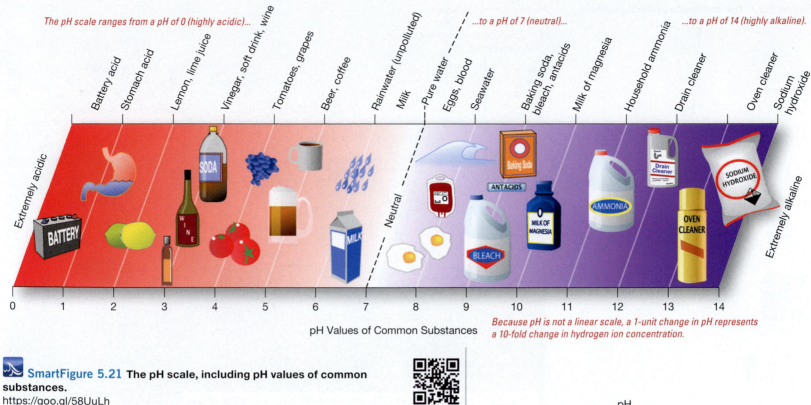

The pH scale ranges from a pH of 0 (highly acidic)... ...to a pH of 7 (neutral)... ...to a pH of 14 (highly alkaline).

Battery acid · Stomach acid · Lemon, lime juice · Vinegar, soft drink, wine · Tomatoes, grapes · Beer, coffee · Rainwater (unpolluted) · Milk · Pure water · Eggs, blood · Seawater · Baking soda, bleach, antacids · Milk of magnesia · Household ammonia · Drain cleaner · Oven cleaner · Sodium hydroxide

Extremely acidic

Neutral

Extremely alkaline

0 1 2 3 4 5 6 7 8 9 10 11 12 13 14

pH Values of Common Substances

Because pH is not a linear scale, a 1-unit change in pH represents a 10-fold change in hydrogen ion concentration.

SmartFigure 5.21 The pH scale, including pH values of common substances.
https://goo.gl/58UuLh

This reaction would seem to make the ocean slightly acidic. Carbonic acid, however, keeps the ocean slightly alkaline through the process of *buffering*.

The Carbonate Buffering System

The chemical reactions in **Figure 5.23** show that carbon dioxide (CO_2) combines with water (H_2O) to form carbonic acid (H_2CO_3). Carbonic acid can then lose a hydrogen ion (H^+) to form the negatively charged bicarbonate ion (HCO_3^-). The bicarbonate ion can lose its hydrogen ion, too, though it does so less readily than carbonic acid. When the bicarbonate ion loses its hydrogen ion, it forms the double-charged negative carbonate ion (CO_3^{2-}), some of which combines with calcium ions to form calcium carbonate ($CaCO_3$). Some of the calcium carbonate is precipitated by various inorganic and organic means, and then it sinks and cycles back into the ocean by dissolving at depth.

The equations below Figure 5.23 show how these chemical reactions involving carbonate minimize changes in the pH of the ocean in a process called **buffering**. Buffering protects the ocean from getting too acidic or too basic, similarly to how buffered aspirin protects sensitive stomachs. For example, if the pH of the ocean increases (becomes too basic), it causes H_2CO_3 to release H^+, and pH drops. Conversely, if the pH of the ocean decreases (becomes too acidic), HCO_3^- combines with H^+ to remove it, causing pH to rise. In this way, buffering prevents large swings of ocean water pH and allows the ocean to stay within a limited range of pH values. Recently, however, increasing amounts of carbon dioxide from human emissions are beginning to enter the ocean and change the ocean's pH, making it more acidic. For more details on this process, see Chapter 16, "The Oceans and Climate Change."

Climate Connection

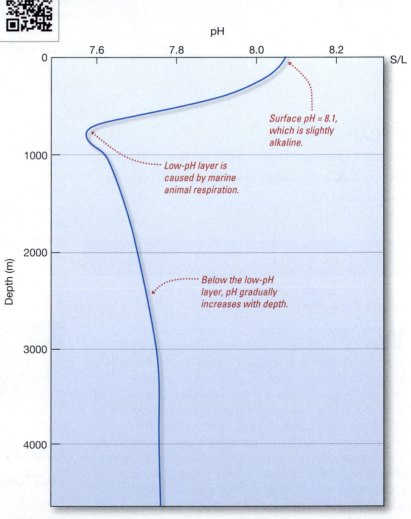

pH

7.6 7.8 8.0 8.2

S/L

Surface pH = 8.1, which is slightly alkaline.

Low-pH layer is caused by marine animal respiration.

Below the low-pH layer, pH gradually increases with depth.

Depth (m)

Figure 5.22 Seawater pH varies with ocean depth.

Figure 5.23 The carbonate buffering system.

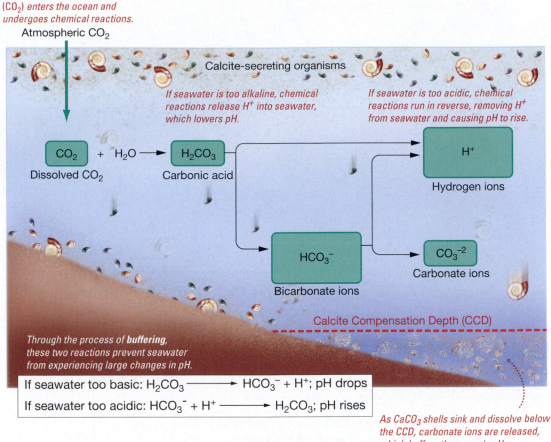

Atmospheric carbon dioxide (CO₂) enters the ocean and undergoes chemical reactions.

Atmospheric CO₂

Calcite-secreting organisms

If seawater is too alkaline, chemical reactions release H⁺ into seawater, which lowers pH.

If seawater is too acidic, chemical reactions run in reverse, removing H⁺ from seawater and causing pH to rise.

CO₂ + H₂O → H₂CO₃

Dissolved CO₂ Carbonic acid

H⁺

Hydrogen ions

HCO₃⁻

Bicarbonate ions

CO₃⁻²

Carbonate ions

Calcite Compensation Depth (CCD)

*Through the process of **buffering**, these two reactions prevent seawater from experiencing large changes in pH.*

If seawater too basic: $H_2CO_3 \longrightarrow HCO_3^- + H^+$; pH drops

If seawater too acidic: $HCO_3^- + H^+ \longrightarrow H_2CO_3$; pH rises

As CaCO₃ shells sink and dissolve below the CCD, carbonate ions are released, which buffers the ocean's pH.

Web Animation
The Carbonate Buffering System
https://goo.gl/ZsCngq

RECAP

Reactions involving carbonate chemicals serve to buffer the ocean and help maintain its average pH at 8.1 (slightly alkaline, or basic).

STUDENTS SOMETIMES ASK ...

Why do carbonated beverages burn my throat when I drink them?

When carbon dioxide gas (CO₂) dissolves in water (H₂O), its molecules often cling to water molecules and form carbonic acid (H₂CO₃). Carbonic acid is a weak acid, in which most molecules are intact at any given moment. However, some of those molecules naturally break apart and exist as two fragments: a negatively charged HCO₃⁻ ion and a positively charged H⁺ ion. The H⁺ ions are responsible for acidity—the higher their concentration in a solution, the more acidic that solution. The presence of carbonic acid in carbonated water makes that water acidic—the more carbonated, the more acidic. What you're feeling when you drink a carbonated beverage is the moderate acidity of that beverage irritating your throat.

Deep-ocean water contains more carbon dioxide than surface water because deep water is cold and has the ability to dissolve more gases. Also, the higher pressures of the deep ocean further aid the dissolution of gases in seawater. If carbon dioxide combines with water to form carbonic acid, why isn't the cold water of the deep ocean highly acidic? When microscopic marine organisms that make their shells out of calcium carbonate (calcite) die and sink into the deep ocean, they neutralize the acid through buffering. In essence, these organisms act as an "antacid" for the deep ocean, analogous to the way commercial antacids use calcium carbonate to neutralize excess stomach acid. As explained in Chapter 4, these shells are readily dissolved below the calcite (calcium carbonate) compensation depth (CCD).

Interdisciplinary

Relationship

| **CONCEPT CHECK 5.5** | Discuss the acid/base properties of seawater. |

1 Explain the difference between an acidic substance and an alkaline (basic) substance.

2 How does the ocean's buffering system work?

5.6 How Does Seawater Salinity Vary at the Surface and with Depth?

From the surface to the ocean depths, the ocean undergoes variations in salinity, temperature, and density that create a layered ocean. This layering affects the mixing of ocean water, the movement of currents, and the distribution of marine life. In

this section and the next, we'll explore the variations of properties both at the surface and with depth that cause the ocean to be layered.

Surface Salinity Variation

Although seawater salinity at the ocean surface averages 35‰, surface salinity varies depending on the latitude (**Figure 5.24**). The red curve in Figure 5.24 shows temperature, which is low in the high latitudes but steadily increases with latitude all the way to the equator. The green curve in the figure shows salinity, which is lowest at high latitudes, peaks in the lower latitudes near the Tropics of Cancer and Capricorn, and dips near the equator.

Why does surface salinity vary in the pattern shown in Figure 5.24? At high latitudes, abundant precipitation and runoff and the melting of freshwater icebergs all decrease salinity. In addition, cool temperatures limit the amount of evaporation that takes place (which would increase salinity). The formation and melting of sea ice balance each other out in the course of a year and are not a factor in changes in salinity.

At low latitudes, the pattern of Earth's atmospheric circulation (see Chapter 6, "Air–Sea Interaction") causes warm, dry air to descend, so near the Tropics of Cancer and Capricorn evaporation rates are high and salinity increases. In addition, little precipitation and runoff occur to decrease salinity. As a result, the regions near the Tropics of Cancer and Capricorn are the continental *and* maritime deserts of the world.

Temperatures are warm near the equator, so evaporation rates are high enough to increase salinity. Increased precipitation and runoff partially offsets the high salinity, though. For example, daily rain showers are common along the equator, adding water to the ocean and lowering its salinity.

Figure 5.25 is a map of satellite-collected data that shows how ocean surface salinity varies worldwide. Notice how the overall pattern of the satellite image matches the graph in Figure 5.24. For example, both the graph and the satellite image show high salinity in the subtropics (Figure 5.25, *orange*) and lower salinity in rainy polar regions and equatorial belts (Figure 5.25, *blue*). In addition, notice that the Atlantic Ocean has higher salinity values than the Pacific. The Atlantic Ocean's higher overall salinity is caused by its proximity to land and the associated continental effect. This causes high rates of evaporation in the narrower Atlantic Ocean, particularly in the tropics. The satellite image also reveals an expanse of low-salinity water from the Amazon River's outflow (Figure 5.25, *purple*).

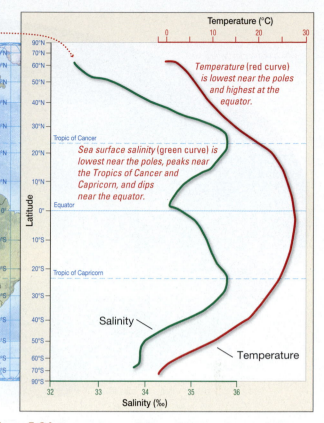

The presence of large amounts of runoff from land in far northern latitudes causes salinity to be lower there compared to equivalent latitudes in the Southern Hemisphere.

Temperature (red curve) is lowest near the poles and highest at the equator.

Sea surface salinity (green curve) is lowest near the poles, peaks near the Tropics of Cancer and Capricorn, and dips near the equator.

Salinity

Temperature

Figure 5.24 Sea surface salinity and temperature variation with latitude. Graph showing variation with latitude in sea surface salinity (*green curve*) along with sea surface temperature (*red curve*).

Salinity Variation with Depth

Figure 5.26 shows how seawater salinity varies with depth. The graph displays data for the open ocean far from land and shows one curve for high-latitude regions and one for low-latitude regions.

The curve to the right in Figure 5.26 shows the salinity change with depth for low-latitude regions (such as in the tropics). This curve shows *increased* salinity at the surface because of the reasons discussed in the preceding section. Note that even along the equator where surface salinity dips (Figure 5.24), the salinity value is still relatively high. Then with increasing depth, the curve changes to a *lower* salinity value for the remainder of the water column.

The curve to the left in Figure 5.26 shows the salinity change with depth for high-latitude regions (such as near Antarctica or in the Gulf of Alaska). This curve shows *decreased* salinity at the surface also because of the reasons discussed in

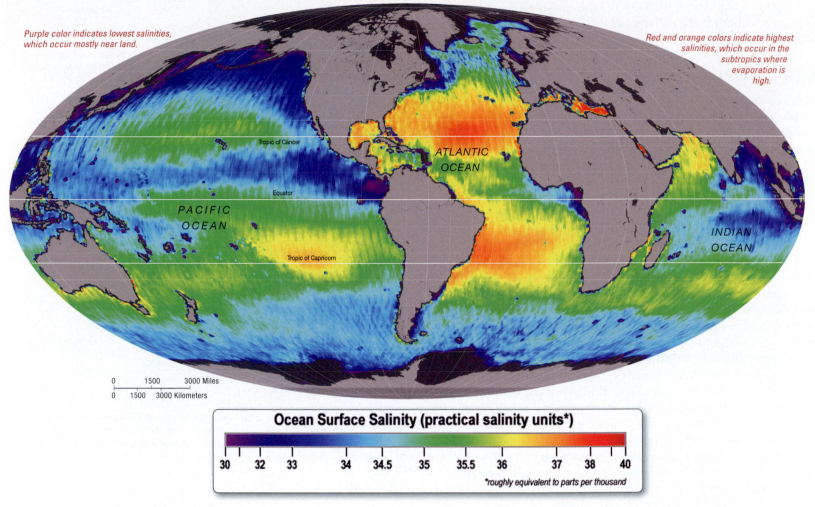

Purple color indicates lowest salinities, which occur mostly near land.

Red and orange colors indicate highest salinities, which occur in the subtropics where evaporation is high.

Tropic of Cancer

ATLANTIC OCEAN

Equator

PACIFIC OCEAN

INDIAN OCEAN

Tropic of Capricorn

| 0 | 1500 | 3000 Miles |
| 0 | 1500 | 3000 Kilometers |

Ocean Surface Salinity (practical salinity units*)

| 30 | 32 | 33 | 34 | 34.5 | 35 | 35.5 | 36 | 37 | 38 | 40 |

roughly equivalent to parts per thousand

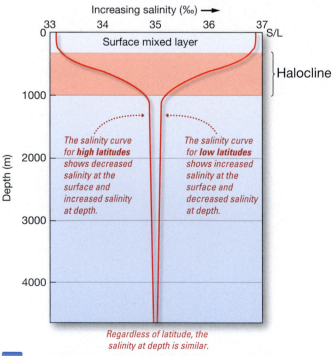

Increasing salinity (‰) →

Surface mixed layer

Halocline

*The salinity curve for **high latitudes** shows decreased salinity at the surface and increased salinity at depth.*

*The salinity curve for **low latitudes** shows increased salinity at the surface and decreased salinity at depth.*

Depth (m)

Regardless of latitude, the salinity at depth is similar.

 SmartFigure 5.26 Salinity variation with depth.
Graph showing a vertical profile of high- and low-latitude salinity variation with depth. Horizontal scale is in ‰; vertical scale is depth in meters, with sea level at the top. The layer of rapidly changing salinity is called the halocline. https://goo.gl/9b1dP7

Figure 5.25 **Satellite-derived surface salinity of the oceans.** Map of ocean surface salinity from data collected by the Aquarius satellite during January, 2015. Values are in practical salinity units, which are roughly equivalent to parts per thousand (‰); black regions indicate no data; the north-south color striations are an artifact from the satellite's orbital path.

the preceding section. Then with increasing depth, the curve changes to a *higher* salinity value for the remainder of the water column. Note that both curves on the graph show that regardless of latitude, there is a similar, intermediate value of salinity at depth.

These two curves, which together resemble the outline of a wide Champagne glass, show that salinity varies widely at the surface but hardly varies at all in the deep ocean. Why is this so? It occurs because all the processes that affect seawater salinity (precipitation, runoff, melting icebergs, melting sea ice, sea ice forming, and evaporation) occur at the *surface* and thus have no effect on deep water below.

Halocline

Both curves in Figure 5.26 show a rapid change in salinity between the depths of about 300 meters (980 feet) and 1000 meters (3300 feet). For the low-latitude curve, the change is a *decrease* in salinity. For the high-latitude curve, the change is an *increase* in salinity. In both cases, this layer of rapidly changing salinity with depth is called a **halocline** (*halo* = salt, *cline* = slope). Haloclines separate layers of different salinity in the ocean.

CONCEPT CHECK 5.6 | Specify how seawater salinity varies at the surface and with depth.

1 Why is there low surface salinity in the high latitudes, and why is there higher surface salinity in the low latitudes?

2 Explain the dip in surface salinity values near the equator that is shown in Figure 5.24.

3 Describe the halocline, including where it occurs in the ocean.

RECAP

A halocline is a layer of rapidly changing salinity that occurs in both high- and low-latitude regions.

5.7 How Does Seawater Density Vary with Depth?

The density of pure water is 1.000 gram per cubic centimeter (g/cm^3) at 4°C (39°F). This value serves as a standard against which the density of all other substances can be measured. Seawater contains various dissolved substances that increase its density. In the open ocean, seawater density averages between 1.022 and 1.030 g/cm^3 (depending on its salinity). Thus, the density of seawater is 2 to 3% greater than that of pure water. Unlike freshwater, seawater continues to increase in density until it freezes at a temperature of –1.9°C (28.6°F). (Recall that below 4°C [39°F], the density of freshwater actually *decreases*; see Figure 5.12.) At its freezing point, however, seawater behaves in a similar fashion to freshwater: Its density decreases dramatically, which is why sea ice floats, too.

Density is an important property of ocean water because density differences determine the vertical position of ocean water and cause water masses to float or sink, thereby creating deep-ocean currents. For example, if seawater with a density of 1.030 g/cm^3 were added to freshwater with a density of 1.000 g/cm^3, the denser seawater would sink below the freshwater, initiating a deep current.

Factors Affecting Seawater Density

The ocean, like Earth's interior, is layered according to density. Low-density water exists near the surface, and higher-density water occurs below. Except for some shallow inland seas with high rates of evaporation that create high-salinity water, the highest-density water is found at the deepest ocean depths. Let's examine how temperature, salinity, and pressure influence seawater density by expressing the relationships using arrows (up arrow = increase; down arrow = decrease):

- As temperature increases (↑), seawater density decreases (↓)[11] (due to thermal expansion).
- As salinity increases (↑), seawater density increases (↑) (due to the addition of more dissolved material).
- As pressure increases (↑), seawater density increases (↑) (due to the compressive effects of pressure).

Of these three factors, only temperature and salinity influence the density of surface water. Pressure influences seawater density only when very high pressures are encountered, such as in deep-ocean trenches. Still, the density of seawater in the deep ocean is only about 5% greater than at the ocean surface, showing that despite tons of pressure per square centimeter, water is nearly incompressible. Unlike air, which can be compressed and put in a tank for use in scuba diving, the molecules in liquid water are already close together and cannot be compressed much

[11] A relationship where one variable *decreases* as a result of another variable's *increase* is known as an inverse relationship, in which the two variables are *inversely proportional*.

RECAP

Differences in ocean density cause the ocean to be layered. Seawater density increases with decreased temperature, increased salinity, and increased pressure. Temperature has the greatest influence on seawater density.

more. Therefore, pressure has the least effect on influencing the density of surface water and can largely be ignored.

Temperature, on the other hand, has the greatest influence on surface seawater density because the range of surface seawater temperature is greater than that of salinity. In fact, only in the extreme polar areas of the ocean, where temperatures are low and remain relatively constant, does salinity significantly affect density. Cold water that also has high salinity is some of the highest-density water in the world. The density of seawater—the result of its salinity and temperature—influences currents in the deep ocean because high-density water sinks below less-dense water.

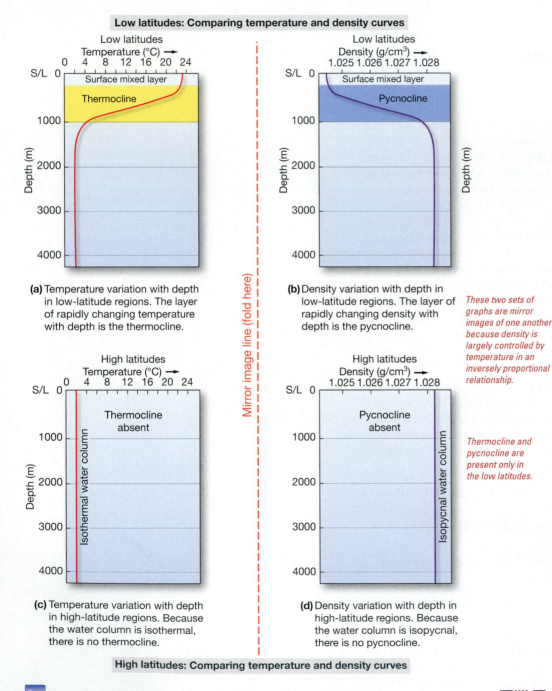

Low latitudes: Comparing temperature and density curves

(a) Temperature variation with depth in low-latitude regions. The layer of rapidly changing temperature with depth is the thermocline.

(b) Density variation with depth in low-latitude regions. The layer of rapidly changing density with depth is the pycnocline.

These two sets of graphs are mirror images of one another because density is largely controlled by temperature in an inversely proportional relationship.

Mirror image line (fold here)

(c) Temperature variation with depth in high-latitude regions. Because the water column is isothermal, there is no thermocline.

(d) Density variation with depth in high-latitude regions. Because the water column is isopycnal, there is no pycnocline.

Thermocline and pycnocline are present only in the low latitudes.

High latitudes: Comparing temperature and density curves

🌊 **SmartFigure 5.27 Comparing vertical profile curves for temperature and density in the low and high latitudes.** Paired graphs comparing temperature and density curves in the low latitudes (*a* and *b, above*) and temperature and density curves in the high latitudes (*c* and *d, below*). https://goo.gl/1zC18l

Temperature and Density Variation with Depth

The four graphs in **Figure 5.27** compares the vertical profile curves for temperature and density in both low- and high-latitude regions. Let's examine each graph individually.

Figure 5.27a shows how temperature varies with depth in low-latitude regions, where surface waters are warmed by high Sun angles and constant length of days. However, the Sun's energy does not penetrate very far into the ocean. Surface water temperatures remain relatively constant until a depth of about 300 meters (980 feet) because of good surface mixing mechanisms such as surface currents, waves, and tides. Below about 300 meters (980 feet), the temperature decreases rapidly until a depth of about 1000 meters (3300 feet). Below 1000 meters, the water's low temperature again remains constant down to the ocean floor.

The density curve for low-latitude regions in **Figure 5.27b** shows that density is relatively low at the surface. Density is low because surface water temperatures are high. (Remember that temperature has the greatest influence on density and temperature is inversely proportional to density.) Below the surface, density remains constant also until a depth of about 300 meters (980 feet) because of good surface mixing. Below about 300 meters (980 feet), the density increases rapidly until a depth of about 1000 meters (3300 feet). Below 1000 meters, the water's low density again remains constant down to the ocean floor.

Figure 5.27c shows how temperature varies with depth in high-latitude regions, where surface waters remain cool year-round and deep-water temperatures are about the same as the surface. The temperature curve for high-latitude regions, therefore, is a straight vertical line, which indicates uniform conditions at the surface and at depth.

The density curve for high-latitude regions (**Figure 5.27d**) also shows hardly any variation with depth. Density is relatively high at the

surface because surface water temperatures are low. Density is high below the surface, too, because water temperature is also low. The density curve for high-latitude regions, therefore, is also represented by a straight vertical line, which indicates uniform conditions at the surface and at depth. These conditions allow cold high-density water to form at the surface, sink, and initiate deep-ocean currents.

One of the most important things to notice about Figure 5.27 is that the two graphs shown in the top portion of Figure 5.27 are related to each other, as are the two graphs shown in the bottom portion of Figure 5.27. If you can imagine folding Figure 5.27 along its vertical dashed line and overlaying the two sets of graphs, you would notice they are identical mirror images of each other. For example, the low latitude temperature graph (Figure 5.27a) is a mirror image of its corresponding density graph (Figure 5.27b). Similarly, the high latitude temperature graph (Figure 5.27c) is a mirror image of its corresponding density graph (Figure 5.27d). Why are the curves mirror images of each other? As discussed previously, temperature is the most important factor that influences seawater density and it operates as an inversely proportional relationship. This is exactly the relationship that is illustrated by the mirror images of these two sets of graphs.

Thermocline and Pycnocline

Analogous to the halocline (the layer of rapidly changing salinity shown in Figure 5.26), the low-latitude temperature graph in Figure 5.27a displays a curving line that indicates a layer of rapidly changing temperature called a **thermocline** (*thermo* = heat, *cline* = slope). Similarly, the low-latitude density graph in Figure 5.27b displays a curving line that indicates a **pycnocline** (*pycno* = density, *cline* = slope), which is a layer of rapidly changing density. Note that the corresponding high-latitude graphs of temperature (Figure 5.27c) and density (Figure 5.27d) both lack a thermocline and a pycnocline, respectively, because these lines show a constant value with depth (they are straight vertical lines that don't curve). Like a halocline, a thermocline and a pycnocline typically occur between about 300 meters (980 feet) and 1000 meters (3300 feet) below the surface. The temperature difference between water above and below the thermocline can be used to generate electricity (see MasteringOceanography **Web Diving Deeper 5.1**).

When a pycnocline is established in an area, it presents an incredible barrier to mixing between low-density water above and high-density water below. A pycnocline has a high gravitational stability and thus physically isolates adjacent layers of water.[12] The pycnocline results from the combined effect of the thermocline and the halocline because temperature and salinity influence density. The interrelationship of these three layers determines the degree of separation between the upper-water and deep-water masses.

The ocean is layered into three distinct water masses based on density. The **mixed surface layer** occurs above a strong permanent thermocline (and corresponding pycnocline; see Figure 5.27). The water is uniform because it is well mixed by surface currents, waves, and tides. The thermocline and pycnocline occur in a relatively low-density layer called the **upper water,** which is well developed throughout the low and middle latitudes. Denser and colder **deep water** extends from below the thermocline/pycnocline to the deep-ocean floor.

Thermoclines (and corresponding pycnoclines) can occur in other locations, too. Scuba divers, for example, often experience minor thermoclines as they descend into the ocean. Thermoclines can also develop in swimming pools, ponds, and lakes. During the spring and fall, when nights are cool but days can be quite warm, the Sun heats the surface water of the pool, yet the water below the surface can be quite cold. If the pool has not been mixed, a thermocline isolates the warm

[12]This is similar to a temperature inversion in the atmosphere, which traps cold (high-density) air underneath warm (low-density) air.

RECAP

A halocline is a layer of rapidly changing salinity, a thermocline is a layer of rapidly changing temperature, and a pycnocline is a layer of rapidly changing density.

surface layer from the deeper cold water. The cold water below the thermocline can be quite a surprise for anyone who dives into the pool!

In high-latitude regions, the temperature of the surface water remains cold year round, so there is very little difference between the temperature at the surface and in deep water below. Thus, a thermocline and corresponding pycnocline rarely develop in high-latitude regions. Only during the short summer when the days are long does the Sun begin to heat surface waters. Even then, the water does not heat up very much. Nearly all year, then, the water column in high latitudes is **isothermal** (*iso* = same, *thermo* = heat) and **isopycnal** (*iso* = same, *pycno* = density), allowing good vertical mixing between surface and deeper waters.

CONCEPT CHECK 5.7 | Specify how seawater density varies with depth.

1 What are the three factors that affect seawater density? Describe how each factor influences seawater density, including which one is the most important.

2 Describe the thermocline, including where it occurs in the ocean.

3 Describe the pycnocline, including where it occurs in the ocean.

4 Why is there such a close association between (a) the curve showing seawater density variation with ocean depth and (b) the curve showing seawater temperature variation with ocean depth?

5.8 What Methods Are Used to Desalinate Seawater?

More than a third of the world's population already suffers from shortages of drinkable water—with a rise to 50% expected by 2025. Because the human consumption of freshwater is growing even as its supply is dwindling, several countries have begun to use the ocean as a source of freshwater. **Desalination,** or salt removal from seawater, can provide freshwater for business, home, and agricultural use.

Although seawater is mostly just water molecules, its ability to form hydrogen bonds, easily dissolve so many substances, and resist changes in temperature and state makes seawater difficult to desalinate. As a result, desalination is energy intensive and expensive. The high cost of desalination, however, is only one issue. Recent studies, for example, indicate that desalination can negatively affect marine life by entrapping them in intake pipes and by releasing highly salty leftover brine back to the ocean. Still, using the sea as a source of freshwater is attractive to many coastal communities that have few other sources.

Currently, there are more than 13,000 desalination plants worldwide, the majority of them very small and located in arid regions of the Middle East, Caribbean, and Mediterranean. These plants produce more than 45 billion liters (12 billion gallons) of freshwater daily. The United States produces only about 10% of the world's desalted water, primarily in Florida. To date, only a limited number of desalination plants have been built along the California coast, primarily because the cost of desalination is generally higher than the costs of other water supply alternatives available in California (such as water transfers and groundwater pumping) but also because the extensive permitting process is an impediment to building desalination facilities. However, as drought conditions occur and concern over water availability increases, desalination projects are being proposed at numerous locations in the state.

Because desalinated water requires a lot of energy and thus is expensive to produce, most desalination plants are small-scale operations. In fact, desalination plants provide less than 0.5% of human water needs. More than half of the world's desalination plants use *distillation* to purify water, while most of the remaining plants use *membrane processes*.

Distillation

The process of **distillation** (*distillare* = to trickle) is shown schematically in **Figure 5.28.** In distillation, saltwater is boiled, and the resulting water vapor is passed through a cooling condenser, where it condenses and is collected as freshwater. This simple procedure is very efficient at purifying seawater. For instance, distillation of 35‰ seawater produces freshwater with a salinity of only 0.03‰, which is about 10 times fresher than bottled water, so it needs to be mixed with less pure water to make it taste better. Distillation is expensive, however, because it requires large amounts of heat energy to boil the saltwater. Because of water's high latent heat of vaporization, it takes 540 calories to convert only 1 gram (0.035 ounce) of water at the boiling point to the vapor state.[13] Increased efficiency, such as using the waste heat from a power plant, is required to make distillation practical on a large scale.

Solar distillation, which is also known as **solar humidification,** does not require supplemental heating and has been used successfully in small-scale agricultural experiments in arid regions such as Israel, West Africa, and Peru. Solar humidification is similar to distillation in that saltwater is evaporated in a covered container, but the water is heated by direct sunlight instead (Figure 5.28). Saltwater in the container evaporates, and the water vapor that condenses on the cover runs into collection trays. The major difficulty lies in effectively concentrating the energy of sunlight into a small area to speed evaporation.

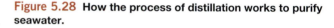

In this laboratory setup, water vapor condenses on a plastic sheet...

...and is captured, producing very fresh water (0.03‰).

(1) Solar Distillation

*The process of distillation requires using the Sun's energy to evaporate seawater (called **solar distillation**), or...*

H_2O vapor from **(1)** evaporation or **(2)** boiling

35‰ seawater

(2) Heat distillation

*...boiling seawater (called **heat distillation**). In either case, water vapor is produced.*

Figure 5.28 **How the process of distillation works to purify seawater.**

Membrane Processes

Electrolysis can be used to desalinate seawater, too. In this method, two electrodes—one positive electrode and one negative electrode—are situated in a container of seawater. When an electrical current is applied to the electrodes, positive ions such as sodium ions are attracted to the negative electrode, and negative ions such as chloride ions are attracted to the positive electrode. Then membranes are used to trap the ions. In time, enough ions are removed to convert seawater to freshwater. The major drawback to electrolysis is that it requires large amounts of energy, and so this method is more suited for desalinating brackish water than seawater.

Reverse osmosis (*osmos* = to push) may have potential for large-scale desalination. In osmosis, water molecules naturally pass through a thin, semipermeable membrane from a freshwater solution to a saltwater solution. In reverse osmosis, water on the salty side is highly pressurized to drive water molecules—but not salt and other impurities—through the membrane to the freshwater side (**Figure 5.29**). A significant problem with reverse osmosis is that the membranes are flimsy, become clogged, and must be replaced frequently. Advanced composite materials may help eliminate these problems because they are sturdier, provide better filtration, and last up to 10 years.

Worldwide, at least 30 countries are operating reverse osmosis units. Saudi Arabia—where energy from oil is cheap but water is scarce—has the world's largest reverse osmosis plant, which produces 485 million liters (128 million gallons) of desalinated water daily. The largest plant in the United States opened in 2008 in Tampa Bay, Florida, and produces up to 95 million liters (25 million gallons) of freshwater per day, which provides about 10% of the drinking water supply of the Tampa Bay region. Once permits are obtained, a new facility in Carlsbad, California, is designed to produce twice as much freshwater as the Tampa Bay plant. Reverse osmosis is also used in many household water purification units and aquariums.

[13]Even at 100% efficiency, it still requires a whopping *540,000 calories* of heat energy to produce 1 liter (about 1 quart) of distilled water.

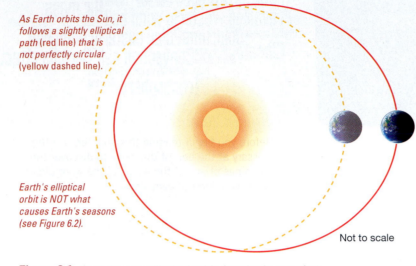

As Earth orbits the Sun, it follows a slightly elliptical path (red line) that is not perfectly circular (yellow dashed line).

Earth's elliptical orbit is NOT what causes Earth's seasons (see Figure 6.2).

Not to scale

Figure 6.1 Earth has an elliptical orbit, but that's not what causes Earth's seasons. Comparison between a slightly elliptical orbit (*red line*) and a perfectly circular orbit (*yellow dashed line*). View is directly above Earth's plane of the ecliptic; Earth's elliptical orbit is exaggerated for clarity (not to scale). Note that Earth's elliptical orbit is not the cause of Earth's seasons.

6.1 What Causes Variations in Solar Radiation on Earth?

A variety of factors cause changes in the amount of solar radiation (solar energy) that Earth receives. One of the most striking examples is the daytime–nighttime cycle: The side of Earth facing the Sun (the daytime side) receives a tremendous dose of intense solar radiation, while the nighttime side receives none. An example of a longer-term cycle is the change in seasons.

What Causes Earth's Seasons?

This seemingly simple question is the source of a common misconception: Even though Earth does revolve around the Sun in an elliptical orbit that varies only slightly from a perfect circle (**Figure 6.1**), Earth's seasons are not caused by Earth's changing distance from the Sun. As will be explained below, Earth's seasons are actually caused by the tilt of Earth's axis.

The surface connecting all points in Earth's orbit is called the **plane of the ecliptic** (**Figure 6.2**). More importantly, Figure 6.2 shows

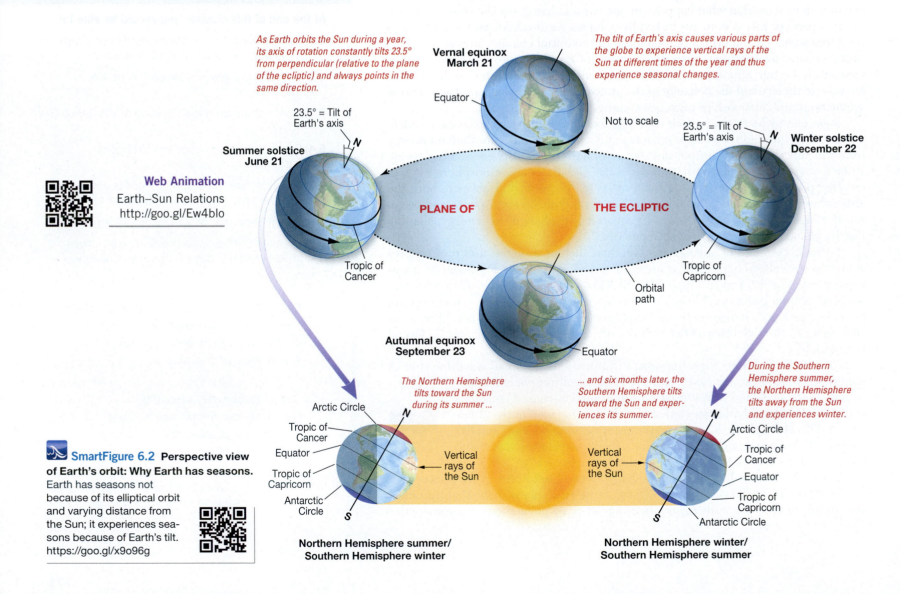

Web Animation
Earth–Sun Relations
http://goo.gl/Ew4blo

As Earth orbits the Sun during a year, its axis of rotation constantly tilts 23.5° from perpendicular (relative to the plane of the ecliptic) and always points in the same direction.

Vernal equinox
March 21

Equator

The tilt of Earth's axis causes various parts of the globe to experience vertical rays of the Sun at different times of the year and thus experience seasonal changes.

23.5° = Tilt of Earth's axis

Not to scale

23.5° = Tilt of Earth's axis

Winter solstice
December 22

Summer solstice
June 21

PLANE OF THE ECLIPTIC

Tropic of Cancer

Tropic of Capricorn

Orbital path

Autumnal equinox
September 23

Equator

SmartFigure 6.2 Perspective view of Earth's orbit: Why Earth has seasons. Earth has seasons not because of its elliptical orbit and varying distance from the Sun; it experiences seasons because of Earth's tilt. https://goo.gl/x9o96g

The Northern Hemisphere tilts toward the Sun during its summer ...

... and six months later, the Southern Hemisphere tilts toward the Sun and experiences its summer.

During the Southern Hemisphere summer, the Northern Hemisphere tilts away from the Sun and experiences winter.

Arctic Circle
Tropic of Cancer
Equator
Tropic of Capricorn
Antarctic Circle

Vertical rays of the Sun

Vertical rays of the Sun

Arctic Circle
Tropic of Cancer
Equator
Tropic of Capricorn
Antarctic Circle

**Northern Hemisphere summer/
Southern Hemisphere winter**

**Northern Hemisphere winter/
Southern Hemisphere summer**

that Earth's axis of rotation is not perpendicular ("upright") relative to the plane of the ecliptic; rather, it tilts at an angle of 23.5 degrees. As a result, *different hemispheres on Earth are tilted more directly toward or away from the Sun* during Earth's yearly orbit (Figure 6.2, *insets*), which is the cause of Earth's seasons (not Earth's elliptical orbit). An interesting consequence of Earth's tilt is that throughout its yearly cycle, Earth's axis *always points in the same direction*, which is toward Polaris, the North Star.

Interdisciplinary

Relationship

So, the tilt of Earth's rotational axis—not its elliptical orbit—is what causes Earth to have seasons. Let's examine a yearly progression of the seasons from spring, summer, fall, to winter:

- At the **vernal equinox** (*vernus* = spring; *equi* = equal, *noct* = night), which occurs on or about March 21, the Sun is directly overhead along the equator. During this time, all places in the world experience equal lengths of night and day (hence the name *equinox*). In the Northern Hemisphere, the vernal equinox is also known as the *spring equinox*.

- At the **summer solstice** (*sol* = the Sun, *stitium* = a stoppage), which occurs on or about June 21, the Sun reaches its most northerly point in the sky, directly overhead along the **Tropic of Cancer**, at 23.5 degrees north latitude (Figure 6.2, *left inset*). To an observer on Earth, the noonday Sun reaches its northernmost or southernmost position in the sky at this time and appears to pause—hence the term *solstice*—before beginning its next six-month cycle.

- At the **autumnal equinox** (*autumnus* = fall), which occurs on or about September 23, the Sun is directly overhead along the equator again. In the Northern Hemisphere, the autumnal equinox is also known as the *fall equinox*.

- At the **winter solstice**, which occurs on or about December 22, the Sun is directly overhead along the **Tropic of Capricorn**, at 23.5 degrees south latitude (Figure 6.2, *right inset*). In the Southern Hemisphere, the seasons are reversed. Thus, the winter solstice is the time when the Southern Hemisphere is most directly facing the Sun, which is the beginning of the Southern Hemisphere summer.

Because Earth's rotational axis is tilted 23.5 degrees, the Sun's **declination** (angular distance from the equatorial plane) varies between 23.5 degrees north and 23.5 degrees south of the equator on a yearly cycle. As a result, the region between these two latitudes (called the **tropics**) receives much greater annual radiation than polar areas.

Seasonal changes in the angle of the Sun and the length of day profoundly influence Earth's climate. In the Northern Hemisphere, for example, the longest day occurs on the summer solstice and the shortest day on the winter solstice.

Daily heating of Earth also influences climate in most locations. Exceptions to this pattern occur north of the **Arctic Circle** (66.5 degrees north latitude) and south of the **Antarctic Circle** (66.5 degrees south latitude), which at certain times of the year do not experience daily cycles of daylight and darkness. For instance, during the Northern Hemisphere winter, the area north of the Arctic Circle receives no direct solar radiation at all and experiences up to six months of darkness. At the same time, the area south of the Antarctic Circle receives continuous radiation ("midnight Sun"), so it experiences up to six months of light. Half a year later, during the Northern Hemisphere summer (the Southern Hemisphere winter), the situation is reversed.

RECAP

Earth's axis is tilted at an angle of 23.5 degrees, which causes the Northern and Southern Hemispheres to take turns "leaning toward" the Sun every six months, and results in the change of seasons.

How Latitude Affects the Distribution of Solar Radiation

If Earth were a flat plate in space, with its flat side directly facing the Sun, sunlight would fall equally on all parts of Earth. Earth is spherical, however, so the amount and intensity of solar radiation received at higher latitudes are much less than at lower latitudes. The following factors influence the amount of radiation received at low and high latitudes:

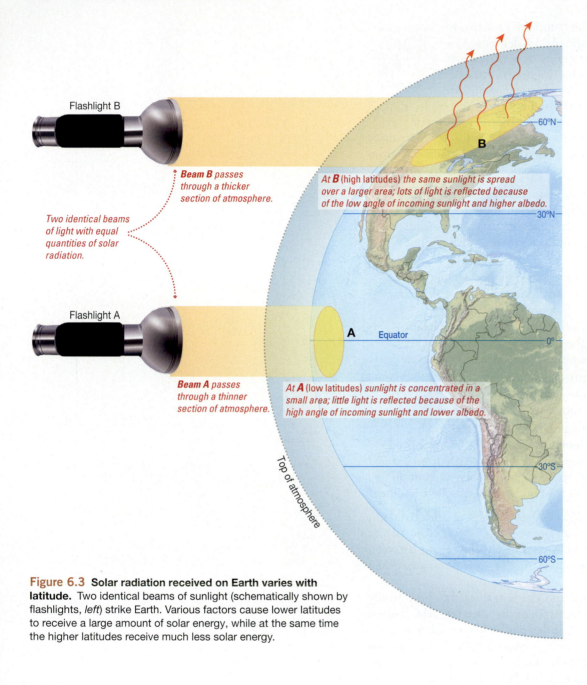

- **Solar footprint.** Most of the time in the equatorial region, the Sun is directly overhead, and so at low latitudes, sunlight strikes at a high angle. This means solar radiation is concentrated in a relatively small area (area *A* in **Figure 6.3**). Closer to the poles, sunlight strikes at a low angle, so in high latitudes, the same amount of radiation is spread over a larger area (area *B* in Figure 6.3).
- **Atmospheric absorption.** Earth's atmosphere absorbs some radiation, so less radiation reaches Earth's surface at high latitudes, compared to low latitudes, because sunlight must pass through more atmosphere at high latitudes.
- **Albedo.** The **albedo** (*albus* = white) of various Earth materials, defined as the percentage of incident radiation that is reflected back to space, varies depending on the material considered. For example, thick sea ice covered by snow reflects back into space as much as 90% of incoming solar radiation, and so has a high albedo. This is one of the reasons a larger proportion of radiation is reflected back into space in ice-covered high latitudes as compared to low latitudes, which lack substantial amounts of ice. Other Earth materials such as ocean, soil, vegetation, sand, and rock have much lower albedo values than ice; the average albedo of Earth's surface is about 30%.
- **Reflection of incoming sunlight.** The angle at which sunlight strikes the ocean surface determines how much is absorbed and how much is reflected. If the Sun shines down on a smooth sea from directly overhead, only 2% of the radiation is reflected, but if the Sun is only 5 degrees above the horizon, 40% is reflected back into the atmosphere (**Table 6.1**). Thus, the ocean reflects more radiation at high latitudes than at low latitudes.

Because of all these reasons, the intensity of radiation at high latitudes is greatly decreased compared with the intensity of radiation received in equatorial regions.

Figure 6.3 Solar radiation received on Earth varies with latitude. Two identical beams of sunlight (schematically shown by flashlights, *left*) strike Earth. Various factors cause lower latitudes to receive a large amount of solar energy, while at the same time the higher latitudes receive much less solar energy.

TABLE 6.1	REFLECTION AND ABSORPTION OF SOLAR ENERGY RELATIVE TO THE ANGLE OF INCIDENCE ON A FLAT SEA				
Elevation of the Sun above the horizon	**90°**	**60°**	**30°**	**15°**	**5°**
Reflected radiation (%)	2	3	6	20	40
Absorbed radiation (%)	98	97	94	80	60

Other factors influence the amount of solar energy that reaches Earth. For example, the amount of radiation received at a particular location on Earth's surface varies *daily* because Earth rotates on its axis, so the surface experiences daylight and darkness each day. In addition, the amount of radiation varies *annually* due to Earth's seasons, as discussed in the previous section.

Oceanic Heat Flow

Close to the poles, most incoming solar radiation strikes Earth's surface at low angles. In addition, ice has a high albedo, so more energy is reflected back into space than is absorbed. In contrast, between about 35 degrees north latitude and 40 degrees south latitude,[1] sunlight strikes Earth at much higher angles, and more energy is absorbed than is reflected back into space. The graph in **Figure 6.4** shows how incoming sunlight and outgoing heat combine on a daily basis for a net heat gain in low-latitude oceans and a net heat loss in high-latitude oceans.

Based on Figure 6.4, you might expect that over time the equatorial zone grows progressively warmer and the polar regions grow progressively cooler. The polar regions are always considerably colder than the equatorial zone, but the temperature *difference* remains the same because excess heat is transferred from the equatorial zone to the poles. How is this accomplished? Circulation in both the oceans and the atmosphere transfers the heat.

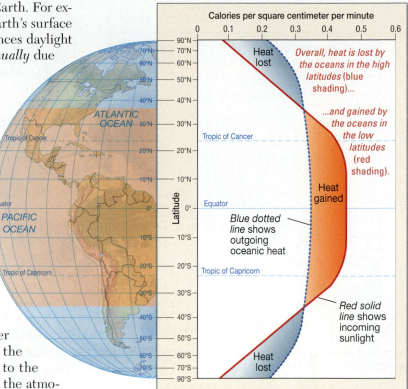

Figure 6.4 Graph showing the balance between heat gained and heat lost by the oceans. On average, heat gained and heat lost by the oceans balance each other on a global scale, whereby the excess heat from low latitudes is transferred to heat-deficient high latitudes by both oceanic and atmospheric circulation.

CONCEPT CHECK 6.1 | Explain variations in solar radiation on Earth, including the cause of Earth's seasons.

1 Sketch a labeled diagram to explain the cause of Earth's seasons.

2 Along the Arctic Circle, how would the Sun appear during the summer solstice? During the winter solstice?

3 If there is a net annual heat loss at high latitudes and a net annual heat gain at low latitudes, why does the temperature difference between these regions not increase?

RECAP

Low-latitude regions receive more solar radiation than high-latitude regions, but oceanic and atmospheric circulation transfer heat around the globe.

6.2 What Physical Properties Does the Atmosphere Possess?

The atmosphere transfers heat and water vapor from place to place on Earth. Within the atmosphere, complex relationships exist among air composition, temperature, density, water vapor content, and pressure. Before we apply these relationships, let's examine the atmosphere's composition and some of its physical properties.

Composition of the Atmosphere

Figure 6.5 lists the composition of dry air and shows that the atmosphere consists almost entirely of nitrogen and oxygen. Other gases include argon (an inert

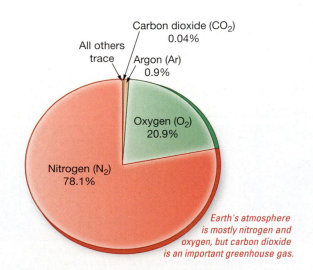

Figure 6.5 Composition of dry air. Pie chart showing the composition of dry air (without any water vapor) by volume. Nitrogen and oxygen gas comprise 99% of the total composition of Earth's atmosphere.

[1] Note that this latitudinal range extends farther into the Southern Hemisphere because the Southern Hemisphere has more ocean surface area in the middle latitudes than the Northern Hemisphere does.

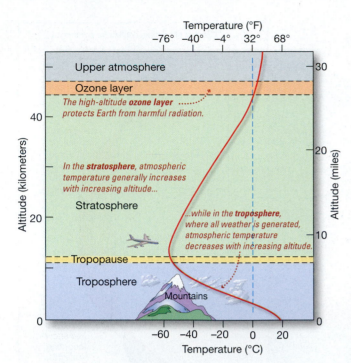

Figure 6.6 **Temperature profile of the atmosphere including the names of atmospheric layers.**

gas), carbon dioxide, and others in trace amounts. Although these gases are present in very small amounts, they can trap significant amounts of heat within the atmosphere. For more about how these gases trap heat in the atmosphere, see Chapter 16, "The Oceans and Climate Change."

Climate

Connection

Temperature Variation in the Atmosphere

Intuitively, it seems logical that the higher one goes in the atmosphere, the warmer it should be since it's closer to the Sun. However, as unusual as it seems, the atmosphere is actually heated from *below*. That is because the Sun's energy passes through the Earth's atmosphere and warms Earth's surface (both land and water), which in turn reradiates this energy back into the atmosphere as heat. This process is one of the mechanisms underlying the *greenhouse effect* and will be discussed in more detail in Chapter 16, "The Oceans and Climate Change."

Climate

Connection

Figure 6.6 shows a temperature profile of the atmosphere. The lowermost portion of the atmosphere, which extends from the surface to about 12 kilometers (7 miles), is called the **troposphere** (*tropo* = turn, *sphere* = a ball) and is where all weather is produced. The troposphere gets its name because of the abundance of mixing that occurs within this layer of the atmosphere, mostly as a result of being heated from below. Within the troposphere, temperature gets cooler with altitude to the point that at high altitudes, the air temperature is well below freezing. If you have ever flown in a jet airplane, for instance, you may have noticed that any water on the wings or inside your window freezes during a high-altitude flight.

Density Variation in the Atmosphere

It may seem surprising that air has density, but since air is composed of molecules, it certainly does. Temperature has a dramatic effect on the density of air. At higher temperatures, for example, air molecules move more quickly, take up more space, and density is decreased. Thus, the general relationship between density and temperature is as follows:

- Warm air is less dense, so it rises; this is commonly expressed as "heat rises."
- Cool air is more dense, so it sinks.

Figure 6.7 shows how a radiator (heater) uses convection to heat a room. The heater warms the nearby air and causes it to expand. This expansion makes the air less dense, causing it to rise. Conversely, a cold window cools the nearby air and causes it to contract, thereby becoming more dense, which causes it to sink. A **convection cell** (*con* = with, *vect* = carried) forms, composed of the rising and sinking air moving in a circular fashion, similar to the convection in Earth's mantle discussed in Chapter 2.

Figure 6.7 **How a convection cell in a room is created by a hot radiator and a cold window.**

Atmospheric Water Vapor Content

The amount of water vapor in air depends in part on the air's temperature. Warm air, for instance, can hold more water vapor than cold air because the air molecules are moving more quickly and come into contact with more water vapor. Thus, warm air is typically moist, and, conversely, cool air is typically dry. As a result, a warm, breezy day speeds evaporation when you hang your laundry outside to dry.

Water vapor influences the density of air. The addition of water vapor decreases the density of air because water vapor has a lower density than air. Thus, humid air is less dense than dry air.

Atmospheric Pressure

Atmospheric pressure is 1.0 atmosphere[2] (14.7 pounds per square inch) at sea level and decreases with increasing altitude. Atmospheric pressure depends on the weight of the column of air above. For instance, a tall column of air produces higher atmospheric pressure than a short column of air. An analogy to this is water pressure in a swimming pool: The taller the column of water above, the higher the water pressure. Thus, the highest pressure in a pool is at the bottom of the deep end.

Similarly, the tall column of air at sea level means air pressure is high at sea level and decreases with increasing elevation. When sealed bags of potato chips or pretzels are taken to a high elevation, there is a shorter column air overhead and the atmospheric pressure is much lower than where the bags were sealed. This may cause the bags to swell and sometimes burst. You may also have experienced this change in pressure when your ears "popped" during the takeoff or landing of an airplane, or while driving on steep mountain roads.

Changes in atmospheric pressure cause air movement as a result of changes in the molecular density of the air. The general relationship is shown in **Figure 6.8**, which indicates that:

- A column of cool, dense air causes high pressure at the surface, which will lead to sinking air (movement *toward* the surface and compression).
- A column of warm, less dense air causes low pressure at the surface, which will lead to rising air (movement *away from* the surface and expansion).

In addition, sinking air tends to warm because of its compression, while rising air tends to cool due to expansion. Note that there are complex relationships among air composition, temperature, density, water vapor content, and pressure.

Movement of the Atmosphere

Air *always* moves from high-pressure regions toward low-pressure regions. This moving air is called **wind**. If a balloon is inflated and let go, what happens to the air inside the balloon? It rapidly escapes, moving from a high-pressure region inside the balloon (caused by the balloon pushing on the air inside) to the lower-pressure region outside the balloon.

An Example: A Nonspinning Earth

Imagine for a moment that Earth is not spinning on its axis but that the Sun rotates around Earth, with the Sun directly above Earth's equator at all

STUDENTS SOMETIMES ASK . . .

Why is there so much nitrogen in the atmosphere?

To understand the abundance of nitrogen in the atmosphere, it's useful to compare it to oxygen, the next most abundant element in the atmosphere. Figure 6.5, for example, shows that nitrogen is about four times more abundant in the atmosphere than oxygen. However, if we consider the relative abundances of oxygen and nitrogen associated with the entire Earth (both within and above it), oxygen is about 10,000 times more abundant. This abundance of oxygen reflects the composition of the material from which Earth originally formed and the process of Earth's accretion. Oxygen is a major component of the solid Earth, along with silicon and elements such as magnesium, calcium, and sodium. Nitrogen does not readily react with these elements to form solids, so it is not incorporated into the solid Earth. This is one reason why nitrogen is so enriched in the atmosphere relative to oxygen. The other primary reason is that, unlike oxygen, nitrogen is very stable in the atmosphere and is not involved to a great extent in chemical reactions that occur there. Thus, over geologic time, it has built up in the atmosphere to a much greater extent than oxygen.

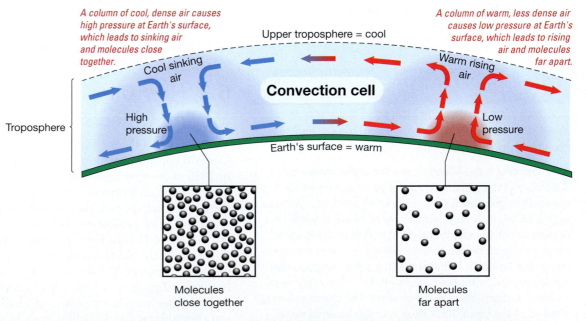

Figure 6.8 **Characteristics of high and low atmospheric pressure zones.**

[2]The *atmosphere* is a unit of pressure; 1.0 atmosphere is the average pressure exerted by the overlying atmosphere at sea level and is equivalent to 760 millimeters of mercury, 101,300 Pascal, or 1013 millibars.

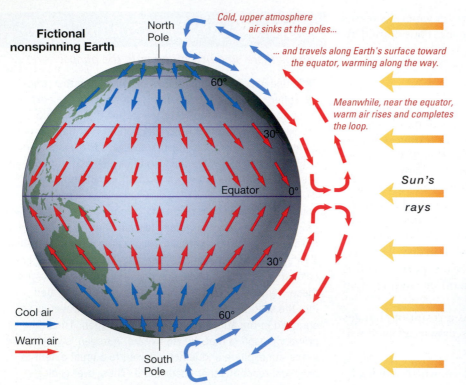

Figure 6.9 **Hypothetical atmospheric circulation on a fictional nonspinning earth.** Diagram showing a fictional nonspinning Earth with the Sun directly above Earth's equator. Arrows show the pattern of winds that would develop due to uneven solar heating on Earth (profile view of winds shown by arrows at *right*). Note the large atmospheric circulation cell that develops in each hemisphere that spans from the equator to each pole.

RECAP

The atmosphere is heated from below; its changing temperature, density, water vapor content, and pressure cause atmospheric movement, which is called wind.

times (**Figure 6.9**). Because more solar radiation is received along the equator than at the poles, the air at the equator in contact with Earth's surface is warmed. This warm, moist air rises, creating low pressure at the surface. This rising air cools (see Figure 6.6) and releases its moisture as rain. Thus, a zone of low pressure and much precipitation occurs along the equator.

As the air along the equator rises, it reaches the top of the troposphere and begins to move toward the poles. Because the temperature is much lower at high altitudes, the air cools, and its density increases. This cool, dense air sinks at the poles, creating high pressure at the surface. The sinking air is quite dry because cool air cannot hold much water vapor. Thus, the poles experience high pressure and clear, dry weather.

In a nonspinning Earth, which way will surface winds blow? Air always moves from high pressure to low pressure, so air travels from the high pressure at the poles toward the low pressure at the equator. Thus, there are strong northerly winds in the Northern Hemisphere and strong southerly winds in the Southern Hemisphere.[3] The air warms as it makes its way back to the equator, completing the loop (called a *convection cell* or *circulation cell*; see Figure 6.7).

Is this fictional case of a nonspinning Earth a good analogy for what is really happening on Earth? Actually, it is not, even though the *principles* that drive the physical movement of air remain the same whether Earth is spinning or not. Let's now examine how Earth's spin influences atmospheric circulation.

CONCEPT CHECK 6.2 | Describe the physical properties of the atmosphere.

1 Describe the physical properties of the atmosphere, including its composition, temperature, density, water vapor content, pressure, and movement.

2 Is Earth's atmosphere heated from above or below? Explain.

6.3 How Does the Coriolis Effect Influence Moving Objects?

The **Coriolis effect** changes the intended path of a moving body. Named after Gaspard Gustave de Coriolis, the French engineer who first calculated its influence in 1835, it is often incorrectly called the Coriolis *force*. It does not accelerate the moving body, so it does not influence the body's speed. As a result, it is an effect and not a true force. In fact, the Coriolis effect is often called a "fictitious" force.

The Coriolis effect causes moving objects on Earth to follow curved paths. In the Northern Hemisphere, an object will follow a path to the *right* of its intended direction; in the Southern Hemisphere, an object will follow a path to the *left* of its intended direction. The directions right and left are the *viewer's perspective looking in the direction in which the object is traveling*. For example, the Coriolis effect very slightly influences the movement of a ball thrown between two people. In the Northern Hemisphere, the ball will veer slightly to its right *from the thrower's perspective.*

[3]Notice that winds are named based on the direction *from which they are moving.*

The Coriolis effect acts on all moving objects. However, it is much more pronounced on objects traveling long distances, especially north or south. This is why the Coriolis effect has a dramatic effect on atmospheric circulation and the movement of ocean currents.

Interdisciplinary

Relationship

The Coriolis effect is a result of Earth's rotation toward the east. More specifically, the *difference* in the speed of Earth's rotation at different latitudes causes the Coriolis effect. In reality, objects travel along straight-line paths,[4] but Earth rotates underneath them, making the object's path appear to curve. Let's look at two examples to help clarify this.

Example 1: Perspectives and Frames of Reference on a Merry-Go-Round

A merry-go-round is a useful experimental apparatus with which to test some of the concepts of the Coriolis effect. A merry-go-round is a large circular wheel that rotates around its center. It has bars that people hang onto while the merry-go-round spins, as shown in **Figure 6.10**.

Imagine that you are on a merry-go-round that is spinning counterclockwise, as viewed from above (Figure 6.10). As you are spinning, what will happen to you if you let go of the bar? If you guessed that you would fly off along a straight-line path perpendicular to the merry-go-round (Figure 6.10, *Path A*), that's not quite right. Your angular momentum would propel you in a straight line *tangent* to your circular path on the merry-go-round at the point where you let go (Figure 6.10, *Path B*). The law of inertia states that a moving object will follow a straight-line path until it is compelled to change that path by other forces. Thus, you would follow a straight-line path (*Path B*) until you collide with some object such as other playground equipment or the ground. From the perspective of another person on the merry-go-round, your departure along *Path B* would *appear* to curve to the right due to the merry-go-round's rotation.

Imagine that you are back on the merry-go-round, spinning counterclockwise, but you are now joined by another person who is facing you directly but on the opposite side of the merry-go-round. If you were to toss a ball to the other person, what path would it appear to follow? Even though you threw the ball straight at the other person (Figure 6.10, *Path C*), from *your perspective* the ball's path would appear to curve to the right (*Path D*). That's because the frame of reference (in this example, the merry-go-round) has rotated during the time that it took the ball to reach where the other person had been (Figure 6.10). A person viewing the merry-go-round from directly overhead would observe that the ball did indeed travel along a straight-line path (Figure 6.10, *Path C*), just as your path was straight when you let go of the merry-go-round bar. Similarly, the perspective of being on the rotating Earth causes objects to appear to travel along curved paths. This is the Coriolis effect. The merry-go-round spinning in a counterclockwise direction is analogous to the Northern Hemisphere because, as viewed from above the North Pole, Earth is spinning counterclockwise. Thus, moving objects appear to follow curved paths to the *right* of their intended direction in the Northern Hemisphere.

If the other person on the merry-go-round had thrown a ball toward you, it would also appear to have curved. From the perspective of the other person, the ball would appear to curve to its right, just as the ball you threw curved to the right. From your perspective, however, the ball thrown toward you would appear to curve to its *left*. When considering the Coriolis effect, the perspective to keep in mind is the one *looking in the same direction that the object is moving.*

[4]Newton's first law of motion (the law of inertia) states that an object at rest remains at rest, and a moving object continues to move in a straight line unless an external force changes its state of motion.

Web Video
Coriolis Effect on a Merry-Go-Round
http://goo.gl/x3iOeX

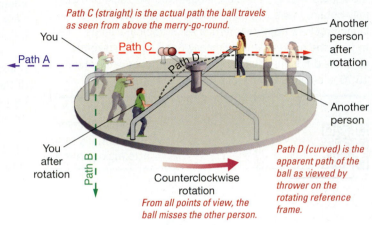

Path C (straight) is the actual path the ball travels as seen from above the merry-go-round.

You

Path C

Path A

Path D

Another person after rotation

You after rotation

Path B

Counterclockwise rotation
From all points of view, the ball misses the other person.

Another person

Path D (curved) is the apparent path of the ball as viewed by thrower on the rotating reference frame.

SmartFigure 6.10 A merry-go-round spinning counterclockwise as viewed from above illustrates some concepts about the Coriolis effect. See text for description of *Paths A, B, C,* and *D.*
https://goo.gl/UE3t03

STUDENTS SOMETIMES ASK . . .

If Earth is spinning so fast, why don't we feel it?

Despite Earth's constant rotation, we have the illusion that Earth is still. The reason that we don't feel the motion is because Earth rotates smoothly and quietly, and the atmosphere moves along with us. Thus, all sensations we receive tell us there is no motion and the ground is comfortably at rest—even though most of the United States is continually moving at speeds greater than 800 kilometers (500 miles) per hour!

To simulate the Southern Hemisphere, the merry-go-round would need to rotate in a *clockwise* direction, which is analogous to Earth when viewed from above the South Pole. Thus, moving objects appear to follow curved paths to the *left* of their intended direction in the Southern Hemisphere.

Example 2: A Tale of Two Missiles

The distance that a point on Earth travels in the course of a day is shorter with increasing latitude. Someone standing near the pole, for example, travels in a circle that is not nearly as large one traveled by a person near the equator. Because people at both locations travel their respective distances in one day, the velocity of the two points on which they stand must not be the same. **Figure 6.11a** shows that as Earth rotates on its axis, the velocity decreases with latitude, ranging from more than 1600 kilometers (1000 miles) per hour at the equator to 0 kilometers per hour at the poles. *This change in velocity with latitude is the true cause of the Coriolis effect.* The following example illustrates how velocity changes with latitude.

Imagine that we have two missiles that fly in straight lines toward their destinations. For simplicity, assume that the flight of each missile takes one hour regardless of the distance flown. The first missile is launched from the North Pole toward New Orleans, Louisiana, which is at 30 degrees north latitude (**Figure 6.11b**). Does the missile land in New Orleans? Actually, no. Earth rotates eastward at 1400 kilometers (870 miles) per hour along the 30 degrees latitude line (**Figure 6.11a**), so the missile lands somewhere near El Paso, Texas, 1400 kilometers west of its target. From your perspective at the North Pole, the path of the missile appears to curve *to its*

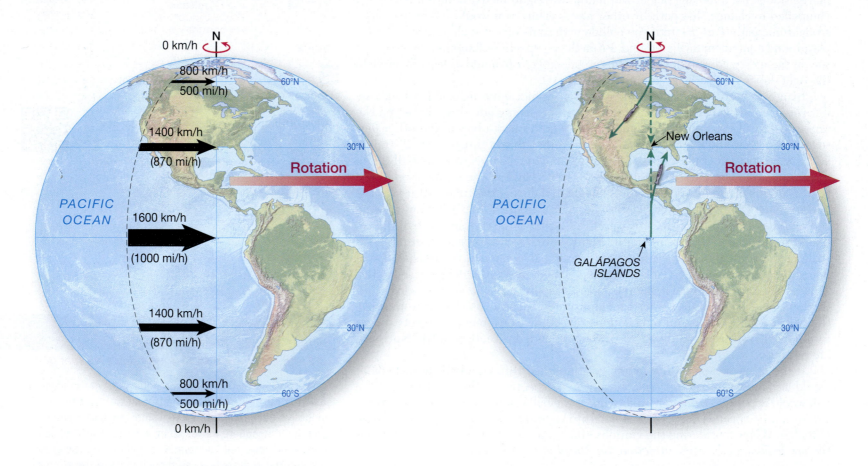

(a) The velocity of any point on Earth varies with latitude from about 1600 kilometers (1000 miles) per hour at the equator to 0 kilometers per hour at either pole.

(b) The path of missiles shot towards New Orleans from the North Pole and from the Galápagos Islands on the equator. Dashed lines indicate intended paths; solid lines indicate paths that the missiles would travel as viewed from Earth's surface.

Figure 6.11 The Coriolis effect and missile paths.

right in accordance with the Coriolis effect. In reality, New Orleans has moved out of the line of fire due to Earth's rotation.

The second missile is launched toward New Orleans from the Galápagos Islands, which are directly south of New Orleans along the equator (**Figure 6.11b**). From their position on the equator, the Galápagos Islands are moving east at 1600 kilometers (1000 miles) per hour, 200 kilometers (124 miles) per hour faster than New Orleans (**Figure 6.11a**). At takeoff, therefore, the missile is also moving toward the east 200 kilometers per hour faster than New Orleans. Thus, when the missile returns to Earth one hour later at the latitude of New Orleans, it will land offshore of Alabama, 200 kilometers east of New Orleans. Again, from your perspective on the Galápagos Islands, the missile appears to curve *to its right*. Keep in mind that both of these missile examples ignore friction, which would greatly reduce the amount the missiles deflect to the right of their intended courses.

Changes in the Coriolis Effect with Latitude

The first missile (shot from the North Pole) missed the target by 1600 kilometers (1000 miles), while the second missile (shot from the Galápagos Islands) missed its target by only 200 kilometers (124 miles). What was responsible for the difference? Not only does the rotational velocity of points on Earth range from 0 kilometers per hour at the poles to more than 1600 kilometers (1000 miles) per hour at the equator, but the *rate of change* of the rotational velocity (per degree of latitude) increases as the pole is approached from the equator.

For example, the rotational velocity differs by 200 kilometers (124 miles) per hour between the equator (0 degrees) and 30 degrees north latitude. From 30 degrees north latitude to 60 degrees north latitude, however, the rotational velocity differs by 600 kilometers (372 miles) per hour. Finally, from 60 degrees north latitude to the North Pole (where the rotational velocity is zero), the rotational velocity differs by more than 800 kilometers (500 miles) per hour.

Thus, the maximum Coriolis effect is at the poles, and there is no Coriolis effect at the equator. The magnitude of the Coriolis effect depends much more, however, on the length of time the object (such as an air mass or ocean current) is in motion. Even at low latitudes, where the Coriolis effect is small, a large Coriolis deflection is possible if an object is in motion for a long time. In addition, because the Coriolis effect is caused by the *difference* in velocity of different latitudes on Earth, there is no Coriolis effect for those objects moving due east or due west along the equator.

For a summary of the Coriolis effect, see MasteringOceanography **Web Table 6.1**.

RECAP

The Coriolis effect causes moving objects to curve to the right in the Northern Hemisphere and to the left in the Southern Hemisphere. It is at its maximum at the poles and is nonexistent at the equator

CONCEPT CHECK 6.3 | Demonstrate an understanding of the Coriolis effect.

1 Describe the Coriolis effect in both the Northern and Southern Hemisphere.

2 What is the underlying cause of the Coriolis effect?

3 Explain how the strength of the Coriolis effect changes with latitude.

6.4 What Global Atmospheric Circulation Patterns Exist?

Figure 6.12 shows atmospheric circulation and the corresponding wind belts on a spinning Earth, which presents a more complex pattern than that of the fictional nonspinning Earth (Figure 6.9).

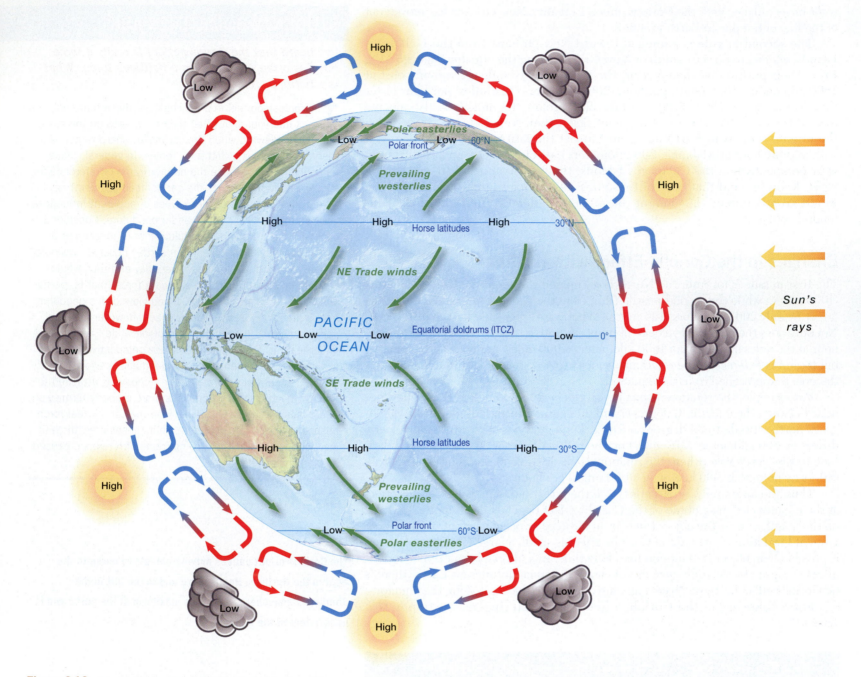

Figure 6.12 **Atmospheric circulation and wind belts of the world.** The three-cell model of atmospheric circulation creates the major wind belts of the world, shown on the globe as green arrows and labeled in green lettering. A profile view of atmospheric circulation cells is shown by red and blue arrows around the edges of the globe. Also shown are the names of the boundaries between the wind belts (*blue lettering*), surface atmospheric pressures (*high* or *low*), and resulting typical weather (*Sun* or *clouds*).

Web Animation
Global Wind Patterns
http://goo.gl/kW1MCv

Circulation Cells

The greater heating of the atmosphere over the equator causes the air to expand, to decrease in density, and to rise. As the air rises, it cools by expansion because the pressure is lower, and the water vapor it contains condenses and falls as rain in the equatorial zone. The resulting dry air mass travels north or south of the equator. Around 30 degrees north and south latitude, the air cools off enough to become denser than the surrounding air, so it begins to descend, completing the loop (Figure 6.12). These circulation cells are called **Hadley cells** after noted English meteorologist George Hadley (1685–1768).

In addition to Hadley cells, each hemisphere has a **Ferrel cell** between 30 and 60 degrees latitude and a **polar cell** between 60 and 90 degrees latitude. The Ferrel cell—named after American meteorologist William Ferrel (1817–1891),

who invented the three-cells-per-hemisphere model for atmospheric circulation—is not driven solely by differences in solar heating; if it were, air within it would circulate in the opposite direction. Similarly to the movement of interlocking gears, the Ferrel cell moves in the direction that coincides with the movement of the two adjoining circulation cells.

Pressure

A column of cool, dense air moves *toward* the surface and creates high pressure. The descending air at about 30 degrees north and south latitude creates high-pressure zones called the **subtropical highs**. Similarly, descending air at the poles creates high-pressure regions called the **polar highs**.

What kind of weather is experienced in these high-pressure areas? Descending air is quite dry, and it tends to warm under its own weight, so these areas typically experience dry, clear, fair conditions. The conditions are not necessarily warm (such as at the poles)—just dry and associated with clear skies.

A column of warm, low-density air rises *away* from the surface and creates low pressure. Thus, rising air creates a band of low pressure at the equator—the **equatorial low**—and at about 60 degrees north and south latitude—the **subpolar low**. The weather in areas of low pressure is dominated by cloudy conditions with lots of precipitation, because rising air cools and cannot hold its water vapor.

Wind Belts

The lowermost portion of the circulation cells—that is, the part that is closest to the surface—generates the major wind belts of the world. The masses of air that move across Earth's surface from the subtropical high-pressure belts toward the equatorial low-pressure belt constitute the **trade winds**. These steady winds are named from the term *to blow trade*, which means to blow in a regular course. If Earth did not rotate, these winds would blow in a north–south direction. In the Northern Hemisphere, however, the **northeast trade winds** curve to the right due to the Coriolis effect and blow *from northeast to southwest*. In the Southern Hemisphere, on the other hand, the **southeast trade winds** curve to the left due to the Coriolis effect and blow *from southeast to northwest*.

Some of the air that descends in the subtropical regions moves along Earth's surface to higher latitudes as the **prevailing westerly wind belts**. Because of the Coriolis effect, the prevailing westerlies blow from southwest to northeast in the Northern Hemisphere and from northwest to southeast in the Southern Hemisphere.

Air moves away from the high pressure at the poles, too, producing the **polar easterly wind belts**. The Coriolis effect is maximized at high latitudes, so these winds are deflected strongly. The polar easterlies blow from the northeast in the Northern Hemisphere, and from the southeast in the Southern Hemisphere. When the polar easterlies come into contact with the prevailing westerlies near the subpolar low pressure belts (at 60 degrees north and south latitude), the warmer, less dense air of the prevailing westerlies rises above the colder, more dense air of the polar easterlies.

Boundaries

The boundary between the two trade wind belts along the equator is known as the **doldrums** (*doldrum* = dull) because, long ago, sailing ships were becalmed there by the lack of winds. Sometimes stranded for days or weeks, the situation was unfortunate but not life-threatening: Daily rain showers supplied sailors with plenty of freshwater. Today, meteorologists refer to this globe-circling region as the **Intertropical Convergence Zone (ITCZ)** because it is the region between the tropics where the Northern and Southern Hemisphere trade winds converge (Figure 6.12).

6.1 Squidtoons

I was sailing across the sea before it was cool.

https://goo.gl/pb9ydB

What is the origin of the name horse latitudes?

The term *horse latitudes* supposedly originates from the days when Spanish sailing vessels transported horses across the Atlantic to the West Indies. Ships would often become becalmed in mid-ocean due to the light winds in these latitudes, thus severely prolonging the voyage; the resulting water shortages would make it necessary for crews to dispose of their horses overboard (see the chapter-opening quote). Alternatively, the term might also have originated by seamen who were paid an advance, called the "dead horse," before a long voyage. A few months into the voyage, the "dead horse" was officially worked off; this was also about the same time sailing vessels were stuck in the middle of the ocean without wind, so these regions became known as the *horse latitudes*.

The boundary between the trade winds and the prevailing westerlies (centered at 30 degrees north or south latitude) is known as the **horse latitudes**. Sinking air in these regions causes high atmospheric pressure (associated with the *subtropical high pressure*) and results in clear, dry, and fair conditions. Because the air is sinking, the horse latitudes are known for surface winds that are light and variable.

The boundary between the prevailing westerlies and the polar easterlies at 60 degrees north or south latitude is known as the **polar front**. This is a battleground for different air masses, so cloudy conditions and lots of precipitation are common here.

Clear, dry, fair conditions are associated with the high pressure at the poles, so precipitation is minimal. The poles are often classified as cold deserts because the annual precipitation is so low.

Table 6.2 summarizes the characteristics of global wind belts and boundaries.

Circulation Cells: Idealized or Real?

The three-cell model of atmospheric circulation first proposed by Ferrel provides a simplified model of the general circulation pattern on Earth. This circulation model is idealized and does not always match the complexities observed in nature, particularly for the location and direction of motion of the Ferrel and polar cells. Nonetheless, it generally matches the pattern of major wind belts of the world and provides a general framework for understanding why they exist.

Further, the following factors significantly alter the idealized wind, pressure, and atmospheric circulation patterns illustrated in Figure 6.12:

1. The tilt of Earth's rotation axis, which produces seasons
2. The lower heat capacity of continental rock compared to seawater,[5] which makes the air over continents colder in winter and warmer in summer than the air over adjacent oceans
3. The uneven distribution of land and ocean over Earth's surface, which particularly affects patterns in the Northern Hemisphere

During winter, therefore, the continents usually develop atmospheric high-pressure cells from the weight of cold air centered over them and, during the summer, they usually develop low-pressure cells (**Figure 6.13**). In fact, such seasonal shifts in atmospheric pressure over Asia cause *monsoon winds*, which have a dramatic effect on Indian Ocean currents, and will be discussed in Chapter 7, "Ocean Circulation." In general, however, the patterns of atmospheric high- and

SmartTable 6.2 **Characteristics of wind belts and boundaries.**
https://goo.gl/nEIsTK

SMARTTABLE 6.2 CHARACTERISTICS OF WIND BELTS AND BOUNDARIES

Region (north or south latitude)	Name of wind belt or boundary	Atmospheric pressure	Characteristics
Equatorial (0–5 degrees)	Doldrums (boundary)	Low	Light, variable winds. Abundant cloudiness and much precipitation. Breeding ground for hurricanes.
5–30 degrees	Trade winds (wind belt)	—	Strong, steady winds, generally from the east.
30 degrees	Horse latitudes (boundary)	High	Light, variable winds. Dry, clear, fair weather with little precipitation. Major deserts of the world.
30–60 degrees	Prevailing westerlies (wind belt)	—	Winds generally from the west. Brings storms that influence weather across the United States.
60 degrees	Polar front (boundary)	Low	Variable winds. Stormy, cloudy weather year round.
60–90 degrees	Polar easterlies (wind belt)	—	Cold, dry winds generally from the east.
Poles (90 degrees)	Polar high pressure (boundary)	High	Variable winds. Clear, dry, fair conditions, cold temperatures, and minimal precipitation. Cold deserts.

[5]An object that has low heat capacity heats up quickly when heat energy is applied. Recall from Figure 5.7 that water has one of the highest specific heat capacities of common substances.

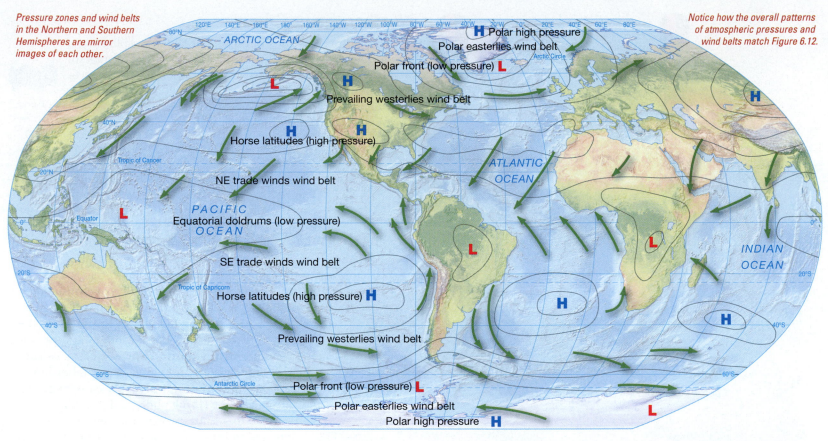

Pressure zones and wind belts in the Northern and Southern Hemispheres are mirror images of each other.

Notice how the overall patterns of atmospheric pressures and wind belts match Figure 6.12.

Figure 6.13 **January sea-level atmospheric pressures and global winds.** Average atmospheric pressure pattern for January. High (*H*) and low (*L*) atmospheric pressure zones correspond closely to those shown in Figure 6.12 but are modified by seasonal changes and the distribution of continents. Green arrows show the direction of winds, which move from high-pressure regions toward low-pressure regions but are modified by the Coriolis effect.

low-pressure zones shown in Figure 6.13 corresponds closely to those shown in Figure 6.12.

Global wind belts have had a profound effect on ocean explorations (**Diving Deeper 6.1**). The world's wind belts also closely match the pattern of ocean surface currents, which are discussed in Chapter 7, "Ocean Circulation."

Web Animation

Seasonal Pressure and Precipitation Patterns.
http://goo.gl/zVaUTq

RECAP

The major wind belts in each hemisphere are the trade winds, the prevailing westerlies, and the polar easterlies. The boundaries between these wind belts include the doldrums, the horse latitudes, the polar front, and the polar high.

CONCEPT CHECK 6.4 | Explain global atmospheric circulation patterns.

1 Sketch the pattern of surface wind belts on Earth, showing atmospheric circulation cells, zones of high and low pressure, the names of the wind belts, and the names of the boundaries between the wind belts.

2 Why are there high-pressure caps at each pole and a low-pressure belt in the equatorial region?

3 Discuss the patterns or trends that you notice about the global wind belts and boundaries as shown in Figure 6.12 and delineated in Table 6.2.

6.5 How Does the Ocean Influence Global Weather Phenomena and Climate Patterns?

Because of the ocean's huge extent over Earth's surface and also because of water's unusual thermal properties, the ocean dramatically influences global weather phenomena and climate patterns.

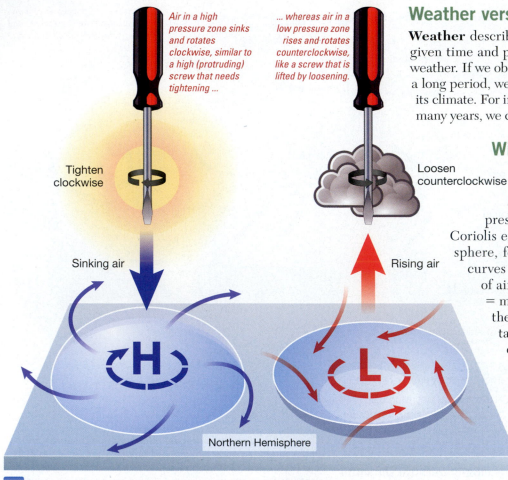

Air in a high pressure zone sinks and rotates clockwise, similar to a high (protruding) screw that needs tightening ...

... whereas air in a low pressure zone rises and rotates counterclockwise, like a screw that is lifted by loosening.

Tighten clockwise

Loosen counterclockwise

Sinking air

Rising air

H

L

Northern Hemisphere

SmartFigure 6.14 **High- and low-pressure regions and resulting air flow in the Northern Hemisphere.**
https://goo.gl/BwyVus

Climate

Connection

Weather versus Climate

Weather describes the conditions of the atmosphere at a given time and place. **Climate** is the long-term average of weather. If we observe the weather conditions in an area over a long period, we can begin to draw some conclusions about its climate. For instance, if the weather in an area is dry over many years, we can say that the area has an arid climate.

Winds

Recall that air always moves from high pressure toward low pressure and that the movement of air is called *wind*. However, as air moves away from high-pressure regions and toward low-pressure regions, the Coriolis effect modifies its direction. In the Northern Hemisphere, for example, air moving from high to low pressure curves to the right and results in a counterclockwise[6] flow of air around low-pressure cells (called **cyclonic** [*kyklon* = moving in a circle] flow). Similarly, as the air leaves the high-pressure region and curves to the right, it establishes a clockwise flow of air around high-pressure cells (called **anticyclonic flow**). **Figure 6.14** shows how a screwdriver can help you remember how air moves around high- and low-pressure regions in the Northern Hemisphere: High pressures are similar to a high (protruding) screw that needs to be tightened, so a screwdriver would be turned clockwise; low pressures are similar to a tightened screw that needs to be loosened (lifted), so a screwdriver would be turned counterclockwise. In addition, Figure 6.14 shows that high pressure is typically associated with fair, dry weather (*sun icon*) whereas low pressure is typically associated with cloudy, rainy weather (*cloud icon*).

Weather maps show the pattern of how winds flow in accordance with high- and low- pressure regions. **Figure 6.15** is a simplified U.S. weather map showing atmospheric pressures in millibars (the lines are called *isobars* [*iso* = same, *baros* = weight]) and associated winds (shown by *green arrows*). In general, winds move from high-pressure toward low-pressure regions in response to the *pressure gradient*, which is the slope the isobars represent. The winds, however, are modified by the Coriolis effect, which causes the winds to end up flowing roughly parallel to the isobars. In comparing Figure 6.15 with Figure 6.14, notice how the pattern of winds matches in the two figures. Because winter high-pressure cells are typically replaced by summer

Web Animation
Cyclones and Anticyclones
http://goo.gl/oKHd3E

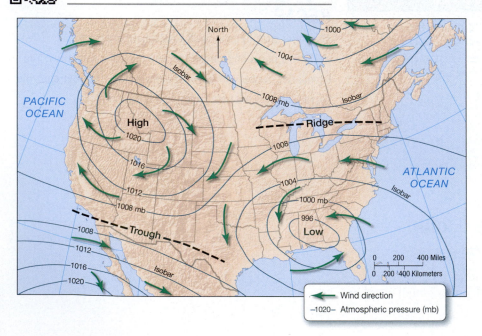

North

1000
1004
Isobar
1008 mb Isobar
PACIFIC OCEAN
High
1020
1016
1012
1008 mb
1008
Trough
1012
1016
1020
Isobar
Ridge
1008
1004
1000 mb
996
Low
ATLANTIC OCEAN
Isobar

0 200 400 Miles
0 200 400 Kilometers

Wind direction
–1020– Atmospheric pressure (mb)

Figure 6.15 U.S. weather map showing atmospheric pressures and associated winds.
Simplified U.S. weather map showing atmospheric pressure in millibars (*mb*), lines of equal pressure called *isobars*, and associated winds (*green arrows*). In general, winds move from high-pressure toward low-pressure regions but are modified by the Coriolis effect, which causes the winds to flow roughly parallel to the isobars.
Note that a trough occurs between two high pressure systems and a ridge occurs between two low pressure systems.

[6]These directions are reversed in the Southern Hemisphere.

WHY CHRISTOPHER COLUMBUS NEVER SET FOOT ON NORTH AMERICA

The Italian navigator and explorer **Christopher Columbus** is widely credited by Europeans with the accomplishment of discovering North America in the year 1492. However, America was already populated with many natives, and the Vikings made a voyage to North America that predated Columbus's by about 500 years. The truth is, Columbus never set foot on North America because the pattern of the major wind belts of the world prevented his sailing ships from reaching continental North America during his four voyages.

Rather than sailing east, Columbus was determined to reach the East Indies (today the country of Indonesia) by sailing west across the Atlantic Ocean. An astronomer in Florence, Italy, named Toscanelli was the first to suggest such a route in a letter to the king of Portugal. Columbus later contacted Toscanelli and was told how far he would have to sail west to reach India. Today, we know that this distance would have carried him just west of North America.

After years of difficulties in initiating the voyage, Columbus received the financial backing of the Spanish monarchs Ferdinand V and Isabella I. He set sail from Spain with 88 men and three ships (the *Niña*, the *Pinta*, and the *Santa María*) on August 3, 1492, and made a stop to resupply in the Canary Islands off Africa (**Figure 6A**). The Canary Islands are located at 28 degrees north latitude and are within the northeast trade winds, which blow steadily from the northeast to the southwest. Instead of sailing directly west, which would have allowed Columbus to reach central Florida, the map in Figure 6A shows that Columbus sailed a more southerly route.

During the morning of October 12, 1492, the first land was sighted; this is generally believed to have been Watling Island in the present-day Bahamas, southeast of Florida. Based on the inaccurate information he had been given, Columbus was convinced that he had arrived in the East Indies and was somewhere near India. Consequently, he called the inhabitants "Indians," and the area is known today as the *West Indies*. Later during this voyage, he explored the coasts of Cuba and Hispaniola (the island comprising modern-day Haiti and the Dominican Republic).

On his return journey, Columbus sailed to the northeast and picked up the prevailing westerlies, which transported him away from North America and toward Spain. Upon his return to Spain and the announcement of his discovery, additional voyages were planned. Columbus made three more trips across the Atlantic Ocean, following similar paths through the Atlantic. Thus, his ships were controlled by the trade winds on the outbound voyage and the prevailing westerlies on the return trip. During his next voyage, in 1493, Columbus explored Puerto Rico and the Leeward Islands and established a colony on Hispaniola. In 1498, he explored Venezuela and landed on South America, unaware that it was a new continent to Europeans. On his last voyage in 1502, he reached Central America.

Although he is today considered a master mariner, Columbus died in neglect in 1506, still convinced that he had explored islands near India. Even though he never set foot on the North American mainland, his journeys inspired other Spanish and Portuguese navigators to explore the "New World," including the coasts of North and South America.

GIVE IT SOME THOUGHT

1. What unique atmospheric conditions prevented Christopher Columbus from setting foot on the continent of North America?

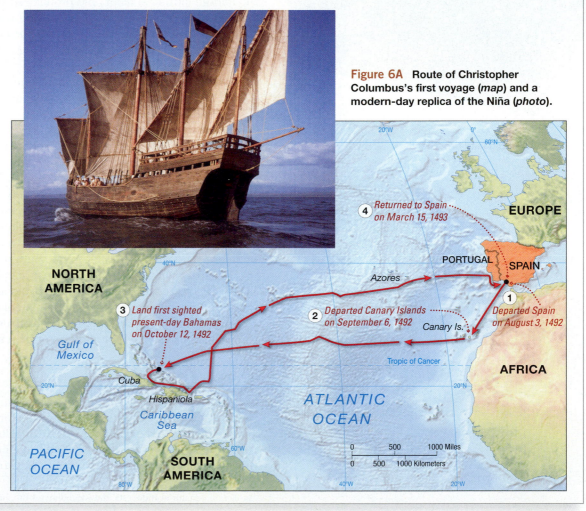

Figure 6A Route of Christopher Columbus's first voyage (*map*) and a modern-day replica of the Niña (*photo*).

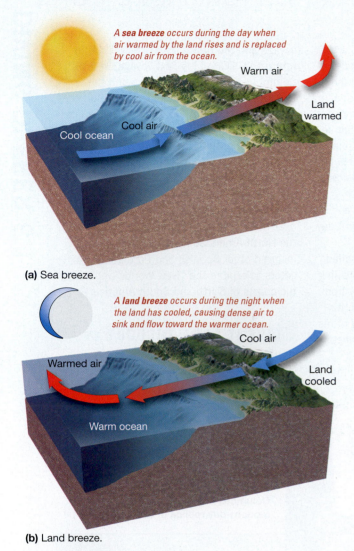

*A **sea breeze** occurs during the day when air warmed by the land rises and is replaced by cool air from the ocean.*

Warm air

Land warmed

Cool air

Cool ocean

(a) Sea breeze.

*A **land breeze** occurs during the night when the land has cooled, causing dense air to sink and flow toward the warmer ocean.*

Cool air

Warmed air

Land cooled

Warm ocean

(b) Land breeze.

Figure 6.16 **Sea and land breezes.**

low-pressure cells over the continents, wind patterns associated with continents often reverse themselves seasonally.

SEA AND LAND BREEZES Other factors that influence regional winds, especially in coastal areas, are **sea breezes** and **land breezes** (**Figure 6.16**). When an equal amount of solar energy is applied to both land and ocean, the land heats up about five times more due to its lower heat capacity. The land heats the air around it and, during the afternoon, the warm, low-density air over the land rises. Rising air creates a low-pressure region over the land, pulling the cooler air over the ocean toward land, creating what is known as a *sea breeze*. At night, the land surface cools about five times more rapidly than the ocean and cools the air around it. This cool, high-density air sinks, creating a high-pressure region that causes the wind to blow from the land. This is known as a *land breeze*, and it is most prominent in the late evening and early morning hours.

Storms and Fronts

At very high and very low latitudes, there is little daily and minor seasonal change in weather.[7] Equatorial regions are usually warm, damp, and typically calm because the dominant direction of air movement in the doldrums is upward. Midday rains are common, even during the supposedly "dry" season. It is within the *middle latitudes* between 30 and 60 degrees north or south latitude where storms are common.

Storms are atmospheric disturbances characterized by strong winds, precipitation, and often thunder and lightning. Due to the seasonal change of pressure systems over continents, air masses from the high and low latitudes may move into the middle latitudes, meet, and produce severe storms. **Air masses** are large volumes of air that have a definite area of origin and distinctive characteristics. Several air masses influence the United States, including polar air masses and tropical air masses (**Figure 6.17**). Some air masses originate over land (c = continental) and are therefore dryer, but most originate over the sea (m = maritime) and are moist. Some are colder (P = polar; A = Arctic) and some are warm (T = tropical). Typically, the United States is influenced more by polar air masses during the winter and more by tropical air masses during the summer.

As polar and tropical air masses move into the middle latitudes, they also move gradually in an easterly direction. A **warm front** is the contact between a warm air mass moving into an area occupied by cold air. A **cold front** is the contact between a cold air mass moving into an area occupied by warm air (**Figure 6.18**).

These confrontations are brought about by the movement of the **jet stream**, which is a narrow, fast-moving, easterly flowing air mass. It exists above the middle latitudes just below the top of the troposphere, centered at an altitude of about 10 kilometers (6 miles). It usually follows a wavy path and may cause unusual weather by steering a polar air mass far to the south or a tropical air mass far to the north.

Regardless of whether a warm front or cold front is produced, the warmer, less-dense air always rises above the denser cold air. The warm air cools as it rises, so its water vapor condenses as precipitation. A cold front is usually steeper, and the temperature difference across it is greater than a warm front. Therefore, rainfall along a cold front is usually heavier and briefer than rainfall along a warm front.

Tropical Cyclones (Hurricanes)

Tropical cyclones (*kyklon* = moving in a circle) are huge rotating masses of low pressure characterized by strong winds and torrential rain. They are the largest storm systems on Earth, though they are not associated with any fronts. In North and South America, tropical cyclones are called **hurricanes** (*Huracan* = Taino god of wind); in the western North Pacific Ocean, they are called **typhoons** (*tai-fung* = great

[7]In fact, in equatorial Indonesia, the vocabulary of Indonesians doesn't include the word *seasons*.

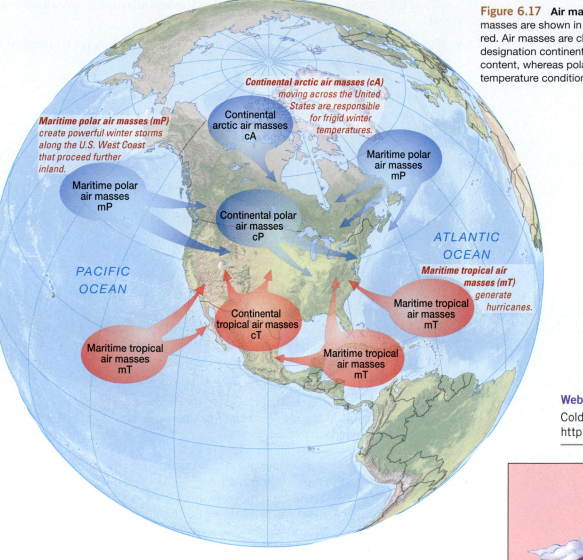

Figure 6.17 **Air masses that affect U.S. weather.** Polar air masses are shown in blue, and tropical air masses are shown in red. Air masses are classified based on their source region: The designation continental (c) or maritime (m) indicates moisture content, whereas polar (P), Arctic (A), and tropical (T) indicate temperature conditions.

Maritime polar air masses (mP) create powerful winter storms along the U.S. West Coast that proceed further inland.

Continental arctic air masses (cA) moving across the United States are responsible for frigid winter temperatures.

Maritime tropical air masses (mT) generate hurricanes.

Maritime polar air masses mP

Continental arctic air masses cA

Maritime polar air masses mP

Maritime polar air masses mP

Continental polar air masses cP

PACIFIC OCEAN

ATLANTIC OCEAN

Maritime tropical air masses mT

Continental tropical air masses cT

Maritime tropical air masses mT

Maritime tropical air masses mT

Web Animation
Cold Fronts and Warm Fronts
http://goo.gl/Xp8yOD

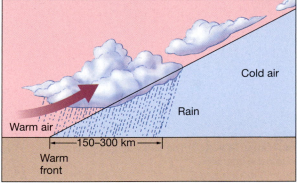

(a) Profile view of a gradually rising warm front.

Cold air

Rain

Warm air

Warm front

|←150–300 km→|

wind); and in the Indian Ocean, they are called **cyclones**. No matter what they are called, tropical cyclones can be highly destructive. In fact, the energy contained in a *single* hurricane is greater than that generated by all energy sources in the United States over the past 20 years.

ORIGIN Remarkably, what powers tropical storms is the release of vast amounts of *latent heat of* condensation[8] that is carried within water vapor and is released as water condenses to form clouds in a hurricane. A tropical cyclone begins as a low-pressure cell that breaks away from the equatorial low-pressure belt and grows as it picks up heat energy in the following manner. Surface winds feed moisture (in the form of water vapor) into the storm. When water evaporates, it stores tremendous amounts of heat in the form of latent heat of evaporation. When water vapor condenses into a liquid (in this case, clouds and rain), it releases this stored heat—latent heat of condensation—into the surrounding atmosphere, which causes the atmosphere to warm and the air to rise. This rising air causes surface pressure to decrease, drawing additional warm moist surface air into the storm. This air, as it rises and cools, condenses into clouds and releases even more latent heat, further powering the storm and continuously repeating itself as a feedback loop, each time intensifying the storm.

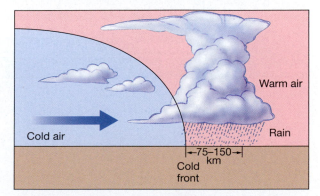

Cold air

Warm air

Rain

|←75–150 km→|

Cold front

(b) Profile view of a steeper cold front.

Figure 6.18 **Warm and cold fronts.** Profile (side) view through **(a)** a gradually rising warm front and **(b)** a steeper cold front. With both fronts, warm air rises, causing precipitation.

[8]For a discussion of water's latent heats, see Section 5.2 in Chapter 5.

Tropical storms are classified according to their maximum sustained wind speed:

- If winds are less than 61 kilometers (38 miles) per hour, the storm is classified as a *tropical depression*.
- If winds are between 61 and 120 kilometers (38 and 74 miles) per hour, the storm is called a *tropical storm*.
- If winds exceed 120 kilometers (74 miles) per hour, the storm is a *tropical cyclone*.

The **Saffir-Simpson Scale** of hurricane intensity (**Table 6.3**) further divides tropical cyclones into categories based on wind speed and damage. In some cases, in fact, the wind in tropical cyclones attains speeds as high as 400 kilometers (250 miles) per hour!

Worldwide, about 100 storms grow to hurricane status each year. The conditions needed to create a hurricane are as follows:

- Ocean water with a temperature greater than 25°C (77°F), which provides an abundance of water vapor to the atmosphere through evaporation.
- Warm, moist air, which supplies vast amounts of latent heat as the water vapor in the air condenses and fuels the storm.
- The Coriolis effect, which causes the hurricane to spin counterclockwise in the Northern Hemisphere and clockwise in the Southern Hemisphere. Generally, hurricanes cannot occur directly on the equator because the Coriolis effect is nonexistent there.

These conditions are found during the late summer and early fall, when the tropical and subtropical oceans are at their maximum temperature. That's why the official Atlantic basin hurricane season each year is from June 1 to November 30, although in rare cases hurricanes form earlier or later.

MOVEMENT When hurricanes are initiated in the low latitudes, they are affected by the trade winds and generally move from east to west across ocean basins. Hurricanes typically last from 5 to 10 days and sometimes migrate into the middle latitudes (**Figure 6.19**). In rare cases, hurricanes have done considerable damage to the northeastern United States and have even affected Nova Scotia, Canada. Figure 6.19 also shows how hurricanes are affected by the Coriolis effect: In the Northern Hemisphere, they curve to the right and in the Southern Hemisphere, they curve to the left. Moreover, this serves to carry them out of the tropics and into the middle latitudes, where they are steered towards the east by the prevailing

SmartTable 6.3 **The Saffir-Simpson scale of hurricane intensity.**
https://goo.gl/FICvaJ

	SMARTTABLE 6.3	THE SAFFIR-SIMPSON SCALE OF HURRICANE INTENSITY			
	Wind speed		**Typical storm surge (sea level height above normal)**		
Category	**km/hr**	**mi/hr**	**meters**	**feet**	**Damage**
1	120–153	74–95	1.2–1.5	4–5	Minimal: Minor damage to buildings
2	154–177	96–110	1.8–2.4	6–8	Moderate: Some roofing material, door, and window damage; some trees blown down
3	178–209	111–130	2.7–3.7	9–12	Extensive: Some structural damage and wall failures; foliage blown off trees and large trees blown down
4	210–249	131–155	4.0–5.5	13–18	Extreme: More extensive structural damage and wall failures; most shrubs, trees, and signs blown down
5	>250	>155	>5.8	>19	Catastrophic: Complete roof failures and entire building failures common; all shrubs, trees, and signs blown down; flooding of lower floors of coastal structures

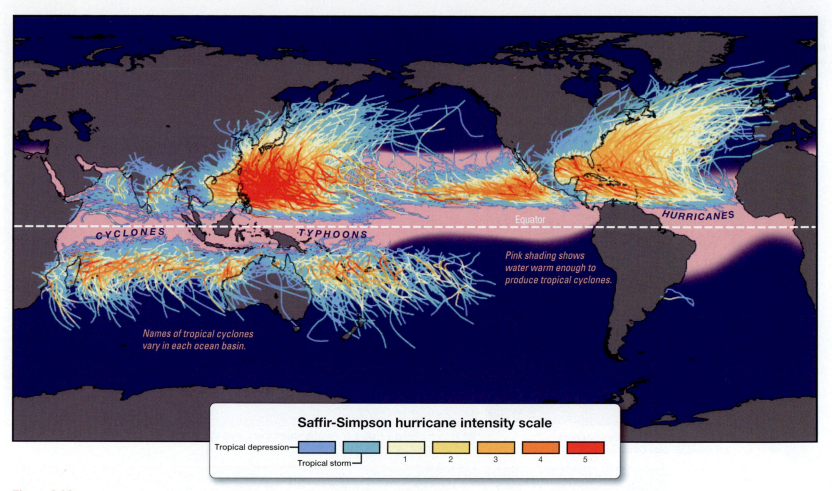

CYCLONES

TYPHOONS

Equator

HURRICANES

Pink shading shows water warm enough to produce tropical cyclones.

Names of tropical cyclones vary in each ocean basin.

Saffir-Simpson hurricane intensity scale

Tropical depression

Tropical storm

1 2 3 4 5

Figure 6.19 Historic tropical cyclone tracks over the past 150 years. Color-coded map showing the intensity and paths of historic tropical cyclones (which, depending on the area, can also be called *hurricanes* or *typhoons*). Tropical cyclones originate in low-latitude regions that have warm ocean surface temperatures (*pink shading*) and initially move from east to west but then curve into higher latitudes because of the Coriolis effect. Note the lack of tropical cyclones along the equator.

westerlies (**Figure 6.20b**). Once hurricanes move over cooler water or land, their energy source is cut off, which causes hurricanes to dissipate.

The diameter of a typical hurricane is less than 200 kilometers (124 miles) (**Figure 6.20a**), although extremely large hurricanes can exceed diameters of 800 kilometers (500 miles). As air moves across the ocean surface toward the low-pressure center, it is drawn up around the **eye of the hurricane** (**Figure 6.20c**). The air in the vicinity of the eye spirals upward, so horizontal wind speeds may be less than 15 kilometers (25 miles) per hour. The eye of the hurricane, therefore, is usually calm. Hurricanes are composed of spiral rain bands where intense rainfall caused by severe thunderstorms can produce tens of centimeters (several inches) of rainfall per hour.

IMPACT OF OTHER FACTORS A host of factors influences the development and strength of hurricanes. For example, warmer sea surface temperatures tend to favor the development of hurricanes, while strong wind shear in the upper atmosphere can ventilate heat away from a developing hurricane and thus interfere with hurricane formation. Other factors that can either enhance or disrupt the development and intensification of hurricanes include the amount of atmospheric convective instability, air humidity, the degree of rotation of spinning winds, and even El Niño/La Niña events (which are discussed in Chapter 7, "Ocean Circulation").

New research combined with careful analysis of historical data shows an out-of-phase relationship between tropical cyclone variability in the North Atlantic and the eastern North Pacific, meaning that when one basin has high storm occurrence, the other has low storm occurrence. The recognition of this pattern has helped improve hurricane forecasting.

STUDENTS SOMETIMES ASK . . .

Figure 6.19 shows a noticeable gap in tropical cyclones along the equator. Has a tropical cyclone ever existed at the equator?

Remarkably, yes. An unusual confluence of weather conditions in 2001 created the first-ever documented instance of a tropical cyclone almost directly over the equator in the eastern Pacific Ocean. Tropical storm Vamei drifted in from the north, existed at the equator briefly and started to lose its spin (remember, there is no Coriolis effect at the equator), then moved off the equator to the north and regained its spin. Statistical models indicate that such an event occurs only once every 300–400 years.

SmartFigure 6.20 Typical North Atlantic hurricane storm track and detail of internal structure.
https://goo.gl/DcEaEV

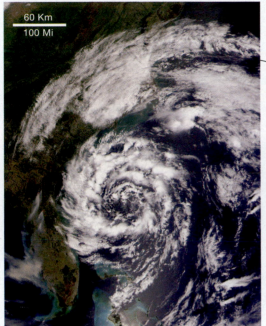

60 Km
100 Mi

(a) Satellite photo of Hurricane Andrea off the U.S. East Coast in 2007.

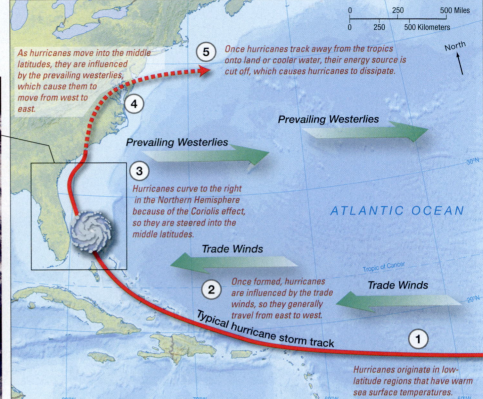

0 250 500 Miles
0 250 500 Kilometers

North

⑤ Once hurricanes track away from the tropics onto land or cooler water, their energy source is cut off, which causes hurricanes to dissipate.

④ As hurricanes move into the middle latitudes, they are influenced by the prevailing westerlies, which cause them to move from west to east.

Prevailing Westerlies

Prevailing Westerlies

③ Hurricanes curve to the right in the Northern Hemisphere because of the Coriolis effect, so they are steered into the middle latitudes.

ATLANTIC OCEAN

30°N

Trade Winds

Tropic of Cancer

② Once formed, hurricanes are influenced by the trade winds, so they generally travel from east to west.

Trade Winds

20°N

Typical hurricane storm track

①

Hurricanes originate in low-latitude regions that have warm sea surface temperatures.

80°W 70°W 60°W 50°W

(b) Map showing a typical hurricane storm track, including the steps involved in its origin, movement, and dissipation.

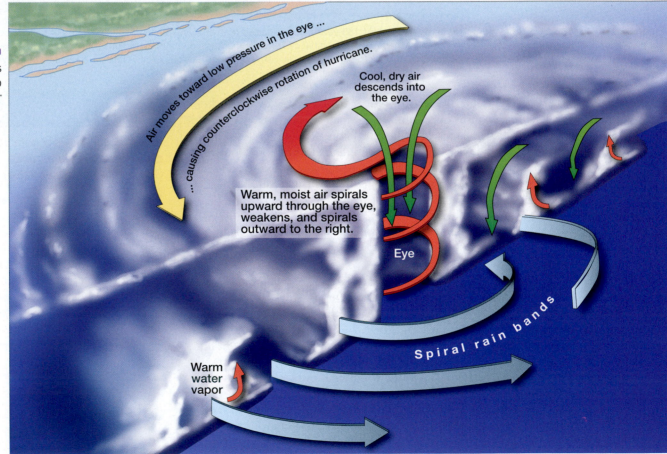

Air moves toward low pressure in the eye ...

... causing counterclockwise rotation of hurricane.

Cool, dry air descends into the eye.

Warm, moist air spirals upward through the eye, weakens, and spirals outward to the right.

Eye

Spiral rain bands

Warm water vapor

(c) Enlarged cut-away view of a hurricane showing its components, internal structure, and winds.

Human-caused climate change has been linked to the documented warming of ocean surface waters, which fuels hurricanes. As a result, several recent studies show that hurricane severity is expected to increase. In fact, climate models suggest that, while typical tropical storms will likely decrease in number, the risk of Category 4 and 5 storms will likely increase. For more details about climate change and its impact on hurricanes, see Chapter 16, "The Oceans and Climate Change."

TYPES OF DESTRUCTION Destruction from hurricanes is caused by high winds and flooding from intense rainfall. **Storm surge**, however, causes the majority of a hurricane's coastal destruction. In fact, storm surge is responsible for 90% of the deaths associated with hurricanes.

(b) Photograph showing the storm surge in Milford, Connecticut caused by Hurricane Sandy in 2012.

When a hurricane develops over the ocean, its low-pressure center produces a low "hill" of water (**Figure 6.21**). As the hurricane migrates across the open ocean, the hill moves with it. As the hurricane approaches shallow water nearshore, the portion of the hill over which the wind is blowing shoreward produces a mass of elevated, wind-driven water. This mass of water—the storm surge—can be as high as 12 meters (40 feet), resulting in a dramatic increase in sea level at the shore, large storm waves, and tremendous destruction to low-lying coastal areas (particularly if it occurs at high tide). In addition, the area of the coast that is hit with the right front quadrant of the hurricane—where onshore winds further pile up water—experiences the most severe storm surge (Figure 6.21). Table 6.3 shows typical storm surge heights associated with Saffir-Simpson hurricane intensities.

The area within the hurricane's right front quadrant (orange shading) experiences the most severe storm surge.

Path of hurricane

Right front quadrant

Wind

L

Low pressure

Land

Northern Hemisphere

Ocean

(a) As a hurricane in the Northern Hemisphere moves ashore, the low-pressure center around which the storm winds blow, combined with strong onshore winds, produces a high-water storm surge that floods the coast.

Figure 6.21 **A hurricane's storm surge batters the coast.**

HISTORIC DESTRUCTION ON THE U.S. MAINLAND Periodic destruction from hurricanes occurs along the East Coast and the Gulf Coast regions of the United States. In fact, the most deadly natural disaster in U.S. history was caused by a hurricane that struck Galveston Island, Texas, in September 1900. Galveston Island is a thin strip of sand called a *barrier island* located in the Gulf of Mexico off Texas (**Figure 6.22**). In 1900, it was a popular beach resort that averaged only 1.5 meters (5 feet) above sea level. At least 6000 people in and around Galveston were killed when the hurricane's 6-meter (20-foot)-high storm surge completely submerged the island, accompanied by heavy rainfall and winds of 160 kilometers (100 miles) per hour.

Category 4 hurricanes, like the one in 1900 that devastated Galveston, have been surpassed by Category 5 hurricanes making landfall only three times in the United States: (1) in 1935, an unnamed hurricane[9] flattened the Florida Keys; (2) in 1969, Hurricane Camille struck Mississippi; and (3) in 1992, Hurricane Andrew came ashore in southern Florida, with winds as high as 258 kilometers (160 miles) per hour, ripping down every tree in its path as it crossed the Everglades. Hurricane Andrew did more than $26.5 billion of damage in Florida and along the Gulf Coast. In the aftermath of Hurricane Andrew, more than 250,000 people were left homeless and although most people heeded the warnings to evacuate, 54 were killed.

In October 1998, Hurricane Mitch proved to be one of the most devastating tropical cyclones to affect the Western Hemisphere. At its peak, it was estimated to have winds of 290 kilometers (180 miles) per hour—a strong Category 5 hurricane. It hit Central America with winds of 160 kilometers (100 miles) per hour

[9]Prior to 1950, Atlantic hurricanes were not named, but this hurricane is often referred to as the "Labor Day Hurricane" because it came ashore then. Today, hurricanes are named by forecasters using an alphabetized list of female and male names.

Figure 6.22 Destruction from the Galveston hurricane of 1900. Photo showing destruction from the 1900 hurricane at Galveston (*above*) and location map of Galveston, Texas (*right*). At least 6000 people died as a result of the Galveston hurricane, which completely submerged Galveston Island and still stands as the single deadliest U.S. natural disaster.

and as much as 130 centimeters (51 inches) of total rainfall, causing widespread flooding and mudslides in Honduras and Nicaragua that destroyed entire towns. The hurricane resulted in more than 11,000 deaths, left more than 2 million homeless, and caused more than $10 billion in damage across the region.

In September 2008, Hurricane Ike reached Category 4 in the Gulf of Mexico and made landfall near Galveston in low-lying Gilchrist, Texas, as a Category 2 hurricane. Ike resulted in 146 deaths and $24 billion in damages, making it the third costliest U.S. hurricane of all time, behind only Hurricane Katrina (2005) and Hurricane Andrew (1992). In August 2011, Hurricane Irene achieved Category 3 status in the Caribbean and wreaked havoc as it moved along the eastern seaboard from Florida to New England. In all, Irene caused severe flooding that was responsible for 56 deaths and more than $10 billion in damages.

In October 2012, Hurricane Sandy, a large Category 1 storm, affected the Caribbean and the eastern seaboard from Florida to Maine. Hurricane Sandy was the largest Atlantic hurricane on record, with winds spanning an enormous area that was over 1800 kilometers (1100 miles) wide. When Hurricane Sandy came ashore in the United States, its peak wave heights and storm surge coincided with peak high tides, with the largest waves and storm surge focused along the heavily populated New York and New Jersey coasts. Hurricane Sandy caused extensive wave damage (**Figure 6.23**), severe coastal erosion, and extreme flooding that destroyed thousands of homes and left millions without electric service throughout the Mid-Atlantic states. In all, the storm was responsible for 233 deaths and more than $68 billion in damages, making it the second costliest hurricane in U.S. history, behind only Hurricane Katrina.

Figure 6.23 Damage to a New Jersey amusement pier from Hurricane Sandy in 2012. Hurricane Sandy, which was the widest Atlantic hurricane on record, caused flooding and destruction from the Caribbean to the U.S. East Coast from Florida to Maine, particularly in New York and New Jersey.

THE RECORD-BREAKING 2005 ATLANTIC HURRICANE SEASON: HURRICANES KATRINA, RITA, AND WILMA

Although the official Atlantic hurricane season extends each year from June 1 to November 30, the 2005 Atlantic hurricane season persisted into January 2006 and was the most active season on record, shattering numerous records. For example, a record 27 named tropical storms formed, of which a record 15 became hurricanes. Of these, seven strengthened into major hurricanes, a record-tying five became Category 4 hurricanes and a record four reached Category 5 strength, the highest categorization for hurricanes on the Saffir-Simpson Scale of hurricane intensity (see Table 6.3). For the first time ever, NOAA's National Hurricane Center, which oversees the naming of Atlantic hurricanes, ran out of the usual names for storms and resorted to naming storms using the Greek alphabet.

The most notable storms of the 2005 season were the five Category 4 and Category 5 hurricanes: Dennis, Emily, Katrina, Rita, and Wilma. These storms made

a combined 12 landfalls as major hurricanes (Category 3 strength or higher) throughout Cuba, Mexico, and the Gulf Coast of the United States, causing over $100 billion in damage and more than 2000 deaths.

Hurricane Katrina, the sixth-strongest Atlantic hurricane ever recorded, was the costliest and one of the deadliest hurricanes in U.S. history. Katrina formed over the Bahamas on August 23 and crossed southern Florida as a moderate Category 1 hurricane before passing over the warm Loop Current and strengthening rapidly in the Gulf of Mexico, becoming one of the strongest hurricanes ever recorded in the Gulf. The storm weakened considerably before making its second landfall as a Category 3 storm on the morning of August 29 in southeast Louisiana (**Figure 6.24a**). Still, Katrina was the largest hurricane of its strength to make landfall in the United States in recorded history; its sheer size caused devastation over a radius of 370 kilometers (230 miles). Katrina's 9-meter (30-foot) storm surge—the highest ever recorded in the United States—caused severe damage along the coasts of Mississippi, Louisiana, and Alabama.

As forecasters watched these events unfold, they recognized a potential catastrophe—Katrina was on a collision course with New Orleans. This scenario was considered particularly disastrous because nearly all of the New Orleans metropolitan area is below sea level along Lake Pontchartrain. Even without a direct hit, the storm surge from Katrina was forecast to be greater than the height of the levees protecting New Orleans. This risk of devastation was well known; several previous studies warned that a direct hurricane strike on New Orleans could lead to massive flooding, which would lead to thousands of drowning deaths, as well as many more suffering from disease and dehydration after the hurricane passed. Although Katrina passed to the east of New Orleans, levees separating Lake Pontchartrain from New Orleans were breached by Katrina's high winds, storm surge, and heavy rains, ultimately flooding roughly 80% of the city and many neighboring areas (**Figure 6.24b**). Damages from Katrina exceeded $100 billion, easily making it the costliest hurricane in U.S. history. The storm also left hundreds of thousands homeless and killed approximately 1800 people, making it the deadliest U.S. hurricane since the 1928 Okeechobee Hurricane, which killed as many as 2500 people. Responders in the aftermath of Katrina attributed many deaths to drowning because of rising water that trapped residents in the attics of single-story homes. In addition, many observers were surprised that a disaster like Katrina could befall a wealthy, technologically advanced nation like the United States. The lack of adequate disaster response by the Federal Emergency Management Agency (FEMA) led to a U.S. Senate investigation in 2006 that recommended disbanding the agency and creating a new National Preparedness and Response Agency. Ten years after the disaster, the population of New Orleans remains below prestorm levels and some areas remain so damaged that they are unusable.

Following on the heels of Katrina, Hurricane Rita set records as the fourth most intense Atlantic hurricane ever recorded and the most intense tropical cyclone observed in the Gulf of Mexico, breaking the record set by Katrina just three weeks earlier. Rita reached its maximum intensity on September 21, with sustained winds of 290 kilometers (180 miles) per hour and an estimated minimum pressure of 89,500 Pascal (895 millibars, or 0.884 atmosphere). Hurricane Rita's unusually rapid intensification in the Gulf can likely be attributed to its passage over the warm Loop Current, as well as higher-than-normal sea surface temperatures in the Gulf. Rita made landfall on September 24 near the Texas–Louisiana border as a Category 3 hurricane. Rita's 6-meter (20-foot) storm surge caused extensive damage along the coasts of Louisiana and extreme southeastern Texas, completely destroying some coastal communities and causing $10 billion in damage.

Later during the same season, Hurricane Wilma set numerous records for both strength and seasonal activity. Wilma was only the third Category 5 ever to develop during the month of October, and its extremely low pressure of 88,200 Pascal (882

(a) Satellite view of Hurricane Katrina coming ashore along the Gulf Coast on August 29, 2005, showing the hurricane's counterclockwise direction of spin and prominent central eye. Hurricane Katrina, which had a diameter of about 670 kilometers (415 miles), was the largest hurricane of its strength to make landfall in the United States in recorded history.

(b) Hurricane Katrina breached levees and flooded New Orleans, Louisiana, causing damages of more than $75 billion and claiming at least 1600 lives.

Figure 6.24 **Hurricane Katrina, the most destructive hurricane in U.S. history.**

RECAP

Hurricanes are intense—and sometimes destructive—tropical storms that form where water temperatures are high, where there is an abundance of warm moist air, and where the Coriolis effect influences their spin.

millibars, or 0.871 atmosphere) ranked it as the most intense hurricane ever recorded in the Atlantic basin. Its maximum sustained near-surface wind speed reached 282 kilometers (175 miles) per hour, with gusts up to 320 kilometers (200 miles) per hour. Wilma made several landfalls, with the most destructive effects felt in the Yucatán Peninsula of Mexico, Cuba, and southern Florida. At least 62 deaths were reported, and damage was estimated at $16 billion to $20 billion ($12.2 billion in the United States), ranking Wilma among the top 10 costliest hurricanes ever recorded in the Atlantic and the sixth costliest storm in U.S. history. Wilma also affected 11 countries with winds or rainfall, more than any other hurricane in recent history.

HISTORIC DESTRUCTION IN OTHER REGIONS The majority of the world's tropical cyclones are formed in the waters north of the equator in the western Pacific Ocean. These storms, called *typhoons*, do enormous damage to coastal areas and islands in Southeast Asia (see Figure 6.19).

Other areas of the world such as Bangladesh, which borders the Indian Ocean, experience tropical cyclones on a regular basis. Bangladesh is particularly vulnerable because it is a highly populated and low-lying country, much of it only 3 meters (10 feet) above sea level. In 1970, a 12-meter (40-foot)-high storm surge from a tropical cyclone killed an estimated 1 million people. Another tropical cyclone hit the area in 1972 and caused up to 500,000 deaths. In 1991, Hurricane Gorky's winds of 233 kilometers (145 miles) per hour and large storm surge caused extensive damage, killing over 200,000 people.

Even islands near the centers of ocean basins can be struck by hurricanes. The Hawaiian Islands, for example, were hit hard by Hurricane Dot in August 1959 and by Hurricane Iwa in November 1982. Hurricane Iwa hit very late in the hurricane season and produced winds up to 130 kilometers (81 miles) per hour. Damage of more than $100 million occurred on the islands of Kauai and Oahu. Niihau, a small island that is inhabited by only a few hundred native Hawaiians, was directly in the path of the storm and suffered severe property damage but no serious injuries. Hurricane Iniki roared across the islands of Kauai and Niihau in September 1992, with 210-kilometer (130-mile)-per-hour winds. It was the most powerful hurricane to hit the Hawaiian Islands in the past 100 years, with property damage that approached $1 billion.

FUTURE THREAT TO LIFE AND PROPERTY Each year, tropical cyclones and hurricanes leave millions homeless worldwide and account for, on average, over $100 billion of damage in the United States alone. Hurricanes will continue to be a threat to life and property around the globe. Because of increasingly accurate forecasts and prompt evacuation, however, the loss of life has been decreasing. Property damage, on the other hand, has been increasing because increasing coastal populations have resulted in more and more construction along the coast. Inhabitants of areas subject to a hurricane's destructive force must be made aware of the danger so that they can be prepared for its eventuality. The impact of human-caused climate change on the inevitable economic losses from tropical cyclones is a major concern, too. In fact, new research shows that human-caused climate change may double the global economic losses caused by tropical cyclones and hurricanes.

Climate

Connection

The Ocean's Climate Patterns

Just as land areas have climate patterns, so do regions of the oceans. The open ocean is divided into climatic regions that run generally east–west (parallel to lines of latitude) and have relatively stable boundaries that are somewhat modified by ocean surface currents (**Figure 6.25**).

Climate

Connection

The **equatorial** region spans the equator, which gets an abundance of solar radiation. As a result, the major air movement is upward because heated air rises. Surface winds, therefore, are weak and variable, which is why this region is called the *doldrums*. Surface waters are warm and the air is saturated with water vapor. Daily rain showers are common, which keeps surface salinity relatively low. The equatorial regions just north or south of the equator are also the breeding grounds for tropical cyclones.

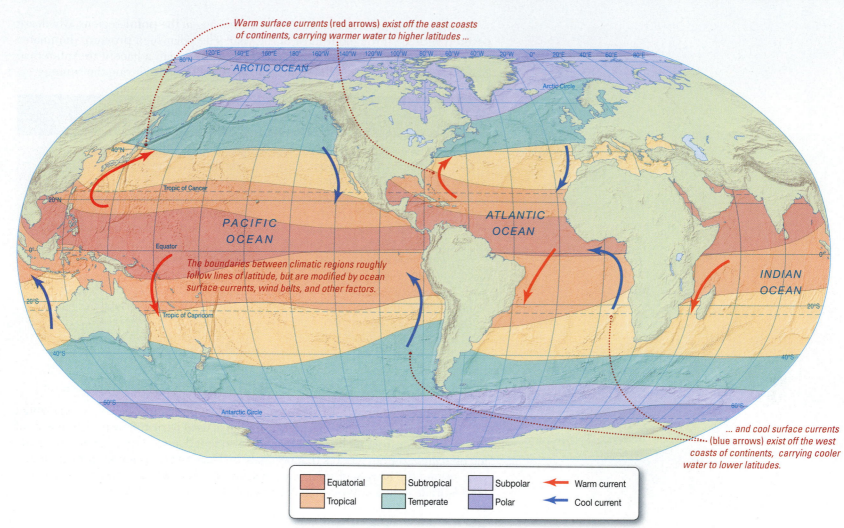

Warm surface currents (red arrows) *exist off the east coasts of continents, carrying warmer water to higher latitudes ...*

The boundaries between climatic regions roughly follow lines of latitude, but are modified by ocean surface currents, wind belts, and other factors.

... and cool surface currents (blue arrows) exist off the west coasts of continents, carrying cooler water to lower latitudes.

| Equatorial | Subtropical | Subpolar | Warm current |
| Tropical | Temperate | Polar | Cool current |

Figure 6.25 **The ocean's climatic regions.**

Tropical regions extend north or south of the equatorial region up to the Tropic of Cancer and the Tropic of Capricorn, respectively. They are characterized by strong trade winds, which blow from the northeast in the Northern Hemisphere and from the southeast in the Southern Hemisphere. These winds push the equatorial currents and create moderately rough seas. Relatively little precipitation falls at higher latitudes within tropical regions, but precipitation increases toward the equator. Once tropical cyclones form, they gain energy here as large quantities of heat are transferred from the ocean to the atmosphere.

Beyond the tropics are the **subtropical** regions. Belts of high pressure are centered there, so the dry, descending air produces little precipitation and a high rate of evaporation, resulting in the highest surface salinities in the open ocean (see the surface salinity map in Chapter 5, Figure 5.25). Winds are weak and currents are sluggish, typical of the horse latitudes. However, strong boundary currents (along the boundaries of continents) flow north and south, particularly along the western margins of the subtropical oceans.

The **temperate** regions (also called the *middle latitudes* or *midlatitudes*) are characterized by strong westerly winds (the prevailing westerlies) that blow from the southwest in the Northern Hemisphere and from the northwest in the Southern Hemisphere (see Figure 6.12). Severe storms are common, especially during winter, and precipitation is heavy. In fact, the North Atlantic is noted for fierce storms, which have claimed many ships and numerous lives over the centuries.

The **subpolar** region experiences extensive precipitation due to the subpolar low. Sea ice covers the subpolar ocean in winter, but it melts away, for the most part, in summer. Icebergs are common, and the surface temperature seldom exceeds 5°C (41°F) in the summer months.

Surface temperatures remain at or near freezing in the **polar** regions, which are covered with ice throughout most of the year. The polar high pressure dominates the area, which includes the Arctic Ocean and the ocean adjacent to Antarctica. There is no sunlight during the winter and constant daylight during the summer.

CONCEPT CHECK 6.5 | Describe how the ocean influences global weather phenomena and climate patterns.

1 Describe the difference between cyclonic and anticyclonic flow and show how the Coriolis effect is important in producing both clockwise and counterclockwise flow patterns.

2 How do sea breezes and land breezes form? During a hot summer day, which one would be most common and why?

3 Name the polar and tropical air masses that affect U.S. weather. Describe the pattern of movement across the continent and patterns of precipitation associated with warm and cold fronts.

4 What are the conditions needed for the formation of a tropical cyclone? Why do most middle latitude areas only rarely experience a hurricane? Why are there no hurricanes at the equator?

5 Describe the types of destruction caused by hurricanes. Which one causes the majority of fatalities and destruction?

6.6 How Do Sea Ice and Icebergs Form?

Low temperatures in high-latitude regions cause a permanent or nearly permanent ice cover on the sea surface. The term **sea ice** is used to distinguish such masses of frozen seawater from **icebergs**, which are also found at sea but originate by breaking off (*calving*) from glaciers that originate on land. Sea ice is found throughout the year around the margin of Antarctica, within the Arctic Ocean, and in the extreme high-latitude region of the North Atlantic Ocean.

Formation of Sea Ice

Sea ice is ice that forms directly from seawater (**Figure 6.26**). It begins as small, needle-like, hexagonal (six-sided) crystals, which eventually become so numerous that a *slush* develops. As the slush begins to form into a thin sheet, it is broken by wind stress and wave action into disk-shaped pieces called **pancake ice** (**Figure 6.26a**). As further freezing occurs, the pancakes merge together to form larger **ice floes** (*flo* = layer; **Figure 6.26b**). Over time, ice floes merge and create large sheets of ice, which are moved by winds and currents to produce **pressure ridges** along their margins (**Figure 6.26c**).

The rate at which sea ice forms is closely tied to temperature conditions. Large quantities of ice form in relatively short periods when the temperature falls to extremely low levels (such as temperatures below –30°C [–22°F]). Even at these low temperatures, the rate of ice formation slows as sea ice thickens because the ice (which has poor heat conduction) effectively insulates the underlying water from freezing. In addition, calm water enables pancake ice to join together more easily, which aids the formation of sea ice.

The process of sea ice formation tends to be a self-perpetuating process. As sea ice forms at the surface, only a small percentage of the dissolved components can be accommodated into the crystalline structure of ice. As a result, most of the dissolved substances remain in the surrounding seawater, which causes its salinity to increase. Recall from Chapter 5 that increasing the amount of dissolved materials decreases the freezing point of water, which doesn't appear to enhance ice formation. However, also recall that increasing the salinity of water increases its density and its tendency to sink. As it sinks below the surface, it is replaced by lower-salinity (and lower-density) water from below, which will freeze more readily than the high-salinity water it replaced, thereby establishing a circulation pattern that enhances the formation of sea ice.

Figure 6.26 **Stages of formation of sea ice.** Sea ice, which forms directly from the freezing of seawater, begins as **(a)** pancake ice. As further freezing occurs, pancake ice thickens into **(b)** ice floes, eventually forming large sheets of ice that contain **(c)** pressure ridges.

1 m
3.3 ft

(a) Pancake ice, which is the initial stage of sea ice formation, is a frozen slush that is broken by wind stress and wave action into disk-shaped pieces.

(b) Ice floes form with additional freezing, causing pancake ice to merge together and become thicker.

(c) Ridged ice is created over time when large sheets of sea ice collide, forming thick pressure ridges.

(a) Icebergs, such as this small North Atlantic berg, are formed when pieces of ice calve from glaciers that extend to the sea.

(b) Map showing North Atlantic currents (*blue arrows*), typical iceberg distribution (*white triangles*), and the site of the 1912 *Titanic* sinking (*black X*).

(c) Aerial view of part of a large tabular Antarctic iceberg, which extends beyond the horizon.

(d) Satellite view of iceberg C-19, which broke off from Antarctica's Ross Ice Shelf in May 2002. Also shown is iceberg B-15A, which is part of the larger B-15 iceberg that was the size of Connecticut when it calved in March 2000.

Figure 6.27 **Icebergs.** Icebergs form when land-based freshwater glacial ice breaks off into the sea. The world's two main iceberg-producing regions are Greenland (*a* and *b*) and Antarctica (*c* and *d*).

Recent satellite analyses of the extent of Arctic Ocean sea ice shows that it has decreased dramatically in the past few decades. This accelerated melting appears to be linked to shifts in Northern Hemisphere atmospheric circulation patterns that have caused the region to experience anomalous warming. For more on this topic, see Chapter 16, "The Oceans and Climate Change."

Climate Connection

Formation of Icebergs

An *iceberg* is a body of floating ice that has broken away from a glacier (**Figure 6.27a**) and so is quite distinct from sea ice. Icebergs are formed by vast ice sheets on land, which grow from the accumulation of snow and slowly flow outward to the sea. Once at sea, the ice either breaks up and produces icebergs there or, because it is less dense than water, floats on top of the water, often extending a great distance away from shore before breaking up under the stress of current, wind, and wave action. Most calving occurs during the summer months, when temperatures are highest.

In the Arctic, icebergs originate primarily by calving from glaciers that extend to the ocean along the western coast of Greenland (**Figure 6.27b**). Icebergs are also produced by glaciers along the eastern coasts of Greenland, Ellesmere Island, and other Arctic islands. In all, about 10,000 or so icebergs are calved off these glaciers each year, and the number of icebergs has been increasing recently. Many of these icebergs are carried by currents in and around the Labrador Sea (Figure 6.27b, *blue arrows*) into North Atlantic shipping lanes, where they become navigational hazards. In recognition of this fact, the area is called *Iceberg Alley*; it is here that the luxury liner RMS *Titanic* hit an iceberg and sank (see *black x* on Figure 6.27b; see also MasteringOceanography **Web Diving Deeper 6.1**). Because of their large size, some of these icebergs take several years to melt, and, in that time, they may be carried as far south as 40 degrees north latitude, which is the same latitude as Philadelphia, Pennsylvania.

SHELF ICE In Antarctica, where glaciers cover nearly the entire continent, the edges of glaciers form thick floating sheets of ice called **shelf ice** that break off and produce vast plate-like icebergs (**Figures 6.27c** and **6.27d**). In March 2000, for example, a Connecticut-sized iceberg (11,000 square kilometers [4250 square miles]) known as *B-15* and nicknamed "Godzilla" broke loose from the Ross Ice Shelf into the Ross Sea. Since its origin more than a decade ago, B-15 has broken up into smaller icebergs, but many of its parts still exist. Even larger icebergs have also been observed in Antarctic waters. For instance, the largest iceberg ever recorded measured an incredible 32,500 square kilometers (12,500 square miles)—nearly three times the size of B-15 or about the same size as Connecticut and Massachusetts combined.

Icebergs from shelf ice have flat tops that may stand as much as 200 meters (650 feet) above the ocean surface, although most rise less than 100 meters (330 feet)

above sea level, and as much as 90% of their mass is below the waterline. Once icebergs are created, ocean currents driven by strong winds carry the icebergs north, where they eventually melt. Because this region is not a major shipping route, the icebergs pose little serious navigation hazard except to supply ships traveling to Antarctica. Officers aboard ships sighting these gigantic bergs have, in some cases, mistaken them for land!

The rate at which Antarctica is producing icebergs—especially large icebergs—has recently increased, most likely as a result of Antarctic warming. In addition, newly discovered troughs on the sea floor surrounding Antarctica could funnel warmer ocean water to the base of glaciers, thereby accelerating their melting. For more information about Antarctic warming and its relationship to climate change, see Chapter 16, "The Oceans and Climate Change."

Climate Connection

RECAP

Sea ice is created when seawater freezes; icebergs form when chunks of ice break off from coastal glaciers that reach the sea.

CONCEPT CHECK 6.6 | Specify how sea ice and icebergs form.

1 Why does the formation of sea ice tend to be a self-perpetuating process?

2 Describe differences between sea ice, icebergs, and shelf ice, including how each is formed.

6.7 Can Power from Wind Be Harnessed as a Source of Energy?

The uneven heating of Earth by the Sun drives various small- and large-scale winds. These winds, in turn, can be harnessed to turn windmills or turbines that generate electricity. At various places on land where the wind blows almost constantly, wind farms have been constructed that consist of hundreds of large turbines mounted on tall towers, thereby taking advantage of this renewable, clean energy source. Similar facilities could be built offshore, where the wind generally blows harder and more steadily than on land. **Figure 6.28** shows the offshore areas where the potential for wind farms exist.

Some offshore wind farms have already been built (**Figure 6.29**), and many more are being planned. In the North Sea, near windswept northern Europe, for example, about 100 sea-based turbines are already operating, with hundreds more planned. In fact, Denmark generates 18% of its power by wind—more than any other country—and hopes to increase its proportion of wind power to 50% by 2030.

One major disadvantage of wind power is that wind strength varies and sometimes it's not windy at all, which is particularly problematic when a high demand for electricity exists. There is also the problem of getting the energy to viable markets such as population centers. On paper, wind and solar power could supply the United States and some other countries with all the electricity they require. In practice, however, both sources are too erratic to supply more than about 20% of a region's total energy capacity, according to the U.S. Department of Energy. What are needed are cheap and efficient ways of storing the generated power, then tapping those supplies when

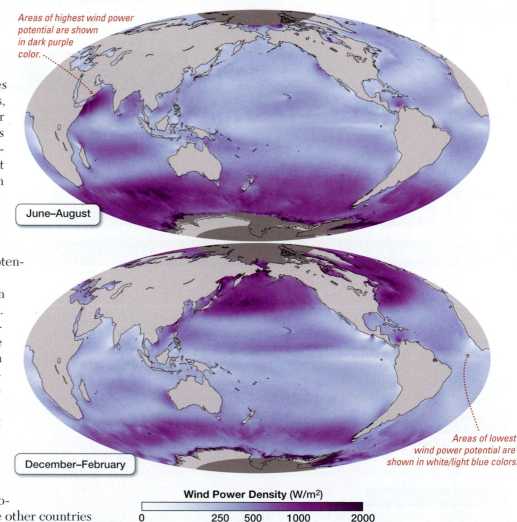

Figure 6.28 **Global ocean wind energy potential.** Average ocean wind intensity maps during 2000–2007 for June–August (*top*) and December–February (*bottom*).

Areas of highest wind power potential are shown in dark purple color.

June–August

December–February

Areas of lowest wind power potential are shown in white/light blue colors.

Wind Power Density (W/m²)

0 250 500 1000 2000

Figure 6.29 **Offshore wind farm.** Offshore wind turbines form part of a wind farm that harnesses wind energy off the west coast of Scotland in the United Kingdom.

RECAP

There is vast potential for developing wind power as a renewable source of energy, although the erratic nature of wind energy is problematic. Several offshore wind farms currently exist.

needed. Some of the best solutions to the storage problem include pumping water to high areas and using the water to turn turbines later, using pumps to store underground air at high pressures, and storing energy in advanced batteries.

CONCEPT CHECK 6.7 | Evaluate the advantages and disadvantages of harnessing winds as a source of energy.

1 Discuss the advantages and disadvantages of building an offshore wind farm.

2 Describe the location, timeframe, and power generating capability of America's first offshore wind farm.

ESSENTIAL CONCEPTS REVIEW

6.1 What causes variations in solar radiation on Earth?

▶ *The atmosphere and the ocean act as one interdependent system,* linked by complex feedback loops. There is a close association between most atmospheric and oceanic phenomena.

▶ *The Sun heats Earth's surface unevenly due to the change of seasons (caused by the tilt of Earth's rotational axis,* which is 23.5 degrees from vertical) and the daily cycle of sunlight and darkness (Earth's rotation on its axis). *Differences in latitude also cause changes in the amount of solar radiation* received on Earth.

Study Resources
MasteringOceanography Study Area Quizzes, MasteringOceanography Web Animation

Critical Thinking Question

Earth's axis of rotation is angled 23.5 degrees from perpendicular relative to the plane of its ecliptic. Specify how the tilt of Earth's axis affects the change of seasons, the length of day, and the angle of sunlight over the timespan of a year using as an example a location from both hemispheres.

Active Learning Exercise

With another student in class, make a list of changes that would occur if Earth's axis of rotation was vertical (not tilted). For example, would seasons still exist?

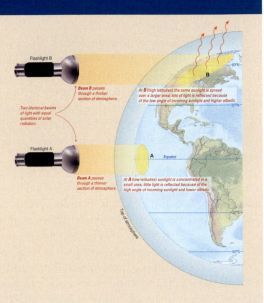

6.2 What physical properties does the atmosphere possess?

▶ The *uneven distribution of solar energy on Earth* influences most of the physical properties of the atmosphere (such as *temperature*, *density*, *water vapor content*, and *pressure differences*) that produce *atmospheric movement*.

Study Resources
MasteringOceanography Study Area Quizzes, MasteringOceanography Web Video

Critical Thinking Question
In a nonspinning Earth, describe the basic atmospheric circulation pattern that would exist.

Active Learning Exercise
With another student in class, use the Internet to research the ancient Greek story of Icarus. Based on the temperature profile of the atmosphere shown in Figure 6.6, is the tragedy that befalls Icarus based in physical fact? Explain.

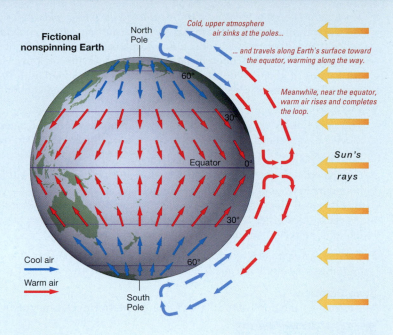

6.3 How does the Coriolis effect influence moving objects?

▶ *The Coriolis effect influences the paths of moving objects on Earth and is caused by Earth's rotation.* Because Earth's surface rotates at different velocities at different latitudes, *objects in motion tend to veer to the right in the Northern Hemisphere and to the left in the Southern Hemisphere.* The Coriolis effect is *nonexistent at the equator but increases with latitude*, reaching a maximum at the poles.

Study Resources
MasteringOceanography Study Area Quizzes, MasteringOceanography Web Animation, MasteringOceanography Web Table 6.1, MasteringOceanography Web Video; Web Video

Critical Thinking Question
How does the Coriolis effect influence the *direction* of moving objects? How does it affect the *speed* of moving objects? Explain.

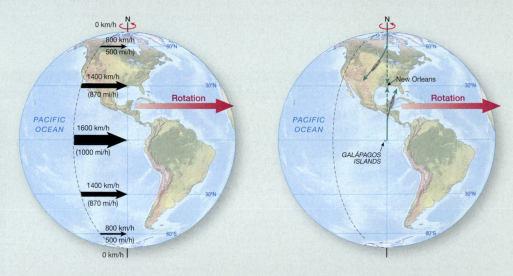

(a) The velocity of any point on Earth varies with latitude from about 1600 kilometers (1000 miles) per hour at the equator to 0 kilometers per hour at either pole.

(b) The path of missiles shot towards New Orleans from the North Pole and from the Galápagos Islands on the equator. Dashed lines indicate intended paths; solid lines indicate paths that the missiles would travel as viewed from Earth's surface.

Active Learning Exercise
With another student in class, explain why the Coriolis effect is strongest at the poles. Then, switch roles and explain why the Coriolis effect is nonexistent at the equator.

6.4 What global atmospheric circulation patterns exist?

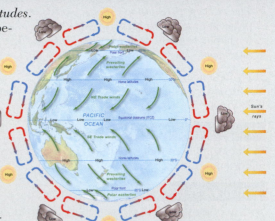

▶ *More solar energy is received than is radiated back into space at low latitudes than at high latitudes.* On the spinning Earth, this creates *three circulation cells in each hemisphere*: a *Hadley cell* between 0 and 30 degrees latitude, a *Ferrel cell* between 30 and 60 degrees latitude, and a *polar cell* between 60 and 90 degrees latitude. High-pressure regions, where dense air descends, are located at about 30 degrees north or south latitude and at the poles. Belts of low pressure, where air rises, are generally found at the equator and at about 60 degrees latitude.

▶ *The movement of air within the circulation cells produces the major wind belts of the world.* The air at Earth's surface that is moving away from the subtropical highs produces *trade winds* moving toward the equator and *prevailing westerlies* moving toward higher latitudes. The air moving along Earth's surface from the polar high to the subpolar low creates the *polar easterlies*.

▶ *Calm winds characterize the boundaries between the major wind belts of the world.* The boundary between the two trade wind belts is called the *doldrums*, which coincides with the Intertropical Convergence Zone (ITCZ). The boundary between the trade winds and the prevailing westerlies is called the *horse latitudes*. The boundary between the prevailing westerlies and the polar easterlies is called the *polar front*.

▶ *The tilt of Earth's axis of rotation,* the *lower heat capacity of rock material* compared to seawater, and the *distribution of continents modify the wind and pressure belts of the idealized three-cell model.* However, the three-cell model closely matches the pattern of the major wind belts of the world.

Study Resources
MasteringOceanography Study Area Quizzes, MasteringOceanography Web Animations, Web Video

Critical Thinking Question

To help reinforce your knowledge of atmospheric circulation patterns, draw from memory the pattern of surface wind belts on Earth, showing atmospheric circulation cells in the upper/lower atmosphere, zones of high and low atmospheric pressure, the names of the wind belts, and the names of the boundaries between the wind belts.

Active Learning Exercise

With another student in class, discuss how the world's idealized wind belts shown in Figure 6.12 are modified on the real Earth (Figure 6.13). Use specific examples including wind belts, boundaries, high and low atmospheric pressures, and differences between ocean and land. Report your findings to the class.

6.5 How does the ocean influence global weather phenomena and climate patterns?

(c) Enlarged cut-away view of a hurricane showing its components, internal structure, and winds.

▶ *Weather describes the conditions of the atmosphere at a given place and time, while climate is the long-term average of weather.* Atmospheric motion (*wind*) is always from *high-pressure regions toward low-pressure regions.* In the Northern Hemisphere, therefore, there is a *counterclockwise cyclonic movement* of air around low-pressure cells and a *clockwise anticyclonic movement* around high-pressure cells. Coastal regions commonly experience *sea and land breezes*, due to the daily cycle of heating and cooling.

▶ *Many storms are due to the movement of air masses.* In the middle latitudes, cold air masses from higher latitudes meet warm air masses from lower latitudes and create *cold and warm fronts* that move from west to east across Earth's surface. *Tropical cyclones (hurricanes) are large, powerful storms that mostly affect tropical regions of the world.* Destruction caused by hurricanes is caused by storm surge, high winds, and intense rainfall.

▶ *The ocean's climate patterns are closely related to the distribution of solar energy and the wind belts of the world.* Ocean surface currents somewhat modify oceanic climate patterns.

Study Resources
MasteringOceanography Study Area Quizzes, MasteringOceanography Web Animations, Web Video

Critical Thinking Question

Specify differences between *weather* and *climate*. Then answer this question: When it rains in a region that experiences an arid climate, does it mean the region's climate has changed from dry to wet? Explain.

Active Learning Exercise

Pair up with another student in class. Using the Internet, have each student determine the latitude and today's offshore surface water temperature for one of the following two locations: San Diego, California, and Charleston, South Carolina. Working together, mark each location on the map shown as Figure 6.25. Using information from Figure 6.25, explain why the surface water temperature is so different at these two locations.

6.6 How do sea ice and icebergs form?

▶ *In high latitudes, low temperatures freeze seawater and produce sea ice*, which forms as a slush and breaks into pancakes that ultimately grow into ice floes and, over time, large sheets of ice with pressure ridges. *Icebergs form when chunks of ice break off glaciers* that form on Antarctica, Greenland, and some Arctic islands. Floating sheets of ice called *shelf ice* near Antarctica produce the largest icebergs.

(b) Ice floes form with additional freezing, causing pancake ice to merge together and become thicker.

Study Resources

MasteringOceanography Study Area Quizzes, MasteringOceanography Web Diving Deeper 6.1

Critical Thinking Question

Specify the places in the world where most icebergs form. What hazards do icebergs present? Explain.

Active Learning Exercise

Using the Internet and working with another student in class, find three instances within the past 10 years where icebergs have been involved in oceanic shipping or transportation accidents (or near accidents). Specify where the incidents took place.

6.7 Can power from wind be harnessed as a source of energy?

▶ *Winds can be harnessed as a source of power.* There is vast potential for developing this clean, renewable resource, but *problems exist* related to generating power when needed, supplying it to consumers, and storing it. Several offshore wind farm systems currently exist.

Study Resources

MasteringOceanography Study Area Quizzes

Critical Thinking Question

Specify some negative environmental factors that might inhibit the development of large offshore wind farms.

Active Learning Exercise

Working as a group and using Figure 6.28, identify a specific location on Earth that would allow wind turbines to work at maximum capacity year-round (not just during a particular season). Present your findings to the class, including the location and an explanation of your reasoning.

MasteringOceanography™

www.masteringoceanography.com

Looking for additional review and test prep materials? With individualized coaching on the toughest topics of the course, MasteringOceanography offers a wide variety of ways for you to move beyond memorization and deeply grasp the underlying processes of how the oceans work. Visit the Study Area in **www.masteringoceanography.com** to find practice quizzes, study tools, and multimedia that will improve your understanding of this chapter's content. Sign in today to enjoy the following features: Self Study Quizzes, SmartFigures, SmartTables, Oceanography Videos, Squidtoons, Geoscience Animation Library, RSS Feeds, Digital Study Modules, and an optional Pearson eText.

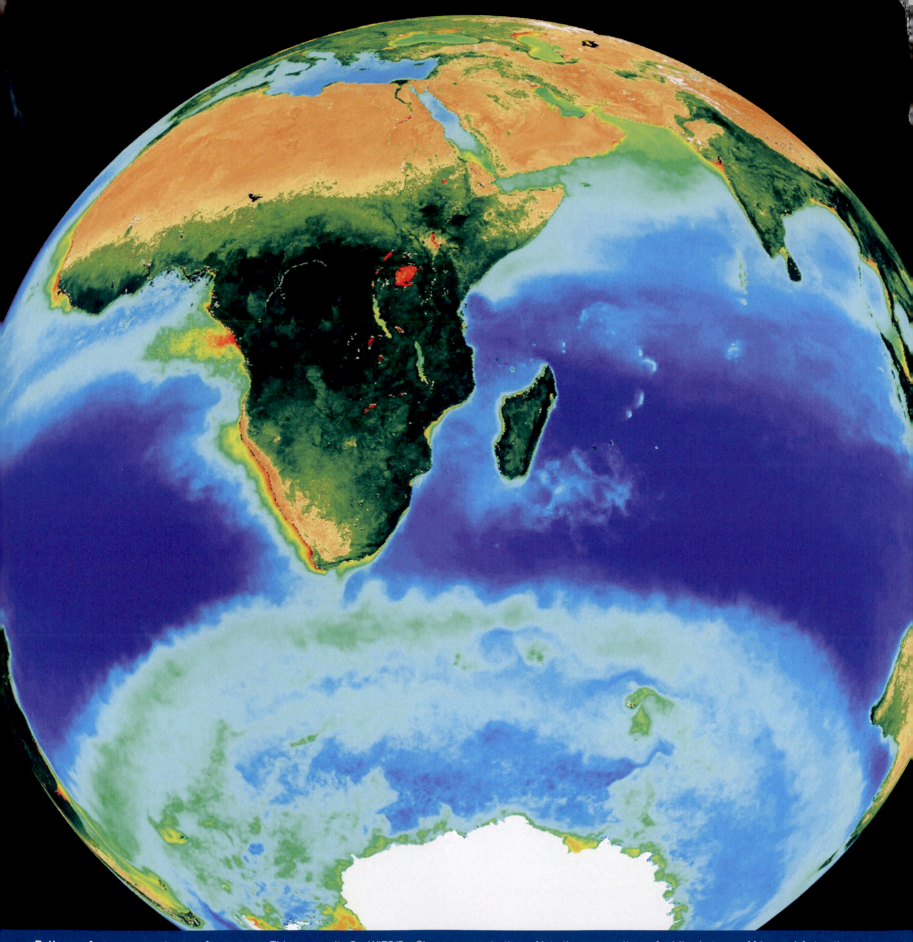

Patterns of ocean currents seen from space. This composite SeaWiFS/SeaStar satellite view during the austral summer highlights ocean circulation patterns. The deep blue color represents low chlorophyll (phytoplankton) concentrations, and the orange and red colors represent high chlorophyll (phytoplankton) concentrations. Note the wavy pattern of eddies between Africa and Antarctica, where the Agulhas Current meets the Antarctic Circumpolar Current and is turned to the east, creating the Agulhas Retroflection. Off the west coast of Africa, coastal upwelling is shown in bright red colors.

Thermohaline circulation
North Atlantic Gyre Sverdrup (Sv)
Conveyer-belt circulation
South Atlantic Gyre Indian Ocean Gyre
Upwelling El Niño–Southern Oscillation (ENSO)
Ekman transport**Ocean current**
Pacific warm pool **Subtropical Gyre South Pacific Gyre**
North Pacific Gyre Western intensification
Ekman spiral Geostrophic current Monsoon
La Niña Subpolar gyre

Ocean Circulation

Before you begin reading this chapter, use the glossary at the end of this book to discover the meanings of any of the words in the word cloud above you don't already know.

Ocean currents are masses of ocean water that flow from one place to another. The amount of water can be large or small, currents can be at the surface or deep below, and the phenomena that create them can be simple or quite complex. Simply put, currents are *water masses in motion*.

Huge current systems dominate the surfaces of the major oceans. These currents transfer heat from warmer to cooler areas on Earth, just as the major wind belts of the world do. Wind belts transfer about two-thirds of the total amount of heat from the tropics to the poles; ocean surface currents transfer the other third. Ultimately, energy from the Sun drives surface currents, and they closely follow the pattern of the world's major wind belts. As a result, the movement of currents has aided the travel of prehistoric people across ocean basins. Ocean currents also influence the abundance of life in surface waters by affecting the growth of microscopic algae, which are the basis of most oceanic food webs.

More locally, surface currents affect the climates of coastal continental regions. Cold currents flowing toward the equator on the western sides of continents produce arid conditions. Conversely, warm currents flowing poleward on the eastern sides of continents produce warm, humid conditions. Ocean currents, for example, contribute to the mild climate of northern Europe and Iceland, whereas conditions at similar latitudes along the Atlantic coast of North America (such as Labrador, Canada) are much colder. In addition, water sinks in high-latitude regions, initiating deep currents that help regulate the planet's climate.

Climate

Connection

ESSENTIAL LEARNING CONCEPTS

At the end of this chapter, you should be able to:

7.1 Demonstrate an understanding of how ocean currents are measured.

7.2 Explain the origin of ocean surface currents and how surface circulation patterns are organized globally.

7.3 Describe the conditions that produce upwelling.

7.4 Specify the main surface circulation patterns in each ocean basin.

7.5 Explain the origin and characteristics of deep-ocean currents.

7.6 Evaluate the advantages and disadvantages of harnessing currents as a source of energy.

"The coldest winter I ever spent was a summer in San Francisco."

—Anonymous, but often attributed to Mark Twain; said in reference to San Francisco's cool summer weather caused by coastal upwelling

7.1 How Are Ocean Currents Measured?

Ocean currents are either *wind driven* or *density driven*. Moving air masses—particularly the major wind belts of the world—set wind-driven currents in motion. Wind-driven currents move water horizontally and occur primarily in the ocean's surface waters, so these currents are called **surface currents**. Density-driven circulation, on the other hand, moves water vertically and accounts for the thorough mixing of the deep masses of ocean water. Some surface waters become high in density—through low temperature and/or high salinity—and so sink beneath the surface. This dense water sinks and spreads slowly beneath the surface, so these currents are called **deep currents**.

Surface Current Measurement

Because surface currents are driven by the wind, they rarely flow in the same direction and at the same rate for very long, so measuring average flow rates can

be difficult. Some consistency, however, exists in the *overall* surface current pattern worldwide. Surface currents can be measured directly or indirectly.

DIRECT METHODS Two main methods are used to measure surface currents *directly*. In one, a floating device is released into the current and its position is tracked through time. Typically, radio-transmitting float bottles or other devices are used (**Figure 7.1a**), but other accidentally released items also make good drift meters (**Diving Deeper 7.1**). The other method uses a current-measuring device, such as the propeller flow meter shown in **Figure 7.1b**, that is deployed from a fixed position, such as a pier or a stationary ship. A propeller device can also be towed behind a ship, with the ship's speed subtracted to determine the current's true flow rate.

INDIRECT METHODS Three different methods can be used to measure surface currents *indirectly*. The first method involves *pressure gradients*, which are the slopes caused by large-scale bulges and depressions in the ocean's surface (note that pressure gradients are also used to determine the movement of winds based on high and low atmospheric pressure; see, for example, the weather map shown in Figure 6.15). Water flows parallel to a pressure gradient (that is, downhill), so this method determines the internal distribution of density and the corresponding pressure gradient across an area of the ocean. A second method uses radar altimeters—such as those launched aboard Earth-observing satellites today—to determine the lumps and bulges at the ocean surface, which are a result of the shape of the underlying sea floor[1] as well as current flow. From these data, *dynamic topography* maps can be produced that show the speed and direction of surface currents (**Figure 7.2**). A third method uses a *Doppler flow meter* to

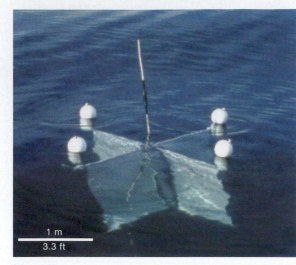

(a) A **drift current meter** afloat in the ocean.

Figure 7.1 Examples of direct-method current-measuring devices.

(b) A **propellor-type flow meter** being brought back aboard a research vessel.

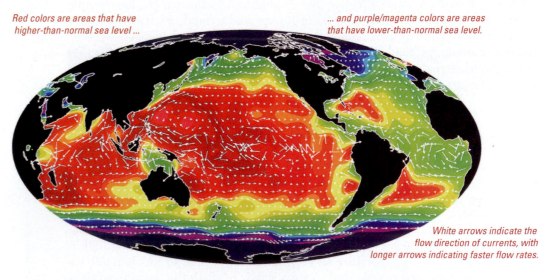

Red colors are areas that have higher-than-normal sea level ...

... and purple/magenta colors are areas that have lower-than-normal sea level.

White arrows indicate the flow direction of currents, with longer arrows indicating faster flow rates.

Colors represent sea surface height above/below average

-120 -80 -40 0 40 80

Ocean dynamic topography, centimeters

Arrows represent current speed

→ 10 centimeters per second

Figure 7.2 Satellite view of ocean dynamic topography. Map of TOPEX/Poseidon radar altimeter data showing variation of sea surface height, in centimeters, from September 1992 to September 1993.

[1]Note that this technique is described in Chapter 3, Section 3.1: "Using Satellites to Map Ocean Properties from Space."

RUNNING SHOES AS DRIFT METERS: JUST DO IT

Any floating object can serve as a makeshift drift meter, as long as it is known where the object entered the ocean and where it was retrieved. The path of the object can then be inferred, providing information about the movement of surface currents. If the time of release and retrieval are known, the speed of currents can also be determined. Oceanographers have long used *drift bottles* (floating "messages in a bottle" or radio-transmitting devices set adrift in the ocean) to track the movement of currents.

Many objects have inadvertently become drift meters when ships lose cargo at sea. Worldwide, in fact, as many as 10,000 shipping containers are lost overboard each year (**Figure 7A**, *top* and *middle photos*). In this way, Nike athletic shoes and colorful floating bathtub toys (Figure 7A, *inset* and *bottom photos*) have advanced the understanding of current movement in the North Pacific Ocean.

In May 1990, the container vessel *Hansa Carrier* was en route from Korea to Seattle, Washington, when it encountered a severe North Pacific storm. The ship was transporting 12.2-meter (40-foot)-long rectangular metal shipping containers, many of which were lashed to the ship's deck for the voyage. During the storm, the ship lost 21 deck

containers overboard, including 5 that held Nike athletic shoes. The shoes floated, and those that were released from their containers were carried east by the North Pacific Current. Within six months, thousands of the shoes began to wash up along the beaches of Alaska, Canada, Washington, and Oregon (Figure 7A, *map*), more than 2400 kilometers (1500 miles) from the site of the spill. A few shoes were found on beaches in northern California, and over two years later, shoes from the spill were even recovered from the north end of the Big Island of Hawaii!

Even though the shoes had spent considerable time drifting in the ocean, they were in good shape and wearable (after barnacles and oil were removed). Because the shoes were

Continued on next page...

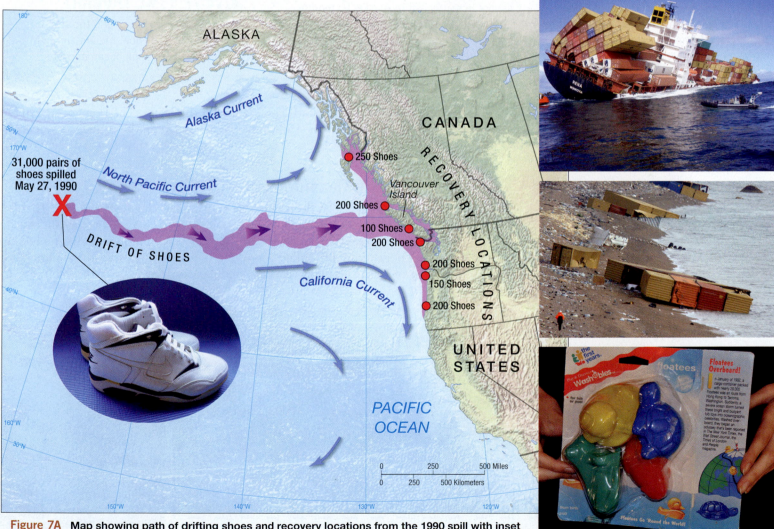

Figure 7A Map showing path of drifting shoes and recovery locations from the 1990 spill with inset photo of basketball shoes (*map*), spilled shipping containers from a cargo ship (*top photo*), shipping containers and their contents washed up at a beach (*middle photo*), and a package of tub toys similar to those involved in the 1992 shipping container spill (*bottom photo*).

...Continued from previous page.

not tied together, many beachcombers found individual shoes or pairs that did not match. Many of the shoes retailed for around $100, so people interested in finding matching pairs placed ads in newspapers or attended local swap meets.

With help from the beachcombing public (as well as lighthouse operators), information on the location and number of shoes collected was compiled during the months following the spill. Serial numbers inside the shoes were traced to individual containers, and they indicated that only four of the five containers had released their shoes; evidently, one entire container sank without opening. Thus, a maximum of 30,910 pairs of shoes (61,820 individual shoes) were released. The almost instantaneous release of such a large number of drift items helped oceanographers refine computer models of North Pacific circulation. Before the shoe spill, the largest number of drift bottles purposefully released at one time by

oceanographers had been about 30,000. Although only 2.6% of the shoes were recovered, this compares favorably with the 2.4% recovery rate of drift bottles released by oceanographers conducting research.

In January 1992, another cargo ship lost 12 containers during a storm to the north of where the shoes had previously spilled. One of these containers held 29,000 packages of small, floatable, colorful plastic bathtub toys in the shapes of blue turtles, yellow ducks, red beavers, and green frogs (Figure 7A, *bottom photo*). Even though the toys were housed in plastic packaging glued to a cardboard backing, studies showed that after 24 hours in seawater, the glue deteriorated, and more than 100,000 of the toys were released.

The floating bathtub toys began to come ashore in southeastern Alaska 10 months later, verifying the computer models. Models suggest that as long as the bathtub toys are able to float, they will continue to be carried

by the Alaska Current, eventually dispersing throughout the North Pacific Ocean. For example, some of the toys have made it all the way to South America. Others have even found their way into the Arctic Ocean, where they became entrapped and transported within the Arctic Ocean's floating ice pack, later released into the northern Atlantic Ocean. Remarkably, the tub toys have been found along beaches in the United Kingdom and eastern North America—entire ocean basins from where they were accidentally released into the ocean.

Oceanographers continue to study ocean currents by tracking these and other floating items spilled by cargo ships (see Mastering-Oceanography **Web Table 7.1**).

GIVE IT SOME THOUGHT

1. How did running shoes and tub toys became inadvertent float meters and help oceanographers track ocean currents?

transmit low-frequency sound signals through the water. The flow meter remains stationary and measures the shift in frequency between the sound waves emitted and those backscattered by particles in the water to determine current movement.

Deep Current Measurement

The great depth at which deep currents exist makes them even more difficult to measure than surface currents. Most often, they are mapped using underwater floats that are carried within deep currents. One such unique oceanographic program that began in 2000, called **Argo**, is a global array of free-drifting profiling floats (**Figure 7.3b**) that move vertically and measure the temperature, salinity, and

Figure 7.3 The Argo system of free-drifting submersible floats

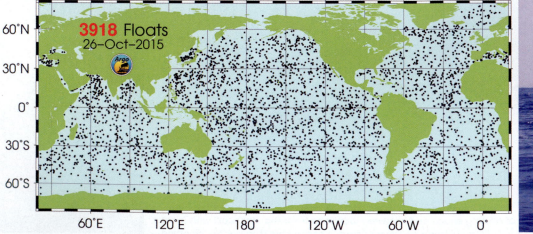

(a) Map showing the location of Argo floats, which can dive to 2000 meters (6600 feet) and collect data on the ocean properties before resurfacing and transmitting their data.

(b) Argo floats are deployed from research or cargo vessels.

other water characteristics of the upper 2000 meters (6600 feet) of the ocean. Once deployed, each float sinks to a particular depth, drifts for up to 10 days collecting data, then resurfaces and transmits data on its location and ocean variables, which are made publically available within hours. Each float then sinks back down to a programmed depth and drifts for up to another 10 days, collecting more data, before resurfacing and repeating the cycle. In 2007, the goal of the program was achieved with the launch of the 3000th Argo float; currently, nearly 4000 floats are operating worldwide (**Figure 7.3a**). The program will allow oceanographers to develop a forecasting system for the oceans analogous to weather forecasting on land and also track changes in ocean properties as a result of human-caused climate change.

Climate

Connection

 Other techniques used for measuring deep currents include identifying the distinctive temperature and salinity characteristics of a deep-water mass and tracking telltale chemical tracers. Some tracers are naturally absorbed into seawater, while others are intentionally added. Some useful tracers that have inadvertently been added to seawater include tritium (a radioactive isotope of hydrogen produced by nuclear bomb tests in the 1950s and early 1960s) and chlorofluorocarbons (Freon and other gases that deplete Earth's ozone layer).

RECAP

Wind-induced surface currents are measured with floating objects, by satellites, and by other techniques. Density-induced deep currents are measured using submerged floats, water properties, and chemical tracers.

CONCEPT CHECK 7.1 | Demonstrate an understanding of how ocean currents are measured.

1 Compare the forces that are directly responsible for creating horizontal and deep vertical circulation in the oceans. What is the ultimate source of energy that drives both circulation systems?

2 Describe the different ways in which currents are measured.

7.2 What Creates Ocean Surface Currents and How Are They Organized?

Surface currents occur within and above the *pycnocline* (layer of rapidly changing density) to a depth of about 1 kilometer (0.6 mile) and affect only about 10% of the world's ocean water. Worldwide, the pattern of ocean surface currents is affected mostly by the major wind belts of the world but also by a variety of factors including the Coriolis effect, seasonal changes, and the geometry of each ocean basin.

Origin of Surface Currents

In a simplistic case, surface currents develop from friction between the ocean and the wind that blows across its surface. Only about 2% of the wind's energy is transferred to the ocean surface, so a 50-knot[2] wind will create a 1-knot current. You can simulate this on a tiny scale simply by blowing gently and steadily across a cup of coffee.

 If there were no continents on Earth, the surface currents would generally follow the major wind belts of the world. In each hemisphere, therefore, a current would flow between 0 and 30 degrees latitude as a result of the trade winds, a second would flow between 30 and 60 degrees latitude as a result of the prevailing westerlies, and a third would flow between 60 and 90 degrees latitude as a result of the polar easterlies.

 In reality, however, ocean surface currents are driven by more than just the wind belts of the world. The distribution of continents on Earth is one factor that influences the nature and direction of flow of surface currents in each ocean basin.

[2]A *knot* is a speed of 1 nautical mile per hour. A nautical mile is defined as the distance of 1 minute of latitude and is equivalent to 1.15 statute (land) miles, or 1.85 kilometers.

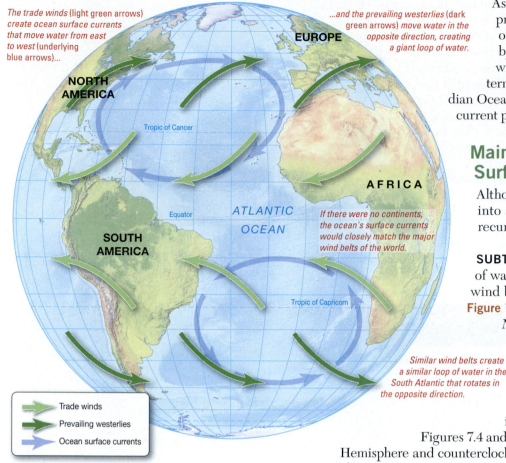

The trade winds (light green arrows) create ocean surface currents that move water from east to west (underlying blue arrows)...

...and the prevailing westerlies (dark green arrows) move water in the opposite direction, creating a giant loop of water.

If there were no continents, the ocean's surface currents would closely match the major wind belts of the world.

Similar wind belts create a similar loop of water in the South Atlantic that rotates in the opposite direction.

EUROPE

NORTH AMERICA

Tropic of Cancer

AFRICA

Equator

ATLANTIC OCEAN

SOUTH AMERICA

Tropic of Capricorn

→ Trade winds
→ Prevailing westerlies
→ Ocean surface currents

Figure 7.4 **How major wind belts affect the movement of surface currents in the Atlantic Ocean.**

Web Animation
Ocean Circulation
http://goo.gl/bZqQBD

As an example, **Figure 7.4** shows how the trade winds and prevailing westerlies create large circular-moving loops of water in the Atlantic Ocean basin, which is bounded by the irregular shape of continents. These same global wind belts affect the other ocean basins, so a similar pattern of surface current flow also exists in the Pacific and Indian Oceans. As we shall see, other factors that influence surface current patterns include gravity, friction, and the Coriolis effect.

Main Components of Ocean Surface Circulation

Although ocean water continuously flows from one current into another, ocean surface currents have a predictable, recurring pattern within each ocean basin.

SUBTROPICAL GYRES The large, circular-moving loops of water shown in Figure 7.4 that are driven by the major wind belts of the world are called **gyres** (*gyros* = a circle). **Figure 7.5** shows the world's five **subtropical gyres**: (1) the *North Pacific Gyre*, (2) the *South Pacific Gyre*, (3) the *North Atlantic Gyre*, (4) the *South Atlantic Gyre*, and (5) the *Indian Ocean Gyre* (which is mostly within the Southern Hemisphere). The reason they are called subtropical gyres is because the center of each gyre coincides with the subtropics at 30 degrees north or south latitude. As shown in Figures 7.4 and 7.5, subtropical gyres rotate clockwise in the Northern Hemisphere and counterclockwise in the Southern Hemisphere. Studies of floating objects (see Diving Deeper 7.1) indicate that the average drift time in a smaller subtropical gyre, such as the North Atlantic Gyre, is about three years, whereas in larger subtropical gyres, such as the North Pacific Gyre, it is about six years.

Generally, *each subtropical gyre is composed of four main currents* that flow progressively into one another (**Table 7.1**). The North Atlantic Gyre, for instance, is composed of the North Equatorial Current, the Gulf Stream, the North Atlantic Current, and the Canary Current (Figure 7.5). Let's examine each of the four main currents that comprise subtropical gyres.

Equatorial Currents The trade winds, which blow from the southeast in the Southern Hemisphere and from the northeast in the Northern Hemisphere, set in motion the water masses between the tropics. The resulting currents are called **equatorial currents**, which travel westward along the equator and form the equatorial boundary current of subtropical gyres (Figure 7.5). They are called *north equatorial currents* or *south equatorial currents*, depending on their position relative to the equator.

Western Boundary Currents When equatorial currents reach the western portion of an ocean basin, they must turn because they cannot cross land. The Coriolis effect deflects these currents away from the equator as **western boundary currents**, which comprise the western boundaries of subtropical gyres. Western boundary currents are so named because they travel along the western boundaries of their respective ocean basins.[3] For example, the Gulf Stream and the Brazil Current, which are shown in Figure 7.5, are western boundary currents. They come from equatorial regions, where water temperatures are warm, so they carry warm water to higher latitudes. Note that Figure 7.5 shows warm currents as red arrows.

[3]Notice that western boundary currents are off the *eastern* coasts of adjoining continents. It's easy to be confused about this because we have a land-based perspective. From an *oceanic perspective*, however, the western side of the ocean basin is where western boundary currents reside.

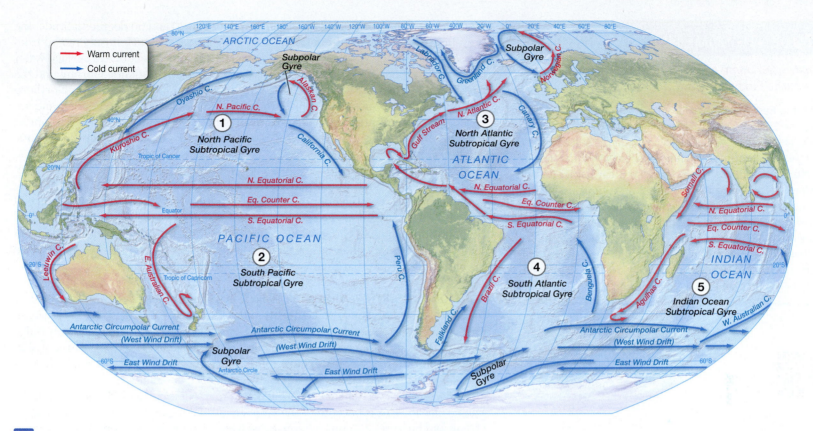

🌊 **SmartFigure 7.5** **Wind-driven surface currents.** Major wind-driven surface currents of the world's oceans during February–March. The five major subtropical gyres are ① the North Pacific Gyre, ② the South Pacific Gyre, ③ the North Atlantic Gyre, ④ the South Atlantic Gyre, and ⑤ the Indian Ocean Gyre. Smaller subpolar gyres exist in higher latitudes and rotate in the reverse direction of their adjacent subtropical gyres. https://goo.gl/Gnj9zw

🌊 **SmartTable 7.1** **Subtropical gyres and surface currents.** https://goo.gl/DHJGM6

SmartTable 7.1 SUBTROPICAL GYRES AND SURFACE CURRENTS

Pacific Ocean	Atlantic Ocean	Indian Ocean
North Pacific Gyre	**North Atlantic Gyre**	**Indian Ocean Gyre**
North Pacific Current	North Atlantic Current	South Equatorial Current
California Current[a]	Canary Current[a]	Agulhas Current[b]
North Equatorial Current	North Equatorial Current	West Wind Drift
Kuroshio (Japan) Current[b]	Gulf Stream[b]	West Australian Current[a]
South Pacific Gyre	**South Atlantic Gyre**	**Other Major Currents**
South Equatorial Current	South Equatorial Current	Equatorial Countercurrent
East Australian Current[b]	Brazil Current[b]	North Equatorial Current
West Wind Drift	West Wind Drift	Leeuwin Current
Peru (Humboldt) Current[a]	Benguela Current[a]	Somali Current
Other Major Currents	**Other Major Currents**	
Equatorial Countercurrent	Equatorial Countercurrent	
Alaskan Current	Florida Current	
Oyashio Current	East Greenland Current	
	Labrador Current	
	Falkland Current	

[a]Denotes an eastern boundary current of a gyre, which is relatively *slow*, *wide*, and *shallow* (and is also a *cold-water* current).
[b]Denotes a western boundary current of a gyre, which is relatively *fast*, *narrow*, and *deep* (and is also a *warm-water* current).

What is the name of the current that's mentioned in the movie Finding Nemo?

It's the East Australian Current, which is called the "EAC" in the 2003 Disney animated film and is even geographically correct (if only all movies could do the same!). The EAC is a western-intensified ocean surface current that helps Nemo's dad and Dory travel from the Great Barrier Reef to Sydney Harbor along the east coast of Australia. While being swept along in the EAC, they meet Crush the sea turtle, who famously asks, "What brings you to the EAC?" In the storyline, Crush and other sea turtles help Nemo's dad and Dory navigate the EAC and after many close calls (...spoiler alert...), Nemo is successfully rescued. In fact, a sequel called *Finding Dory* is set to be released in 2016.

Web Video
Perpetual Ocean by NASA
https://goo.gl/q1Vo4i

RECAP

The principal ocean surface current pattern on Earth consists of large subtropical gyres and smaller subpolar gyres, both of which are big, circular-moving loops of water powered by the major wind belts of the world.

Northern or Southern Boundary Currents Between 30 and 60 degrees latitude, the prevailing westerlies blow from the northwest in the Southern Hemisphere and from the southwest in the Northern Hemisphere. These winds direct ocean surface water in an easterly direction across an ocean basin (see the North Atlantic Current and the Antarctic Circumpolar Current [West Wind Drift] in Figure 7.5). In the Northern Hemisphere, these currents comprise the northern parts of subtropical gyres and are called **northern boundary currents**; in the Southern Hemisphere, they comprise the southern parts of subtropical gyres and are called **southern boundary currents**.

Eastern Boundary Currents When currents flow back across the ocean basin, the Coriolis effect and continental barriers turn them toward the equator, creating **eastern boundary currents** of subtropical gyres along the eastern boundary of the ocean basins. Examples of eastern boundary currents include the Canary Current and the Benguela Current,[4] which are shown in Figure 7.5. They come from high-latitude regions where water temperatures are cool, so they carry cool water to lower latitudes. Note that Figure 7.5 shows cold currents as blue arrows.

EQUATORIAL COUNTERCURRENTS A large volume of water is driven westward by the north and south equatorial currents and piles up water on the western side of an ocean basin near the equator, creating higher sea level there. As a result, this bulge of water flows downhill toward the east under the influence of gravity. This current, called the **equatorial countercurrent**, is a narrow, easterly flow of water that occurs *counter to* and *between* the adjoining equatorial currents.

Figure 7.5 shows that an equatorial countercurrent is particularly apparent in the Pacific Ocean. This is because of the large equatorial region that exists in the Pacific Ocean and because of a dome of equatorial water that becomes trapped in the island-filled embayment between Australia and Asia. Continual influx of water from equatorial currents builds the dome and creates an eastward countercurrent that stretches across the Pacific toward South America. The equatorial countercurrent in the Atlantic Ocean, on the other hand, is not nearly as well defined because of the shapes of the adjoining continents, which limit the equatorial area that exists in the Atlantic Ocean. The presence of an equatorial countercurrent in the Indian Ocean is strongly influenced by the monsoons, which will be discussed later in this chapter.

SUBPOLAR GYRES Northern or southern boundary currents that flow eastward as a result of the prevailing westerlies eventually move into subpolar latitudes (about 60 degrees north or south latitude). Here, they are driven in a westerly direction by the polar easterlies, producing **subpolar gyres** that rotate opposite the adjacent subtropical gyres. Subpolar gyres are smaller and fewer than subtropical gyres. Two examples include the subpolar gyre in the Atlantic Ocean between Greenland and Europe and in the Weddell Sea off Antarctica (Figure 7.5).

Other Factors Affecting Ocean Surface Circulation

Several other factors influence circulation patterns in subtropical gyres, including Ekman spiral and Ekman transport, geostrophic currents, and western intensification of subtropical gyres.

EKMAN SPIRAL AND EKMAN TRANSPORT During the voyage of the *Fram* (see MasteringOceanography Web Diving Deeper 7.1), Norwegian explorer Fridtjof Nansen (1861–1930) observed that Arctic Ocean ice moved 20 to 40 degrees to the right of the wind blowing across its surface (**Figure 7.6**). Not only ice but surface waters in the Northern Hemisphere were observed to move to the right of the wind direction; in the Southern Hemisphere, surface waters move to the *left* of the wind direction.

[4]Currents are often named for a prominent geographic location near where they pass. For instance, the Canary Current passes the Canary Islands, and the Benguela Current is named for the Benguela Province in Angola, Africa.

Why does surface water move in a direction different than the wind? **V. Walfrid Ekman** (1874–1954), a Swedish physicist, developed an ocean circulation model in 1905 called the **Ekman spiral (Figure 7.7)** that explains Nansen's observations as a balance between the Coriolis effect, which causes objects to curve from their intended path, and frictional effects, which reduce the water's speed with depth.

The Ekman spiral describes the speed and direction of flow of surface waters at various depths. Ekman's model assumes that a uniform column of water is set in motion by wind blowing across its surface (Figure 7.7, *large green wind arrow*). Under ideal conditions in the Northern Hemisphere, the Coriolis effect causes surface water in contact with the wind to move in a direction 45 degrees to the right of the wind direction (Figure 7.7, *purple arrow*). In the Southern Hemisphere, where Coriolis curvature is to the left, the surface layer moves 45 degrees to the left of the wind direction. The surface water moves as a thin layer on top of deeper layers of water. As the surface layer moves, other layers beneath it are set in motion, thus passing the energy of the wind down through the water column. This is similar to how a deck of cards can be fanned out by pressing on and rotating only the top card in the deck.

Current speed decreases with increasing depth, however, and the Coriolis effect increases curvature to the right (like a spiral). Thus, each successive layer of water is set in motion at a progressively slower speed and in a direction progressively to the right of the one above it. In Figure 7.7, for example, the purple surface current arrow sets the water in motion beneath it, which is curved more to the right but moves more slowly and is represented by the shorter pink arrow. In turn, the pink arrow sets the water in motion beneath it, which is curved more to the right but moves more slowly and is represented by the even shorter gray arrow, and so on down the water column. Deeper in the ocean, a layer of water actually exists that moves in a direction *exactly opposite from the wind direction that initiated it!* (See Figure 7.7, *small orange arrow.*) If the water is deep enough, friction will consume

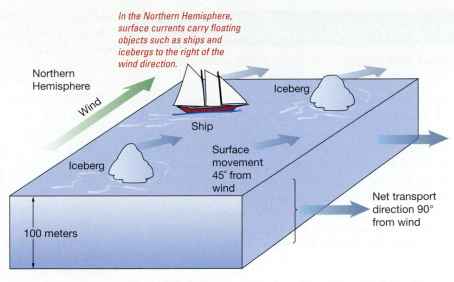

In the Northern Hemisphere, surface currents carry floating objects such as ships and icebergs to the right of the wind direction.

Figure 7.6 Transport of floating objects is to the right of the wind direction in the Northern Hemisphere.

Web Animation

Ekman Spiral and Ekman Transport
http://goo.gl/GmBbWto

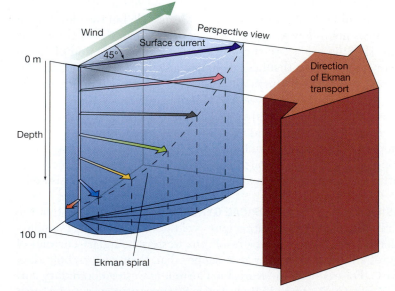

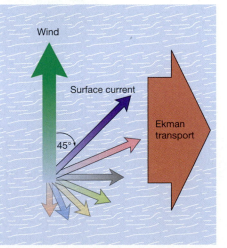

(a) Wind drives surface water in a direction 45 degrees to the right of the wind in the Northern Hemisphere. Deeper water continues to deflect to the right and moves at a slower speed with increased depth, causing the Ekman spiral.

(b) Ekman transport, which is the average water movement for the entire column, is at a right angle (90 degrees) to the wind direction.

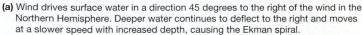

 SmartFigure 7.7 The Ekman spiral produces Ekman transport. (a) Perspective view and (b) top view of the same area showing how the Ekman spiral produces Ekman transport.
https://goo.gl/Tn7Upl

What does an Ekman spiral look like at the surface? Is it strong enough to disturb ships?

The Ekman spiral creates different layers of surface water that move in slightly different directions at slightly different speeds. It is too weak to create eddies or whirlpools (vortexes) at the surface and so presents no danger to ships. In fact, the Ekman spiral is unnoticeable at the surface. It can be observed, however, by lowering oceanographic equipment over the side of a vessel. At various depths, the equipment can be observed to drift at various angles from the wind direction according to the Ekman spiral.

the energy imparted by the wind, and no motion will occur below that depth. Although it depends on wind speed and latitude, this stillness normally occurs at a depth of about 100 meters (330 feet).

Figure 7.7 shows the spiral nature of this movement with increasing depth from the ocean's surface. The length of each colored arrow in Figure 7.7 is proportional to the speed of each individual thin layer, and the direction of each colored arrow indicates the direction it moves.[5] Under ideal conditions, therefore, the surface layer should move at an angle of 45 degrees from the direction of the wind (Figure 7.7, *purple arrow*). All the layers combine, however, to create a net water movement that is 90 degrees from the direction of the wind. This average movement, called **Ekman transport**, is 90 degrees to the *right* in the Northern Hemisphere and 90 degrees to the *left* in the Southern Hemisphere.

"Ideal" conditions rarely exist in the ocean, so the actual movement of surface currents deviates slightly from the angles shown in Figure 7.7. Generally, surface currents move at an angle somewhat less than 45 degrees from the direction of the wind and Ekman transport in the open ocean is typically about 70 degrees from the wind direction. In shallow coastal waters, Ekman transport may be very nearly the same direction as the wind.

GEOSTROPHIC CURRENTS Ekman transport deflects surface water to the right in the Northern Hemisphere, so a clockwise rotation develops within an ocean basin and produces the **Subtropical Convergence** of water in the middle of the gyre, causing water literally to pile up in the center of the subtropical gyre. Thus, there is a hill of water within all subtropical gyres that is as much as 2 meters (6.6 feet) high.

Surface water in the Subtropical Convergence tends to flow downhill in response to gravity. The Coriolis effect opposes gravity, however, deflecting the water to the right in a curved path (**Figure 7.8a**) into the hill again. When these two factors balance, the net effect is a **geostrophic current** (*geo* = earth, *strophio* = turn) that moves in a circular path around the hill and is shown in Figure 7.8a as the *path of ideal geostrophic flow*.[6] Friction between water molecules, however, causes the water to move gradually down the slope of the hill as it flows around it. This is the path of *actual geostrophic flow* labeled in Figure 7.8a.

If you reexamine the satellite image of sea surface elevation in Figure 7.2, you'll see that the hills of water within the subtropical gyres of the Atlantic Ocean are clearly visible. The hill in the North Pacific is visible as well, but the elevation of the equatorial Pacific is not as low as expected because the map shows conditions during a moderate El Niño event,[7] so there is a well-developed region of warm water across the equatorial Pacific that has an anomalously high sea surface height. Figure 7.2 also shows very little distinction between the North and South Pacific Gyres. Moreover, the South Pacific Gyre hill is less pronounced than in other gyres because (1) it covers such a large area, (2) it lacks confinement by continental barriers along its western margin, and (3) it is interfered with by numerous islands (really the tops of tall sea floor mountains). The southern Indian Ocean hill is rather well developed in the figure, although its northeastern boundary stands high because of the influx of warm Pacific Ocean water through the East Indies islands.

WESTERN INTENSIFICATION OF SUBTROPICAL GYRES Figure 7.8a shows that the apex (top) of the hill formed within a rotating gyre is closer to the western boundary than the geographic center of the gyre. As a result, the western boundary currents of the subtropical gyres are faster, narrower, and deeper than their eastern boundary current counterparts. For example, the Kuroshio Current (a western boundary current) of the North Pacific Gyre is up to 15 times faster, 20 times narrower, and 5 times deeper than the California Current (an eastern boundary current). This phenomenon

[5]The name Ekman *spiral* refers to the spiral observed by connecting the tips of the arrows shown in Figure 7.7.
[6]The term *geostrophic* for these currents is appropriate, since the currents behave as they do because of Earth's rotation.
[7]El Niño events are discussed later in this chapter, under "Pacific Ocean Circulation."

Figure 7.8 Geostrophic current and western intensification.
(a) Perspective view and (b) map view of a subtropical gyre showing how the center of rotation of the gyre is shifted to the west by Earth's rotation and creates the western intensification of currents.

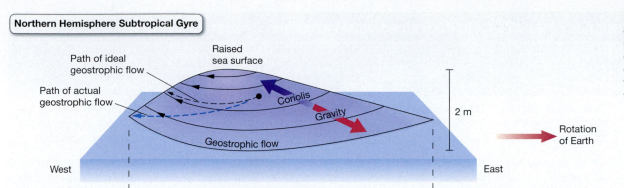

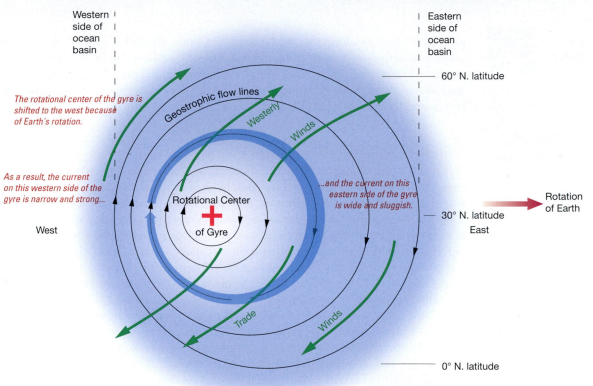

(a) Perspective view of a subtropical gyre showing how water literally piles up in the center, forming a hill up to 2 meters (6.6 feet) high. Ideally, gravity and the Coriolis effect balance each other to create an ideal geostrophic current that flows in equilibrium around the hill. However, friction makes the current gradually run downslope *(path of actual geostrophic flow)*.

(b) Corresponding map view of the same subtropical gyre, showing that the flow pattern is restricted (lines are closer together) on the western side of the gyre, resulting in western intensification.

is called **western intensification**, and currents affected by this phenomenon are said to be *western intensified*. Note that the western boundary currents of *all* subtropical gyres are western intensified, *even in the Southern Hemisphere.*

A number of factors cause western intensification, including the Coriolis effect. The Coriolis effect increases toward the poles, so eastward-flowing high-latitude water turns toward the equator more strongly than westward-flowing equatorial water turns toward higher latitudes. This causes a wide, slow, and shallow flow of water toward the equator across most of each subtropical gyre, leaving only a narrow band through which the poleward flow can occur along the western margin of the ocean basin. If a constant volume of water rotates around the apex of the hill in **Figure 7.8b**, then the velocity of the water along the western margin will be much greater than the velocity around the eastern side.[8] In Figure 7.8b, the lines

[8]A good analogy for this phenomenon is a funnel: In the narrow end of a funnel, the flow rates are speeded up (such as in western intensified currents); in the wide end, the flow rates are sluggish (such as in eastern boundary currents).

SMARTTABLE 7.2 CHARACTERISTICS OF WESTERN AND EASTERN BOUNDARY CURRENTS OF SUBTROPICAL GYRES

Current type	Examples	Width	Depth	Speed	Transport volume (millions of cubic meters per second[a])	Comments
Western boundary current	Gulf Stream, Brazil Current, Kuroshio Current	*Narrow:* usually less than 100 kilometers (60 miles)	*Deep:* to depths of 2 kilometers (1.2 miles)	*Fast:* hundreds of kilometers per day	*Large:* as much as 100 Sv[a]	Waters derived from low latitudes and are warm; little or no upwelling
Eastern boundary current	Canary Current, Benguela Current, California Current	*Wide:* up to 1000 kilometers (600 miles)	*Shallow:* to depths of 0.5 kilometer (0.3 mile)	*Slow:* tens of kilometers per day	*Small:* typically 10 to 15 Sv[a]	Waters derived from middle latitudes and are cool; coastal upwelling common

[a]One million cubic meters (35.3 million cubic feet) per second is a flow rate equal to one Sverdrup (Sv).

 SmartTable 7.2 Characteristics of western and eastern boundary currents of subtropical gyres. https://goo.gl/P2G56w

RECAP

Western intensification is a result of Earth's rotation and causes the western boundary currents of all subtropical gyres to be fast, narrow, and deep.

are close together along the western margin, indicating the faster flow. The end result is a high-speed western boundary current that flows along the hill's steeper westward slope and a slow drift of water toward the equator along the more gradual eastern slope. Table 7.2 summarizes the differences between western and eastern boundary currents of subtropical gyres.

Ocean Currents and Climate

Ocean surface currents directly influence the climate of adjoining landmasses. For instance, warm ocean currents warm the nearby air. This warm air can hold a large amount of water vapor, which puts more moisture (high humidity) in the atmosphere. When this warm, moist air travels over a continent, it releases its water vapor in the form of precipitation. Continental margins that have warm ocean currents offshore (Figure 7.9, *red arrows*) typically have a humid climate. The presence of a warm current off the East Coast of the United States helps explain why the area experiences such high humidity, especially in the summer.

 Climate Connection

Conversely, cold ocean currents cool the nearby air, which is more likely to have low water vapor content. When the cool, dry air travels over a continent, it results in very little precipitation. Continental margins that have cool ocean currents offshore (Figure 7.9, *blue arrows*) typically have a dry climate. The presence of a cold current off California is part of the reason the climate there is so arid.

CONCEPT CHECK 7.2 | Explain the origin of ocean surface currents and how surface circulation patterns are organized globally.

1 How many subtropical gyres exist worldwide? How many main currents exist within each subtropical gyre?

2 On a base map of the world, plot and label the major currents involved in the surface circulation gyres of the oceans. Use colors to represent warm versus cool currents and indicate which currents are western intensified. On an overlay, superimpose the major wind belts of the world on the gyres and describe the relationship between wind belts and currents.

3 Explain why the subtropical gyres in the Northern Hemisphere move in a clockwise fashion while the subpolar

gyres rotate in a counterclockwise pattern.

4 Diagram and discuss how Ekman transport produces the "hill" of water within subtropical gyres that causes geostrophic current flow. As a starting place on the diagram, use the wind belts (the trade winds and the prevailing westerlies). What causes the apex of the geostrophic "hills" to be offset to the west of the center of the ocean gyre systems?

5 Describe western intensification, including the characteristics of western and eastern boundary currents of subtropical gyres.

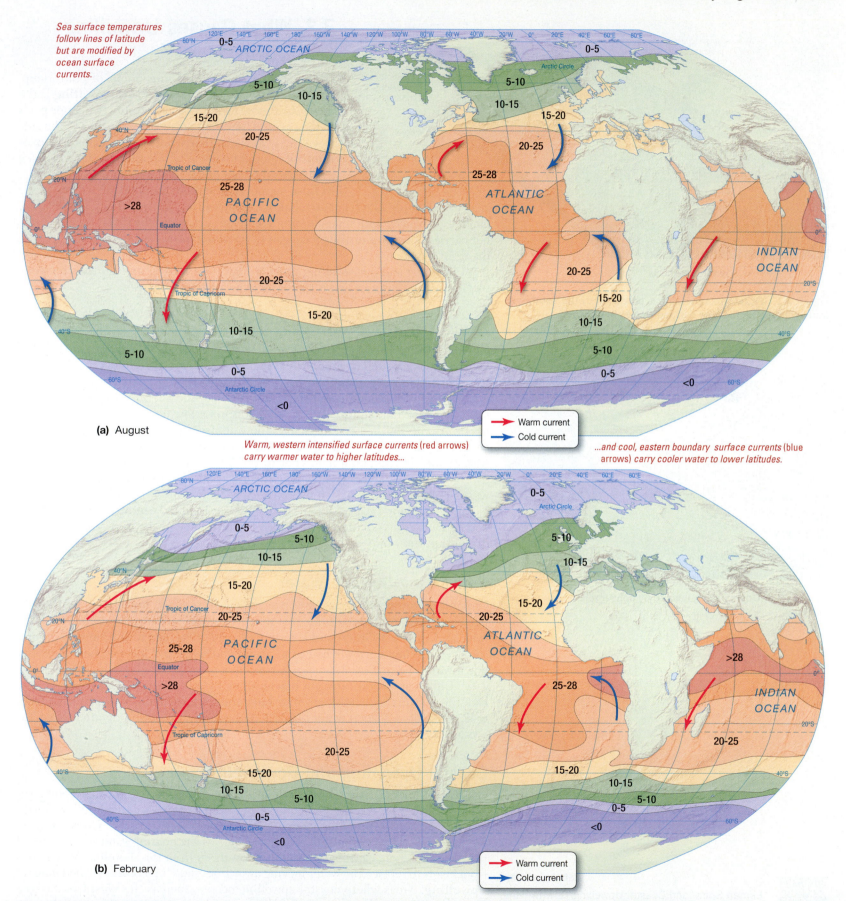

Sea surface temperatures follow lines of latitude but are modified by ocean surface currents.

(a) August

➡ Warm current
➡ Cold current

Warm, western intensified surface currents (red arrows) *carry warmer water to higher latitudes...*

...and cool, eastern boundary surface currents (blue arrows) *carry cooler water to lower latitudes.*

(b) February

➡ Warm current
➡ Cold current

Figure 7.9 Sea surface temperature of the world ocean. Paired maps showing average sea surface temperature in degrees centigrade for **(a)** August and **(b)** February. A comparison of the maps shows that sea surface temperatures migrate north–south with the seasons.

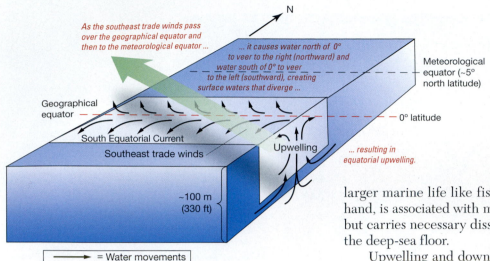

As the southeast trade winds pass over the geographical equator and then to the meteorological equator ...

... it causes water north of 0° to veer to the right (northward) and water south of 0° to veer to the left (southward), creating surface waters that diverge ...

N

Meteorological equator (~5° north latitude)

Geographical equator

0° latitude

South Equatorial Current

Southeast trade winds

Upwelling

... resulting in equatorial upwelling.

~100 m (330 ft)

→ = Water movements

Figure 7.10 Equatorial upwelling is produced by the southeast trade winds, which cause surface waters to diverge.

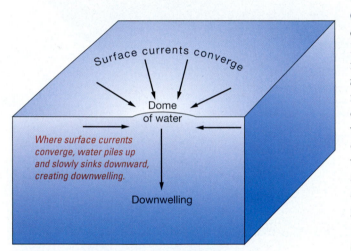

Surface currents converge

Dome of water

Where surface currents converge, water piles up and slowly sinks downward, creating downwelling.

Downwelling

Figure 7.11 Downwelling caused by convergence of surface currents.

Web Animation
Ekman Spiral and Coastal Upwelling/Downwelling
http://goo.gl/UIFbuF

7.3 What Causes Upwelling and Downwelling?

Upwelling is the upward movement of cold, deep, nutrient-rich water to the surface; **downwelling** is the downward movement of surface water to deeper parts of the ocean. Upwelling hoists chilled water to the surface. This cold water, rich in nutrients, creates high **productivity** (an abundance of microscopic algae), which establishes the base of the food web and, in turn, supports incredible numbers of larger marine life like fish and whales. Downwelling, on the other hand, is associated with much lower amounts of surface productivity but carries necessary dissolved oxygen to those organisms living on the deep-sea floor.

Interdisciplinary

Relationship

Upwelling and downwelling provide important mixing mechanisms between surface and deep waters and is the result of a variety of processes.

Diverging Surface Water

Current divergence occurs when surface waters move *away from* an area on the ocean's surface, such as along the equator. As shown in **Figure 7.10**, the South Equatorial Current occupies the area along the *geographical equator* (most notably in the Pacific Ocean; see Figure 7.5), while the *meteorological equator* (where the doldrums exist) typically occurs a few degrees of latitude to the north. As the southeast trade winds blow across this region, Ekman transport causes surface water north of the equator to veer to the right (northward) and water south of the equator to veer to the left (southward). The net result is a divergence of surface currents along the geographical equator, which causes upwelling of cold, nutrient-rich water. Because this type of upwelling is common along the equator—especially in the Pacific—it is called **equatorial upwelling**, and it creates areas of high productivity that are some of the most prolific fishing grounds in the world.

Converging Surface Water

Current convergence occurs when surface waters move *toward* each other. In the North Atlantic Ocean, for instance, the Gulf Stream, the Labrador Current, and the East Greenland Current all come together in the same region of the ocean. When currents converge, water stacks up and has no place to go but downward. The surface water slowly sinks in a process called *downwelling* (**Figure 7.11**). Unlike upwelling, areas of downwelling are not associated with prolific marine life because the necessary nutrients are not continuously replenished from cold, nutrient-rich deep water below. Consequently, downwelling areas have low productivity.

Coastal Upwelling and Downwelling

Coastal winds can cause upwelling or downwelling due to Ekman transport. **Figure 7.12** shows a coastal region along the west coast of a continent in the Northern Hemisphere with winds moving parallel to the coast. If the winds are from the north (Figure 7.12a), Ekman transport moves the coastal water to the right of the wind direction, causing the water to flow *away from* the shoreline. Water rises from below to replace the water moving away from shore in a process called **coastal upwelling**. Areas where coastal upwelling occurs, such as the West Coast of the United States, are characterized by high concentrations of nutrients, resulting in high productivity and rich marine life. This coastal upwelling also creates low water

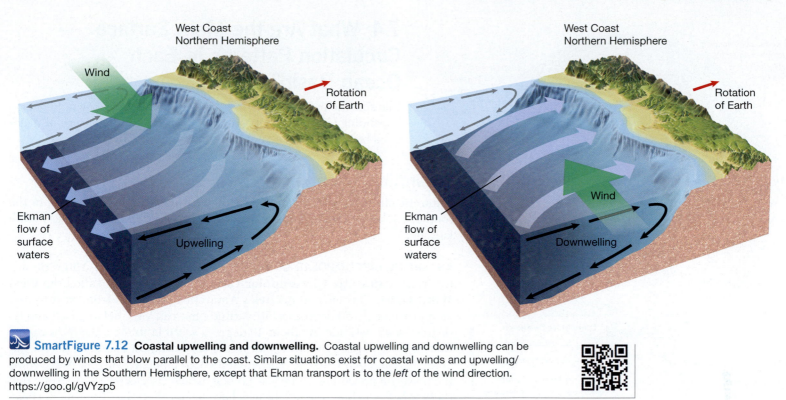

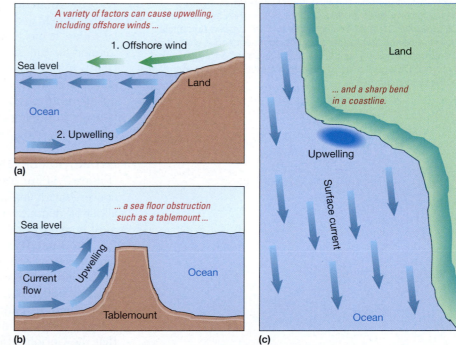

SmartFigure 7.12 **Coastal upwelling and downwelling.** Coastal upwelling and downwelling can be produced by winds that blow parallel to the coast. Similar situations exist for coastal winds and upwelling/downwelling in the Southern Hemisphere, except that Ekman transport is to the *left* of the wind direction. https://goo.gl/gVYzp5

temperatures in areas such as San Francisco, providing a natural form of air conditioning (and much cool weather and fog) in the summer.

If the winds are from the south, Figure 7.12b shows that Ekman transport still moves the coastal water to the right of the wind direction but, in this case, the water flows *toward* the shoreline. The water stacks up along the shoreline and has nowhere to go but down, in a process called **coastal downwelling**. Areas where coastal downwelling occurs have low productivity and a lack of marine life. If the winds reverse, areas that are typically associated with coastal downwelling can experience upwelling.

A similar situation exists for coastal winds and upwelling/downwelling in the Southern Hemisphere, except that Ekman transport is to the *left* of the wind direction.

Other Causes of Upwelling

Figure 7.13 shows how upwelling can be created by offshore winds, sea floor obstructions, or a sharp bend in a coastline. Upwelling also occurs in high-latitude regions, where there is no pycnocline (a layer of rapidly changing density). The absence of a pycnocline allows significant vertical mixing between high-density cold surface water and high-density cold deep water below. Thus, both upwelling and downwelling are common in high latitudes.

Figure 7.13 **Examples of other types of upwelling.**

RECAP

Upwelling and downwelling cause vertical mixing between surface and deep water. Upwelling brings cold, deep, nutrient-rich water to the surface, which results in high productivity.

CONCEPT CHECK 7.3 | Describe the conditions that produce upwelling.

1 Draw and describe several different oceanographic conditions that produce upwelling.

2 Explain why upwelling areas are associated with an abundance of marine life.

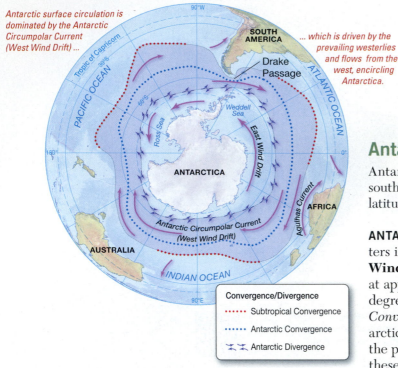

Antarctic surface circulation is dominated by the Antarctic Circumpolar Current (West Wind Drift) ...

... which is driven by the prevailing westerlies and flows from the west, encircling Antarctica.

Convergence/Divergence

· · · · · Subtropical Convergence

· · · · · Antarctic Convergence

⊃⊂ Antarctic Divergence

Figure 7.14 South polar view of Earth showing Antarctic surface circulation. The East Wind Drift is driven by the polar easterlies and flows around Antarctica from the east. The Antarctic Circumpolar Current (West Wind Drift) flows around Antarctica from the west but is further from the continent and is a result of the strong prevailing westerlies. The Antarctic Convergence and Antarctic Divergence are caused by interactions at the boundaries of these two currents.

7.4 What Are the Main Surface Circulation Patterns in Each Ocean Basin?

The specific pattern of surface currents varies from ocean to ocean, depending on the geometry of the ocean basin, the pattern of major wind belts, seasonal factors, and other periodic changes.

Antarctic Circulation

Antarctic circulation is dominated by the movement of water masses in the southern Atlantic, Indian, and Pacific Oceans south of about 50 degrees south latitude.

ANTARCTIC CIRCUMPOLAR CURRENT The main current in Antarctic waters is the **Antarctic Circumpolar Current**, which is also called the **West Wind Drift**. This current encircles Antarctica and flows from west to east at approximately 50 degrees south latitude but can vary between 40 and 65 degrees south latitude. At about 40 degrees south latitude is the *Subtropical Convergence* (**Figure 7.14**), which forms the northernmost boundary of the Antarctic Circumpolar Current. The Antarctic Circumpolar Current is driven by the powerful prevailing westerly wind belt, which creates winds so strong that these Southern Hemisphere latitudes have been called the "Roaring Forties," "Furious Fifties," and "Screaming Sixties."

The Antarctic Circumpolar Current is the only current that completely circumscribes Earth, and it is allowed to do so because of the lack of land at high southern latitudes. It meets its greatest restriction as it passes through the *Drake Passage* (named for the famed English sea captain and ocean explorer Sir Francis Drake [1540–1596]) between the Antarctic Peninsula and the southern islands of South America, which is about 1000 kilometers (600 miles) wide. Although the current is not speedy (its maximum surface velocity is about 2.75 kilometers [1.65 miles] per hour), it transports on average about 130 Sverdrups[9] (130 million cubic meters [4.6 billion cubic feet] per second), which is more than any other surface current.

ANTARCTIC CONVERGENCE AND DIVERGENCE The **Antarctic Convergence** (Figure 7.14), or *Antarctic Polar Front*, is where colder, denser Antarctic waters converge with (and sink sharply below) warmer, less-dense sub-Antarctic waters at about 50 degrees south latitude. The Antarctic Convergence marks the northernmost boundary of the Southern, or Antarctic, Ocean.

The **East Wind Drift**, a surface current propelled by the polar easterlies, moves from an easterly direction around the margin of the Antarctic continent. The East Wind Drift is most extensively developed to the east of the Antarctic Peninsula in the Weddell Sea region and in the area of the Ross Sea (Figure 7.14). As the East Wind Drift and the Antarctic Circumpolar Current (West Wind Drift) flow around Antarctica in opposite directions, they create a surface divergence. Recall that the Coriolis effect deflects moving masses to the left in the Southern Hemisphere, so the East Wind Drift is deflected toward the continent, and the Antarctic Circumpolar Current is deflected away from it. This creates a divergence of currents along a boundary called the **Antarctic Divergence**. The Antarctic Divergence has abundant marine life in the Southern Hemisphere summer because of the mixing of these two currents, which supplies nutrient-rich water to the surface through upwelling.

[9]One million cubic meters (35.3 million cubic feet) per second is a useful flow rate for describing ocean currents, so it has become a standard unit, named the **Sverdrup (Sv)**, after Norwegian meteorologist and physical oceanographer Harald Sverdrup (1888-1957).

Atlantic Ocean Circulation

Figure 7.15 shows Atlantic Ocean surface circulation, which consists of two large subtropical gyres: the North Atlantic Gyre and the South Atlantic Gyre.

THE NORTH AND SOUTH ATLANTIC SUBTROPICAL GYRES The **North Atlantic Subtropical Gyre** rotates clockwise, and the **South Atlantic Subtropical Gyre** rotates counterclockwise, due to the combined effects of the trade winds, the prevailing westerlies, and the Coriolis effect. Figure 7.15 shows that each gyre consists of a poleward-moving warm current (*red*) and an equatorward-moving cold "return" current (*blue*). The two gyres are partially offset by the shapes of the surrounding continents, and the **Atlantic Equatorial Countercurrent** moves in between them.

In the South Atlantic Gyre, the **South Equatorial Current** reaches its greatest strength just below the equator, where it encounters the coast of Brazil and splits in two. Part of the South Equatorial Current moves off along the northeastern coast of South America toward the Caribbean Sea and the North Atlantic. The rest is turned southward as the **Brazil Current**, which ultimately merges with the Antarctic Circumpolar Current (West Wind Drift) and moves eastward across the South Atlantic. The Brazil Current is much smaller than its Northern Hemisphere counterpart, the Gulf Stream, due to the splitting of the South Equatorial Current. The **Benguela Current**, slow moving and cold, flows toward the equator along Africa's western coast, completing the gyre.

Outside the gyre, the *Falkland Current* (Figure 7.15), which is also called the *Malvinas Current*, moves a significant amount of cold water along the coast of Argentina as far north as 25 to 30 degrees south latitude, wedging its way between the continent and the southbound Brazil Current.

THE GULF STREAM The **Gulf Stream** is the world's best studied ocean current. It moves northward along the East Coast of the United States, warming coastal states and moderating winters in these and northern European regions.

Figure 7.16 shows the network of currents in the North Atlantic Ocean that contribute to the flow of the Gulf Stream. The **North Equatorial Current** moves parallel to the equator in the Northern Hemisphere, where it is joined by the portion of the South Equatorial Current that turns northward along the South American coast. This flow then splits into the **Antilles Current**, which passes along the Atlantic side of the West Indies, and the **Caribbean Current**, which passes through the Yucatán Channel into the Gulf of Mexico. These water masses reconverge to form the **Florida Current**.

The Florida Current flows close to shore over the continental shelf at a rate that at times exceeds 35 Sverdrups. As it moves off North Carolina's Cape Hatteras and flows across the deep ocean in a northeasterly direction, it is called the *Gulf Stream*. The Gulf Stream is a western boundary current, so it is subject to western intensification. Thus, it is only 50 to 75 kilometers (31 to 47 miles) wide, but it reaches depths of 1.5 kilometers (1 mile) and speeds from 3 to 10 kilometers (2 to 6 miles) per hour, making it the world's fastest ocean current.

The western boundary of the Gulf Stream is usually abrupt, but it periodically migrates closer to and farther away from the shore. Its eastern boundary is very difficult to identify because it is usually masked by circular and meandering water flows that continuously change their positions.

The Sargasso Sea The Gulf Stream gradually merges eastward with the water of the **Sargasso Sea**. The Sargasso Sea is the water that circulates around the North Atlantic Gyre's center of rotation, which is shifted to the west because of Earth's rotation. The Sargasso Sea can be thought of as the stagnant eddy on the western side of the North Atlantic Gyre. Its name is derived from a type of floating marine alga called *sargassum* (*sargassum* = grapes) that abounds on its surface.

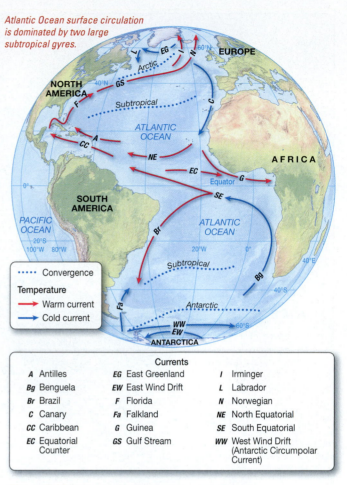

Atlantic Ocean surface circulation is dominated by two large subtropical gyres.

Currents

A Antilles	*EG* East Greenland	*I* Irminger
Bg Benguela	*EW* East Wind Drift	*L* Labrador
Br Brazil	*F* Florida	*N* Norwegian
C Canary	*Fa* Falkland	*NE* North Equatorial
CC Caribbean	*G* Guinea	*SE* South Equatorial
EC Equatorial Counter	*GS* Gulf Stream	*WW* West Wind Drift (Antarctic Circumpolar Current)

Figure 7.15 Atlantic Ocean surface currents.

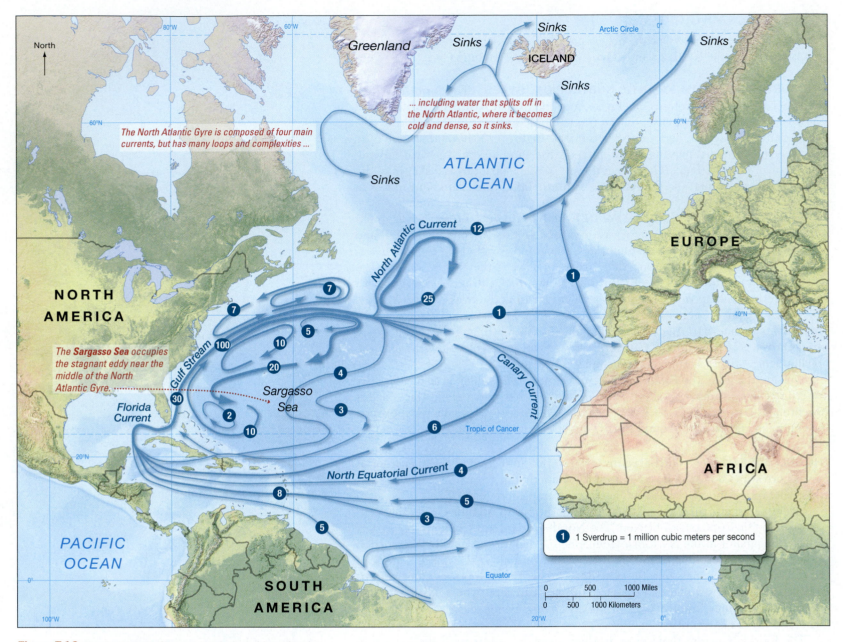

North

The North Atlantic Gyre is composed of four main currents, but has many loops and complexities …

… including water that splits off in the North Atlantic, where it becomes cold and dense, so it sinks.

Greenland Sinks Sinks ICELAND Sinks Sinks Sinks

Arctic Circle

Sinks

ATLANTIC OCEAN

North Atlantic Current 12

EUROPE

25

7

7

1

5

The Sargasso Sea occupies the stagnant eddy near the middle of the North Atlantic Gyre.

100

NORTH AMERICA

10

20

Gulf Stream

4

Canary Current

30

Florida Current

2

Sargasso Sea

3

10

6

Tropic of Cancer

AFRICA

North Equatorial Current 4

8

5

5

1 1 Sverdrup = 1 million cubic meters per second

3

PACIFIC OCEAN

5

Equator

0 500 1000 Miles
0 500 1000 Kilometers

SOUTH AMERICA

Figure 7.16 North Atlantic Ocean circulation. Map of major currents in the North Atlantic Gyre, showing average flow rates in *Sverdrups* (1 Sverdrup = 1 million cubic meters [35.3 million cubic feet] per second). The four major currents include the western intensified Gulf Stream, the North Atlantic Current, the Canary Current, and the North Equatorial Current, but there are many complex flow patterns.

The transport rate of the Gulf Stream off Chesapeake Bay is about 100 Sverdrups,[10] which suggests that a large volume of water from the Sargasso Sea has combined with the Florida Current to produce the Gulf Stream. By the time the Gulf Stream nears Newfoundland, however, the transport rate is only 40 Sverdrups, which suggests that a large volume of water has returned to the diffuse flow of the Sargasso Sea.

Warm- and Cold-Core Rings The mechanisms that produce the dramatic loss of water as the Gulf Stream moves northward are yet to be determined. Meanders, however, may cause much of it. **Meanders** (*Menderes* = a river in Turkey that has a very sinuous course) are snakelike bends in the current that often disconnect from the Gulf Stream and form large rotating masses of water called *vortexes* (*vortex* = to turn), which are more commonly known as *eddies* or *rings*.

[10]The Gulf Stream's flow of 100 Sverdrups equates to a volume of about 100 major league sport stadiums passing by the southeast U.S. coast *each second* and is more than 100 times greater than the combined flow of *all* the world's rivers!

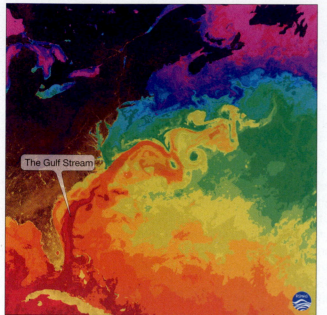

(a) The Gulf Stream is shown flowing along the U.S. East Coast in this NOAA satellite false-color image of sea surface temperature (warm waters = red and orange; cool waters = green, blue, purple, pink).

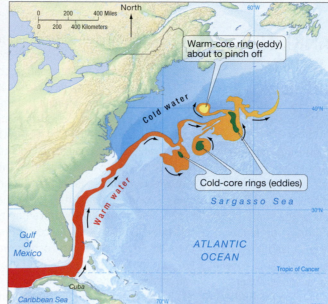

(b) Matching map of the same area as part a. As the Gulf Stream flows northward, some of its meanders pinch off and form either warm-core or cold-core rings.

Figure 7.17 The Gulf Stream and sea surface temperatures. **(a)** A NOAA satellite false-color image of sea surface temperature and **(b)** a matching map of the same area, showing the development of warm- and cold-core rings (eddies).

Figure 7.17 shows several of these rings, which are noticeable near the center of each image. The figure also shows that meanders along the north boundary of the Gulf Stream pinch off and trap warm Sargasso Sea water in eddies that rotate clockwise, creating **warm-core rings** (*yellow*) surrounded by cooler (*blue and green*) water. These warm rings contain shallow, bowl-shaped masses of warm water about 1 kilometer (0.6 mile) deep, with diameters of about 100 kilometers (60 miles). Warm-core rings remove large volumes of water as they disconnect from the Gulf Stream.

Cold nearshore water spins off to the south of the Gulf Stream as counterclockwise-rotating **cold-core rings** (*green*) surrounded by warmer (*yellow and red-orange*) water (Figure 7.17). The cold rings consist of spinning cone-shaped masses of cold water that extend over 3.5 kilometers (2.2 miles) deep. These rings may exceed 500 kilometers (310 miles) in diameter at the surface. The diameter of the cone increases with depth and sometimes reaches all the way to the sea floor, where cones have a tremendous impact on sea floor sediment. Cold rings move southwest at speeds of 3 to 7 kilometers (2 to 4 miles) per day toward Cape Hatteras, where they often rejoin the Gulf Stream.

Because both warm- and cold-core rings maintain unique temperature characteristics, they often contain distinct populations of marine life. For example, studies of rings have found they are isolated habitats for either warm-water organisms in a cold ocean or, conversely, cold-water organisms in a warmer ocean. The organisms can survive as long as the ring does; in some cases, rings have been documented to last up to two years. In addition, cold-core rings are typically associated with high nutrient levels and an abundance of marine life, while warm-core rings are zones of downwelling that lack nutrients and are deficient in marine life.

Interdisciplinary

Relationship

OTHER NORTH ATLANTIC CURRENTS Southeast of Newfoundland, the Gulf Stream continues in an easterly direction across the North Atlantic (Figure 7.16). Here the Gulf Stream breaks into numerous branches, many of which become cold and dense enough to sink beneath the surface. As shown in Figure 7.15, one major branch

Web Animation
The Gulf Stream: Its Meanders and Cold and Warm Core Eddies
https://goo.gl/LwqnV4

STUDENTS SOMETIMES ASK . . .

Is the Gulf Stream rich in life?

The Gulf Stream *itself* isn't rich in life, but its *boundaries* often are. The oceanic areas that have abundant marine life are typically associated with cool water—either in high-latitude regions, or in any region where upwelling occurs. These areas are constantly resupplied with oxygen- and nutrient-rich water, which results in high productivity. Warm-water areas develop a prominent thermocline that isolates the surface water from colder, nutrient-rich water below. Nutrients used up in warm waters tend not to be resupplied. The Gulf Stream, which is a western intensified, warm-water current, is therefore associated with low productivity and an absence of marine life. New England fishers knew about the Gulf Stream (see **Diving Deeper 7.2**) because they sought their catch along the sides of the current, where mixing and upwelling occur.

Actually, all western intensified currents are warm and are associated with low productivity. The Kuroshio Current in the North Pacific Ocean, for example, is named for its conspicuous absence of marine life. In Japanese, *Kuroshio* means "black current," in reference to its clear, lifeless waters.

Interdisciplinary

Relationship

DIVING DEEPER 7.2

BENJAMIN FRANKLIN: THE WORLD'S MOST FAMOUS PHYSICAL OCEANOGRAPHER

Benjamin Franklin (**Figure 7B**, *inset*) is well known as a scientist, an inventor, an economist, a statesman, a diplomat, a writer, a poet, an international celebrity, and one of the founding fathers of the United States. He even held the position of deputy postmaster general of the colonies from 1753 to 1774. Remarkably, he also became known as one of the first physical oceanographers because he contributed greatly to the understanding of the Gulf Stream, a North Atlantic Ocean surface current. Why would a postmaster general be interested in an ocean current?

Franklin became interested in North Atlantic Ocean circulation patterns because he needed to explain why mail ships coming from Europe to New England took about two weeks less time when they took a longer, more southerly route than when they took a more direct, northerly route. In about 1769 or 1770, Franklin mentioned this dilemma to his cousin, a Nantucket sea captain named Timothy Folger. Folger told Franklin that a strong current with which the mail ships were unfamiliar was impeding their journey because it flowed against them. The whaling ships were familiar with the current because they often hunted whales along its boundaries. The whalers often met the mail ships within the current and told their crews they would make swifter progress if they avoided the current. The British captains of the mail ships, however, would not accept advice from simple American fishers, so they continued to make slow progress within the current. If the winds were light, their ships were actually carried *backward*!

Folger sketched the current for Franklin, including directions for avoiding it by taking a more southerly route when sailing from Europe to North America. Franklin then asked other ship captains for information concerning the movement of surface waters in the North Atlantic Ocean. Franklin inferred that there was a significant current moving northward along the eastern coast of the United States, which then headed east across the North Atlantic. He concluded that this current was responsible for aiding the progress of ships traveling through the North Atlantic to Europe and slowing ships traveling in the reverse direction. This strong current is named the *Gulf Stream* because it carries warm water from the Gulf of Mexico and because it is narrow and well defined—similar to a stream, but in the ocean. Franklin subsequently measured the temperature of the current himself.

In 1777, Franklin published his first chart of the Gulf Stream based on his collected data and distributed it to the captains of the mail ships, who initially ignored it. It wasn't until later that the ship captains, using their own experience, verified that Franklin's chart was accurate. Franklin's revised 1786 chart is shown in Figure 7B and includes a rather accurate inset showing surface current circulation in the North Atlantic Ocean.

In 1969, six scientists studied the Gulf Stream aboard a submersible vessel that was allowed to float for a month underwater wherever the current took her. During the vessel's 2640-kilometer (1650-mile) journey, the scientists observed and measured the properties of and cataloged marine life within the Gulf Stream. Appropriately enough, the vessel was named the *Ben Franklin*.

GIVE IT SOME THOUGHT

1. What dilemma was Deputy Postmaster General Benjamin Franklin trying to solve that led him to map ocean surface currents and become one of the world's most famous physical oceanographers?

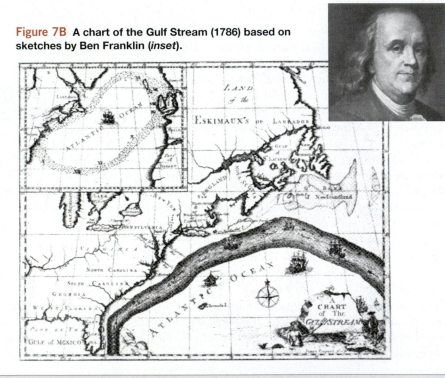

Figure 7B A chart of the Gulf Stream (1786) based on sketches by Ben Franklin (*inset*).

combines the cold water of the **Labrador Current** with the warm Gulf Stream, producing abundant fog in the North Atlantic. This branch eventually breaks into the **Irminger Current**, which flows along Iceland's west coast, and the **Norwegian Current**, which moves northward along Norway's coast. The other major branch crosses the North Atlantic as the **North Atlantic Current** (also called the *North Atlantic Drift*, emphasizing its sluggish nature), which turns southward to become the cool **Canary Current**. The Canary Current is a broad, diffuse southward flow that eventually joins the North Equatorial Current, thus completing the gyre.

CLIMATIC EFFECTS OF NORTH ATLANTIC CURRENTS The warming effects of the Gulf Stream are far ranging. The Gulf Stream moderates temperatures not only along the East Coast of the United States but also in northern Europe (in conjunction with heat transferred by the atmosphere). Thus, the temperatures across the Atlantic at different latitudes are much higher in Europe than in North America because of the effects of heat transfer from the Gulf Stream to Europe. For example, Spain and Portugal have warm climates, even though they are at the same latitude as the New England states, which are known for severe winters. The warming that northern Europe experiences because of the Gulf Stream is as much as 9°C (20°F), which is enough to keep high-latitude Baltic ports ice-free throughout the year.

Climate Connection

The warming effects of western boundary currents in the North Atlantic Ocean can be seen on the average sea surface temperature map for February shown in Figure 7.9b. Off the east coast of North America from latitudes 20 degrees north (the latitude of Cuba) to 40 degrees north (the latitude of Philadelphia), for example, there is a 20°C (36°F) difference in sea surface temperatures. On the eastern side of the North Atlantic, on the other hand, there is only a 5°C (9°F) difference in temperature between the same latitudes, indicating the moderating effect of the Gulf Stream.

The average sea surface temperature map for August (see Figure 7.9a) also shows how the North Atlantic and Norwegian Currents (branches of the Gulf Stream) warm northwestern Europe, compared with the same latitudes along the North American coast. On the western side of the North Atlantic, the southward-flowing Labrador Current—which is cold and often contains icebergs from western Greenland—keeps Canadian coastal waters much cooler. During the Northern Hemisphere winter (Figure 7.9b), North Africa's coastal waters are cooled by the southward-flowing Canary Current and are much cooler than waters near Florida and the Gulf of Mexico.

Indian Ocean Circulation

Because of the shape and location of India, the Indian Ocean exists mostly in the Southern Hemisphere. From November to March, equatorial circulation in the Indian Ocean is similar to that in the Atlantic Ocean, with two westward-flowing equatorial currents (North and South Equatorial Currents) separated by an eastward-flowing Equatorial Countercurrent. As compared to circulation in the Atlantic, however, the Equatorial Countercurrent in the Indian Ocean lies in a more southerly position because most of the Indian Ocean lies in the Southern Hemisphere. The shape of the Indian Ocean basin and its proximity to the high mountains of Asia cause it to experience strong seasonal changes.

MONSOONS The winds of the northern Indian Ocean have a seasonal pattern called **monsoon** (*mausim* = season) winds. During winter, air over the Asian mainland rapidly cools, creating high atmospheric pressure, which causes the wind to blow from southwest Asia off the continent and out over the ocean (**Figure 7.19a**, *green arrows*). These northeast trade winds are called the *northeast monsoon*. During this season, there is little precipitation because the air associated with the high pressure over land is so dry.

What is the Loop Current, and how are hurricanes affected by it?

The *Loop Current* is a warm ocean surface current in the Gulf of Mexico that flows northward between Cuba and the Yucatán Peninsula, moves north into the Gulf of Mexico, loops east and south before exiting to the east through the Florida Straits and eventually merging with other waters to form the Gulf Stream (**Figure 7.18**). In the Gulf of Mexico, the warmest waters are associated with the Loop Current and the rings that spin off from the Loop Current, which are commonly called *Loop Current eddies*. As explained in Chapter 6, warm water provides energy that fuels hurricanes. As a result, a hurricane typically intensifies when it passes over the warm waters of the Loop Current or its associated warm-core eddies.

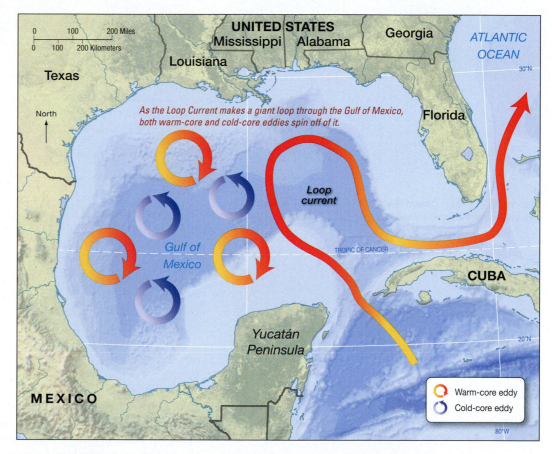

Figure 7.18 The Loop Current and its eddies.

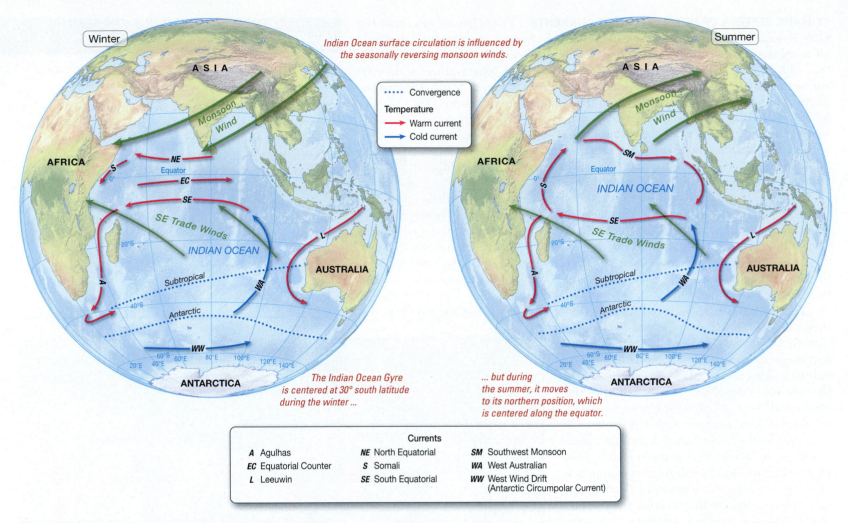

Indian Ocean surface circulation is influenced by the seasonally reversing monsoon winds.

The Indian Ocean Gyre is centered at 30° south latitude during the winter ...

... but during the summer, it moves to its northern position, which is centered along the equator.

Currents		
A Agulhas	**NE** North Equatorial	**SM** Southwest Monsoon
EC Equatorial Counter	**S** Somali	**WA** West Australian
L Leeuwin	**SE** South Equatorial	**WW** West Wind Drift (Antarctic Circumpolar Current)

Figure 7.19 **Indian Ocean surface currents are influenced by the seasonal monsoon winds.**

During summer, the winds reverse. Because of the lower heat capacity of rocks and soil compared with water, the Asian mainland warms faster than the adjacent ocean, creating low atmospheric pressure over the continent. As a result, the winds blow strongly from the Indian Ocean onto the Asian landmass (**Figure 7.19b**, *green arrows*), giving rise to the *southwest monsoon*, which may be thought of as a continuation of the southeast trade winds across the equator. During this season, there is heavy precipitation on land because the air brought in from the Indian Ocean is warm and full of moisture.

Not only does this seasonal cycle influence weather patterns that affect millions of people on land, it also affects surface current circulation in the Indian Ocean. In fact, the northern Indian Ocean is the only place in the world where reversing seasonal winds actually cause major ocean surface currents to switch direction. During the wintertime northwest monsoon (Figure 7.19a), offshore winds cause the North Equatorial Current to flow from east to west and its extension, the **Somali Current**, flows south along the coast of Africa. An *Equatorial Countercurrent* is also established. During the summertime southwest monsoon (Figure 7.19b), the winds reverse, causing the North Equatorial Current to be replaced by the *Southwest Monsoon Current*, which flows in the opposite direction. The winds cause the Somali Current to reverse as well, which flows rapidly northward with velocities approaching 4 kilometers (2.5 miles) per hour and feeds the Southwest Monsoon Current. By October, the northeast trade winds are reestablished, and the North Equatorial Current reappears (Figure 7.19a).

The movement of winds during the summertime southwest monsoon also affects sea surface temperatures, which cool near the Arabian Peninsula because of upwelling as water is drawn away from shore. This cool water also supports large populations of phytoplankton during the

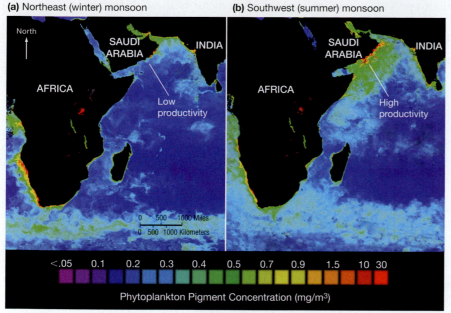

During the northeast (winter) monsoon, lack of upwelling conditions result in low concentrations of phytoplankton along the coast of Saudi Arabia.

(a) Northeast (winter) monsoon

During the southwest (summer) monsoon, strong winds generate upwelling of nutrient-rich waters, leading to an increase in the concentration of phytoplankton along the coast of Saudi Arabia.

(b) Southwest (summer) monsoon

Figure 7.20 Seasonal variations in phytoplankton concentration in the Indian Ocean. Paired satellite images for the two seasonal monsoons showing oceanic phytoplankton pigment in mg/m³, where orange and red colors indicate higher amounts of phytoplankton and thus higher productivity.

summer southwest monsoon (**Figure 7.20**). Studies of productivity in the Indian Ocean show that upwelling has increased in recent years due to stronger winds caused by warming of the Eurasian landmass, resulting in higher-than-normal summer productivity in the Arabian Sea.

INDIAN OCEAN SUBTROPICAL GYRE Surface circulation in the southern Indian Ocean (the **Indian Ocean Subtropical Gyre**) is similar to subtropical gyres observed in other southern oceans. When the northeast trade winds blow, the South Equatorial Current provides water for the Equatorial Countercurrent and the **Agulhas Current**,[11] which flows southward along Africa's east coast and joins the Antarctic Circumpolar Current (West Wind Drift). The *Agulhas Retroflection* is created when the Agulhas Current makes an abrupt turn as it meets the strong Antarctic Circumpolar Current (see the chapter-opening satellite image). Turning northward out of the Antarctic Circumpolar Current is the **West Australian Current**, an eastern boundary current that merges with the South Equatorial Current, completing the gyre.

LEEUWIN CURRENT Eastern boundary currents in other subtropical gyres are cold drifts toward the equator that produce arid coastal climates (that is, they receive less than 25 centimeters [10 inches] of rain per year). In the southern Indian Ocean, however, the **Leeuwin Current** displaces the West Australian Current offshore. The Leeuwin Current is driven southward along the Australian coast from the warm-water dome piled up in the East Indies by the Pacific equatorial currents.

The Leeuwin Current produces a mild climate in southwestern Australia, which receives about 125 centimeters (50 inches) of rain per year. During El Niño events, however, the Leeuwin Current weakens, so the cold Western Australian Current brings drought instead.

Pacific Ocean Circulation

Two large subtropical gyres dominate the circulation pattern in the Pacific Ocean, resulting in surface water movement and climatic effects similar to those found in the Atlantic. However, the Equatorial Countercurrent is much better developed in the Pacific Ocean than in the Atlantic (**Figure 7.21**), largely because the Pacific Ocean basin is larger and more unobstructed than the Atlantic Ocean.

[11]The Agulhas Current is named for Cape Agulhas, which is the southernmost tip of Africa.

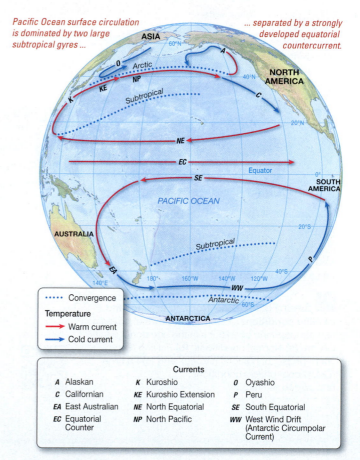

Pacific Ocean surface circulation is dominated by two large subtropical gyres ...

... separated by a strongly developed equatorial countercurrent.

Convergence		
Temperature		
→ Warm current		
→ Cold current		

Currents		
A Alaskan	**K** Kuroshio	**O** Oyashio
C Californian	**KE** Kuroshio Extension	**P** Peru
EA East Australian	**NE** North Equatorial	**SE** South Equatorial
EC Equatorial Counter	**NP** North Pacific	**WW** West Wind Drift (Antarctic Circumpolar Current)

Figure 7.21 Pacific Ocean surface currents.

Web Animation

El Niño and La Niña
http://goo.gl/1YAdt9

NORMAL CONDITIONS "Normal" conditions in the Pacific Ocean are a bit of a misnomer because they are experienced so infrequently. As we'll see, various atmospheric and oceanic disturbances dominate conditions in the Pacific. Still, normal conditions provide a baseline from which to measure these disturbances.

North Pacific Subtropical Gyre Figure 7.21 shows that the **North Pacific Subtropical Gyre** includes the North Equatorial Current, which flows westward into the western intensified **Kuroshio Current**[12] near Asia. The warm waters of the Kuroshio Current make Japan's climate warmer than would be expected for its latitude. This current flows into the **North Pacific Current**, which connects to the cool-water **California Current**. The California Current flows south along the coast of California to complete the loop. Some North Pacific Current water also flows to the north and merges into the **Alaskan Current** in the Gulf of Alaska.

South Pacific Subtropical Gyre Figure 7.21 also shows how the **South Pacific Subtropical Gyre** includes the South Equatorial Current, which flows westward into the western intensified **East Australian Current**.[13] From there, it joins the Antarctic Circumpolar Current (West Wind Drift) and completes the gyre as the **Peru Current** (also called the *Humboldt Current*, after German naturalist Friedrich Heinrich Alexander von Humboldt [1769–1859]).

FISHERIES AND THE PERU CURRENT The cool waters of the Peru Current have historically been one of Earth's richest fishing grounds. What conditions produce such an abundance of fish? **Figure 7.22a** shows that along the west coast of South America, coastal winds create Ekman transport that moves water away from shore, causing upwelling of cool, nutrient-rich water. This upwelling increases productivity and results in an abundance of marine life, including small silver-colored fish called *anchovetas* (anchovies) that become particularly plentiful near Peru and Ecuador. Anchovies provide a food source for many larger marine organisms and also supply Peru's commercial fishing industry, which was established in the 1950s. Anchovies had been so abundant in the waters off South America that by 1970 Peru was the largest producer of fish from the sea in the world, with a peak production of 12.3 million metric tons (27 billion pounds), accounting for about one-quarter of *all* fish from the sea worldwide.

Interdisciplinary

Relationship

WALKER CIRCULATION Figure 7.22a shows that high pressure and sinking air dominate the coastal region of South America, resulting in clear, fair, and dry weather. On the other side of the Pacific, a low-pressure region and rising air create cloudy conditions with plentiful precipitation in Indonesia, New Guinea, and northern Australia. This pressure difference causes the strong southeast trade winds to blow across the equatorial South Pacific. The resulting atmospheric circulation cell in the equatorial South Pacific Ocean is named the **Walker Circulation Cell** (*green arrows*) after Sir Gilbert T. Walker (1868–1958), the British meteorologist who first described the effect in the 1920s.

(a) Normal conditions

(b) El Niño conditions (strong)

(c) La Niña conditions

SmartFigure 7.22 Normal, El Niño, and La Niña conditions. Perspective views of oceanic and atmospheric conditions in the equatorial Pacific Ocean. **(a)** Normal conditions; **(b)** El Niño (ENSO warm phase) conditions (strong); and **(c)** La Niña (ENSO cool phase) conditions. https://goo.gl/FhkBlE

[12]Kuroshio is pronounced "kuhr-ROH-shee-oh." Because of its proximity to Japan, the Kuroshio Current is also called the *Japan Current*.

[13]Note that the western intensified East Australian Current was named because it lies off the *east coast* of Australia, even though it occupies a position along the *western* margin of the Pacific Ocean basin.

PACIFIC WARM POOL The southeast trade winds set ocean water in motion across the Pacific from east to west. The water warms as it flows in the equatorial region and creates a wedge of warm water on the western side of the Pacific Ocean, called the **Pacific Warm Pool** (see Figure 7.9). Due to the movement of equatorial currents to the west, the Pacific Warm Pool is thicker along the western side of the Pacific than along the eastern side. The *thermocline* beneath the Warm Pool in the western equatorial Pacific occurs below 100 meters (330 feet) depth. In the eastern Pacific, however, the thermocline is within 30 meters (100 feet) of the surface. The difference in depth of the thermocline can be seen by the sloping boundary between the warm surface water and the cold deep water in Figure 7.22a.

EL NIÑO–SOUTHERN OSCILLATION (ENSO) CONDITIONS Peru's residents have known for generations that every few years, a current of warm water reduces the population of anchovies in coastal waters. The decrease in anchovies causes a dramatic decline not only in the fishing industry, but also in marine life such as sea birds, sea lions, and seals that depend on anchovies for food. The warm current also brings about changes in the weather—usually intense rainfall—and even brings such interesting items as floating coconuts from tropical islands near the equator. At first, these events were called *anos de abundancia* (years of abundance) because the additional rainfall dramatically increased plant growth on the normally arid land. What was once thought of as a joyous event, however, soon became associated with the ecological and economic disaster that is now a well-known consequence of the phenomenon.

The Pacific warming was first described in the late 1880s by a Peruvian Navy captain who reported on an unusually warm *"corriente del Niño"* (ocean current of the Christ Child), so named because it appeared around Christmas time. Thus, this warm-water current was given the name **El Niño**, Spanish for "the child," in reference to baby Jesus. In the 1920s, Walker was the first to recognize that an east–west atmospheric pressure seesaw accompanied the warm current and he called the phenomenon the **Southern Oscillation**. Today, the combined oceanic and atmospheric effects are called **El Niño–Southern Oscillation (ENSO)**, which periodically alternates between warm and cold phases and causes dramatic environmental changes.

ENSO Warm Phase (El Niño) **Figure 7.22b** shows the atmospheric and oceanic conditions during an ENSO warm phase, which is known as *El Niño*. The high pressure along the coast of South America weakens, reducing the difference between the high- and low-pressure regions of the Walker Circulation Cell. This, in turn, causes the southeast trade winds to diminish. In very strong El Niño events, the trade winds actually blow in the *reverse* direction.

Without the trade winds, the Pacific Warm Pool that has built up on the western side of the Pacific begins to flow back across the ocean toward South America, creating a band of warm water that stretches across the equatorial Pacific Ocean (**Figure 7.23a**). The warm water usually begins to move in September of an El Niño year and reaches South America by December or January. During strong to very strong El Niños, the water temperature off Peru can be up to 10°C (18°F) higher than normal. In addition, the average sea level can increase as much as 20 centimeters (8 inches), simply due to thermal expansion of the warm water along the coast.

As the warm water increases sea surface temperatures across the equatorial Pacific, temperature-sensitive corals are decimated in Tahiti, the Galápagos, and other tropical Pacific islands. In addition, many other organisms are affected by the warm water (see MasteringOceanography **Web Diving Deeper 7.2**). Once the warm water reaches South America, it moves north and south along the west coast of the Americas, increasing average sea level and the number of tropical hurricanes formed in the eastern Pacific.

The flow of warm water across the Pacific also causes the sloped thermocline boundary between warm surface waters and the cooler waters below to flatten out and become more horizontal (Figure 7.22b). Near Peru, upwelling brings warmer, nutrient-depleted water to the surface instead of cold, nutrient-rich water. In fact, *downwelling* can sometimes occur as the warm water stacks up along coastal South America. Productivity diminishes and most types of marine life in the area are dramatically reduced.

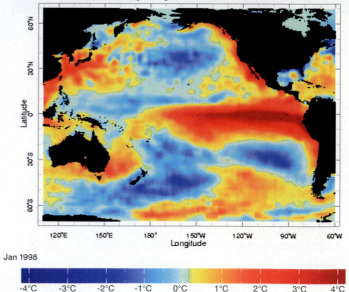

(a) Map of the Pacific Ocean in January 1998, showing the anomalous warming during the 1997–1998 El Niño.

Jan 1998

-4°C -3°C -2°C -1°C 0°C 1°C 2°C 3°C 4°C
Sea surface temperature anomaly

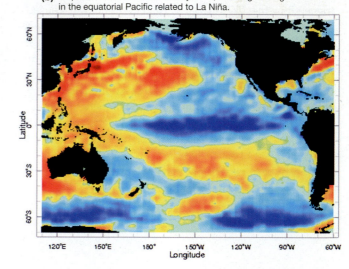

(b) Map of the same area in January 2000, showing cooling in the equatorial Pacific related to La Niña.

Figure 7.23 **Sea surface temperature anomaly maps during El Niño and La Niña.** Maps showing satellite-derived sea surface temperature anomalies, which represent departures from normal conditions. Values are in °C; red colors indicate water warmer than normal and blue colors represent water cooler than normal.

Web Video
Weekly Pacific Sea Surface Temperature Anomalies
http://goo.gl/9jBBnr

The amount of anchovies produced by Peru is impressive! Besides a topping for pizza, what are some other uses of anchovies?

Anchovies are an ingredient in certain dishes, hors d'oeuvres, sauces, and salad dressing, and fishers also use them as bait. Historically, however, most of the *anchoveta* caught in Peruvian waters were exported and used as fishmeal (consisting of ground anchovies). The fishmeal, in turn, was used largely in pet food and as a high-protein chicken feed. As unbelievable as it may seem, El Niños affected the price of eggs! Prior to the collapse of the Peruvian *anchoveta* fishing industry in 1972–1973, El Niño events significantly reduced the availability of *anchovetas*. This drastically cut the export of anchovies from Peru, causing U.S. farmers to pursue more expensive options for chicken feed. Thus, egg prices typically increased.

Interdisciplinary

Relationship

The collapse of the *anchoveta* fishing industry in Peru was triggered by the 1972–1973 El Niño event but was caused by chronic overfishing in prior years (see MasteringOceanography **Web Diving Deeper 13.2**). Interestingly, the shortage of fishmeal after 1972–1973 led to an increased demand for soyameal, an alternative source of high-quality protein. Increased demand for soyameal increased the price of soy commodities, thereby encouraging U.S. farmers to plant soybeans instead of wheat. Reduced production of wheat, in turn, caused a major global food crisis—and all this was triggered by an El Niño event.

As the warm water moves to the east across the Pacific, the low-pressure zone also migrates. In a strong to very strong El Niño event, the low pressure can move across the entire Pacific and remain over South America. The low pressure substantially increases precipitation along coastal South America. Conversely, high pressure replaces the Indonesian low, bringing dry conditions or, in strong to very strong El Niño events, drought conditions to Indonesia and northern Australia.

ENSO Cool Phase (La Niña) In some instances, conditions opposite of El Niño prevail in the equatorial South Pacific; these events are known as *ENSO cool phase* or **La Niña** (Spanish for "the female child"). Figure 7.22c shows La Niña conditions, which are similar to normal conditions but more intensified because there is a larger pressure difference across the Pacific Ocean. This larger pressure difference creates stronger Walker Circulation and stronger trade winds, which in turn cause more upwelling, a shallower thermocline in the eastern Pacific, and a band of cooler than normal water that stretches across the equatorial South Pacific (**Figure 7.23b**).

La Niña conditions commonly occur following an El Niño. For instance, the 1997–1998 El Niño was followed by several years of persistent La Niña conditions. The alternating pattern of El Niño–La Niña conditions since 1950 is shown by the multivariate **ENSO index** (**Figure 7.24**), which is calculated using a weighted average of atmospheric and oceanic factors, including atmospheric pressure, winds, and sea surface temperatures. Positive ENSO index numbers indicate El Niño conditions, whereas negative numbers reflect La Niña conditions. Normal conditions are indicated by a value near zero, and the greater the index value differs from zero (either negative or positive), the stronger the respective condition.

For a summary and comparison of atmospheric and oceanic conditions during normal, El Niño, and La Niña conditions, see MasteringOceanography **Web Tables 7.2** and **7.3**.

How Often Do El Niño Events Occur? Records of sea surface temperatures over the past 100 years reveal that throughout the 20th century, El Niño conditions occur on average about every 2 to 10 years, but in a highly irregular pattern. In some decades, for instance, there has been an El Niño event every few years, while in others there may have been only one. Figure 7.24 shows the pattern since 1950, revealing that the equatorial Pacific fluctuates between El Niño and La Niña conditions, with only a few years that could be considered normal conditions (represented by an ENSO index value close to zero). Typically, El Niño events last for 12 to 18 months and are followed by La Niña conditions that usually exist for a similar length of time. However, some El Niño or La Niña conditions can last for several years.

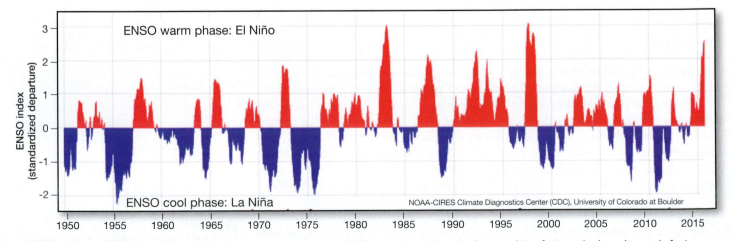

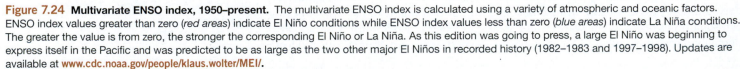

Figure 7.24 Multivariate ENSO index, 1950–present. The multivariate ENSO index is calculated using a variety of atmospheric and oceanic factors. ENSO index values greater than zero (*red areas*) indicate El Niño conditions while ENSO index values less than zero (*blue areas*) indicate La Niña conditions. The greater the value is from zero, the stronger the corresponding El Niño or La Niña. As this edition was going to press, a large El Niño was beginning to express itself in the Pacific and was predicted to be as large as the two other major El Niños in recorded history (1982–1983 and 1997–1998). Updates are available at **www.cdc.noaa.gov/people/klaus.wolter/MEI/.**

Recently recovered sediments from a South American lake provide a continuous 10,000-year record of the frequency of El Niño events. The sediments indicate that between 10,000 and 7000 years ago, no more than five strong El Niños occurred each century. The frequency of El Niños then increased, peaking at about 1200 years ago (and coinciding with the early Middle Ages in Europe), when they occurred every three years or so. If the pattern observed in the lake sediments continues, researchers predict that there should be an increase in El Niños in the early part of the 22nd century.

El Niño events—especially severe ones—may occur more frequently as a result of increased global warming. For instance, the two most severe El Niño events in the 20th century occurred in 1982–1983 and 1997–1998. Presumably, increased ocean temperatures could trigger more frequent and more severe El Niños, although a recent study of the past 7000 years of ocean temperature data from Pacific corals suggests that there is only a weak correlation between El Niño events and ocean warming episodes. However, the recent pattern of El Niños could also be a part of a long-term natural climate cycle. For example, oceanographers have discovered a phenomenon called the **Pacific Decadal Oscillation (PDO)**, which lasts 20 to 30 years and appears to influence Pacific sea surface temperatures. Analysis of satellite data suggests that the Pacific Ocean was in the warm phase of the PDO from 1977 to 1999 and that it is now in its cool phase, which may suppress the initiation of El Niño events during the next few decades.

Climate

Connection

Effects of El Niños and La Niñas Mild El Niño events influence only the equatorial South Pacific Ocean, while strong to very strong El Niño events can influence worldwide weather patterns. Typically, stronger El Niños alter the atmospheric jet stream and produce unusual weather in most parts of the globe. Sometimes the weather is drier than normal; at other times, it is wetter. The weather may also be warmer or cooler than normal. It is still difficult to predict exactly how a particular El Niño will affect any region's weather.

Figure 7.25 shows how very strong El Niño events can result in flooding, erosion, droughts, fires, tropical storms, and effects on marine life worldwide.

Figure 7.25 Effects of severe El Niños. Map showing the locations of flooding, erosion, droughts, fires, tropical storms, and effects on marine life that are associated with severe El Niño events.

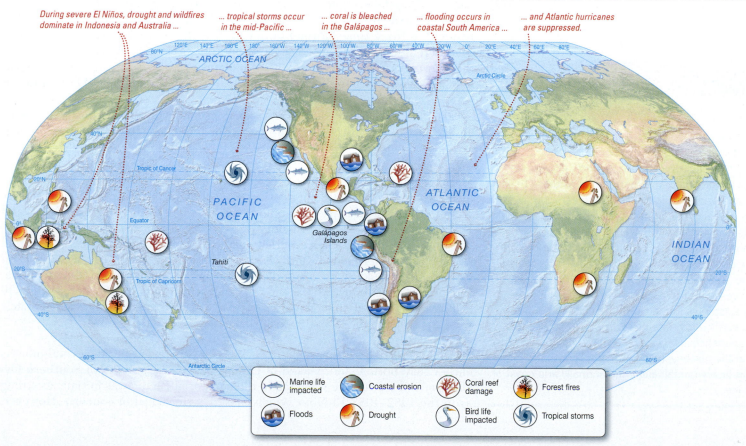

During severe El Niños, drought and wildfires dominate in Indonesia and Australia ...

... tropical storms occur in the mid-Pacific ...

... coral is bleached in the Galápagos ...

... flooding occurs in coastal South America ...

... and Atlantic hurricanes are suppressed.

Marine life impacted

Coastal erosion

Coral reef damage

Forest fires

Floods

Drought

Bird life impacted

Tropical storms

(a) Sea surface temperature anomaly map for January 1998, an El Niño year.

(b) Sea surface temperature anomaly map for the same region a year later (January 1999), during a La Niña event.

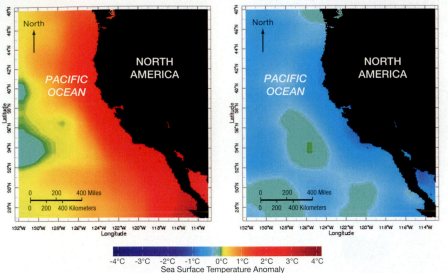

-4°C -3°C -2°C -1°C 0°C 1°C 2°C 3°C 4°C
Sea Surface Temperature Anomaly

Figure 7.26 **Sea surface temperatures off western North America during El Niño and La Niña.** Satellite-derived sea surface temperature anomaly maps (in °C) along the west coast of North America. Red color represents water that is warmer than normal; blue color represents water that is colder than normal.

STUDENTS SOMETIMES ASK . . .

Do El Niño events occur in other ocean basins?

Yes, the Atlantic and Indian Oceans both experience events similar to the Pacific's El Niño. These events are not nearly as strong, however, nor do they influence world-wide weather phenomena to the same extent as those that occur in the equatorial Pacific Ocean. The great width of the Pacific Ocean in equatorial latitudes is the main reason that El Niño events occur more strongly in the Pacific.

In the Atlantic Ocean, this phenomenon is related to the North Atlantic Oscillation (NAO), which is a periodic change in atmospheric pressure between Iceland and the Azores Islands. This pressure difference determines the strength of the prevailing westerlies in the North Atlantic, which in turn affects ocean surface currents there. The Atlantic Ocean periodically experiences NAO events, which sometimes cause intense cold in the northeast United States, unusual weather in Europe, and heavy rainfall along the normally arid coast of southwest Africa.

These weather perturbations also affect the production of corn, cotton, and coffee. More locally, the satellite images in **Figure 7.26** show that sea surface temperatures off western North America are significantly higher during an El Niño year.

Even though severe El Niños are typically associated with vast amounts of destruction, they can be beneficial in some areas. Tropical hurricane formation, for instance, is generally suppressed in the Atlantic Ocean because of greater wind shear in the upper atmosphere, some desert regions receive much-needed rain, and organisms adapted to warm-water conditions thrive in the Pacific.

La Niña events are associated with sea surface temperatures and weather phenomena opposite those of El Niño. Indian Ocean monsoons, for instance, are typically drier than usual in El Niño years but wetter than usual in La Niña years.

Examples from Recent El Niños Recent El Niños provide an indication of the variability of the effects of El Niño events. For instance, in the winter of 1976, a moderate El Niño event coincided with northern California's worst drought of the 20th century, showing that El Niño events don't always bring torrential rains to the western United States. During that same winter, the eastern United States experienced record cold temperatures.

THE 1982–1983 EL NIÑO The 1982–1983 El Niño was the strongest ever recorded, causing far-ranging effects around the globe. Not only was there anomalous warming in the tropical Pacific, but the warm water also spread along the west coast of North America, influencing sea surface temperatures as far north as Alaska. Sea level was higher than normal (due to thermal expansion of the water), which, when high surf was experienced, caused damage to coastal structures and increased coastal erosion. In addition, the jet stream swung much farther south than normal across the United States, bringing a series of powerful storms that resulted in three times normal rainfall across the southwestern United States. The increased rainfall caused severe flooding and landslides as well as higher-than-normal snowfall in the Rocky Mountains. Alaska and western Canada had a relatively warm winter, and the eastern United States had its mildest winter in 25 years.

The full strength of El Niño was experienced in western South America. Normally arid Peru was drenched with more than 3 meters (10 feet) of rain, causing extreme flooding and landslides. Sea surface temperatures were so high for so long that temperature-sensitive coral reefs across the equatorial Pacific were decimated. Marine mammals and sea birds, which depend on the food normally available in the highly productive waters along the west coast of South America, went elsewhere or died. In the Galápagos Islands, for example, over half of the island's fur seals and sea lions died of starvation during the 1982–1983 El Niño.

French Polynesia had not experienced a hurricane in 75 years; in 1983, it endured six. The Hawaiian Island of Kauai also experienced a rare hurricane. Meanwhile, in Europe, severe cold weather prevailed. Elsewhere, droughts occurred in Australia, Indonesia, China, India, Africa, and Central America. Worldwide, more than 2000 deaths and at least $10 billion in property damage ($2.5 billion in the United States) were attributed to the 1982–1983 El Niño event.

THE 1997–1998 EL NIÑO The 1997–1998 El Niño event began several months earlier than normal and peaked in January 1998. The amount of Southern Oscillation and sea surface warming in the equatorial Pacific was initially as strong as in the 1982–1983 El Niño, which caused a great deal of concern. However,

the 1997–1998 El Niño weakened in the last few months of 1997 before reintensifying in early 1998. The impact of the 1997–1998 El Niño was felt mostly in the tropical Pacific, where surface water temperatures in the eastern Pacific averaged more than 4°C (7°F) warmer than normal, and, in some locations, reached up to 9°C (16°F) above normal (see Figure 7.23a). High pressure in the western Pacific brought drought conditions that caused wildfires to burn out of control in Indonesia. Also, the warmer than normal water along the west coast of Central and North America increased the number of hurricanes off Mexico.

In the United States, the 1997–1998 El Niño caused killer tornadoes in the Southeast, massive blizzards in the upper Midwest, and flooding in the Ohio River Valley. Most of California received twice the normal rainfall, which caused flooding and landslides in many parts of the state. The lower Midwest, the Pacific Northwest, and the eastern seaboard, on the other hand, had relatively mild weather. In all, the 1997–1998 El Niño caused 2100 deaths and $33 billion in property damage worldwide.

Predicting El Niño Events The 1982–1983 El Niño event was not predicted, nor was it recognized until it was near its peak. Because it affected weather worldwide and caused such extensive damage, the **Tropical Ocean–Global Atmosphere (TOGA)** program was initiated in 1985 to study how El Niño events develop. The goal of the TOGA program was to monitor the equatorial South Pacific Ocean during El Niño events to enable scientists to model and predict future El Niño events. The 10-year program studied the ocean from research vessels, analyzed surface and subsurface data from radio-transmitting sensor buoys, monitored oceanic phenomena by satellite, and developed computer models.

These models have made it possible to predict El Niño events since 1987 as much as one year in advance. After the completion of TOGA, the **Tropical Atmosphere and Ocean (TAO)** project (sponsored by the United States, Canada, Australia, and Japan) has continued to monitor the equatorial Pacific Ocean with a series of 70 moored buoys, providing real-time information about the conditions of the tropical Pacific that is available on the Internet. Although monitoring has improved, the trigger mechanisms of El Niño events are still not fully understood.

RECAP

El Niño is a combined oceanic–atmospheric phenomenon that occurs periodically in the tropical Pacific Ocean, bringing warm water to the east. La Niña describes conditions opposite those of El Niño.

CONCEPT CHECK 7.4 | Specify the main surface circulation patterns in each ocean basin.

1 Explain why Gulf Stream eddies that develop northeast of the Gulf Stream rotate clockwise and have warm-water cores, whereas those that develop to the southwest rotate counterclockwise and have cold-water cores.

2 Describe changes in atmospheric pressure, precipitation, winds, and ocean surface currents during the two monsoon seasons of the Indian Ocean.

3 Describe changes in atmospheric and oceanographic phenomena that occur during El Niño/La Niña events, including changes in atmospheric pressure, winds, Walker Circulation, weather, equatorial surface currents, coastal upwelling/downwelling and the abundance of marine life, sea surface temperature and the Pacific Warm Pool, sea surface elevation, and position of the thermocline.

4 How often do El Niño events occur? Using Figure 7.24, determine how many years since 1950 have been El Niño years. Has the pattern of El Niño events occurred at regular intervals?

5 How is La Niña different from El Niño? Describe the pattern of La Niña events in relation to El Niños since 1950 (see Figure 7.24).

6 Describe the global effects of severe El Niños.

7.5 How Do Deep-Ocean Currents Form?

Deep currents occur in the deep zone below the pycnocline, so they influence about 90% of all ocean water. Density differences create deep currents. Although these density differences are usually small, they are large enough to cause denser waters to sink. Deep-water currents move larger volumes of water and are much slower than surface currents. Typical speeds of deep currents range from 10 to 20 kilometers (6 to 12 miles) per year. Thus, it takes a deep current an *entire year* to travel the same distance that a western intensified surface current can move in *one hour*.

Because the density variations that cause deep-ocean circulation are the result of differences in temperature and salinity, deep-ocean circulation is also referred to as **thermohaline circulation** (*thermo* = heat, *haline* = salt).

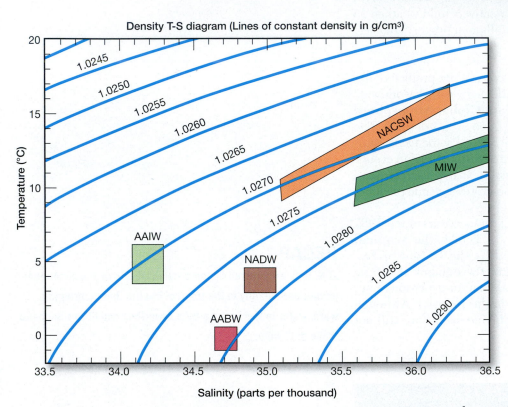

Origin of Thermohaline Circulation

Recall from Chapter 5 that an increase in seawater density can be caused by a *decrease* in temperature or an *increase* in salinity. Temperature, though, has the greater influence on density. Density changes due to salinity are important only in very high latitudes, where water temperature remains low and relatively constant.

Most water involved in deep-ocean currents (thermohaline circulation) originates in high latitudes *at the surface*. In these regions, surface water becomes cold and its salinity increases as sea ice forms. When this surface water becomes dense enough, it sinks, initiating deep-ocean currents. Once this water sinks, it is removed from the physical processes that increased its density in the first place, so its temperature and salinity remain largely unchanged for the duration it spends in the deep ocean. Thus, a **temperature–salinity (T–S) diagram** can be used to identify deep-water masses based on their characteristic temperature, salinity, and resulting density. **Figure 7.27** shows a T–S diagram for the North Atlantic Ocean.

As these surface water masses become dense and are sinking (downwelling) in high-latitude areas, deep-water masses are also rising to the surface (upwelling). Because the water temperature in high-latitude regions is the same at the surface as it is down below, the water column is isothermal, there is no thermocline or associated pycnocline (see Chapter 5), and upwelling and downwelling can easily occur.

North Atlantic Water Masses:

- (AAIW) Antarctic Intermediate Water
- (AABW) Antarctic Bottom Water
- (NADW) North Atlantic Deep Water
- (NACSW) North Atlantic Central Surface Water
- (MIW) Mediterranean Intermediate Water

Figure 7.27 Temperature–salinity (T–S) diagram. A density T–S diagram for the North Atlantic Ocean. Lines of constant density are in grams/cm³. After various deep-water masses sink below the surface, they can be identified based on their characteristic temperature, salinity, and resulting density.

Sources of Deep Water

In southern subpolar latitudes, huge masses of deep water form beneath sea ice along the margins of the Antarctic continent. Here, rapid winter freezing produces very cold, high-density water that sinks down the continental slope of Antarctica and becomes **Antarctic Bottom Water**, the densest water in the open ocean (**Figure 7.28**). Antarctic Bottom Water slowly sinks beneath the surface and spreads into all the world's ocean basins, eventually returning to the surface perhaps 1000 years later.

In the northern subpolar latitudes, large masses of deep water form in the Norwegian Sea. From there, the deep water flows as a subsurface current into the North Atlantic, where it becomes part of the **North Atlantic Deep Water**. North Atlantic Deep Water also comes from the margins of the Irminger Sea off southeastern Greenland, the Labrador Sea, and the dense, salty Mediterranean Sea. Like Antarctic Bottom Water, North Atlantic Deep Water spreads throughout the ocean basins. It is less dense, however, so it layers on top of the Antarctic Bottom Water (Figure 7.28).

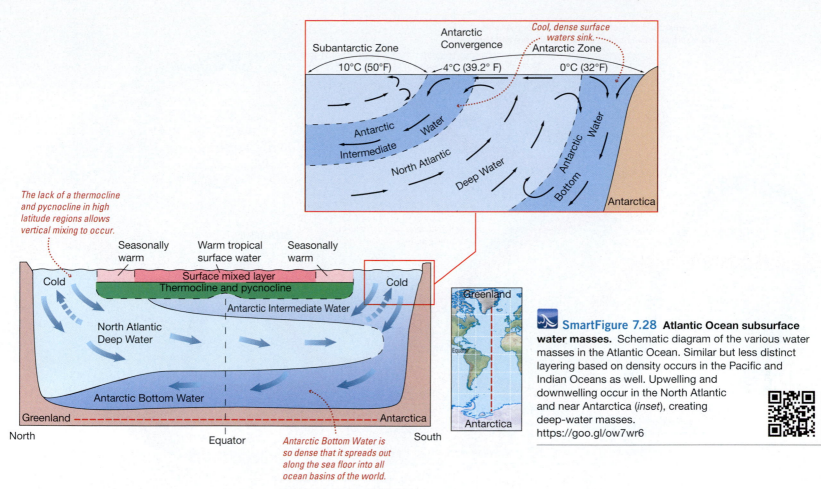

The lack of a thermocline and pycnocline in high latitude regions allows vertical mixing to occur.

Antarctic Bottom Water is so dense that it spreads out along the sea floor into all ocean basins of the world.

SmartFigure 7.28 Atlantic Ocean subsurface water masses. Schematic diagram of the various water masses in the Atlantic Ocean. Similar but less distinct layering based on density occurs in the Pacific and Indian Oceans as well. Upwelling and downwelling occur in the North Atlantic and near Antarctica (*inset*), creating deep-water masses. https://goo.gl/ow7wr6

Surface water masses converge within the subtropical gyres and in the Arctic and Antarctic. Subtropical Convergences do not produce deep water, however, because the density of warm surface waters is too low for them to sink. Major sinking does occur, however, along the **Arctic Convergence** and Antarctic Convergence (Figure 7.28, *inset*). The deep-water mass formed from sinking at the Antarctic Convergence is called the **Antarctic Intermediate Water** mass (Figure 7.28), which remains one of world's most poorly studied water masses.

Figure 7.28 also shows that the highest-density water is found along the ocean bottom, with less-dense water above. In low-latitude regions, the boundary between the warm surface water and the deeper cold water is marked by a prominent thermocline and corresponding pycnocline that prevent vertical mixing. There is no pycnocline in high-latitude regions, so substantial vertical mixing (upwelling and downwelling) occurs.

This same general pattern of layering based on density occurs in the Pacific and Indian Oceans as well. These oceans have no source of Northern Hemisphere deep water, however, so they lack a deep-water mass. In the northern Pacific Ocean, the low salinity of surface waters prevents them from sinking into the deep ocean. In the northern Indian Ocean, surface waters are too warm to sink. **Oceanic Common Water**, which is created when Antarctic Bottom Water and North Atlantic Deep Water mix, lines the bottoms of these basins.

Worldwide Deep-Water Circulation

For every liter of water that sinks from the surface into the deep ocean, a liter of deep water must return to the surface somewhere else. However, it is difficult to identify specifically *where* this vertical flow to the surface is occurring. It is generally believed that it occurs as a gradual, uniform upwelling throughout the ocean

Web Animation

North Atlantic Deep-Water Circulation
https://goo.gl/ojdtTq

7.1 Squidtoons

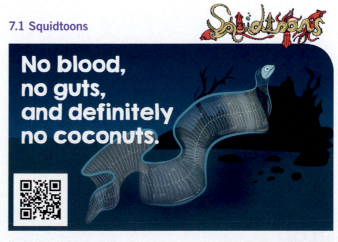

No blood, no guts, and definitely no coconuts.

https://goo.gl/sXd731

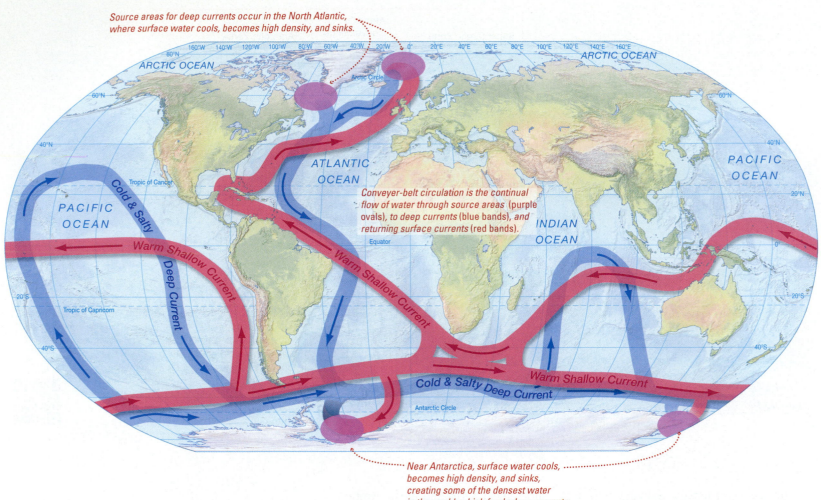

Source areas for deep currents occur in the North Atlantic, where surface water cools, becomes high density, and sinks.

Conveyer-belt circulation is the continual flow of water through source areas (purple ovals), to deep currents (blue bands), and returning surface currents (red bands).

Near Antarctica, surface water cools, becomes high density, and sinks, creating some of the densest water in the world, which feeds deep currents.

Figure 7.29 Idealized conveyer-belt circulation. Schematic map of conveyer-belt circulation of the world ocean. Source areas for deep water (*purple ovals*) exist in high-latitude regions where surface water cools, becomes high density, and sinks. These source areas feed the flow of deep, high-density waters (*blue bands*), which slowly drifts into all oceans. Deep water returns to the surface in localized areas of upwelling and also as gradual, uniform upwelling throughout the ocean basins. Surface currents (*red bands*) complete the conveyer by returning water to the source areas.

Web Animation
Deep-Ocean Conveyer-Belt Circulation
http://goo.gl/jzGmDh

basins and that it may be somewhat greater in low-latitude regions, where surface temperatures are higher. Alternatively, scientific studies on turbulent mixing rates between deep-ocean and surface waters in the Southern Ocean suggest that deep water traveling across rugged bottom topography is a major factor in producing the upwelling that returns deep water toward the surface.

CONVEYER-BELT CIRCULATION An integrated model combining deep thermohaline circulation and surface currents is shown in **Figure 7.29**. Because the overall circulation pattern resembles a large conveyer belt, the model is called **conveyer-belt circulation**. Beginning in the North Atlantic, surface water carries heat to high latitudes via the Gulf Stream. During the cold winter months, this heat is transferred to the overlying atmosphere, warming northern Europe.

Climate

Connection

Cooling in the North Atlantic increases the density of this surface water to the point where it sinks to the bottom and flows southward, initiating the lower limb of the "conveyor." Here, seawater flows downward at a rate equal to 100 Amazon Rivers and begins its long journey into the deep basins of all the world's oceans. This limb extends all the way to the southern tip of Africa, where it joins the deep water that encircles Antarctica. The deep water that encircles Antarctica includes deep water that descends along the margins of the Antarctic continent. This mixture of deep waters flows northward into the deep Pacific and Indian Ocean basins, where it eventually surfaces and completes the conveyer belt by flowing west and then north again into the North Atlantic Ocean.

Does this simple conveyer-belt model of ocean circulation adequately reflect the movement of both surface and deep-ocean currents? Satellite and deep sampling of the oceans confirms the basic conveyer-belt movement of warm waters poleward at the surface and cold waters equatorward at depth. However, the conveyer-belt model ignores some crucial complex components of the ocean's circulation system—including small-scale eddies and oceanic fronts—that affect climate change and are now being studied at ever-finer scales.

DISSOLVED OXYGEN IN DEEP WATER Cold water can dissolve more oxygen than warm water. Thus, deep-water circulation brings dense, cold, oxygen-enriched water from the surface to the deep ocean. During its time in the deep ocean, deep water becomes enriched in nutrients as well, due to decomposition of dead organisms and the lack of organisms using nutrients there.

At various times in the geologic past, warmer water probably constituted a larger proportion of deep oceanic waters. As a result, the oceans had a lower oxygen concentration than today because warm water cannot hold as much oxygen. Moreover, the oxygen content of the oceans has probably fluctuated widely throughout time.

Climate
Connection

If high-latitude surface waters did not sink and eventually return from the deep sea to the surface, the distribution of life in the sea would be considerably different. There would be very little life in the deep ocean, for instance, because there would be no oxygen for organisms to breathe. In addition, life in surface waters might be significantly reduced without the circulation of deep water that brings nutrients to the surface.

Interdisciplinary
Relationship

CONVEYER-BELT CIRCULATION AND CLIMATE CHANGE Conveyer-belt circulation is an important part of global ocean circulation; as such, it can profoundly affect global climate. For example, the part of the conveyer in the North Atlantic that includes both surface and deep currents is called the **Atlantic meridional overturning circulation (AMOC)**. Studies of the AMOC reveal that it is weaker now than it has been in the past 1000 years because glacial ice from Greenland has melted, thereby releasing low-density freshwater into the North Atlantic, which acts as a cap and prohibits deep water from sinking. Scientists who study the phenomenon predict that the AMOC will continue to weaken in the next few decades as global warming causes more glacial ice to melt in Greenland, which would increase the size of the freshwater wedge in the North Atlantic and may cause the complete shutdown of the AMOC. If this happens, the formation of deep water will be prevented and global circulation patterns will be reorganized. For example, the freshwater wedge may cause the Gulf Stream in the North Atlantic to loop back towards the equator instead of reaching higher latitudes. If the warming waters of the Gulf Stream don't reach high latitudes, climate would likely be dramatically altered in Europe, North America, and many other places around the world. For more details about how the reorganization of ocean circulation patterns can alter Earth's climate, see Chapter 16, "The Oceans and Climate Change."

Climate
Connection

see Chapter 16, "The Oceans and Climate Change."

STUDENTS SOMETIMES ASK ...

The movie The Day After Tomorrow *depicts a freezing superstorm as a result of alterations in deep-water circulation. Could this really happen?*

Although Hollywood movies are well known for dramatization, what is interesting about this blockbuster 2004 film is that its story line is based on recent scientific findings: The ocean's deep-water circulation helps drive ocean currents around the globe and is important to world climate. In fact, strong evidence suggests that North Atlantic deep-water circulation has already weakened and may further weaken during this century, resulting in unpleasant effects on our climate—but certainly not as rapid or as cataclysmic as the situation portrayed in the film. Computer models suggest that the continued weakening of the North Atlantic deep-water circulation will result in some long-term cooling, particularly over parts of northern Europe.

RECAP

Thermohaline circulation describes the movement of deep currents, which is initiated when surface water in high latitudes becomes cold and dense, and as a result, sinks.

CONCEPT CHECK 7.5 | Explain the origin and characteristics of deep-ocean currents.

1 Discuss the origin of thermohaline vertical circulation. Why do deep currents form only in high-latitude regions?

2 What are the two major deep-water masses? Where do they form at the ocean's surface?

3 Explain why no matter where you are in the ocean, if you go deep enough, you will encounter Oceanic Common Water.

4 Describe how the distribution of life in the ocean would be different if there were very little dissolved oxygen in deep-water currents.

Figure 7.30 **Power from ocean currents.** A prototype ocean current power system in Strangford Lough, Ireland, which generates power with underwater blades that can be raised for maintenance. As currents flow past the tower, they turn the 16-meter (52-foot)-long propellers, which rotate internal rotors and generate electricity. The propellers can rotate in either direction, so this system can generate power from bi-directional tidal flows.

RECAP

In spite of problems with placing mechanical devices in the ocean, there is vast potential for developing power from currents as a renewable source of energy.

7.6 Can Power from Currents Be Harnessed as a Source of Energy?

The movement of ocean currents has often been considered to be capable of providing a source of renewable, clean energy similar to that of wind farms (see Chapter 6)—but underwater. Even though currents move much more slowly than winds, currents carry much more energy because water has about 800 times the density of air. As a result, currents have the potential to generate even more power than wind farms do. In theory, oceans could power the entire globe without adding any pollution to the atmosphere. Power from currents could also provide a more dependable source of electricity than the wind or Sun because ocean currents flow all day and night.

One location that has received much consideration as a site for harnessing power from ocean currents is the Florida–Gulf Stream Current System, which you may recall is a fast, western intensified surface current that runs along the East Coast of the United States. In fact, researchers have determined that at least 2000 megawatts[14] of electricity could be recovered from this ocean current system along the southeastern coast of Florida alone.

Various devices have been proposed to extract energy from ocean currents. All of them involve some sort of mechanism for converting the movement of water into electrical energy. **Figure 7.30** shows one solution. Here, underwater turbines similar to windmills (but anchored to the ocean floor) are placed in waters that experience strong currents. As these currents flow past the tower, they turn the propellers, which rotate internal rotors and generate electricity. These systems must also work in locations where bi-directional tidal flows occur, so in some cases the propellers can rotate in either direction; in other cases, the entire turbine can swivel on its anchor to face oncoming currents. Such a system of six turbines has been successfully tested in the East River near New York City. Once this system is expanded to its capacity of 300 turbines, it will have the ability to generate about 10 megawatts of electricity.

Systems that generate power from currents have some significant hurdles to overcome. For example, current systems are expensive, difficult to maintain, and potentially hazardous to ship traffic. Further, the moving devices that harness current power can potentially disturb, injure, or kill marine life, although detailed environmental studies suggest that marine organisms are not harmed by the devices.[15] Also, underwater deployment of complex machines with moving parts exposed to seawater for long periods of time presents challenges associated with corrosion and *biofouling*, which is the accumulation of algae and other sea life on machinery. In addition, the variability of current flow causes irregular power generation, which is problematic. Still, similar turbine systems powered by currents are in use in Strangford Lough, Ireland (Figure 7.30), and are planned for Canada's Bay of Fundy and offshore South Korea.

Interdisciplinary

Relationship

CONCEPT CHECK 7.6 | Evaluate the advantages and disadvantages of harnessing currents as a source of energy.

1 Explain why ocean currents have the potential to generate even more power than wind farms, which are much more common today.

2 Discuss the advantages and disadvantages of building an offshore current power system.

[14]Each megawatt of electricity is enough to serve the energy needs of about 800 average U.S. homes.

[15]Interestingly, the same environmental studies suggest that these sea floor devices may, in fact, act as makeshift marine sanctuaries because fishing practices do not occur at these locations.

ESSENTIAL CONCEPTS REVIEW

7.1 How are ocean currents measured?

▶ *Ocean currents are masses of water that flow* from one place to another and can be divided into *surface currents that are wind driven* and *deep currents that are density driven*. Currents can be measured directly or indirectly by various methods.

Study Resources

MasteringOceanography Study Area Quizzes, MasteringOceanography Web Table 7.1

Critical Thinking Question

Compare and contrast the characteristics and origins of surface currents and deep currents.

Active Learning Exercise

Working with another student in class, compile a list of the ways in which both surface and deep currents are measured.

7.2 What creates ocean surface currents and how are they organized?

▶ *Surface currents occur within and above the pycnocline.* They consist of circular-moving loops of water called gyres, set in motion by the major wind belts of the world. They are modified by the positions of the continents, the Coriolis effect, seasonal changes, and other factors. There are *five major subtropical gyres in the world*; they rotate *clockwise in the Northern Hemisphere* and *counterclockwise in the Southern Hemisphere*. Water is pushed toward the center of the gyres, forming low "hills" of water.

▶ The *Ekman spiral* influences shallow surface water and is *caused by winds and the Coriolis effect*. The average net flow of water affected by the Ekman spiral causes the water to move at *90-degree angles to the wind direction*. At the center of a gyre, the Coriolis effect deflects the water so that it tends to move into the hill, whereas gravity moves the water down the hill. When gravity and the Coriolis effect balance, a *geostrophic current* flowing parallel to the contours of the hill is established.

▶ As a result of Earth's rotation, *the apex (top) of the hill is located to the west of the geographical center of the gyre*. A phenomenon called *western intensification* occurs in which *western boundary currents of subtropical gyres are faster, narrower, and deeper* than their eastern boundary counterparts.

Study Resources

MasteringOceanography Study Area Quizzes, MasteringOceanography Web Diving Deeper 7.1, MasteringOceanography Web Animations, MasteringOceanography Web Video, Web Video

Critical Thinking Question

From memory, construct a map showing the world's five subtropical gyres, including nearby continents. For each gyre, name the main currents involved, including any that experience western intensification.

Active Learning Exercise

With another student in class, describe what the pattern of ocean surface currents would look like if there were no continents on Earth.

7.3 What causes upwelling and downwelling?

▶ *Upwelling and downwelling help vertically mix deep and surface waters.* Upwelling—the movement of *cold, deep, nutrient-rich water to the surface*—stimulates productivity and creates a large amount of marine life. Upwelling and downwelling can occur in a variety of ways.

Study Resources

MasteringOceanography Study Area Quizzes, MasteringOceanography Web Animation

Critical Thinking Question

From a practical standpoint, give several reasons why upwelling is a much more intensely studied process than downwelling. For example, why are the world's major fisheries associated with areas of upwelling?

Active Learning Exercise

Working with another student in class, discuss this question: If Earth's rotation on its axis were to suddenly stop, how would it affect the processes that cause both upwelling and downwelling?

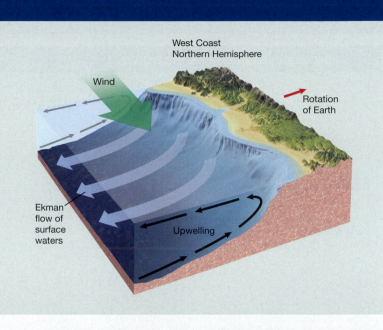

7.4 What are the main surface circulation patterns in each ocean basin?

▶ *Antarctic circulation is dominated by a single large current, the Antarctic Circumpolar Current* (West Wind Drift), which flows in a clockwise direction around Antarctica and is driven by the Southern Hemisphere's prevailing westerly winds. Between the Antarctic Circumpolar Current and the Antarctic continent is a current called the *East Wind Drift*, which is powered by the polar easterly winds. The two currents flow in opposite directions, and the Coriolis effect deflects them away from each other, creating the *Antarctic Divergence*, an area of abundant marine life due to upwelling and current mixing.

▶ *The North Atlantic Gyre and the South Atlantic Gyre dominate circulation in the Atlantic Ocean.* A poorly developed equatorial countercurrent separates these two subtropical gyres. The highest-velocity and best-studied ocean current is the *Gulf Stream*, which carries warm water along the southeastern U.S. Atlantic coast. Meanders of the Gulf Stream produce *warm- and cold-core rings*. The warming effects of the Gulf Stream extend along its route and reach as far away as northern Europe.

▶ *The Indian Ocean consists of one gyre, the Indian Ocean Gyre*, which exists mostly in the Southern Hemisphere. The *monsoon wind system*, which changes direction with the seasons, dominates circulation in the Indian Ocean. The monsoons blow from the northeast in the winter and from the southwest in the summer.

▶ *Circulation in the Pacific Ocean consists of two subtropical gyres: the North Pacific Gyre and the South Pacific Gyre*, which are separated by a well-developed equatorial countercurrent.

▶ *A periodic disruption of normal sea surface and atmospheric circulation patterns in the Pacific Ocean is called El Niño–Southern Oscillation (ENSO).* The *warm phase of ENSO (El Niño)* is associated with the eastward movement of the Pacific Warm Pool, halting or reversal of the trade winds, a rise in sea level along the equator, a decrease in productivity along the west coast of South America, and, in very strong El Niños, worldwide changes in weather. El Niños fluctuate with the *cool phase of ENSO (La Niña conditions)*, which are associated with cooler than normal water in the eastern tropical Pacific.

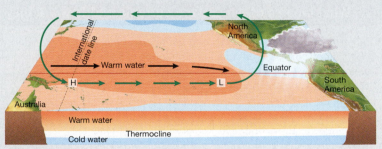

(b) El Niño conditions (strong)

Study Resources

MasteringOceanography Study Area Quizzes, MasteringOceanography Web Diving Deeper 7.2, MasteringOceanography Web Tables 7.2 and 7.3, MasteringOceanography Web Animations, Web Video

Critical Thinking Question

Using Figure 7.22, specify the atmospheric and oceanic changes that occur between normal and El Niño conditions. Also, compare normal and La Niña conditions.

Active Learning Exercise

During flood stage, the largest river in the world—the mighty Amazon River—dumps 200,000 cubic meters of water into the Atlantic Ocean each second. With another student in class, compare its flow rate with the volume of water transported by (1) the West Wind Drift and (2) the western-intensified Gulf Stream. How many times larger than the Amazon is each of these two ocean currents?

7.5 How do deep-ocean currents form?

▶ *Deep currents occur below the pycnocline.* They affect much larger amounts of ocean water and move much more slowly than surface currents. Changes in temperature and/or salinity at the surface create slight increases in density, which set deep currents in motion. Deep currents, therefore, are called *thermohaline circulation*.

▶ *The deep ocean is layered based on density.* Antarctic Bottom Water, the densest deep-water mass in the oceans, forms near Antarctica and sinks along the continental shelf into the South Atlantic Ocean. Farther north, at the Antarctic Convergence, the low-salinity *Antarctic Intermediate Water* sinks to an intermediate depth dictated by its density. Sandwiched between these two masses is the *North Atlantic Deep Water*, which has high nutrient levels after remaining below the surface for hundreds of years. Layering in the Pacific and Indian Oceans is similar, except there is no source of Northern Hemisphere deep water.

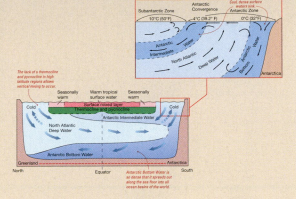

▶ *Worldwide circulation models that include both surface and deep currents resemble a conveyer belt.* Deep currents carry oxygen into the deep ocean, which is extremely important for life on the planet.

Study Resources
MasteringOceanography Study Area Quizzes, MasteringOceanography Web Animations, MasteringOceanography Web Video

Critical Thinking Question
Antarctic Intermediate Water can be identified throughout much of the South Atlantic based on its temperature, salinity, and dissolved oxygen content. Why is it colder and less salty—and why does it contain more oxygen—than the surface water mass above it and the North Atlantic Deep Water below it?

Active Learning Exercise
The density T–S graph shown as Figure 7.27 shows that the Antarctic Intermediate Water (AAIW) has about the same density as the North Atlantic Central Surface Water (NACSW). With another student in class, discuss how you can differentiate the two water masses based on physical properties (not just their locations). As a group, explain how it is possible that these two distinct water masses have nearly the same density.

7.6 Can power from currents be harnessed as a source of energy?

▶ *Ocean currents can be harnessed as a source of power.* Although there is vast potential for developing this clean, renewable resource, significant problems must be overcome to make this a practical source of energy.

Study Resources
MasteringOceanography Study Area Quizzes

Critical Thinking Question
What are some environmental factors that could inhibit the development of offshore ocean current power systems?

Active Learning Exercise
With another student in class, discuss the advantages of current power devices that are attached to the sea floor and capable of operating in two or more directions.

MasteringOceanography™

www.masteringoceanography.com

Looking for additional review and test prep materials? With individualized coaching on the toughest topics of the course, MasteringOceanography offers a wide variety of ways for you to move beyond memorization and deeply grasp the underlying processes of how the oceans work. Visit the Study Area in **www.masteringoceanography.com** to find practice quizzes, study tools, and multimedia that will improve your understanding of this chapter's content. Sign in today to enjoy the following features: Self Study Quizzes, SmartFigures, SmartTables, Oceanography Videos, Squidtoons, Geoscience Animation Library, RSS Feeds, Digital Study Modules, and an optional Pearson eText.

CONCEPT CHECK 8.2 | Describe the characteristics of waves.

1 Can a wave with a wavelength of 14 meters ever be more than 2 meters high? Why or why not?

2 What physical feature of a wave is related to the depth of the wave base? What is the difference between the *wave base* and *still water level*?

3 Calculate the speed (*S*) in meters per second for deep-water waves with the following characteristics:

a. *L* = 351 meters, *T* = 15 seconds

b. *T* = 12 seconds

c. *f* = 0.125 wave per second

4 For each of the following statements about deep-water waves, determine if the statement is *true* or *false*, then explain why:

a. The longer the wave, the deeper the wave base.

b. The greater the wave height, the deeper the wave base.

c. The longer the wave, the faster the wave travels.

d. The greater the wave height, the faster the wave travels.

e. The faster the wave, the greater the wave height.

8.3 How Do Wind-Generated Waves Develop?

Most ocean waves are generated by the wind and so are termed *wind-generated waves*.

Wave Development

The "lifecycle" of a wind-generated wave includes its origin in a windy region of the ocean, its movement across great expanses of open water without subsequent aid of wind, and its termination when it breaks and releases its energy, either in the open ocean or against the shore.

CAPILLARY WAVES, GRAVITY WAVES, AND THE "SEA" As the wind blows over the ocean surface, it pushes down on and parallel to the surface, causing water to pile up into small wavelets called **capillary waves**, which are more commonly called *ripples*. These capillary waves are small, rounded waves with V-shaped troughs and have wavelengths less than 1.74 centimeters (0.7 inch) (**Figure 8.9**, *left*). The name comes from *capillarity*, a property that results from the surface tension of water.[3]

As capillary wave development increases, the sea surface takes on a rougher appearance. The water "catches" more of the wind, allowing the wind and ocean surface to interact more efficiently. As more energy is transferred to the ocean, **gravity waves** develop. These are symmetric waves that have wavelengths exceeding 1.74 centimeters (0.7 inch) (Figure 8.9, *middle*).

The length of gravity waves is generally 15 to 35 times their height. As additional energy is gained, wave height increases more rapidly than wavelength. The crests become pointed and the troughs are rounded, resulting in a *trochoidal* (*trokhos* = wheel) waveform (Figure 8.9, *right*).

Energy imparted by the wind increases the height, length, and speed of the wave. When wave speed equals wind speed, neither wave height nor wavelength can change because there is no net energy exchange and the wave has reached its maximum size.

The area where wind-driven waves are generated is called the **sea**, or *sea area*. It is characterized by

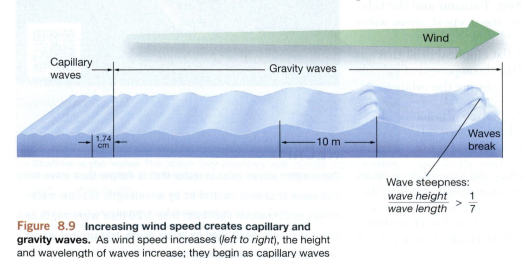

Figure 8.9 Increasing wind speed creates capillary and gravity waves. As wind speed increases (*left to right*), the height and wavelength of waves increase; they begin as capillary waves and progress to gravity waves. When the wave steepness exceeds a 1:7 ratio, the waves become unstable and break. Diagram is schematic and not to scale.

[3]See Section 5.1 in Chapter 5 for a description of water's surface tension.

choppiness and waves moving in many directions. The waves have a variety of periods and wavelengths (most of them short) due to frequently changing wind speed and direction.

FACTORS AFFECTING WAVE ENERGY Figure 8.10 shows the factors that determine the amount of energy in waves. These factors are (1) the *wind speed*, (2) the *duration*—the length of time during which the wind blows in one direction, and (3) the *fetch*—the distance over which the wind blows in one direction.

Wave height is directly related to the energy in a wave. Wave heights in a sea area are usually less than 2 meters (6.6 feet), but waves with heights of 10 meters (33 feet) and periods of 12 seconds are not uncommon. As "sea" waves gain energy, their steepness increases. When steepness reaches a critical value of 1/7, open ocean breakers—called *whitecaps*—form. The **Beaufort Wind Scale** and the state of the sea (Table 8.1), which was initially devised by Admiral Sir Francis Beaufort (1774–1857) of the British Navy, describes the appearance of the sea surface from dead calm conditions to hurricane-force winds.

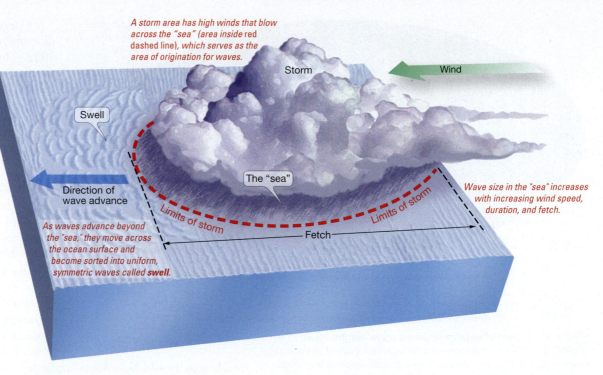

A storm area has high winds that blow across the "sea" (area inside red dashed line), which serves as the area of origination for waves.

Storm

Wind

Swell

The "sea"

Direction of wave advance

*As waves advance beyond the "sea," they move across the ocean surface and become sorted into uniform, symmetric waves called **swell**.*

Limits of storm

Limits of storm

Fetch

Wave size in the "sea" increases with increasing wind speed, duration, and fetch.

Figure 8.10 Factors that contribute to creating large waves in the "sea" and the origination of swell.

SmartTable 8.1 Beaufort wind scale and the state of the sea.
https://goo.gl/5JCG1R

SMARTTABLE 8.1	BEAUFORT WIND SCALE AND THE STATE OF THE SEA			
Beaufort number	**Descriptive term**	**Wind speed**		**Appearance of the sea**
		km/hr	mi/hr	
0	Calm	<1	<1	Like a mirror
1	Light air	1–5	1–3	Ripples with the appearance of scales, no foam crests
2	Light breeze	6–11	4–7	Small wavelets; crests of glassy appearance, no breaking
3	Gentle breeze	12–19	8–12	Large wavelets; crests begin to break, scattered whitecaps
4	Moderate breeze	20–28	13–18	Small waves, becoming longer; numerous whitecaps
5	Fresh breeze	29–38	19–24	Moderate waves, taking longer form; many whitecaps, some spray
6	Strong breeze	39–49	25–31	Large waves begin to form, whitecaps everywhere, more spray
7	Near gale	50–61	32–38	Sea heaps up and white foam from breaking waves begins to be blown in streaks
8	Gale	62–74	39–46	Moderately high waves of greater length, edges of crests begin to break into spindrift, foam is blown in well-marked streaks
9	Strong gale	75–88	47–54	High waves, dense streaks of foam and sea begins to roll, spray may affect visibility
10	Storm	89–102	55–63	Very high waves with overhanging crests; foam is blown in dense white streaks, causing the sea to appear white; the rolling of the sea becomes heavy; visibility reduced
11	Violent storm	103–117	64–72	Exceptionally high waves (small and medium-sized ships might for a time be lost from view behind the waves), the sea is covered with white patches of foam, everywhere the edges of the wave crests are blown into froth, visibility further reduced
12	Hurricane	118+	73+	The air is filled with foam and spray, sea completely white with driving spray, visibility greatly reduced

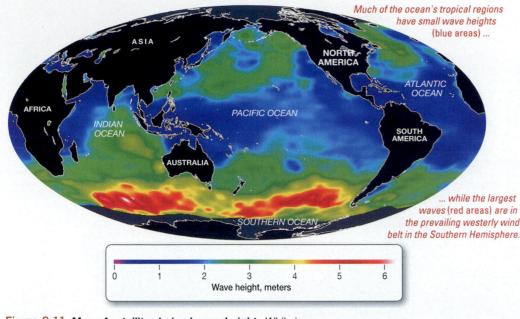

Much of the ocean's tropical regions have small wave heights (blue areas) ...

... while the largest waves (red areas) are in the prevailing westerly wind belt in the Southern Hemisphere.

Wave height, meters
0 1 2 3 4 5 6

Figure 8.11 Map of satellite-derived wave height. While in operation, the TOPEX/Poseidon satellite received a return of stronger radar signals from calm seas and weaker radar signals from seas with large waves. Based on these data, a map of average wave height can be produced. Data was collected during October 3–12, 1992; scale is in meters.

STUDENTS SOMETIMES ASK . . .

I know that swell is what surfers hope for. Is swell always big?

Not necessarily. Swell is defined as waves that have moved out of their area of origination, so these waves do not have to be a certain wave height to be classified as swell. It is true, however, that the uniform and symmetrical shape of most swell delights surfers.

Figure 8.11 is a map based on satellite data of average wave heights during October 3–12, 1992. The waves in the Southern Hemisphere are particularly large because the prevailing westerlies between 40 and 60 degrees south latitude reach the highest average wind speeds on Earth, creating the latitudes called the "Roaring Forties," "Furious Fifties," and "Screaming Sixties."

HOW HIGH CAN WAVES BE? According to a U.S. Navy Hydrographic Office bulletin published in the early 1900s, the theoretical maximum height of wind-generated waves should be no higher than 18.3 meters (60 feet); this became known as the "60-foot rule." Although there were some isolated eyewitness accounts of larger waves, the U.S. Navy considered any sightings of waves over 60 feet to be exaggerations. Certainly, embellishment of reported wave height under conditions of extremely rough seas would be understandable. For many years, the "60-foot rule" was accepted as fact.

Careful observations made aboard the 152-meter- (500-foot-) long U.S. Navy tanker USS *Ramapo* in 1933 proved the "60-foot rule" incorrect. The ship was caught in a typhoon in the western Pacific Ocean and encountered 108-kilometer-per-hour (67-mile-per-hour) winds en route from the Philippines to San Diego. The resulting waves were symmetrical, uniform, and had a period of 14.8 seconds. Because the *Ramapo* was traveling with the waves, the vessel's officers were able to measure the waves accurately. The officers used the dimensions of the ship, including the *eye height* of an observer on the ship's bridge (**Figure 8.12**). Geometric relationships revealed that the waves were 34 meters (112 feet) high, which is taller than an 11-story building! These waves proved to be a record that still stands today for the largest authentically recorded wind-generated waves, shattering the "60-foot rule." Although the *Ramapo* was largely undamaged, other ships traveling in rough seas aren't always so lucky (**Figure 8.13**). In fact, several large ships disappear at sea every year simply due to enormous waves.

FULLY DEVELOPED SEA For a given wind speed, **Table 8.2** lists the minimum fetch and duration of wind beyond which the waves cannot grow. Waves cannot grow because an equilibrium condition, called a **fully developed sea**, has been achieved. Waves can grow no further in a fully developed sea because they lose as much energy breaking as whitecaps under the force of gravity as they receive from the wind. Table 8.2 also lists the average characteristics of waves resulting from a fully developed sea, including the height of the highest 10% of the waves.

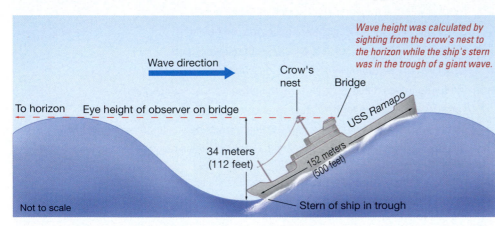

Wave direction

Crow's nest

Bridge

USS Ramapo

Wave height was calculated by sighting from the crow's nest to the horizon while the ship's stern was in the trough of a giant wave.

To horizon Eye height of observer on bridge

34 meters (112 feet)

152 meters (500 feet)

Stern of ship in trough

Not to scale

Figure 8.12 The U.S. Navy tanker USS *Ramapo* measured the largest authentically recorded waves (1933). Diagrammatic view of the record-setting wave height of 34 meters (112 feet) that was calculated based on geometric relationships of the vessel in the giant waves. Note that the record has not been exceeded since.

SWELL As waves generated in a sea area move toward its margins, wind speeds diminish and the waves eventually move faster than the wind. When this occurs, wave steepness decreases and waves become long-crested waves called **swells** (*swellan* = swollen), which are uniform, symmetrical waves that have traveled out of their area of origination.

Figure 8.13 **Wave** *damage on the aircraft* **carrier** *Bennington* **(1945).** The *Bennington* returns from heavy seas encountered in a typhoon off Okinawa in 1945, with part of its reinforced steel flight deck bent down over its bow. Damage to the flight deck, which is 16.5 meters (54 feet) above still water level, was caused by large waves.

Swells move with little loss of energy over large stretches of the ocean surface, transporting energy away from one sea area and depositing it in another. The movement of swells to distant areas is the reason why there can be waves at a shoreline even though there is no wind.

Waves with longer wavelengths travel faster and thus leave the sea area first. They are followed by slower, shorter **wave trains**, or groups of waves. The progression from long, fast waves to short, slow waves illustrates the principle of **wave dispersion** (*dis* = apart, *spargere* = to scatter)—the sorting of waves by their wavelength. Waves of many wavelengths are present in the generating area. Wave speed depends on wavelength in deep water (see Figure 8.8), however, so the longer waves "outrun" the shorter ones. The distance over which waves change from a choppy "sea" to uniform swell is called the **decay distance**, which can be up to several hundred kilometers.

As a group of waves leaves a sea area and becomes a *swell wave train*, the leading wave keeps disappearing. However, the same number of waves always remains in the group because as the leading wave disappears, a new wave replaces

STUDENTS SOMETIMES ASK . . .

What is the difference between "groundswell" and "wind swell"?

The term *groundswell* was originally a sailor's word (and is now a surfer's term) for deep-ocean swell, such as might be generated by a distant storm or earthquake. In its original sense, groundswell referred to waves so huge that their troughs bared the "ground" of the sea bottom. Groundswell is essentially the same thing as *wind swell*, although groundswell often refers to very large waves from a distant origin, whereas wind swell refers to smaller, locally produced waves.

TABLE 8.2	CONDITIONS NECESSARY TO PRODUCE A FULLY DEVELOPED SEA AT VARIOUS WIND SPEEDS AND THE CHARACTERISTICS OF THE RESULTING WAVES						
These conditions . . .			. . . produce these waves				
Wind speed in km/hr (mi/hr)	Fetch in km (mi)	Duration in hours	Average height in m (ft)	Average wave-length in m (ft)	Average period in seconds	Highest 10% of waves in m (ft)	
20 (12)	24 (15)	2.8	0.3 (1.0)	10.6 (34.8)	3.2	0.8 (2.5)	
30 (19)	77 (48)	7.0	0.9 (2.9)	22.2 (72.8)	4.6	2.1 (6.9)	
40 (25)	176 (109)	11.5	1.8 (5.9)	39.7 (130.2)	6.2	3.9 (12.8)	
50 (31)	380 (236)	18.5	3.2 (10.5)	61.8 (202.7)	7.7	6.8 (22.3)	
60 (37)	660 (409)	27.5	5.1 (16.7)	89.2 (292.6)	9.1	10.5 (34.4)	
70 (43)	1093 (678)	37.5	7.4 (24.3)	121.4 (398.2)	10.8	15.3 (50.2)	
80 (50)	1682 (1043)	50.0	10.3 (33.8)	158.6 (520.2)	12.4	21.4 (70.2)	
90 (56)	2446 (1517)	65.2	13.9 (45.6)	201.6 (661.2)	13.9	28.4 (93.2)	

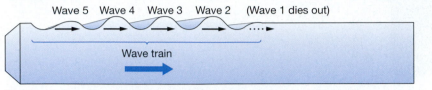

(a) Energy in the leading waves (*Waves 1 and 2*) is transferred into circular orbital motion.

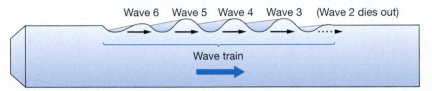

(b) Wave 1 dies out and is replaced by Wave 2; note new Wave 5 behind.

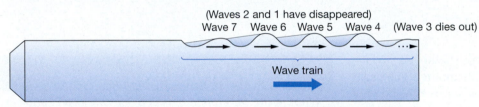

(c) Wave 2 dies out and is replaced by Wave 3; note new Wave 6 behind.

(d) Wave 3 dies out and is replaced by Wave 4; note new Wave 7 behind. Even though new waves take up the lead, the length of the wave train and the total number of waves remain the same. This causes the group speed to be one-half that of the individual wave.

Figure 8.14 Movement of a wave train. Diagrammatic sequence showing the movement of a wave train. Note how waves die out in the front but are replaced in the back, causing the length of the wave train and the total number of waves to remain the same even as the wave train progresses.

RECAP

Constructive interference results from in-phase overlapping of waves and creates larger waves, while destructive interference results from waves overlapping out of phase, reducing wave height.

it at the back of the group (**Figure 8.14**). For example, if four waves are generated, the lead wave keeps dying out as the wave train travels, but one is created in the back, so the wave train stays four waves. Because of the progressive dying out and creation of new waves, the group moves across the ocean surface at only *half* the velocity of an individual wave in the group.

Interference Patterns

When swells from different storms run together, the waves clash, or *interfere* with one another, giving rise to **interference patterns**. An interference pattern produced when two or more wave systems collide is the sum of the disturbance that each wave would have produced individually. **Figure 8.15** shows that the result may be a larger or smaller trough or crest, depending on conditions.

CONSTRUCTIVE INTERFERENCE Constructive interference occurs when wave trains having the same wavelength come together *in phase*, meaning crest to crest and trough to trough. If the displacements from each wave are added together, the interference pattern results in a wave with the same wavelength as the two overlapping wave systems but with a wave height equal to the sum of the individual wave heights (Figure 8.15, *top*).

DESTRUCTIVE INTERFERENCE Destructive interference occurs when wave trains having the same wavelength come together *out of phase*, meaning the crest from one wave coincides with the trough from a second wave. If the waves have identical heights, the sum of the crest of one and the trough of another is zero, so the energies of these waves cancel each other (Figure 8.15, *middle*).

MIXED INTERFERENCE In most ocean areas, it is likely that two or more swells of different heights and lengths will come together and produce a mixture of constructive and destructive interference. In this scenario, a more complex **mixed interference** pattern develops (Figure 8.15, *bottom*). Mixed interference patterns explain the varied sequence of higher and lower waves called **surf beat** that most people notice at the beach as well as other irregular wave patterns that occur when two or more swells approach the shore. In the open ocean, several swell systems often interact, creating complex wave patterns (**Figure 8.16**) and, sometimes, large waves that can be hazardous to ships.

Rogue Waves

Rogue waves are massive, solitary, spontaneous ocean waves that can reach enormous height and often occur at times when normal ocean waves are not unusually large. In a sea of 2-meter (6.5-foot) waves, for example, a 20-meter (65-foot) rogue wave may suddenly appear. *Rogue* means "unusual" and, in this case, the waves are unusually large. Rogue waves—sometimes called *superwaves, monster waves, sleeper waves,* or *freak waves*—are defined as individual waves of exceptional height or abnormal shape that are more than twice the average of the highest one-third of all the wave heights in a given wave record. Mariners often describe rogue wave crests as "mountains of water" and their troughs as "holes in the sea," creating a roller-coaster ride that can be disastrous.

Because of their size and destructive power, rogue waves have been popularized in literature and movies such as *The Perfect Storm* and the disaster film remake *Poseidon*. The impressive size of some rogue waves makes them quite hazardous to oil-drilling platforms and vessels at sea. In 1966, for example, the Italian luxury liner *Michelangelo* received extensive damage from a rogue wave it encountered during a storm in the North Atlantic (**Figure 8.17**). In a nearby area in 1995, the giant luxury liner *Queen Elizabeth II*, which carries 1500 passengers, plowed through a 29-meter (95-foot) wave generated by Hurricane Luis.

In the open ocean, 1 wave in 23 will be over twice the height of the wave average, 1 in 1175 will be three times as high, and 1 in 300,000 will be four times as high. The chances of a truly monstrous wave, therefore, are only one in several billion. Nevertheless, rogue waves do occur, and satellite observations over a three-week period in 2001 confirmed that rogue waves occur more frequently than was previously thought: The study documented more than 10 individual giant waves around the globe of over 25 meters (82 feet) in height. Even with satellite monitoring of ocean waves (see MasteringOceanography **Web Table 3.1**), it remains difficult to forecast specifically when or where rogue waves will arise. For instance, the 17-meter (56-foot) NOAA research vessel R/V *Ballena* was flipped and sunk in 2000 by a 4.6-meter (15-foot) rogue wave off the California coast while conducting a survey in shallow, calm water. Fortunately, the three people on board survived the incident.

Worldwide each year, about 10 large ships such as supertankers or containerships are reported missing without a trace; as many as 1000 vessels of all sizes are lost each year, and rogue waves are the suspected cause of many of these sinkings. Still, scientists lack detailed shipboard measurements of rogue waves because they tend to

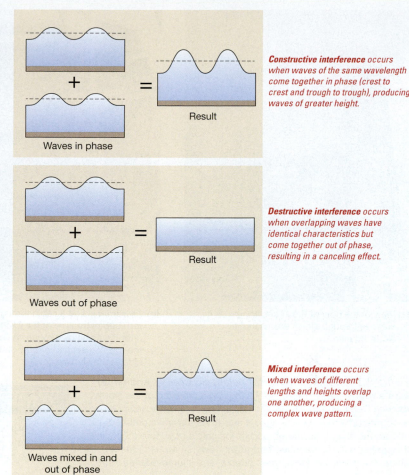

Constructive interference occurs when waves of the same wavelength come together in phase (crest to crest and trough to trough), producing waves of greater height.

Destructive interference occurs when overlapping waves have identical characteristics but come together out of phase, resulting in a canceling effect.

Mixed interference occurs when waves of different lengths and heights overlap one another, producing a complex wave pattern.

SmartFigure 8.15 Constructive, destructive, and mixed interference produce a variety of wave patterns.
https://goo.gl/0TbyY2

Web Animation
Interference Patterns in Waves
http://goo.gl/0rQ7Vq

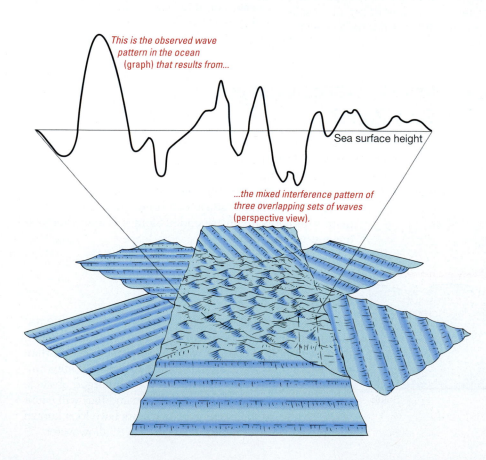

This is the observed wave pattern in the ocean (graph) that results from...

Sea surface height

...the mixed interference pattern of three overlapping sets of waves (perspective view).

Figure 8.16 An example of a mixed interference pattern that develops from three overlapping sets of waves.

(a) View of the bow of the *Michelangelo* as it crashes through storm waves in the North Atlantic.

(b) View from inside a cabin through the torn superstructure of the *Michelangelo* after it was hit by a rogue wave. Note the missing section of bow on the right side of the ship *(red oval)*.

Figure 8.17 Damage to the *Michelangelo* from its encounter with a rogue wave (1966). While making a crossing of the North Atlantic Ocean in April 1966, the Italian luxury liner *Michelangelo* encountered a rogue wave amid a storm. As the wave crashed aboard the ship, it ripped off part of the ship's bow, tore a hole in its superstructure, smashed bridge windows 24 meters (80 feet) above the waterline, killed three people, and injured dozens more.

appear without warning, and bobbing ships make poor observation platforms.

The main cause of rogue waves is theorized to be an extraordinary case of constructive wave interference where multiple waves overlap in-phase to produce an extremely large wave. Rogue waves also tend to occur more frequently near weather fronts and downwind from islands or shoals. Recent modeling of the wave conditions in the Pacific that caused a Japanese fishing vessel to capsize in 2008 suggests that rogue waves can be created when the low- and high-frequency components of ordinary ocean waves interact and channel their energy into a narrow frequency band.

Another way in which rogue waves can be created is when strong ocean currents focus and amplify opposing swells. Such conditions exist along the "Wild Coast" off the southeast coast of Africa, where the Agulhas Current flows directly against large Antarctic storm waves, creating rogue waves that can crash onto a ship's bow, overcome its structural capacity, and sink even large, well-constructed ships (**Figure 8.18**).

CONCEPT CHECK 8.3 | Discuss how wind-generated waves develop.

1 Define *swell*. Does swell necessarily imply a particular wave size? Why or why not?

2 Waves from separate sea areas move away as swell and produce an interference pattern when they come together. If Sea A has wave heights of 1.5 meters (5 feet) and Sea B has wave heights of 3.5 meters (11.5 feet), what would be the height of waves resulting from constructive interference and destructive interference? Illustrate your answer (see Figure 8.15).

8.4 How Do Waves Change in the Surf Zone?

Most waves generated in the sea area by storm winds move across the ocean as swell. These waves then release their energy along the margins of continents in the **surf zone**, which is the zone of breaking waves. Breaking waves exemplify power and persistence, sometimes moving objects weighing several tons. In doing so, energy from a distant storm can travel thousands of kilometers until it is finally expended along a distant shoreline in a few wild moments.

Physical Changes as Waves Approach Shore

As deep-water waves of swell move toward continental margins over gradually **shoaling** (*shold* = shallow) water, they eventually encounter water depths that are less than one-half of their wavelength (**Figure 8.19**) and become transitional waves. Actually, any shallowly submerged obstacle (such as a coral reef, sunken wreck, or sand bar) will cause waves to release some energy. Navigators have long known that breaking waves indicate dangerously shallow water.

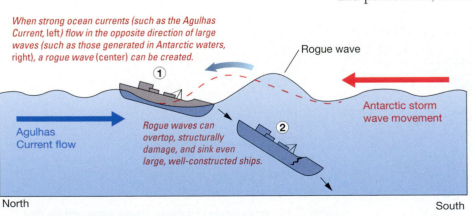

When strong ocean currents (such as the Agulhas Current, left) flow in the opposite direction of large waves (such as those generated in Antarctic waters, right), a rogue wave (center) can be created.

①

Rogue wave

Antarctic storm wave movement

Agulhas Current flow

Rogue waves can overtop, structurally damage, and sink even large, well-constructed ships.

②

North

South

Figure 8.18 The creation of a rogue wave along Africa's "Wild Coast."

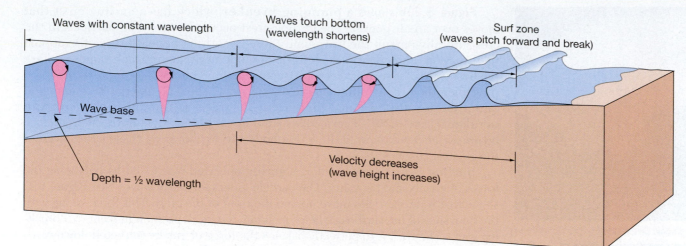

Waves with constant wavelength

Waves touch bottom
(wavelength shortens)

Surf zone
(waves pitch forward and break)

Wave base

Depth = ½ wavelength

Velocity decreases
(wave height increases)

Figure 8.19 **Physical changes of a wave in the surf zone.** As waves approach the shore and encounter water depths of less than one-half wavelength, the waves "feel bottom." The *wave speed decreases*, and waves stack up against the shore, causing the *wavelength to decrease*. This results in an *increase in wave height* to the point where the wave *steepness is increased* beyond the 1:7 ratio, causing the wave to *pitch forward and break* in the surf zone.

Many physical changes occur to a wave as it encounters shallow water, becomes a shallow-water wave, and eventually breaks. The progressively shallower depths interfere with water particle movement at the base of the wave, so the *wave speed decreases*. As one wave slows, the following waveform, which is still moving at its original speed, moves closer to the wave that is being slowed, causing a *decrease in wavelength*. Although some wave energy is lost due to friction, the wave energy that remains must go somewhere, so *wave height increases*. This increase in wave height combined with the decrease in wavelength causes an *increase in wave steepness (H/L)*. When the wave steepness reaches the 1:7 ratio, the *waves break as surf* (Figure 8.19).

If the surf is swell that has traveled from distant storms, breakers will develop relatively near shore, in shallow water. The horizontal motion characteristic of shallow-water waves moves water alternately toward and away from the shore as an oscillation. The surf will be characterized by parallel lines of relatively uniform breakers.

If the surf consists of waves generated by local winds, the waves may not have been sorted into swell. The surf may be mostly unstable, deep-water, high-energy waves with steepness already near the 1:7 ratio. In this case, the waves will break shortly after feeling bottom some distance from shore, and the surf will be rough, choppy, and irregular.

When the water depth is about one and one-third times the wave height, the crest of the wave breaks, producing surf.[4] When the water depth becomes less than $1/20$ the wavelength, waves in the surf zone begin to behave like shallow-water waves (see Figure 8.7). Particle motion is greatly impeded by the bottom, and a significant transport of water toward the shoreline occurs (Figure 8.19).

Waves break in the surf zone because particle motion near the bottom of the wave is severely restricted, slowing the waveform. At the surface, however, individual orbiting water particles have not yet been slowed because they have no contact with the bottom. In addition, the wave height increases in shallow water. The difference in speed between the top and bottom parts of the wave causes the top part of the wave to overrun the lower part, which results in the wave toppling over and breaking. Breaking waves are analogous to a person who leans too far forward. If you don't catch yourself, you may also "break" something when you fall.

Breakers and Surfing

There are three main types of breakers. **Figure 8.20a** shows a **spilling breaker**, which is a turbulent mass of air and water that runs down the front slope of the wave as it breaks. Spilling breakers result from a gently sloped ocean bottom, which gradually extracts energy from the wave over an extended distance and produces breakers with low overall energy. As a result, spilling breakers have a longer life span and give surfers a long—but somewhat less exciting—ride than other breakers.

[4]Using this relationship provides a handy way of estimating water depth in the surf zone: The depth of the water where waves are breaking is one and one-third times the breaker height.

Web Animation

Wave Motion and Wave Refraction
http://goo.gl/55FMau

(a) Spilling breaker, resulting from a gradual beach slope.

(b) Plunging breaker, resulting from a steep beach slope; these are the best waves for surfing.

(c) Surging breaker, resulting from an abrupt beach slope.

Figure 8.20 **Types of breakers.** Photos of the three types of breakers, which are a result of different beach slopes.

Web Animation
Three Types of Breakers
https://goo.gl/7D8t9S

Figure 8.20b shows a **plunging breaker**, which has a curling crest that moves over an air pocket. The curling crest occurs because the particles in the crest literally outrun the wave, and there is nothing beneath them to support their motion. Plunging breakers form on moderately steep beach slopes and are the best waves for surfing (see the chapter-opening photo).

When the ocean bottom has an abrupt slope, the wave energy is compressed into a shorter distance, and the wave will surge forward, creating a **surging breaker** (**Figure 8.20c**). These waves build up and break right at the shoreline, so board surfers tend to avoid them. For body surfers, however, these waves present the greatest challenge.

Surfing is analogous to riding a gravity-operated water sled by balancing the forces of gravity and buoyancy. The particle motion of ocean waves (see Figure 8.4) shows that water particles move up into the front of the crest. This force, along with the buoyancy of the surfboard, helps maintain a surfer's position in front of a breaking wave. The trick is to perfectly balance the force of gravity (directed downward) with the buoyant force (directed perpendicular to the wave face) to enable a surfer to be propelled forward by the wave's energy. A skillful surfer, by positioning the board properly on the wave front, can regulate the degree to which the propelling gravitational forces exceed the buoyancy forces, and the surfer can obtain speeds up to 40 kilometers (25 miles) per hour while moving along the face of a breaking wave. When the wave passes over water that is too shallow to allow the upward movement of water particles to continue, the wave has expended its energy, and the ride is over.

Wave Refraction

Waves seldom approach a shore at a perfect right angle (90 degrees). Instead, some segment of the wave will "feel bottom" first and will slow before the rest of the wave. This results in the **refraction** (*refringere* = to break up), or *bending*, of each *wave crest* (also called a *wave front*) as waves approach the shore.

Figure 8.21a shows how waves coming toward a straight shoreline are refracted and tend to align themselves *nearly* parallel to the shore. This explains why all waves come almost straight in toward a beach, no matter what their original orientation was. **Figure 8.21b** shows how waves coming toward an irregular shoreline refract so that they, too, nearly align with the shore. **Figure 8.21c** shows a classic example of wave refraction around Rincon Point in California.

The refraction of waves along an irregular shoreline distributes wave energy unevenly along the shore. To help illustrate how this works, notice the long black arrows in Figure 8.21b, which are called **orthogonal lines** (*ortho* = straight, *gonia* = angle), or *wave rays*. Orthogonal lines are always oriented perpendicular to the wave crests, so they indicate the direction that waves travel. More importantly, orthogonals are spaced so that the energy between lines is equal at all times and can thus be used to show variations in wave energy. Far from shore, for example, notice how the orthogonals in Figure 8.21b are evenly spaced, which indicates that all parts of the wave have the same amount of energy. As the waves approach the shore and refract, however, notice how the orthogonals *converge* on headlands that jut into the ocean and *diverge* in bays. This means that wave energy is focused against the headlands but dispersed in bays. As a result, large waves occur at headlands, which are areas of good surfing[5] and sites of erosion. Conversely, smaller waves occur in bays, which often provide areas for boat anchorages and are also associated with sediment deposition. Interestingly, the same waves break against both headlands and in nearby bays, but their energy is different because wave refraction causes the spacing of orthogonals (and thus energy) to change. In addition, waves approaching shore are also influenced by sea floor features such as shallow banks or submarine canyons.

[5]Sailors have long known that "the points draw the waves." Surfers also know how wave refraction causes good "point breaks."

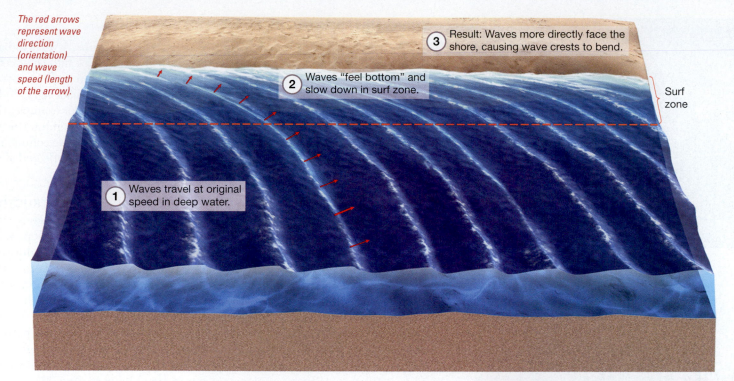

The red arrows represent wave direction (orientation) and wave speed (length of the arrow).

③ Result: Waves more directly face the shore, causing wave crests to bend.

② Waves "feel bottom" and slow down in surf zone.

Surf zone

① Waves travel at original speed in deep water.

(a) Aerial view of wave refraction along a straight shoreline.

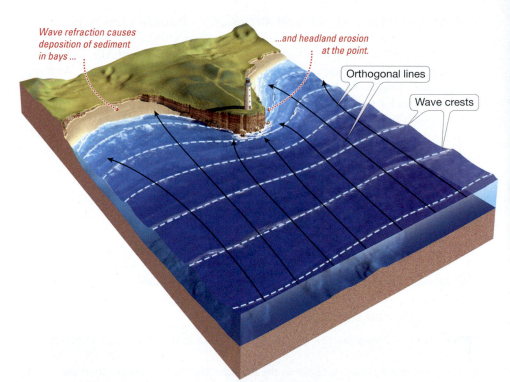

Wave refraction causes deposition of sediment in bays ...

...and headland erosion at the point.

Orthogonal lines

Wave crests

(b) Perspective view of wave refraction along an irregular shoreline.

Original wave angle

Waves refract (bend) around point

(c) Photo of wave refraction at Rincon Point, California (looking west).

SmartFigure 8.21 Wave refraction. Wave refraction is the bending of waves, as shown here along **(a)** a straight shoreline and **(b and c)** an irregular shoreline.
https://goo.gl/mJrwgb

Web Animation
Wave Motion and Wave Refraction
http://goo.gl/1wEScd

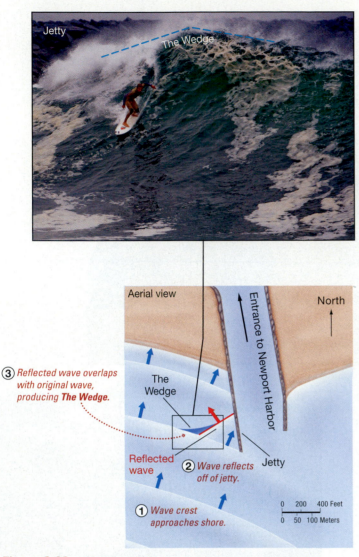

Figure 8.22 **Wave reflection and constructive interference at *The Wedge*, Newport Harbor, California.** As waves approach the shore, ① some of the wave energy is reflected off the long jetty at the entrance to the harbor. ② The reflected wave overlaps and constructively interferes with the original wave, ③ resulting in a wedge-shaped wave (*dark blue triangle*) that may reach heights exceeding 8 meters (26 feet). Photo (*inset*) shows a surfer riding *The Wedge*; note the jetty in the background.

Web Video

Wave Dispersion, Wave Reflection, and Wave Interference
https://goo.gl/s6jfkv

RECAP

Wave refraction is the bending of waves caused when waves slow in shallow water; wave reflection is the bouncing back of wave energy caused when waves strike a hard barrier.

Wave Reflection

Not all wave energy is expended as waves rush onto the shore. A vertical barrier, such as a seawall or a rock ledge, can reflect waves back into the ocean with little loss of energy—a process called **wave reflection** (*reflecten* = to bend back), which is similar to how a mirror reflects (bounces) back light. If the incoming wave strikes the barrier at a right (90-degree) angle, for example, the wave energy is reflected back parallel to the incoming wave, often interfering with the next incoming wave and creating unusual waveforms. More commonly, waves approach the shore at an angle, causing wave energy to be reflected at an angle equal to the angle at which the wave approached the barrier.

THE WEDGE: A CASE STUDY OF WAVE REFLECTION AND CONSTRUCTIVE INTERFERENCE An outstanding example of wave reflection and constructive interference occurs in an area called *The Wedge*, which develops west of the jetty that protects the harbor entrance at Newport Harbor, California (**Figure 8.22**). The jetty is a solid human-made object that extends into the ocean 400 meters (1300 feet) and has a near-vertical side facing the waves. As incoming waves strike the vertical side of the jetty at an angle, they are reflected at an equivalent angle. Because the original waves and the reflected waves have the same wavelength, a constructive interference pattern develops, creating plunging breakers that may exceed 8 meters (26 feet) in height (Figure 8.22, *inset*). Too dangerous for board surfers, these waves present a fierce challenge to the most experienced body surfers. The Wedge has crippled and even killed many who have come to challenge it.

STANDING WAVES, NODES, AND ANTINODES **Standing waves** (also called *stationary waves*) can be produced when waves are reflected at right angles to a barrier. Standing waves are the sum of two waves with the same wavelength moving in opposite directions, resulting in no net movement. Although the water particles continue to move vertically and horizontally, there is none of the circular motion that is characteristic of a progressive wave.

Figure 8.23 shows the movement of water during the wave cycle of a standing wave. Lines along which there is no vertical movement are called *nodes* (*nodus* = knot), or *nodal lines*. *Antinodes*, which are crests that alternately become troughs, are the points of greatest vertical movement within a standing wave.

When water sloshes back and forth in a basin, the maximum vertical displacements are at the antinodes. When water moves from crest to trough at

STUDENTS SOMETIMES ASK . . .

Why is surfing so much better along the West Coast of the United States than along the East Coast?

There are three main reasons the U.S. West Coast has better surfing conditions:

1. The waves are generally bigger in the Pacific. The Pacific is larger than the Atlantic, so the fetch is larger, allowing bigger waves to develop in the Pacific.
2. The beach slopes are generally steeper along the West Coast. Along the East Coast, the gentle slopes often create spilling breakers, which are not as favorable for surfing. Steeper beach slopes along the West Coast cause plunging breakers, which are better for surfing.
3. The wind is more favorable. Most of the United States is influenced by the prevailing westerlies, which blow toward shore and enhance waves along the West Coast. Along the East Coast, the wind generally blows away from shore.

tsunami that have struck since 1990—with the majority occurring along the Pacific Ring of Fire.

THE ERUPTION OF KRAKATAU (1883) One of the most destructive historical tsunami ever generated came from the eruption of the volcanic island of Krakatau[7] on August 27, 1883. Approximately the size of a small Hawaiian island in what is now Indonesia, Krakatau exploded with the greatest release of energy from Earth's interior ever recorded in historic times. The island, which stood 450 meters (1500 feet) above sea level, was nearly obliterated. The sound of the explosion was heard throughout the Indian Ocean, up to 4800 kilometers (3000 miles) away, and remains the loudest noise on human record. Dust from the explosion ascended into the atmosphere and circled Earth on high-altitude winds, producing brilliant red sunsets worldwide for nearly a year.

Not many people were killed by the outright explosion of the volcano because the island was uninhabited. However, the displacement of water from the energy released during the explosion was enormous, creating a tsunami that exceeded 35 meters (116 feet)—as high as a 12-story building. It devastated the coastal region of the Sunda Strait between the nearby islands of Sumatra and Java, drowning more than 1000 villages and taking more than 36,000 lives. The energy carried by this wave reached every ocean basin and was even detected by tide-recording stations as far away as London and San Francisco.

Like most of the other approximately 130 active volcanoes in Indonesia, Krakatau was formed along the Sunda Arc, a 3000-kilometer (1900-mile) curving chain of volcanoes associated with the subduction of the Australian Plate beneath the Eurasian Plate. Where these two sections of Earth's crust meet, earthquakes and volcanic eruptions are common.

THE SCOTCH CAP, ALASKA/HILO, HAWAII TSUNAMI (1946) A strong and damaging tsunami hit the Hawaiian Islands on April 1, 1946, with north-facing shores including the port city of Hilo receiving the majority of damage. The tsunami was from a magnitude $M_w = 7.3$ earthquake in the Aleutian Trench off the island of Unimak, Alaska, more than 3000 kilometers (1850 miles) away. From that direction, the offshore bathymetry in Hilo Bay focused the tsunami's energy directly toward town, so the city of Hilo experienced tsunami surges of tremendous heights. In this case, the tsunami expressed itself as a strong recession followed by a surge of water nearly 17 meters (55 feet) above normal high tide, causing more than $25 million in damage and killing 159 people. Remarkably, it stands as Hawaii's worst natural disaster (Figure 8.26).

Closer to the source of the earthquake, the tsunami was considerably larger. The tsunami struck Scotch Cap, Alaska, on Unimak Island, where a two-story reinforced concrete lighthouse stood 14 meters (46 feet) above sea level at its base. The lighthouse was destroyed by a wave that is estimated to have reached 36 meters (118 feet), killing all five people inside the lighthouse at the time. Vehicles on a nearby mesa 31 meters (103 feet) above water level were also moved by the onrush of water.

PAPUA NEW GUINEA (1998) In July 1998, off the north coast of Papua New Guinea in the western part of the Pacific Ring of Fire, a magnitude $M_w = 7.1$ earthquake was followed shortly thereafter by a 15-meter (49-foot) tsunami, which was up to five times larger than expected for a tsunami created by an earthquake that size. The tsunami completely overtopped a heavily populated low-lying sand bar, destroying three entire villages, and resulting in at least 2200 deaths. Researchers who mapped the sea floor after the tsunami discovered the remains of a huge underwater landslide, which was apparently triggered by the shaking and caused the unusually large tsunami.

STUDENTS SOMETIMES ASK ...

If a tsunami warning is issued, what is the best thing to do?

The *smartest* thing to do is to stay out of coastal areas, but people often want to observe the tsunami first-hand. For instance, after the February 2010 magnitude $M_w = 8.8$ Chilean earthquake, tsunami warnings were issued in 53 countries. Across the Pacific Ocean in Australia, widespread media broadcasts warned people of the impending danger, yet live television coverage showed hundreds of people standing on low-lying beaches waiting for the tsunami to arrive. Worse, some of them were deliberately swimming into the incoming tsunami, despite the efforts of volunteer lifeguards to prevent this.

If you *must* go to the beach to observe a tsunami, expect crowds, road closings, and general mayhem. It would be a good idea to stay at least 30 meters (100 feet) above sea level. If you happen to be at a remote beach and notice the water suddenly withdrawing, evacuate immediately to higher ground (**Figure 8.28**). And, if you happen to be at a beach when an earthquake occurs and shakes the ground so hard that you can't stand up, then *RUN*—don't walk—for high ground as soon as you *can* stand up!

After the first surge of the tsunami, stay out of low-lying coastal areas for several hours because several more surges (and withdrawals) can be expected. There are many documented cases of curious people being killed when they were caught in the third or fourth (. . . or ninth) surge of a tsunami.

Figure 8.28 Tsunami warning sign. This tsunami warning sign in coastal Oregon advises residents to evacuate low-lying areas during a tsunami.

[7]The island Krakatau (which is *west* of Java) is also called Krakatoa.

(a) Before: January 10, 2003

North

Town mosque

Lhoknga
INDIAN
OCEAN
Earthquake
Epicenter

0.6 Mi
1 Km

Town mosque

(b) After: December 29, 2004

Figure 8.29 Satellite views of tsunami destruction in Indonesia. Paired satellite images of Lhoknga, in the province of Aceh, west coast of Sumatra, Indonesia. **(a)** Before the tsunami, on January 10, 2003 and **(b)** the same area on December 29, 2004, three days after the tsunami that inundated the city with a 15-meter (50-foot) surge of seawater. The town mosque, indicated in both images, was one of the few buildings that remained standing.

INDIAN OCEAN (2004) Although most tsunami occur in the Pacific Ocean, they do sometimes occur in other ocean basins. On December 26, 2004, an enormous earthquake struck in a subduction zone off the west coast of Sumatra in Indonesia on the shores of the Indian Ocean. Known as the Sumatra–Andaman Earthquake, it was the second biggest earthquake recorded during the past century and the largest to be recorded by modern seismograph equipment. The earthquake was so large, in fact, that it triggered small earthquakes as far away as Alaska, altered Earth's rotation, and by some calculations may have even changed Earth's gravity field. Initially, it was deemed a magnitude $M_w = 9.0$, but it was subsequently upgraded to magnitude $M_w = 9.2$. The earthquake occurred about 30 kilometers (19 miles) beneath the sea floor, near the Sunda Trench, where the Indian Plate is being subducted beneath the Eurasian Plate. Moreover, about 1200 kilometers (750 miles) of sea floor was ruptured along the interface of these two tectonic plates, thrusting sea floor upward and generating about 10 meters (33 feet) of vertical displacement. This abrupt vertical movement of the sea floor is what generated the deadliest tsunami in history.

Once generated, the tsunami spread out at jetliner speeds across the Indian Ocean. Only 15 minutes after the earthquake, the tsunami hit the shores of Sumatra with a series of alternating rapid withdrawals and strong surges up to 35 meters (115 feet) high. Many coastal villages were completely washed away (**Figure 8.29**), causing several hundred thousand deaths. Sites farther from the quake zone experienced smaller but nevertheless deadly waves. Particularly affected were Thailand (see Figure 8.25b), which was struck by the tsunami about 75 minutes after the quake occurred, and Sri Lanka and India, which were pounded by devastating waves about three hours after the earthquake. After seven hours and at a distance of more than 5000 kilometers (3000 miles), the tsunami hit the east coast of Africa, where it still had enough power to kill more than a dozen people. Although much smaller, the tsunami was also detected in the Atlantic, Pacific, and Arctic Oceans.

In a remarkable coincidence, the Jason-1 satellite happened to be passing over the Indian Ocean two hours after the tsunami originated (**Figure 8.30**). The satellite's radar altimeter, which is designed to accurately measure the elevation of the ocean surface (see Diving Deeper 3.1), was able to detect the crests and troughs of the tsunami as it radiated out across the Indian Ocean with a wavelength of about 500 kilometers (300 miles). Although this sighting occurred about an hour before the first waves struck Sri Lanka and India, the satellite data couldn't have been used to warn tsunami victims because scientists needed several hours to analyze the information. However, these satellite data are particularly valuable because they have allowed scientists to check the accuracy of open-ocean tsunami travel models, which are based on seismic data, bathymetric information, and coastal tide gauges.

In all, between 230,000 and 300,000 people in 11 countries were killed by the tsunami, ranking this tragedy as the most deadly tsunami in recorded history. The tsunami also caused billions of dollars of damage throughout the region and

DIVING DEEPER 8.1

WAVES OF DESTRUCTION: THE 2011 JAPANESE TSUNAMI

On March 11, 2011, the nation of Japan and seismologists around the world received a terrible surprise: A huge earthquake, significantly stronger than people had anticipated or prepared for in the region, struck off the northeastern shore of Japan. The great 2011 $M_w = 9.0$ Tohoku Earthquake—named for the region it struck—occurred along the Japan Trench where the Pacific Plate is being subducted underneath Japan (**Figure 8A**). It shifted the coast of Japan up to 8 meters (26 feet) eastward and also uplifted the sea floor by as much as 5 meters (16.4 feet) over an area comparable in size to the state of Connecticut; this sudden vertical movement of the sea floor imparted energy into the Pacific Ocean and generated one of the largest and best-studied tsunami in history.

Although Japan prides itself in disaster preparedness, many residents did not get accurate tsunami warnings, and many chose to stay in dangerous locations in part because they misunderstood the risks. Based on the initial estimates of the earthquake, the Japan Meteorological Agency (JMA) issued a tsunami warning within three minutes from the start of the shaking. The agency's warning predicted a tsunami surge of only 3 to 6 meters (10 to 20 feet), and unfortunately, many people did not take immediate action because of their belief that Japan's extensive network of tsunami walls would protect them. In addition, some residents ignored the warning because of previous false alarms. Within the next few minutes, JMA reissued its warning for a 10-meter (33-foot) tsunami, but many people did not receive the updated warning because of the destruction of the regional power supply system.

About 20 minutes after the earthquake began, the tsunami swept onto the coastline of northeastern Japan, where most fishing villages and towns were built at the ends of long, narrow bays because these bays offer protection from wind waves. However, these long narrow bays amplified the height of the tsunami. The initial surge of the tsunami reached heights of about 15 meters (49 feet) and easily overtopped harbor-protecting tsunami walls and coastal margins (**Figure 8B**), penetrating as far as 10 kilometers (6 miles) inland. In one location, the tsunami reached a record height of 40 meters (131 feet) because of amplification by offshore topography.

The Pacific Tsunami Warning Center, with its network of buoys that had been greatly expanded since the 2004 Indian Ocean tsunami, sent out tsunami warnings across the Pacific and accurately predicted its arrival, allowing coastal communities to prepare. The tsunami struck Hawaii seven hours after the earthquake—and the coast of California three hours later still—with a surge up to 2 meters (7 feet) that damaged harbors, marinas, and coastal resorts. Minor damage occurred in Peru and Chile when the wave struck their coastlines 20 to 22 hours after the event. The Philippines, Indonesia, and New Guinea were also affected by the tsunami.

In Japan, shaking from the earthquake caused few fatalities and did minimal damage due largely to Japan's strict building codes, thorough earthquake preparation, and advance warning systems. However, the tsunami destroyed many small towns and villages, killed 19,508 people, and initially displaced nearly half a million people; nearly two years later, thousands were still living in temporary shelters such as high school gymnasiums. The tsunami also disrupted the power supply needed to maintain water circulation for cooling the reactors at the Fukushima Daiichi nuclear power plant. Three reactors exploded, releasing radioactivity that continues to plague central Japan. Radioactivity was also released into the ocean, and studies are now under way to determine its effect on marine life.

The destructive power of the tsunami also generated a huge amount of floating debris, including boats, cars, building parts, household items, and even entire houses. Japanese authorities estimate that 5 million metric tons (11 billion pounds) of debris was swept out to sea by the tsunami, with about 70% sinking to the sea floor and 1.5 million metric tons (3 billion pounds) left floating. The tsunami debris is being monitored as it slowly spreads out and moves across the Pacific Ocean, transported by currents in the North Pacific Gyre. Although some of the floating debris will eventually sink, disperse, or biodegrade, the wreckage may impact Hawaiian coral reefs. Along the U.S. West Coast, a derelict Japanese fishing vessel and other tsunami debris has already arrived, and more floating debris is expected in the coming years.

Figure 8B The 2011 tsunami surges over a protective seawall in Miyako, northeastern Japan.

GIVE IT SOME THOUGHT

1. Why did many Japanese residents not heed the tsunami warnings that were issued immediately after the 2011 Tohoku Earthquake?

2. How much floating debris was released into the Pacific Ocean by the tsunami? Where did this debris end up?

Figure 8A The great 2011 Tohoku Earthquake.
Location of the $M_w = 9.0$ Tohoku Earthquake, which was the most powerful earthquake ever to have hit Japan. It occurred along the Japan Trench, uplifted the sea floor, and created a large, devastating tsunami.

left millions homeless. Why was there such a large loss of life during the tsunami? One factor was the lack of a warning system in the Indian Ocean, like the network of buoys and deep-sea instruments that monitors earthquake and wave activity in the Pacific Ocean, where most tsunami have historically occurred. Another is the lack of an emergency response system to warn high-population-density coastal communities and beachgoers. Still another was the lack of good public awareness in recognizing the signs of an impending tsunami, such as when a rapid withdrawal of water is observed at the shore, which is caused by the trough of the tsunami arriving first and indicates that an equally strong surge of water will soon follow.

Studies of the effect of the tsunami on different coastlines suggest that those areas lacking protective coral reefs or mangroves received stronger surges. In many cases, the shape of the coastline as well as the geometry of the offshore sea floor affected the height of the tsunami and the amount of coastal destruction. Sediment cores reveal that several other large tsunami have occurred in the region during the past 1000 years.

On July 17, 2006, another strong earthquake—this time an $M_w = 7.7$ earthquake associated with faulting in the Java Trench off the southern coast of Java, Indonesia—triggered a 3-meter (10-foot) tsunami that killed 668 people, injured at least 600, and destroyed many coastal structures. This destruction near the area that experienced even greater damage from the 2004 tsunami emphasized the fact that the region still lacked a comprehensive tsunami warning system. In 2010, however, the Indian Ocean Tsunami Warning and Mitigation System became fully operational, with its network of deep-ocean pressure sensors, buoys, land seismographs, tidal gauges, data centers, and communications upgrades. The system will continue to be upgraded with more detection buoys in the coming years. With this new warning system in place and enhanced public education about what to do in the event of a tsunami, all countries bordering the Indian Ocean will be much better prepared for the destructive power of future tsunami.

JAPAN (2011) On March 11, 2011, an $M_w = 9.0$ earthquake—the fourth largest ever to be recorded—occurred in the Japan Trench offshore of northeastern Japan. Known as the Tohoku Earthquake, it uplifted the sea floor by several meters, moved the island of Japan itself 4 meters (13 feet), and created a large, devastating tsunami that was felt throughout the Pacific Ocean (**Diving Deeper 8.1**). In all, damages have been estimated at $235 billion, making it the most expensive natural disaster in world history.

Tsunami Warning System

In response to the tsunami that struck Hawaii in 1946, a tsunami warning system was established throughout the Pacific Ocean. It led to what is now the **Pacific Tsunami Warning Center (PTWC)**, which coordinates information from 25 Pacific Rim countries and is headquartered in Ewa Beach (near Honolulu), Hawaii. The tsunami warning system uses seismic waves—some of which travel through Earth at speeds 15 times faster than tsunami—to forecast destructive tsunami. In addition, oceanographers have recently established a network of sensitive pressure sensors on the deep-ocean floor of the Pacific. The program, called **Deep-ocean Assessment and Reporting of Tsunamis (DART)**, utilizes sea floor sensors that are capable of picking up the small yet distinctive pressure pulse from a tsunami passing above. The pressure sensors relay information to a buoy at the surface that transmits the data via satellite,

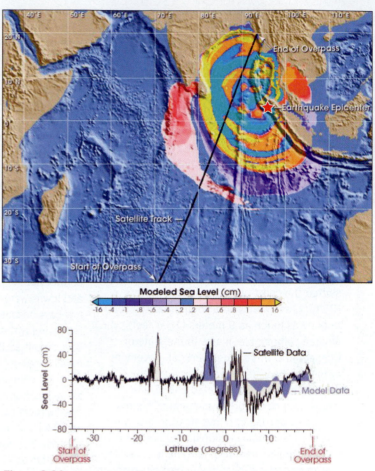

Figure 8.30 Jason-1 satellite detects the Indian Ocean Tsunami. The Indian Ocean Tsunami was initiated by a large earthquake offshore Sumatra on December 26, 2004 (*red star*). By a fortuitous circumstance, the Jason-1 satellite passed over the Indian Ocean (*black line*) two hours after the tsunami was generated. Its radar altimeter detected the crests and troughs of the tsunami (*colors*), which showed a wave height of about 1 meter (3.3 feet) in the open ocean. The graph of the satellite's overpass (*below*) shows the difference between the measured sea level from satellite data (*black line*) and the modeled wave height (*blue curve*).

STUDENTS SOMETIMES ASK …

Where and when will the next major tsunami hit?

Although most tsunami occur in the Pacific Ocean, a tsunami can affect any coastal region worldwide. Tsunami prediction is still in its infancy, yet seismologists who examine the history of large earthquakes (*paleoseismologists*) can produce forecasts of future tsunami based on evidence such as sediment records that indicate the size and frequency of past tsunami. Using this technique, paleoseismologists have issued forecasts for large, damaging tsunami in two regions: (1) In the Indian Ocean, where in 2004 the deadliest tsunami in history killed 300,000 people, a similar tsunami has been forecast to occur sometime within the next 30 years and (2) off the U.S. Pacific Northwest coast, where there is a subduction zone just offshore and a large tsunami generated by faulting in the offshore trench occurred in 1700; another tsunami is expected in the next 250 to 500 years. For both these locations, it's only a matter of time.

(a) Before: January 10, 2003

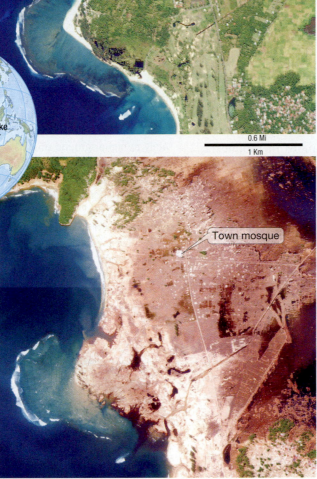

North

Town mosque

Lhoknga
INDIAN OCEAN
Earthquake Epicenter

0.6 Mi
1 Km

Town mosque

(b) After: December 29, 2004

Figure 8.29 Satellite views of tsunami destruction in Indonesia. Paired satellite images of Lhoknga, in the province of Aceh, west coast of Sumatra, Indonesia. **(a)** Before the tsunami, on January 10, 2003 and **(b)** the same area on December 29, 2004, three days after the tsunami that inundated the city with a 15-meter (50-foot) surge of seawater. The town mosque, indicated in both images, was one of the few buildings that remained standing.

INDIAN OCEAN (2004) Although most tsunami occur in the Pacific Ocean, they do sometimes occur in other ocean basins. On December 26, 2004, an enormous earthquake struck in a subduction zone off the west coast of Sumatra in Indonesia on the shores of the Indian Ocean. Known as the Sumatra–Andaman Earthquake, it was the second biggest earthquake recorded during the past century and the largest to be recorded by modern seismograph equipment. The earthquake was so large, in fact, that it triggered small earthquakes as far away as Alaska, altered Earth's rotation, and by some calculations may have even changed Earth's gravity field. Initially, it was deemed a magnitude $M_w = 9.0$, but it was subsequently upgraded to magnitude $M_w = 9.2$. The earthquake occurred about 30 kilometers (19 miles) beneath the sea floor, near the Sunda Trench, where the Indian Plate is being subducted beneath the Eurasian Plate. Moreover, about 1200 kilometers (750 miles) of sea floor was ruptured along the interface of these two tectonic plates, thrusting sea floor upward and generating about 10 meters (33 feet) of vertical displacement. This abrupt vertical movement of the sea floor is what generated the deadliest tsunami in history.

Once generated, the tsunami spread out at jetliner speeds across the Indian Ocean. Only 15 minutes after the earthquake, the tsunami hit the shores of Sumatra with a series of alternating rapid withdrawals and strong surges up to 35 meters (115 feet) high. Many coastal villages were completely washed away (**Figure 8.29**), causing several hundred thousand deaths. Sites farther from the quake zone experienced smaller but nevertheless deadly waves. Particularly affected were Thailand (see Figure 8.25b), which was struck by the tsunami about 75 minutes after the quake occurred, and Sri Lanka and India, which were pounded by devastating waves about three hours after the earthquake. After seven hours and at a distance of more than 5000 kilometers (3000 miles), the tsunami hit the east coast of Africa, where it still had enough power to kill more than a dozen people. Although much smaller, the tsunami was also detected in the Atlantic, Pacific, and Arctic Oceans.

In a remarkable coincidence, the Jason-1 satellite happened to be passing over the Indian Ocean two hours after the tsunami originated (**Figure 8.30**). The satellite's radar altimeter, which is designed to accurately measure the elevation of the ocean surface (see Diving Deeper 3.1), was able to detect the crests and troughs of the tsunami as it radiated out across the Indian Ocean with a wavelength of about 500 kilometers (300 miles). Although this sighting occurred about an hour before the first waves struck Sri Lanka and India, the satellite data couldn't have been used to warn tsunami victims because scientists needed several hours to analyze the information. However, these satellite data are particularly valuable because they have allowed scientists to check the accuracy of open-ocean tsunami travel models, which are based on seismic data, bathymetric information, and coastal tide gauges.

In all, between 230,000 and 300,000 people in 11 countries were killed by the tsunami, ranking this tragedy as the most deadly tsunami in recorded history. The tsunami also caused billions of dollars of damage throughout the region and

tsunami that have struck since 1990—with the majority occurring along the Pacific Ring of Fire.

THE ERUPTION OF KRAKATAU (1883)

One of the most destructive historical tsunami ever generated came from the eruption of the volcanic island of Krakatau[7] on August 27, 1883. Approximately the size of a small Hawaiian island in what is now Indonesia, Krakatau exploded with the greatest release of energy from Earth's interior ever recorded in historic times. The island, which stood 450 meters (1500 feet) above sea level, was nearly obliterated. The sound of the explosion was heard throughout the Indian Ocean, up to 4800 kilometers (3000 miles) away, and remains the loudest noise on human record. Dust from the explosion ascended into the atmosphere and circled Earth on high-altitude winds, producing brilliant red sunsets worldwide for nearly a year.

Not many people were killed by the outright explosion of the volcano because the island was uninhabited. However, the displacement of water from the energy released during the explosion was enormous, creating a tsunami that exceeded 35 meters (116 feet)—as high as a 12-story building. It devastated the coastal region of the Sunda Strait between the nearby islands of Sumatra and Java, drowning more than 1000 villages and taking more than 36,000 lives. The energy carried by this wave reached every ocean basin and was even detected by tide-recording stations as far away as London and San Francisco.

Like most of the other approximately 130 active volcanoes in Indonesia, Krakatau was formed along the Sunda Arc, a 3000-kilometer (1900-mile) curving chain of volcanoes associated with the subduction of the Australian Plate beneath the Eurasian Plate. Where these two sections of Earth's crust meet, earthquakes and volcanic eruptions are common.

THE SCOTCH CAP, ALASKA/HILO, HAWAII TSUNAMI (1946)

A strong and damaging tsunami hit the Hawaiian Islands on April 1, 1946, with north-facing shores including the port city of Hilo receiving the majority of damage. The tsunami was from a magnitude $M_w = 7.3$ earthquake in the Aleutian Trench off the island of Unimak, Alaska, more than 3000 kilometers (1850 miles) away. From that direction, the offshore bathymetry in Hilo Bay focused the tsunami's energy directly toward town, so the city of Hilo experienced tsunami surges of tremendous heights. In this case, the tsunami expressed itself as a strong recession followed by a surge of water nearly 17 meters (55 feet) above normal high tide, causing more than $25 million in damage and killing 159 people. Remarkably, it stands as Hawaii's worst natural disaster (Figure 8.26).

Closer to the source of the earthquake, the tsunami was considerably larger. The tsunami struck Scotch Cap, Alaska, on Unimak Island, where a two-story reinforced concrete lighthouse stood 14 meters (46 feet) above sea level at its base. The lighthouse was destroyed by a wave that is estimated to have reached 36 meters (118 feet), killing all five people inside the lighthouse at the time. Vehicles on a nearby mesa 31 meters (103 feet) above water level were also moved by the onrush of water.

PAPUA NEW GUINEA (1998)

In July 1998, off the north coast of Papua New Guinea in the western part of the Pacific Ring of Fire, a magnitude $M_w = 7.1$ earthquake was followed shortly thereafter by a 15-meter (49-foot) tsunami, which was up to five times larger than expected for a tsunami created by an earthquake that size. The tsunami completely overtopped a heavily populated low-lying sand bar, destroying three entire villages, and resulting in at least 2200 deaths. Researchers who mapped the sea floor after the tsunami discovered the remains of a huge underwater landslide, which was apparently triggered by the shaking and caused the unusually large tsunami.

[7]The island Krakatau (which is *west* of Java) is also called Krakatoa.

STUDENTS SOMETIMES ASK ...

If a tsunami warning is issued, what is the best thing to do?

The *smartest* thing to do is to stay out of coastal areas, but people often want to observe the tsunami first-hand. For instance, after the February 2010 magnitude $M_w = 8.8$ Chilean earthquake, tsunami warnings were issued in 53 countries. Across the Pacific Ocean in Australia, widespread media broadcasts warned people of the impending danger, yet live television coverage showed hundreds of people standing on low-lying beaches waiting for the tsunami to arrive. Worse, some of them were deliberately swimming into the incoming tsunami, despite the efforts of volunteer lifeguards to prevent this.

If you *must* go to the beach to observe a tsunami, expect crowds, road closings, and general mayhem. It would be a good idea to stay at least 30 meters (100 feet) above sea level. If you happen to be at a remote beach and notice the water suddenly withdrawing, evacuate immediately to higher ground (**Figure 8.28**). And, if you happen to be at a beach when an earthquake occurs and shakes the ground so hard that you can't stand up, then *RUN*—don't walk—for high ground as soon as you *can* stand up!

After the first surge of the tsunami, stay out of low-lying coastal areas for several hours because several more surges (and withdrawals) can be expected. There are many documented cases of curious people being killed when they were caught in the third or fourth (. . . or ninth) surge of a tsunami.

Figure 8.28 Tsunami warning sign. This tsunami warning sign in coastal Oregon advises residents to evacuate low-lying areas during a tsunami.

an antinode, the displaced volume of water has to move horizontally to raise the water level of the adjacent antinode from trough to crest. As a result, there is no vertical motion at the node; instead, there is only horizontal motion. At the antinodes, however, the movement of water particles is entirely vertical.

We'll consider standing waves further when tidal phenomena are discussed in Chapter 9, "Tides." Under certain conditions, the development of standing waves significantly affects the tidal character in coastal regions.

> ### CONCEPT CHECK 8.4 | Explain how waves change in the surf zone.
>
> **1** What is the 1:7 ratio? What happens to a wave when the 1:7 ratio is exceeded?
>
> **2** Describe the physical changes that occur to a wave's wave speed (S), wavelength (L), height (H), and wave steepness (H/L) as the wave moves across progressively shallower water to break on the shore.
>
> **3** Describe the three different types of breakers and indicate the slope of the beach that produces each of the three types. How is the energy of the wave distributed differently within the surf zone by the three types of breakers?
>
> **4** Using examples, explain how wave refraction is different from wave reflection.
>
> **5** Using orthogonal lines, illustrate how wave energy is distributed along a shoreline with headlands and bays. Identify areas of high- and low-energy release.

8.5 How Are Tsunami Created?

The Japanese term for the large, sometimes destructive waves that occasionally roll into their harbors is **tsunami** (*tsu* = harbor, *nami* = wave[s]). Tsunami originate from sudden changes in the topography of the sea floor caused by such events as slippage along underwater faults, underwater landslides such as those that form turbidity currents or from the collapse of large oceanic volcanoes, and underwater volcanic eruptions. Many people mistakenly call them "tidal waves," but tsunami are unrelated to the tides. The mechanisms that trigger tsunami are typically seismic events, so tsunami are more accurately called *seismic sea waves*.

The majority of tsunami are caused by vertical *fault movement*. Underwater fault movement displaces Earth's crust, generates earthquakes, and, if it ruptures the sea floor, produces a sudden change in water level at the ocean surface (**Figure 8.24**). Faults that produce *vertical* displacements (the uplift or downdropping of ocean floor) change the volume of the ocean basin, which affects the entire water column and generates tsunami. Conversely, faults that produce *horizontal* displacements (such as the lateral movement associated with transform faulting) generally do not generate tsunami because the side-to-side movement of these faults does not change the volume of the ocean basin. Much less common events, such as underwater avalanches triggered by shaking or underwater volcanic eruptions—which create the largest waves—also produce tsunami. In addition, large objects that splash into the ocean, such as above-water coastal landslides (see MasteringOceanography **Web Diving Deeper 8.1**) or meteorite impacts, produce **splash waves**, which are a type of tsunami.

Because tsunami generally originate from energy imparted to the ocean basin, they affect the entire water column of the ocean, which

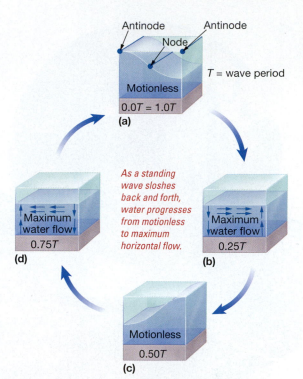

Figure 8.23 Sequence of motion in a standing wave. In a standing wave, water is motionless when antinodes reach maximum displacement (**a** and **c**). Water movement is at a maximum (*blue arrows*) when the water level is horizontal (**b** and **d**). Movement is vertical beneath the antinodes, and maximum horizontal movement occurs beneath the node.

... while wind-generated waves exist only in surface waters.

A triggering event such as a sea floor earthquake imparts energy throughout the water column and generates a tsunami ...

Figure 8.24 How a tsunami differs from wind-generated waves.

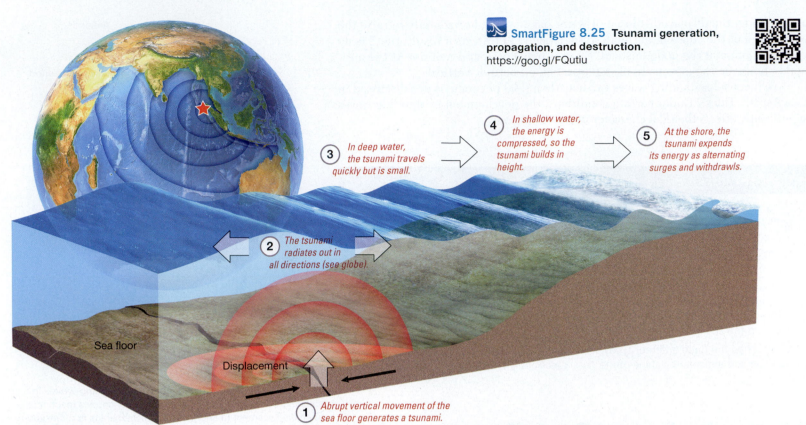

SmartFigure 8.25 **Tsunami generation, propagation, and destruction.**
https://goo.gl/FQutiu

③ In deep water, the tsunami travels quickly but is small.

④ In shallow water, the energy is compressed, so the tsunami builds in height.

⑤ At the shore, the tsunami expends its energy as alternating surges and withdrawls.

② The tsunami radiates out in all directions (see globe).

Sea floor

Displacement

① Abrupt vertical movement of the sea floor generates a tsunami.

(a) How a tsunami is generated, propagated, and surges to extreme heights at the shore.

(b) Sequence of photos of the 2004 Indian Ocean Tsunami inundating the Chedi Resort in Phuket, Thailand, on December 26, 2004.

Web Animation

Tsunami
http://goo.gl/H6j8jK

on average is 4 kilometers (2.5 miles) deep and in the deepest trenches the depth can exceed 11 kilometers (7 miles). Wind-generated surface waves, on the other hand, affect only the topmost waters of the ocean (Figure 8.24) and do not contain nearly as much energy as tsunami. When tsunami come into shallow water, the energy from the entire oceanic water column is compressed into a shallow region and tsunami expend their energy at the shore as a series of strong alternating surges and withdrawals of seawater that can cause considerable damage (Figure 8.25).

During the 20th century, 498 measurable tsunami occurred worldwide, 66 of them resulting in fatalities. The source events were earthquakes (86%), volcanic activity (5%), landslides (4%), and combinations of those processes (5%). Tsunami generated by meteorite or asteroid impacts occur much less frequently.

The wavelength of a typical tsunami exceeds 200 kilometers (125 miles), so it is a shallow-water wave everywhere in the ocean.[6] Because tsunami are shallow-water waves, their speed is determined only by water depth. In the open ocean, tsunami move at well over 700 kilometers (435 miles) per hour—they can easily keep pace with a jet airplane—and have heights of only about 1 meter (3.3 feet) or less. Even though they are fast, tsunami are small in the open ocean and pass unnoticed in deep water until they reach shore, where they slow in the shallow water and undergo physical changes (just as wind-generated waves do) that cause tsunami to increase greatly in wave height.

Coastal Effects

Contrary to popular belief, a tsunami does not form a huge breaking wave at the shoreline. Instead, it is a strong flood or surge of water that causes the ocean to advance (or, in certain cases, retreat) dramatically. In fact, a tsunami resembles a sudden, *extremely* high tide, which is why they are misnamed "tidal waves." It may take many minutes for the crest of the tsunami to arrive on shore, during which time sea level can rise up to 40 meters (131 feet) above normal, with normal waves superimposed on top of the higher sea level. The strong surge of water can rush into low-lying areas and cause extensive destruction (Figure 8.25b), including loss of life.

As the trough of the tsunami approaches the shore, the water rapidly drains off the land. In coastal areas, it looks like a sudden and *extremely* low tide, where sea level is many meters lower than even the lowest low tide. Because tsunami are typically a series of waves, there is often an alternating series of dramatic surges and withdrawals of water that are widely separated in time. The first surge is only rarely the largest; instead, the third, fourth—or even seventh—surge may be the largest and can occur several hours later.

Depending on the geometry of sea floor motion that creates a tsunami, the trough can sometimes arrive at a coast first. For example, on the side of a fault where the sea floor drops down, the first part of the wave to propagate outward will be the trough, followed by the crest. Conversely, on the side of the fault that moves upward, the crest leads the trough. At a coast where the trough arrives first, the water drains out and exposes parts of the lowermost shoreline that are rarely seen. For people at the shoreline, the temptation is to explore these newly exposed areas and catch stranded organisms. Within a few minutes, however, a strong surge of water (the crest of the tsunami) is due to arrive.

The alternating surges and retreats of water by tsunami can severely damage coastal structures and can injure and kill people as well. The speed of the advance—up to 4 meters (13 feet) per second—is faster than any human can run. Those who are caught in a tsunami are often drowned or crushed by floating debris (Figure 8.26).

Figure 8.26 Tsunami damage in Hilo, Hawaii.
Flattened parking meters in Hilo, Hawaii, caused by the 1946 tsunami that resulted in more than $25 million in damage and 159 deaths.

STUDENTS SOMETIMES ASK ...

What is the record height of a tsunami?

Japan, which is in close proximity to several subduction zones and endures more tsunami than any other place on Earth (followed by Chile and Hawaii), holds the record. The largest documented high water from a tsunami occurred in the Ryukyu Islands of southern Japan in 1971, when normal sea level was raised by an astonishing 85 meters (278 feet). In low-lying coastal areas, such an enormous vertical rise can send seawater many kilometers inland, causing flooding and widespread damage.

[6]Recall that the depth of the wave base is equal to one-half a wave's wavelength. Thus, tsunami can typically be felt to depths of 100 kilometers (62 miles), which is deeper than even the deepest ocean trenches.

Figure 8.27 **Significant tsunami since 1990.**
Global locations and corresponding table showing significant tsunami (those with large tsunami height and/or responsible for a large number of fatalities) since 1990. Tsunami have claimed more than 300,000 lives worldwide since 1990. These killer waves are most often generated by earthquakes along colliding tectonic plates of the Pacific Rim, although the most deadly tsunami in history was the 2004 Indian Ocean Tsunami (*number 8*). Locations of ocean trenches are shown by dark red lines; Pacific Ring of Fire is shown by red shading.

Some Examples of Historic and Recent Tsunami

Although many small tsunami are created each year, most go unnoticed. On average, about 50 noticeable tsunami occur every decade, with a large tsunami occurring somewhere in the world every 2 to 3 years and an extremely large and damaging one occurring every 15 to 20 years. Remarkably, two of the largest and most deadly tsunami have occurred within the past decade.

Where do most tsunami occur? About 86% of all great waves are generated in the Pacific Ocean because large-magnitude earthquakes occur along the series of trenches that ring its ocean basin where oceanic plates are subducted along convergent plate boundaries. Volcanic activity is also common along the *Pacific Ring of Fire*, and the large earthquakes that occur along its margin are capable of producing extremely large tsunami. **Figure 8.27** shows significant

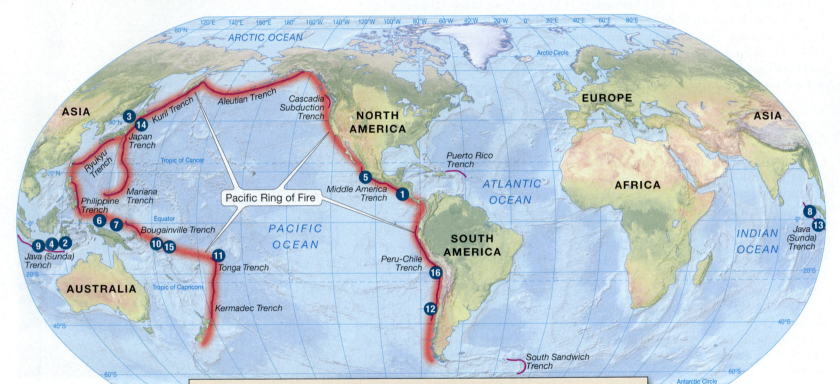

Map number	Date	Location	Max. height (m)	(ft)	Fatalities
1	Sep. 2, 1992	Nicaragua	10	33	170
2	Dec. 12, 1992	Flores Island, Indonesia	26	85	>1,000
3	Jul. 12, 1993	Okushiri, Japan	31	102	239
4	Jun. 2, 1994	East Java, Indonesia	14	46	238
5	Oct. 9, 1995	Jalisco, Mexico	11	36	1
6	Feb. 17, 1996	Irian Jaya, Indonesia	8	26	161
7	Jul. 17, 1998	Papua New Guinea	15	49	>2,200
8	Dec. 26, 2004	Sumatra, Indonesia	35	115	300,000
9	Jul. 17, 2006	Central Java, Indonesia	3	10	668
10	Apr. 1, 2007	Solomon Islands	5	16	52
11	Sep. 29, 2009	Samoa	14	46	189
12	Feb. 27, 2010	Chile	3	10	550
13	Oct. 25, 2010	Pagai Island, Indonesia	3	10	435
14	Mar. 11, 2011	Tohoku, Japan	40	131	19,508
15	Feb. 16, 2013	Solomon Islands	2	7	9
16	Apr. 1, 2014	Northern Chile	2	7	7

allowing oceanographers to detect the passage of a tsunami in the open ocean (**Figure 8.31**). DART buoys, which are essential components of tsunami warning systems, have now been deployed in all oceans.

TSUNAMI WATCHES AND TSUNAMI WARNINGS When a seismic disturbance occurs beneath the ocean surface that is large enough to be tsunamigenic (capable of producing a tsunami), a *tsunami watch* is issued. At this point, a tsunami may or may not have been generated, but the potential for one exists.

The PTWC is linked to a series of sea floor pressure sensors, ocean buoys, and tide-measuring stations throughout the Pacific, so the recording stations nearest the earthquake are closely monitored for any indication of unusual wave activity. If unusual wave activity is verified, the tsunami watch is upgraded to a *tsunami warning*. Generally, earthquakes smaller than magnitude $M_w = 6.5$ are not typically tsunamigenic because they lack the duration of ground shaking necessary to initiate a tsunami. In addition, transform faults do not usually produce tsunami because lateral movement does not offset the ocean floor and impart energy to the water column in the same way that vertical fault movements do.

When a tsunami is detected, warnings are sent to all the coastal regions that might encounter the destructive wave, along with its estimated time of arrival. This warning, usually just a few hours in advance of the tsunami, makes it possible to evacuate people from low-lying areas and remove ships from harbors before the waves arrive. If the disturbance is nearby, however, there is not enough time to issue a warning because a tsunami travels so rapidly. Unlike hurricanes, whose high winds and waves threaten ships at sea and send them to the protection of a coastal harbor, a tsunami washes ships from their coastal moorings into the open ocean or onto shore. The best strategy during a tsunami warning is to move ships out of coastal harbors and into deep water, where tsunami are not easily felt.

EFFECTIVENESS OF TSUNAMI WARNINGS Since the PTWC was established in 1948, it has effectively prevented loss of life due to tsunami when people have heeded the evacuation warnings. Property damage, however, has increased as more buildings have been constructed close to shore. To combat the damage caused by tsunami, countries that are especially prone to tsunami, such as Japan, have invested in shoreline barriers, seawalls, and other coastal fortifications.

Perhaps one of the best strategies to limit tsunami damage and loss of life is to restrict construction projects in low-lying coastal regions where tsunami have frequently struck in the past. However, the long time interval between large tsunami—typically 200 years or more—often causes people to let down their guard about remembering past disasters.

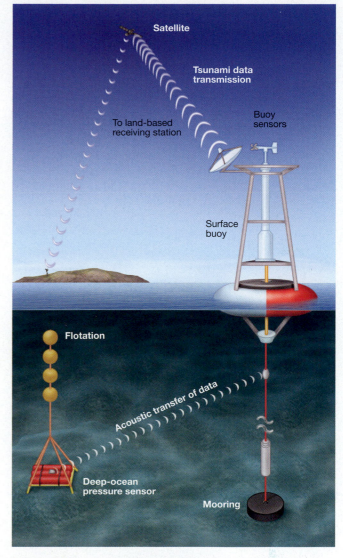

Figure 8.31 Deep-ocean Assessment and Reporting of Tsunamis (DART). The DART system consists of a deep-ocean pressure sensor that can detect a tsunami passing above. The pressure sensor relays information to a buoy at the surface that transmits the data via satellite, allowing oceanographers to detect the passage of a tsunami in the open ocean.

RECAP

Most tsunami are generated by underwater fault movement, which transfers energy to the entire water column. When these fast and long waves surge ashore, they can do considerable damage.

CONCEPT CHECK 8.5 | Specify the origin and characteristics of tsunami.

1 Why is it more likely that a tsunami will be generated by the vertical movement of sea floor faults rather than the horizontal movement of sea floor faults?

2 Describe the differences between wind-generated waves and tsunami. Which waves transmit more energy?

3 Explain what it would look like at the shoreline when the trough of a tsunami arrives there first. What is the impending danger?

4 Explain how the tsunami warning system in the Pacific Ocean works. Why must the tsunami be verified at the closest tide recording station?

8.6 Can Power from Waves Be Harnessed as a Source of Energy?

Moving water has a huge amount of energy, which is why there are so many hydroelectric power plants on rivers. Even greater energy exists in ocean waves, but significant problems must be overcome for the power to be harnessed efficiently. For example, a serious obstacle to the use of any device to harness wave energy is the monumental engineering problem of preventing the devices from being destroyed by the wave force they are built to harness. Also, deployment of complex machines with exposed moving parts to the marine environment presents challenges associated with corrosion and *biofouling*, which is the accumulation of algae and other sea life on machinery.

Another key disadvantage of wave energy is that the system produces significant power only when large storm waves break against it, so the system can't continuously provide reliable power and thus would serve only as a power supplement. In addition, a series of 100 or more of these structures along the shore would be required. Structures of this type could have a significant impact on the environment, with negative effects on marine organisms that rely on wave energy for dispersal, transporting food supplies, or removing wastes. Also, harnessing wave energy might alter the transport of sand along the coast, causing erosion in areas deprived of sediment.

Still, the immense power contained in waves could be used for generating electricity. Offshore wave-generating plants would be able to tap into the higher wave energy found offshore, but they are more likely to be damaged in large waves and more difficult to maintain. The most promising locations for coastal power generation from waves are where waves refract (bend) and converge, such as at headlands, which tend to focus wave energy (see Figure 8.21b). Using this advantage, an array of wave power plants might extract up to 10 megawatts of power[8] per 1 kilometer (0.6 mile) of shoreline.

Internal waves are a potential source of energy, too. Along shores that have favorable sea floor shape for focusing wave energy, internal waves may be effectively concentrated by refraction and thus could power an energy-conversion device that generates electricity.

Wave Power Plants and Wave Farms

In 2000, the world's first commercial-scale wave power plant began generating electricity. Built by Voith Hydro Wavegen and called **LIMPET 500** (*Land Installed Marine Powered Energy Transformer*), the plant is located on Islay, a small island off the west coast of Scotland that experiences high wave energy potential. The plant consists of a partially submerged chamber facing the sea (**Figure 8.32a**). As a wave approaches, the water level inside the structure rises, compressing air in the top of the chamber that is expelled through a turbine, which rotates to generate electricity (**Figure 8.32b**). As the wave recedes, the water level falls in the chamber and air is drawn back into the structure through the turbine. Even though there are two directions of air flow through the turbine, it is designed to continually produce power throughout the wave cycle. LIMPET 500 was constructed as a research and test facility that was originally configured with a turbo-generator of only 500 kilowatts (capable of supplying the energy needs of about 125 average U.S. homes) but has since been upgraded with a more efficient turbine that can generate more power and is continually being improved.

Wavegen's technology has recently been deployed in the 16-unit Mutriku breakwater power plant. Located in the Basque Country in northern Spain and developed by the Basque Energy Board EVE, this power plant is the world's only commercially operated wave energy plant. Another wave power plant using

[8]Each megawatt of electricity is enough to serve the energy needs of about 250 average U.S. homes.

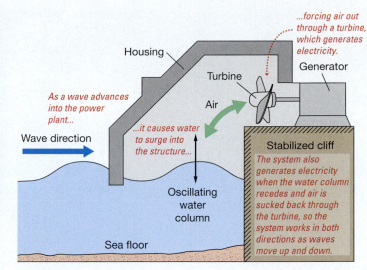

...forcing air out through a turbine, which generates electricity.

Housing

Turbine

Generator

As a wave advances into the power plant...

Air

Wave direction

...it causes water to surge into the structure...

Stabilized cliff
The system also generates electricity when the water column recedes and air is sucked back through the turbine, so the system works in both directions as waves move up and down.

Oscillating water column

Sea floor

(a) Photo of the exterior of LIMPET 500, the world's first commercial wave power plant.

(b) Schematic view of the interior of a wave power plant showing how it generates electricity from waves.

SmartFigure 8.32 How a wave power plant works.
https://goo.gl/Vx6PJT

Wavegen technology will also be installed in Siadar on the Isle of Lewis off the west coast of Scotland. This installation will have a peak operating capacity of 30 megawatts and will be one of the world's first large-scale developments of wave energy.

In 2008, Pelamis Wave Power completed the world's first wave farm off the coast of northern Portugal. This project uses three 150-meter-long (500-foot-long) devices that resemble giant segmented snakes and float half-submerged in the ocean (**Figure 8.33**). As each segment surges up or down with the crest of an oncoming wave, its hydraulic power plant pumps a biodegradable hydraulic fluid through a turbine, thus generating electricity. These wave-energy devices are already supplying up to 2.25 megawatts of power to Portugal's electrical grid, and up to 26 more devices are planned for the future.

Currently, about 50 wave-energy projects are in development at various sites around the world. These projects use different methods to harness wave power, including floats or submerged pistons that move up and down with each passing wave, tethered paddles that oscillate back and forth, and collection of water from breaking waves that overtops coastal structures and then using the weight of that water to turn turbines as it moves downhill and returns to the ocean. In 2013, the London-based consultancy Bloomberg New Energy Finance projected that up to 22 tidal projects and 17 wave projects generating more than one megawatt of electricity—enough to power around 250 average U.S. homes—could be installed by 2020.

Figure 8.33 Harnessing the power of ocean waves. This wave energy device resembles a large segmented floating snake and is designed to flex as waves pass, generating electricity in the process. A wave farm of three of these devices is currently generating electricity off northern Portugal, and up to 26 more are planned.

Global Coastal Wave Energy Resources

Leading estimates suggest that the global resource for wave energy lies between 1 and 10 terawatts; the world currently produces about 12 terawatts from all sources. Where are the best places to develop additional wave power plants and wave farms? **Figure 8.34** shows the average wave heights experienced along coastal regions and indicates the sites most favorable for wave-energy generation (*red areas*). The map shows that west-to-east movement of storm systems in the middle latitudes between 30 and 60 degrees north or south latitude causes the western coasts of continents to be struck by larger waves than eastern coasts. Thus, more wave energy is generally available along western than eastern shores. Furthermore, some of the largest waves (and greatest potential for wave power) are associated with the prevailing westerly wind belt in the Southern Hemisphere middle latitudes.

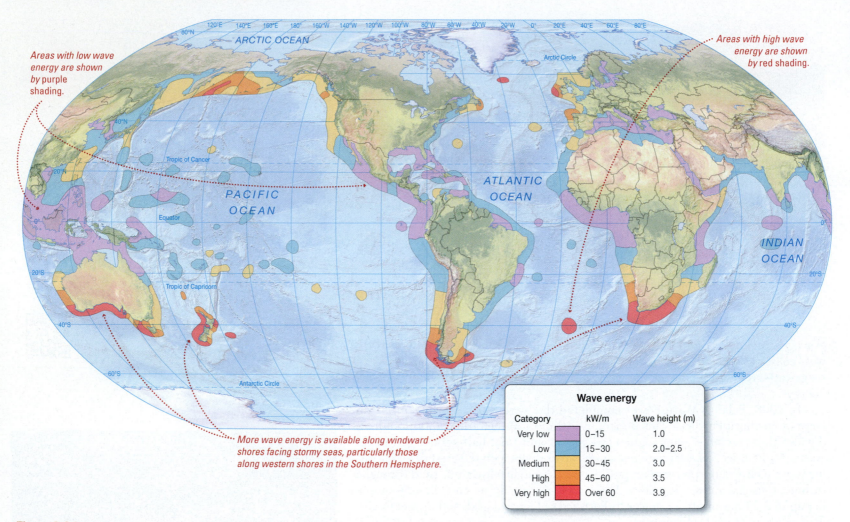

Areas with low wave energy are shown by purple shading.

Areas with high wave energy are shown by red shading.

More wave energy is available along windward shores facing stormy seas, particularly those along western shores in the Southern Hemisphere.

Wave energy

Category	kW/m	Wave height (m)
Very low	0–15	1.0
Low	15–30	2.0–2.5
Medium	30–45	3.0
High	45–60	3.5
Very high	Over 60	3.9

Figure 8.34 Global coastal wave energy resources. In this map of coastal wave energy, areas of highest wave energy are shown in red. kW/m is kilowatts per meter (for example, every meter of "red" shoreline has the potential of generating over 60 kilowatts of electricity); corresponding average wave height is in meters.

RECAP

Ocean waves produce large amounts of energy. Although significant problems exist in harnessing wave energy effectively, several types of devices are extracting wave energy today.

CONCEPT CHECK 8.6 | Evaluate the advantages and disadvantages of harnessing waves as a source of energy.

1 Discuss some problems that might result from developing facilities for conversion of wave energy to electrical energy.

2 Describe the locations and power-generating capability of existing offshore wave power plants and wave farms.

3 Worldwide, where are the best locations for new wave farms? Explain the oceanographic conditions that make these locations favorable.

ESSENTIAL CONCEPTS REVIEW

8.1 How are waves generated, and how do they move?

▶ *All ocean waves begin as disturbances that release energy into the ocean*. Energy sources include wind, the movement of fluids of different densities (which create internal waves), catastrophic releases of energy into the ocean by landslides and meteor impacts, underwater sea floor movements, the gravitational pull of the Moon and the Sun on Earth, and human activities in the ocean.

▶ Once initiated, *waves transmit energy through matter by setting up patterns of oscillatory motion* in the particles that make up the matter. Progressive waves are longitudinal, transverse, or orbital, depending on the pattern of particle oscillation. Particles in ocean waves move primarily in orbital paths.

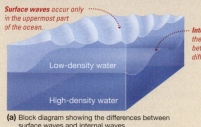

Surface waves occur only in the uppermost part of the ocean.

Internal waves occur within the ocean along the boundary between water masses of different density (the pycnocline).

Low-density water

High-density water

(a) Block diagram showing the differences between surface waves and internal waves.

Study Resources

MasteringOceanography Study Area Quizzes, MasteringOceanography Web Diving Deeper 8.1

Critical Thinking Question

Compare and contrast the properties of the three types of progressive waves: longitudinal, transverse, and orbital waves.

Active Learning Exercise

Working in groups of three, have each student in class pick one of the following types of progressive waves: (1) longitudinal wave, (2) transverse wave, and (3) orbital wave. For each type of wave, use your own words to explain how the wave moves and whether the wave can transmit through the three states of matter: (a) solid, (b) liquid, and (c) gas. Then recombine as a group and have each student take turns explaining their wave type.

8.2 What characteristics do waves possess?

▶ *Waves are described according to their wavelength* (L), *wave height* (H), *wave steepness* (H/L), *wave period* (T), *frequency* (f), *and wave speed* (S). As a wave travels, the water passes the energy along by moving in a circle, called *circular orbital motion*. This motion advances the waveform, not the water particles themselves. *Circular orbital motion decreases with depth*, ceasing entirely at wave base, which is equal to one-half the wavelength measured from still water level.

▶ If water depth is greater than one-half the wavelength, a progressive wave travels as a *deep-water wave with a speed that is directly proportional to wavelength*. If water depth is less than $1/20$ wavelength (L/20), the wave moves as a *shallow-water wave with a speed that is directly proportional to water depth. Transitional waves* have wavelengths between deep- and shallow-water waves, with speeds that depend on both wavelength and water depth.

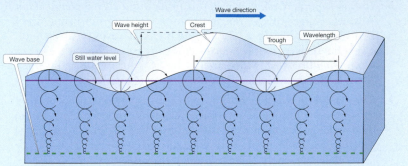

Wave direction

Wave height Crest Trough Wavelength

Wave base Still water level

(a) Diagrammatic view of an idealized progressive ocean wave showing wave characteristics and terminology.

Study Resources

MasteringOceanography Study Area Quizzes, MasteringOceanography Web Animation

Critical Thinking Question

To help reinforce your knowledge of wave terminology, draw a diagram of a simple progressive wave from memory. Include wave orbitals and label the crest, trough, wavelength, wave height, wave base, and still water level.

Active Learning Exercise

Working in pairs, have each student in class use Figure 8.8 to consider one of the following two deep-water waves: (1) a wave with a wavelength of 200 meters and (2) a wave with a wavelength of 400 meters. Determine the corresponding wave period and wave speed for each wave and compare your results. Then discuss how you determined your answers and what general relationships exist between wavelength, wave period, and wave speed.

8.3 How do wind-generated waves develop?

▶ As wind-generated waves form in a sea area, *capillary waves* with rounded crests and wavelengths less than 1.74 centimeters (0.7 inch) *form first. As the energy of the waves increases, gravity waves form*, with increased wave speed, wavelength, and wave height. Factors that influence the size of wind-generated waves include wind *speed*, *duration* (time), and *fetch* (distance). An equilibrium condition called a *fully developed sea* is reached when the maximum wave height is achieved for a particular wind speed, duration, and fetch.

▶ *Energy is transmitted from the sea area across the ocean by uniform, symmetrical waves called swell*. Different wave trains of swell can create either *constructive*, *destructive*, or *mixed interference* patterns. Constructive interference produces unusually large waves called rogue waves or superwaves.

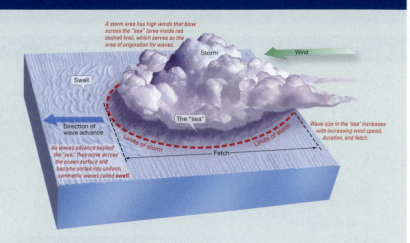

A storm area has high winds that blow across the "sea" (area inside red dashed line), which serves as the area of origination for waves.

Storm

Wind

Swell

The "sea"

Direction of wave advance

Limits of storm

Limits of storm

Fetch

As waves advance beyond the "sea," they move across the ocean surface and become sorted into uniform, symmetric waves called **swell**.

Wave size in the "sea" increases with increasing wind speed, duration, and fetch.

Study Resources
MasteringOceanography Study Area Quizzes, MasteringOceanography Web Animation, MasteringOceanography Web Video

Critical Thinking Question
Using the information about the record-breaking waves experienced by the USS *Ramapo* in 1933, determine the waves' wavelength and speed.

Active Learning Exercise
Working with another student in class, use Table 8.2 to determine the characteristics of the average height, wavelength, period, and highest 10% of waves created by the following two conditions: (1) a wind speed of 40 kilometers (25 miles) per hour, and (2) a wind speed of 80 kilometers (50 miles) per hour. With a doubling of wind speed, are waves created that have double the properties? Explain.

8.4 How do waves change in the surf zone?

▶ *As waves approach shallow water near the shore, they undergo many physical changes*. Waves release their energy in the surf zone when their steepness exceeds a 1:7 ratio and break. If waves break on a relatively flat surface, they produce *spilling breakers*. The curling crests of *plunging breakers*, which are the best for surfing, form on steep slopes, and abrupt beach slopes create *surging breakers*.

▶ When swell approaches the shore, *segments of the waves that first encounter shallow water are slowed* whereas other segments of the wave in deeper water move at their original speed, *causing each wave to refract, or bend*. Refraction concentrates wave energy on headlands, while low-energy breakers are characteristic of bays.

▶ *Reflection of waves off seawalls or other barriers* can cause an interference pattern called a *standing wave*. The crests of standing waves do not move laterally as in progressive waves but alternate with troughs at antinodes. Between the antinodes are nodes, where there is no vertical movement of the water.

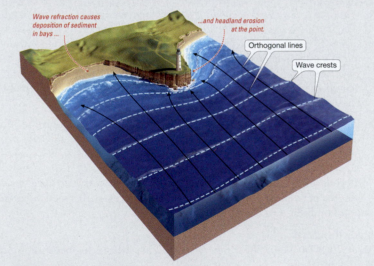

Wave refraction causes deposition of sediment in bays ...

...and headland erosion at the point.

Orthogonal lines

Wave crests

(b) Perspective view of wave refraction along an irregular shoreline.

Study Resources
MasteringOceanography Study Area Quizzes, MasteringOceanography Web Animations, MasteringOceanography Web Video, Web Video

Critical Thinking Question
As a wave comes into shallow water, it experiences five physical changes that cause the wave to break at the shore. Name each physical change and explain why each occurs.

Active Learning Exercise
Working with another student in class and using the Internet, explain the basic physics of surfing. Your answer should include important properties of waves as well as important properties of surfboards.

8.5 How are tsunami created?

▶ *Sudden changes in the elevation of the sea floor, such as from fault movement or volcanic eruptions, generate tsunami, or seismic sea waves.* These waves often have lengths exceeding 200 kilometers (125 miles) and travel across the open ocean with undetectable heights of about 0.5 meter (1.6 feet) at speeds in excess of 700 kilometers (435 miles) per hour. Upon approaching shore, a tsunami produces a *series of rapid withdrawals and surges*, some of which may increase the height of sea level by 40 meters (131 feet) or more.

▶ *Most tsunami occur in the Pacific Ocean*, where they have caused billions of dollars of coastal damage and taken tens of thousands of lives. A recent example is the *2011 Japanese Tohoku Earthquake and resulting tsunami.* However, the *2004 Indian Ocean tsunami killed as many as 300,000*, making it the most deadly tsunami in history. The *Pacific Tsunami Warning Center (PTWC)* has dramatically reduced fatalities by successfully predicting tsunami using real-time seismic information and a network of deep-ocean pressure sensors. A new tsunami warning system has become operational in the Indian Ocean.

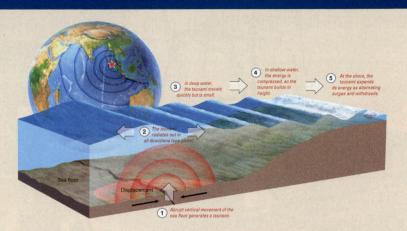

Study Resources
MasteringOceanography Study Area Quizzes, MasteringOceanography Web Diving Deeper 8.1, MasteringOceanography Web Animation

Critical Thinking Question
What ocean depth would be required for a tsunami with a wavelength of 220 kilometers (136 miles) to travel as a deep-water wave? Is it possible that such a wave could become a deep-water wave anyplace in the world ocean? Explain.

Active Learning Exercise
Working with another student in class, analyze this scenario: While shopping in a surf shop, you overhear some surfing enthusiasts mention that they would really like to ride the curling wave of a tidal wave at least once in their life because it is a single breaking wave of enormous height. What would you say to these surfers?

8.6 Can power from waves be harnessed as a source of energy?

▶ *Ocean waves can be harnessed to produce hydroelectric power*, but significant problems must be overcome to make this a practical source of energy. Some of these problems include the *constant battering* of any structure that exists in such a high-energy environment, *corrosion and/or biofouling* of machinery exposed to ocean waters, and the *varying amount of wave energy* at different times associated with natural waves.

Study Resources
MasteringOceanography Study Area Quizzes

Critical Thinking Question
What are some negative environmental factors that could inhibit the development of ocean wave power systems? What are your ideas for how these issues could be overcome?

Active Learning Exercise
Working as a group and using Figure 8.34, identify a specific location on Earth that has high wave power potential *and* is close to a major

population center so that it is easy to distribute the generated electricity. Present your findings to the class, including the location and an explanation of your reasoning.

MasteringOceanography™
www.masteringoceanography.com

Looking for additional review and test prep materials? With individualized coaching on the toughest topics of the course, MasteringOceanography offers a wide variety of ways for you to move beyond memorization and deeply grasp the underlying processes of how the oceans work. Visit the Study Area in **www.masteringoceanography.com** to find practice quizzes, study tools, and multimedia that will improve your understanding of this chapter's content. Sign in today to enjoy the following features: Self Study Quizzes, SmartFigures, SmartTables, Oceanography Videos, Squidtoons, Geoscience Animation Library, RSS Feeds, Digital Study Modules, and an optional Pearson eText.

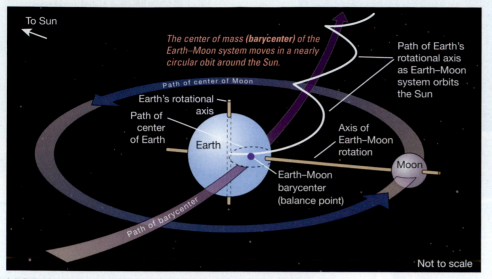

(a) Schematic diagram showing the movement of the barycenter (balance point) of the Earth–Moon system around the Sun.

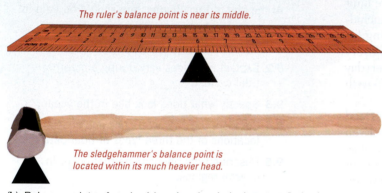

(b) Balance points of a ruler *(above)* and a sledgehammer *(below)*.

Figure 9.1 Earth–Moon system barycenter. The Earth–Moon barycenter is similar to a sledgehammer flung into space, with Earth represented by the hammer's head and the Moon represented by the end of the handle.

STUDENTS SOMETIMES ASK . . .

Are there also tides in other objects, such as lakes and swimming pools?

The Moon's and Sun's gravity act on all objects, so anything that has the ability to flow will exhibit measurable tides. For example, there are tides in lakes, wells, and swimming pools. In fact, there are even extremely tiny tidal bulges in a glass of water! However, tidal effects become negligible the smaller the body of water is, and other effects dominate them so they are not observable for the most part. On the other hand, tides in the atmosphere and the "solid" Earth are much larger. Tides in the atmosphere—called *atmospheric tides*—can be miles high and are also affected by solar heating. The tides inside Earth's interior—called *solid-body tides*, or *Earth tides*—are up to 50 centimeters (1.6 feet) high and cause the daily stretching and contraction of Earth. Interestingly, Earth tides have been identified as a trigger mechanism for tremors along certain weak faults.

Why isn't the barycenter halfway between the two bodies? It's because Earth's mass is so much greater than that of the Moon. This can be visualized by imagining Earth and its Moon as ends of an object that is much heavier on one end than the other. Visualize a sledgehammer, for example, which has a lighter handle and a much heavier head. If you were to position your finger to balance the sledgehammer with its handle sticking out to one side, you'd find that its balance point is within the head of the hammer (**Figure 9.1b**). Now imagine that the sledgehammer is flung into space, tumbling slowly end over end about its balance point. This is exactly the situation that describes the movement of the Earth–Moon system. The purple arrow in Figure 9.1a shows the smooth, nearly circular path of the Earth–Moon barycenter around the Sun.

If the Moon and Earth are attracted to one another, why don't the two collide? The Earth–Moon system is in a mutually stable orbit held together by a balance between centripetal (gravitational) and inertial (motion) forces, which prevents the Moon and Earth from colliding or flying apart. This is how orbits are established that keep objects at more or less fixed distances.

Newton's work also allowed an understanding of why the tides behave as they do. Just as gravity and motion serve to keep bodies in mutual orbits, they also exert an influence on every particle of water in the oceans, thus creating the tides.

GRAVITATIONAL AND CENTRIPETAL FORCES IN THE EARTH–MOON SYSTEM To understand how *tide-generating forces* influence the oceans, let's examine how *gravitational forces* and *centripetal forces* affect objects on Earth within the Earth–Moon system. (We'll ignore the influence of the Sun for the moment.)

The **gravitational force** is derived from **Newton's law of universal gravitation**, which states that *every object that has mass in the universe is attracted to every other object*. An object can be as small as a sub-atomic particle or as large as a sun. The basic equation for this relationship is

$$F_g = \frac{Gm_1m_2}{r^2} \qquad (9.1)$$

This equation states that the gravitational force (F_g) is directly proportional to the product of the masses of the two bodies (m_1, m_2) and is inversely proportional to the square of the distance between the two masses (r^2). Note that G is the gravitational constant, so it does not change.

Let's simplify Newton's law of universal gravitation and examine the effect of both mass and distance on the gravitational force, which can be expressed with arrows (up arrow = increase, down arrow = decrease):

If mass increases ($\uparrow$), then gravitational force increases ($\uparrow$).

A practical example of this can be seen in an object with a large mass (such as the Sun), which produces a large gravitational attraction (**Figure 9.2a**).

Looking at how distance influences gravitational force, the relationship is as follows:

If distance increases ($\uparrow$), then gravitational force greatly decreases ($\downarrow\downarrow$).

Equation 9.1 shows that the gravitational attraction varies with the *square* of distance, so even a small *increase* in the distance between two objects significantly *decreases* the gravitational force between them—hence the double arrows in the

distance relationship illustrated above. This means that when an object is twice as far away, the gravitational attraction is only one-quarter as strong. As a practical example, this is why astronauts can experience zero gravity (weightlessness) in space if they get far enough away from Earth's gravitational pull (**Figure 9.2b**). In summary, then, the *greater* the mass of the objects and (especially) the *closer* they are together, the greater their gravitational attraction.

Figure 9.3 shows how gravitational forces for points on Earth (caused by the Moon) vary depending on their distances from the Moon. The greatest gravitational attraction (the longest arrow) is at Z, the **zenith** (*zenith* = a path over the head), which is the point closest to the Moon. The gravitational attraction is weakest at N, the **nadir** (*nadir* = opposite the zenith), which is the point farthest from the Moon. The direction of the gravitational attraction between most particles and the center of the Moon is at an angle relative to a line connecting the center of Earth and the Moon (Figure 9.3). This angle causes the force of gravitational attraction between each particle and the Moon to be slightly different.

The **centripetal force**[1] (*centri* = the center, *pet* = seeking) required to keep planets in their orbits is provided by the gravitational attraction between the center-of-mass of each of the planets and the Sun. Centripetal force connects the center-of-mass of an orbiting body to its parent, pulling the object *inward* toward the parent, "seeking the center" of its orbit. For example, if you tie a string to a ball and swing the ball around your head (**Figure 9.4**), the string pulls the ball toward your hand. The string exerts a *centripetal force* on the ball, forcing the ball to *seek the center* of its orbit. If the string breaks, the force is gone, and the ball can no longer maintain its circular orbit. The ball flies off in a *straight* line,[2] *tangent* (*tangent* = touching) to the circle (Figure 9.4).

Earth and the Moon are interconnected, too, but by gravity rather than by a string. Gravity provides the centripetal force that holds the Moon in its orbit around Earth. If all gravity in the solar system could be shut off, centripetal force would vanish, and the momentum of the celestial bodies would send them flying off into space along straight-line paths, tangent to their orbits.

RESULTANT FORCES Particles of identical mass rotate in identical-sized paths due to the Earth–Moon rotation system (**Figure 9.5**). Each particle requires an identical centripetal force to maintain it in its circular path. Gravitational attraction between the particle and the Moon supplies the centripetal force, but the *supplied* force is different than the *required* force (because gravitational attraction varies with distance from the Moon) except at the center of Earth. This difference creates tiny **resultant forces**, which are the mathematical difference between the two sets of arrows shown in Figures 9.3 and 9.5.

Figure 9.6 combines Figures 9.3 and Figure 9.5 to show that resultant forces are produced by the difference between the required centripetal (C) and supplied gravitational (G) forces. However, do not think that both of these forces are being applied to the points because C is a force that would be required to keep the particles in a perfectly circular path, while G is the force actually provided for this purpose by gravitational attraction between the particles and the Moon. The resultant forces (*blue arrows*) are established by constructing an arrow from the tip of the centripetal (*red*) arrow to the tip of the gravity (*black*) arrow and located where the red and black arrows begin.

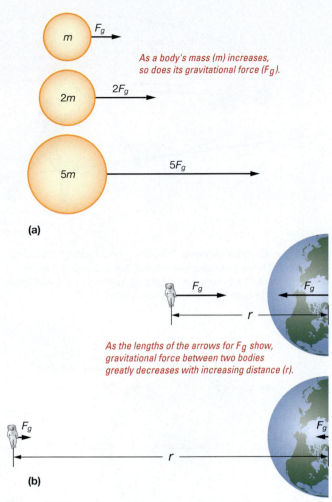

(a)

As a body's mass (m) increases, so does its gravitational force (Fg).

(b)

As the lengths of the arrows for Fg show, gravitational force between two bodies greatly decreases with increasing distance (r).

Figure 9.2 **The relationship of gravitational force to both mass (a) and distance (b).**

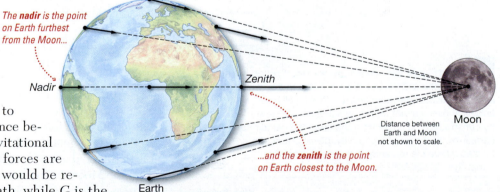

*The **nadir** is the point on Earth furthest from the Moon...*

Nadir

Zenith

Moon

Distance between Earth and Moon not shown to scale.

*...and the **zenith** is the point on Earth closest to the Moon.*

Earth

Figure 9.3 **Gravitational forces on Earth due to the Moon.** Black arrows represent the gravitational forces on objects located at different places on Earth due to the Moon. The length and orientation of the arrows indicate the strength and direction of the gravitational force. Notice the length and angular differences of the arrows for different points on Earth.

[1]This is not to be confused with the so-called *centrifugal force*, (*centri* = the center, *fug* = flee) an apparent or fictitious force that is oriented outward.

[2]At the moment that the string breaks, the ball will continue along a straight-line path, obeying Newton's first law of motion (the *law of inertia*), which states that moving objects follow straight-line paths until they are compelled to change that path by other forces.

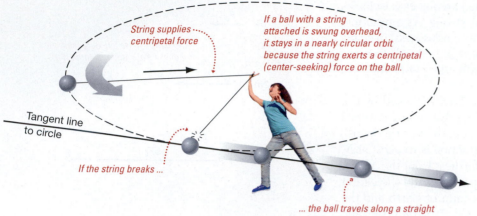

If a ball with a string attached is swung overhead, it stays in a nearly circular orbit because the string exerts a centripetal (center-seeking) force on the ball.

String supplies centripetal force

Tangent line to circle

If the string breaks ...

... the ball travels along a straight path that is tangent to the circle.

Figure 9.4 Centripetal force. The string on a rotating ball exerts a centripetal force that holds the rotating ball in a nearly circular orbit. This is similar to how Earth's gravity exerts a centripetal force on the Moon to hold the Moon in a nearly circular orbit.

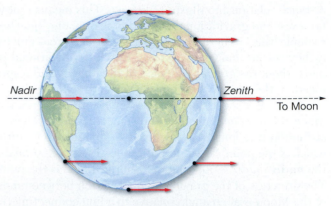

Nadir Zenith

To Moon

Figure 9.5 Required centripetal (center-seeking) forces. Red arrows represent the centripetal forces required to keep identical-sized particles in identical-sized orbits as a result of the rotation of the Earth–Moon system around its barycenter. Notice that the arrows are all the same length and are oriented in the same direction for all points on Earth.

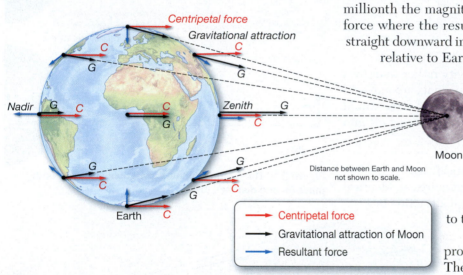

Centripetal force
Gravitational attraction

Nadir Zenith

Moon

Distance between Earth and Moon not shown to scale.

Earth

→ Centripetal force
→ Gravitational attraction of Moon
→ Resultant force

SmartFigure 9.6 Resultant forces. Red arrows indicate centripetal forces (*C*), which are not equal to the black arrows that indicate gravitational attraction (*G*). The small blue arrows show resultant forces, which are established by constructing an arrow from the tip of the centripetal (*red*) arrow to the tip of the gravity (*black*) arrow and located where the red and black arrows begin. https://goo.gl/eBCMYZ

Web Animation
Tidal Forces
http://goo.gl/tdDbEx

RECAP

The tides are caused by an imbalance between the required centripetal and the provided gravitational forces acting on Earth. This difference produces residual forces, the horizontal component of which creates two equal tidal bulges on opposite sides of Earth.

TIDE-GENERATING FORCES Resultant forces are small, averaging about one-millionth the magnitude of Earth's gravity. Moreover, there is no tide-generating force where the resultant forces either point straight upward toward the sky or straight downward into Earth. In these cases, the resultant force is oriented *vertical* relative to Earth's surface (**Figure 9.7**), and no tides occur here. These conditions occur at three locations on Earth: (1) at the zenith, (2) at the nadir, and (3) along an "equator" connecting all points halfway between the zenith and nadir. However, if the resultant force has a significant *horizontal component*—that is, tangential or parallel to Earth's surface—it produces tidal bulges on Earth, creating what are known as the **tide-generating forces**. These tide-generating forces are quite small but reach their maximum value at points on Earth's surface at a "latitude" of 45 degrees relative to the "equator" between the zenith and nadir (Figure 9.7).

As previously discussed, gravitational attraction is inversely proportional to the *square* of the distance between two masses. The tide-generating force, however, is inversely proportional to the *cube* of the distance between each point on Earth and the *center* of the tide-generating body (Moon or Sun). Although the tide-generating force is derived from the gravitational force, it is not linearly proportional to it. As a result, distance is a more highly weighted variable for tide-generating forces.

The tide-generating forces create two simultaneous bulges: one on the side of Earth directed *toward* the Moon (the zenith) and the other on the side directed *away from* the Moon (the nadir) (**Figure 9.8**). On the side directly facing the Moon, the bulge is created because the provided gravitational force is greater than the required centripetal force. Conversely, on the side facing away from the Moon, the bulge is created because the required centripetal force is greater than the provided gravitational force. Although the forces are oriented in opposite directions on the two sides of Earth, the resultant forces are equal in magnitude, so the bulges are equal, too.

Tidal Bulges: The Moon's Effect

It is easier to understand how tides on Earth are created if we consider an ideal Earth and an ideal ocean. The ideal Earth has two tidal bulges, one toward the Moon and one away from the Moon (called the **lunar bulges**), as shown in Figure 9.8. The ideal ocean has a uniform depth, with no friction between the seawater

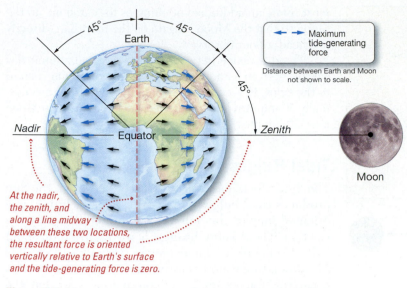

Figure 9.7 Tide-generating forces. Schematic diagram showing the tide-generating forces on Earth (*arrows*) that are caused by the Moon. The maximum tide-generating forces (*blue arrows*) occur where the resultant forces have a significant horizontal component. See the text for a description.

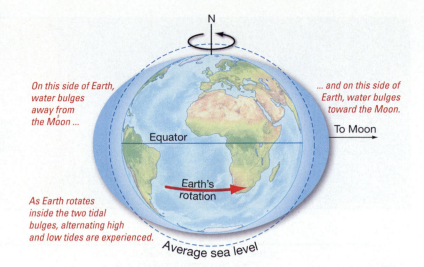

Figure 9.8 Idealized tidal bulges. In an idealized case, the Moon creates two bulges in the ocean surface: one that extends *toward* the Moon and the other *away from* the Moon. As Earth rotates, it carries various locations into and out of the two tidal bulges so that all points on its surface (except the poles) experience two alternating high and low tides daily.

and the sea floor. Newton made these same simplifications when he first explained Earth's tides.

If the Moon is stationary and aligned with the ideal Earth's equator, the maximum bulge will occur on the equator on opposite sides of Earth. If you were standing on the equator, you would experience two high tides each day. The time between high tides, which is the **tidal period**, would be 12 hours. If you moved to any latitude north or south of the equator, you would experience the same tidal period, but the high tides would be less high because you would be at a lower point on the bulge.

In most places on Earth, however, high tides occur every 12 hours 25 minutes because tides depend on the lunar day, not the solar day. The **lunar day** (also called a *tidal day*) is measured from the time the Moon is on the meridian of an observer—that is, directly overhead—to the next time the Moon is on that meridian and is 24 hours 50 minutes long.[3] The **solar day** is measured from the time the Sun is on the meridian of an observer to the next time the Sun is on that meridian and is 24 hours long. Why is the lunar day 50 minutes longer than the solar day? During the 24 hours it takes Earth to make a full rotation, the Moon has continued moving another 12.2 degrees to the east in its orbit around Earth (**Figure 9.9**). Thus, Earth

RECAP

A solar day (24 hours) is shorter than a lunar day (24 hours and 50 minutes). The extra 50 minutes is the result of the Moon's movement in its orbit around Earth.

Web Animation

The Lunar Day
https://goo.gl/ElkpUH

 SmartFigure 9.9 **The lunar day.** A lunar day is the time that elapses between when the Moon is directly overhead and the next time the Moon is directly overhead. During one complete rotation of Earth (the 24-hour solar day), the Moon moves eastward 12.2 degrees, and Earth must rotate an additional 50 minutes for the Moon to be in the exact same position overhead. Thus, a lunar day is 24 hours 50 minutes long.
https://goo.gl/bbYGx0

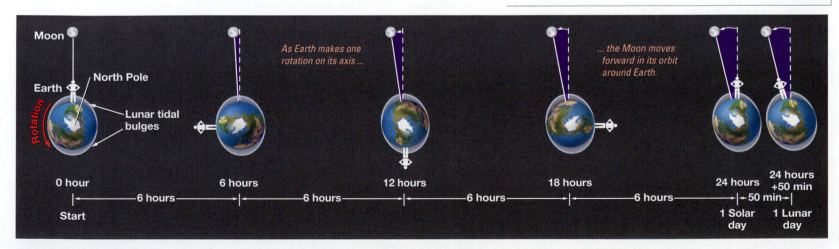

[3]A lunar day is exactly 24 hours, 50 minutes, 28 seconds long.

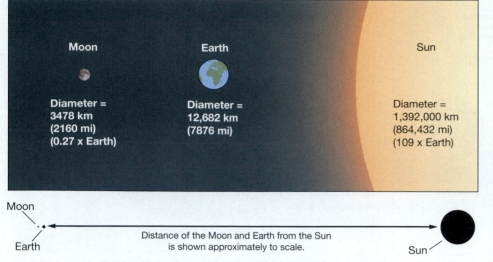

Figure 9.10 **Relative sizes and distances of the Moon, Earth, and Sun.** *Top*: The relative sizes of the Moon, Earth, and Sun are shown to scale. Note that the diameter of the Moon is roughly one-fourth that of Earth, while the diameter of the Sun is 109 times the diameter of Earth. *Bottom*: The relative distances of the Moon, Earth, and Sun are shown to scale.

RECAP

Although the Moon is much smaller than the Sun, it controls tides far more than the Sun because the Moon is much closer to Earth. As a result, the Moon creates lunar tidal bulges that are twice the size of the solar tidal bulges.

RECAP

In an idealized case, the rise and fall of the tides are caused by Earth's rotation carrying various locations into and out of the tidal bulges.

must rotate an additional 50 minutes to "catch up" to the Moon so that the Moon is again on the meridian (directly overhead) of our observer.

The difference between a solar day and a lunar day can be seen in some of the natural phenomena related to the tides. For example, alternating high tides are normally 50 minutes *later* each successive day, and the Moon rises 50 minutes *later* each successive night.

Tidal Bulges: The Sun's Effect

The Sun affects the tides, too. Like the Moon, the Sun produces tidal bulges on opposite sides of Earth, one oriented *toward* the Sun and one oriented *away from* the Sun. These **solar bulges**, however, are only about half the size of the lunar bulges. Although the Sun is 27 million times more massive than the Moon, its tide-generating force is not 27 million times greater than the Moon's. This is because the Sun is 390 times farther from Earth than the Moon (**Figure 9.10**). Moreover, tide-generating forces vary inversely as the *cube* of the distance between objects. Thus, the tide-generating force is reduced by the cube of 390, or about 59 million times compared with that of the Moon. These conditions result in the Sun's tide-generating force being $^{27}/_{59}$ that of the Moon, or 46% (about one-half). Consequently, the solar bulges are only 46% the size of the lunar bulges and, as a result, the Moon exerts over two times the gravitational pull of the Sun on the tides.

Even though the Moon exerts over two times the gravitational pull of the Sun on Earth's tides, note that the Sun does not exert a smaller gravitational force on Earth as compared to the Moon. In fact, the Sun's total "pull" on all points on Earth is much greater than that of the Moon, but the *difference* across Earth is small because the diameter of Earth is very small in relation to the distance from the Sun. In contrast, the diameter of Earth is quite large in relation to the distance to the center of the Moon. In summary, the reason the Moon controls tides far more than the Sun is that the Moon is much closer to Earth, although it is much smaller in size and mass compared to the Sun.

Earth's Rotation and the Tides

The tides appear to move water in toward shore (the **flood tide**) and to move water away from shore (the **ebb tide**). However, according to the nature of the idealized tides presented so far, *Earth's rotation carries various locations into and out of the tidal bulges*, which are in fixed positions relative to the Moon and the Sun. In essence, alternating high and low tides are created as Earth constantly rotates inside fluid bulges that are supported by the Moon and the Sun.

CONCEPT CHECK 9.1 | Demonstrate an understanding of the forces that cause ocean tides.

1 Why are there tidal bulges on both sides of Earth (for example, not just the side of Earth that faces the Moon or the Sun)?

2 Explain why the Sun's influence on Earth's tides is only 46% that of the Moon, even though the Sun is so much more massive than the Moon.

3 Why is a lunar day 24 hours 50 minutes long, while a solar day is 24 hours long?

4 If Earth did not have the Moon orbiting it, would there still be tides? Why or why not?

9.2 How Do Tides Vary during a Monthly Tidal Cycle?

The monthly tidal cycle is 29-and-a-half days because that's how long it takes the Moon to complete an orbit around Earth.[4] During its orbit around Earth, the Moon's changing position influences tidal conditions on Earth.

The Monthly Tidal Cycle

During the monthly tidal cycle, the phase of the Moon changes dramatically. When the Moon is between Earth and the Sun, it cannot be seen at night; this phase is called **new moon**. When the Moon is on the side of Earth opposite the Sun, its entire disk is brightly visible; this phase is called **full moon**. A **quarter moon**—a moon that is half lit and half dark as viewed from Earth—occurs when the Moon is at right angles to the Sun relative to Earth.

Figure 9.11 shows the positions of Earth, the Moon, and the Sun at various points during the 29-and-a-half-day lunar cycle. When the Sun and Moon are aligned, either with the Moon between Earth and the Sun (new moon; Moon in *conjunction*) or with the Moon on the side opposite the Sun (full moon; Moon in *opposition*), the tide-generating forces of the Sun and Moon combine (Figure 9.11, *top*). At this time, the **tidal range** (the vertical difference between high and low tides) is large (very *high* high tides and quite *low* low tides) because there is *constructive interference*[5] between the lunar and solar tidal bulges. The maximum tidal range is called a **spring tide**[6] (*springen* = to rise up) because the tide is extremely large, or "springs forth." Note that spring tides have nothing to do with the seasons and happen twice a month during all times of the year. When the Earth–Moon–Sun system is aligned, the Moon is said to be in **syzygy** (*syzygia* = union).

When the Moon is in either the first- or third-quarter[7] phase (Figure 9.11, *bottom*), the tide-generating force of the Sun is working at right angles to the tide-generating force of the Moon. The tidal range is small (*lower* high tides and *higher* low tides) because there is *destructive interference*[8] between the lunar and solar tidal bulges. This is called a **neap tide** (*nep* = scarcely or barely touching),[9] and the Moon is said to be in **quadrature** (*quadra* = four).

The time between successive spring tides (full moon and new moon) or neap tides (first quarter and third quarter) is one-half the monthly lunar cycle, which is about two weeks. The time between a spring tide and a successive neap tide is one-quarter the monthly lunar cycle, which is about one week.

Figure 9.12 shows the appearance of the Moon as it moves through its monthly cycle. As the Moon progresses from new moon to first-quarter phases, the Moon is a **waxing crescent** (*waxen* = to increase; *crescere* = to grow). In between the first-quarter and full moon phases, the Moon is a **waxing gibbous** (*gibbus* = hump). Between the Moon's full and third-quarter phases, it is a

[4]The 29-and-a-half-day monthly tidal cycle is also called a *lunar cycle*, a *lunar month*, or a *synodic* (*synod* = meeting) *month*. In fact, the word "month" is derived from the word "moon."

[5]As mentioned in Chapter 8, *constructive interference* occurs when two waves (or, in this case, two tidal bulges) overlap crest to crest and trough to trough.

[6]Spring tides have no connection with the spring season; they occur twice a month during the time when the Earth–Moon–Sun system is aligned.

[7]The third-quarter moon is often called the last-quarter moon, which is not to be confused with certain sports that have a fourth or last quarter.

[8]*Destructive interference* occurs when two waves (or, in this case, two tidal bulges) match up crest to trough and trough to crest.

[9]To help remember a *neap tide*, think of it as one that has been "*nipped in the bud*," thus indicating a small tidal range.

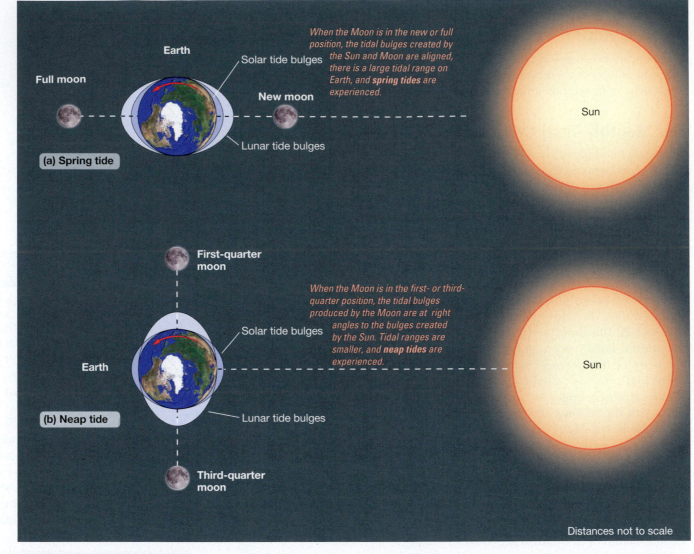

Distances not to scale

 SmartFigure 9.11 Earth–Moon–Sun positions and the tides. Schematic diagram showing **(a)** spring tide and **(b)** neap tide. Note in both cases that there is only one moon in orbit around Earth. https://goo.gl/DmuidP

Web Animation
Monthly Tidal Cycle
http://goo.gl/cg3NoK

RECAP

Spring tides occur during the full moon and new moon phases, when the lunar and solar tidal bulges constructively interfere, producing a large tidal range. Neap tides occur during the quarter moon phases, when the lunar and solar tidal bulges destructively interfere, producing a small tidal range.

waning gibbous (*wanen* = to decrease). And, in between the third-quarter and new moon phases, the Moon is a **waning crescent**. The Moon has identical periods of rotation on its axis and orbit around Earth (a property called *synchronous rotation*). What this means is that during the Moon's early history, it became "locked" in Earth's gravitational "tractor beam." As a result, the same side of the Moon always faces Earth.

Why does the Moon appear to change phase? The answer is that, unlike the Sun and other stars, the Moon emits no light of its own. Instead, it shines brightly only because it reflects light from the Sun. As shown in Figure 9.12, the Moon's circular disk is always present but only half of the Moon's surface is illuminated by the Sun at all times. However, not all of the Moon's sunlit face can be seen because of the Moon's position with respect to Earth and the Sun. When the Moon is in full moon phase, for example, we see the entire "daylit" face because the Sun and the Moon are in opposite directions from Earth in the sky (although they do not exist on the same plane; as a result, sunlight is not blocked). In the case of a new moon, the Moon and the Sun are in almost the same part of the sky, and the sunlit side of the Moon is oriented away from our view (Figure 9.12).

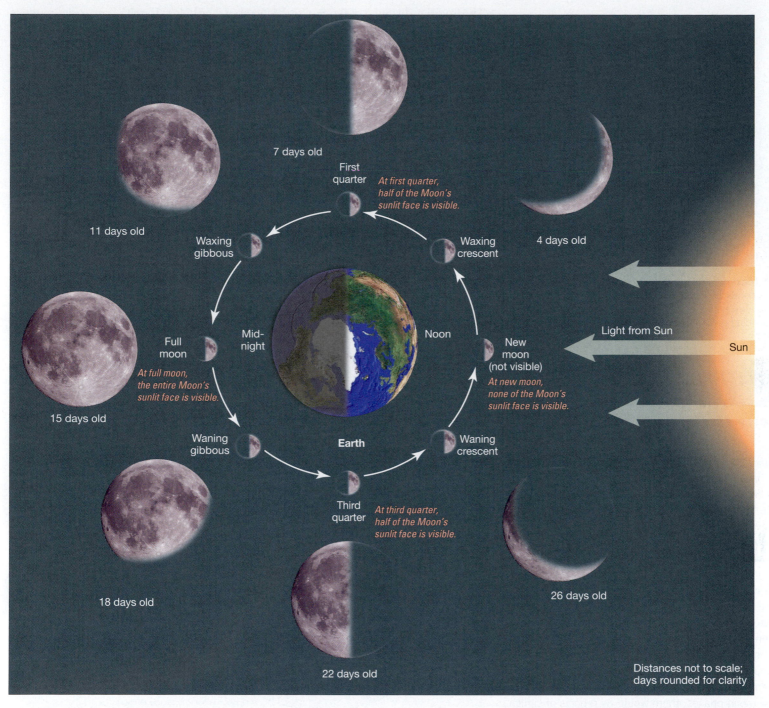

7 days old

First quarter · *At first quarter, half of the Moon's sunlit face is visible.*

11 days old

Waxing gibbous

Waxing crescent

4 days old

Light from Sun

Sun

Full moon · *At full moon, the entire Moon's sunlit face is visible.*

Mid-night

Noon

New moon (not visible) · *At new moon, none of the Moon's sunlit face is visible.*

15 days old

Waning gibbous

Earth

Waning crescent

18 days old

Third quarter · *At third quarter, half of the Moon's sunlit face is visible.*

26 days old

22 days old

Distances not to scale; days rounded for clarity

Figure 9.12 Phases of the Moon. As the Moon moves around Earth during its 29-and-a-half-day lunar cycle, its phase changes depending on its position relative to the Sun and Earth. Names of Moon phases are shown (*inner circle*) along with what the Moon looks like as viewed from Earth (*outer circle*).

Complicating Factors

Besides Earth's rotation and the relative positions of the Moon and the Sun, there are many other factors that influence tides on Earth. Two of the most prominent of these factors are the declination of the Moon and Sun and the elliptical shapes of Earth's and the Moon's orbits. Let's examine both of these factors.

DECLINATION OF THE MOON AND SUN Up to this point, we have assumed that the Moon and Sun have remained directly overhead at the equator, but this is not usually the case. Most of the year, in fact, they are either north or south of the equator.

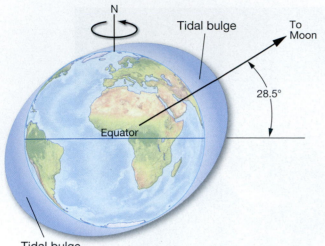

Figure 9.13 Maximum declination of tidal bulges from the equator. The center of the tidal bulges may lie at any latitude from the equator to a maximum of 28.5 degrees on either side of the equator, depending on the season of the year (solar angle) and the Moon's position.

STUDENTS SOMETIMES ASK . . .

Do the phases of the Moon influence human behavior?

All kidding about werewolves aside, tidal and gravitational effects have been used to explain the apparent increase in people's bizarre behavior when the Moon is full. The belief is that since the phases of the Moon affect ocean tides, they may have similar effects on the bodily fluids of human beings, whose composition is about 65% water. A number of scientists have tried to determine whether the Moon is in fact related to human behavior by calculating the precise number of births, crimes, and unusual behaviors that purportedly occur during various phases of the Moon. A few such investigators have indeed found an association between a full moon and unintentional poisonings, absenteeism, and crime. More often, however, researchers have found no relationship between lunar cycles and human biology or behavior. And it would seem that since tidal forces are so similar during full and new moon phases, there would be reports of similar unusual behavior during both full and new moons. It may simply be that certain people expect to be influenced by a full moon, so they may be more attentive and responsive to their internal sensations or desires at that time. It's no wonder these people are called *lunatics*!

Web Animation
Effects of Elliptical Orbits
https://goo.gl/wOSc4Q

The angular distance of the Sun or Moon above or below Earth's equatorial plane is called **declination** (*declinare* = to turn away).

Earth orbits the Sun along an invisible ellipse in space. The imaginary plane that contains this ellipse is called the **ecliptic** (*ekleipein* = to fail to appear). Recall from Chapter 6 that Earth's axis of rotation is tilted 23.5 degrees with respect to the ecliptic and that this tilt causes Earth's seasons. It also means the maximum declination of the Sun relative to Earth's equator is 23.5 degrees.

To complicate matters further, the plane of the Moon's orbit is tilted 5 degrees with respect to the ecliptic. Thus, the maximum declination of the Moon's orbit relative to Earth's equator is 28.5 degrees (5 degrees plus the 23.5 degrees of Earth's tilt). The declination changes from 28.5 degrees south to 28.5 degrees north and back to 28.5 degrees south of the equator during the multiple lunar cycles within one year. As a result, tidal bulges are rarely aligned with the equator. Instead, they occur mostly north and south of the equator. The Moon affects Earth's tides more than the Sun, so tidal bulges follow the Moon, ranging from a maximum of 28.5 degrees north to a maximum of 28.5 degrees south of the equator (**Figure 9.13**).

EFFECTS OF ELLIPTICAL ORBITS Earth orbits the Sun in an elliptical orbit (**Figure 9.14**) such that Earth is 148.5 million kilometers (92.2 million miles) from the Sun during the Northern Hemisphere winter and 152.2 million kilometers (94.5 million miles) from the Sun during summer. Thus, the distance between Earth and the Sun varies by 2.5% over the course of a year. Tidal ranges are largest when Earth is near its closest point, called **perihelion** (*peri* = near, *helios* = Sun) and smallest near its most distant point, called **aphelion** (*apo* = away from, *helios* = Sun). Thus, the greatest tidal ranges typically occur in January each year.

The Moon orbits Earth in an elliptical orbit, too. The Earth–Moon distance varies by 8% (between 375,000 kilometers [233,000 miles] and 405,800 kilometers [252,000 miles]). Tidal ranges are largest when the Moon is closest to Earth, called **perigee** (*peri* = near, *geo* = Earth), and smallest when most distant, called **apogee** (*apo* = away from, *geo* = Earth) (Figure 9.14, *top*). As viewed from Earth, the Moon appears about 14% larger during perigee as compared to its smaller size during

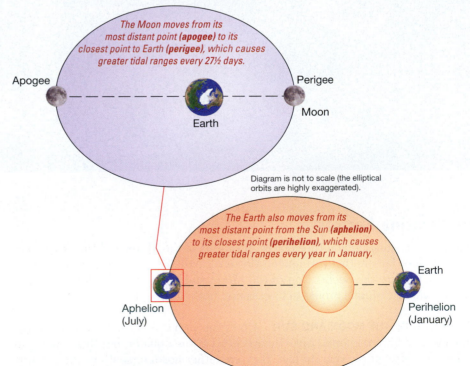

Figure 9.14 Effects of elliptical orbits.

During **apogee**, the Moon appears smaller than normal.

During **perigee**, the Moon appears about 14% larger.

When a full moon occurs at perigee, the Moon is 30% brighter and is called a **supermoon**.

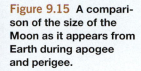

Figure 9.15 A comparison of the size of the Moon as it appears from Earth during apogee and perigee.

apogee (**Figure 9.15**); these larger moons, when they happen to coincide with full moon phase, are called *supermoons*, which are also 30% brighter.

Note that the Moon cycles between perigee, apogee, and back to perigee every 27-and-a-half days.[10] Spring tides happen to coincide with perigee about every one-and-a-half years, producing **proxigean** (*proximus* = nearest, *geo* = Earth), or "closest of the close moon," tides. During this time, the tidal range is especially large and often results in the flooding of low-lying coastal areas; if a storm occurs simultaneously, damage can be extreme. In 1962, for example, a winter storm that occurred at the same time as a proxigean tide caused widespread damage along the entire U.S. East Coast.

The elliptical orbits of Earth around the Sun and the Moon around Earth change the distances between Earth, the Moon, and the Sun, thus affecting Earth's tides. The net result is that spring tides have greater ranges during the Northern Hemisphere winter than in the summer, and spring tides have greater ranges when they coincide with perigee.

Idealized Tide Prediction

The declination of the Moon determines the position of the tidal bulges. The example illustrated in **Figure 9.16** shows that the Moon is directly overhead at 28 degrees north latitude when its declination is 28 degrees north of the equator. Imagine standing at 28 degrees north latitude and experiencing tidal conditions during a day, which is the sequence shown in **Figure 9.16a–e**:

- With the Moon directly overhead, the tidal condition experienced will be high tide (Figure 9.16a).
- Low tide occurs 6 lunar hours later (6 hours 12½ minutes solar time) (Figure 9.16b).
- Another high tide, but one much lower than the first, occurs 6 lunar hours later (Figure 9.16c).
- Another low tide occurs 6 lunar hours later (Figure 9.16d).

[10]Note that this 27-and-a-half-day perigee–apogee–perigee cycle is not to be confused with the 29-and-a-half-day monthly tidal cycle; both cycles occur simultaneously and are ongoing but are independent of one another.

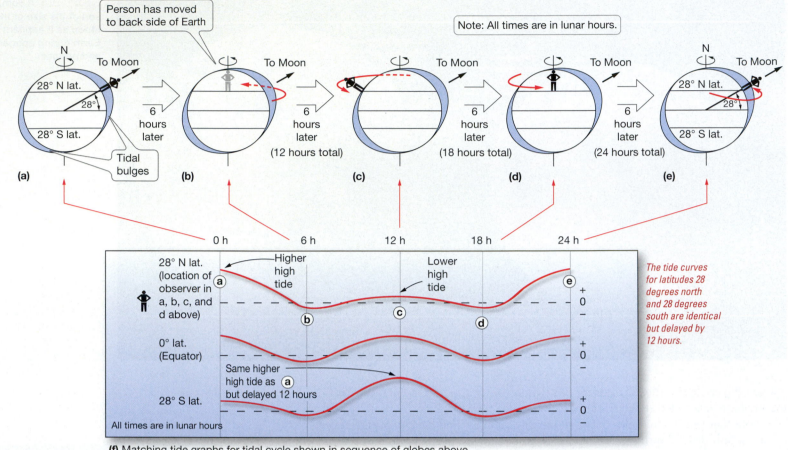

(f) Matching tide graphs for tidal cycle shown in sequence of globes above.

 SmartFigure 9.16 **Predicted idealized tides.** **(a)–(e)** Sequence showing the tide experienced every 6 lunar hours at 28 degrees north latitude when the declination of the Moon is 28 degrees north. **(f)** Corresponding tide curves for the lunar day shown in the sequence above for the latitudes 28 degrees north, 0 degrees (equator), and 28 degrees south. https://goo.gl/CFdvNW

- Six lunar hours later, at the end of a 24-lunar-hour period (24 hours 50 minutes solar time), you will have passed through a complete lunar-day cycle of two high tides and two low tides (Figure 9.16e).

The graphs in **Figure 9.16f** show the heights of the tides observed during the same lunar day at 28 degrees north latitude, the equator, and 28 degrees south latitude when the declination of the Moon is 28 degrees north of the equator. Tide curves for 28 degrees north and 28 degrees south latitude have identically timed highs and lows, but the *higher* high tides and *lower* low tides occur 12 hours later. The reason they occur out of phase by 12 hours is because the bulges in the two hemispheres are on opposite sides of Earth in relation to the Moon. MasteringOceanography **Web Table 9.1** summarizes the characteristics of the tides on the idealized Earth.

STUDENTS SOMETIMES ASK . . .

What are tropical tides?

Differences between successive high tides and successive low tides occur each lunar day (see, for example, Figure 9.16f). Because these differences occur within a period of one day, they are called *diurnal* (*daily*) *inequalities*. These inequalities are at their greatest when the Moon is at its maximum declination relative to the equator, and such tides are called *tropical tides* because at maximum declination the Moon is over one of Earth's tropics. When the Moon is over the equator (*equatorial tides*), the difference between successive high tides and low tides is minimal.

CONCEPT CHECK 9.2 | Explain how tides vary during a monthly tidal cycle.

1 To help reinforce your knowledge of how tides work, draw the positions of the Earth–Moon–Sun system during a complete monthly tidal cycle from memory. Indicate the tide conditions experienced on Earth, the phases of the Moon, the time between those phases, and syzygy and quadrature.

2 Explain why the maximum tidal range (spring tide) occurs during new and full moon phases and the minimum tidal range (neap tide) at first-quarter and third-quarter moons.

3 What is declination? Discuss the degree of declination of the Moon and Sun relative to Earth's equator. What are the effects of declination of the Moon and Sun on the tides?

4 Diagram the Earth–Moon system's orbit about the Sun. Label the positions on the orbit at which the Moon and Sun are closest to and farthest from Earth, stating the terms used to identify them. Discuss the effects of the Moon's and Earth's positions on Earth's tides.

9.3 What Do Tides Look Like in the Ocean?

The idealized tidal bulges discussed so far have been treated as *freely propagating waves* with the crests of the waves (the peaks of the tidal bulges) separated by a distance of one-half Earth's circumference—about 20,000 kilometers (12,420 miles). However, the tidal bulges themselves are continually pulled on by astronomical forces, so they actually exist on Earth as *forced waves*. What this means is they are continuously pulled along by a driving force, which is the tidal force caused mostly by the Moon. As a result, the tidal bulges move across Earth's oceans at about 1600 kilometers (1000 miles) per hour to keep up with the tidal force of the Moon. Note that it's the tidal bulge waveform—not the water itself—that moves at such great speeds.

The idealized tidal bulge model also assumes no continents on Earth, oceans that are infinitely deep, and no frictional losses. These conditions clearly don't exist on Earth, so the idealized tidal bulge model can only take us so far in explaining the tides. In reality, a variety of conditions on the real Earth cause ocean tides to break up into distinct, large circulation units in each ocean basin called *cells*.

Amphidromic Points and Cotidal Lines

In the open ocean, the crests and troughs of the tide wave rotate around an **amphidromic point** (*amphi* = around, *dromus* = running) near the center of each cell. There is essentially no tidal range at amphidromic points, but radiating from each point are **cotidal lines** (*co* = with, *tidal* = tide), which connect all nearby locations where high tide occurs simultaneously. The labels on the cotidal lines in Figure 9.17 indicate the time of high tide in hours as they rotate around the cell.

The times in Figure 9.17 indicate that the tide wave in each cell generally rotates counterclockwise in the Northern Hemisphere and clockwise in the Southern Hemisphere. The tide wave in each cell must complete one rotation during the tidal period (usually 12 lunar hours), so this limits the size of the cells.

Low tide occurs six hours after high tide in an amphidromic cell. If high tide is occurring along the cotidal line labeled "10," for example, then low tide is occurring along the cotidal line labeled "4."

Effect of the Continents

The continents affect tides, too, because they interrupt the free movement of the tidal bulges across the ocean surface. Tides are expressed in each ocean basin as freestanding waves that are affected by the position and shape of the continents that ring the ocean basin. In fact, two of the most important factors that influence tidal conditions along a coast are coastline shape and offshore depth.

Just like surface waves that undergo physical changes as they move into shallow water (such as slowing down and increasing in height; see Section 8.4 in Chapter 8), tides experience similar physical changes as they enter the shallow water of continental shelves. These changes tend to amplify the tidal range as compared to the deep ocean, where the maximum tidal range is only about 45 centimeters (18 inches).

In addition, increased turbulent mixing rates in deep water over areas of rough bottom topography (as discussed in Chapter 7) are associated with internal waves created by tides breaking on this rough topography and against continental slopes. These tide-generated internal waves have recently been observed along the chain of Hawaiian Islands, have heights of up to 300 meters (1000 feet), and contribute to increased turbulence and mixing, which strongly affect the tides.

STUDENTS SOMETIMES ASK . . .

How often are conditions right to produce the maximum tide-generating force?

Maximum tides occur when Earth is closest to the Sun (at perihelion), the Moon is closest to Earth (at perigee), and the Earth–Moon–Sun system is aligned (at syzygy) with both the Sun and Moon at zero declination. This rare condition—which creates an absolute *maximum* spring tidal range—occurs once every 1600 years. Fortunately, the next occurrence is predicted for the year 3300.

However, there are other times when conditions produce large tide-generating forces. During early 1983, for example, large, slow-moving low-pressure cells developed in the North Pacific Ocean and caused strong northwest winds. In late January, the winds produced a near fully developed 3-meter (10-foot) swell that affected the U.S. West Coast from Oregon to Baja California. The large waves would have been trouble enough under normal conditions, but there were also unusually high spring tides of 2.25 meters (7.4 feet) because Earth was near perihelion at the same time that the Moon was at perigee. In addition, a strong El Niño had raised sea level by as much as 20 centimeters (8 inches). When the waves hit the coast during these unusual conditions, they caused more than $100 million in damage, including the destruction of 25 homes, damage to 3500 others, the collapse of several commercial and municipal piers, and at least a dozen deaths.

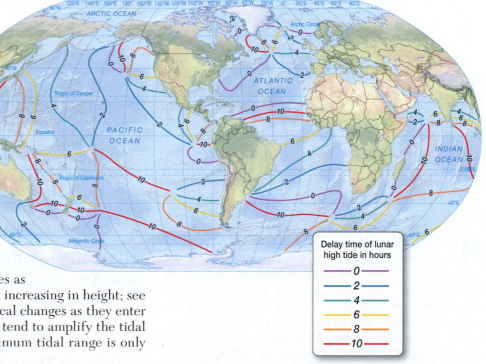

Figure 9.17 Cotidal map of the world. Cotidal lines indicate times of the main lunar daily high tide in lunar hours after the Moon has crossed the Greenwich Meridian (0 degrees longitude). Tidal ranges generally increase with increasing distance along cotidal lines away from the amphidromic points (center of the cell). Where cotidal lines terminate at both ends in amphidromic points, maximum tidal range will be near the midpoints of the lines.

Other Considerations

A detailed analysis of all the variables that affect the tides at any particular coast reveals that nearly 400 factors are involved—far more than can adequately be addressed here. The combination of all these factors creates some conditions that are unexpected based on a simple tidal model. For example, high tide rarely occurs when the Moon is at its highest point in the sky. Instead, the time between the Moon crossing the meridian and a corresponding high tide varies from place to place.

Because of the complexity of the tides, a completely mathematical model of the tides is beyond the limits of marine science. Instead, a combination of mathematical analysis and observation is required to adequately model the tides. Moreover, successful models must take into account at least 37 independent factors related to tides (the two most important are forces caused by the Moon and the Sun). Tidal models that take into account the most dominant tide-producing factors are usually quite successful in predicting future tides.

RECAP

Many factors influence the tides, and tidal prediction is more complicated than what is predicted based on a simple lunar and solar tidal bulge model.

CONCEPT CHECK 9.3 | Specify what tides look like in the ocean.

1 Are tides considered deep-water waves anywhere in the ocean? Why or why not?

2 What are amphidromic points and cotidal lines?

3 Discuss reasons why the tides don't follow a simple tidal bulge model.

9.4 What Types of Tidal Patterns Exist?

In theory, most coastal regions on Earth should experience two high tides and two low tides of unequal heights during a lunar day. In practice, however, the various depths, sizes, and shapes of ocean basins modify tides so they exhibit three different patterns in different parts of the world. The three types of tidal patterns, which are illustrated in **Figure 9.18**, are *diurnal* (*diurnal* = daily), *semidiurnal* (*semi* = half, *diurnal* = daily), and *mixed*.[11]

Diurnal Tidal Pattern

A **diurnal tidal pattern** has one high tide and one low tide each lunar day. These tides are common in shallow inland seas such as the Gulf of Mexico and along the coast of Southeast Asia. Diurnal tides have a tidal period of 24 hours 50 minutes.

Semidiurnal Tidal Pattern

A **semidiurnal tidal pattern** has two high ties and two low tides each lunar day. The heights of successive high tides and successive low tides are approximately the same.[12] Semidiurnal tides are common along the Atlantic coast of the United States. The tidal period is 12 hours 25 minutes.

Mixed Tidal Pattern

A **mixed tidal pattern** may have characteristics of both diurnal and semidiurnal tides. Successive high tides and/or low tides will have significantly different heights, a condition called *diurnal inequality*. Mixed tides commonly have a tidal period of 12 hours 25 minutes, but they may also exhibit diurnal periods. Mixed tidal patterns are commonly found along the Pacific coast of North America.

Figure 9.19 shows examples of monthly tidal curves for various coastal locations. Even though a tide at any particular location follows a single tidal pattern, it still may

Web Animation
Tidal Patterns
http://goo.gl/tdDbEx

STUDENTS SOMETIMES ASK . . .

I noticed that Figure 9.18 shows negative tides. How can there ever be a negative tide?

Negative tides occur because the *datum*—the starting point or reference point from which tides are measured—is an average of the tides over many years. Along the West Coast of the United States, for instance, the datum is mean lower low water (MLLW), which is the average of the *lower* of the two low tides that occur daily in a mixed tidal pattern. Because the datum is an average, there will be some days when the tide is less than the average (similar to the distribution of exam scores, some of which will be below the average). These lower-than-average tides are given negative values, occur only during spring tides, and are often the best times to visit local tide pool areas.

[11]Sometimes a *mixed* tidal pattern is referred to as *mixed semidiurnal*.
[12]Because tides are always growing higher or lower at any location due to the spring tide–neap tide sequence, successive high tides and successive low tides can never be *exactly* the same at any location.

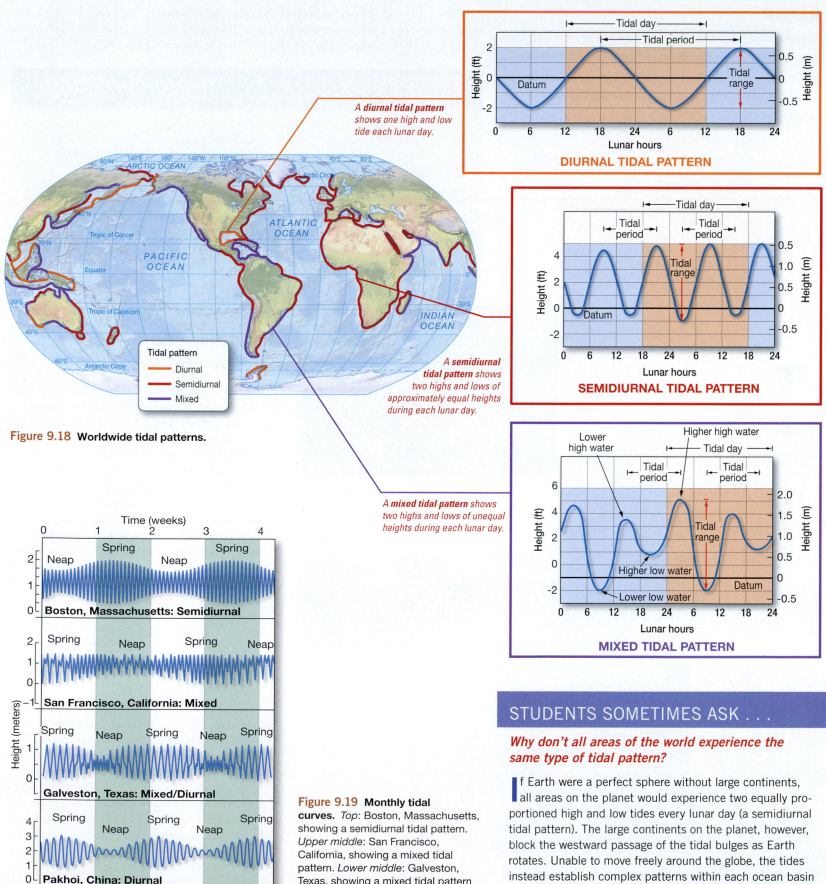

Figure 9.18 Worldwide tidal patterns.

A diurnal tidal pattern shows one high and low tide each lunar day.

DIURNAL TIDAL PATTERN

A semidiurnal tidal pattern shows two highs and lows of approximately equal heights during each lunar day.

SEMIDIURNAL TIDAL PATTERN

A mixed tidal pattern shows two highs and lows of unequal heights during each lunar day.

MIXED TIDAL PATTERN

Figure 9.19 **Monthly tidal curves.** *Top:* Boston, Massachusetts, showing a semidiurnal tidal pattern. *Upper middle:* San Francisco, California, showing a mixed tidal pattern. *Lower middle:* Galveston, Texas, showing a mixed tidal pattern with strong diurnal tendencies. *Bottom:* Pakhoi, China, showing a diurnal tidal pattern.

STUDENTS SOMETIMES ASK . . .

Why don't all areas of the world experience the same type of tidal pattern?

If Earth were a perfect sphere without large continents, all areas on the planet would experience two equally proportioned high and low tides every lunar day (a semidiurnal tidal pattern). The large continents on the planet, however, block the westward passage of the tidal bulges as Earth rotates. Unable to move freely around the globe, the tides instead establish complex patterns within each ocean basin that often differ greatly from tidal patterns of adjacent ocean basins or even other regions of the same ocean basin.

RECAP

A diurnal tidal pattern exhibits one high and one low tide each lunar day, a semidiurnal tidal pattern exhibits two high and two low tides daily of about the same height, and a mixed tidal pattern usually has two high and two low tides of different heights daily but may also exhibit diurnal characteristics.

pass through stages of one or both of the other tidal patterns. Typically, however, the tidal pattern for a location remains the same throughout the year. Also, the tidal curves in Figure 9.19 clearly show the weekly switching of the spring tide–neap tide cycle.

CONCEPT CHECK 9.4 | Compare the characteristics and coastal locations of the three types of tidal patterns.

1 What do the terms *diurnal*, *semidiurnal*, and *mixed* mean, as related to tidal patterns?

2 Describe the number of high and low tides in a lunar day, the period, and any inequality of the following tidal patterns: diurnal, semidiurnal, and mixed.

3 Using the Internet, find a coastal tidal prediction for your birthday this year (or, if your birthday has already happened, find next year's tide prediction). What tidal pattern is displayed?

4 Of the three tidal patterns, which one is most common along the U.S. East Coast? The U.S. West Coast? Worldwide?

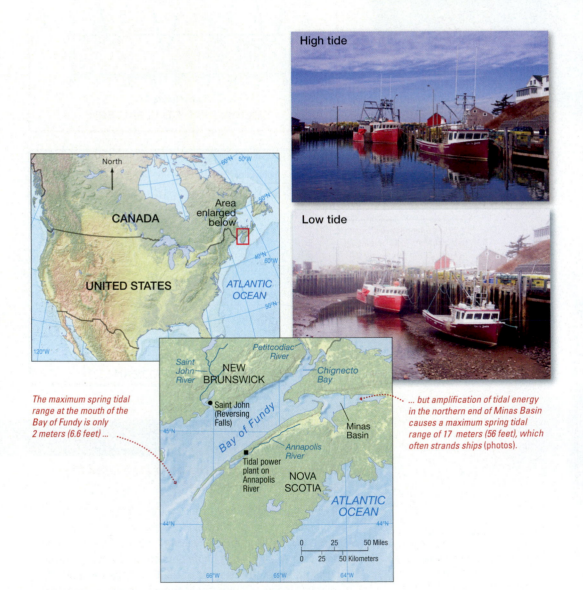

High tide

Low tide

North

CANADA

Area enlarged below

UNITED STATES

ATLANTIC OCEAN

The maximum spring tidal range at the mouth of the Bay of Fundy is only 2 meters (6.6 feet) ...

Petitcodiac River

Saint John River

NEW BRUNSWICK

Chignecto Bay

Saint John (Reversing Falls)

Bay of Fundy

Minas Basin

Annapolis River

Tidal power plant on Annapolis River

NOVA SCOTIA

ATLANTIC OCEAN

... but amplification of tidal energy in the northern end of Minas Basin causes a maximum spring tidal range of 17 meters (56 feet), which often strands ships (photos).

0 25 50 Miles
0 25 50 Kilometers

Figure 9.20 The Bay of Fundy, site of the world's largest tidal range.

9.5 What Tidal Phenomena Occur in Coastal Regions?

Remember that tides are fundamentally very long-wavelength waves. When tide waves enter coastal waters, they are subject to reflection and amplification similar to what wind-generated waves experience. In certain locations, reflected wave energy causes water to slosh around in a bay, producing *standing waves*.[13] As a result, interesting tidal phenomena are sometimes experienced in coastal waters.

Large lakes and coastal rivers experience tidal phenomena, too. In some low-lying rivers, for instance, an incoming high tide produces a *tidal bore* (**Diving Deeper 9.1**). And, as we'll see later in this section, the tides profoundly affect the spawning behavior of a certain type of marine fish.

An Example of Tidal Extremes: The Bay of Fundy

The largest tidal range in the world is found in Nova Scotia's **Bay of Fundy**. With a length of 258 kilometers (160 miles), the Bay of Fundy has a wide opening into the Atlantic Ocean. At its northern end, however, it splits into two narrow basins, Chignecto Bay and Minas Basin (**Figure 9.20**). The period of free oscillation in the bay—the oscillation that occurs when a body is displaced and then released—is very nearly that of the tidal period. The resulting constructive interference—along with the narrowing and

[13]See Chapter 8 for a discussion of standing waves, including the terms *node* and *antinode*.

TIDAL BORES: BORING WAVES THESE ARE NOT!

A tidal bore (*bore* = crest, or wave) is a wall of water that moves up certain low-lying rivers due to an incoming tide. Because it is a wave created by the tides, it is a *true* tidal wave. When an incoming tide rushes up a river, it develops a steep forward slope because the flow of the river resists the advance of the tide (**Figure 9A**). This creates a tidal bore, which may reach heights of 5 meters (16.4 feet) or more and move at speeds up to 24 kilometers (15 miles) per hour.

Tidal bores develop only in coastal regions where the following conditions are present: (1) a large spring tidal range of at least 6 meters (20 feet); (2) a tidal cycle that has a very abrupt rise of the flood tide phase and an elongated ebb tide phase; (3) a low-lying river with a persistent seaward current during the time when an incoming high tide begins; (4) a progressive shallowing of the sea floor as the basin progresses inland; and (5) a progressive narrowing of the basin toward its upper reaches. Because of these unique circumstances, only about 60 places on Earth experience tidal bores.

Although tidal bores do not commonly attain the size of waves in the surf zone, tidal bores have successfully been rafted, kayaked, and even surfed (**Figure 9B**). They can give a surfer a very long ride because the bore travels many kilometers upriver. If you miss the bore, though, you have to wait about half a day before the next one comes along because the incoming high tide occurs only twice a day.

The Amazon River is probably the longest estuary affected by oceanic tides: Tides can be measured as far as 800 kilometers (500 miles) from the river's mouth, although the effects are quite small at this distance. Tidal bores near the mouth of the Amazon River can reach heights up to 5 meters (16.4 feet) and are locally called *pororocas*, meaning "mighty noise." Other rivers that have notable tidal bores include the Qiantang River in China (which has the largest tidal bores in the world, often reaching 8 meters [26 feet] high); the Petitcodiac River in New Brunswick, Canada; the River Seine in France; the Trent and Severn Rivers in England;

and Cook Inlet near Anchorage, Alaska (where the largest tidal bore in the United States can be found). Although the Bay of Fundy has the world's largest tidal range, its tidal bore rarely exceeds 1 meter (3.3 feet), mostly because the bay is so wide.

GIVE IT SOME THOUGHT

1. Compare and contrast a tidal bore with a tsunami (see Section 8.5 in Chapter 8). Which of the two is a *true* tidal wave?

Figure 9A How a tidal bore forms (figure) and a tidal bore moving quickly upriver near Chignecto Bay, New Brunswick, Canada (photo).

Figure 9B Brazilian surf star Alex "Picuruta" Salazar tidal bore surfing on the Amazon River.

shoaling of the bay to the north—causes a buildup of tidal energy in the northern end of the bay. In addition, the bay curves to the right, so the Coriolis effect in the Northern Hemisphere adds to the extreme tidal range.

During maximum spring tide conditions, the tidal range at the mouth of the bay (where it opens to the ocean) is only about 2 meters (6.6 feet). However, the tidal range increases progressively from the mouth of the bay northward. In the northern end of Minas Basin, the maximum spring tidal range is 17 meters (56 feet), which leaves boats high and dry during low tide (Figure 9.20, *insets*).

Coastal Tidal Currents

The current that accompanies the slowly turning tide crest in a Northern Hemisphere basin rotates counterclockwise, producing a **rotary current** in the open portion of the basin. Friction increases in nearshore shoaling waters, so the rotary current changes to an alternating current, or **reversing current**, that moves into and out of restricted passages along a coast.

The velocity of rotary currents in the open ocean is usually well below 1 kilometer (0.6 mile) per hour. Reversing currents, however, can reach velocities up to 44 kilometers (28 miles) per hour in restricted channels such as between islands of coastal waters.

Reversing currents also exist in the mouths of bays (and some rivers) due to the daily flow of tides. **Figure 9.21** shows that a **flood current** is produced when water rushes into a bay (or river) with an incoming high tide. Conversely, an **ebb current** is produced when water drains out of a bay (or river) because a low tide is approaching. No currents occur for several minutes during either **high slack water** (which occurs at the peak of each high tide) or **low slack water** (at the peak of each low tide).

Reversing currents in bays can sometimes reach speeds of 40 kilometers (25 miles) per hour, creating a navigation hazard for ships. On the other hand, the daily flow of

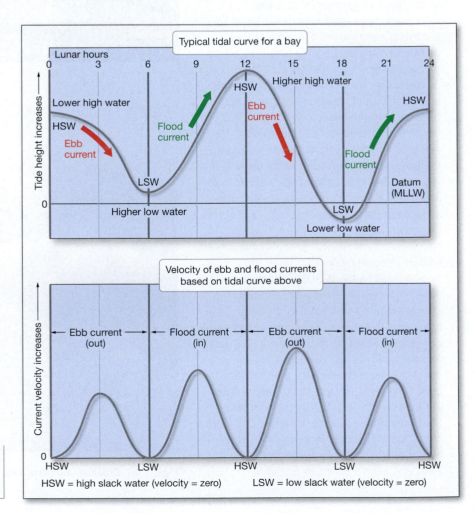

SmartFigure 9.21 **Reversing tidal currents in a bay.**
Top: Typical tidal curve for a bay that experiences a mixed tidal pattern, showing alternating ebb currents (approaching low tide) and flood currents (approaching high tide). No currents occur during either high slack water (*HSW*) or low slack water (*LSW*). The datum *MLLW* stands for *mean lower low water*, which is the average of the lower of the two low tides that occur daily in a mixed tidal pattern. *Bottom*: Corresponding chart showing the velocity of ebb and flood currents based on the tidal curve above.
https://goo.gl/g2XtZD

these currents often keeps sediment from closing off the bay and replenishes the bay with seawater and ocean nutrients.

Tidal currents can be significant even in deep-ocean waters. For example, tidal currents were encountered shortly after the discovery of the remains of the *Titanic* at a depth of 3795 meters (12,448 feet) on the continental slope south of Newfoundland's Grand Banks in 1985. These tidal currents were so strong that they forced researchers to abandon the use of the camera-equipped, tethered, remotely operated vehicle *Jason Jr.*

Whirlpools: Fact or Fiction?

A **whirlpool**—a rapidly spinning body of water, which is also termed a *vortex* (*vertere* = to turn)—can be created in some restricted coastal passages due to reversing tidal currents. Whirlpools most commonly occur in shallow passages connecting two large bodies of water that have different tidal cycles. The different tidal heights of the two bodies cause water to move vigorously through the passage. As water rushes through the passage, it is affected by the shape of the shallow sea floor, causing turbulence, which, along with spin due to opposing tidal currents, creates whirlpools. The larger the tidal difference between the two bodies of water and the smaller the passage, the greater the vortex caused by the tidal currents. Because whirlpools can have high flow rates of up to 16 kilometers (10 miles) per hour, they can cause ships to spin out of control for a short time.

One of the world's most famous whirlpools is the *Maelstrom* (*malen* = to grind in a circle, *strom* = stream), which occurs in a passage off the west coast of Arctic Norway (**Figure 9.22**). This and another famous whirlpool in the Strait of Messina, which separates mainland Italy from Sicily, are probably the source of ancient legends of huge churning funnels of water that destroy ships and carry mariners to their deaths, although they are not nearly as deadly as legends suggest. Other notable whirlpools occur off the west coast of Scotland, in the Bay of Fundy at the border between Maine and the Canadian province of New Brunswick, and off Japan's Shikoku Island.

Grunion: Doing What Comes Naturally on the Beach

From March through September, shortly after the maximum spring tide has occurred, **grunion** (*Leuresthes tenuis*) come ashore along sandy beaches of Southern California and Baja California to bury their eggs. Grunion—slender, silvery fish up to 15 centimeters (6 inches) long—are the only marine fish in the world that come completely out of water to spawn. The name *grunion* comes from the Spanish *gruñón*, which means "grunter" and refers to the faint noise they make during spawning.

A mixed tidal pattern occurs along Southern California and Baja California beaches. On most lunar days (24 hours and 50 minutes), there are two high tides and two low tides. There is usually a significant difference in the heights of the two high tides that occur each day. During the summer months, the higher high tide occurs at night. The night high tide becomes higher each night as the maximum spring tide range is approached, causing sand to be eroded from the beach (**Figure 9.23**, *graph*). After the maximum spring tide has occurred, the night high tide diminishes each night. As neap tide is approached, sand is deposited on the beach.

Grunion spawn only after each night's higher high tide has peaked on the three or four nights following the night of the highest spring high tide. This ensures that

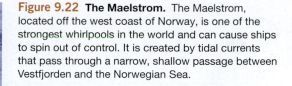

Figure 9.22 **The Maelstrom.** The Maelstrom, located off the west coast of Norway, is one of the strongest whirlpools in the world and can cause ships to spin out of control. It is created by tidal currents that pass through a narrow, shallow passage between Vestfjorden and the Norwegian Sea.

9.1 Squidtoons

Born only when sun, moon, & earth align.

https://goo.gl/TRduZw

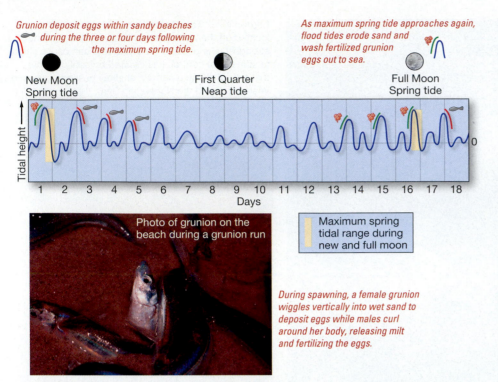

Grunion deposit eggs within sandy beaches during the three or four days following the maximum spring tide.

As maximum spring tide approaches again, flood tides erode sand and wash fertilized grunion eggs out to sea.

New Moon
Spring tide

First Quarter
Neap tide

Full Moon
Spring tide

Tidal height

0

1 2 3 4 5 6 7 8 9 10 11 12 13 14 15 16 17 18

Days

Photo of grunion on the beach during a grunion run

Maximum spring tidal range during new and full moon

During spawning, a female grunion wiggles vertically into wet sand to deposit eggs while males curl around her body, releasing milt and fertilizing the eggs.

Figure 9.23 The tidal cycle and spawning grunion. During summer months and for 3 or 4 days after the highest spring tides (*graph*), grunion deposit their eggs on sandy beaches (*photo*). The successively lower high tides during the approaching neap tide conditions won't wash the eggs from the sand until they are ready to hatch about 10 days later. As the next spring tide is approached, successively higher high tides wash the eggs free and allow them to hatch. The spawning cycle begins a few days later, after the peak of spring tide conditions, with the next cycle of successively lower high tides.

RECAP

Coastal tidal phenomena include large tidal ranges (the largest of which—17 meters [56 feet]—occurs in the Bay of Fundy), tidal currents, rapidly spinning vortices called whirlpools, and grunion, which time their spawning cycles with the tides.

their eggs will be covered deeply in sand deposited by the receding higher high tides each succeeding night. The fertilized eggs buried in the sand are ready to hatch about 10 days after spawning. By this time, another spring tide is approaching, so the night high tide is getting progressively higher each night again. The beach sand is eroding again, too, which exposes the eggs to the waves that break ever higher on the beach. The eggs hatch about three minutes after being freed in the water. Tests done in laboratories have shown that the grunion eggs will not hatch until agitated in a manner that simulates that of the eroding waves.

The spawning begins as the grunion come ashore immediately following an appropriate high tide, and it may last from one to three hours. Spawning usually peaks about an hour after it starts and may last an additional 30 minutes to an hour. During this time, the beach may be littered with thousands of fish. Females, which are larger than males, catch waves and swim high onto the beach. If no males are near, a female may return to the water without depositing her eggs. In the presence of males, she drills her tail into the semifluid sand until only her head is visible. The female continues to twist, depositing her eggs 5 to 7 centimeters (2 to 3 inches) below the surface.

The male curls around the female's body and deposits his milt against it (Figure 9.23, *photo*). The milt runs down the body of the female to fertilize the eggs. When the spawning is completed, both fish return to the water with the next wave.

Larger females are capable of producing up to 3000 eggs for each series of spawning runs, which are separated by the two-week period between spring tides. As soon as the eggs are deposited, another group of eggs begins to form within the female. These eggs will be deposited during the next spring tide run. Early in the season, only older fish spawn. By May, however, even the one-year-old females are in spawning condition.

Young grunion grow rapidly and are about 12 centimeters (5 inches) long when they are a year old and ready for their first spawning. They usually live two or three years, but four-year-olds have been recovered. The age of a grunion can be determined by its scales. After growing rapidly during the first year, they grow very slowly. In fact, there is no growth at all during the six-month spawning season, which causes marks to form on each scale that can be used to identify a grunion's age.

It is not known exactly how grunion are able to time their spawning behavior so precisely with the tides. Research suggests that grunion are somehow able to sense very small changes in hydrostatic pressure caused by rising and falling sea level due to changing tides. Certainly, a very dependable detection mechanism keeps the grunion accurately informed of the tidal conditions because their survival depends on a spawning behavior precisely tuned to the tides.

Interdisciplinary

Relationship

CONCEPT CHECK 9.5 | Describe tidal phenomena that occur in coastal regions.

1 Discuss factors that help produce the world's largest tidal range in the Bay of Fundy.

2 Specify the differences between rotary and reversing tidal currents.

3 Of flood current, ebb current, high slack water, and low slack water, when is the best time to enter a bay

by boat? When is the best time to navigate in a shallow, rocky harbor? Explain.

4 Describe the spawning cycle of grunion, indicating the relationship among tidal phenomena, where grunion lay their eggs, and the movement of sand on the beach.

9.6 Can Tidal Power Be Harnessed as a Source of Energy?

Throughout history, ocean tides have been used as a source of power. During high tide, water can be trapped in a basin and then harnessed to do work as it flows back to the sea. In the 12th century, for example, water wheels driven by the tides were used to power gristmills and sawmills. During the 17th and 18th centuries, much of Boston's flour was produced at a tidal mill.

Today, tidal power is considered a clean, renewable resource with vast potential. The initial cost of building a tidal power electricity-generating plant may be higher than the cost of building a conventional thermal power plant, but the operating costs are lower because it does not use fossil fuels or radioactive substances to generate electricity.

One disadvantage of tidal power, however, is the periodicity of the tides, which allows power to be generated only during a portion of a 24-hour day. People operate on a solar period, but tides operate on a lunar period, so the energy available from the tides would coincide with need only part of the time. Power would have to be distributed to the point of need at the moment it was generated, which could be a great distance away, resulting in an expensive transmission problem. The power could be stored, but even this alternative presents a large and expensive technical problem.

To generate electricity effectively, electrical turbines (generators) need to run at a constant speed, which is difficult to maintain when generated by the variable flow of tidal currents in two directions (flood tide and ebb tide). Specially designed turbines that allow both advancing and receding water to spin their blades are necessary to solve the problem of generating electricity from the tides.

Other potential disadvantages of tidal power include environmental concerns such as change in habitat and harm to wildlife. For example, most tidal power plants involve a dam that alters the ecology of estuaries where they are located. Also, tidal power plants disturb the normal flow of tidal currents and negatively affect marine organisms that depend on these currents for bringing food or for aiding migrations. Another example is the harm done to marine organisms entrapped or injured by moving tidal power devices. Underwater turbines produce noise, too, that could disturb marine organisms or have long-term negative effects. Lastly, a tidal power plant would likely interfere with many traditional human uses of estuaries, such as transportation and fishing.

Tidal Power Plants

Tidal power can be harnessed in one of two ways: (1) Tidal water trapped behind coastal barriers in bays and estuaries during high tide can be released at a later time to turn turbines and generate electrical energy and (2) tidal currents that pass through narrow channels can be used to turn underwater pivoting turbines, which produce energy (see Section 7.6 in Chapter 7). Although the first type is much more commonly employed, Norway, the United Kingdom, and the United States have recently installed offshore turbines that harness swift coastal tidal currents and plan to expand these devices into tidal energy farms.

One example of a successful tidal power plant that uses water trapped behind a coastal barrier and has been in operation since 1966 is at Saint-Malo, France, on the estuary of the La Rance River (**Figure 9.24**). The La Rance estuary has a surface area of approximately 23 square kilometers (9 square miles) and its tidal range is 13.4 meters (44 feet). A general rule of tidal power is that usable tidal energy increases as the area of the basin increases and as the tidal range increases.

The power-generating barrier was built across the La Rance estuary a little over 3 kilometers (2 miles) upstream to protect it from storm waves. The barrier is 760 meters (2500 feet) wide and supports a two-lane road (Figure 9.24). Water passing through the barrier powers 24 electricity-generating units that operate beneath the power plant. At peak operating capacity, each unit can generate 10 megawatts of electricity, for a total of 240 megawatts.[14]

[14]Each megawatt of electricity is enough to serve the energy needs of about 250 average U.S. homes.

9.2 How do tides vary during a monthly tidal cycle?

▶ Tides would be easy to predict if Earth were a uniform sphere covered with an ocean of uniform depth. For most places on Earth, *the time between successive high tides would be 12 hours 25 minutes (half a lunar day)*. The *29-and-a-half-day monthly tidal cycle* would consist of tides with maximum tidal range (spring tides) and minimum tidal range (neap tides). *Spring tides would occur each new moon and full moon, and neap tides would occur each first- and third-quarter phases of the Moon.*

▶ The *declination of the Moon* varies between 28.5 degrees north or south of the equator during the lunar month, and the *declination of the Sun* varies between 23.5 degrees north or south of the equator during the year, so *the location of tidal bulges usually creates two high tides and two low tides of unequal height per lunar day*. Tidal ranges are greatest when Earth is nearest the Sun and Moon.

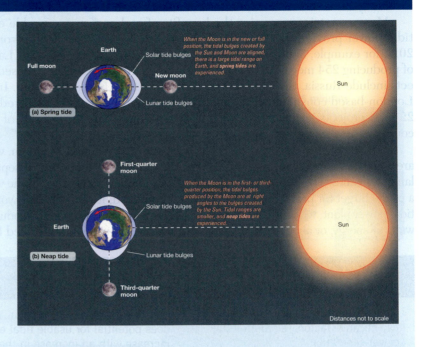

Study Resources

MasteringOceanography Study Area Quizzes, MasteringOceanography Web Table 9.1, MasteringOceanography Web Animations

Critical Thinking Question

Specify differences between these two astronomical cycles that occurs simultaneously: the 27-and-a-half-day perigee–apogee–perigee cycle and the 29-and-a-half-day monthly tidal cycle.

Active Learning Exercise

This is a take-home assignment. Observe the Moon from a reference location every night at about the same time for two weeks. Keep track of your observations about the shape (phase) of the Moon and its position in the sky. Then compare your observations to the reported tides in your area (or a coastal area you visit frequently) and report your findings to the class.

9.3 What do tides look like in the ocean?

▶ *When friction and the true shape of ocean basins are considered, the dynamics of tides become more complicated.* Moreover, the two bulges on opposite sides of Earth cannot exist because they cannot keep up with the rotational speed of Earth. Instead, the bulges are broken up into *several tidal cells that rotate around an amphidromic point*—a point of zero tidal range. Rotation is counterclockwise in the Northern Hemisphere and clockwise in the Southern Hemisphere. *Many other factors influence tides on Earth*, too, such as the positions of the continents, the varying depth of the ocean, and coastline shape.

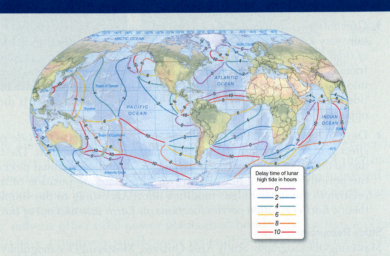

Study Resources

MasteringOceanography Study Area Quizzes

Critical Thinking Question

Explain amphidromic points and cotidal lines. How do they influence tides in the open ocean?

Active Learning Exercise

Assume that the tides are free waves and are therefore governed by the conditions and equations for ocean waves in Chapter 8. Consider the tidal bulges as wave crests with a wavelength of one-half Earth's circumference—about 20,000 kilometers (12,420 miles)—and assume that the average height of the lunar bulges is 3 meters (10 feet). Working with another student in class and using information from Chapter 8, determine if tides are deep-water or shallow-water waves. Note that the average depth of the oceans is 3.7 kilometers (2.3 miles) and the deepest ocean depth is 11 kilometers (6.8 miles).

CONCEPT CH

1 Explain the
shore and the

2 What speci
included in a t

3 How does th
beach face?

10.2 How

The movement
shoreline (*towar*
(often referred to

Movement F

Sand on the bea
breaking waves.

MECHANISM A
ward the berm. S
returns to the oc
as **backwash**, tho
swash over the top

While standin
swash and backwa
pendicular to the
mines whether sa

LIGHT VERSUS HE
acterized by less
beach, so backwa
tem, therefore c
toward the berm,

During *heavy*
the beach is satu
of the swash soak
system, therefore
which erodes the
swash comes *on to*
ing the beach fro
backwash.

During heavy
The orbital motio
shore. Thus, the sa
forms one or more

SUMMERTIME AND
alternate seasonall
they produce chang
duces a wide sandy
beach—at the ex

Figure 10.3 Summer
(a) summertime and **(b)**

9.4 What types of tidal patterns exist?

▶ The *three types of tidal patterns* observed on Earth are *diurnal* (a single high and low tide each lunar day), *semidiurnal* (two high and two low tides each lunar day), and *mixed* (characteristics of both). Mixed tidal patterns usually consist of semidiurnal periods with significant diurnal inequality. Mixed tidal patterns are the most common type in the world.

Study Resources
MasteringOceanography Study Area Quizzes, MasteringOceanography Web Animation

Critical Thinking Question
Assume that there are two moons in orbit around Earth that are on the same orbital plane but always on opposite sides of Earth and that each moon is the same size and mass of our Moon. How would this affect the tidal range during spring and neap tide conditions? Also, how would this affect the tidal pattern observed?

Active Learning Exercise
Working with another student in class, draw a monthly tidal curve for a fictional location. On the horizontal axis, use time in hours to show the length of two days; on the vertical axis, use tidal height. Also label the high and low tides. Present it to the class and indicate if it represents diurnal, semidiurnal, or mixed tide conditions.

9.5 What tidal phenomena occur in coastal regions?

▶ There are many types of *observable tidal phenomena in coastal areas. Tidal bores are true tidal waves* (a wave produced by the tides) that occur in certain rivers and bays due to an incoming high tide. The effects of constructive interference together with the shoaling and narrowing of coastal bays creates the *largest tidal range in the world—17 meters (56 feet)—at the northern end of Nova Scotia's Bay of Fundy. Tidal currents follow a rotary pattern* in open-ocean basins but are converted to *reversing currents* along continental margins. The maximum velocity of reversing currents occurs during flood and ebb currents, when the water is halfway between high slack water and low slack water. *Whirlpools* can be created in some restricted coastal passages due to reversing tidal currents. *The tides are also important to many marine organisms.* For instance, *grunion*—small silvery fish that inhabit waters along the West Coast of North America—time their spawning cycle to match the pattern of the tides.

Study Resources
MasteringOceanography Study Area Quizzes

Critical Thinking Question
Examine a current monthly tide calendar for a coastal location. When is the best time for a field trip to the tide pools? If this were a grunion spawning month, when would the conditions be right for the next grunion spawning run?

Active Learning Exercise
Working with another student in class and using the Internet, find a monthly tidal chart for the Bay of Fundy. Discuss the tidal information that is displayed, including spring and neap tide conditions, the corresponding phases of the Moon, and the amount of tidal range.

9.6 Can tidal power be harnessed as a source of energy?

▶ *Tides can be used to generate power* without need for fossil or nuclear fuel. There are some *significant drawbacks*, however, to creating successful tidal power plants. Still, many sites worldwide have the *potential for tidal power generation.*

Study Resources
MasteringOceanography Study Area Quizzes

Critical Thinking Question
What are some negative environmental factors that could inhibit the development of ocean tidal power systems?

Active Learning Exercise
Split the class into groups of two. In each group, have each student pick one of the two methods in which tidal power is harnessed. Debate the relative advantages and disadvantages of each method. For example, which method has fewer negative environmental concerns?

MasteringOceanography™

www.masteringoceanography.com

Looking for additional review and test prep materials? With individualized coaching on the toughest topics of the course, MasteringOceanography offers a wide variety of ways for you to move beyond memorization and deeply grasp the underlying processes of how the oceans work. Visit the Study Area in **www.masteringoceanography. com** to find practice quizzes, study tools, and multimedia that will improve your understanding of this chapter's content. Sign in today to enjoy the following features: Self Study Quizzes, SmartFigures, SmartTables, Oceanography Videos, Squidtoons, Geoscience Animation Library, RSS Feeds, Digital Study Modules, and an optional Pearson eText.

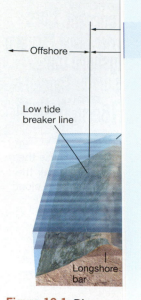

- Offshore -

Low tide
breaker line

Longshore
bar

Figure 10.1 Diagram
showing beach termin
fined as the entire active
waves; it extends from th
far end of the berm (*righ*
can be found at any bea

Figure 10.2 Photo of a
a typical beach are seen i
the dry, flat area close to
wet, gently-sloping area (*l*

Emerging shorelines include ancient
sea cliffs and marine terraces
with stranded beach
deposits.

Present sea level

Submerging shorelines
include wave-cut benches with
drowned beach deposits.

Uplift

Figure 10.17 Features of ancient emerging and submerging
shorelines.

Features of Emerging Shorelines

Marine terraces (**Figure 10.17**; see also Figures 10.8
and 10.9) are one feature characteristic of emerging
shorelines. Marine terraces are flat platforms backed
by cliffs, which form when a wave-cut bench is exposed
above sea level. **Stranded beach deposits** and other
evidence of marine processes such as ancient sea cliffs
may exist many meters above the present shoreline, in-
dicating that the former shoreline has risen above sea
level (Figure 10.17).

Features of Submerging Shorelines

Features characteristic of submerging shorelines include
wave-cut benches below sea level that contain **drowned**
beaches (Figure 10.17). Other features of submerging
shorelines include **submerged dune topography** and
drowned river valleys along the present shoreline.

Changes in Sea Level

What causes the changes in sea level that produce submerging and emerging shore-
lines? One mechanism is the raising or lowering of the land surface relative to sea
level through the movement of Earth's crust. Another mechanism is the alteration
of the level of the sea itself through worldwide changes in sea level.

MOVEMENT OF EARTH'S CRUST The elevation of Earth's crust relative to sea level
can be affected by tectonic movements and by isostatic adjustment.[2] These are called
changes in relative sea level, because it's the land that has changed, not the sea.

Tectonic Movements The most dramatic changes in sea level during the past
3000 years have been caused by *tectonic movements*, which affect the elevation
of the land. These changes include uplift or subsidence of major portions of conti-
nents or ocean basins, as well as localized folding, faulting, or tilting of the conti-
nental crust.

Most of the U.S. Pacific coast, for example, is an emerging shoreline because
continental margins where plate boundaries occur are tectonically active, produc-
ing earthquakes, volcanoes, and mountain chains paralleling the coast. Most of the
U.S. Atlantic coast, on the other hand, is a submerging shoreline. When a continent
moves away from a spreading center (such as the Mid-Atlantic Ridge), its trailing
edge subsides because of cooling and the additional weight of accumulating sedi-
ment. Passive margins experience only a low level of tectonic deformation, earth-
quakes, and volcanism, making the Atlantic coast far more quiet and stable than the
Pacific coast.

Isostatic Adjustment Earth's crust also undergoes *isostatic adjustment*: It sinks
under the accumulation of heavy loads of ice, vast piles of sediment, or outpour-
ings of lava, and it rises when heavy loads are removed (**Figure 10.18**).

For example, at least four major accumulations of glacial ice—and dozens of
smaller ones—have occurred in high-latitude regions over the past 3 million years.
Although Antarctica is still covered by a very large, thick ice cap, much of the ice
that once covered northern Asia, Europe, and North America has melted.

[2]Recall that isostatic adjustment of Earth's crust is also discussed in Chapter 1.

The weight of ice sheets as much as 3 kilometers (2 miles) thick caused the crust beneath to sink (Figure 10.18). Today, these areas are still slowly rebounding, 18,000 years after the ice began to melt. The floor of Hudson Bay, for example, which is now about 150 meters (500 feet) deep, will be close to or above sea level by the time it stops isostatically rebounding. Another example is the Gulf of Bothnia (between Sweden and Finland), which has isostatically rebounded 275 meters (900 feet) during the past 18,000 years.

Generally, tectonic and isostatic changes in sea level are confined to a segment of a continent's shoreline. For a *worldwide* change in sea level, there must be a change in seawater volume or ocean basin capacity.

WORLDWIDE (EUSTATIC) CHANGES IN SEA LEVEL Changes in sea level that are experienced worldwide due to changes in seawater volume or ocean basin capacity are called **eustatic sea level changes** (*eu* = good, *stasis* = standing).[3] The formation or destruction of large inland lakes, for example, causes small eustatic changes in sea level. When lakes form, they trap water that would otherwise run off the land into the ocean, so sea level is lowered worldwide. When lakes are drained and release their water back to the ocean, sea level rises.

Another example of a eustatic change in sea level is through changes in sea floor spreading rates, which can change the capacity of the ocean basin and affect sea level worldwide. Fast spreading, for instance, produces larger rises, such as the East Pacific Rise, which displace more water than slow-spreading ridges such as the Mid-Atlantic Ridge. Thus, fast spreading raises sea level, whereas slower spreading lowers sea level worldwide. Significant changes in sea level due to changes in spreading rate typically take hundreds of thousands to millions of years and may have changed sea level by 1000 meters (3300 feet) or more in the geologic past.

Changes to Sea Level during Ice Ages Ice ages cause eustatic sea level changes, too. As glaciers form, they tie up vast volumes of water on land, eustatically lowering sea level. An analogy to this effect is a sink of water representing an ocean basin. To simulate an ice age, some of the water from the sink is removed and frozen, causing the water level of the sink to be lower. In a similar fashion, worldwide sea level is lower during an ice age. During interglacial stages (such as the one we are in at present), the glaciers melt and release great volumes of water that drain to the sea, eustatically raising sea level. This would be analogous to putting a frozen chunk of ice on the counter near the sink and letting the ice melt, causing the water to drain into the sink and raise "sink level."

During the *Pleistocene Epoch*,[4] glaciers advanced and retreated many times on land in middle- to high-latitude regions, causing sea level to fluctuate considerably. The thermal contraction and expansion of the ocean as its temperature decreased and increased, respectively, affected sea level, too. The thermal contraction and expansion of seawater work much like a mercury thermometer: As

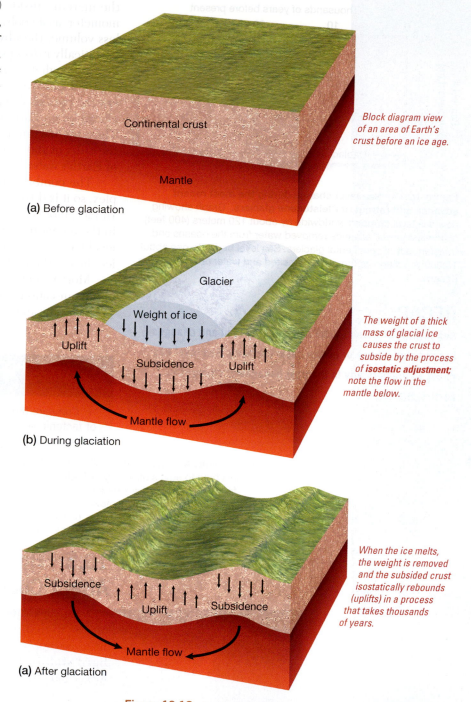

(a) Before glaciation

Block diagram view of an area of Earth's crust before an ice age.

Continental crust

Mantle

(b) During glaciation

*The weight of a thick mass of glacial ice causes the crust to subside by the process of **isostatic adjustment**; note the flow in the mantle below.*

Glacier

Weight of ice

Uplift Subsidence Uplift

Mantle flow

(a) After glaciation

When the ice melts, the weight is removed and the subsided crust isostatically rebounds (uplifts) in a process that takes thousands of years.

Subsidence

Uplift Subsidence

Mantle flow

Figure 10.18 **Isostatic adjustment caused by glacial ice.**

Web Animation
Glacial Isostasy
http://goo.gl/vz3ZDT

[3]The term *eustatic* refers to a highly idealized situation in which all of the continents remain static (in *good standing*), while only the sea rises or falls.

[4]The Pleistocene Epoch of geologic time, which is also called the "Ice Age," occurred 2.6 million to 10,000 years ago (see the Geologic Time Scale, Figure 1.31).

Land

Jetties

Another type
similar to a g
structed of rip
from waves, a

More groins ar
is created (Figu

Does a gr
Sand eventual
tional sand on
gineering and
seasonal wave
sand to move a
the last groin.
many areas be
excessive use

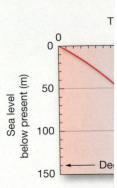

Figure 10.19 Sea le
advance and retreat
how sea level dropped
as the last glacial adva
transferred it to contin
18,000 years ago as th
oceans.

RECAP

Sea level is affected

in seawater volume

changed dramaticall

Earth's climate.

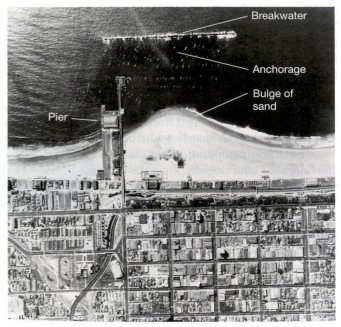

(a) The shoreline and pier at Santa Monica as it appeared in September 1931, before the breakwater was constructed in 1933. Note that the pier is on stilts and thus does not affect longshore transport.

(b) The same area in 1949, showing that the construction of the breakwater to create a boat anchorage disrupted the longshore transport of sand and caused a bulge of sand on the beach in the wave shadow behind the breakwater.

Figure 10.26 Breakwater at Santa Monica, California. Paired aerial photos of the Santa Monica pier and shoreline **(a)** before a breakwater was built and **(b)** after the breakwater, showing how the breakwater caused a bulge of sand on the beach. After the breakwater was destroyed by waves in 1983, the bulge disappeared and the shoreline returned to a straight shoreline.

STUDENTS SOMETIMES ASK . . .

I have the opportunity to live in a house at the edge of a coastal cliff where there is an incredible view along the entire coast. Is it safe from coastal erosion?

Based on what you've described, most certainly not! Geologists have long known that cliffs are naturally unstable. Even if cliffs appear to be stable (or have been stable for a number of years), they can be severely damaged during just one significant storm.

The most common cause of coastal erosion is direct wave attack, which undermines the support and causes the cliff to fail. You might want to check the base of the cliff and examine the local bedrock to determine for yourself if you think it will withstand the pounding of powerful storm waves that can move rocks weighing several tons. Other dangers include drainage runoff, weaknesses in the bedrock, slumps and landslides, seepage of water through the cliff, and even burrowing animals. Although all states enforce a setback from the edge of the cliff for all new buildings, sometimes that isn't enough because large sections of "stable" cliffs can fail all at once. For instance, several city blocks of real estate have been eroded from the edge of cliffs during the past 100 years in some areas of Southern California. Even though the view sounds outstanding, you may find out the hard way that the house is built a little *too* close to the edge of a cliff!

Seawalls

One of the most destructive types of hard stabilization is a **seawall** (**Figure 10.27**), which is built parallel to the shore, along the landward side of the berm. The purpose of a seawall is to armor the coastline and protect landward developments from ocean waves.

Once waves begin breaking against a seawall, however, turbulence generated by the abrupt release of wave energy quickly erodes the sediment on its seaward side, which can eventually cause it to collapse into the surf (Figure 10.27). In many cases where seawalls have been used to protect property on barrier islands, the seaward slope of the island beach has steepened and the rate of erosion has increased, causing the destruction of the recreational beach.

A well-designed seawall may last for many decades, but the constant pounding of waves eventually takes its toll (**Figure 10.28**). In the long run, the cost of repairing or replacing seawalls will be more than the property is worth, and the sea will claim more of the coast through the natural processes of erosion. It's just a matter of time for homeowners who live too close to the coast, many of whom are gambling that their houses won't be destroyed in their lifetimes.

Alternatives to Hard Stabilization

Is it a good idea to preserve the houses of a few people who have built too close to the shore by armoring the coast with hard stabilization even though it destroys the recreational beach? If you own coastal property, your response would probably be different from the response of the general beachgoing public. Because hard stabilization has been shown to have negative environmental consequences, alternatives have been sought.

CONSTRUCTION RESTRICTIONS One of the simplest alternatives to the use of hard stabilization is to restrict construction in areas prone to coastal erosion. Unfortunately, this is becoming less and less an option as coastal regions experience population increases and governments increase the risk of damage and injuries because of programs like the *National Flood Insurance Program (NFIP)*. Since its inception in 1968, NFIP has paid out billions of dollars in federal subsidies to repair or replace high-risk coastal structures. As a result, NFIP has actually *encouraged* construction in exactly the unsafe locations it was designed to prevent![5] Further, many homeowners spend large amounts of money rebuilding structures and fortifying their property.

BEACH REPLENISHMENT Another alternative to hard stabilization is **beach replenishment** (also called **beach nourishment**), in which sand is added to the beach to replace lost sediment (**Figure 10.29**). Although rivers naturally supply sand to most beaches, dams on rivers restrict the sand supply that would normally arrive at beaches. When inland dams are built, their effects on beaches far downstream are rarely considered. It's not until beaches begin disappearing that the rivers are seen as parts of much larger systems that operate along the coast.

Beach replenishment is expensive, however, because huge volumes of sand must be continually supplied to the beach. The cost of beach replenishment depends on the type and quantity of material placed on the beach, how far the material must be transported, and how it is to be distributed on the beach. Most sand used for replenishment comes from offshore areas, but sand that is dredged from nearby rivers, drained dams, harbors, and lagoons is also used.

The average cost of sand used to replenish beaches is between $5 and $10 per 0.76 cubic meter (1 cubic yard). In comparison, a typical top-loading trash dumpster holds about 2.3 cubic meters (3 cubic yards) of material, and a typical dump truck holds a volume of about 45 cubic meters (60 cubic yards). The drawbacks of beach replenishment projects are that a huge volume of sand must be moved and that new sand must be supplied on a regular basis. These problems often cause replenishment projects to exceed the monetary limits of what can be reasonably accomplished. For example, a small beach replenishment project of several hundred cubic meters can cost around $10,000 per year. Larger projects—requiring several thousand cubic meters of sand—cost several million dollars per year.

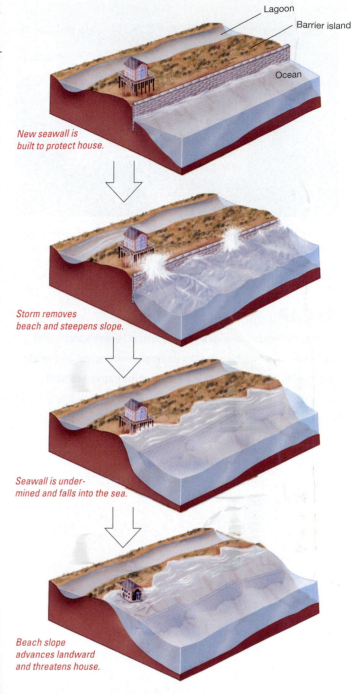

New seawall is
built to protect house.

Storm removes
beach and steepens slope.

Seawall is undermined and falls into the sea.

Beach slope
advances landward
and threatens house.

Figure 10.27 Seawalls and beaches. Diagrammatic sequence of the negative consequences that can occur when a seawall is built along a barrier island beach to protect beachfront property.

Figure 10.28 Seawall damage. A seawall in Solana Beach, California, that has been damaged by waves and needs repair. Although seawalls appear to be sturdy, they can be destroyed by the continual pounding of high-energy storm waves. In addition, high energy waves sometimes carry driftwood or logs that pound seawalls with the strength of battering rams.

[5]Changes in regulations of the Federal Emergency Management Agency (FEMA), which oversees NFIP, are intended to curb this practice.

Figure 10.29 Beach replenishment. Beach replenishment projects, such as this one in Carlsbad, California, are used to widen beaches. Beach replenishment involves dredging sand from offshore or coastal locations, pumping it through a pipe (*lower right*), and spreading it across the beach.

> Dredged sand exiting pipe

RECAP

Hard stabilization includes groins, jetties, breakwaters, and seawalls, all of which alter the coastal environment and result in changes in the shape of the beach. Alternatives to hard stabilization include construction restrictions, beach replenishment, and relocation.

RELOCATION U.S. coastal policy has recently shifted from defending coastal property in high-hazard areas to removing structures and letting nature reclaim the beach. This approach, called **relocation**, involves moving structures to safer locations as they become threatened by erosion. One example of the successful use of this technique is the relocation of the Cape Hatteras Lighthouse in North Carolina (see MasteringOceanography Web Diving Deeper 10.1). Relocation, if done wisely, can allow humans to live in balance with the natural processes that continually modify beaches.

CONCEPT CHECK 10.5	Describe the types of hard stabilization and evaluate various alternatives.

1 List the types of hard stabilization and describe what each is intended to do.

2 Overall, does a groin add any additional sand to the beach? Explain.

3 Why do groins often multiply to form a groin field?

4 When a breakwater was built in Santa Monica, what unexpected problem occurred? What was done to alleviate the problem (before the breakwater was destroyed by waves)?

5 Describe alternatives to hard stabilization, including potential drawbacks of each.

10.6 What are the Characteristics and Types of Coastal Waters?

Just offshore of beaches are **coastal waters**, which are the relatively shallow-water areas that adjoin continents or islands. If the continental shelf is broad and shallow, coastal waters can extend several hundred kilometers from land. If the continental shelf has significant relief or drops rapidly onto the deep-ocean basin, on the other hand, coastal waters will occupy a relatively thin band near the margin of the land. Beyond coastal waters lies the *open ocean*.

Coastal waters are important for many reasons. This section first describes the unique characteristics of coastal waters, then examines the various types of coastal waters, including estuaries, lagoons, and marginal seas.

Characteristics of Coastal Waters

Because of their proximity to land, coastal waters are directly influenced by processes that occur on or near land. River runoff and tidal currents, for example, have a far more significant effect on coastal waters than on the open ocean.

SALINITY Freshwater is less dense than seawater, so river runoff does not mix well with seawater along the coast. Instead, the freshwater forms a wedge at the surface, which creates a well-developed **halocline**[6] (**Figure 10.30a**). When water is shallow enough, however, tidal mixing causes freshwater to mix with seawater, thus reducing the salinity of the water column (**Figure 10.30c**). There is no halocline here; instead, the water column is **isohaline** (*iso* = same, *halo* = salt).

Freshwater runoff from the continents generally lowers the salinity of coastal regions compared to the open ocean. Where precipitation on land is mostly rain, river runoff peaks in the rainy season. Where runoff is due mainly to melting snow and ice, on the other hand, runoff always peaks in summer.

Prevailing offshore winds can increase the salinity in some coastal regions. As winds travel over a continent, they usually lose most of their moisture. When these

[6]Recall that a halocline (*halo* = salt, *cline* = slope) is a layer of rapidly changing salinity, as discussed in Chapter 5.

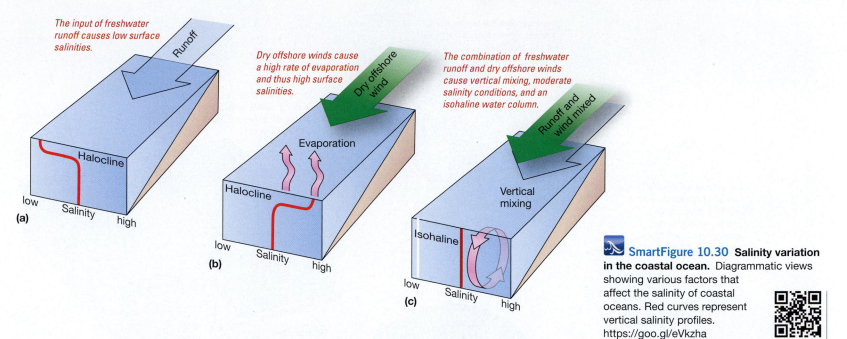

The input of freshwater runoff causes low surface salinities.

Runoff

Halocline

low Salinity high

(a)

Dry offshore winds cause a high rate of evaporation and thus high surface salinities.

Dry offshore wind

Evaporation

Halocline

low Salinity high

(b)

The combination of freshwater runoff and dry offshore winds cause vertical mixing, moderate salinity conditions, and an isohaline water column.

Runoff and wind mixed

Vertical mixing

Isohaline

low Salinity high

(c)

SmartFigure 10.30 **Salinity variation in the coastal ocean.** Diagrammatic views showing various factors that affect the salinity of coastal oceans. Red curves represent vertical salinity profiles. https://goo.gl/eVkzha

dry winds reach the ocean, they typically evaporate considerable amounts of water as they move across the surface of the coastal waters. The increased evaporation rate increases surface salinity, creating a halocline (**Figure 10.30b**). The gradient of the halocline, however, is reversed compared to the one developed from the input of freshwater (Figure 10.30a).

TEMPERATURE In low-latitude coastal regions, where circulation with the open ocean is restricted, surface waters are prevented from mixing thoroughly, so sea surface temperatures may approach 45°C (113°F) (**Figure 10.31a**). Alternatively, sea ice forms in many high-latitude coastal areas where water temperatures are uniformly cold—generally lower than –2°C (28.4°F) (**Figure 10.31b**). In both low- and high-latitude coastal waters, **isothermal** (*iso* = same, *thermo* = heat) conditions prevail.

Surface temperatures in middle-latitude coastal regions are coolest in winter and warmest in late summer. A strong **thermocline**[7] may develop from surface water being warmed during the summer (**Figure 10.31c**) and cooled during the winter (**Figure 10.31d**). In summer, very high-temperature surface water may form a relatively thin layer. Vertical mixing reduces the surface temperature by distributing the heat through a greater volume of water, thus pushing the thermocline deeper and making it less pronounced. In winter, cooling increases the density of surface water, which causes it to sink.

Prevailing offshore winds can significantly affect surface water temperatures. These winds are relatively warm during the summer, so they increase the ocean surface temperature and seawater evaporation. During winter, they are much cooler than the ocean surface, so they absorb heat and cool surface water near shore. Mixing from strong winds may drive the thermoclines in Figures 10.31c and 10.31d deeper and even mix the entire water column, producing isothermal conditions. Tidal currents can also cause considerable vertical mixing in shallow coastal waters.

COASTAL GEOSTROPHIC CURRENTS Recall from Chapter 7 that *geostrophic* (*geo* = earth, *strophio* = turn) *currents* move in a circular path around the middle of a current gyre. Wind and runoff create geostrophic currents in coastal waters, too, where they are called **coastal geostrophic currents**.

[7]Recall that a *thermocline* (*thermo* = heat, *cline* = slope) is a layer of rapidly changing temperature, as discussed in Chapter 5.

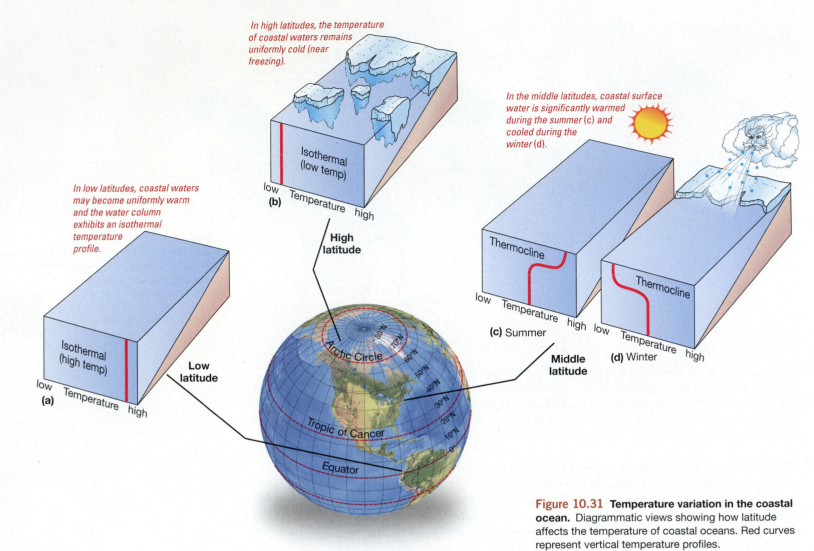

In high latitudes, the temperature of coastal waters remains uniformly cold (near freezing).

Isothermal (low temp)

low Temperature high
(b)

High latitude

In low latitudes, coastal waters may become uniformly warm and the water column exhibits an isothermal temperature profile.

Isothermal (high temp)

low Temperature high
(a)

Low latitude

In the middle latitudes, coastal surface water is significantly warmed during the summer (c) and cooled during the winter (d).

Thermocline

low Temperature high
(c) Summer

Thermocline

low Temperature high
(d) Winter

Middle latitude

Arctic Circle

80°N 70°N 60°N 50°N 40°N 30°N 20°N 10°N 0°

Tropic of Cancer

Equator

Figure 10.31 Temperature variation in the coastal ocean. Diagrammatic views showing how latitude affects the temperature of coastal oceans. Red curves represent vertical temperature profiles.

RECAP

The shallow coastal ocean adjoins land and experiences changes in salinity and temperature that are more dramatic than the open ocean. Coastal geostrophic currents can also develop.

Where winds blow in a certain direction parallel to a coastline, they transport water toward the coast, where it piles up along the shore. Gravity eventually pulls this water back toward the open ocean. As it runs downslope away from the shore, the Coriolis effect causes it to curve to the right in the Northern Hemisphere and to the left in the Southern Hemisphere. Thus, in the Northern Hemisphere, the coastal geostrophic current curves *northward* on the western coast and *southward* on the eastern coast of continents. These currents are reversed in the Southern Hemisphere.

A high-volume runoff of freshwater produces a surface wedge of freshwater that slopes away from the shore (**Figure 10.32**). This causes a surface flow of low-salinity water toward the open ocean, which curves to the right because of the Coriolis effect in the Northern Hemisphere and to the left in the Southern Hemisphere.

Coastal geostrophic currents are variable because they depend on the wind and the amount of runoff for their strength. If the wind is strong and the volume of runoff is high, then the currents are relatively strong. They are bounded on the ocean side by the steadier eastern or western boundary currents of subtropical gyres.

An example of a coastal geostrophic current is the **Davidson Current**, which develops along the coast of Washington and Oregon (Figure 10.32). Although the

current is present year-round, it is more strongly developed during the rainy winter season when high volumes of runoff combine with strong southwesterly winds to produce a relatively strong northward-flowing current. It flows between the shore and the southward-flowing California Current.

Estuaries

An **estuary** (*aestus* = tide) is a partially enclosed coastal body of water in which freshwater runoff from a river dilutes the input of salty ocean water. Estuaries are marine environments whose pH, salinity, temperature, and water levels vary, depending on the mixing between the river that feeds the estuary and the ocean from which it derives its salinity. The most common example of an estuary is a river mouth, where a river empties into the sea. Other coastal bodies of water such as bays, inlets, gulfs, and sounds are considered estuaries, too.

The mouths of large rivers form the most economically significant estuaries because many are seaports, centers of ocean commerce, and important commercial fisheries. Examples include Baltimore, New York, San Francisco, Buenos Aires, London, Tokyo, and many others.

ORIGIN OF ESTUARIES The estuaries of today exist because sea level has risen approximately 120 meters (400 feet) since major continental glaciers began melting about 18,000 years ago. As described in Section 10.4, these glaciers covered portions of North America, Europe, and Asia during the Pleistocene Epoch, which is also referred to as the *Ice Age*.

Four major types of estuaries can be identified based on their geologic origin (**Figure 10.33**):

1. A **coastal plain estuary** forms as sea level rises and floods existing river valleys. These estuaries, such as the Chesapeake Bay in Maryland and Virginia, are called *drowned river valleys* (Figure 10.33a).

2. A **fjord**[8] forms as sea level rises and floods a glaciated valley. Water-carved valleys have V-shaped profiles, but fjords are U-shaped valleys with steep walls. Commonly, a shallowly submerged glacial deposit of debris (called a *moraine*) is located near the ocean entrance, marking the farthest extent of the glacier. Fjords are common along the coasts of Alaska, Canada, New Zealand, Chile, and Norway (Figure 10.33b).

3. A **bar-built estuary** is shallow and is separated from the open ocean by sand bars that are deposited parallel to the coast by wave action. Lagoons that separate *barrier islands* from the mainland are bar-built estuaries. They are very common along the U.S. Gulf Coast and East Coast (see Figure 10.13). Examples include Laguna Madre in Texas and Pamlico Sound in North Carolina (Figure 10.33c).

4. A **tectonic estuary** forms when faulting or folding of rocks creates a restricted downdropped area into which the sea has flooded. California's San Francisco Bay is in part a tectonic estuary (Figure 10.33d), formed by movement along faults, including the San Andreas Fault.

WATER MIXING IN ESTUARIES Because freshwater from a river is less dense than seawater, the basic flow pattern in an estuary is a surface flow of less dense freshwater toward the ocean and an opposite flow below the surface of salty seawater into the estuary. Mixing takes place where these two water masses are in contact with one another.

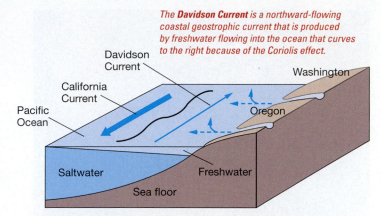

*The **Davidson Current** is a northward-flowing coastal geostrophic current that is produced by freshwater flowing into the ocean that curves to the right because of the Coriolis effect.*

Figure 10.32 Davidson coastal geostrophic current. Block diagram of the Pacific Northwest coast showing that runoff from Oregon and Washington produces a freshwater wedge (*light blue*) that thins away from shore. This causes a surface flow of low-salinity water toward the open ocean, which is acted upon by the Coriolis effect and curves to the right, producing the Davidson Current that flows close to shore and in the opposite direction of the California Current. Note that the Davidson Current is more strongly developed during the winter rainy season when there is a large amount of river runoff.

[8]The Norwegian term *fjord* is pronounced "FEE-yord" and means a long, narrow sea inlet bordered by steep cliffs.

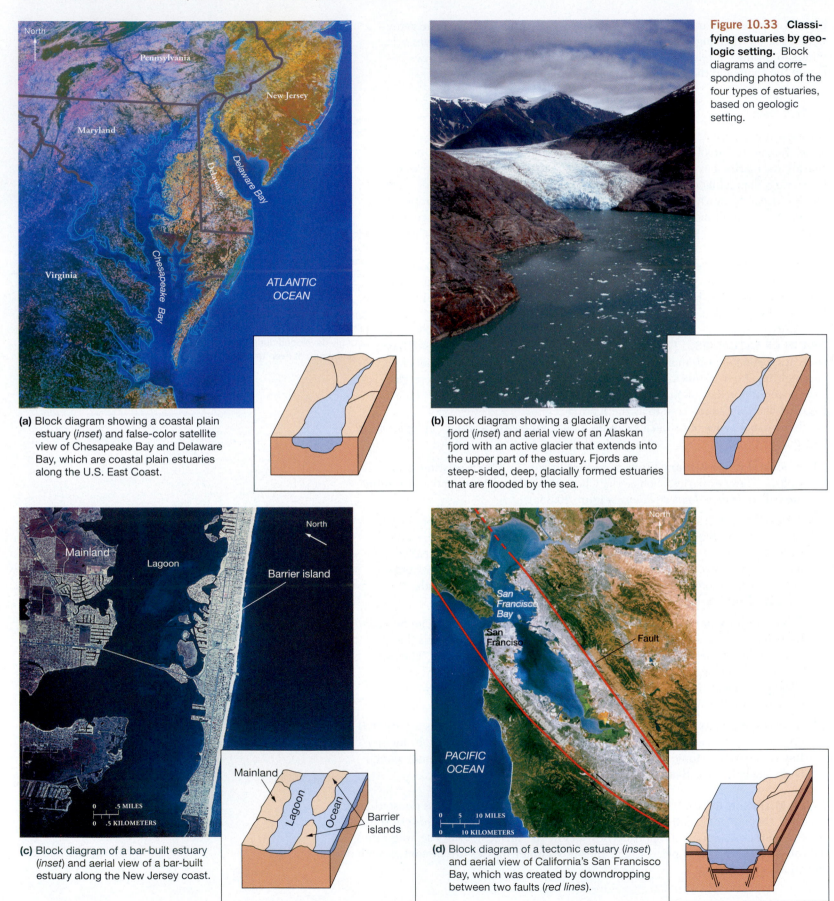

Figure 10.33 **Classifying estuaries by geologic setting.** Block diagrams and corresponding photos of the four types of estuaries, based on geologic setting.

(a) Block diagram showing a coastal plain estuary (*inset*) and false-color satellite view of Chesapeake Bay and Delaware Bay, which are coastal plain estuaries along the U.S. East Coast.

(b) Block diagram showing a glacially carved fjord (*inset*) and aerial view of an Alaskan fjord with an active glacier that extends into the upper part of the estuary. Fjords are steep-sided, deep, glacially formed estuaries that are flooded by the sea.

(c) Block diagram of a bar-built estuary (*inset*) and aerial view of a bar-built estuary along the New Jersey coast.

(d) Block diagram of a tectonic estuary (*inset*) and aerial view of California's San Francisco Bay, which was created by downdropping between two faults (*red lines*).

Based on the physical characteristics of the estuary and the resulting mixing of freshwater and seawater, estuaries are classified into one of four main types, as shown in **Figure 10.34**:

1. A **vertically mixed estuary** is a shallow, low-volume estuary where the net flow always proceeds from the head of the estuary toward its mouth. Salinity at any point in the estuary is uniform from surface to bottom because river water mixes evenly with ocean water at all depths. Salinity simply increases from the head to the mouth of the estuary, as shown in Figure 10.34a. Salinity lines curve at the edge of the estuary because the Coriolis effect influences the inflow of seawater.

2. A **slightly stratified estuary** is a somewhat deeper estuary in which salinity increases from the head to the mouth at any depth, as in a vertically mixed estuary. However, two water layers can be identified. One is the less-saline, less-dense upper water from the river, and the other is the more-saline, more-dense deeper water from the ocean. These two layers are separated by a zone of mixing. The circulation that develops in slightly stratified estuaries is a net surface flow of low-salinity water toward the ocean and a net subsurface flow of seawater toward the head of the estuary (Figure 10.34b), which is called an **estuarine circulation pattern**.

3. A **highly stratified estuary** is a deep estuary in which upper-layer salinity increases from the head to the mouth, reaching a value close to that of open-ocean water. The deep-water layer has a rather uniform open-ocean salinity at any depth throughout the length of the estuary. An estuarine circulation pattern is well developed in this type of estuary (Figure 10.34c). Mixing at the interface of the upper water and the lower water creates a net movement from the deep-water mass into the upper water. Less-saline surface water simply moves from the head toward the mouth of the estuary, growing more saline as water from the deep mass mixes with it. Relatively strong haloclines develop at the contact between the upper and lower water masses.

4. A **salt wedge estuary** is an estuary in which a wedge of salty water intrudes from the ocean beneath the river water. This kind of estuary is typical of the mouths of deep, high-volume rivers. No horizontal salinity gradient exists at the surface because surface water is essentially fresh throughout the length of—and even beyond—the estuary (Figure 10.34d). There is, however, a *horizontal* salinity gradient at depth and a very pronounced vertical salinity gradient (a halocline) at any location throughout the length of the estuary. This halocline is shallower and more highly developed near the mouth of the estuary.

Within all estuaries, the predominant mixing pattern may vary with location, season, or tidal conditions. In addition, mixing patterns in real estuaries are rarely as simple as the models presented here.

ESTUARIES AND HUMAN ACTIVITIES Estuaries are important breeding grounds and protective nurseries for many marine animals, so the ecological well-being of estuaries is vital to fisheries and coastal environments worldwide. Nevertheless, estuaries support shipping, logging, manufacturing, waste disposal, and other activities that can potentially damage the environment.

Estuaries are most threatened where human population is large and expanding, but they can be severely damaged where populations are still modest, too. Development in the sparsely populated Columbia River estuary, for example, demonstrates how human activities can damage an estuary.

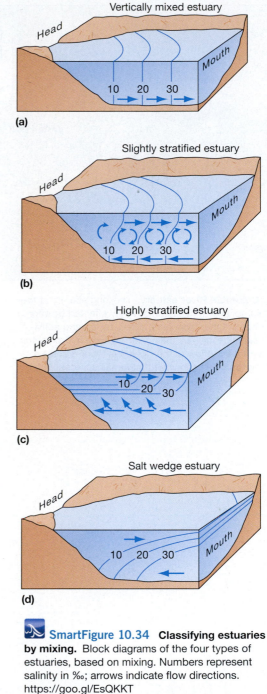

(a) Vertically mixed estuary
(b) Slightly stratified estuary
(c) Highly stratified estuary
(d) Salt wedge estuary

SmartFigure 10.34 Classifying estuaries by mixing. Block diagrams of the four types of estuaries, based on mixing. Numbers represent salinity in ‰; arrows indicate flow directions. https://goo.gl/EsQKKT

RECAP

Estuaries were formed by the rise in sea level after the last ice age. They can be classified based on geologic origin as coastal plain, fjord, bar-built, or tectonic estuaries. Estuaries can also be classified based on mixing as vertically mixed, slightly stratified, highly stratified, or salt wedge.

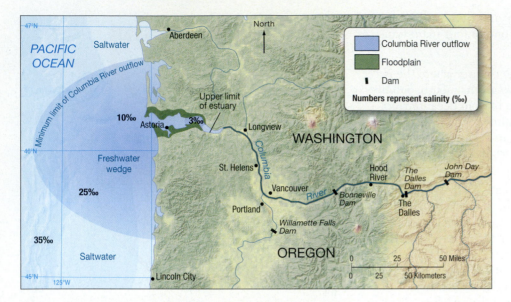

Figure 10.35 Columbia River estuary. The long estuary at the mouth of the Columbia River has been severely affected by interference with floodplains that have been diked, by logging activities, and—most significantly—by the construction of hydroelectric dams. The tremendous outflow of the Columbia River creates a large wedge of low-density freshwater that remains traceable far out at sea.

Columbia River Estuary The Columbia River, which forms most of the border between Washington and Oregon, has a long salt-wedge estuary at its entrance to the Pacific Ocean (**Figure 10.35**). The strong flow of the river and tides drive a salt wedge as far as 42 kilometers (26 miles) upstream and raise the river's water level more than 3.5 meters (12 feet). When the tide falls, the huge flow of freshwater (up to 28,000 cubic meters [1,000,000 cubic feet] per second) creates a freshwater wedge that can extend hundreds of kilometers into the Pacific Ocean.

Most rivers create floodplains along their lower courses, which have rich soil that can be used for growing crops. In the late 19th century, farmers moved onto the floodplains to establish agriculture along the Columbia River. Eventually, protective dikes were built to prevent agricultural damage done by annual flooding. Flooding brings new nutrients, however, so the dikes deprived the floodplain of the nutrients necessary to sustain agriculture.

The river has been the principal conduit for the logging industry, which has dominated the region's economy through most of its modern history. Fortunately, the river's ecosystem has largely survived the additional sediment caused by clear cutting by the logging industry. The construction of more than 250 dams along the river and its tributaries, on the other hand, has permanently altered the river's ecosystem. Many of these dams, for example, do not have salmon ladders, which help fish "climb" in short vertical steps around the dams to reach their spawning grounds at the headwaters of their home streams.

Even though the dams have caused a multitude of problems, they do provide flood control, electrical power, and a dependable source of water, all of which have become necessary to the region's economy. To aid shipping operations, the river receives periodic dredging of sediment, which brings an increased risk for pollution. If these kinds of problems have developed in such sparsely populated areas as the Columbia River estuary, then larger environmental effects must exist in more highly populated estuaries, such as the Chesapeake Bay.

Chesapeake Bay Estuary Chesapeake Bay is about 320 kilometers (200 miles) long and 56 kilometers (35 miles) wide at its widest point, making it the largest (and best studied) estuary in the United States (**Figure 10.36**). It drains a watershed of about 166,000 square kilometers (64,000 square miles) spread over six states that includes a population of over 15 million people. The length of the bay's shoreline is an astonishing 17,700 kilometers (11,000 miles) because of all the inlets created by the 19 major rivers and 400 creeks and tributaries that flow into it. The bay formed when the lower parts of the Susquehanna River were drowned by rising sea level after the most recent ice age.

Chesapeake Bay is a slightly stratified estuary that experiences large seasonal changes in salinity, temperature, and dissolved oxygen. Figure 10.36a shows the estuary's average surface salinity, which increases oceanward. The salinity lines are oriented virtually north–south in the middle of the bay because of the Coriolis effect. Recall that the Coriolis effect causes flowing water to curve to the right in the Northern Hemisphere, so seawater entering the bay tends to hug the bay's *eastern*

[9]Recall that a *pycnocline* (*pycno* = density, *cline* = slope) is a layer of rapidly changing density, as discussed in Chapter 5. A pycnocline is caused by a change in temperature and/or salinity with depth.

side, and freshwater flowing through the bay toward the ocean tends to hug its *western* side.

With maximum river flow in the spring, a strong halocline (and *pycnocline*[9]) develops, preventing the fresh surface water and saltier deep water from mixing. Beneath the pycnocline, which can be as shallow as 5 meters (16 feet), waters may become **anoxic** (*a* = without, *oxic* = oxygen) from May through August, as dead organic matter decays in the deep water (Figure 10.36b). Major kills of commercially important blue crab, oysters, and other bottom-dwelling organisms occur during this time.

The degree of stratification and extent of mortality of bottom-dwelling animals have increased since the early 1950s. Increased nutrients from sewage and agricultural fertilizers have been added to the bay during this time, too, which has increased the productivity of microscopic algae (algal blooms). When these organisms die, their remains accumulate as organic matter at the bottom of the bay and promote the development of anoxic conditions. In drier years with less river runoff, however, anoxic conditions aren't as widespread or severe in bottom waters (Figure 10.36c) because fewer nutrients are supplied.

Lagoons

Landward of barrier islands lie protected, shallow bodies of water called **lagoons** (see Figure 10.33c). Lagoons form in a bar-built type of estuary. Because of restricted circulation between lagoons and the ocean, three distinct zones can usually be identified within lagoons (**Figure 10.37**): (1) A *freshwater zone* that lies near the head of the lagoon where rivers enter, (2) a *transitional zone* of brackish[10] water that occurs near the middle of the lagoon, and (3) a *saltwater zone* that lies close to the lagoon's mouth.

Salinity within a lagoon is highest near the entrance and lowest near the head (Figure 10.37b). In latitudes that have seasonal variations in temperature and precipitation, ocean water flows through the entrance during a warm, dry summer to compensate for the volume of water lost through evaporation, thus increasing the salinity in the lagoon. Lagoons actually may become hypersaline[11] in arid regions, where evaporation rates are extremely high. Even though water flows into the lagoon from the open ocean to replace water lost by evaporation, the dissolved components do not evaporate and sometimes accumulate to extremely high levels. During the rainy season, the lagoon becomes much less saline as freshwater runoff increases.

Tidal effects are greatest near the entrance to the lagoon (Figure 10.37c) and diminish inland from the saltwater zone until they are nearly undetectable in the freshwater zone.

LAGUNA MADRE Laguna Madre is located along the Texas coast between Corpus Christi and the mouth of the Rio Grande (**Figure 10.38**). This long, narrow body of water is protected from the open ocean by Padre Island, a barrier island 160 kilometers (100 miles) long. The lagoon probably formed about 6000 years ago, as sea level approached its present height.

The tidal range of the Gulf of Mexico in this area is about 0.5 meter (1.6 feet). The inlets at each end of Padre Island are quite narrow (Figure 10.38), so there is very little tidal interchange between the lagoon and the open sea.

Laguna Madre is a hypersaline lagoon, and much of it is less than 1 meter (3.3 feet) deep. As a result, there are large seasonal changes in temperature and

[10]Brackish water is water with salinity between that of freshwater and seawater.
[11]Hypersaline conditions are created when water becomes excessively salty.

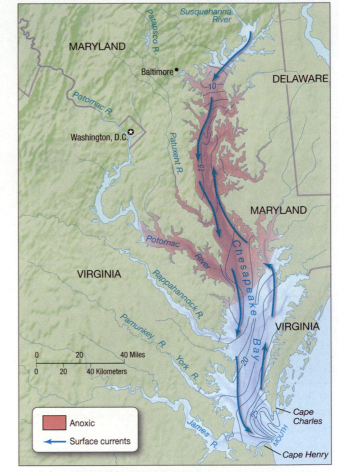

(a) Map of Chesapeake Bay, showing average surface salinity (*blue lines*) in ‰. The red area in the middle of the bay represents anoxic (oxygen-depleted) waters.

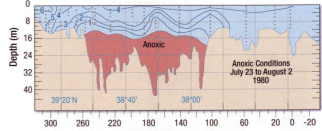

(b) Profile along length of Chesapeake Bay showing dissolved oxygen concentration (in ppm) during July–August 1980, indicating deep anoxic waters (*dark red*).

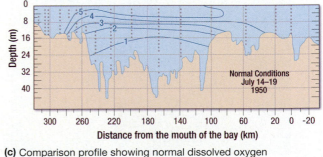

(c) Comparison profile showing normal dissolved oxygen concentration (in ppm) during July 1950.

Figure 10.36 **Chesapeake Bay salinity and dissolved oxygen.**

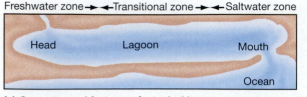

Freshwater zone → ← Transitional zone → ← Saltwater zone

(a) Geometry and features of a typical lagoon as seen from above.

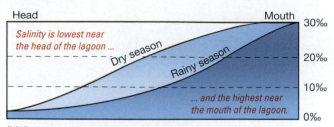

(b) Salinity profile of a typical lagoon, which is affected by seasonal changes in freshwater input.

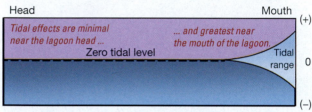

(c) Tidal effects from the head to the mouth of a typical lagoon.

Figure 10.37 Lagoons. Diagrammatic representations of the general features of a typical lagoon.

Figure 10.38 Laguna Madre summer surface salinity. Map showing geometry of Laguna Madre, Texas, and typical summer surface salinity (in ‰).

salinity. Water temperatures reach 32°C (90°F) in the summer and can dip below 5°C (41°F) in winter. Salinities range from 2‰ when infrequent local storms provide large volumes of freshwater to over 100‰ during dry periods. High evaporation generally keeps salinity well above 50‰.[12]

Because even salt-tolerant marsh grasses cannot withstand such high salinities, the marsh has been replaced by an open sand beach on Padre Island. At the inlets, ocean water flows in as a surface wedge *over* the denser water of the lagoon and water from the lagoon flows out as a *subsurface* flow, which is the exact opposite of a typical estuarine circulation pattern.

Marginal Seas

At the margins of the ocean are relatively large semi-isolated bodies of water called **marginal seas**. Most of these seas result from tectonic events that have isolated low-lying pieces of ocean crust between continents, such as the Mediterranean Sea, or are created behind volcanic island arcs, such as the Caribbean Sea. These waters are shallower than and have varying degrees of exchange with the open ocean, depending on climate and geography; as a result, salinities and temperatures are substantially different from those of typical open ocean seawater.

A CASE STUDY: THE MEDITERRANEAN SEA The **Mediterranean Sea** (*medi* = middle, *terra* = land) is actually a number of small seas connected by narrow necks of water into one larger sea. It is the remnant of the ancient Tethys Sea that existed when all the continents were combined about 200 million years ago. It is more than 4300 meters (14,100 feet) deep and is one of the few inland seas in the world underlain by oceanic crust. Thick salt deposits and other evidence on the floor of the Mediterranean suggest that it nearly dried up about 6 million years ago, only to refill with a large saltwater waterfall (see Diving Deeper 4.1).

The Mediterranean is bounded by Europe and Asia Minor on the north and east and Africa on the south (**Figure 10.39a**). It is surrounded by land except for very shallow and narrow connections to the Atlantic Ocean through the Strait of Gibraltar (about 14 kilometers [9 miles] wide), and to the Black Sea through the Bosporus (roughly 1.6 kilometers [1 mile] wide). In addition, the Mediterranean Sea has a human-made passage to the Red Sea via the Suez Canal, a waterway 160 kilometers (100 miles) long that was completed in 1869. The Mediterranean Sea has a very irregular coastline, which divides it into subseas such as the Aegean Sea and Adriatic Sea, each of which has a separate circulation pattern.

An underwater ridge called a **sill**, which extends from Sicily to the coast of Tunisia at a depth of 400 meters (1300 feet), separates the Mediterranean into two major basins. This sill restricts the flow between the two basins, resulting in strong currents that run between Sicily and the Italian mainland through the Strait of Messina (Figure 10.39a).

Mediterranean Circulation The Mediterranean Sea has a unique circulation pattern. This circulation is caused by the dry, intense heat of the Middle East, where a huge volume of water evaporates from the eastern Mediterranean and causes a tremendous surface inflow of Atlantic Ocean water through the Strait of Gibraltar to replace the evaporated water. In fact, the water level in the eastern Mediterranean is generally 15 centimeters (6 inches) lower than at the Strait of Gibraltar. The surface flow follows the northern coast of Africa throughout the length of the Mediterranean and spreads northward across the sea (Figure 10.39a).

[12]Recall that normal salinity in the open ocean averages 35‰.

Figure 10.39 **Mediterranean Sea bathymetry and circulation.**

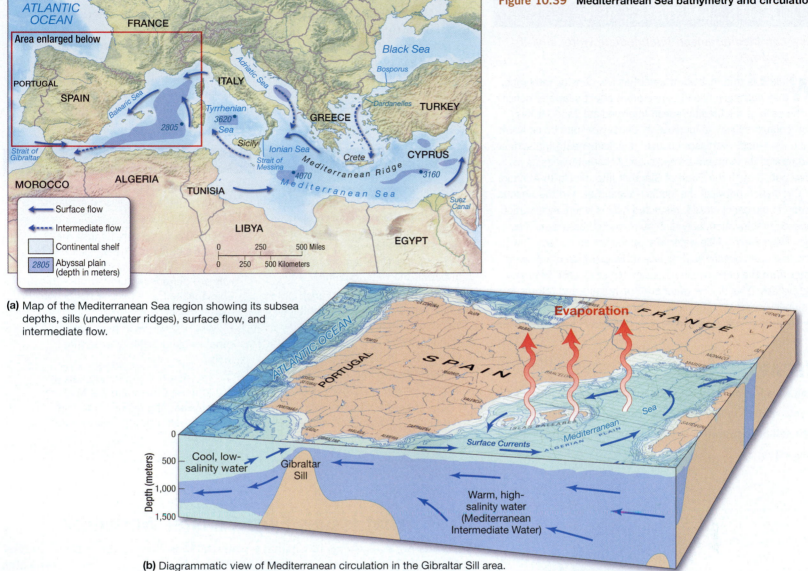

(a) Map of the Mediterranean Sea region showing its subsea depths, sills (underwater ridges), surface flow, and intermediate flow.

(b) Diagrammatic view of Mediterranean circulation in the Gibraltar Sill area.

The remaining Atlantic Ocean water continues eastward to Cyprus. During winter, it sinks to form what is called the *Mediterranean Intermediate Water*, which has a temperature of 15°C (59°F) and a salinity of 39.1‰. This water flows westward at a depth of 200 to 600 meters (660 to 2000 feet) and returns to the North Atlantic as a *subsurface* flow through the Strait of Gibraltar (Figure 10.39b). During World War II, German submarines routinely escaped detection when crossing through the Strait of Gibraltar by switching off their engines and taking advantage of the currents flowing into and out of the Mediterranean Sea. The submarine captains would adjust the buoyancy of the submarine so that the sub would be transported either into the Mediterranean Sea with its surface current or out of the sea within its intermediate waters.

By the time the Mediterranean Intermediate Water passes through Gibraltar, its temperature has dropped to 13°C (55°F) and its salinity to 37.3‰. It is still denser than even Antarctic Bottom Water and much denser than water at this depth in the Atlantic Ocean, so it moves down the continental slope. While descending, it mixes with Atlantic Ocean water and becomes less dense. At a depth of about 1000 meters (3300 feet), its density equals that of the surrounding

How can Mediterranean Intermediate Water sink if it's so warm?

While it is true that warm water has low density, remember that *both* salinity and temperature affect seawater density. In the case of the Mediterranean Intermediate Water, it has high enough salinity to increase its density despite being warm. Once its density increases enough, it sinks beneath the surface and retains its temperature and salinity characteristics as it flows out through the Strait of Gibraltar into the North Atlantic.

Circulation between the Mediterranean Sea and the Atlantic Ocean is typical of closed, restricted basins where evaporation exceeds precipitation. Low-latitude restricted basins such as this always rapidly lose water to evaporation, so surface flow from the open ocean must replace it. Evaporation of inflowing water from the open ocean increases the sea's salinity to very high values. This denser water eventually sinks and returns to the open ocean as a subsurface flow.

RECAP

High evaporation rates in the Mediterranean Sea cause it to have a shallow inflow of surface seawater and a subsurface high-salinity outflow—a circulation pattern opposite that of most estuaries.

Atlantic Ocean, so it spreads in all directions (Figure 10.39b), sometimes forming deep-ocean eddies that last for more than two years and can be detected by satellite as far north as Iceland.

This circulation pattern, which is called **Mediterranean circulation**, is opposite that of most estuaries, which experience estuarine circulation where freshwater flows at the surface into the open ocean and salty water flows below the surface into the estuary. In estuaries, however, freshwater input exceeds water loss to evaporation, whereas evaporation exceeds input in the Mediterranean.

CONCEPT CHECK 10.6 | Compare the various types of coastal waters.

1 For coastal oceans where deep mixing does not occur, describe the effect that offshore winds and freshwater runoff have on salinity distribution. How will the winter and summer seasons affect the temperature distribution in the water column?

2 Describe how coastal runoff of low-salinity water produces a coastal geostrophic current and give a specific location where a coastal geostrophic current can be found.

3 Describe the four main types of estuaries, based on geologic origin.

4 Describe the difference between vertically mixed and salt wedge estuaries in terms of salinity distribution, depth, and volume of river flow. Which displays the more classical estuarine circulation pattern?

5 Discuss factors that cause the surface salinity of Chesapeake Bay to be greater along its east side. Also, why are periods of summer anoxia in Chesapeake Bay's deep water becoming increasingly worse?

6 What factors lead to a wide seasonal range of salinity in Laguna Madre?

7 Describe the circulation between the Atlantic Ocean and the Mediterranean Sea, and explain how and why it differs from typical estuarine circulation.

10.7 What Issues Face Coastal Wetlands?

Wetlands are ecosystems in which the water table is close to the surface, so they are typically saturated most of the time. Wetlands can border either freshwater or coastal environments. Coastal wetlands occur along the margins of coastal waters such as estuaries, lagoons, and marginal seas; they include swamps, tidal flats, coastal marshes, and bayous.

Types of Coastal Wetlands

The two most important types of coastal wetlands are **salt marshes** and **mangrove swamps**. Both experience intermittent submergence by ocean water and contain salt-adapted plants, oxygen-depleted muds, and accumulations of organic matter called *peat deposits*.

Salt marshes generally occur between about 30 and 65 degrees latitude (**Figures 10.40a** and **10.40b**) and support a variety of salt-tolerant grasses and other low-lying plants that are termed *halophytic* (*halo* = salt, *phyto* = plant). Examples of halophytic grasses include cordgrass and salt-meadow cordgrass, both of which belong to the genus *Spartina* and have the ability to get rid of excess salt by producing exterior salt crystals. Other plants that live in this habitat, like pickleweed (*Salicornia*), accumulate salts in their tissues and dispose of excess salts by breaking off the tissues once they become highly salty. Well-developed salt marsh habitats are found along most coasts of the continental United States and also along the coasts of Europe, Japan, and eastern South America.

10.2 Squidtoons

HOW CAN MANGROVES PROTECT THE CORAL REEFS?

https://goo.gl/WHfOoR

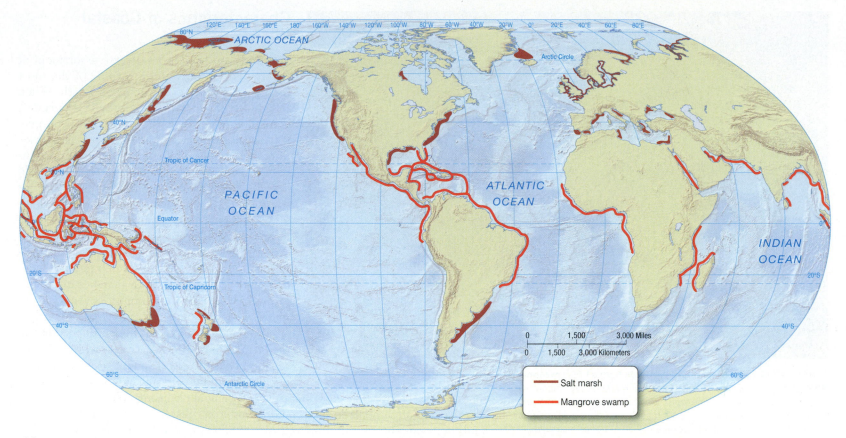

(a) Map showing the distribution of salt marshes in the higher latitudes and mangrove swamps in the lower latitudes.

(b) A typical salt marsh in Morro Bay, California.

(c) A dense mangrove swamp bordering a seaway in the Florida Keys.

Figure 10.40 Salt marshes and mangrove swamps. (a) Map showing the distribution of salt marshes and mangrove swamps, with accompanying photos (**b** and **c**).

Mangrove swamps are restricted to tropical regions (below 30 degrees latitude; **Figures 10.40a** and **10.40c**) and support various species of salt-tolerant mangrove trees, shrubs, and palms. To live in these salty conditions, some mangroves produce tall tripod-like root systems to stay above the salty water; others crystalize excess salt on their leaves. Mangrove swamps occur throughout the Caribbean and Florida; the most extensive mangroves in the world are found throughout Southeast Asia.

Figure 10.41 **Marine wetlands provide habitat and protection for many species of fish.** Both salt marshes and mangrove swamps are types of marine wetlands that are important nursery areas for many species of fish, such as these Atlantic silversides (*Menidia menidia*) that seek protection within mangrove roots.

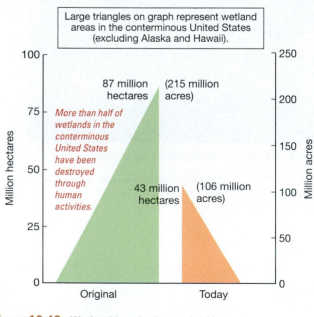

Figure 10.42 **Wetland loss in the conterminous United States.**

Characteristics of Coastal Wetlands

Wetlands are home to a diverse assortment of plants and animals and are some of the most highly productive ecosystems on Earth. When left undisturbed, wetlands provide enormous economic benefits. Salt marshes, for example, serve as nurseries for more than half the species of commercially important fishes in the southeastern United States (**Figure 10.41**). Other fish, such as flounder and bluefish, use marshes for feeding and protection during the winter. Fisheries of oysters, scallops, clams, eels, and smelt are located directly in marshes, too. Mangrove ecosystems are important nursery areas and habitats for commercially valuable shrimp, prawn, shellfish, and fish species. Both marshes and mangroves also serve as important stopover points for many species of waterfowl and migrating birds.

Wetlands also soak up the nutrients that run off farmlands and down rivers, which, if they reached coastal waters, could fuel harmful algal blooms and create marine oxygen-free dead zones. In essence, wetlands are amazingly efficient at cleansing polluted water; that's why they are often referred to as "nature's kidneys." Just 0.4 hectare (1 acre) of wetlands, for example, can filter up to 2,760,000 liters (730,000 gallons) of water each year, cleaning agricultural runoff, toxins, and other pollutants long before they reach the ocean. Wetlands remove inorganic nitrogen compounds (from sewage and fertilizers) and metals (from groundwater polluted by land sources), which become attached to clay-sized particles in wetland mud. Some nitrogen compounds trapped in sediment are decomposed by bacteria that release the nitrogen to the atmosphere as gas, and many of the remaining nitrogen compounds fertilize plants, further increasing the productivity of wetlands. As marsh plants die, their remains either accumulate as peat deposits or are broken up to become food for bacteria, fungi, and fish.

Interdisciplinary

Relationship

In addition, wetlands protect shorelines from erosion and serve as a first line of defense against hurricanes and tsunami by dissipating wave energy and absorbing excess water. The 2004 Indian Ocean tsunami, for example, devastated some coastal regions, yet others with protective offshore coral reefs or coastal mangroves experienced much less damage. As another example, the loss of protective coastal wetlands in the Mississippi River Delta contributed to the extensive flooding associated with the storm surge caused by Hurricane Katrina in 2005 (see Section 6.5 in Chapter 6). During Hurricane Sandy in 2012, the complete lack of protective wetlands around New York City caused much more severe flooding there than in neighboring regions that retained just remnants of wetlands.

Serious Loss of Valuable Wetlands

Despite all the benefits wetlands provide, more than half of the nation's wetlands have vanished. Of the original 87 million hectares (215 million acres) of wetlands that once existed in the conterminous United States, only about 43 million hectares (106 million acres) remain (**Figure 10.42**). Wetlands have been filled in and developed for housing, industry, and agriculture because people want to live near the oceans and because they often view wetlands as unproductive, useless land that harbors diseases. In many places, wetland loss is compounded by the lack of fresh sediment from regular river floods. Instead, flooding rivers and their sediment are channeled away from wetland areas.

Louisiana's coastal wetlands, for example, are among those that are steadily disappearing. Over time, the soil in wetlands naturally compresses under its own weight, in a process called *subsidence*. Normally, the growth of plants and the infusion of fresh sediment from river floodwaters offset subsidence. With these factors reduced or eliminated, many wetlands are sinking into the ocean faster than they are building up. For example, scientists estimate that subsidence of the Mississippi River Delta—along with rising sea level—will cause about 10% of Louisiana to sink beneath the ocean surface by the end of the century.

Other countries have experienced similar losses of wetlands, too. In fact, scientists estimate that 50% of wetlands worldwide have been destroyed in the past century. Mangroves, for example, are already critically endangered or approaching extinction in 26 out of the 120 countries that have mangroves. Indonesia has lost over 50% of its mangroves in the past three decades, and the Philippines has reported losing 70% of its original mangrove cover. Worldwide, 3.6 million hectares (8.9 million acres) of mangroves have been lost since 1980. Of the mangrove swamps that remain, many are in critical condition or seriously damaged. At the current rate of mangrove loss, there is increasing concern that all mangrove ecosystems worldwide will be destroyed within the next 100 years.

To help prevent the loss of remaining wetlands, the U.S. Environmental Protection Agency (EPA) established an Office of Wetlands Protection (OWP) in 1986. At that time, wetlands were being lost to development at a rate of 121,000 hectares (300,000 acres) per year! In 1997, the rate of coastal wetland loss had slowed to about 8100 hectares (20,000 acres) per year. The goal of the OWP is to reduce the loss of wetlands in the United States to zero by actively enforcing regulations against wetlands pollution and identifying the most valuable wetlands to be protected or restored.

In spite of these global, long-term trends, recent documentation of U.S. wetlands shows that there has been an overall *increase* in wetlands during this century. In fact, a study from 1998 to 2004 revealed that the conterminous United States gained an estimated 13,000 hectares (32,000 acres) of wetlands each year of the study. This gain—although small—was primarily due to an increase in freshwater wetlands; coastal wetlands were still decreasing, but at a slower rate than previously reported. The fact that coastal regions were losing wetlands despite the national trend of a net gain in wetlands points to the need for more research on the natural and human forces behind these trends and to an expanded effort on conservation of wetlands, particularly in coastal areas.

Future sea level rise is predicted to exacerbate the loss of wetlands. Even using a conservative estimate of sea level rise over the next 100 years of 50 centimeters (20 inches), it is estimated that as much as 61% of existing U.S. coastal wetlands will be lost. Human-caused global warming could cause additional sea level rise and thus even more wetland loss. Some of this wetland loss, however, would be partially offset by new wetland formation on former upland areas, although even under ideal circumstances, not all lost wetlands would be replaced.

Climate

Connection

RECAP

Coastal wetlands such as salt marshes and mangrove swamps are highly productive areas that serve as important nurseries for many marine organisms and act as filters for polluted runoff.

CONCEPT CHECK 10.7 | Specify the issues that face coastal wetlands.

1 Name the two types of coastal wetland environments and the latitude ranges where each will likely develop.

2 How do wetlands contribute to the biology of the oceans and the cleansing of polluted river water?

3 From the information in Figure 10.42, determine how large an area of wetlands has been lost in the conterminous United States. What percentage of original wetlands still remains? What efforts are being made in the United States to reverse this trend?

Figure 11.1 Pollution washed up on a beach.

Is dilution the solution to ocean pollution?

That's certainly a catchy phrase, but the implications are controversial. It suggests that the oceans can be used as a repository for society's wastes, as long as the wastes are diluted to the point that they no longer threaten marine organisms (which is often difficult to determine). Because the oceans are vast and consist of a good solvent (water), they appear ideally suited to this disposal strategy. In addition, the oceans have good mixing mechanisms (currents, waves, and tides) that dilute many forms of pollution.

Air pollution was once viewed in a similar manner. Disposal of pollutants into the atmosphere was thought to be acceptable, as long as they were dispersed widely and high enough—so tall smokestacks were constructed. Over time, however, pollutants, such as nitric acid and acid sulfates, increased in the atmosphere to the point that acid rain is now a problem. The ocean, like the atmosphere, has a finite *holding capacity* for pollutants, although experts disagree on exactly how much that is.

As disposal sites on land begin to fill, the ocean is increasingly evaluated as an area for disposal of society's wastes. One thing that we can all do is to limit the amount of waste we generate, alleviating some of the problem of where to put waste. It is likely, however, that the ocean will continue to be used as a dumping ground in the foreseeable future. Despite many new disposal techniques, a long-term *solution* to ocean pollution hasn't yet appeared.

marine pollution has increased as well (**Figure 11.1**). But what exactly is marine pollution?

Marine Pollution: A Definition

Pollution can broadly be defined as *any harmful substance*, but how do scientists determine which substances are harmful? For example, a substance may be esthetically unappealing to people yet not be harmful to the environment. Conversely, certain types of pollution cannot easily be detected by humans, yet they can do harm to the environment. A substance may not immediately be harmful, but it may cause harm years, decades, or even centuries later. Also, to whom must this harm be done? For instance, some marine species thrive when exposed to a particular compound that is quite toxic to other species. Interestingly, natural conditions in coastal waters, such as dead seaweed on the beach, may be considered *pollution* by some people. It should be remembered, however, that although nature may produce conditions we dislike, it does not pollute. The amount of a pollutant is also important: If a substance that causes pollution is present in extremely tiny amounts, can it still be characterized as a pollutant? All of these questions are difficult to answer.

The World Health Organization defines pollution of the marine environment as follows:

> *The introduction by man, directly or indirectly, of substances or energy into the marine environment, including estuaries, which results or is likely to result in such deleterious effects as harm to living resources and marine life, hazards to human health, hindrance to marine activities, including fishing and other legitimate uses of the sea, impairment of quality for use of sea water, and reduction of amenities.*

It is often difficult to determine the degree to which pollution affects the marine environment. Most areas were not studied sufficiently before they were polluted, so scientists do not have an adequate baseline from which to determine how pollutants have altered the marine environment. The marine environment is affected by decade- to century-long cycles, too, so it is difficult to determine whether a change is due to a natural biological cycle or any number of introduced pollutants, many of which combine to produce new compounds.

Environmental Bioassay

One of the most widely used techniques for determining the concentration of pollutants that negatively affect the living resources of the ocean is to conduct a carefully controlled experiment to assess how a particular pollutant impacts marine organisms. Such an experiment is called an *environmental biological assay*, or **environmental bioassay** (*bio* = biologic, *essaier* = to weigh out). For example, regulatory agencies such as the U.S. Environmental Protection Agency (EPA) use an environmental bioassay to determine the concentration of a pollutant that causes 50% mortality among a specific group of test organisms within a prescribed period of time. If a pollutant exceeds a

Interdisciplinary

Relationship

50% mortality rate, then concentration limits are established for the discharge of the pollutant into coastal waters.

There are several drawbacks to using a specific environmental bioassay to draw general conclusions about a pollutant. One shortcoming is that it does not predict the long-term effect of pollution on marine organisms. Another is that it does not take into account how pollutants may combine with other substances to create new types of pollutants. In addition, environmental bioassays are often time consuming, laborious, and organism-specific and so may result in data that isn't applicable to other species.

The Issue of Waste Disposal in the Ocean

Waste disposal facilities on land (such as landfills) have limited capacities that are already being exceeded in many cases. Should additional waste be discarded in the open ocean? Unlike coastal areas, the open ocean has mixing mechanisms (waves, tides, and currents) that distribute pollutants over a wide area—including an entire ocean basin. Diluting pollutants often renders them less harmful. On the other hand, do we really want to distribute a pollutant across an entire ocean without knowing what its long-term effects might be?

Some experts believe that we should not dump *anything* in the ocean, while others believe that the ocean can continue to be a repository for many of society's wastes, as long as proper monitoring is conducted. Unfortunately, there are no easy answers and the issues are complex. What is clear is that more research is needed to determine how various types of pollution are affecting the ocean and its inhabitants.

11.1 Squidtoons

Tuna, the ~~chicken~~ cheetah of the sea.

https://goo.gl/9SvLVo

RECAP

The ocean has become a global repository for much of the waste humans generate. Marine pollution is difficult to define but includes any substance introduced by humans that is harmful to the marine environment.

CONCEPT CHECK 11.1 | Explain how pollution is defined.

1 Examine the World Health Organization's definition of pollution. Why does it need to be so specific?

2 What is an environmental bioassay? What are some drawbacks to using an environmental bioassay to determine whether or not a substance should be classified as a pollutant?

3 From memory, define pollution. Then consider these items and

determine whether each one is a pollutant based on your definition (and refine your definition as necessary):

a. Dead seaweed on a beach
b. Natural oil seeps
c. A small amount of sewage
d. Warm water from a power plant dumped into the ocean
e. Sound from boats in the ocean

11.2 What Marine Environmental Problems Are Associated with Petroleum Pollution?

Petroleum (*petra* = rock, *oleum* = oil), which is more commonly called *oil*, is a naturally occurring liquid composed of hydrocarbons and other organic compounds. Underground petroleum deposits are highly valued for their energy content and are recovered mostly through drilling. Once oil is collected, it is sent to a refinery via pipeline or tanker. As a result, oil spills into the ocean are a fact of our modern oil-powered economy. Some oil spills result from oil extraction. Other spills result from loading or unloading accidents or tanker collisions. Still others are caused by oil tankers running aground, as in the case of the 1989 oil spill from the *Exxon Valdez* tanker in Prince William Sound, Alaska. Let's examine some of these examples.

Figure 11.11 The *New Carissa* on fire off the Oregon coast. When the freighter M/V *New Carissa* ran aground in 1999 in shallow water offshore Coos Bay, Oregon, and began leaking oil, it was intentionally set on fire to prevent further oil from spilling into the ocean.

RECAP

Petroleum pollution enters the ocean naturally through seeps but also through human activities. Large oil spills are an inevitable part of our modern society but the ocean's bacteria and other microbes naturally biodegrade oil.

In February 1999, the Japanese-owned freighter M/V *New Carissa* ran aground just offshore of Coos Bay, Oregon, with nearly 1.5 million liters (400,000 gallons) of tarlike fuel oil aboard that began leaking through cracks in its hull. When the ship washed into the surf zone and an approaching storm threatened to tear it apart, federal and state authorities decided to ignite the vessel and its fuel rather than risk a larger oil spill (**Figure 11.11**). This was the first time that oil on a ship in U.S. waters was intentionally burned to prevent an oil spill. Eventually, the ship split in two, and about half of its oil burned, limiting the amount of oil spilled into the ocean. Most of the remaining oil was sunk with the wrecked ship a month later, when it was towed offshore and sunk in water 3 kilometers (1.9 miles) deep by U.S. Naval gunfire and a torpedo.

In addition to the chemicals that comprise petroleum, a variety of other chemical compounds—including sewage sludge—is released into the ocean. In sufficient quantities, these materials are often responsible for marine pollution incidents. We'll explore these substances next.

CONCEPT CHECK 11.2	Specify the marine environmental problems associated with petroleum pollution.

1 Explain why many marine pollution experts consider oil among the *least* damaging pollutants in the ocean.

2 Discuss techniques used to clean oil spills. Why is it important to begin the cleanup immediately?

3 Describe the world's largest accidental and intentionally released oil spills. How many times larger was each than the *Exxon Valdez* oil spill?

11.3 What Marine Environmental Problems Are Associated with Non-Petroleum Chemical Pollution?

There are many other types of pollution besides petroleum that are considered types of marine chemical pollution. Examples include sewage sludge, DDTs, PCBs, mercury, and even chemicals contained in prescription and non-prescription drugs. Let's examine some of these types of chemical pollutants, including how these substances get into the ocean.

Sewage Sludge

One of the main types of marine chemical pollution is **sewage sludge**. Sewage treated at a facility typically undergoes **primary treatment**, where solids are allowed to settle and separate from the liquid, and **secondary treatment**, where it is exposed to bacteria-killing chlorine. Sewage sludge is the semisolid material that remains after such treatment. It contains a toxic brew of human waste, oil, zinc, copper, lead, silver, mercury, pesticides, and other chemicals. Since the 1960s, at least 500,000 metric tons (1.1 billion pounds) of sewage sludge have been dumped into the coastal waters of Southern California, and more than 8 million metric tons (18 billion pounds) of sewage sludge has been dumped into the New York Bight between Long Island and the New Jersey shore.

Although the Clean Water Act of 1972 prohibited the dumping of sewage into the ocean after 1981, the high cost of treating and disposing of sewage sludge on land resulted in extension waivers being granted to many municipalities. In the summer of 1988, however, non-biodegradable debris including medical waste—probably carried by heavy rains into the ocean through storm drains—washed up on Atlantic coast beaches and adversely affected the tourist business. Although this event was

completely unrelated to sewage disposal at sea, it focused public awareness on ocean pollution and helped pass legislation that makes it illegal to dump sewage sludge at sea.

NEW YORK'S SEWAGE SLUDGE DISPOSAL AT SEA Sewage sludge from New York and Philadelphia has traditionally been transported offshore by barge and dumped in the ocean at sites totaling 150 square kilometers (58 square miles) within the New York Bight sludge site and the Philadelphia sludge site (**Figure 11.12**).

The water depth is about 29 meters (95 feet) at the New York Bight sludge site and about 40 meters (130 feet) at the Philadelphia sludge site. The water column in such shallow water is relatively uniform, so even the smallest sludge particles reach the bottom without undergoing much horizontal transport, and the ecology of the dump site can be severely affected. At the very least, such a concentration of organic and inorganic matter seriously disrupts the chemical cycling of nutrients. Greatly reduced species diversity results, and in some locations, the result is an overabundance of algae that causes dissolved oxygen to be reduced to very low levels.[3]

In 1986, the shallow-water sites were abandoned, and sewage was subsequently transported to a deep-water site 171 kilometers (106 miles) out to sea (Figure 11.12). The deep-water site is beyond the continental shelf break, so there is usually a well-developed density gradient that separates low-density, warmer surface water from high-density, colder deep water. Internal waves moving along this density gradient can horizontally transport particles at rates 100 times faster than they sink.

Local fishermen reported adverse effects on their fisheries soon after deep-water dumping began. Also, concern was expressed that the sewage could be transported great distances in eddies of the Gulf Stream (see Chapter 7), even as far as the coast of the United Kingdom. This program was terminated in 1993, and municipalities must now dispose of their sewage on land.

BOSTON HARBOR SEWAGE PROJECT Prior to the 1980s, some 48 different communities that comprise the greater Boston area used an antiquated sewage system to dump sludge and partially treated sewage at the entrance to Boston Harbor. Tidal currents often swept the sewage back into the bay, and at other times, the system became overloaded and dumped raw sewage directly into the bay, making Boston Harbor one of the most polluted bays in the country.

A court-ordered cleanup of Boston Harbor in the 1980s resulted in the construction of a new waste treatment facility at Deer Island, which came online in 1998. The facility treats all sewage with bacteria-killing chlorine and carries it through a tunnel 15.3 kilometers (9.5 miles) long into deeper waters offshore (**Figure 11.13a**), which prevents it from returning to the bay. Since the cleanup of Boston Harbor, beaches have reopened, clammers are digging again, and marine life—including harbor seals, porpoises, and even whales—has returned. To pay for the $3.8 billion sewage system, however, the average annual sewage bill for a Boston-area household was increased to about $1200, more than five times what it had been.

Despite the benefits, at the time it was proposed, some opponents feared that the project would degrade the environment in Cape Cod Bay and Stellwagen Bank (**Figure 11.13b**), which is an important whale habitat. In 1992, six years before the facility went online, this area was designated as a U.S. National Marine Sanctuary, which restricts the amount of sewage that can be dumped there.

DDT and PCBs

The pesticide **DDT** and the industrial chemicals called **PCBs** are now found throughout the marine environment. They are persistent, biologically active chemicals that have been introduced into the oceans entirely as a result of human activities. Because of their toxicity, persistence, and propensity for accumulating in

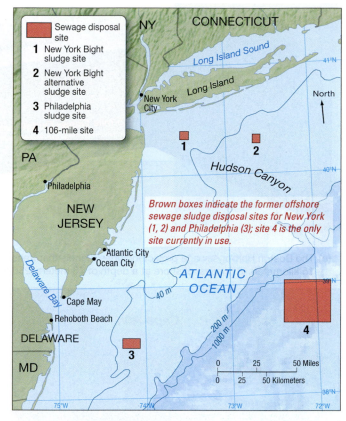

Figure 11.12 Atlantic sewage sludge disposal sites. Prior to 1986, more than 8,000,000 metric tons (18 billion pounds) of sewage sludge was dumped each year by barge at the New York Bight sludge sites (*1* and *2*) and the Philadelphia sludge site (*3*). In 1986, the dump site was moved to the larger and deeper-water 171-kilometer (106-mile) site (*4*).

[3]Water with very low dissolved oxygen creates *hypoxic* (*hypo* = under, *oxic* = oxygen) conditions that can kill marine life; for details, see Section 13.2 in Chapter 13 "Biological Productivity and Energy Transfer."

living and nonliving things are composed of the same basic building blocks: atoms, which move continuously in and out of living and nonliving systems. This free exchange of identical components between life and nonlife is one of the factors that complicates attempts at formally defining life.

A simple definition of life is that it consumes energy from its environment. Using this definition, a car engine could probably be classified as alive. An engine, however, cannot self-replicate or otherwise reproduce itself, which is another key component of life.

Several other qualities are crucial in defining life. Water probably needs to be a part of a living organism because living things need a solvent for biochemical reactions—though ammonia or sulfuric acid might also work. A living thing probably has to have some sort of a membrane to distinguish itself from its environment. In addition, most living things tend to respond to stimuli or adapt to their environment. Finally, life as we know it is carbon based, since carbon is so useful in making chemical compounds. Because NASA's definition of life must encompass the potential of extraterrestrial life, they use the following simple working definition of life: "Life is a self-sustained chemical system capable of undergoing Darwinian evolution."[1] However, even this definition is problematic in that it would likely require observation of several successive generations over a considerable length of time to verify evolution in a life-form.

A good working definition of life, then, should incorporate most of these ideas: that living things can capture, store, and convert energy; they are capable of reproduction; they can adapt to their environment; and they change through time.

The Three Domains of Life

All living things belong to one of three domains, or "superkingdoms," of life: Bacteria (or Eubacteria), Archaea, and Eukarya (**Figure 12.1a**). Domain **Bacteria** (*bakterion* = a rod) includes simple life-forms with cells that usually lack a nucleus,

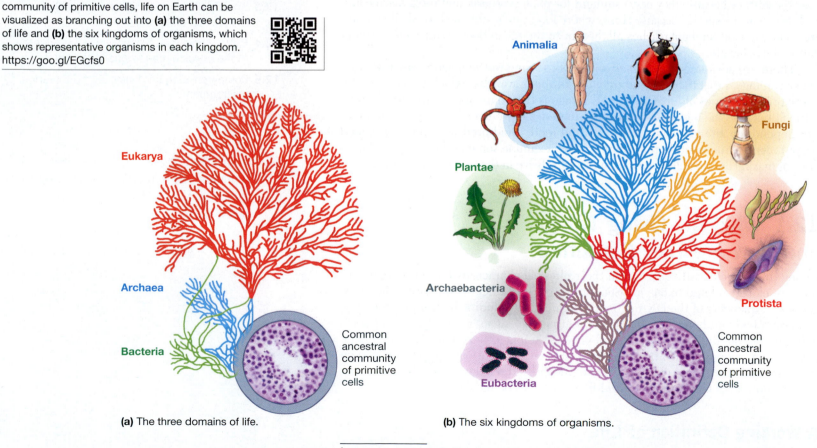

SmartFigure 12.1 **The three domains of life and the six kingdoms of organisms.** Starting with a common ancestral community of primitive cells, life on Earth can be visualized as branching out into **(a)** the three domains of life and **(b)** the six kingdoms of organisms, which shows representative organisms in each kingdom. https://goo.gl/EGcfs0

(a) The three domains of life.

(b) The six kingdoms of organisms.

[1] See Diving Deeper 1.3 for a description of Darwinian evolution.

including purple bacteria, green nonsulfur bacteria, and cyanobacteria (blue-green algae). Domain **Archaea** (*achaeo* = ancient) is a group of simple, microscopic, bacterialike creatures that includes methane producers and sulfur oxidizers that inhabit deep-sea vents and seeps,[2] as well as other forms—many of which prefer environments with extreme temperatures and/or pressures. Domain **Eukarya** (*eu* = good, *karuon* = nut) includes complex organisms: multicellular plants; multicellular animals; fungi; and protists, which are a diverse array of mostly microscopic organisms that don't fit into any other group. The main component of eukaryotes, DNA, is housed in a discrete nucleus, and their cells contain structures that supply energy to build the cell and maintain its functions.

What were the ancestors of these three domains of life? According to modern evolutionary theory (and first suggested by Charles Darwin over 150 years ago), all organisms on Earth share a common genetic heritage, each being the genealogical descendant of a single primitive species from the distant past. This idea, called *universal common ancestry*, has been proved to be valid in rigorous statistical analyses using a set of commonly retained proteins found in a wide range of living organisms. This means that the ancestors of the three domains of life are thought to have consisted of a community of early primitive cells, some of which apparently acquired new genetic material by engulfing microbial neighbors, including their genetic codes. Through *symbiosis* (*sym* = together, *bios* = life),[3] these groups of organisms helped each other coexist for their mutual benefit, and as a result evolved into new organisms that contained merged genes. Today, in fact, it is well established that such adoption of foreign genetic material is common among single-celled bacteria and other microbes but also occurs in some fungi, plants, insects, worms, and other animals. For example, a scientific study of venomous jellies has revealed that one of the genes necessary for jellies to sting is identical to a gene in bacteria, suggesting that the ancestors of jellies picked up the gene from microbes. This process, called *lateral gene transfer*, was likely quite important in the evolution of some of the first complex cells.

The Six Kingdoms of Organisms

Within the three domains of life, a system of five kingdoms of organisms was first proposed by ecologist and biologist Robert H. Whittaker in 1969. In 1977, microbiologist and biophysicist Carl R. Woese and his colleagues enlarged the groupings into six kingdoms of organisms based on biochemical differences. Although there is some debate by biologists about the validity of this organizational grouping, the six kingdoms of organisms that are most widely accepted are Eubacteria, Archaebacteria, Plantae, Animalia, Fungi, and Protista (**Figure 12.1b**). More recently, some biologists now group the Eubacteria and Archaebacteria together as kingdom Bacteria and recognize additional divisions within the kingdom Protista.[4]

Kingdom **Eubacteria** (*eu* = good, *bakterion* = a rod) includes some of the simplest organisms. These organisms are single celled but lack discrete nuclei and the internal organelles present in all other organisms. Included in this kingdom are cyanobacteria (blue-green algae) and heterotrophic bacteria. Recent discoveries have shown bacteria to be a much more important part of marine ecology than previously believed. These organisms are found throughout the breadth and depth of the oceans.

Kingdom **Archaebacteria** (*achaeo* = ancient, *bakterion* = a rod) is a group of simple, microscopic, bacterialike creatures that includes methane producers and sulfur oxidizers that inhabit deep-sea vents and seeps, as well as other forms—many of which prefer environments with extreme temperatures and/or pressures. Genetic analyses suggest that these organisms are some of the most ancient life-forms on Earth.

What is coevolution?

The term *coevolution* refers to the way species adapt reciprocally to one another as they evolve. Organism interactions actually form a large web, but we tend to think of pairs of interacting species. The evolutionary dance between a predator and its prey is a good example. It is a kind of arms race, with each improvement in the prey's defense—such as better armor or camouflage—challenging the predator to come up with better adaptations that counter the advantage—such as stronger jaws or more sensitive eyes. And it also works vice versa. Because it's a mutual evolution of adaptations, it's hard to tell if the chicken or the egg came first (so to speak).

[2]A *seep* is an area where various fluids trickle out of the sea floor.

[3]For more on the various types of symbiosis, see Section 14.3 in Chapter 14, "Animals of the Pelagic Environment."

[4]The divisions of kingdom Protista that are recognized by some biologists are kingdom Protozoa (*proto* = first, *zoa* = animal) and kingdom Chromista (*chromo* = color).

Kingdom **Plantae** (*planta* = plant) comprises the multicelled plants, all of which photosynthesize. Only a few species of true plants—such as surf grass (*Phyllospadix*) and eelgrass (*Zostera*)—inhabit shallow coastal environments. In the ocean, photosynthetic marine algae occupy the ecological niche of land plants. However, certain plants are vital parts of coastal ecosystems, including mangrove swamps and salt marshes.

Kingdom **Animalia** (*anima* = breath) comprises the multicelled animals. Organisms from kingdom Animalia range in complexity from the simple sponges to complex vertebrates (animals with backbones), which also includes humans.

Kingdom **Fungi** (*fungus*; probably from the Greek *sp(h)ongos* = sponge) includes more than 100,000 species of molds and lichens, though less than 0.5% of them are sea dwellers. Fungi exist in specialized places throughout the marine environment, and they are most commonly found in the intertidal zone, where they live symbiotically with cyanobacteria or green algae to form lichen. Other fungi remineralize organic matter and function primarily as decomposers in the marine ecosystem.

Kingdom **Protista** (*proto* = first, *ktistos* = to establish) includes a diverse collection of single-celled and multicelled organisms that have a nucleus. Examples of organisms in kingdom Protista include various types of marine *algae* (aquatic photosynthetic organisms that can be either microscopic single-celled or macroscopic multicelled) and single-celled organisms called **protozoa** (*proto* = first, *zoa* = animal).

Linnaeus and Taxonomic Classification

In an effort to determine the relationships of all living things on Earth, Swedish botanist Carl von Linné—who Latinized his name to **Carolus Linnaeus** (1707–1778) (**Figure 12.2**)—created a system in 1758 that is the basis of the modern scientific system of classification used today. Linnaeus developed a system similar to the social hierarchy of his day, with its kingdoms, countries, provinces, parishes, and villages. Linnaeus was a gifted observationist and tireless collector who spent much of his time naming species and organizing them into groups. In fact, Linnaeus's original names for the roughly 12,000 organisms he examined over the course of his life became the starting point for all biological classification. Linnaeus's organizational scheme can be visualized as a series of nested boxes (**Figure 12.3**). Today, the systematic classification of organisms—called **taxonomy** (*taxix* = arrangement, *nomia* = a law)—involves using physical characteristics as well as genetic information to recognize organism similarities and then grouping them into the following increasingly specific categories:

- Kingdom (Less specific grouping)
- Phylum[5]
- Class
- Order
- Family
- Genus
- Species (More specific grouping)

Figure 12.2 Carolus Linnaeus. An 1805 engraving of Swedish botanist and the father of taxonomic classification, Carolus Linnaeus, who is wearing traditional Lapland clothing.

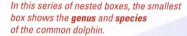

*In this series of nested boxes, the smallest box shows the **genus** and **species** of the common dolphin.*

Genus • *Delphinus*
Species • *delphis*
(Common dolphin)

FAMILY • DELPHINIDAE
ORDER • CETACEA
CLASS • MAMMALIA
PHYLUM • CHORDATA
KINGDOM • ANIMALIA

As the boxes become larger, they represent groups of organisms that the dolphin is more distantly related to.

Figure 12.3 A visualization of Linnaeus's classification scheme. The Linnaean system of classification can be visualized as a series of nested boxes, with genus and species as the most specific groupings (smallest boxes). The larger boxes represent more distant genealogical connections, grouping organisms with fewer shared characteristics and evolutionary similarities.

[5]Note that for plants, the term "division" is used instead of "phylum"; the current International Code of Botanical Nomenclature, however, allows the use of either term.

SmartTable 12.1 TAXONOMIC CLASSIFICATION OF SELECTED ORGANISMS

Category	Human	Common dolphin	Killer whale	Bat star	Giant kelp
Kingdom	Animalia	Animalia	Animalia	Animalia	Protista
Phylum	Chordata	Chordata	Chordata	Echinodermata	Phaeophyta
Subphylum	Vertebrata	Vertebrata	Vertebrata		
Class	Mammalia	Mammalia	Mammalia	Asteroidea	Phaeophycae
Order	Primates	Cetacea	Cetacea	Valvatida	Laminariales
Family	Hominidae	Delphinidae	Delphinidae	Oreasteridae	Lessoniaceae
Genus	*Homo*	*Delphinus*	*Orcinus*	*Asterina*	*Macrocystis*
Species	*sapiens*	*delphis*	*orca*	*miniata*	*pyrifera*

All organisms that share a common category (for instance, a *family*—such as the cat or dolphin family) have certain characteristics and evolutionary similarities. In some cases, subdivisions of these categories are also used, such as subphylum (Table 12.1). The categories assigned to an individual species must be agreed upon by an international panel of experts.

The fundamental unit of taxonomic classification is the **species** (*species* = a kind). Species consist of populations of genetically similar, interbreeding (or potentially interbreeding) individuals that share a collection of inherited characteristics whose combination is unique. Sometimes species may also be defined in other ways, such as individuals coexisting in a population that are similar in appearance. Although species is a useful concept for classifying both present and past life (as represented by fossils), there is still debate amongst biologists about what defines a species.

As an offshoot of his classification scheme, Linnaeus also invented *binomial nomenclature*, where every living thing is known by just two Latin names (previously, organisms were known by a combination of as many as a dozen Latin names). In this way, every type of organism has a unique two-word scientific name composed of its genus and species, which is italicized with the first letter of the generic (genus) name capitalized—for example, *Delphinus delphis*. Most organisms also have one or more common names. *Delphinus delphis*, for instance, is the common dolphin, and *Orcinus orca* is the killer whale or orca. If the scientific name is referred to repeatedly within a document, it is often shortened by abbreviating the genus name to its first letter. Thus, *Delphinus delphis* becomes *D. delphis*.

The Linnaean system, even in its modern form, is far from perfect. New evidence routinely requires taxonomists to move species from one genus to another, or even to an entirely different order. Linnaeus, however, is credited with forging a single, flexible, universally applicable scientific language that is still in use over 250 years after he invented it.

SmartTable 12.1 **Taxonomic classification of selected organisms.**
https://goo.gl/jFfjkM

Web Animation

A Visualization of Taxonomic Classification
https://goo.gl/SzqOAU

STUDENTS SOMETIMES ASK . . .

Why are there scientific names for all organisms? Wouldn't it be easier just to know the common name of an organism?

Each individual species has a unique two-word scientific name, which identifies a particular species more clearly than the common name. Common names are often used for more than one species of organisms, which can be confusing. "Dolphin," for example, is used to describe dolphins, porpoises, and even a type of fish! Most people would strongly object to being served dolphin in a restaurant, but dolphin *fish* (also called mahi-mahi) is often a featured menu item.

Common names can also be confusing because there can be more than one for the same species and because they vary from language to language. Scientific names, on the other hand, are Latin based, so they are the same in all languages. This allows a Chinese scientist, for example, to communicate effectively with a Greek scientist about a particular organism. Thus, scientific names are useful, descriptive (if you know a bit about Latin terms and word roots), and unambiguous.

CONCEPT CHECK 12.1 | Discuss the characteristics of life and how living things are classified.

1 What characteristics should be included in a good working definition of life?

2 List the three major domains of life and the six kingdoms of organisms.

Describe the fundamental criteria used in assigning organisms to these divisions.

3 What are thought to be the ancestors of the three domains of life?

RECAP

Living organisms use energy, reproduce, adapt, and change through time. Living things can be classified into one of three domains and six kingdoms, each of which is split into increasingly specific groupings of phylum, class, order, family, genus, and species.

12.2 How Are Marine Organisms Classified?

Marine organisms can be classified according to where they live (their habitat) and how they move (their mobility). Organisms that inhabit the water column can be classified as either *plankton* (drifters) or *nekton* (swimmers). All other organisms are *benthos* (bottom dwellers).

in temperature, salinity, viscosity, and availability of nutrients effectively limit their lateral range. The deaths of large numbers of fish, for example, can be caused by temporary horizontal shifts of water masses in the ocean. Changes in water pressure normally limit the vertical range of nekton.

Fish may appear to exist everywhere in the oceans, but they are most abundant near continents and islands and in colder waters. Some fish, such as salmon, ascend freshwater rivers to spawn. Many eels do just the reverse, growing to maturity in freshwater and then descending the streams to breed in the great depths of the ocean.

Benthos (Bottom Dwellers)

The term **benthos** (*benthos* = bottom) describes organisms living on or in the ocean bottom. **Epifauna** (*epi* = upon, *fauna* = animal) live on the surface of the sea floor, either attached to rocks or moving along the bottom. **Infauna** (*in* = inside, *fauna* = animal) live buried in the sand, in discarded shells, or within the mud that exists on the sea bottom. Some benthos, called **nektobenthos** (*nektos* = swimming, *benthos* = bottom), live on the bottom yet also have the ability to swim or crawl through the water above the ocean floor (such as flat-fish, octopuses, crabs, and sea urchins). Examples of benthos are shown in **Figure 12.7**. The shallow coastal ocean floor contains a wide variety of physical and nutritive conditions, which have allowed a great number of animal species to develop. Moving across the bottom from the shore into deeper water, the *number* of benthos species per square meter may remain relatively constant, but the

Figure 12.7 Benthos (bottom dwellers): Representative intertidal and shallow subtidal forms. Schematic drawing of various benthos organisms. Organisms include ① sponges, ② sand dollars, ③ crinoid, ④ sea anemones (open and closed), ⑤ barnacles, ⑥ mussels, ⑦ sea urchin, ⑧ sea cucumber, ⑨ sea hare, ⑩ shore crab, ⑪ sea star, ⑫ abalone, ⑬ ghost crab, ⑭ lug worm, ⑮ annelid worm, and ⑯ clam.

biomass of benthos organisms decreases. This is because these shallow sea floor areas receive sufficient sunlight, so they can support many species of large marine algae (often called *seaweeds*) that are attached to the bottom.

Throughout most of the deeper parts of the sea floor, animals live in perpetual darkness, where photosynthetic production cannot occur. They must feed on each other or on whatever outside nutrients fall from the highly productive upper sunlit surface zone of the ocean.

The deep-sea bottom is an environment of coldness, stillness, and darkness. Under these conditions, life progresses slowly, and organisms that live in the deep sea usually are widely distributed because physical conditions vary little on the deep-ocean floor, even over great distances.

HYDROTHERMAL VENT BIOCOMMUNITIES Prior to the discovery of deep-sea hydrothermal vent biocommunities in 1977, marine scientists believed that only sparse and small life existed on the deep-ocean floor. Then the first biocommunity at a hydrothermal vent was discovered in waters 2500 meters (8200 feet) deep in the Galápagos Rift off South America, demonstrating for the first time that high concentrations of abundant and large deep-ocean benthos are possible. Because the primary factor that limits life on the deep-ocean floor is sparse food supply, scientists wondered how the hydrothermal vent organisms were able to obtain enough food to exist.

It turns out that bacterialike archaea, which produce food not by photosynthesis (for no sunlight is available) but instead thrive on sea floor chemicals, are the base of this marine food web. As a result, the size of individuals and the total biomass in hydrothermal communities far exceed those previously known for deep-ocean benthos. These biocommunities are discussed in Chapter 15, "Animals of the Benthic Environment."

| **CONCEPT CHECK 12.2** | Demonstrate an understanding of how marine organisms are classified. |

1 Describe the lifestyles of plankton, nekton, and benthos. Why does plankton account for a much larger percentage of the ocean's biomass than benthos and nekton combined?

2 List the subdivisions of plankton and benthos and the criteria used for assigning individual species to each.

3 For the following marine organisms, determine if they are plankton, nekton, or benthos: (a) sharks, (b) octopuses, (c) clams, (d) diatoms, (e) corals, (f) crabs, (g) giant kelps, (h) jellies, (i) dolphins.

12.3 How Many Marine Species Exist?

The total number of cataloged species on Earth in both marine and terrestrial environments is currently 1.8 million; the figure is constantly increasing as new species are discovered. Experts suggest that there must be many marine species (maybe millions) that have not yet been identified due in large part to the difficulty and expense of exploring the deep sea. Overall, as many as 2000 new marine and terrestrial species are discovered each year. Therefore, the total number of species on Earth, known and still undiscovered, has been estimated to be between 3 million and 100 million. Although estimates range widely, the most likely total number of species on Earth is probably between 6 million and 12 million.[9] Recently, a sophisticated analysis that takes advantage of a natural mathematical pattern in the biodiversity produced by evolution proposed that there must be 8.7 million eukaryotic species on Earth (give or take 1 million). Remarkably, newly discovered species are not all microbes or small invertebrates. In

[9]Note that this figure does not include the millions of species throughout geologic time that were once living but are now extinct.

RECAP

Marine organisms can be classified according to their habitat and mobility as plankton (drifters), nekton (swimmers), or benthos (bottom dwellers).

12.1 Squidtoons

Ironically, this yeti is found in the deep, dark sea.

https://goo.gl/XPCrID

2.5 cm
1 in

Figure 12.8 Yeti crab. In 2005, Census of Marine Life (CoML) researchers on a voyage to study the deep sea floor in the South Pacific discovered a white crab with long hairy arms, which they dubbed the yeti crab (*Kiwa hirsuta*). The crab uses its hairs to grow bacteria, which it eats or uses to detoxify poisonous minerals from the water emitted by hydrothermal vents where the crab lives.

fact, new species of frogs, lizards, birds, fish, mammals, dolphins, and even primates previously unknown to science have been recently discovered in remote areas of the world.

Some of the difficulties of describing the extent to which the marine environment is inhabited include the immense size of the marine habitat and its inaccessibility. In addition, many marine organisms have not been studied in any detail, relationships between organisms are poorly known, and some populations fluctuate greatly each season. To help address these shortcomings, an ambitious $650 million, 10-year program called the *Census of Marine Life (CoML)* was completed in 2010. The CoML included ocean surveys conducted by thousands of researchers in more than 80 countries that assessed the diversity, distribution, and abundance of marine life. These organisms included not only fish but also sea birds, marine mammals, invertebrates, and microscopic sea life. The census produced thousands of scientific publications and led to the creation of the Ocean Biogeographic Information System database, which holds the millions of records generated by the surveys. The CoML discovered at least 1200 new marine species, such as the yeti crab (**Figure 12.8**), and will undoubtedly add more species as the surveys are fully analyzed.

In 2015, taxonomists at the World Register of Marine Species published an updated list based on the work of CoML and other researchers, cataloguing a total of 228,445 known marine species. The team eliminated 190,400 previous listed species because they were duplicate identities. The total number of 228,445 marine species represents only about 13% of the 1.8 million known species on Earth (**Figure 12.9**). The team also acknowledged that there are many more marine species to be discovered. In fact, experts estimate that the world's oceans probably contain somewhere between 700,000 and 1 million eukaryotic species.

Why Are There So Few Marine Species?

If the ocean is such a prime habitat for life and if life originated there, why do so few of the world's species live in the ocean? The disparity is likely a result of the fact that the marine environment is more stable than the terrestrial environment. A major factor that leads to the creation of different species is the variability of the environment: The more variable the environment, the more species are generally present. For example, the higher environmental variability that exists on land presents many opportunities for natural selection to produce new species to inhabit a variety of new niches. This is one reason tropical rainforests have such a high biodiversity and a resulting large number of species. On the other hand, the relatively uniform conditions of the open ocean do not pressure organisms to adapt, so there are fewer species there. For example, sharks have existed almost unchanged in Earth's oceans for more than 400 million years. In addition, ocean temperatures are not only stable but are also relatively low below the sunlit surface waters. The rate of chemical reaction is slowed, which may further reduce the tendency for speciation to occur.

Climate

Connection

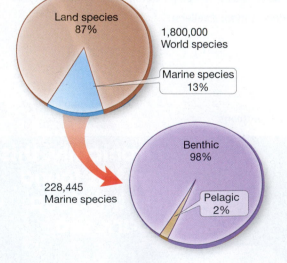

Figure 12.9 Distribution of species on Earth. Of the 1.8 million known species on Earth, 87% inhabit land environments and only 13% inhabit the ocean. Of the 228,445 known marine species, 98% inhabit the benthic environment and live in or on the ocean floor, while only 2% inhabit the pelagic environment and live within the water column as either plankton or nekton.

Species in Pelagic and Benthic Environments

Figure 12.9 shows that only 2% or about 5000 of the 228,445 known marine species inhabit the **pelagic** (*pelagios* = of the sea) **environment** and live within the water column. The other 98% inhabit the **benthic** (*benthos* = bottom) **environment** and live either in or on the sea floor. These numbers are minimums, however, because recent discoveries indicate that many more species may inhabit the benthic environment than previously thought.

Why do most marine species inhabit the benthic environment? The ocean floor contains numerous benthic environments (such as rocky, sandy, muddy, flat, sloped, irregular, and mixed bottoms) that create different habitats to which organisms have adapted. On the other hand, most of the pelagic environment—especially that below the sunlit surface waters—is a watery world that is quite uniform from one region to the next and does not experience extreme environmental variability to which organisms need to be adapted in order to survive.

RECAP

Marine species represent only 13% of the total number of known species on Earth. The benthic environment, which has large environmental variability, is home to 98% of the 228,445 known marine species.

CONCEPT CHECK 12.3 | Specify the number of marine species that exist.

1 How many total species have been cataloged on Earth? How many marine species exist (both number and percentage of total)? Of marine species, how many are benthic versus pelagic?

2 Why do scientists think there are many more undiscovered species on Earth?

3 What factors account for the fact that most marine species inhabit the benthic environment?

12.4 How Are Marine Organisms Adapted to the Physical Conditions of the Ocean?

For organisms to survive, they must be able to adapt to conditions of their environment. The ocean's physical conditions provide benefits but also present challenges to anything living within it.

For example, the marine environment—particularly its temperature—is far more stable than the terrestrial environment. As a result, ocean-dwelling organisms have not developed highly specialized regulatory systems to adjust to sudden changes that might occur within their environment. Marine organisms can, therefore, be affected adversely by very small changes in temperature, salinity, turbidity, pressure, or other environmental conditions.

Climate

Connection

Water constitutes more than 80% of the mass of **protoplasm** (*proto* = first, *plasm* = something molded), which is the substance of living matter. In fact, more than 65% of your body's weight—and 95% of a jelly's weight—is water (**Figure 12.10**). Water carries dissolved within it the gases and minerals organisms need to survive. Water is also a raw material in the photosynthesis of food by marine phytoplankton. Land plants and animals have developed complex "plumbing systems" to retain water and to distribute it throughout their bodies. The inhabitants of the open ocean do not risk atmospheric *desiccation* (drying out), however, because they live in an environment of abundant water.

Need for Physical Support

One basic need of all plants and animals is for simple physical support. Land plants, for example, have vast root systems that anchor the plants securely to the ground. Land animals have skeletons and combinations of appendages—legs, arms, fingers, and toes—to support their entire weight.

In the ocean, water physically supports marine plants and animals. Organisms such as photosynthetic phytoplankton, which must live in the upper surface waters of the ocean, depend primarily upon buoyancy and frictional resistance to sinking in order to maintain their desired position. Still, maintaining position can be difficult, and some organisms have developed special adaptations to increase their efficiency. These adaptations are discussed in this and succeeding chapters.

STUDENTS SOMETIMES ASK . . .

Which group of organisms has the most species on Earth?

Although you might expect that bacteria or other microbes comprise the largest number of species on Earth, it's actually ... (wait for it) ... insects, which comprise 56% of total species on Earth (more than 1 million species identified). Of insect species, nearly half—or one out of every four species on Earth—are beetles, such as ladybugs, fireflies, rhinoceros beetles, soldier beetles, leaf beetles, rove beetles, click beetles, dung beetles, bark beetles, scarabs, and weevils. Even though insect species are incredibly abundant on land, only 1400 species—less than one-quarter of 1%—inhabit the ocean. Of those, just five species (sea skaters of the genus *Halobates*) live in the open ocean; the rest are coastal. Apparently, the lack of solid surfaces in the open ocean is what has kept insects at bay (literally!).

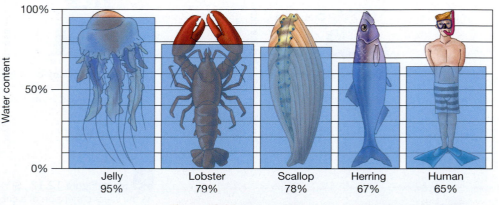

100%

Water content

50%

0%

| Jelly | Lobster | Scallop | Herring | Human |
| 95% | 79% | 78% | 67% | 65% |

Figure 12.10 Percentage water content of selected organisms. Bar graph showing the water content of selected organisms. Jellies are 95% water, and humans are 65% water.

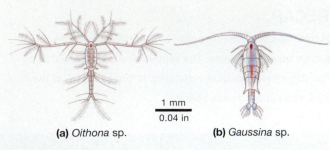

(a) *Oithona* sp. **(b)** *Gaussina* sp.

Figure 12.11 **Water temperature and appendages.** Similar species display different adaptations based on water temperature. For example, **(a)** the copepod *Oithona* displays the ornate plumage characteristic of warm-water varieties and **(b)** the copepod *Gaussina* displays the less ornate appendages found on temperate and cold-water forms.

Water's Viscosity

Viscosity (*viscos* = sticky) is a substance's *internal resistance to flow*. Recall from Chapter 1 that a substance that has high resistance to flow (high viscosity)—such as toothpaste—does not flow easily. Conversely, a substance that has low viscosity—such as water—flows more readily. Viscosity is strongly affected by temperature. Tar, for example, must be heated to decrease its viscosity before it can be spread onto roofs or roads.

The viscosity of ocean water increases as salinity increases and temperature decreases. Thus, single-celled organisms that float in colder, higher-viscosity waters have less need for extensions to help them maintain their positions near the surface. **Figure 12.11** shows, for example, that a warm-water floating crustacean has ornate, featherlike appendages, whereas a cold-water variety does not.

THE IMPORTANCE OF ORGANISM SIZE The basic requirements of phytoplankton are that they (1) stay in the upper portion of ocean water where solar radiation is available, (2) have available necessary nutrients, (3) efficiently take in these nutrients from surrounding waters, and (4) expel waste materials. Their small size and ingenious shapes help single-celled phytoplankton satisfy these requirements without needing specialized multiple cells.

Phytoplankton cannot propel themselves, so they use frictional resistance to maintain their general position near the surface of the water. Frictional resistance to sinking increases as an organism's ratio of surface area to volume (mass) increases. For example, **Figure 12.12** shows the surface-area-to-volume ratio of three different cubes and illustrates that the ratio increases as an organism's size decreases. In the figure, Cube a has twice the surface area per unit of volume of Cube b and four times the surface area per unit of volume as Cube c. If the cubes were plankton, Cube a would have four times the resistance to sinking per unit of mass as Cube c, so Cube a would need to exert far less energy to stay afloat. Single-celled organisms, which make up the bulk of photosynthetic marine life, clearly benefit from being as small as possible. They are so small, in fact, that one needs a microscope to see them!

Small size also serves the other basic requirements of phytoplankton. Photosynthetic cells take in nutrients from surrounding water and expel waste through their cell membranes, and the efficiency of both functions increases with a higher surface area-to-volume ratio. Thus, if Cubes a and c in Figure 12.12 were planktonic algae, Cube a could take in nutrients and dispose of waste four times more efficiently than

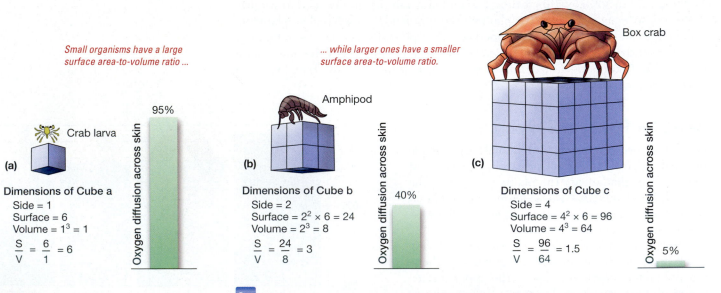

Small organisms have a large surface area-to-volume ratio ...

... while larger ones have a smaller surface area-to-volume ratio.

Box crab

Crab larva

Amphipod

Oxygen diffusion across skin

95%

40%

5%

(a)

Dimensions of Cube a
Side = 1
Surface = 6
Volume = 1^3 = 1
$\dfrac{S}{V} = \dfrac{6}{1} = 6$

(b)

Dimensions of Cube b
Side = 2
Surface = $2^2 \times 6$ = 24
Volume = 2^3 = 8
$\dfrac{S}{V} = \dfrac{24}{8} = 3$

(c)

Dimensions of Cube c
Side = 4
Surface = $4^2 \times 6$ = 96
Volume = 4^3 = 64
$\dfrac{S}{V} = \dfrac{96}{64} = 1.5$

SmartFigure 12.12 **Surface area-to-volume ratio of cubes of different sizes.** As the linear dimension of a cube increases, the ratio of surface area-to-volume decreases. Thus, smaller bodies have a higher surface area-to-volume ratio, which allows them to stay afloat more easily, exchange nutrients and wastes more efficiently, and diffuse oxygen across their skin more effectively. https://goo.gl/K46zXC

Cube c. This is why cells in all plants and animals are microscopic, regardless of the overall size of the organism.

Diatoms—one of the most important groups of phytoplankton—often have unusual appendages, needlelike extensions, or even rings (**Figure 12.13**) to increase their surface area, thus preventing them from sinking below sunlit surface waters. Other planktonic marine organisms—particularly warm-water species—use similar strategies to stay afloat. Some small organisms produce a tiny droplet of oil, which lowers their overall density and increases buoyancy. Interestingly, accumulations of vast amounts of these organisms in sea floor sediment can produce offshore oil deposits. This happens when the droplets of oil combine, and, over millions of years, become buried deep below the sea floor, which exposes the oil to the natural heat and high pressure of Earth. Once the oil has undergone chemical changes, it can migrate and become trapped in an oil reservoir.

Despite adaptations to remain in the upper layers of the ocean, organisms still have a higher density than seawater, so they tend to sink, if ever so slowly. This is not a serious handicap, however, because wind causes considerable mixing and turbulence near the surface. Mixing returns phytoplankton to surface waters, keeping these organisms positioned to bask in the solar radiation needed to photosynthesize and produce the energy used by essentially all other members of the marine community.

VISCOSITY AND STREAMLINING As organisms increase in size, viscosity ceases to enhance survival and instead becomes an obstacle. This is particularly true of large organisms that swim freely in the open ocean. They must pursue prey or flee predators, yet the faster they swim, the more the viscosity of water impedes their progress. Not only must water be displaced ahead of the swimmer, but water also must move in behind it to occupy the space that the animal has vacated.

Figure 12.14 shows the advantage of **streamlining**, which is having a shape that offers the least resistance to fluid flow. Streamlining allows marine organisms to overcome water's viscosity and move more easily through water. A streamlined shape usually consists of a flattened body, which presents a small cross section at the front end and a gradual tapering at the back end to reduce the wake created by eddies. It is exemplified in the shape of free-swimming fish (and in marine mammals such as whales and dolphins).

REPRODUCTION Marine organisms also take advantage of water's high viscosity to enhance chances of reproduction and to populate new habitats. For example, many organisms use the reproductive strategy called **broadcast spawning**, which involves releasing eggs and sperm directly into seawater, like pollen adrift in the wind. In some instances, broadcast spawning occurs during large assemblages when organisms of the opposite sex are in close proximity to one another. In other cases, large quantities of reproductive materials are simply released into the water under the assumption that at least a few cells of the right type will come together to successfully produce offspring. In addition, larvae are sometimes released in great numbers into ocean currents, which carry juvenile organisms into new habitats.

Temperature

Figure 12.15 compares extremes in land and ocean surface temperatures and shows that ocean temperatures have a far narrower range than temperatures on land. The minimum surface temperature of the open ocean is seldom much below −2°C (28.4°F), and the maximum surface temperature seldom exceeds 32°C (89.6°F), except in some shallow-water coastal regions, where the temperature may reach 40°C (104°F). On land, however, extremes in temperatures have ranged from −88°C (−127°F) to 58°C (136°F), which represents a temperature range more than

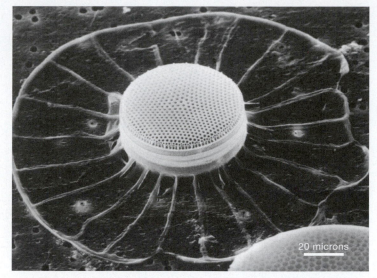

Figure 12.13 Warm-water diatom. Scanning electron micrograph of the warm-water diatom *Planktoniella sol*, which has a prominent marginal ring and spokes to increase its surface area, thus preventing it from sinking.

20 microns

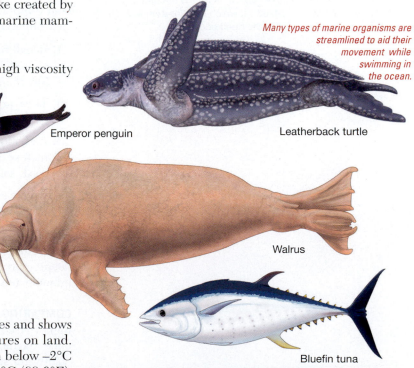

Many types of marine organisms are streamlined to aid their movement while swimming in the ocean.

Emperor penguin

Leatherback turtle

Walrus

Bluefin tuna

Figure 12.14 Streamlining. Schematic drawing showing examples of marine organisms that have a streamlined body shape, which enables them to efficiently move through and displace water with minimum resistance, creating only a small wake.

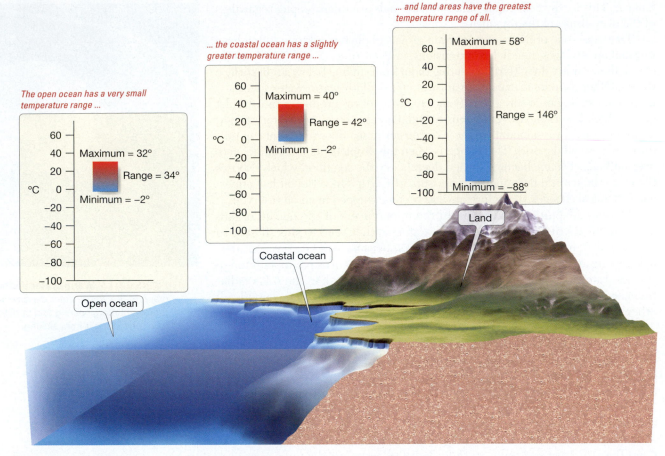

The open ocean has a very small temperature range ...

... the coastal ocean has a slightly greater temperature range ...

... and land areas have the greatest temperature range of all.

Open ocean — Maximum = 32°, Range = 34°, Minimum = −2°

Coastal ocean — Maximum = 40°, Range = 42°, Minimum = −2°

Land — Maximum = 58°, Range = 146°, Minimum = −88°

Figure 12.15 Comparison of extremes in ocean and land surface temperatures. In the open ocean, the maximum temperature range is limited to only 34°C (61°F); it increases to 42°C (76°F) in the coastal ocean. On land, the maximum temperature range is 146°C (263°F), which is more than four times the temperature range in the open ocean.

four times greater than that experienced by the ocean. This *continental effect* was discussed in Chapter 5. Further, the ocean has a smaller daily, seasonal, and annual temperature range than that experienced on land, which provides a stable environment for marine organisms. The reasons for this are fourfold:

1. Recall from Chapter 5 that the heat capacity of water is much higher than that of land, which causes land to heat up by a greater amount and much more rapidly than the ocean.

2. The warming of the ocean is reduced substantially because of evaporation, a cooling process that stores excess heat as latent heat.

3. Radiation received at the surface of the ocean can penetrate several tens of meters deep and distribute its energy throughout a very large mass. In contrast, solar radiation absorbed by land heats only a very thin surface layer.

4. Unlike solid land surfaces, water has good mixing mechanisms, such as currents, waves, and tides that allow heat from one area to be transported to other areas.

In addition, the small daily and seasonal temperature variations are confined to ocean surface waters and decrease with depth, becoming insignificant throughout the deeper parts of the ocean. At ocean depths that exceed 1.5 kilometers (0.9 mile), for example, temperatures hover around 3°C (37.4°F) year-round, regardless of latitude.

COMPARING COLD- AND WARM-WATER SPECIES Cold water is denser and has a higher viscosity than warm water. These factors, among others, profoundly influence marine life, resulting in the following differences between warm-water and cold-water species in the marine environment:

- Floating organisms are physically smaller in warm waters than in colder waters. Small organisms expose more surface area per unit of body mass, which helps them maintain their position in the lower viscosity and density of warm seawater more easily.

- Warm-water species often have ornate plumage to increase surface area, which is strikingly absent in the larger cold-water species (see Figures 12.11 and 12.13).
- Warmer temperatures increase the rate of biological activity, which more than doubles with an increase of 10°C (18°F). Tropical organisms apparently grow faster, have a shorter life expectancy, and reproduce earlier and more frequently than those in colder water.
- There are more *species* in warm waters, but the total *biomass* of plankton in colder, high-latitude waters greatly exceeds that of the warmer tropics. Note that the high biomass of plankton in high-latitude regions is not directly caused by temperature and viscosity; it is only that these conditions are associated with the upwelling of nutrients, which in turn supports a vast biomass of phytoplankton.

Some animal species can live only in cooler waters, whereas others can live only in warmer waters. Many of these organisms can withstand only very small temperature changes and are called **stenothermal** (*steno* = narrow, *thermo* = temperature). Stenothermal organisms are found predominantly in the open ocean, at depths where large temperature ranges do not occur.

Other species are little affected by different temperatures and can withstand large and even rapid changes in temperature. These organisms are called **eurythermal** (*eury* = wide, *thermo* = temperature) and are predominantly found in shallow coastal waters—where the largest temperature ranges are found—and in surface waters of the open ocean.

Salinity

The sensitivity of marine animals to changes in their environment varies from organism to organism. Those that inhabit estuaries, for example, such as oysters, must be able to withstand considerable fluctuations in salinity. The daily rise and fall of the tides forces salty ocean water into river mouths and draws it out again, changing the salinity considerably. During floods, the salinity in estuaries can reach extremely low levels. The coastal organisms that can tolerate large changes in salinity are known as **euryhaline** (*eury* = wide, *halo* = salt).

Marine organisms that inhabit the open ocean, on the other hand, are seldom exposed to a large variation in salinity. They have adapted to a constant salinity and can tolerate only very small changes. These organisms are called **stenohaline** (*steno* = narrow, *halo* = salt).

EXTRACTION OF SALINITY COMPONENTS Some organisms extract minerals from ocean water—particularly *silica* (SiO_2) and *calcium carbonate* ($CaCO_3$)—to construct the hard parts of their bodies, which serve as protective coverings. In doing so, they reduce the amount of dissolved material in ocean water. For example, phytoplankton (including diatoms) and microscopic protozoans such as radiolarians and silicoflagellates extract silica from seawater. Coccolithophores, foraminifers, most mollusks, corals, and some algae that secrete a calcium carbonate skeletal structure extract calcium carbonate from seawater.

Interdisciplinary

Relationship

DIFFUSION Molecules of soluble substances, such as nutrients, move through water from areas of *high* concentration to areas of *low* concentration until the distribution of the substance is uniform (**Figure 12.16a**). This process, called **diffusion** (*diffuse* = dispersed), is caused by random motion of molecules. The outer membrane of a living cell is permeable to many molecules. Organisms may take in nutrients they need from the surrounding water by diffusion of the nutrients through their cell walls. Because nutrients are usually plentiful in seawater, they pass through the cell wall into the interior, where nutrients are less concentrated (**Figure 12.16b**). In addition to passive diffusion, organisms also import nutrients to cells by active transport.

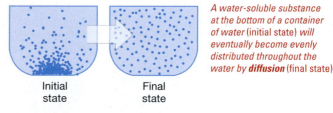

A water-soluble substance at the bottom of a container of water (initial state) *will eventually become evenly distributed throughout the water by* **diffusion** (final state).

Initial state Final state

(a) Diffusion in a container of water.

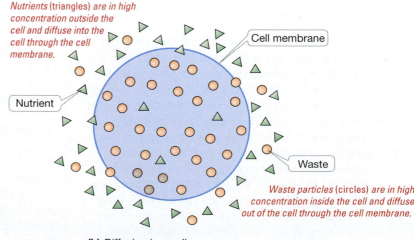

Nutrients (triangles) *are in high concentration outside the cell and diffuse into the cell through the cell membrane.*

Cell membrane

Nutrient

Waste

Waste particles (circles) *are in high concentration inside the cell and diffuse out of the cell through the cell membrane.*

(b) Diffusion in a cell.

Figure 12.16 Diffusion. Diffusion is the process by which materials move through fluids from higher to lower concentrations and become more evenly distributed.

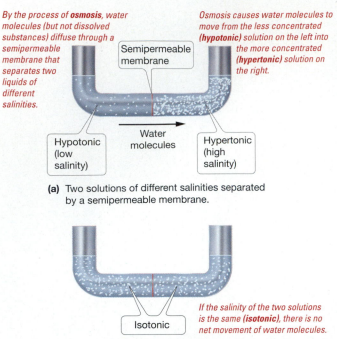

By the process of **osmosis**, water molecules (but not dissolved substances) diffuse through a semipermeable membrane that separates two liquids of different salinities.

Semipermeable membrane

Osmosis causes water molecules to move from the less concentrated (**hypotonic**) solution on the left into the more concentrated (**hypertonic**) solution on the right.

Water molecules

Hypotonic (low salinity)

Hypertonic (high salinity)

(a) Two solutions of different salinities separated by a semipermeable membrane.

Isotonic

If the salinity of the two solutions is the same (**isotonic**), there is no net movement of water molecules.

(b) Two solutions of the same salinity separated by a semipermeable membrane.

Figure 12.17 Osmosis. Osmosis is the process by which water molecules move through a semipermeable membrane from higher water molecule concentration (lower salinity) to lower water molecule concentration (higher salinity).

Web Animation
Osmosis
https://goo.gl/Ei4Lpa

After a cell uses the energy stored in nutrients, it must dispose of waste. Waste passes out of a cell by diffusion, too. As the concentration of waste materials becomes greater within the cell than in the water surrounding it, these materials pass from within the cell into the surrounding fluid. The waste products are then carried away by circulating fluid that services cells in higher animals or by the water that surrounds simple one-celled organisms.

OSMOSIS When water solutions of unequal salinity are separated by a semipermeable membrane (such as skin or the membrane around a living cell), water molecules (but not dissolved ions) diffuse through the membrane. To equalize the salt concentration on both sides of the membrane, water molecules always move from the *less* concentrated solution into the *more* concentrated solution in a process called **osmosis** (*osmos* = to push) (**Figure 12.17a**). **Osmotic pressure** is the pressure that must be applied to the more concentrated solution to prevent water molecules from passing into it. Osmosis causes water to move through an organism's skin (its semipermeable membrane) and affects both marine and freshwater organisms. If the salinity of an organism's body fluid equals that of the ocean, it is **isotonic** (*iso* = same, *tonos* = tension) and has equal osmotic pressure, and so no net transfer of water will occur through the membrane in either direction (**Figure 12.17b**).

Interdisciplinary

Relationship

If seawater has a lower salinity than the fluid within an organism's cells, seawater will pass through the cell walls into the cells (always toward the more concentrated solution). This organism is **hypertonic** (*hyper* = over, *tonos* = tension), which means it is saltier than the surrounding seawater.

If the salinity within an organism's cells is less than that of the surrounding seawater, water from the cells will pass through the cell membranes out into the seawater (toward the more concentrated solution). This organism is **hypotonic** (*hypo* = under, *tonos* = tension) relative to the water outside its body.

In essence, osmosis is diffusion that produces a net transfer of water molecules through a semipermeable membrane from the side with the *greatest concentration* of water molecules to the side with the *lesser concentration* of water molecules.

During osmosis, three things can occur simultaneously across the cell membrane:

1. Water molecules move through the semipermeable membrane toward the side with the lower concentration of water.

2. Nutrient molecules or ions move from where they are more concentrated into the cell, where they are used to maintain the cell.

3. Waste molecules move from within the cell to the surrounding seawater.

Molecules or ions of all the substances in the system are passing through the membrane in both directions. A net transport of molecules of a given substance always occurs from the side on which they are most highly concentrated to the side where the concentration is less, until equilibrium is attained.

The body fluids of marine invertebrates (those without backbones) such as worms, mussels, and octopuses, and the seawater in which they live, are nearly isotonic. As a result, these organisms have not had to evolve special mechanisms to maintain their body fluids at a proper concentration. This gives them an advantage over their freshwater relatives, whose body fluids are hypertonic.

AN EXAMPLE OF OSMOSIS: SALTWATER VERSUS FRESHWATER FISH Saltwater fish have body fluids that are only slightly more than one-third as saline as ocean water, possibly because they evolved in low-salinity coastal waters. They are, therefore, hypotonic (less salty) compared to the surrounding seawater.

This salinity difference means that saltwater fish, without some means of regulation, would lose water from their body fluids into the surrounding ocean and eventually dehydrate. This loss is counteracted, however, because saltwater fish drink ocean water and excrete the salts through special chloride-releasing cells located in their gills. These fish

Interdisciplinary

Relationship

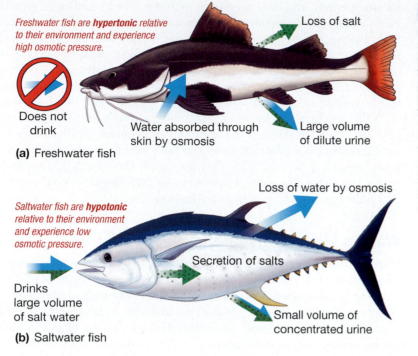

*Freshwater fish are **hypertonic** relative to their environment and experience high osmotic pressure.*

Loss of salt

Does not drink

Water absorbed through skin by osmosis

Large volume of dilute urine

(a) Freshwater fish

Loss of water by osmosis

*Saltwater fish are **hypotonic** relative to their environment and experience low osmotic pressure.*

Secretion of salts

Drinks large volume of salt water

Small volume of concentrated urine

(b) Saltwater fish

SmartFigure 12.18 Salinity adaptations of freshwater and saltwater fish. Osmotic processes cause freshwater and saltwater fish to have different adaptations to their environment. As a result, freshwater fish are hypertonic relative to their environment, and saltwater fish are hypotonic relative to theirs.
https://goo.gl/o9YA04

also help maintain their body water by discharging a very small amount of very highly concentrated urine (**Figure 12.18a**). Freshwater fish are hypertonic (internally more saline) compared to the very dilute water in which they live. The osmotic pressure of the body fluids of such fish may be 20 to 30 times greater than that of the freshwater that surrounds them, so freshwater fish risk rupturing cell walls from taking in excessive quantities of water through osmosis. To prevent this, freshwater fish do not drink water, and their cells have the capacity to absorb salt. They also excrete large volumes of very dilute urine to reduce the amount of water in their cells (**Figure 12.18b**).

RECAP

Osmosis produces a net transfer of water molecules through a semipermeable membrane, from the side with the greater concentration of water molecules to the side with the lesser concentration.

Dissolved Gases

The amount of gases that dissolve in seawater increases as the temperature of seawater decreases, so cold water dissolves more gas than warm water. This helps vast phytoplankton communities develop in high latitudes during summer, when solar energy becomes available for photosynthesis. These cold waters contain an abundance of dissolved gases, specifically carbon dioxide (which phytoplankton need for photosynthesis) and oxygen (which all organisms need to metabolize their food). In addition, the cold, oxygen-rich water of high-latitude regions sinks and flows along the ocean bottom, supplying deep-sea organisms with an abundant supply of dissolved oxygen.

Most animals that live in the ocean—except air-breathing marine mammals and certain fishes—must extract dissolved oxygen from seawater. How do they do this? Most marine animals have specially designed fibrous respiratory organs called **gills** that exchange oxygen and carbon dioxide directly with seawater. Most fish, for instance, take water in through their mouths (which gives them the appearance of "breathing" underwater), pass it through their gills to extract oxygen, and then expel it through the gill slits on the sides of their bodies (**Figure 12.19**). Most fish need at least 4.0 parts per million (ppm by weight) of dissolved oxygen in seawater to survive

Interdisciplinary

Relationship

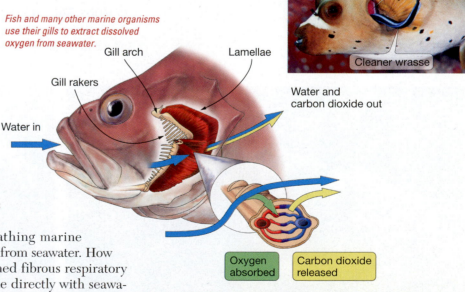

Fish and many other marine organisms use their gills to extract dissolved oxygen from seawater.

Photo

Gills

Cleaner wrasse

Gill arch

Lamellae

Gill rakers

Water in

Water and carbon dioxide out

Oxygen absorbed

Carbon dioxide released

Figure 12.19 Gills on fish. Water is taken in through the mouth and passes through the gills, which extract dissolved oxygen. Afterward, water and carbon dioxide are expelled through the gill slits. Photo (*inset*) shows a blue-streaked cleaner wrasse (*Labroides dimidiatus*) cleaning the gills of a blackspotted puffer (*Arothron nigropunctatus*).

Figure 12.20 Moon jellies. Most jellies such as these moon jellies (*Aurelia aurita*) are nearly transparent, making them very difficult for predators to see. To enhance their appearance in this photograph, they are illuminated by a strong white light from above.

for long periods, and they need even more for activity and rapid growth. That's why an aquarium needs an aerator to continually resupply oxygen to the water in the tank. During low-oxygen conditions, most marine animals with gills cannot simply breathe air at the surface. Their adaptations allow them to use only oxygen that is dissolved in water. If dissolved oxygen levels become low enough (such as after an algae bloom, when decomposition consumes dissolved oxygen), it may cause many marine organisms to suffocate and die unless they can move to other oxygen-rich areas.

Gill structure and location vary among animals of different groups. In fishes, gills are located at the rear of the mouth and contain capillaries. In higher aquatic invertebrates, they protrude from the body surface and contain extensions of the vascular system. In mollusks, they are inside the mantle cavity. In higher vertebrates (including humans), they occur as rudimentary, nonfunctional gill slits, which disappear during embryonic development.

Water's High Transparency

Water—including seawater—has relatively high transparency compared to many other substances, allowing sunlight to penetrate to a depth of about 1000 meters (3300 feet) in the open ocean. The actual depth depends on the amount of turbidity (suspended sediment) in the water, the amount of plankton in the water, latitude, time of day, and the season. Even though there are few places to hide in the open ocean, marine organisms employ many clever adaptations to prevent being seen.

TRANSPARENCY Because of water's high transparency, many marine organisms have developed large eyes, which help them locate and capture prey in dim light. To combat keen-eyed predators, many marine organisms such as jellies are themselves nearly transparent, which helps them blend into their environment (**Figure 12.20**). In fact, almost all open-ocean animals not otherwise protected by teeth, toxins, speed, or small size have some degree of invisibility. Only at depths where sunlight never penetrates is transparency uncommon. Another strategy used to increase an organism's transparency is to have silver sides that act like a mirror and reflect back weak light. Not only can transparency help organisms elude their predators, it can also help transparent organisms stalk their prey.

CAMOUFLAGE AND COUNTERSHADING Some marine organisms hide by using their coloration pattern as camouflage (**Figure 12.21a**). Still others use **countershading**, which means they are dark colored on top and light colored on bottom (**Figure 12.21b**),

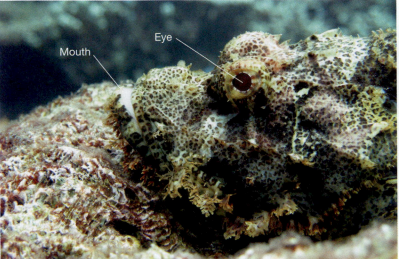

(a) The head and eye of a well-camouflaged rock fish.

(b) Halibut on a dock in Alaska show countershading.

Figure 12.21 Examples of camouflage and countershading.

to blend into their environment. Many fish—especially flat fish—have countershading so they cannot easily be seen against the dark background of deep water or the ocean floor and they blend into the sunlight when viewed from below. Countershading also helps predators sneak up on prey.

DAILY VERTICAL MIGRATION: THE DEEP SCATTERING LAYER (DSL) Many marine organisms undertake a daily vertical migration to deeper, darker parts of the ocean to avoid becoming prey. These organisms create a curious feature called the **deep scattering layer (DSL)**, which was discovered when the U.S. Navy was testing sonar equipment to detect enemy submarines early in World War II. On many of the sonar recordings, a mysterious sound-reflecting surface appeared that was much too shallow to be the ocean floor. It was often referred to as a "false bottom" (see Figure 3.1). What was even more surprising was that the depth of the deep scattering layer changed with time. It was at a depth of about 100 to 200 meters (330 to 660 feet) during the night but was as deep at 900 meters (3000 feet) during the day.

With the help of marine biologists, sonar specialists were able to determine that sonar signals were reflecting off densely packed concentrations of marine organisms. Investigation with plankton nets, submersibles, and detailed sonar revealed that the DSL contains many different organisms, including copepods (which constitute a large proportion of planktonic animals), krill (small crustaceans), and lantern fish (family Myctophidae) (**Figure 12.22**, *enlargement*). The daily movement of the deep scattering layer is caused by the vertical migration of marine organisms that feed in the highly productive surface waters under cover of darkness to protect themselves from being seen by predators. These predators include daytime, **crepuscular** (*crepusculum* = twilight), and nocturnal (*nocturnus* = night) feeders (Figure 12.22). Organisms within the DSL ascend to the surface only at night and then migrate to deeper (darker) water to hide during the day.

Web Animation

Daily Movement of the Deep Scattering Layer (DSL)
http://goo.gl/mV17kD

Figure 12.22 Daily movement and organisms of the deep scattering layer. **(a)** Organisms in the deep scattering layer (DSL) migrate to deep depths during the day to hide from predators and rise at night to feed in surface waters. **(b)** Predators that feed in the DSL include deep-diving daytime, crepuscular (twilight), and nocturnal hunters.

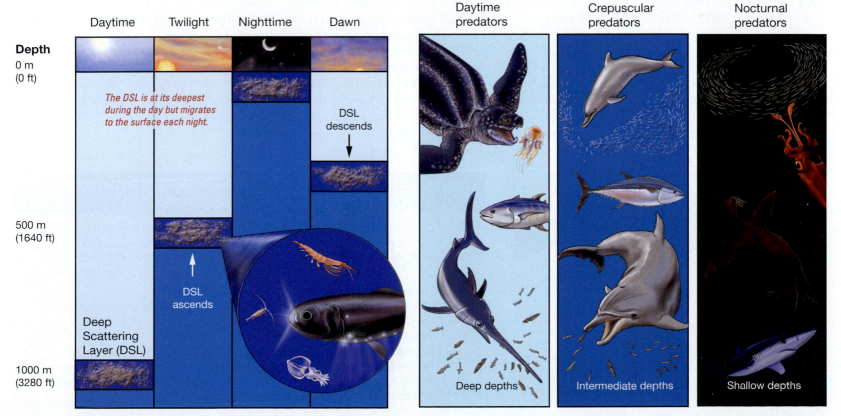

(a) Daily migration and typical organisms associated with the deep scattering layer.

(b) Types of predators that feed on organisms within the deep scattering layer.

Figure 12.23 Use of color by tropical fish. Many tropical fish such as this mandarinfish (*Synchiropus splendidus*) have bright colors and bold patterns, which allow them to blend into the environment by using disruptive coloration. Alternatively, they may also stand out so that they can advertise their identity, sex, or weaponry.

STUDENTS SOMETIMES ASK . . .

If marine animals with large internal air pockets are affected by the extreme pressures at depth, how are sperm whales able to dive so deeply?

All marine mammals have lungs and breathe air, and certain ones have special adaptations that allow them to make extremely deep dives. Sperm whales, for example, can dive to more than 2800 meters (9200 feet) and stay submerged for more than two hours during their search for food! They are able to use small amounts of oxygen very efficiently and have a collapsible rib cage, which forces air out and collapses the lungs, thereby closing the air cavities inside their bodies. Marine mammals and their adaptations are discussed in Chapter 14, "Animals of the Pelagic Environment."

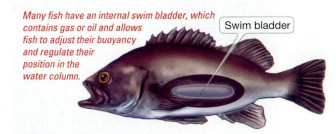

Many fish have an internal swim bladder, which contains gas or oil and allows fish to adjust their buoyancy and regulate their position in the water column.

Swim bladder

Figure 12.24 Swim bladder.

RECAP

The ocean's physical support, viscosity, temperature, salinity, sunlit surface waters, dissolved gases, high transparency, and pressure create conditions to which marine organisms are superbly adapted.

Research reveals that the daily migration of the DSL causes increased vertical mixing of ocean waters. A recent study, in fact, suggests that energy from small animals such as those found in the DSL is a major component of the total contributed to the ocean by all swimming creatures, which is comparable to that of winds and tides.

DISRUPTIVE COLORATION In contrast to species trying to blend into their environment, many species of tropical fish display bright colors (**Figure 12.23**). Why would they be brightly colored if it makes them easily seen by predators? The bright markings of tropical fish may be an example of **disruptive coloration**, where large, bold patterns of contrasting colors tend to make an object blend in when viewed against an equally variable, contrasting background such as a tropical reef. Zebras use this principle to evade predators, tigers use it to conceal themselves from their prey, and military uniforms are camouflaged in a similar way. Even considering disruptive coloration, many tropical fish still don't seem to blend into their environment. Perhaps the bright colors and distinctive markings that make tropical fish more apparent make it easier to advertise their identity, to attract mates, or to display weaponry such as spines or poison. Scientists have yet to agree on why tropical fish have such vivid colors, but it must serve some biological advantage, or the fish wouldn't be so brightly colored.

Pressure

Water pressure increases about 1 kilogram per square centimeter (1 atmosphere, or 14.7 pounds per square inch) with every 10 meters (33 feet) of water depth. Humans are not well adapted to the high pressures that exist below the surface (**Diving Deeper 12.1**). Even when diving to the bottom of the deep end of a swimming pool, one can feel the dramatic increase in pressure in one's ears.

In the deep ocean, water pressure is on the order of several hundred kilograms per square centimeter (several hundred atmospheres, or several tons per square inch). How do deep-water marine organisms withstand pressures that can easily kill humans? Most marine organisms lack large compressible air pockets inside their bodies. They do not have lungs, ear canals, or other passageways as we do, so these organisms don't feel the high pressure pushing in on their bodies. Water is nearly incompressible, so their water-filled bodies have the same amount of pressure pushing outward, and they are unaffected by the high pressures found in deep-ocean environments. However, many fish have a **swim bladder** (**Figure 12.24**) that contains gas or oil and allows fish to adjust their buoyancy, thus helping them regulate their position in the water column.

A few species appear to be extremely tolerant of pressure changes. In fact, some marine species that are found in nearshore areas can also be found at depths of several kilometers.

CONCEPT CHECK 12.4	Explain how marine organisms are adapted to the physical conditions of the ocean.

1 Discuss some adaptations other than size that organisms use to increase their resistance to sinking.

2 List differences between cold- and warm-water species in the marine environment.

3 Describe the process of osmosis. How is it different from diffusion? What three things can occur simultaneously across the cell membrane during osmosis?

4 What is the problem requiring osmotic regulation faced by *hypotonic* fish in the ocean? What adaptation do these animals have to overcome this problem?

5 How does water temperature affect the water's ability to hold gases? How do marine organisms extract dissolved oxygen from seawater?

6 What is the deep scattering layer (DSL), and why does it migrate vertically during the day?

DIVING INTO THE MARINE ENVIRONMENT

Throughout history, humans have submerged themselves in the marine environment to observe it directly for scientific exploration, profit, or adventure (**Figure 12A**). As early as 4500 B.C., brave and skillful divers reached depths of 30 meters (100 feet) on one breath of air to retrieve red coral and mother-of-pearl shells. Later, diving bells (bell-shaped structures full of trapped air) were lowered into the sea to provide passengers or underwater divers with an air supply. In 360 B.C., Aristotle, in his *Problematum*, recorded the use by Greek sponge divers of

kettles full of air lowered into the sea. However, technology to move around freely while breathing underwater was not developed until 1943, when Jacques-Yves Cousteau and Émile Gagnan invented the fully automatic, compressed-air Aqualung. The equipment was later dubbed **scuba**, an acronym for *self-contained underwater breathing apparatus*, and it is used by millions of recreational divers today. By using scuba, divers can experience the ocean firsthand, leading to a fuller appreciation of the wonder and beauty of the marine environment.

Those who venture underwater must contend with many obstacles inherent in ocean diving, such as low temperatures, darkness, and the effects of greatly increased pressure.

To combat low temperatures, specially designed clothing is worn. Waterproof, high-intensity diving lights are used to combat darkness. To combat the deleterious effects of pressure, depth and duration of dives must be limited. As a result, most scuba divers rarely venture below a depth of 30 meters (100 feet)—where the pressure is three times that at the surface—and they stay there less than 30 minutes.

It is relatively dangerous for humans to enter the marine environment because our bodies are adapted to living in the relatively low pressure of the atmosphere. In water, pressure increases rapidly with depth, to which anyone who has been to the bottom of the deep end of a swimming pool can attest. The increased pressure at depth in the ocean can cause problems for divers. For instance, higher pressure causes more

nitrogen to be dissolved in a diver's body and may cause a disorienting condition known as *nitrogen narcosis*, or *rapture of the deep*. Further, if a diver surfaces too rapidly, expanding gases within the body can catastrophically rupture cell membranes (a condition called *barotrauma*).

In addition, when divers return to the surface, they may experience *decompression sickness*, which is also called *caisson disease*, or *the bends*. The bends affects divers who ascend to the lower pressure at the surface too rapidly, causing nitrogen bubbles to form in the bloodstream and other tissues (analogous to the bubbles that form in a carbonated beverage when the container is opened). Various symptoms can result, from nosebleed and joint pain (which causes divers to stoop over, hence the term *the bends*) to permanent neurological injury and even fatal paralysis. To avoid it, divers must ascend slowly, allowing time for excess dissolved nitrogen to be eliminated from the blood via the lungs.

Despite these risks, divers venture to greater and greater depths in the ocean. In 1962, Hannes Keller and Peter Small made an open-ocean dive from a diving bell to a then-record-breaking depth of 304 meters (1000 feet). Although they used a special gas mixture, Small died once they returned to the surface. Presently, the record ocean dive is 534 meters (1750 feet), but researchers who study the physiology of deep divers have simulated a dive to 701 meters (2300 feet) in a pressure chamber using a special mix of oxygen, hydrogen, and helium gases. Researchers believe that humans will eventually be able to stay underwater for extended periods of time at depths below 600 meters (1970 feet).

GIVE IT SOME THOUGHT

1. What causes the bends? How can divers avoid this debilitating condition?

Figure 12A Oceanographer and explorer Willard Bascom in early diving gear.

12.5 What Are the Main Divisions of the Marine Environment?

The oceans can be divided into two main environments. The *ocean water itself is the pelagic environment*, where drifters and swimmers play out their lives in a complex food web. The *ocean bottom is the benthic environment*, where marine algae and animals that do not float or swim (or at least not very well) spend their lives.

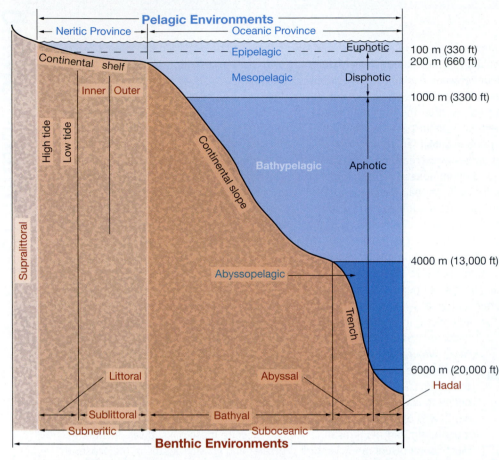

Figure 12.25 Oceanic biozones of the pelagic and benthic environments. Pelagic environments are shown in blue, and benthic environments are shown in brown. Pelagic and benthic environments are both based on depth, not necessarily on distance from shore. Sea floor features and sunlight zones are shown in black lettering.

Pelagic (Open Sea) Environment

The pelagic environment can be divided into distinctive life zones called **biozones** that possess unique physical characteristics, as shown in **Figure 12.25**. The pelagic environment is divided into neritic and oceanic provinces (Figure 12.25). The **neritic province** (*neritos* = of the coast) extends from the shore seaward and includes all water less than 200 meters (660 feet) deep. Seaward of the neritic province is the **oceanic province**, where depth increases beyond 200 meters (660 feet). The oceanic province is further subdivided into four biozones, which are defined according to depth:

1. The **epipelagic zone**, (*epi* = top, *pelagios* = of the sea) from the surface to a depth of 200 meters (660 feet)

2. The **mesopelagic zone**, (*meso* = middle, *pelagios* = of the sea) from 200 to 1000 meters (660 to 3280 feet)

3. The **bathypelagic zone**, (*bathos* = depth, *pelagios* = of the sea) from 1000 to 4000 meters (3300 to 13,000 feet)

4. The **abyssopelagic zone**, (*a* = without, *byssus* = bottom, *pelagios* = of the sea) which includes all the deepest parts of the ocean below 4000 meters (13,000 feet).

The single most important factor that determines the distribution of life in the oceanic province is the availability of sunlight. Thus, in addition to the four biozones, the distribution of life in the ocean is also divided into three zones based on the availability of sunlight as follows:

- The **euphotic zone** (*eu* = good, *photos* = light) extends from the surface to a depth where enough light still exists to support photosynthesis, which is rarely deeper than 100 meters (330 feet). The euphotic zone—often called the thin sunlit surface layer—accounts for only about 2.5% of the marine environment, but as discussed in subsequent chapters, this is where most marine life exists.

- The **disphotic zone** (*dis* = apart from, *photos* = light) has small but measurable quantities of light. It extends from the euphotic zone to a depth where light no longer exists—usually about 1000 meters (3300 feet).

- The **aphotic zone** (*a* = without, *photos* = light) has no light, and it exists below about 1000 meters (3300 feet).

EPIPELAGIC ZONE The upper half of the epipelagic zone is the only place in the ocean where there is sufficient light to support photosynthesis. The boundary between the epipelagic and mesopelagic zones, at 200 meters (660 feet), is also where the level of dissolved oxygen begins to decrease significantly (**Figure 12.26**, *red curve*). Oxygen decreases at this depth because no photosynthetic algae live below about 150 meters (500 feet), and dead organic tissue descending from the biologically productive upper waters is decomposing by bacterial oxidation, which consumes dissolved oxygen and releases nutrients back into the water. Accordingly, nutrient content also increases abruptly below 200 meters (600 feet) (Figure 12.26, *green curve*). The boundary between the epipelagic and mesopelagic zones, then, is the approximate bottom of the mixed layer, seasonal thermocline, and surface water mass.

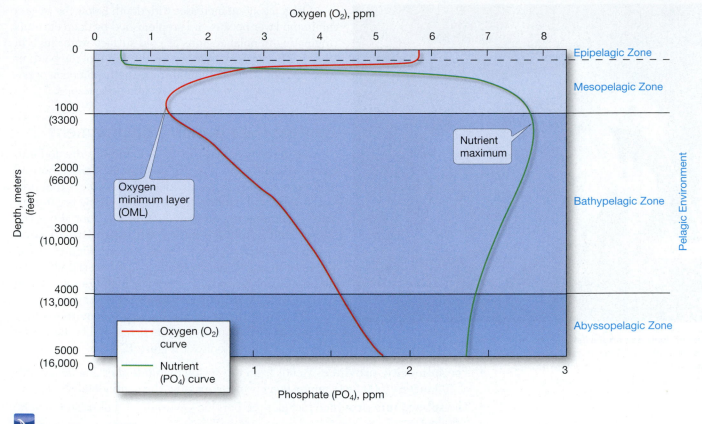

SmartFigure 12.26 Abundance of dissolved oxygen and nutrients with depth.
In surface water, oxygen is abundant due to mixing with the atmosphere and plant photosynthesis, and nutrient content (phosphate) is low due to uptake by algae. At deeper depths, oxygen decreases and produces an oxygen minimum layer (OML), which coincides with a nutrient maximum. Below that, nutrient levels remain high and oxygen increases as it is replenished with high-oxygen cold water from polar regions. Note that ppm = parts per million by weight.
https://goo.gl/EJZe34

MESOPELAGIC ZONE A dissolved **oxygen minimum layer (OML)** occurs at a depth of about 700 to 1000 meters (2300 to 3280 feet) (Figure 12.26). The intermediate-water masses that move horizontally in this depth range often possess the highest levels of nutrients in the ocean.

Organisms capable of **bioluminescence** (*bio* = life, *lumen* = light, *esc* = becoming), which have the ability to produce light biologically and "glow in the dark," are common in the mesopelagic and deeper zones. In these areas below the sunlit surface waters, having the ability to produce light has many advantages, and so the vast majority of organisms are capable of producing light. Examples of bioluminescent organisms include certain species of shrimp, squid, and especially deep-sea fish (**Figure 12.27**). The mechanism involved in allowing organisms to bioluminesce is discussed more fully in Chapter 14, "Animals of the Pelagic Environment."

BATHYPELAGIC AND ABYSSOPELAGIC ZONES The aphotic (lightless) bathypelagic and abyssopelagic zones represent over 75% of the living space in the oceanic province. Many completely blind fish exist in this region of total darkness, and all are small, bizarre-looking, and predaceous.

Many species of shrimp that normally feed on **detritus**[10] become predators at these depths, where the food supply is greatly reduced compared to surface waters. Animals that live in these deep zones feed mostly upon one another. They have evolved impressive warning devices and unusual apparatuses to make them extremely efficient predators (Figure 12.27). Many also have sharp teeth and extremely large mouths relative to their body size.

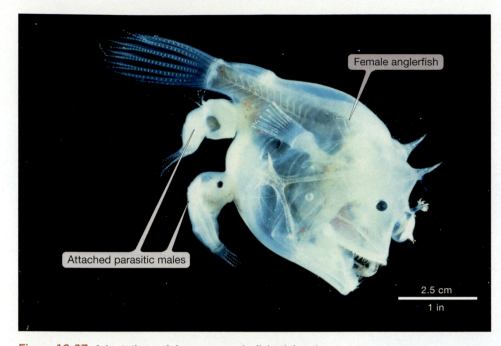

Figure 12.27 Adaptations of deep-sea anglerfish. A female deep-sea anglerfish (*Edriolychnus schmidti*) has a transparent body, small eyes, and sharp teeth. The feathery structure projecting from the front of its head is a bioluminescent lure, which is used to attract prey. Note the two much smaller parasitic males that are attached to the bottom of the female's body.

Oxygen content increases with depth below the oxygen minimum layer because it is replenished by deep currents originating in polar regions as cold surface water high in oxygen. The abyssopelagic zone is the realm of the bottom-water masses, which commonly move in the direction opposite the deep-water masses in the bathypelagic zone.

Benthic (Sea Bottom) Environment

Similar to the way that the water column is divided into zones with different physical conditions, the sea bottom environment can also be divided into provinces that provide a variety of habitats for bottom-dwelling organisms. The transitional region from land to sea floor above the spring high tide line is called the **supralittoral zone** (*supra* = above, *littoralis* = the shore) (see Figure 12.25). Commonly called the spray zone, it is covered with water only during periods of extremely high tides and when tsunami or large storm waves break on the shore.

The rest of the benthic, or sea floor, environment is divided into two main units that correspond to the neritic and oceanic provinces of the pelagic environment (see Figure 12.25):

- The **subneritic province** extends from the spring high tide shoreline to a depth of 200 meters (660 feet), approximately encompassing the continental shelf.
- The **suboceanic province** includes the benthic environment below 200 meters (660 feet).

SUBNERITIC PROVINCE The subneritic province is subdivided into the littoral and sublittoral zones. The *intertidal zone* (the zone between high and low tides) coincides with the **littoral zone** (*littoralis* = the shore). The **sublittoral zone** (*sub* = below, *littoralis* = the shore), or *shallow subtidal zone*, extends from low tide shoreline out to a depth of 200 meters (660 feet).

The sublittoral zone consists of inner and outer regions. The **inner sublittoral zone** extends to the depth at which marine algae no longer grow attached to the ocean bottom (approximately 50 meters [160 feet]), so the seaward boundary varies. All photosynthesis seaward of the inner sublittoral zone is carried out by floating microscopic algae.

The **outer sublittoral zone** extends from the inner sublittoral zone out to a depth of 200 meters (660 feet) or the shelf break, which is the seaward edge of the continental shelf.

SUBOCEANIC PROVINCE The suboceanic province is subdivided into bathyal, abyssal, and hadal zones. The **bathyal zone** (*bathus* = deep) extends from a depth of 200 to 4000 meters (660 to 13,000 feet) and corresponds generally to the continental slope.

The **abyssal zone** (*a* = without, *byssus* = bottom) extends from a depth of 4000 to 6000 meters (13,000 to 20,000 feet) and includes more than 80% of the benthic environment. The ocean floor of the abyssal zone is covered by soft oceanic sediment, primarily abyssal clay. Tracks and burrows of animals that live in this sediment can be seen in **Figure 12.28**.

Figure 12.28 Benthic organisms produce tracks on the ocean floor. As benthic organisms move across or burrow through the ocean bottom, they often leave tracks in the sediment on the ocean floor. Width of view is about 0.6 meter (2 feet).

[10]*Detritus* (*detritus* = to lessen) is a catchall term for dead and decaying organic matter, including waste products.

The **hadal zone** (*hades* = hell)[11] extends below 6000 meters (20,000 feet), so it consists only of deep trenches along the margins of continents. Animal communities that are found in these deep environments have been isolated from each other, often resulting in unique adaptations.

CONCEPT CHECK 12.5 | Compare the main divisions of the marine environment.

1 Construct a table listing the sub-divisions of the pelagic and benthic environments and the physical factors used in assigning their boundaries.

2 Describe the three zones based on the availability of sunlight. Which one is where most marine life exists?

RECAP

The pelagic environment includes the water column and the benthic environment includes the sea bottom. Subdivisions of pelagic and benthic environments are based on depth, which influences the amount of sunlight.

[11]The inhospitable high-pressure environment of the hadal zone is, in fact, aptly named.

ESSENTIAL CONCEPTS REVIEW

12.1 What are living things, and how are they classified?

▶ *A wide variety of organisms lives in the ocean*, ranging in size from microscopic bacteria and algae to blue whales. *All living things belong to one of the three major domains (branches) of life: Archaea*, simple microscopic bacterialike creatures; *Bacteria*, simple life-forms consisting of cells that usually lack a nucleus; and *Eukarya*, complex organisms (including plants and animals) consisting of cells that have a nucleus.

▶ *Organisms are further divided into six kingdoms: Eubacteria*, microscopic single-celled organisms without a nucleus; *Archaebacteria*, ancient bacterialike organisms that live in extreme environments; *Plantae*, many-celled plants; *Animalia*, many-celled animals; *Fungi*, molds and lichens, and *Protista*, single-celled and multicelled organisms with a nucleus. *Classification of organisms involves placing individuals within the kingdoms into increasingly specific groupings of phylum, class, order, family, genus, and species*, the last two of which denote an organism's scientific name. Many organisms also have one or more common names.

Study Resources
MasteringOceanography Study Area Quizzes, MasteringOceanography Web Animation

Critical Thinking Question
Discuss why it is often difficult to differentiate between living and nonliving things.

Active Learning Exercise
With another student in class, construct a nested box diagram similar to Figure 12.3 showing the taxonomic classification of humans. Use the information from Table 12.1 to construct your diagram. Also, use the Internet to find taxonomic information about a marine invertebrate (which lacks a backbone) not listed in Table 12.1 and construct a nested box diagram for it.

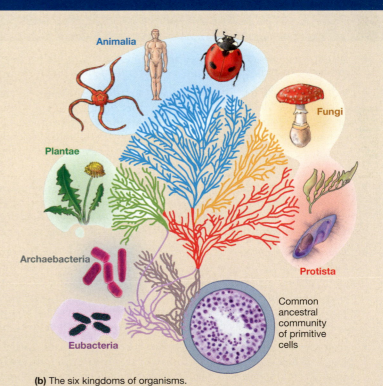

(b) The six kingdoms of organisms.

12.2 How are marine organisms classified?

▶ *Marine organisms can be classified into one of three groups, based on habitat and mobility. Plankton are free-floating forms with little power of locomotion, nekton are swimmers, and benthos are bottom dwellers. Most of the ocean's biomass is planktonic.*

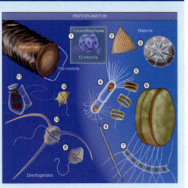

Study Resources
MasteringOceanography Study Area Quizzes

Critical Thinking Question
Explain why most of the ocean's biomass is planktonic.

Active Learning Exercise
Working with another student in class and the Internet, come up with a list of eight different marine organisms in each of the three main categories of marine life: *plankton*, *nekton*, and *benthos*. Do not duplicate any examples of organisms found in the text. Share your list with the class.

12.3 How many marine species exist?

▶ *Only about 13% of all known species inhabit the ocean, and more than 98% of marine organisms are benthic. The marine environment—especially the pelagic environment—is much more stable than the terrestrial environment, so there is less pressure on marine organisms to diversify.*

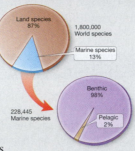

Study Resources
MasteringOceanography Study Area Quizzes

Critical Thinking Question
Explain how environmental variability affects the number of species present.

Active Learning Exercise
Working with another student in class and the Internet, go to the Census of Marine Life Website at **www.coml.org**. Explore the site and make a list of five new species that were discovered during the census. Report your findings to the class.

12.4 How are marine organisms adapted to the physical conditions of the ocean?

▶ *Marine organisms are well adapted to life in the ocean.* Those organisms that have established themselves on land have had to develop complex systems for support and for acquiring and retaining water.

▶ Both *algae*, which must stay in surface waters to receive sunlight, and *small animals* that feed on them *lack an effective means of locomotion.* To keep from sinking below sunlit surface waters, they depend on their *small size* and other adaptations to *increase their ratio of surface area to body mass*, which gives them high frictional resistance to sinking. Their small size also allows them to efficiently absorb nutrients and dispose of wastes. Many nektonic organisms have developed *streamlined bodies so they can overcome the viscosity of seawater* and move through it more easily.

▶ *Surface temperature of the world ocean does not vary* on a daily, seasonally, or yearly basis *as much as on land. Organisms living in warm water tend to be individually smaller, have ornate plumage, comprise a greater number of species, and constitute a much smaller total biomass than organisms living in cold water.* Warm-water organisms also tend to live shorter lives and reproduce earlier and more frequently than cold-water organisms.

▶ *Osmosis is the passing of water molecules through a semipermeable membrane from a region of higher concentration to a region of lower concentration.* If the body fluids of an organism and ocean water are separated by a membrane that allows water molecules to pass through, the organism may become severely dehydrated from osmosis. Many marine invertebrates are essentially *isotonic*: The salinity of their body fluids is similar to that of ocean water. Most marine vertebrates are *hypotonic*: The salinity of their body fluids is lower than that of ocean water, so they tend to lose water through osmosis. Freshwater organisms are essentially all *hypertonic*: The salinity of their body fluids is greater than the water in which they live, so they tend to gain water through osmosis.

▶ *Most marine animals extract oxygen through their gills.* Many marine organisms have well-developed eyesight because water is so transparent. *To avoid being seen and consumed by predators,* many marine organisms are transparent, camouflaged, countershaded, or disruptively colored. Unlike humans, most *marine organisms are unaffected by the high pressure at depth* because they do not have large internal air pockets that can be compressed.

Small organisms have a large surface area-to-volume ratio ...

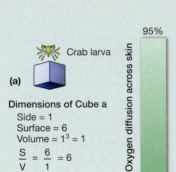

Study Resources
MasteringOceanography Study Area Quizzes, MasteringOceanography Web Animations, Web Video

Critical Thinking Question
Determine the surface-to-volume ratio of an organism whose average linear dimension is (a) 1 centimeter (0.4 inch), (b) 3 centimeters (1.1 inches), and (c) 5 centimeters (2 inches). Which one is better able to resist sinking, and why?

Active Learning Exercise
Working with another student in class, describe how the depth of the deep scattering layer varies over the course of a day. Be sure to include reasons why it does this and which organisms comprise the DSL. Report your findings to the class.

12.5 What are the main divisions of the marine environment?

▶ *The marine environment is divided into pelagic (open sea) and benthic (sea bottom) environments*. These regions are further divided based on depth and have varying physical conditions to which marine life is superbly adapted. One of the most important layers of the pelagic environment is the *euphotic zone*, which includes the sunlit surface waters and *contains enough sunlight to support photosynthesis*.

Study Resources
MasteringOceanography Study Area Quizzes, MasteringOceanography Web Diving Deeper 12.1

Critical Thinking Question
To help reinforce your knowledge of oceanic biozones, construct and label your own diagram similar to Figure 12.25 from memory.

Active Learning Exercise
With another student in class, explain why the two curves in Figure 12.26 have the shape that they do (for example, why the oxygen minimum layer exists, and why it coincides with the nutrient maximum). Share your analysis with the class.

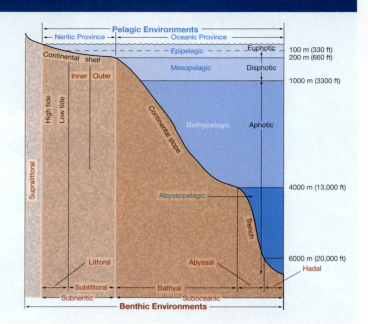

MasteringOceanography™

www.masteringoceanography.com

Looking for additional review and test prep materials? With individualized coaching on the toughest topics of the course, MasteringOceanography offers a wide variety of ways for you to move beyond memorization and deeply grasp the underlying processes of how the oceans work. Visit the Study Area in **www.masteringoceanography.com** to find practice quizzes, study tools, and multimedia that will improve your understanding of this chapter's content. Sign in today to enjoy the following features: Self Study Quizzes, SmartFigures, SmartTables, Oceanography Videos, Squidtoons, Geoscience Animation Library, RSS Feeds, Digital Study Modules, and an optional Pearson eText.

Energy from sunlight

PHOTOSYNTHESIS

CARBON DIOXIDE GAS AND WATER

During photosynthesis, plant cells combine carbon dioxide (CO_2) and water (H_2O) in the presence of sunlight to produce sugar ($C_6H_{12}O_6$) and oxygen gas (O_2).

SUGAR AND OXYGEN GAS

During respiration, animals combine sugar ($C_6H_{12}O_6$) and oxygen gas (O_2) to produce energy, releasing carbon dioxide gas (CO_2) and water (H_2O).

Phytoplankton

100 microns

RESPIRATION

Heat

SmartFigure 13.1 **Photosynthesis and respiration are cyclic and complimentary processes that are fundamental to life on Earth.** Note that this is the same image used in Chapter 1 as Figure 1.27. https://goo.gl/avxu5g

RECAP

Primary productivity is the rate at which carbon (organic matter) is produced by microbes, algae, and plants, mostly through photosynthesis; however, primary productivity also includes microbes that perform chemosynthesis.

supplied by photosynthetic primary productivity as its source of food, and only 0.1% of the ocean's biomass relies on chemosynthesis. Therefore, the discussion of primary productivity presented here focuses on photosynthetic productivity.

Chemically, photosynthesis is a reaction in which energy from the Sun is stored in organic molecules. In photosynthesis (**Figure 13.1**), plant, bacteria, and algae cells capture energy from sunlight and store it as sugars, releasing oxygen gas as a by-product. Alternatively, in cellular **respiration** (*respirare* = to breathe) (Figure 13.1), animals that consume the sugars produced by photosynthesis combine them with oxygen, releasing the stored energy of the sugars to carry on cellular tasks important for various life processes. Note that this is the same figure from Chapter 1, where photosynthesis and respiration were previously discussed as complimentary and cyclic processes.

Measurement of Primary Productivity

Various properties of the ocean can be measured to give an approximation of the amount of primary productivity. One of the most direct at-sea methods is to capture plankton in cone-shaped nylon **plankton nets** (**Figure 13.2**). These fine mesh nets—which resemble windsocks at airports—filter plankton from the ocean as they are towed at a specific depth by research vessels. Analysis of the amounts and types of organisms captured reveals much about the productivity of the area.

Other methods of determining oceanic primary productivity include lowering specially designed bottles into the ocean, analyzing the amount of radioactive carbon in seawater, and even monitoring ocean color using orbiting satellites to determine the presence of chlorophyll. Photosynthetic organisms such as **phytoplankton** (*phyto* = plant, *planktos* = wandering) use the green pigment **chlorophyll** (*khloros* = green, *phylum* = leaf) to capture energy from the Sun and perform photosynthesis. The color of surface waters is strongly affected by the amount of chlorophyll. Hence, ocean color can be used as an approximation of phytoplankton abundance and, in turn, productivity. One such instrument that collected ocean color measurements was the **SeaWiFS** (Sea-viewing Wide Field-of-view Sensor) instrument aboard the *SeaStar* satellite, which operated from 1997 to 2010. It replaced *Nimbus-7*'s Coastal Zone Color Scanner instrument, which operated between 1978 and 1986. During its operational lifespan, SeaWiFS measured the color of Earth's surface with a radiometer and provided global coverage of estimated ocean chlorophyll levels as well as the abundance of land vegetation. Today, ocean color is collected worldwide every two days by MODIS (Moderate Resolution Imaging Spectroradiometer) instruments aboard the *Terra* and *Aqua* satellites. MODIS measures 36 spectral frequencies of light, including ocean fluorescence, which provides a wealth of information about ocean phytoplankton productivity, health, and efficiency.

Factors Affecting Primary Productivity

In the ocean, the two main factors that limit the amount of photosynthetic primary productivity are the availability of solar radiation and the availability of nutrients. Sometimes other variables—such as the amount of carbon dioxide—can also limit primary productivity if they become scarce in seawater. Human-caused climate change can also affect marine productivity, too.

AVAILABILITY OF NUTRIENTS The distribution of life throughout the ocean's breadth and depth depends mainly on the availability of nutrients that phytoplankton need, such as nitrogen, phosphorus, iron, and silica. Marine populations reach their greatest concentration where the physical conditions supply large quantities of nutrients. The sources of nutrients must be considered to understand where these areas are found.

Water in the form of runoff erodes the continents, carrying material to the oceans and depositing it as sediment on the continental margins. Runoff also dissolves and transports compounds such as nitrates and phosphates, which are the main nutrients for phytoplankton. Nitrates and phosphates are also the basic ingredients in all garden and farm fertilizers. When these chemicals reach coastal areas, they can cause **eutrophication** (*eu* = good, *tropho* = nourishment, *ation* = action), which is the enrichment of an ecosystem with chemical nutrients. Eutrophication and its associated problems will be discussed in the next section.

The continents are the major sources of nutrients, so the greatest concentrations of marine life are found along the continental margins. The concentration of marine life decreases as the distance from the continental margins into the open sea increases. Marine life also decreases with increasing depth in the ocean because sunlight doesn't penetrate that far into the ocean, even in the clearest waters. The vast depth of the world's oceans and the great distance between the open ocean and the coastal regions where nutrients are concentrated account for these differences.

Often, the lack of certain nutrients, particularly nitrogen (as nitrates) and phosphorus (as phosphates), can limit productivity. As a result, these compounds are among the most studied in chemical oceanography. Comparatively, nitrogen compounds involved in photosynthesis can be up to 10 times greater than the total measured yearly nitrogen compound concentration. What this implies is that the soluble nitrogen compounds are being completely recycled up to 10 times per year. Similarly, available phosphates may be turned over up to four times per year.

Carbon is an important element in productivity, too, because carbon is the basic component of all organic compounds (including carbohydrates, proteins, and fats). In the ocean, however, various forms of carbon are quite abundant, so there is no scarcity of carbon for photosynthetic production. Thus, carbon does not limit productivity.

When nutrients are not limiting productivity, the ratio of carbon to nitrogen to phosphorus in the tissues of algae is in the proportion of 106:16:1 (C:N:P), which is called the *Redfield ratio*, after American oceanographer Alfred C. Redfield, who first described it in 1963. This ratio is also observed in zooplankton that feed on

This photomicrograph of a plankton sample includes both phytoplankton and zooplankton.

100 microns

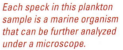

These large, cone-shaped, fine-mesh plankton nets are lowered into the water and towed behind a research vessel to collect plankton.

Each speck in this plankton sample is a marine organism that can be further analyzed under a microscope.

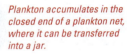

Plankton accumulates in the closed end of a plankton net, where it can be transferred into a jar.

Figure 13.2 Plankton nets collect a plankton sample.

diatoms, and in most ocean water samples taken worldwide. Moreover, phytoplankton take up nutrients in the ratio in which they are available in the ocean water and pass them on to zooplankton in the same ratio. When these plankton and animals die, carbon, nitrogen, and phosphorus are recycled into the water in this same ratio.

Scientific studies in the waters near Antarctica and the Galápagos Islands have revealed that photosynthetic production is low even though the concentration of all nutrients—except iron—is high.[4] Production is high only in regions of shallow water down-current from islands or landmasses where a significant amount of iron from rocks and sediments is dissolved in water. Therefore, the lack of iron can also severely limit primary productivity.

AVAILABILITY OF SOLAR RADIATION Photosynthesis cannot proceed unless light energy (solar radiation) is available. Despite the atmosphere's thickness of more than 80 kilometers (50 miles), its high transparency allows sunlight to penetrate it quite readily, so land-based plants almost always have an abundance of solar radiation to conduct photosynthesis.

In the clearest ocean water, however, solar energy may be detected to depths of only about 1 kilometer (0.6 mile) and, even then, the amount reaching these depths is inadequate for photosynthesis. Photosynthesis in the ocean, therefore, is restricted to the uppermost surface waters and those areas of the sea floor where the water is shallow enough to allow light to penetrate. The water depth at which light is so limited that net photosynthesis becomes zero is called the **compensation depth for photosynthesis**.

The **euphotic zone** (*eu* = good, *photos* = light) extends from the surface down to the compensation depth for photosynthesis, which is approximately 100 meters (330 feet) in the open ocean (see Figure 12.25). Near the coast, the euphotic zone may extend to less than 20 meters (66 feet) because the water contains more suspended inorganic material (turbidity) or microscopic organisms that limit light penetration.

How do the two factors necessary for photosynthesis—the supply of nutrients and the presence of solar radiation—differ between coastal areas and the open ocean? In the open ocean (far from continental margins), solar energy extends deeper into the water column, but concentration of nutrients is low. In coastal regions, on the other hand, light penetration is much less, but the concentration of nutrients is much higher. Because the coastal zone is much more productive, nutrient availability must be the most important factor affecting the distribution of life in the oceans.

Light Transmission in Ocean Water

The graph in **Figure 13.3** shows that most solar energy falls in the range of wavelengths called **visible light**. This radiant energy from the Sun powerfully affects three major components of the oceans:

1. *Ocean winds.* The major wind belts of the world, which produce ocean currents and wind-driven ocean waves, ultimately derive their energy from solar radiation. Wind belts and ocean currents strongly influence world climates.

2. *Ocean stratification.* At the ocean surface, a thin layer of water created by solar heating is warmer than the water below and overlies a great mass of cold water that fills most ocean basins. In most places, this causes the ocean's water column to be stratified into layers.

3. *Primary productivity.* Photosynthesis can occur only where sunlight penetrates the ocean water, so phytoplankton and most animals that eat them must live where the light is, in the relatively thin layer of sunlit surface water, which is the "life layer" where most marine life exists.

STUDENTS SOMETIMES ASK . . .

How is the ocean's primary productivity being affected by climate change?

Human-caused global climate change is expected to produce significant negative impacts on the primary productivity of entire ocean ecosystems. In fact, human-caused climate change is already affecting ocean primary production in many regions. For example, because of changes in wind patterns or water temperatures, areas of the ocean that in the past exhibited strong seasonal phytoplankton blooms now experience weakened blooms or periods in which phytoplankton don't bloom at all. Human-caused climate change is also predicted to produce a decrease in the strength of ocean surface currents, altering the transportation range of fish larvae and plankton, which will ultimately negatively affect primary production.

Another major negative effect of climate change is a shift in the duration or timing of oceanic growing seasons. As explained by the "match-mismatch hypothesis," the growth and survival of predatory species depends on the synchronous production of their main food source. In essence, if a phytoplankton bloom is too early or is delayed, it changes the time when food is available for the organisms that feed on them. In a worst-case scenario, the absence of phytoplankton would cause most zooplankton to die from starvation, which, in turn, would dramatically impact ocean food webs.

Climate

Connection

THE ELECTROMAGNETIC SPECTRUM The Sun radiates a wide range of wavelengths of electromagnetic radiation. Together they comprise the **electromagnetic spectrum**, which is shown in the upper part of Figure 13.3. Only a very narrow portion of the electromagnetic spectrum is visible to humans as visible light. We call it "visible" light because our electromagnetic sensors—our eyes—are adapted to detect only the wavelengths in the visible region. In essence, our eyes "tune into" the visible light wavelengths, just as a radio "tunes into" specific radio waves.

Visible light can be further divided by wavelength into energy levels associated with the colors red, orange, yellow, green, blue, and violet (the acronym used for the color spectrum is ROYGBV). Together, these different wavelengths produce white light. The lower energy, longer wavelengths of light to the left of visible light (for example, infrared, microwaves, and radio waves) are used for heat transfer and communication. The higher energy, shorter wavelengths of light to the right of visible light (for example, X-rays and gamma rays) are capable of damaging tissue, in high enough doses.

THE COLOR OF OBJECTS As Figure 13.3 shows, light from the Sun includes all the visible colors. Most of the light we see is reflected from objects. All objects absorb and reflect different wavelengths of light, and each wavelength represents a color in the visible spectrum. Vegetation, for example, absorbs most wavelengths except green and yellow, which they reflect, so most plants look green. Similarly, a red jacket absorbs all wavelengths of color except red, which is reflected.

The lower part of Figure 13.3 shows how the ocean selectively absorbs the longer-wavelength colors (red, orange, and yellow) of visible light. The true colors of objects can be observed in natural light only in the surface waters because only there can all wavelengths of the visible spectrum be found. Red light is absorbed within the upper 10 meters (33 feet) of the ocean, and yellow is completely absorbed before a depth of 100 meters (330 feet). Thus, the shorter-wavelength portion of the visible spectrum is all that can be transmitted to greater depths (mostly blue light, with some violet and green wavelengths), and even then, their intensity is low. In the open ocean, sunlight strong enough to support photosynthesis occurs only within the euphotic zone

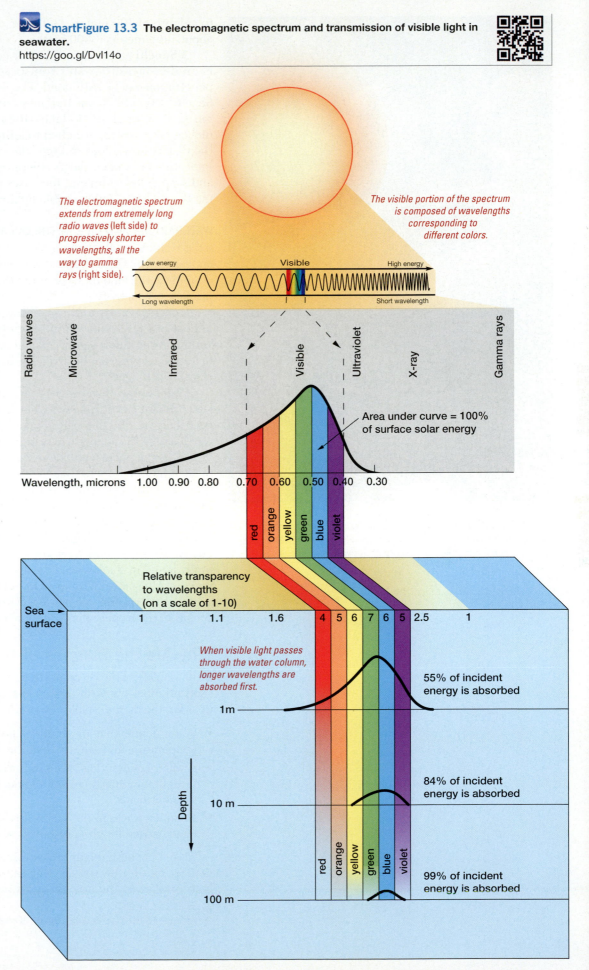

SmartFigure 13.3 The electromagnetic spectrum and transmission of visible light in seawater.
https://goo.gl/Dvl14o

The electromagnetic spectrum extends from extremely long radio waves (left side) to progressively shorter wavelengths, all the way to gamma rays (right side).

The visible portion of the spectrum is composed of wavelengths corresponding to different colors.

Low energy Visible High energy

Long wavelength Short wavelength

Radio waves | Microwave | Infrared | Visible | Ultraviolet | X-ray | Gamma rays

Area under curve = 100% of surface solar energy

Wavelength, microns 1.00 0.90 0.80 0.70 0.60 0.50 0.40 0.30

red | orange | yellow | green | blue | violet

Relative transparency to wavelengths (on a scale of 1-10)

Sea surface

1 1.1 1.6 4 5 6 7 6 5 2.5 1

When visible light passes through the water column, longer wavelengths are absorbed first.

1m —

55% of incident energy is absorbed

Depth

10 m —

84% of incident energy is absorbed

red | orange | yellow | green | blue | violet

100 m —

99% of incident energy is absorbed

Figure 13.4 Using a Secchi disk. A Secchi disk is lowered into the water with a line to measure the depth of penetration of sunlight, which indicates the clarity of the water.

to a depth of 100 meters (330 feet), and no sunlight penetrates below a depth of about 1000 meters (3300 feet).

A **Secchi** (pronounced "SECK-ee") **disk**, such as the one shown in **Figure 13.4**, is used to measure water transparency, and based on that, the depth of light penetration can be estimated. The Secchi disk is named after its inventor, Angelo Secchi (1818–1878), an Italian astronomer who first used the device in 1865 to measure water clarity in lakes. It consists of a disk 20 to 40 centimeters (8 to 16 inches) in diameter attached to a line that is marked off at regular intervals. As the disk is slowly lowered into the ocean, the depth at which it can last be seen indicates the water's clarity. Increased turbidity, which includes microorganisms and suspended sediment, increases the degree of light absorption, thus decreasing the depth to which visible light can penetrate into the ocean.

WATER COLOR AND LIFE IN THE OCEANS The color of the ocean ranges from deep indigo (blue) to yellow-green. Why are some areas of the ocean blue, whereas others appear green? Ocean color is influenced by (1) the amount of turbidity from run-off and (2) the amount of photosynthetic pigment, which increases with increasing primary productivity.

Coastal waters and upwelling areas are biologically very productive and almost always yellow-green in color because they contain large amounts of yellow-green microscopic marine algae and suspended particles. When these materials are present in surface waters, they scatter the wavelengths for greenish or yellowish light.

Water in the open ocean—particularly in the tropics—is less productive and has less turbidity, so it is usually a clear, indigo-blue color. Here, it is water molecules that contribute most to the scattering of light, and they scatter light primarily in the blue wavelengths. The atmosphere scatters blue light, too, which is why clear skies are blue.

Although photosynthetic marine algae and bacteria are microscopic, they occur in such large numbers that they can change the color of the ocean to such a degree that orbiting satellites are able to measure the changes from space. **Figure 13.5**, for example, shows a *SeaStar* satellite/SeaWiFS instrument view of ocean chlorophyll, which is an approximation for productivity. The figure shows high chlorophyll concentrations (highly productive areas) in light green color, which are called **eutrophic** (*eu* = good, *tropho* = nourishment). Generally, eutrophic waters are naturally found in shallow-water coastal regions, areas of upwelling, and high-latitude regions. Alternatively, areas of low chlorophyll concentration (low productivity) are called **oligotrophic** (*oligo* = few, *tropho* = nourishment) and are found in the open oceans of the tropics; they are shown in dark blue colors in the figure.

Why Are the Margins of the Oceans So Rich in Life?

If the stability of the ocean environment is ideal for sustaining life, why are the richest concentrations of marine organisms in the very margins of the oceans, where conditions are the most *un*stable? For example, characteristics of the coastal ocean include:

- Water depths that are shallow, allowing much greater seasonal variations in temperature and salinity than the open ocean.

- A water column that varies in thickness in the nearshore region in response to tides that regularly cover and uncover a thin strip of land along the margins of the continents.

- Breaking waves in the surf zone that release large amounts of energy, which has been carried for great distances across the open ocean.

Each of these conditions stresses organisms. In spite of hardships, however, new species have evolved over the vast expanse of geologic time spanning billions of years by the process of natural selection[5] to fit every imaginable biological

[5]See Diving Deeper 1.3 for a description of evolution and natural selection.

from these coasts, so nu
3300 feet) constantly ris
the process of *equatoria*

13.2 What K
Organisms Ex

Many types of marin
sented by microscopic
seed-bearing plants.

Seed-Bearing Pl

The only members of k
to the seed-bearing r
= plant), which occur
example, is a grasslike
of bays and estuaries f
grass (*Phyllospadix*) (**Fi**
is typically found in th
intertidal zones down t

Other seed-bearin
of the genus *Spartina*),
Rhizophora, *Avicenni*
of food and protection

Macroscopic (L

Various types of marin
waters along the ocean
a few species float. Al
they contain (**Figure 1**
just its color, the divis
of describing the diffe

GREEN ALGAE Alth
phytum = plant) are
resented in the ocean

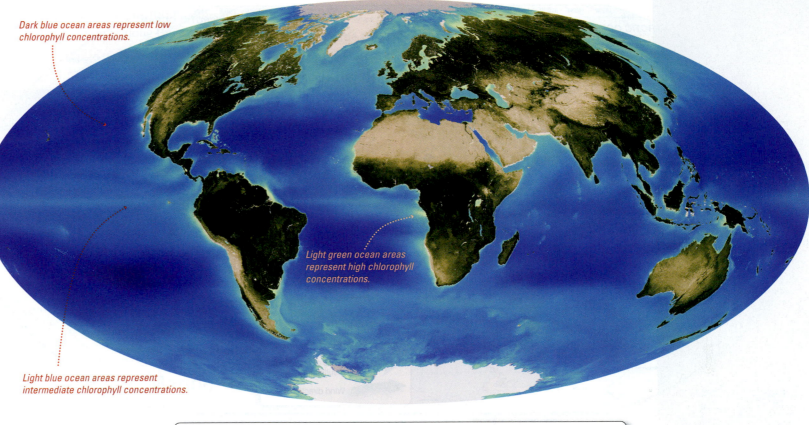

Figure 13.5 Satellite image of ocean chlorophyll.
Satellite data (1998–2010) showing average ocean chlorophyll concentration, which is an approximation for productivity. Data were gathered during the entire 13-year operation of the SeaWiFS instrument aboard the *SeaStar* satellite, which detected changes in seawater color caused by changing concentrations of chlorophyll that varies with photosynthetic productivity. Ocean chlorophyll concentrations are reported in milligrams per cubic meter (mg/m^3). On land, data are depicted using the Normalized Difference Vegetation Index (NDVI), which shows the density of green vegetation.

niche—even in environments that pose difficulties for organisms. In fact, many organisms have adapted to live under adverse conditions—such as coastal environments—as long as nutrients are available.

Along continental margins, some areas have more abundant life than others. What characteristics create such an uneven distribution of life? Again, only the basic requirements for the production of food need be considered. For example, areas that have the lowest water temperatures have the greatest biomass, because cold water contains higher amounts of nutrients and dissolved gases—such as oxygen and carbon dioxide—than warm water. These nutrients and gases stimulate phytoplankton growth, which profoundly affects the distribution of all other life in the oceans.

UPWELLING AND NUTRIENT SUPPLY As discussed in Chapter 7, **upwelling** is a flow of deep water toward the surface that brings water from depths below the euphotic zone. This deep water is rich in nutrients and dissolved gases because there are no phytoplankton at these depths to consume these compounds. When chilled water from below the surface rises, it hoists nutrients from the depths to the surface, where phytoplankton thrive and become food for larger organisms—copepods, fish, and on up to larger organisms such as sharks and whales. However, as will be discussed later, surface warming and the resulting stratification of the ocean's water column can limit upwelling and thereby inhibit primary productivity.

Where does upwelling occur in the oceans? One common location are the highly productive areas of *coastal upwelling* that are found along the western margins of continents, where surface currents are moving toward the equator (**Figure 13.6**). Ekman transport (see Chapter 7) causes surface water to move away

Web Animation
Ekman Spiral and Coastal Upwelling/Downwelling
https://goo.gl/Y4LOD9

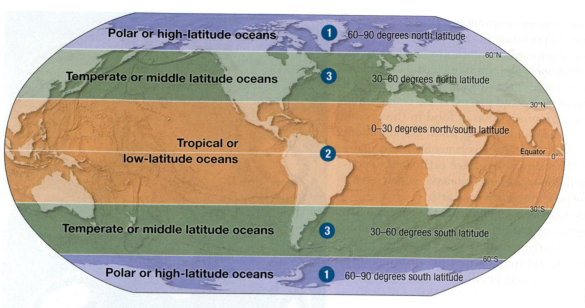

Figure 13.18 Locations of three marine productivity areas. Map showing the locations of three open ocean areas where yearly marine productivity patterns are examined: (1) polar or high-latitude oceans (60 degrees to 90 degrees north and south latitude); (2) tropical or low-latitude oceans (0 degrees to 30 degrees north and south latitude); and (3) temperate or middle latitude oceans (30 degrees to 60 degrees north and south latitude). For full descriptions of the conditions found in each productivity area, see the text.

RECAP

The thermocline acts as an impenetrable lid that inhibits the movement of nutrient-rich deep water to the surface, thereby limiting primary productivity.

nutrients from the upper ocean and concentrates them in deep-sea waters and sea floor sediments.

Surface warming and the resulting stratification of the ocean's water column also affects primary productivity. Throughout much of the subtropical oceans, for example, a permanent **thermocline** (and resulting **pycnocline**[10]) develops. The thermocline forms a barrier to vertical mixing, preventing the resupply of nutrients to the sunlit surface layer. In essence, the thermocline acts as an impenetrable lid that prohibits the movement of nutrient-rich deep water to the surface, thereby inhibiting primary productivity. In middle latitude oceans, a thermocline develops only during the summer season. A thermocline does not usually develop in polar oceans because of the lack of adequate surface warming. As discussed below, the degree to which waters develop a thermocline profoundly affects the patterns of primary productivity in different latitudes.

Let's examine the yearly productivity patterns of three open ocean areas: (1) polar or high-latitude oceans, (2) tropical or low-latitude oceans, and (3) temperate or middle latitude oceans (**Figure 13.18**). Note that in the following discussion, we'll consider only open ocean areas far from land, ensuring that these locations aren't affected by runoff from the continents, which often contain high levels of nutrients that interfere with the seasonal patterns described below.

Productivity in Polar (High-Latitude) Oceans: 60 to 90 degrees North and South Latitude

Polar oceans such as the Arctic Ocean's Barents Sea, which is off the northern coast of Europe, experience continuous darkness for about three months during winter and continuous illumination for about three months during summer. Diatom productivity peaks in the Barents Sea during May (**Figure 13.19a**), when the Sun rises high enough in the sky and there is deep penetration of sunlight into the water. As soon as diatoms develop, zooplankton—mostly small crustaceans such as copepods (**Figure 13.19b**)—begin feeding on them. The zooplankton biomass peaks in June and continues at a relatively high level until winter darkness begins in October.

In the Antarctic region—particularly at the southern end of the Atlantic Ocean—productivity is somewhat greater. This is caused by the upwelling of North Atlantic Deep Water, which forms on the opposite side of the ocean basin, where it sinks and moves southward below the surface. Hundreds of years later, it rises to the surface near Antarctica, carrying with it high concentrations of nutrients (**Figure 13.19c**). When the Sun provides sufficient solar radiation in summer, there is an explosion of biological productivity. A recent study of Antarctic waters, however, documented as much as a 12% decrease in phytoplankton productivity because of increased ultraviolet radiation as a result of the Antarctic ozone hole caused by the use of CFCs (chlorofluorocarbons). For more details about the atmospheric ozone

[10]Recall that a *thermocline* is a layer of rapidly changing temperature, and a *pycnocline* is a layer of rapidly changing density. Development of ocean thermoclines and pycnoclines is discussed in Chapter 5.

(b) SeaWiFS image of chlor
west coast of Africa (Feb
concentrations indicate
caused by coastal upwe
milligrams of chlorophyl

hole and CFCs, which are also strong greenhouse gases, see Chapter 16, "The Oceans and Climate Change."

Blue whales—the largest of all whales (see Figure 14.20)—eat mostly zooplankton and time their migration through middle latitude and polar oceans to coincide with maximum zooplankton productivity. This enables the whales to develop and support calves that can exceed 7 meters (23 feet) in length at birth. The mother blue whale suckles the calf with rich, high-fat milk for six months. By the time the calf is weaned, it is over 16 meters (50 feet) long. In two years, it will be 23 meters (75 feet) long, and after about three years, it will weigh 55 metric tons (121,000 pounds)! This phenomenal growth rate gives some indication of the enormous biomass of small copepods and krill upon which these large mammals feed.[11]

Density and temperature change very little with depth in polar oceans (**Figure 13.19d**), so these waters are **isothermal** (*iso* = same, *thermo* = temperature), and there is no barrier to mixing between surface waters and deeper, nutrient-rich waters. In the summer, however, melting ice creates a thin, low-salinity layer that does not readily mix with the deeper waters. This stratification is crucial to summer primary production because it helps prevent phytoplankton from being carried into deeper, darker waters. Instead, they are concentrated in the sunlit surface waters, where they reproduce continuously.

Nutrient concentrations (mostly nitrates and phosphates) are usually adequate in high-latitude surface waters, so the availability of solar energy limits photosynthetic productivity in these areas more than the availability of nutrients.

Productivity in Tropical (Low-Latitude) Oceans: 0 to 30 degrees North and South Latitude

Perhaps surprisingly, productivity is low in tropical oceans. Because the Sun is more directly overhead, light penetrates much deeper into tropical oceans than in middle latitude and polar waters, and solar energy is available year-round. However, productivity is low in tropical oceans because a permanent thermocline produces a stratification (layering) of water masses. This prevents mixing between surface waters and nutrient-rich deeper waters, effectively eliminating any supply of nutrients from deeper waters below (**Figure 13.20**).

At about 20 degrees north and south latitude, phosphate and nitrate concentrations are commonly less than $1/100$ of their concentrations in middle latitude oceans during winter. In fact, nutrient-rich waters in the tropics lie below 150 meters (500 feet), with the highest concentrations between 500 and 1000 meters (1640 and 3300 feet). So, productivity in tropical oceans is limited by the lack of nutrients (unlike in polar oceans, where productivity is limited by the lack of sunlight).

Generally, primary production in tropical oceans occurs at a steady but rather low rate. The total annual production of tropical oceans is only about half of that found in middle latitude oceans.

[11]As a similar size analogy, consider how many ants you would have to eat as a child to grow to adult size!

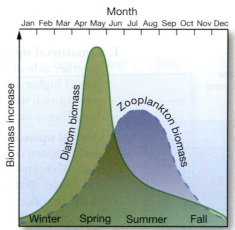

(a) Barents Sea productivity graph, which shows the dramatic springtime increase of diatom biomass that results in an increase in zooplankton abundance.

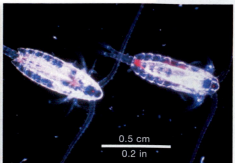

(b) Copepods are an important type of zooplankton in polar oceans. These copepods are of the genus *Calanus*.

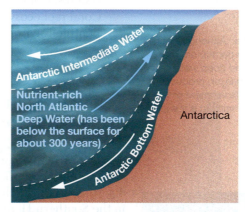

(c) The upwelling of cold, nutrient-rich North Atlantic Deep Water near Antarctica continually supplies Antarctic waters with nutrients.

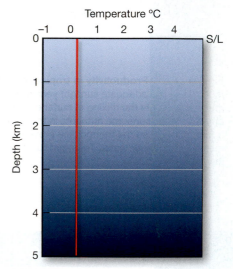

(d) Temperature graph of Antarctic waters showing a nearly uniform water temperature with depth (an isothermal water column).

Figure 13.19 **Productivity in polar oceans.**

Figure 13.34 Fishing bycatch. Underwater photograph of dead or dying fish being discarded over the side of a boat. In commercial fishing operations, about one-fourth of all catch is discarded as unwanted bycatch, which includes birds, turtles, sharks, dolphins, and many species of noncommercial fish.

Figure 13.35 Spotted dolphin (*Stenella attenuata*), which are commonly associated with yellowfin tuna. Eastern Pacific spotted dolphins are often found swimming above yellowfin tuna. This relationship is thought to exist because the dolphins use tuna to help find schools of prey items (squid and small fish), which are eaten by both animals.

species. Incidental catch includes birds, turtles, sharks, and dolphins, as well as many species of noncommercial fish (**Figure 13.34**). In most cases, these animals die before they are thrown back overboard, even though some of them are protected by U.S. and international law. Globally, an estimated 20 million metric tons (44 billion pounds) of bycatch is produced each year by the fishing industry, accounting for nearly one-fourth of the world's total marine fish catch.

TUNA AND DOLPHINS Schools of yellowfin tuna are commonly found swimming beneath spotted and spinner dolphins in the eastern Pacific Ocean (**Figure 13.35**). Fishers commonly use these dolphins to locate tuna and set a *purse seine net* around the entire school. When an underwater line is drawn tight, the net traps the tuna underwater as well as the dolphins at the surface. Unfortunately, dolphins, which are marine mammals, get caught in the net below the water's surface and can't reach the surface to breathe air, so they often drown. **Figure 13.36** illustrates how a purse seine net works; the figure also shows the variety of methods and fishing gear used in modern-day commercial fishing practices.

In 1988, biologist Samuel F. La Budde presented the problem of dolphin deaths caused by tuna fishing through graphic video footage taken of dolphins struggling in tuna fishing nets. In 1990, under intense public outcry and a boycott on tuna, the U.S. tuna canning industry declared it would not buy or sell tuna caught using methods that kill or injure dolphins. In 1992, a special addendum was added to the **Marine Mammals Protection Act**, further protecting dolphins. As a result of these measures, purse seine nets were modified so that dolphins could be released alive. In spite of reducing dolphin mortality as bycatch, dolphin populations have not rebounded accordingly. Research suggests that tuna fishing operations are still having a negative effect on dolphin populations by reducing dolphin survival and birth rates.

DRIFTNETS Another means of netting tuna and other species is by use of **driftnets**, or **gill nets**, which involves capturing fish by their gills (Figure 13.36). Driftnets are made of crisscrossed monofilament fishing line that is virtually invisible and cannot be detected by most marine animals as they swim into them. Depending on the size of the holes in the net, it is highly effective at catching anything large enough to become entangled in it. As a result, driftnets often have high amounts of bycatch.

Up until 1993, Japan, Korea, and Taiwan had the largest driftnet fleets, deploying as many as 1500 fishing vessels into the North Pacific and setting over 48,000 kilometers (30,000 miles) of driftnets in one day. Although driftnetting was supposed to be restricted to specific fisheries, some fishers who claimed to be fishing for squid were involved in illegally taking large quantities of salmon and steelhead trout. Driftnetters were also targeting immature tuna in the South Pacific, which could result in the reduced abundance of South Pacific tuna. In addition, tens of thousands of birds, turtles, dolphins, and other species were killed annually in these nets as bycatch.

In an effort to reduce the wasteful practice of driftnet fishing, the United States in 1989 signed an international treaty that prohibits driftnets greater than 2.5 kilometers (1.5 miles) in the South Pacific and includes a clause that prohibits the import of any fish caught in driftnets. Although long driftnets are banned in international waters, the United States allows limited use of shorter driftnets in some rivers, lakes, and bays. Additionally, some fishers continue to use long driftnets illegally in international waters.

GHOST FISHING Another area of concern for fisheries is called **ghost fishing**, which describes any lost or discarded fishing gear that continues to catch fish, marine mammals, or other organisms after it has been abandoned. Examples of

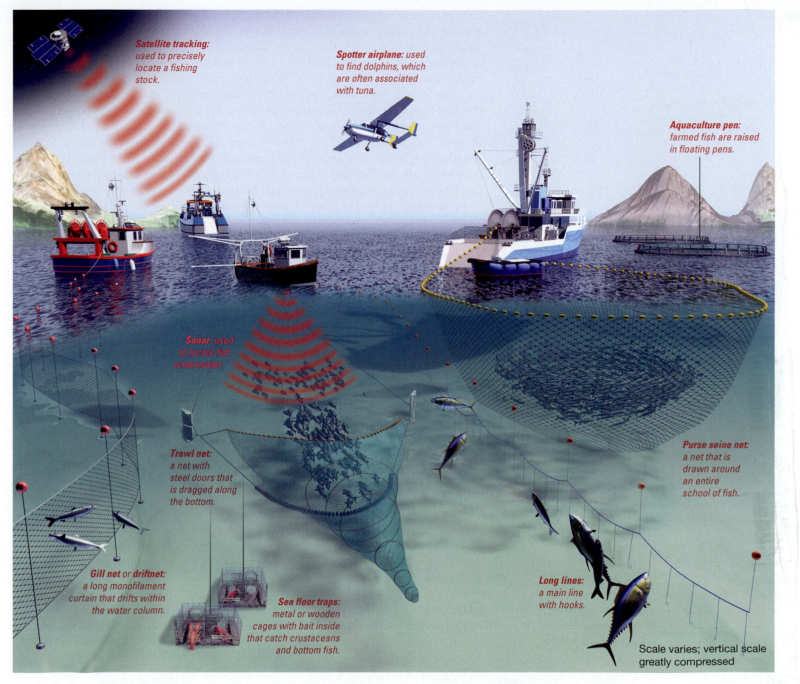

Satellite tracking: used to precisely locate a fishing stock.

Spotter airplane: used to find dolphins, which are often associated with tuna.

Aquaculture pen: farmed fish are raised in floating pens.

Sonar: used to locate fish underwater.

Trawl net: a net with steel doors that is dragged along the bottom.

Purse seine net: a net that is drawn around an entire school of fish.

Gill net or **driftnet:** a long monofilament curtain that drifts within the water column.

Sea floor traps: metal or wooden cages with bait inside that catch crustaceans and bottom fish.

Long lines: a main line with hooks.

Scale varies; vertical scale greatly compressed

SmartFigure 13.36 Methods and gear used in commercial fishing.
https://goo.gl/tc03yN

ghost fishing gear include longlines, gill nets, entangling nets, trammel nets, traps, and even crab and lobster pots. Ghost fishing is detrimental to the environment because anything caught by ghost fishing is killed and wasted. One of the reasons ghost fishing is so deadly is that the abandoned fishing gear continues to entangle and kill marine organisms as long as it remains intact. One solution to this problem, particularly for crab and lobster pots, is the use of biodegradable panels that decompose within a few months after the pots are lost, allowing the captured animals to escape.

Fisheries Management

Fisheries management is the organized effort directed at regulating fishing activity with the goal of maintaining a long-term fishery. Fisheries management practices include assessing ecosystem health, determining fish stocks, analyzing fishing practices (including recommending gear modification), establishing areas closed to

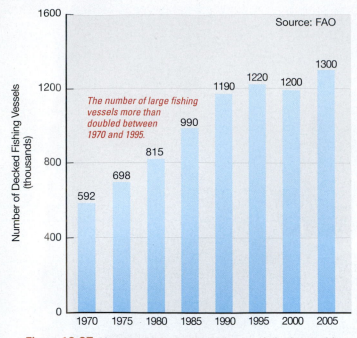

Figure 13.37 Number of decked fishing vessels in the world (in thousands). Bar graph showing that the number of engine-powered, mechanized-decked vessels engaged in commercial fishing worldwide has more than doubled since 1970. Even though the number of vessels has leveled off recently, the existence of so many well-equipped fishing vessels results in an increased fishing effort, which often leads to overfishing. Note that the FAO no longer keeps records of decked fishing vessels.

STUDENTS SOMETIMES ASK . . .

Is tuna labeled as "dolphin-safe" really safe to dolphins?

It depends on where the tuna is from, but the answer is probably yes. Tuna and tuna products harvested in the eastern tropical Pacific Ocean can only be labeled as "dolphin-safe" in the United States if no nets were intentionally set on dolphins during fishing and no dolphins were killed or seriously injured during the set in which the tuna were caught. In the United States, the National Marine Fisheries Service (NMFS) has developed an extensive monitoring, tracking, and verification program to ensure that only tuna and tuna products that meet the definition of dolphin-safe are indeed labeled as "dolphin-safe." This program, along with consumer awareness, has substantially reduced dolphin mortality, which numbered in the hundreds of thousands per year in the 1980s. Recently, however, changes to the definition of "dolphin-safe" have been accepted by the NMFS that address international trade concerns and allow more tuna to be imported from international sources, particularly Mexico. Tuna from international sources caught along with dolphins may be labeled dolphin-safe in the United States as long as observers aboard fishing vessels certify no dolphins were killed or seriously injured during the catch; however, this relies on the honesty of the observers.

fishing, and setting and enforcing catch limits. Unfortunately, however, fisheries management has historically been more concerned with maintaining human employment than preserving a self-sustaining marine ecosystem. For example, fisheries such as anchovy, cod, flounder, haddock, herring, and sardine are suffering from overfishing in spite of being managed (see MasteringOceanography **Web Diving Deeper 13.2**).

One of the most pressing problems facing fisheries management results from the fact that some fisheries encompass the waters of many different countries and involve a variety of ecosystems. Many species of commercial fish, for instance, reproduce in coastal estuaries along the world and migrate long distances across international waters to their preferred environment. Fishing limits are difficult to enforce internationally, and if human interference occurs at any location where the fish reproduce or migrate to, these species can be severely reduced in number. However, in some locations such as Italy and Alaska, fishers feel responsible for managing "their" stock in offshore waters and cooperatively arrive at decisions about fish stocks with the government, resulting in increased fishing stocks and revenue for fishers.

Another problem is the degradation of many ecosystems that sustain fisheries. For example, efforts in the United States to increase populations of the once-abundant Gulf sturgeon (*Acipenser oxyrinchus desotoi*), which lives in the Gulf of Mexico but frequents freshwater streams for breeding, are hampered by the fact that four of the seven major river systems used by the sturgeon are at, or exceed, the capacity for breeding because of the reduction of available habitat. In this case, a recovery plan must include efforts to increase available habitat and may require drastic actions, such as the removal of dams that interfere with fish migrations.

To be successful, fisheries management must also take into account a wide range of factors. For example, a review of historical fish landings indicates recovery plans for many species are not realistic because the plans are based on today's fish capture data, which is compiled after the populations have already reached dangerously low levels. Accordingly, the current levels of MSY—the maximum amount of fish biomass that can be removed yearly and still be sustained by the fishery ecosystem—are severely overestimated because they are based on fish population levels that are a fraction of the levels that existed before commercial fishing started.

Other issues that inhibit good fisheries management include limited scientific analyses, a lack of regulations enforcement (particularly of illegal, private, and international fishers), poaching, misreporting of catch and bycatch, seafood fraud and mislabeling, political barriers, and inadequate guidance for minimally regulated new fisheries.

REGULATION OF FISHING VESSELS A major regulatory failure has been the absence of restrictions on the number of fishing vessels. In 2004, according to the FAO, the world's fishing fleet consisted of about 4 million vessels. Of these, about 1.3 million were large, engine-powered, mechanized vessels with multiple decks called *decked fishing vessels*, many of which are over 24 meters (80 feet) in length. **Figure 13.37** shows that the number of decked fishing vessels in the world more than doubled between 1970 and 1995, and the number continues to increase. Many of these larger vessels use nets that can hold up to 27,000 kilograms (60,000 pounds) of fish in one haul. In addition, there are more than 2.1 million smaller, nondecked fishing vessels owned and operated by subsistence fishers, mostly in Asia, Africa, and the Middle East.

The increase in fishing vessels has resulted in an increased fishing effort, which often leads to overfishing. Additionally, commercial fishers use technology like GPS, depth finders, and spotter airplanes to locate fish populations (Figure 13.36). In some locations, fish are becoming so scarce that the fishing effort costs more than what the catch is worth! In 2003, for instance, the world fishing fleet spent over $120 billion to catch $80 billion worth of fish. To make up for the shortfall, many governments have given fishers subsidies—assistance in the form of cash or other benefits—that can exceed $25 billion yearly. Governments subsidies compound the problem by maintaining an unsustainable number of fishing vessels—or worse, encouraging new fishers to qualify for the subsidy, thus expanding the fishing fleet.

A CASE STUDY: FISHERIES IN THE NORTHWEST ATLANTIC The effects of inadequate fisheries management are illustrated by the history of fisheries in the northwest Atlantic. Under management by the International Commission for the Northwest Atlantic Fisheries, the fishing capacity of the international fleet increased 500% from 1966 to 1976; the total catch, however, rose by only 15%. This was a significant decrease in the catch per unit of effort, which is a good indication that the fishing stocks of the Newfoundland–Grand Banks area were being overexploited. The biologists who recommended the total allowable fishing quotas for the major species within this region complained that enforcement was lacking and that countries' quotas were being bartered in a game of international politics. As a result, fishers exceeded the total allowable catch set by the commission.

The difficulty of enforcing regulations by the international commission was largely responsible for Canada's unilateral decision to extend its right to control fish stocks for a distance of 200 nautical miles (370 kilometers) from its shores beginning January 1, 1977. The United States followed with a similar action only two months later.

Claiming coastal waters, however, proved to have limited effectiveness. After essentially all coastal nations assumed regulatory control over their coastal waters, the situation continued to deteriorate, and overfishing became even more of a problem. The Canadian government, in fact, had to shut down the Grand Banks fishery off Newfoundland in 1992, which resulted in the loss of about 40,000 jobs and subsequent government outlays of more than $3 billion in welfare. This cost far exceeded the value of the fishery, which generated no more than $125 million during one of its best years.

A similar situation has emerged on Georges Bank, which lies in Canadian and U.S. waters. On Georges Bank, however, tough international restrictions and proper enforcement are beginning to bring back haddock and yellowtail flounder fisheries.

In spite of protection, some fish stocks in the North Atlantic have not rebounded as anticipated. Atlantic cod stocks, for example, continue to decrease despite sharply curtailed annual catches. In 2003, to help halt the decline of cod—one of the North Atlantic's most impacted fish species—the Canadian government declared cod completely off-limits to fishing. A similar decline in cod stocks is also occurring in European waters, but to date officials have rejected a proposed cod-fishing ban.

DEEP-WATER FISHERIES One of the results of depleting and/or banning certain fish stocks is that the fishing industry then expands its efforts into the deep sea, where there are few regulations. As compared to surface species, most deep-water species have low metabolic and reproduction rates, which cause them to be severely impacted by fishing. For example, the collapse of Atlantic cod stocks forced fishers to seek the deep-water Greenland halibut as a substitute. Predictably, that species is now in danger of becoming overfished throughout the Atlantic.

The deep-water orange roughy fishery near Australia is another prime example. It was developed to satisfy the middle-American market for a bland white fish. Even the name was picked through careful supermarket research (its original moniker—the slimehead—sounded far less appealing). And when that fishery began its inevitable decline, the industry moved on to another deep-water species, the Patagonian toothfish. It was renamed the Chilean sea bass, even though it is neither a bass nor exclusively Chilean. The precipitous decline of these deep-water fish has qualified them for endangered status.

Deep-sea ecosystems are beginning to feel the effect of this increased fishing effort, too. For example, tall seamounts rise precipitously from the deep sea floor and provide a unique environment that harbors many species of fish and deep-water coral. To effectively catch fish here, fishers use large bottom-dragging trawl nets the size of football fields with two large steel doors that each weighs several tons. As these nets are dragged along the sea floor, they do long-lasting damage and have proven to negatively affect these slow-growing, yet important ecosystems.

STUDENTS SOMETIMES ASK . . .

Does fish farming relieve some of the demand for wild fish?

Fish farming describes the practice of raising fish in enclosed ocean pens or in restricted coastal bodies of water and is similar to raising farm animals on land such as cattle or sheep. Worldwide, nearly 250 species of fish and shellfish are farmed. In fact, the global production of farmed fish has quadrupled in the past 20 years and now provides nearly half of all seafood directly consumed by humans. Although it seems logical that fish farming would help reduce the demand for wild fish, some types of aquaculture have actually *increased* that demand because wild fish are used as feed for carnivorous farmed fish such as salmon, tuna, and seabass. For example, it takes 3 kilograms (6.6 pounds) of wild mackerel or anchovies to produce 1 kilogram (2.2 pounds) of farmed salmon or shrimp.

Some aquaculture systems further reduce wild fish supplies through habitat modification (transforming mangroves and coastal wetlands into fish and shrimp ponds) and by the collection of wild fish for initial stocking of aquaculture operations. Other negative impacts of aquaculture include waste disposal in natural environments, introduction of non-native organisms, inbreeding of wild populations with accidentally released "domesticated" species, and the rapid spread of disease in organisms that live in close quarters, all of which have probably contributed to the collapse of "wild" fish stocks worldwide. If the aquaculture industry is to sustain its contribution to world fish supplies, it must reduce wild fish inputs as food and adopt more ecologically sound management practices.

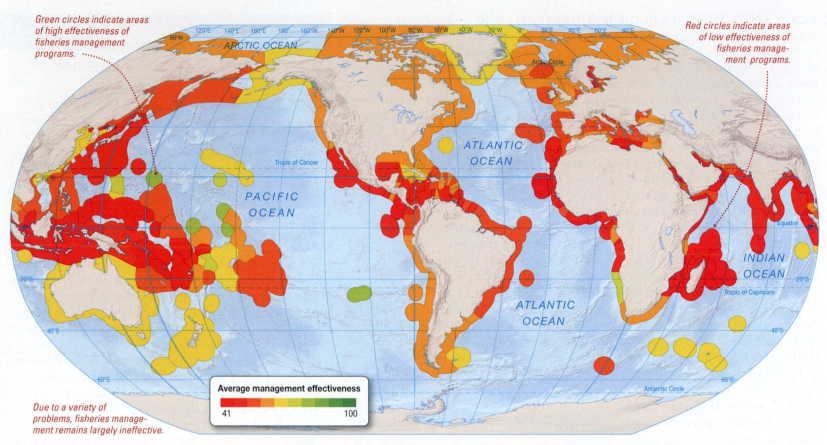

Green circles indicate areas of high effectiveness of fisheries management programs.

Red circles indicate areas of low effectiveness of fisheries management programs.

Due to a variety of problems, fisheries management remains largely ineffective.

Average management effectiveness
41 100

Figure 13.38 Effectiveness of world fisheries management. Map showing the effectiveness of the world's fisheries management programs, based on 6 key parameters (see text for description).

RECAP

For fisheries to be sustainable, they must employ ecosystem-based fisheries management, which considers variables in marine ecosystems and not just individual species. In addition, fishing limits must be upheld despite political factors, unwanted bycatch must be reduced, and critical fish habitats must be protected.

ECOSYSTEM-BASED FISHERY MANAGEMENT The historic record of fishery management shows many examples of attempts to regulate fish stocks that consider only individual species; however, these efforts have been largely unsuccessful. To ensure sustainable fisheries and a healthy marine environment in U.S. waters, there has been a recent push to institute *ecosystem-based fishery management*, which employs a more comprehensive approach to understanding fish stocks including analysis of such variables as fish habitat, migration routes, and predator–prey interactions. Ecosystem-based fishery management essentially reverses the order of management priorities, starting with an ecosystem rather than a target species. It also focuses on rebuilding fisheries on a global basis.

Research has shown that sustainable fisheries can be achieved if management strategies are adopted that remove subsidies and give individual fishers a right to a proportion of the total catch. By removing competition, fishers of all types would be encouraged to act to sustain the entire fishery because, as the common fish stock improves, each individual's quota will increase. In this way, fishers will have the incentive to manage the stock sustainably, thus avoiding overfishing and the eventual collapse of the fishery.

EFFECTIVENESS OF FISHERY MANAGEMENT In 2010, a scientific study was published on the effectiveness of current fisheries management practices. Scientists analyzed six key parameters including the scientific quality of management recommendations, the transparency of converting recommendations into policy, the enforcement of policies, the influence of subsidies, fishing effort, and the extent of fishing by foreign entities. Results of the study (**Figure 13.38**) show that fisheries management remains largely ineffective despite broad acceptance and commitments by governments to initiatives for improvements. However, since 68% of the world's

wild ocean fish are caught by only nine countries and the European Union, fisheries managers remain optimistic that ecosystem-based fisheries management can be attainable.

How can fisheries in danger of collapse be brought back? Fisheries experts agree that the following three key items must be implemented immediately: (1) set scientifically established quotas or limits on the amount of fish caught and enforce those limits, (2) reduce wasteful and unwanted bycatch, and (3) protect vital habitats that fish use as spawning and nursery areas. Humans have the technology and understanding of fisheries ecosystems to achieve these items. But only if all three items can be accomplished will fisheries around the world have a chance of recovering.

Effect of Global Climate Change on Marine Fisheries

According to fisheries scientists, human-caused global climate change is already impacting fisheries in many areas. Researchers used temperature preferences of fish and other marine species as "biological thermometers" to assess the effects of ocean warming on marine species. Marine fish are normally found in specific regions of the ocean, related to the specific water temperature to which they are adapted. If the same species of fish is now found in a region of the ocean outside of its normal area of distribution, this suggests that the water temperature in this region has changed. Between 1970 and 2006, for example, researchers found that ocean warming changed the catch composition of commercial fishers' landings from cooler-water to warmer-water species in most ecosystems (Figure 13.39).

Climate

Connection

RECAP

Higher ocean temperatures are driving species away from where they normally exist into cooler waters, deeper into the oceans, or to high-latitude regions. Species that prefer warmer water, in turn, are populating areas vacated by cool-water species.

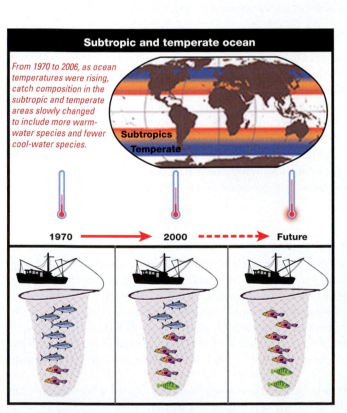

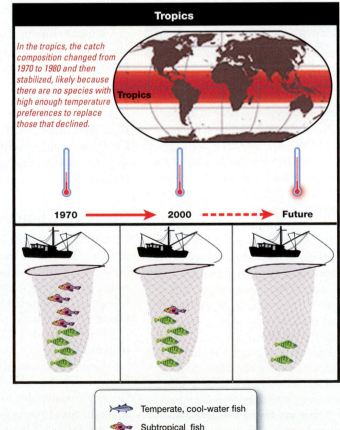

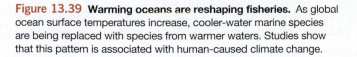

Figure 13.39 Warming oceans are reshaping fisheries. As global ocean surface temperatures increase, cooler-water marine species are being replaced with species from warmer waters. Studies show that this pattern is associated with human-caused climate change.

Best Choices	Good Alternatives	Avoid	Support ocean-friendly seafood
Arctic Char (farmed) Barramundi (US farmed) Catfish (US farmed) Clams (farmed) Cod: Pacific (US non-trawled) Crab: Dungeness, Stone Halibut: Pacific (US) Lobster: California Spiny (US) Mussels (farmed) Oysters (farmed) Saberfish/Black Cod (Alaska & Canada) Salmon (Alaska wild) Sardines: Pacific (US) Scallops (farmed) Shrimp: Pink (OR) Striped Bass (farmed & wild*) Tilapia (US farmed) Trout: Rainbow (US farmed) Tuna: Albacore (Canada & US Pacific, troll/pole caught) Tuna: Skipjack, Yellowfin (US troll/pole caught)	Basa/Pangasius/Swai (farmed) Caviar, Sturgeon (US farmed) Clams (wild) Cod: Atlantic (imported) Cod: Pacific (US trawled) Crab: Blue*, King (US), Snow Flounders, Soles (Pacific) Flounder: Summer (US Atlantic) Grouper: Black, Red (US Gulf of Mexico) Herring: Atlantic Lobster: American/Maine Mahi Mahi (US) Oysters (wild) Pollock: Alaska (US) Saberfish/Black Cod (CA, OR, WA) Salmon (CA, OR, WA*, wild) Scallops (wild) Shrimp (US, Canada) Squid Swai, Basa (farmed) Swordfish (US)* Tilapia (Central & South America farmed) Tuna: Bigeye, Tongoi, Yellowtail (US farmed)	Caviar, Sturgeon* (imported wild) Chilean Seabass/Toothfish* Cobia (imported farmed) Cod: Atlantic (Canada & US) Crab: King (imported) Flounders, Halibut, Soles (US Atlantic except summer flounder) Groupers (US Atlantic)* Lobster: Spiny (Brazil) Mahi Mahi (imported longline) Marlin: Blue, Striped (Pacific)* Monkfish Orange Roughy* Salmon (farmed, including Atlantic)* Sharks* & Skates Shrimp (imported) Snapper: Red (US Gulf of Mexico) Swordfish (imported)* Tilapia (Asia farmed) Tuna: Albacore*, Bigeye*, Skipjack, Tongol, Yellowfin* (except troll/pole caught) Tuna: Bluefin* Tuna: Canned (except troll/pole caught)	**Best Choices** are abundant, well-managed and caught or farmed in environmental friendly ways. **Good Alternatives** are an option, but there are concerns with how they're caught or farmed–or with the health of their habitat due to other human impacts. **Avoid** for now as these items are caught or farmed in ways that harm other marine life or the environment. **Key** CA = California OR = Oregeon WA = Washington *Limit consumption due to concerns about mercury or other contaminants. Visit www.edf.org/seafoodhealth Contaminant information provided by: ENVIRONMENTAL DEFENSE FUND Seafood may appear in more than one column

Figure 13.40 Recommended seafood choices. The Monterey Bay Aquarium Seafood Watch shows recommended seafood choices (*green*), seafood with some concerns (*yellow*), and seafood to avoid (*red*). Printable regional pocket guides and mobile apps are available at **www.seafoodwatch.org**.

STUDENTS SOMETIMES ASK . . .

I've heard of a movement to ban shark fin soup. What is shark fin soup?

Shark fin soup is a Chinese delicacy dating back to the Sung Dynasty (around 960 A.D.). The soup is traditionally made with chicken, ham broth, vegetables, and strands of cartilage from shark fin (and likely some MSG). Interestingly, none of the soup's flavor actually comes from shark fin, so including shark fin is a symbolic gesture. Like many other dishes in traditional cultures, shark fin soup is served as a symbol of class and wealth. The dish has become an ingrained tradition of status, "face," and respect; as such, it has come to be expected at big celebrations, especially weddings. Unfortunately, obtaining shark fins is a wasteful practice. Each year, up to 73 million sharks are killed primarily for their fins, after which their bleeding bodies are discarded, threatening one-third of open-ocean sharks with extinction. In response to this environmental issue, grassroots organizations spearheaded by marine biologists have sprung up to protect sharks and encourage wedding couples to stop serving shark fin soup. Two such examples are the Websites Shark Truth (**www.sharktruth.com**) and Happy Hearts Love Sharks (**www.happyheartslovesharks.org**), which sponsors a fin-free wedding contest each year. These and similar initiatives have recently caused China's demand for shark fin soup to drop by as much as 70%.

Global climate change is also having a severe impact on over-exploited fisheries. In fact, a report from the FAO concluded that global climate change will very likely cause the collapse of several depleted fish stocks. But even healthy fisheries can be affected, too. In areas such as the west coasts of continents, where fish populations are dependent on upwelling, the warming of the oceans and associated reduction in upwelling could devastate healthy fisheries such as salmon, tuna, and mackerel. In addition, rising sea level as a result of ocean warming and ice sheet melting will inundate low-lying coastal areas such as mangroves and marshes, which are vital breeding grounds and nurseries for many commercial fish species.

Seafood Choices

Consumer demand has driven some fish populations to the brink of disappearing. However, consumers can help by making wise choices in the fish they consume by purchasing only fish from healthy, thriving fisheries. Certainly, some types of seafood carry less environmental impact than others because of differences in abundance, how they're caught, and how well fishing is managed. **Figure 13.40** shows some recommendations for seafood (both fish and shellfish) in three categories: best choices (*green*), good alternatives (*yellow*), and seafood to avoid (*red*).

Consumers can also help by purchasing seafood that carries a "sustainable" label, which is issued by the following two organizations that certify fisheries as scientifically sustainable: the Marine Stewardship Council and Friend of the Sea. Such schemes aim to help consumers and retailers support fisheries that are sustainable and not exploited by overfishing. Recent research, however, has indicated that about one-quarter of seafood sold with a "sustainable" label is not meeting the criteria of sustainability.

CONCEPT CHECK 13.5 | Evaluate several issues that affect marine fisheries.

1 Define *overfishing*. When a species is overfished, what changes are there in the standing stock, the size of individuals in the remaining fish population, and the maximum sustainable yield?

2 What impact do lost or abandoned traps, nets, and fishing gear have on the environment?

3 Describe three unintended consequences that have occurred in marine ecosystems when top predators are removed.

4 As an environmentally conscious consumer, what factors should you take into consideration before purchasing seafood at a restaurant or grocery store?

RECAP

The marine fishing industry suffers from overfishing, wasteful practices that produce a large amount of unwanted bycatch, and a lack of effective fisheries management.

ESSENTIAL CONCEPTS REVIEW

13.1 What is primary productivity?

▸ *Microscopic planktonic bacteria and algae that photosynthesize represent the largest biomass in the ocean.* They are the ocean's *primary producers*—the foundation of the ocean's food web. Organic biomass is also produced near deep-sea hydrothermal springs through *chemosynthesis*, in which bacteria-like organisms trap chemical energy by the oxidation of hydrogen sulfide.

▸ *The availability of nutrients and the amount of solar radiation limit the photosynthetic productivity* in the oceans. *Nutrients*—such as *nitrogen, phosphorus, iron,* and *silica*—are most abundant in coastal areas, due to runoff and upwelling. The depth at which net photosynthesis is zero is the *compensation depth for photosynthesis.* Generally, algae cannot live below this depth, which may be less than 20 meters (65 feet) in turbid coastal waters or as much as 100 meters (330 feet) in the open ocean.

▸ *Marine life is most abundant along continental margins,* where nutrients and sunlight are optimal. It decreases with distance from the continents and with increased depth. In addition, *cool water typically supports more abundant life than warm water* because cool water can dissolve more of the gases necessary for life (oxygen and carbon dioxide). *Areas of upwelling bring cold, nutrient-rich water to the surface and have some of the highest productivities.*

▸ *Ocean water selectively absorbs the colors of the visible spectrum. Red and yellow light are absorbed at relatively shallow depths,* whereas *blue and green light are the last to be removed.* Ocean water of *low biological productivity* scatters the short wavelengths of visible light, producing a *blue color.* Turbidity and photosynthetic algae in *more productive ocean water* scatter more green light wavelengths, which produces a *green color.*

(a) Coastal winds (green arrows) cause Ekman transport, which drives surface water away from the west coasts of continents (blue arrows).

(b) SeaWiFS image of chlorophyll concentration along the southwest coast of Africa (February 21, 2000). High chlorophyll concentrations indicate high phytoplankton biomass, which is caused by coastal upwelling. Concentration is reported in milligrams of chlorophyll a per cubic meter of seawater (mg/m³).

(c) Block diagram showing how coastal upwelling in the Southern Hemisphere is caused by coastal winds that cause surface waters to move away from shore due to Ekman transport, thus bringing cold, nutrient-rich water to the surface.

Study Resources

MasteringOceanography Quizzes, MasteringOceanography Web Animation

Critical Thinking Question

As you take a day boat trip in the ocean, you notice the color of the water near the dock is green while the color of the water in deeper waters is blue. What inference can you make about life in the ocean from the water color difference between shallow and deep water?

Active Learning Exercise

With another student in class, analyze how a marine organism's ability to use color to "disappear" in the water column is affected by depth.

13.2 What kinds of photosynthetic marine organisms exist?

▸ *There are many different types of photosynthetic marine organisms.* The seed-bearing Anthophyta are represented by a few genera of *nearshore plants* such as eelgrass (*Zostera*), surf grass (*Phyllospadix*), marsh grass (*Spartina*), and mangrove trees (*Rhizophora* and *Avicennia*). *Macroscopic algae* include *green algae* (Chlorophyta), *red algae* (Rhodophyta), and *brown algae* (Phaetophyta). *Microscopic algae* include *diatoms* and *coccolithophores* (Chrysophyta), and *dinoflagellates* (Pyrrophyta).

▸ *Dinoflagellates sometimes exist in such great abundance they color surface waters red, producing a red tide,* which is more accurately called a *harmful algal bloom (HAB).* Dinoflagellates also *produce powerful biotoxins* that can lead to poisonings. *Ocean eutrophication is the artificial enrichment of waters* by a previously scarce nutrient—usually provided by river runoff—that can *trigger an overabundance of algae* and lead to the *creation of an oxygen-poor dead zone.*

Study Resources
MasteringOceanography Quizzes, MasteringOceanography Web Diving Deeper 13.1

Critical Thinking Question

Compare and contrast the following: red tides, harmful algal blooms, ocean eutrophication, and dead zones.

Active Learning Exercise

With another student in class, compile a list of ways to reduce harmful algae blooms in coastal areas.

13.3 How does regional primary productivity vary?

▸ *The deep ocean acts as a reservoir of nutrients* because the lack of sunlight there limits the uptake of these substances by photosynthetic organisms. When these *deep, cold, nutrient-rich waters are brought to the sunlit surface,* all the right conditions exist to create high productivity and an abundance of marine life. However, *the development of a thermocline acts as an impenetrable lid* that prohibits the movement of nutrient-rich deep water to the surface, thereby *inhibiting productivity.*

▸ In *high-latitude (polar) oceans,* a thermocline is generally absent, so *upwelling can readily occur.* The *availability of solar radiation limits productivity* in polar oceans more than the availability of nutrients.

▸ In *low-latitude (tropical) oceans,* a *strong thermocline* usually exists year-round, so the *absence of upwelling and resulting lack of nutrients in surface water limits productivity.* Productivity can be higher in areas of localized upwelling or near coral reefs, which tend to hold and concentrate nutrients.

▸ In *middle latitude (temperate) oceans, productivity peaks in the spring and fall and is limited by lack of solar radiation in the winter and lack of nutrients in the summer.*

Study Resources
MasteringOceanography Quizzes, MasteringOceanography Web Animation

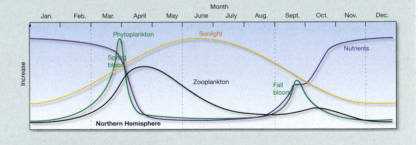

Critical Thinking Question

Compare the biological productivity of polar, middle latitude, and tropical oceans. Consider all factors such as seasonal changes in solar radiation, the development of a thermocline, and the availability of nutrients.

Active Learning Exercise

With another student in class, discuss how the distribution of nutrients in the ocean and the timing of peak productivity affects the migration of marine mammals such as blue whales from their breeding grounds in Baja California to their feeding grounds in the North Pacific.

13.4 How are energy and nutrients passed along in marine ecosystems?

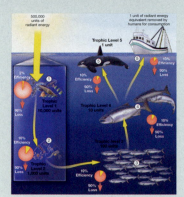

▶ *Radiant energy captured by algae is converted into chemical energy* and passed through the different *trophic levels* of a *biotic community.* It is expended as *mechanical and heat energy* and ultimately becomes biologically useless. Upon death, organisms are *decomposed* to an inorganic form that algae can use again for nutrients.

▶ *Marine ecosystems are composed of populations of organisms called producers* (which photosynthesize or chemosynthesize), *consumers* (which eat producers), and *decomposers* (which break down detritus). Animals can be categorized as *herbivores* (eat plants), *carnivores* (eat animals), *omnivores* (eat both), or *bacteriovores* (eat bacteria). Through *biogeochemical cycles,* the organisms of a biotic community *cycle nutrients and other chemicals* from one form to another.

▶ *Feeding strategies* include *suspension* or *filter feeding* (filtering planktonic organisms from seawater), *deposit feeding* (ingesting sediment and detritus), and *carnivorous feeding* (preying directly upon other organisms). On average, *only about 10% of the mass taken in at one feeding level is passed on to the next.* As a result, the *size of individuals increases* but the *number of individuals decreases* with each *trophic level* of a *food chain* or *food web.* Overall, the *total biomass of populations decreases the higher they are in the biomass pyramid.*

Study Resources
MasteringOceanography Quizzes, Web Video

Critical Thinking Question
The average efficiency of energy transfer between trophic levels is 10%. Use this efficiency to determine how much phytoplankton mass is required to add just 1 gram (0.04 ounce) of new mass to a killer whale, which is a third-level or top carnivore. Create a diagram that summarizes the different trophic levels and the relative size and abundance of organisms at each level. How would your answer change if the efficiency were half the average rate? Twice the average rate?

Active Learning Exercise
With another student in class, evaluate the fate of top carnivores (e.g., sharks and killer whales) as the catch of secondary consumers, such as tuna, increases.

13.5 What issues affect marine fisheries?

▶ *Marine fisheries harvest standing stocks of populations from various ecosystems,* particularly shallow shelf and coastal waters and areas of upwelling.

▶ *Overfishing occurs when adult fish are harvested faster than they can reproduce* and results in the decline of fish populations as well as a reduction of a fishery's *maximum sustainable yield (MSY).*

▶ *Many fishing practices capture unwanted bycatch. Ghost fishing describes any lost or discarded fishing gear* that continues to catch fish, marine mammals, or other organisms after it has been abandoned.

▶ Despite the *management of fisheries, many fish stocks worldwide are still declining. Wise seafood choices can help* reverse the decline in fish populations.

Study Resources
MasteringOceanography Quizzes, MasteringOceanography Web Diving Deeper 13.2

Critical Thinking Question
List several significant problems with current fisheries management practices. How can fisheries management be improved to increase the sustainability of the standing stocks? What changes can be made to fishing gear to reduce the damage to the ecosystem and unwanted bycatch?

Active Learning Exercise
With another student in class, describe three things that can be done to facilitate the recovery, diversity, and abundance of large marine fish in the world ocean.

MasteringOceanography™
www.masteringoceanography.com

Looking for additional review and test prep materials? With individualized coaching on the toughest topics of the course, MasteringOceanography offers a wide variety of ways for you to move beyond memorization and deeply grasp the underlying processes of how the oceans work. Visit the Study Area in **www.masteringoceanography.com** to find practice quizzes, study tools, and multimedia that will improve your understanding of this chapter's content. Sign in today to enjoy the following features: Self Study Quizzes, SmartFigures, SmartTables, Oceanography Videos, Squidtoons, Geoscience Animation Library, RSS Feeds, Digital Study Modules, and an optional Pearson eText.

A superbly streamlined shark glides through the ocean. Sharks like this silky shark (*Carcharhinus falciformis*) have unique adaptations that make them efficient ocean predators.

14

Animals of the Pelagic Environment

Before you begin reading this chapter, use the glossary at the end of this book to discover the meanings of any of the words in the word cloud above you don't already know.

Pelagic organisms live suspended in seawater (not on the ocean floor) and comprise the vast majority of the ocean's **biomass**.[1] Phytoplankton and other photosynthesizing microbes live within the sunlit surface waters of the ocean and are the food source for nearly all other marine life. As a result, many marine animals live in surface waters so they can be close to their food supply. One of the most important challenges facing many marine organisms is to stay afloat and not sink below surface waters into the immense depth of the oceans.

Phytoplankton and other photosynthesizing microbes depend primarily on their small size to provide a high degree of frictional resistance to sinking. Most animals, however, are more dense than ocean water and have less surface area per volume of body mass (they have a smaller surface-area-to-volume ratio). Therefore, they tend to sink more rapidly than phytoplankton.

To remain in surface waters where the food supply is greatest, pelagic marine animals must increase their buoyancy or swim continually. Animals apply one or both of these strategies in amazing ways using a variety of adaptations.

14.1 How Are Marine Organisms Able to Stay above the Ocean Floor?

Some animals increase their buoyancy to remain in near-surface waters. They may have internal structures containing gas, which significantly reduce their average density, or they may have soft bodies void of hard, high-density parts. Larger animals often have the ability to swim, but if their bodies are denser than seawater, they must exert more energy to propel themselves through the water.

Use of Gas Containers

Air is almost 800 times less dense than water at sea level, so even a small amount of air inside an organism can dramatically increase its buoyancy. Generally, animals use either an internal, rigid gas container or a swim bladder to achieve *neutral buoyancy*, using the amount of air in their bodies to regulate their density, so they can remain at a particular depth without expending energy to do so.

Interdisciplinary Relationship

[1]Remember that *biomass* is the mass of living organisms.

"The whale rose even closer. It had a distinct hazel eye that looked directly at me. It studied my hair, looked at my beard, passed its gaze over my nose, and then looked deeply into my eyes. It looked past all those biology classes, between the volumes of whale literature I had studied, and beyond the thousands of gray whales in my memory. It looked into my soul."

—Lindblad Expeditions Naturalist Robert "Pete" Pederson, describing a close encounter with a gray whale (1999).

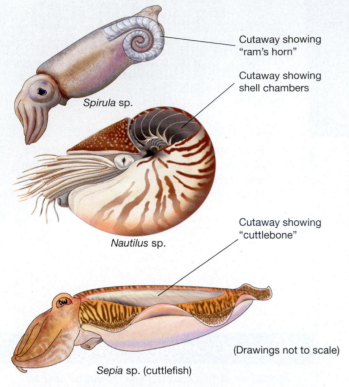

Spirula sp.

Cutaway showing "ram's horn"

Cutaway showing shell chambers

Nautilus sp.

Cutaway showing "cuttlebone"

(Drawings not to scale)

Sepia sp. (cuttlefish)

Figure 14.1 **Gas containers in cephalopods.** The *Nautilus* has an external chambered shell, while *Sepia* and *Spirula* have rigid internal chambered structures that can be filled with gas to provide buoyancy.

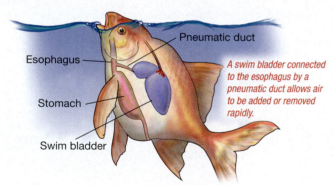

Pneumatic duct

Esophagus

Stomach

Swim bladder

A swim bladder connected to the esophagus by a pneumatic duct allows air to be added or removed rapidly.

(a) Adaptations that allow rapid buoyancy changes.

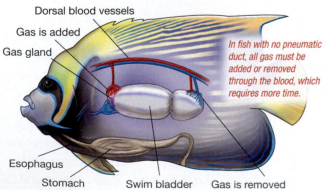

Dorsal blood vessels

Gas is added

Gas gland

In fish with no pneumatic duct, all gas must be added or removed through the blood, which requires more time.

Esophagus

Stomach Swim bladder Gas is removed

(b) Adaptations that only allow slow buoyancy changes.

SmartFigure 14.2 **Swim bladder.** Many bony fish have a swim bladder, which is used to regulate their buoyancy and corresponding position in the water column. Some fish have adaptations that allow for **(a)** rapid buoyancy changes while others **(b)** experience slow buoyancy changes. https://goo.gl/ywFGFq

RIGID GAS CONTAINERS Some animals, such as the cephalopods (*cephalo* = head, *poda* = foot), have rigid gas containers in their bodies. For instance, the genus *Nautilus* has an external many-chambered shell, whereas the cuttlefish *Sepia*[2] and deep-water squid *Spirula* have an internal chambered structure (**Figure 14.1**).

Because the pressure in their air chambers is always 1 kilogram per square centimeter (1 atmosphere or 14.7 pounds per square inch), *Nautilus* must stay above a depth of approximately 500 meters (1640 feet) to prevent collapse of its chambered shell. *Nautilus*, therefore, rarely ventures below about 250 meters (800 feet).

SWIM BLADDER Instead of using rigid gas containers, some slow-moving fish use an internal organ called a **swim bladder** (**Figure 14.2**) to achieve neutral buoyancy and determine their position in the water column. Very active swimmers (such as tuna) or fish that live on the bottom, however, do not usually have a swim bladder because they don't have a problem maintaining their positions in the water column.

A change in depth either expands or contracts the swim bladder, so fish must remove or add gas to the bladder in order to maintain a constant volume. In some fish, a pneumatic duct connects the swim bladder to the esophagus (**Figure 14.2a**), allowing these fish to rapidly add or remove gases through the duct. In other fish without a pneumatic duct (**Figure 14.2b**), the gases of the swim bladder must be added or removed more slowly by an interchange with the blood, so they cannot withstand rapid changes in depth.

The composition of gases in the swim bladders of shallow-water fish is similar to that of the atmosphere. At the surface, the concentration of oxygen in the swim bladder is about 20%, which is the same as the atmosphere. As depth increases, however, the oxygen concentration can increase to as much as 90% or more. This is because chemical reactions within the fish at depth cause oxygen to leave the blood, which then diffuses into the swim bladder. Fish with swim bladders have been captured from as deep as 7000 meters (23,000 feet), where the pressure is 700 kilograms per square centimeter (700 atmospheres, or 10,300 pounds per square inch). Pressure this high compresses the gas to a density of 0.7 gram per cubic centimeter (0.025 pound per cubic inch).[3] This is approximately the same density as fat, and as a result, many deep-water fish have special organs for buoyancy that are filled with fat instead of compressed gas.

Interdisciplinary

Relationship

Ability to Float

Floating animals range in size from microscopic shrimplike organisms to relatively large species, such as the familiar jellies. These floating organisms—collectively called *zooplankton*—comprise the second largest biomass in the ocean after phytoplankton and other photosynthesizing microbes. Microscopic forms of zooplankton usually have a hard shell called a **test** (*testa* = shell). Many larger forms have soft, gelatinous bodies with little if any hard tissue, which reduces their density and allows them to stay afloat.

Microscopic zooplankton are incredibly abundant in the ocean. They are primary consumers because they eat microscopic phytoplankton, the primary producers. Thus, many zooplankton are herbivores. Others are omnivores because they eat other zooplankton in addition to phytoplankton. Most types of microscopic zooplankton have adaptations to increase the surface area of their

[2]Many species of cephalopods have an inking response. In fact, the ink of *Sepia* was used as writing ink (with the brand name Sepia) before alternatives were developed.

[3]For comparison, note that the density of water is 1.0 gram per cubic centimeter (0.04 pound per cubic inch).

bodies (or shells) so they can remain in the sunlit surface waters near their food source.[4]

In addition, some organisms produce low-density fats or oils to stay afloat. Many zooplankton, for example, produce tiny droplets of oil to help maintain neutral or nearly neutral buoyancy. As another example, sharks have a very large, oil-rich liver to help reduce their density and float more easily.

Ability to Swim

Many larger pelagic animals, such as fish and marine mammals, can maintain their position in the water column by swimming and can also swim easily against currents. These organisms are called *nekton* (*nektos* = swimming). Because of their swimming ability, some of these organisms undertake long migrations.

The Diversity of Planktonic Animals

Diversity in planktonic marine animals stems from a shifting balance between competition for food and avoidance of predators. As a result, a wide variety of planktonic animals inhabit the oceans.

EXAMPLES OF MICROSCOPIC ZOOPLANKTON Three of the most important groups of microscopic zooplankton are the radiolarians, foraminifers (both of which are discussed in more detail in Chapter 4, "Marine Sediments"), and copepods.

Radiolarians (*radio* = spoke or ray) are single-celled, microscopic protozoans (*proto* = first, *zoa* = animal) that build their hard shells (*tests*) out of silica (**Figure 14.3**). Their tests have intricate ornamentation, including long projections. Although the spikes and spines appear to be a defense mechanism against predators, they increase the test's surface area so the organism won't sink through the water column.

Foraminifers (*foramen* = an opening) are microscopic to barely macroscopic single-celled protozoans. While the most abundant types of foraminifers are planktonic, the most diverse (in terms of number of species) are benthic. Foraminifers produce a hard test made of calcium carbonate (**Figure 14.4**) that is segmented or chambered, with a prominent opening in one end. The tests of both radiolarians and foraminifers are common components of deep-sea sediment.

Copepods (*kope* = oar, *poda* = foot) are microscopic shrimplike animals of the subphylum Crustacea, which also includes shrimps, crabs, and lobsters. Like other crustaceans, copepods have a hard exoskeleton (*exo* = outside) and a segmented body with jointed legs (**Figure 14.5**). Most copepods have forked tails and distinct and elaborate antennae.

More than 7500 species of copepods exist, with most possessing special adaptations for filtering tiny floating food particles from water. Some copepods are herbivores that eat algae, others are carnivores that eat other zooplankton, and still others are parasitic.

All copepods lay eggs, which are sometimes carried in egg sacs attached to the abdomen but generally are simply released into seawater, where they hatch in about a day. Their rapid reproduction allows great numbers of copepods to occur wherever favorable conditions—generally an abundance of food, which is mostly algae—exist.

RECAP

Marine organisms use a variety of adaptations to stay within the sunlit surface waters, such as rigid gas containers, swim bladders, spines to increase their surface area, soft bodies, and the ability to swim.

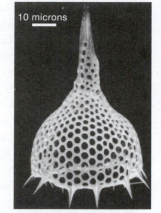

10 microns

(a) *Anthocyrtidium ophirense*

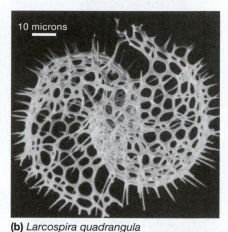
10 microns

(b) *Larcospira quadrangula*

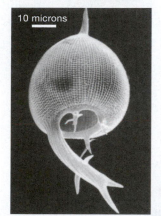

10 microns

(c) *Euphysetta elegans*

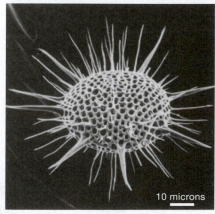

10 microns

(d) *Heliodiscus asteriscus*

Figure 14.3 Radiolarians. Scanning electron micrographs of the tests of various radiolarians.

1.0 mm
0.04 in

Figure 14.4 Foraminifers. Photomicrograph of various tests of pelagic foraminifers that were collected in the Mediterranean Sea.

[4]For a more complete discussion of how increased surface area affects an organism's ability to float, see Section 12.4 in Chapter 12.

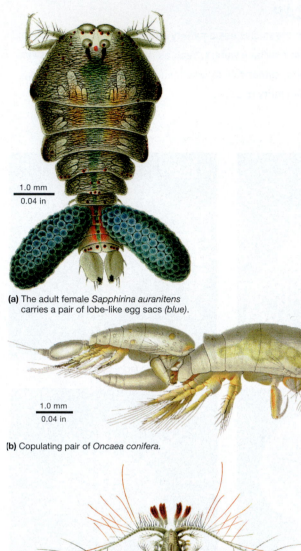

(a) The adult female *Sapphirina auranitens* carries a pair of lobe-like egg sacs *(blue)*.

(b) Copulating pair of *Oncaea conifera*.

(c) *Calocalanus pavo* showing elaborate feathery appendages that are characteristic of warm-water species.

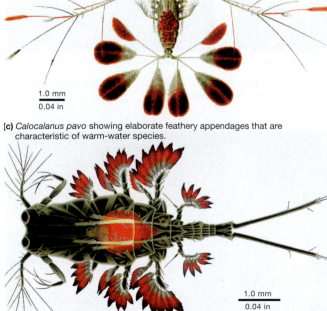

(d) *Copilia vitrea* uses its appendages to cling to large particles in the water column or to larger zooplankton.

Figure 14.5 Copepods. Line drawings of various copepods from Wilhelm Giesbrecht's 1892 book on the flora and fauna of the Gulf of Naples (Mediterranean Sea).

Although copepods are small (only a few species are larger than 1 millimeter [0.04 inch]), they are extremely numerous in the ocean. They are so abundant, in fact, that they are one of the most numerically dominant types of multicelled organisms on the planet. As such, copepods comprise the majority of the ocean's zooplankton biomass and are a vital link in many marine food webs between phytoplankton (producers) and larger species such as plankton-eating fish.

EXAMPLES OF MACROSCOPIC ZOOPLANKTON Many types of zooplankton are large enough to be seen without the aid of a microscope yet don't swim well. Two of the most important groups of macroscopic zooplankton are krill and various types of cnidarians.

Krill, which means "young fry of fish" in Norwegian, are actually in the subphylum Crustacea (genus *Euphausia*) and resemble minishrimp or large copepods (**Figure 14.6**). There are more than 1500 species of krill, most of which achieve a length no longer than 5 centimeters (2 inches). They are abundant near Antarctica and form a critical link in the food web there, supplying food for many organisms from sea birds to the largest whales in the world.

Cnidarians (*cnid* = nematocyst [*nemato* = thread, *cystis* = bladder]), which were formerly known as *coelenterates* (*coel* = hollow, *enteron* = intestines), have soft bodies that are more than 95% water and tentacles armed with stinging cells called *nematocysts*. Macroscopic zooplankton that are cnidarians fall into one of two basic groups: the *hydrozoans* and the *scyphozoans* (jellies).

Hydrozoan (*hydro* = water, *zoa* = animal) *cnidarians* are represented in all oceans by the "Portuguese man-of-war" (genus *Physalia*) and the "by-the-wind sailor" (genus *Velella*). Their gas chambers, called *pneumatophores* (*pneumato* = breath,

Figure 14.6 Krill. Krill that has washed up on a beach in Antarctica and close-up view of *Meganyctiphanes norvegica* (*inset*).

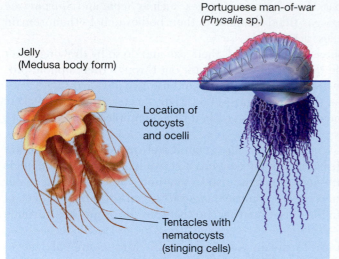

Portuguese man-of-war
(*Physalia* sp.)

Jelly
(Medusa body form)

Location of
otocysts
and ocelli

Tentacles with
nematocysts
(stinging cells)

(a) Line drawings of a typical medusa jelly (*left*) and a Portuguese man-of-war (*Physalia*) (*right*).

10 cm
4 in

(b) Photo of medusa jelly (*Cyanea capillata*).

Figure 14.7 Planktonic cnidarians. Line drawings and a photo of various species of planktonic cnidarians.

phoros = bearing), serve as floats and sails that allow the wind to push them across the ocean surface (**Figure 14.7a**, *left*). Sometimes the wind pushes large numbers of these organisms toward a beach, where they can wash ashore and die. In the living organism, an entire colony of other small organisms that rely on the hydrozoan for habitat can be found within and beneath the float. Portuguese man-of-war tentacles may be many meters long and, because they are armed with nematocysts, they have been known to inflict a painful and occasionally dangerous neurotoxin poisoning in humans.

Jellies, or **scyphozoan** (*skyphos* = cup, *zoa* = animal) *cnidarians*, have a bell-shaped body with a fringe of tentacles and a mouth at the end of a clapper-like extension hanging beneath the bell-shaped float (**Figure 14.7**). Ranging in size from nearly microscopic to 2 meters (6.6 feet) in diameter, most jellies are less than 0.5 meter (1.6 feet) in diameter. The largest jellies can have tentacles as long as 60 meters (200 feet).

Jellies move by muscular contraction of their bell. Water enters the cavity under the bell and is forced out when muscles that circle the bell contract, slowly jetting the animals in random directions as they pulsate.[5] More importantly, the motion swooshes water that contains tiny floating food particles toward their tentacles. To allow the animal to orient itself generally in an upward direction, light-sensitive or gravity-sensitive organs exist around the outer edge of the bell. The ability to orient is important because jellies feed by swimming to the surface and sinking slowly through the rich surface waters. Like hydrozoans, jellies are particularly important in the open ocean, where they provide habitat for a variety of other organisms ranging from juvenile fish to worms and crabs.

Other types of macroscopic zooplankton include tunicates and salps (barrel-shaped, often colonial organisms), ctenophores (comb jellies or sea gooseberries), and chaetognaths (arrowworms).

EXAMPLES OF SWIMMING ORGANISMS Swimming, or *nektonic*, organisms include invertebrate squids, fish, sea turtles, and marine mammals.

Swimming squid include the common squid (genus *Loligo*), flying squid (*Ommastrephes*), and giant squid (*Architeuthis*), all of which are active predators of fish. Most squid possess long, slender bodies with paired fins (**Figure 14.8**) and must

Squid move by trapping water in their mantle cavity between their soft body and penlike shell. They then jettison the water through their siphon for rapid propulsion, as this flying squid (Ommastrephes) does to become airborne.

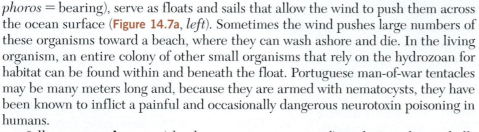

Water jet

Siphon

Mantle

Figure 14.8 Squid motility.

[5]Recall from Chapter 12 that jellies are considered plankton (floaters) and not nekton (swimmers) because they can't control where they move.

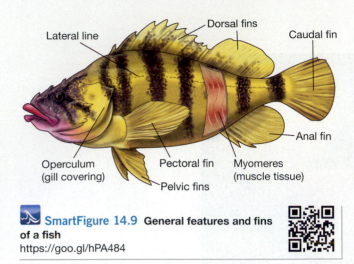

SmartFigure 14.9 **General features and fins of a fish**
https://goo.gl/hPA484

remain active to stay afloat. However, a few species, such as *Sepia* and *Spirula* (see Figure 14.1), have a hollow gas-filled chamber in their bodies to help them remain afloat and so they can be less active.

Squid can swim about as fast as any fish their size and do so by drawing water into their body cavity and then expelling the water out through their siphon, propelling themselves backward. To capture prey, squid use two long arms with pads containing suction cups at the ends (Figure 14.8). Eight shorter arms with suckers convey the prey to the mouth, where it is crushed by a mouthpiece that resembles a parrot's beak.

Most fish are good swimmers, too. The general features and fins of fish are shown in **Figure 14.9**. A fish's lateral line contains a network of sensors that detects changes in water pressure and allows the fish to monitor vibrations in the water around them; the myomeres (*myo* = muscle, *merous* = parted) are packages of muscle tissue attached to the vertebrae that the fish uses for propulsion through the water.

Fish use a variety of fins for swimming (**Figure 14.10**). Most commonly, movement is achieved as a fish alternately contracts and relaxes its myomeres on either side of its body, generating a curvature that travels the length of its body, ending in the tail or *caudal* (*cauda* = tail) *fin* and generating thrust. This type of swimming motion is called *thunniform* (*thunnus* = tuna, *form* = shape). In some cases, fish sacrifice speed for maneuverability, such as in crowded reef environments, or for camouflage by moving only their small transparent fins. As a result, other types of swimming motions involve using various sets of fins as shown in Figure 14.10, such as *thunniform* (*thunnus* = tuna, *form* = shape), *amiiform* (*amia* = a kind of fish, *form* = shape), *labriform* (*labri* = a kind of fish, *form* = shape), and *ostraciform* (*ostracia* = a kind of fish, *form* = shape).

TYPES OF FINS AND THEIR USAGE IN FISH Most active swimming fish use two sets of paired fins—*pelvic fins* and *pectoral* (*pectoralis* = breast) *fins*—to turn, brake, and balance (Figure 14.9). When not in use, these fins can be folded against the body. Vertical fins, both *dorsal* (*dorsum* = back) and *anal*, serve primarily as stabilizers.

The fin that is most commonly used to propel high-speed fish is the *caudal fin*. Caudal fins flare vertically to increase the surface area available to develop thrust against the water.[6] The increased surface area also increases frictional drag. In comparing the caudal fins of various fish, caudal fins have different efficiencies that are related to their size and shape. For example, the larger the size of the caudal fin, the more thrust it generally provides. Caudal fin shapes are organized into one of the five basic forms illustrated in

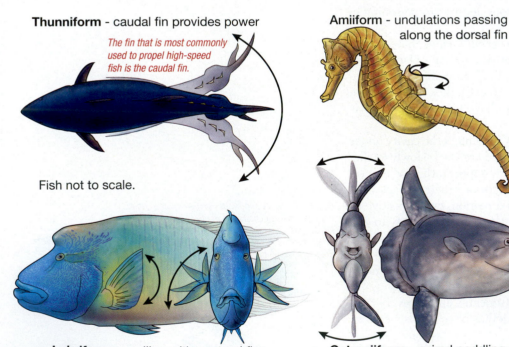

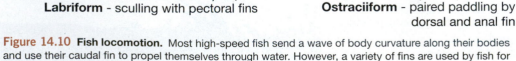

Figure 14.10 Fish locomotion. Most high-speed fish send a wave of body curvature along their bodies and use their caudal fin to propel themselves through water. However, a variety of fins are used by fish for locomotion.

[6]This is similar to humans donning swimming fins on their feet to enable them to swim more efficiently.

Figure 14.11 and keyed by letter to the following descriptions:

a. *Rounded fins* are flexible and useful in accelerating and maneuvering at slow speeds.

b. and **c.** *Truncate fins* (b) and *forked fins* (c) are found on faster fish; the fins are somewhat flexible for better propulsion but are also used for maneuvering.

d. *Lunate fins* are found on fast-cruising fish such as tuna, marlin, and swordfish; the fins are very rigid and useless for maneuverability but very efficient for propulsion.

e. *Heterocercal fins* (*hetero* = uneven, *cercal* = tail) are asymmetrical, with most of their mass and surface area in the upper lobe. Sharks have a heterocercal fin, which produces significant lift and is necessary because sharks have no swim bladder and tend to sink when they stop moving. In fact, the fundamental body shape of sharks has many adaptations to compensate for the shark's negative buoyancy. The pectoral (chest) fins, for example, are large and flat and positioned so they function like airplane wings to lift the front of the shark's body, balancing the rear lift supplied by its heterocercal caudal fin. Although the shark gains tremendous lift from this adaptation of its pectoral fins, it sacrifices maneuverability. This is why sharks tend to swim in broad circles—like a circling airplane—and do not make sharp turns while swimming.

Interdisciplinary

Relationship

(a) Rounded caudal fin on a blue-face angel (other examples: sculpin, flounder).

(b) Truncate caudal fin on a gray angelfish (other examples: salmon, bass).

(c) Forked caudal fin on a yellowfin doctorfish (other examples: herring, goatfish).

(d) Lunate caudal fin on a striped marlin (other examples: bluefish, tuna).

(e) Heterocercal caudal fin on a gray reef shark (other examples: many other types of sharks).

Figure 14.11 Caudal fin shapes. Photos of various fish displaying different caudal fin shapes, including **(a)** rounded, **(b)** truncate, **(c)** forked, **(d)** lunate, and **(e)** heterocercal.

CONCEPT CHECK 14.1 | Compare the various methods by which marine organisms are able to avoid sinking through the water column.

1 Discuss why the rigid gas chamber in cephalopods limits the depth to which they can descend. Why do fish with a swim bladder not have this limitation?

2 Explain why hydrozoans and scyphozoans (jellies) are classified as macroscopic forms of zooplankton. Why are they not considered nekton?

3 From memory, name and describe the different types of fins that fish exhibit. What are the five basic shapes of caudal fins, and what are their uses?

RECAP

Fish use their fins for swimming and staying afloat within the water column. The fin that provides the most thrust is the caudal (tail) fin, which can have a variety of shapes, depending on the lifestyle of the fish.

14.2 What Adaptations Do Pelagic Organisms Possess for Seeking Prey?

Pelagic organisms possess several adaptations that enhance their ability to seek and capture food. These adaptations include mobility (lunging versus cruising), swimming speed, body temperature, and unique circulatory systems. In addition, deep-water nekton have some unique adaptations to help them succeed in capturing prey in their world of darkness.

Mobility: Lungers versus Cruisers

Some fish wait patiently for prey and exert themselves only in short bursts as they lunge at the prey. Others cruise relentlessly through the water, seeking prey. A marked difference occurs in the musculature of fish that use these different styles of obtaining food.

For example, **lungers** (such as the grouper shown in **Figure 14.12a**) sit and wait for prey to come close by. Lungers have truncate caudal fins for speed and maneuverability, and almost all their muscle tissue is white.

On the other hand, **cruisers** (such as the tuna shown in **Figure 14.12b**) actively seek prey. Less than half of a cruiser's muscle tissue is white; most is red.

What is the significance of red versus white muscle tissue? Red muscle tissue contains fibers that are 25 to 50 microns in diameter (0.01 to 0.02 inch), whereas the fibers in white muscle tissue are 135 microns (0.05 inch) and contain much lower concentrations of **myoglobin** (*myo* = muscle, *globus* = sphere), a red pigment with an affinity for oxygen. Moreover, red muscle tissue supplies a much greater amount of oxygen and supports a much higher metabolic rate than white tissue. This is why cruisers have so much red muscle tissue: It allows them to have the endurance needed to support their active lifestyle.

Lungers, on the other hand, need very little red tissue because they do not move continually. Instead, they need white tissue, which fatigues much more rapidly than red tissue, for quick bursts of speed to capture prey. Cruisers use white tissue, too, for short periods of acceleration while on the attack.

Interdisciplinary

Relationship

Figure 14.12 Feeding styles of lungers and cruisers. Photos of **(a)** a typical lunger and **(b)** cruisers.

(a) Lungers, such as this tiger grouper, sit patiently on the bottom and capture prey with quick, short lunges.

(b) Cruisers, such as these yellowfin tuna, swim constantly in search of prey and capture it with short bursts of high-speed swimming.

Swimming Speed

Although rapid swimming consumes much energy, it can help organisms capture prey. Fish normally swim slowly when cruising, fast when hunting for prey, and fastest of all when trying to escape from predators.

Generally, when comparing fish of similar shapes, the larger the fish, the faster it can swim. For tuna, well adapted for sustained cruising and short, high-speed bursts, cruising speed averages about 3 body lengths per second. They can, however, maintain a maximum speed of about 10 body lengths per second, but only for about 1 second. Remarkably, yellowfin tuna (*Thunnus albacares*) have been clocked at 74.6 kilometers (46 miles) per hour! Even though this speed is more than 20 body lengths per second, the tuna could maintain it for only a *fraction* of a second. Theoretically, a 4-meter (13-foot) bluefin tuna (*Thunnus thynnus*) can reach speeds up to about 144 kilometers (90 miles) per hour.[7]

Like fish, many of the toothed whales are fast swimmers, too. For instance, spotted dolphins of the genus *Stenella* have been clocked at 40 kilometers (25 miles) per hour, and killer whales may exceed 55 kilometers (34 miles) per hour during short bursts.

SWIMMING SPEED IN COLD-BLOODED VERSUS WARM-BLOODED FISH The temperature of a fish relative to its environment also affects the speed at which a fish swims. For example, most marine fish are **cold-blooded**, or **poikilothermic** (*poikilos* = varied, *theromos* = heat), so their body temperatures are nearly the same as their environment. Usually, these fish are not fast swimmers. The mackerel (*Scomber*), yellowtail (*Seriola*), and bonito (*Sarda*), on the other hand, are indeed fast swimmers. They also have body temperatures that are 1.3, 1.4, and 1.8°C (2.3, 2.5, and 3.2°F), respectively, higher than the surrounding seawater. Remarkably, mackerel sharks (genera *Lamna* and *Isurus*), tuna (genera *Thunnus*), and the moonfish opah (*Lampris guttatus*) have body temperatures much higher than their environment. Bluefin tuna, for example, can maintain a body temperature of 30 to 32°C (86 to 90°F) regardless of the water temperature, which is characteristic of organisms that are **warm-blooded** or **homeothermic** (*homeo* = alike, *thermos* = heat). As bluefin tuna swim, they use a heat exchange system around swimming muscles along the midsection to generate heat and keep its body warm. Although these tuna are more commonly found in warmer water, where the temperature difference between fish and water is no more than 5°C (9°F), body temperatures of 30°C (86°F) have been measured in bluefin tuna swimming in 7°C (45°F) water. The moonfish opah, which in 2015 was discovered to have a complex heat exchange system in its gills that helps keep its blood and heart warm, can maintain a core body temperature of 3 to 6°C (6 to 11°F) higher than the surrounding water.

Interdisciplinary

Relationship

Why do these fish exert so much energy to maintain their high body temperatures, which is a costly adaptation that is hard to maintain in the ocean, while other fish do quite well with lower body temperatures? Scientific studies suggest that high body temperature in fish is associated with higher metabolic rates, which increases the power output of their muscle tissue and allows them to more effectively seek and capture prey. In addition, having a higher temperature speeds up physiological process within the body, resulting in muscles that contract faster and neurological transmissions that occur more quickly, leading to faster swimming speeds, better vision, and enhanced response times, all of which are greatly advantageous for predators in the marine environment.

[7]Imagine how difficult it would be to clock a bluefin tuna in the ocean at this speed!

SOME MYTHS (AND FACTS) ABOUT SHARKS

"Now we know that almost every attack on a human is an accident: the shark mistakes the human for its normal prey."

—*Peter Benchley (2000), author of* Jaws

Sharks (**Figure 14A**) are the fish humans most fear. Their strength, large size, sharp teeth, and unpredictable nature are enough to keep some people from *ever* entering the ocean. The media hype generated by occasional shark attacks on humans has led to many myths about sharks, including the following:

- *Myth #1: All sharks are dangerous.* Of the world's roughly 400 shark species, 80% are unable to hurt people or rarely encounter people. The shark species that are responsible for most attacks on humans are great white (*Carcharodon carcharias*), tiger (*Galeocerdo cuvier*), and bull sharks (*Carcharhinus leucas*). The world's largest shark is the whale shark (*Rhincodon typus*), which reaches lengths of up to 15 meters (50 feet) but is a filter-feeder that eats only tiny plankton and so is not considered dangerous.

- *Myth #2: Sharks are voracious eaters that must eat continuously.* Like other large animals, sharks eat periodically, depending on their metabolism and the availability of food. Humans are not a primary food source of any shark, and many large sharks prefer the higher fat content of seals and sea lions.

- *Myth #3: Most people attacked by sharks are killed.* Of every 100 people attacked by

sharks, 85 survive. Many large sharks commonly attack by biting their prey to immobilize it before trying to eat it. Consequently, many potential prey escape and survive. The most common targets of shark attacks are surfers (49%), swimmers/waders (29%), divers/snorkelers (15%), and kayakers (6%). Most shark encounters with humans appear to be investigative, not predatory.

- *Myth #4: Many people are killed by sharks each year.* The chances of a person being killed by a shark are quite low (**Table 14A**). Over the past several decades, sharks have killed an average of only 5 to 15 people worldwide each year. Humans, on the other hand, kill as many as 100 million sharks a year, mostly as bycatch from fishing activities. Compared to many other fish, sharks have low reproduction rates and grow slowly, which may result in many sharks being designated as endangered species in the future.

- *Myth #5: The great white shark is a common, abundant species found off most beaches.* Great white sharks are relatively uncommon predators that prefer cooler waters. At most beaches, great whites are rarely encountered.

- *Myth #6: Sharks are not found in freshwater.* A specialized osmoregulatory system enables some species (such as the bull shark) to cope with dramatic changes in salinity, from the high salinity of seawater to the low salinity of freshwater rivers and lakes. For example, bull sharks have been seen as far up the Mississippi River as Illinois and also in Lake Michigan.

- *Myth #7: All sharks need to swim constantly.* Some sharks can remain at rest for long periods on the bottom and obtain enough oxygen by opening and closing their mouths to pump water across their gills. Typically, sharks swim very slowly—cruising speeds are less than 9 kilometers (6 miles) per hour—but they can swim at bursts of over 37 kilometers (23 miles) per hour.

- *Myth #8: Sharks have poor vision.* The lens of a shark's eye is up to seven times more powerful than that of a human's. Sharks can even distinguish color.

- *Myth #9: Eating shark meat makes one aggressive.* There is no indication that eating shark meat will alter a person's temperament. The firm texture, white flesh, low fat content, and mild taste of shark meat have made it a favorite seafood in many countries.

- *Myth #10: No one would ever want to enter water filled with sharks.* Long regarded with fear and suspicion, sharks have more recently been viewed as skilled, highly evolved predators that are vitally important to the health of ocean ecosystems. Diving tours specializing in close encounters with sharks are becoming increasingly popular.

TABLE 14A	TYPES AND NUMBERS OF OCCURRENCES IN THE UNITED STATES
Occurrence to people in the United States	**Average number per year**
Transportation fatalities	42,000
Death involving slipping, tripping, or stumbling	565
Struck by lightning	352
Killed by lightning	50
Bitten by a squirrel in New York City	88
Killed by a dog bite	26
Killed by a snake bite	12
Bitten by a shark	10
Killed by a shark	0.4

"Today I could not, for instance, portray the shark as a villain, especially not as a mindless omnivore that attacks boats and humans with reckless abandon. No, the shark in an updated Jaws *could not be the villain; it would have to be written as the victim, for, worldwide, sharks are much more the oppressed than the oppressors."*

—*Peter Benchley (2005), author of* Jaws

GIVE IT SOME THOUGHT

1. In which freshwater locations have sharks been sighted?

Figure 14A Great white shark (*Carcharodon carcharias*).

Adaptations of Deep-Water Nekton

Living below the surface water but still above the ocean floor are deep-water nektonic species—mostly various species of fish—that are specially adapted to the deep-water environment, where it is very still and completely dark. Their food source is either **detritus** (*detritus* = to lessen)—dead and decaying organic matter including waste products that slowly sink downward from surface waters—or each other. The lack of abundant food limits the *number* of organisms (total biomass) and the *size* of these organisms. As a result, small populations of these organisms exist, and most individuals are less than 30 centimeters (1 foot) long. Many have low metabolic rates to conserve energy, too.

These **deep-sea fish** (Figure 14.13) have special adaptations to efficiently find and collect food. They have good sensory devices, for example, such as long antennae or sensitive lateral lines that can detect movement of other organisms within the water column.

Many marine organisms such as deep-water species of shrimp and squid can **bioluminesce** (*bios* = life, *lumen* = light, *esc* = becoming), which means they can produce light biologically and "glow in the dark." Only a tiny percentage of terrestrial life has bioluminescent capabilities—fireflies and glowworms most famously, but also some millipedes, click beetles, fungus gnats, jack-o-lantern mushrooms, and a few others. In the ocean, however, scientists estimate that 90% of deep-sea marine life is bioluminescent. The vast majority of bioluminescent organisms use light-producing organs called **photophores** (*photo* = light, *phoros* = bearing), which can be simple luminous spots or may be quite complex and equipped with lenses, shutters, color filters, and reflectors.

Bioluminescent light is produced from compounds during the digestion of prey, from specialized cells in the organism, or associated with symbiotic bacteria that is cultured and lives inside the organism. The light is produced when molecules of the biological pigment *luciferin* (*lucifer* = light bringing) are excited and emit photons of light in the presence of oxygen.[8] The light-production process is remarkably efficient: only a 1% loss of energy is required to produce this illumination. Some marine animals control light production in unique ways. For example, scientific research suggests that some sharks control their bioluminescence by using hormones.

Interdisciplinary

Relationship

In a world of darkness, the ability to bioluminesce is useful for a variety of purposes, including:

- Searching for food items in the dark
- Attracting prey (for example, the female deep-sea anglerfish shown in Figures 14.13f and 14.13g use their specially modified dorsal fin as a bioluminescent lure)
- Staking out territory by constantly patrolling an area
- Communicating or seeking a mate by sending signals
- Escaping from predators by using a flash of light to temporarily blind or distract them
- Avoiding predators by use of a "burglar alarm" by attracting unwanted attention with brilliant displays of bioluminescence
- Camouflaging by using belly lights to match the color and intensity of dim filtered sunlight from above and obliterate a telltale shadow to become effectively invisible; this is known as **counterillumination**

To take advantage of bioluminescent light, many deep-sea fish species have large and sensitive eyes—perhaps 100 times more sensitive to light than

[8]This is a bit like the chemical reaction that occurs when you snap a glow stick.

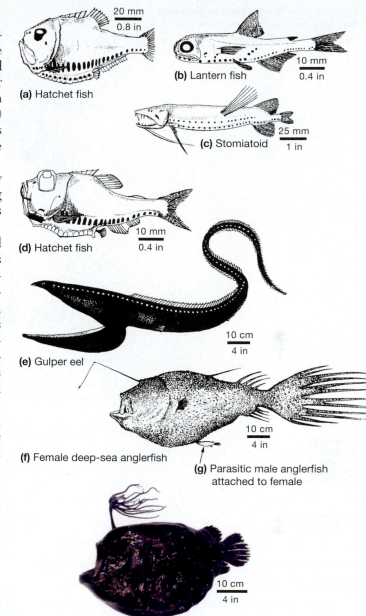

(a) Hatchet fish — 20 mm / 0.8 in

(b) Lantern fish — 10 mm / 0.4 in

(c) Stomiatoid — 25 mm / 1 in

(d) Hatchet fish — 10 mm / 0.4 in

(e) Gulper eel — 10 cm / 4 in

(f) Female deep-sea anglerfish — 10 cm / 4 in

(g) Parasitic male anglerfish attached to female

(h) Photo of female deep-sea anglerfish (*Himantolophus danae*) that washed ashore in Carlsbad, California, in 2001. — 10 cm / 4 in

Figure 14.13 Deep-sea fish. Line drawings and a photo (part *h*) of various deep-sea fish.

Figure 14.14 Adaptations of deep-sea fish. Line drawings showing various adaptions of deep-sea fish.

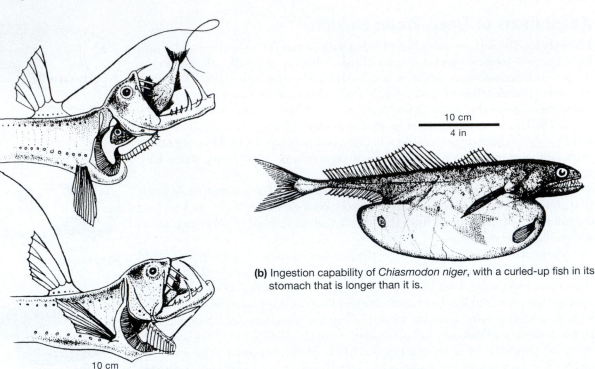

(b) Ingestion capability of *Chiasmodon niger*, with a curled-up fish in its stomach that is longer than it is.

(a) Large teeth, hinged jaw, and swallowing mechanism of the deep-sea viper fish *Chauliodus sloani*.

Web Video
Bioluminescent Organisms
https://goo.gl/Y5s76Y

RECAP

Adaptations of pelagic organisms for seeking prey include mobility (lunging versus cruising), high swimming speed, and high body temperature. Deep-water nekton exhibit a host of unusual adaptations—including bioluminescence—that allow them to survive in deeper waters.

14.1 Squidtoons

Life is a like a game of tag, except I'm never it.

https://goo.gl/w8c9a9

human eyes—that enable them to see potential prey. To avoid being prey, most species are dark in color so that they blend into the environment. Still other species are blind and rely on senses such as smell to track down prey.

Other adaptations that various deep-sea fish species possess include large sharp teeth, expandable bodies to accommodate large food items, hinged jaws that can unlock to open widely, and mouths that are huge in proportion to their bodies (**Figure 14.14**). These adaptations allow deep-sea fish to ingest species that are larger than they are and to process food efficiently whenever it is captured.

CONCEPT CHECK 14.2 | Specify adaptations that pelagic organisms possess for seeking prey.

1 Are most fast-swimming fish cold-blooded or warm-blooded? What advantage does this provide?

2 What are the two food sources of deep-water nekton? List several adaptations of deep-water nekton that allow them to survive in their environment.

3 Describe the mechanism by which bioluminescence is accomplished in deep-sea organisms. What is bioluminescence useful for in the marine environment?

14.3 What Adaptations Do Pelagic Organisms Possess to Avoid Being Prey?

Many animals have unique adaptations to avoid being captured and eaten. Examples of adaptations that organisms use to enhance their survival include schooling and symbiosis.

Schooling

The term **school** refers to large numbers of fish, squid, or shrimp that form well-defined social groupings. Although vast populations of phytoplankton and zooplankton may be highly concentrated in certain areas of the ocean, they are not usually referred to as schools.

The number of individuals in a school can vary from a few larger predaceous fish (such as bluefin tuna) to hundreds of thousands of small filter feeders (such as anchovies). Within the school, individuals move in the same direction and are evenly spaced. Spacing is probably maintained through visual contact and, in the case of fish, by use of the lateral line system (see Figure 14.9) that detects vibrations of swimming neighbors. The school can turn abruptly in spectacular fashion or even reverse direction as individuals at the head or rear of the school assume leadership positions (**Figure 14.15**). Research suggests that each fish decides where to move not based on the behavior of its nearest neighbors, as is often assumed, but on a synthesis of where all the fish in its field of view are headed.

What are the advantages of schooling? One advantage is that during spawning, schooling ensures that there will be males to release sperm to fertilize the eggs released into the water or deposited on the bottom by females. Another advantage is that schools of smaller fish can invade the territory of larger aggressive species and feed there because the "owner" of the territory can never chase away the whole school. The most important function of schooling in small fish, however, is protection from predators.

It may seem illogical that schooling would be protective. For instance, schooling creates tighter groupings of organisms so that any predator lunging into a school would surely catch something, just as land predators run a herd of grazing animals until one weakens and becomes a meal. So, aren't the smaller fish making it easier for the predators by forming a large target? Scientists who study fish behavior suggest that schooling does indeed serve to protect a group of organisms based on strategies that give them "safety in numbers," much like flocks of birds. In many parts of the marine environment, such as the open ocean where there is no place to hide, schooling has the following advantages:

1. When members of a species form schools, they reduce the percentage of ocean volume in which a cruising predator might find one of their kind.

2. When a predator encounters a large school, it is less likely that every fish in the school is consumed, compared to encounters with an individual or even a small school.

3. The school may appear as a single large and dangerous opponent to a potential predator and prevent some attacks.

4. Predators may find the continually changing position and direction of movement of fish within the school confusing, making attack particularly difficult for predators, which can attack only one fish at a time.

In addition, the fact that more than half of all fish species join schools during at least a portion of their lives suggests that schooling enhances survival of species, especially for those with no other means of defense. Schooling may also help fish swim greater distances than individuals because each schooling fish gets a boost from the vortex created by the fish swimming in front of it.

Recently, a new ocean predator has developed a method to take advantage of the schooling behavior of many species of fish. Human fishers have developed nets large enough to encircle whole schools of fish. These nets are very efficient at catching fish, thereby leading to the decline of many fish stocks (see Section 13.5 in Chapter 13).

Figure 14.15 Schooling. A school of blue-lined snapper near a reef in the Maldives, Indian Ocean. Schooling increases chances of survival, and more than half of all fish species are known to join schools during at least a portion of their lives.

(a) Commensalism occurs when an organism benefits without harming its host, such as these remoras attached to a lemon shark.

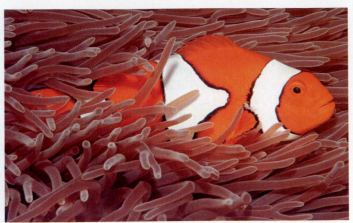

(b) Mutualism occurs when both participants benefit, such as this clown anemonefish and sea anemone.

(c) Parasitism occurs when one participant benefits at the expense of the other, such as this isopod that has attached itself to the head of a whitetip soldierfish.

Figure 14.16 Types of symbiosis. The three main types of symbiosis are **(a)** commensalism, **(b)** mutualism, and **(c)** parasitism.

STUDENTS SOMETIMES ASK . . .

What are the world's largest and smallest fishes?

The world's largest fish is the whale shark (*Rhincodon typus*), which reaches lengths of up to 15 meters (50 feet) and can weigh up to 13.6 metric tons (30,000 pounds). Its mouth is an enormous 1.5 meters (5 feet) wide. It is a slow-moving, wide-ranging, filter-feeding animal that exists almost entirely on plankton. Alternatively, the smallest known fish is *Paedocypris progenetica*, a relative of carp and minnows. Its adult length is only 7.9 millimeters (0.31 inch), which is about as thick as a pencil. It lives exclusively in Indonesian swamps and, remarkably, was only discovered in 2005. There is, however, one other fish that is smaller: the male deep-sea anglerfish *Photocorynus spiniceps*, which has an adult length of only 6.2 millimeters (0.24 inch). It is generally not considered the world's smallest fish because it is not a self-sustaining organism. Instead, the parasitic males bite into larger females and fuse for life (see Figure 14.13g).

Symbiosis

Many marine organisms seek relationships with other organisms to help them survive. One such relationship is **symbiosis** (*sym* = together, *bios* = life), which occurs when two or more organisms associate in a way that benefits at least one of them. There are three main types of symbiotic relationships: commensalism, mutualism, and parasitism.

In **commensalism** (*commensal* = sharing a meal, *ism* = process), a smaller or less dominant participant benefits without harming its host, which affords subsistence or protection to the other. A remora, for example, attaches itself to a shark or another fish to obtain food and transportation, generally without harming its host (**Figure 14.16a**).

In **mutualism** (*mutuus* = borrowed, *ism* = process), both participants benefit. For example, the stinging tentacles of the sea anemone protect the clown fish (**Figure 14.16b**), and the clown fish, which is small but aggressive, chases away any fish that tries to feed on the anemone itself. In addition, the clown fish helps clean the anemone and may even supply scraps of food. Remarkably, clown fish are not stung by the anemone because clown fish have a protective agent in the mucus that coats their bodies.

In **parasitism** (*parasitos* = a person who eats at someone else's table, *ism* = process), one participant (the parasite) benefits at the expense of the other (the host). Many fish are hosts to isopods, which attach to the fish and derive their nutrition from the body fluids of the fish, thereby robbing the host of some of its energy supply (**Figure 14.16c**). Usually, the parasite does not rob enough energy to kill the host because if the host dies, so does the parasite.

Recently, symbiosis had been discovered to be an important component that drives evolution. For example, the sequencing of the genome of a species of diatom reveals that it apparently acquired new genes by engulfing microbial neighbors. The research suggests that early on in the evolution of diatoms, the most significant acquisition was an algal cell that provided the diatom with photosynthetic machinery.

Other Adaptations

Marine animals exhibit a variety of behaviors that serve as defensive mechanisms to help them ward off predators—or to be more successful predators themselves. These include using speed, secreting poisons, and mimicking other poisonous or

distasteful species. Others use transparency, camouflage, or countershading, as discussed in Chapter 12.

14.4 What Characteristics Do Marine Mammals Possess?

Marine mammals include some of the largest, best known, and most charismatic animals in the sea, such as seals, sea lions, manatees, porpoises, dolphins, and whales.

Although all marine mammals have an aquatic existence, their ancestors were land animals. For example, a series of striking fossil discoveries of ancient whales in Pakistan, India, and Egypt provide strong evidence that whales evolved from mammals on land about 50 million years ago. Some whale ancestors had small, unusable hind legs, suggesting that the land mammal predecessor had no need for its hind legs when it developed a large paddle-shaped structure for a tail used to swim through water. Other fossils show a remarkable progression of skeletal adaptations for an increasingly aquatic existence, such as the migration of the blowhole (nostrils) toward the top of the head, upper vertebrae that become increasingly fused, shrinking hip and ankle structures, and jawbone and ear components adapted for underwater hearing. Additional lines of evidence—including DNA analysis of modern whales and a host of anatomical similarities between land and marine mammals—confirm the evolution of whales from a hippopotamus-like land-dwelling ancestor.

The geologic record shows that life on land evolved from marine organisms millions of years ago. Why would a land mammal migrate back to the sea? One hypothesis suggests that they may have returned to the sea because of more abundant food sources. Another is that the extinction of many large marine predators that occurred at the same time as the demise of the dinosaurs allowed mammals to expand into a new environment: the sea. Interestingly, recent research suggests that the rise of diatoms as dominant marine primary producers and global temperature change were key factors that influenced the evolution of modern whales.

Interdisciplinary Relationship

Mammalian Characteristics

All organisms in class Mammalia (including marine mammals) share the following characteristics:

- They are warm-blooded.
- They breathe air.
- They have hair (or fur) during at least some stage of their development.
- They bear live young.[9]
- The females of each species have mammary glands that produce milk for their young.

Marine mammals include at least 117 species within the orders Carnivora, Sirenia, and Cetacea. The major groups of marine mammals are shown in **Figure 14.17** and described below.

9This is true except for a few egg-laying monotreme mammals of Australia from the subclass Prototheria, which includes the duck-billed platypus and the spiny anteater (echidna).

STUDENTS SOMETIMES ASK . . .

What is the fat content of the milk of marine mammals?

Marine mammals produce milk with some of the highest fat content in the animal kingdom. For example, the milk of most dolphins contains about 14% fat (by comparison, whole cow milk contains 4% fat and human breastmilk contains 4.5% fat). Polar bears have milk with 31% fat and baleen whales have milk with 35–41% fat. Pinnipeds have even a higher percentage of milkfat. In fact, the gray seal (*Halichoerus grypus*) produces milk with 53% fat content and the world record for milkfat by animals is the hooded seal (*Cystophora cristata*), which produces milk with 61% fat! The high fat content in milk helps baby marine animals quickly build up a layer of blubber that insulates them from cold water.

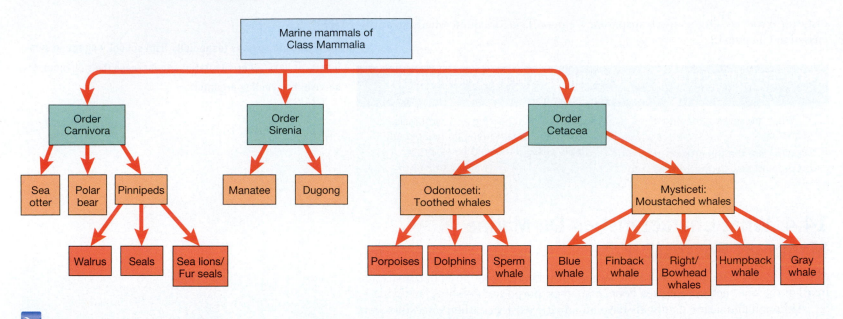

 SmartFigure 14.17 **Major groups of marine mammals.** Organizational chart showing the taxonomic relationships of the various groups of marine mammals, including representative examples. https://goo.gl/8VCFwE

14.2 Squidtoons

NO, THEY'RE NOT THE SAME ANIMALS.

https://goo.gl/w8c9a9

Order Carnivora

All animals within order **Carnivora** (*carni* = meat, *vora* = eat)—such as the familiar cat and dog families on land—have prominent canine teeth. Marine representatives of order Carnivora include sea otters, polar bears, and the **pinnipeds** (*pinna* = feather, *ped* = foot), which include walruses, seals, sea lions, and fur seals. The name *pinniped* describes these organisms' prominent skin-covered flippers, which are well adapted for propelling them through water.

Sea otters (Figure 14.18a) inhabit kelp beds in coastal waters of the eastern North Pacific Ocean. They are some of the smallest marine mammals, with adults reaching lengths up to 1.2 meters (4 feet). Sea otters lack an insulating layer of blubber but have extremely dense fur, which is extremely luxurious and was highly sought for pelts; as a result, they were hunted to the brink of extinction in the late 1800s. Fortunately, they have made a remarkable comeback and now inhabit most areas where they were formerly hunted. Sea otters seem particularly playful because of their habit of continually scratching themselves, which serves to clean their fur and adds an insulating layer of air. Because they lack the insulative benefit of blubber, they have high caloric requirements and are voracious eaters.

Sea otters eat more than 50 kinds of marine life, including sea urchins, crabs, lobsters, sea stars, abalone, clams, mussels, octopuses and fish. They are one of the few types of animals known to use tools. During dives for prey items, they keep a tool—usually a rock—tucked under one arm while they use their dexterous hands to obtain food. When they return to the surface, they use the tool to break open the shells of their food while floating on their backs.

Polar bears (Figure 14.18b) are a type of marine mammal with massive webbed paws that make them excellent swimmers. The polar bear's fur is thick, and each hair is hollow to trap air for better insulation; the hairs also function like fiber optic cables by channeling sunlight to the animal's dark/black skin that absorbs heat from sunlight. Polar bears also have large teeth and sharp claws, which they use for prying and killing. Their diet consists mainly of seals, which they often capture at holes in the Arctic ice when the seals come up for a breath of air. The topic of shrinking Arctic sea ice and its effect on polar bear populations is discussed in Chapter 16, "The Oceans and Climate Change."

Climate Connection

Walruses have large bodies, and adults—both male and female—have ivory tusks up to 1 meter (3 feet) long (**Figure 14.18c**). Their tusks are used for territorial fighting, for hauling themselves onto icebergs, and sometimes for stabbing their prey.

Seals—also called the *earless seals* or *true seals*—differ from the **sea lions** and **fur seals**—also called the *eared seals*—in the following ways:

- Seals lack prominent ear flaps that are specific to sea lions and fur seals. (Look closely and compare **Figures 14.18d** and **14.18e.**)

- Seals have smaller and less-prominent front flippers (called *fore flippers*) than sea lions and fur seals.

- Seals have prominent claws that extend from their fore flippers that sea lions and fur seals lack (**Figure 14.19**).

- Seals have a different hip structure than sea lions and fur seals. As a result, seals cannot move their rear flippers underneath their bodies in the way that sea lions and fur seals do (Figure 14.19).

- Seals, with their smaller front flippers and different hip structure, do not move around on land very well and can only slither along like caterpillars. Sea lions and fur seals, on the other hand, use their large front flippers and their rear flippers, which they can turn under their bodies, to walk easily on land and can even ascend steep slopes, climb stairs, and do other acrobatic tricks.

- Seals propel themselves through the water by using a back-and-forth motion of their rear flippers (similar to a wagging tail), whereas sea lions and fur seals flap their large front flippers.

(a) Sea otter (*Enhydra lutris*).

(b) Polar bear (*Ursus maritimus*).

(c) Walrus (*Odobenus rosmarus*).

(d) Harbor seals (*Phoca vitulina*).

(e) California sea lions (*Zalophus californianus*).

Figure 14.18 Marine mammals of order Carnivora. Photos of various types of marine mammals of order Carnivora, including **(a)** sea otter and **(b)** polar bear. Pinnipeds include **(c)** walrus, **(d)** harbor seals, and **(e)** California sea lions.

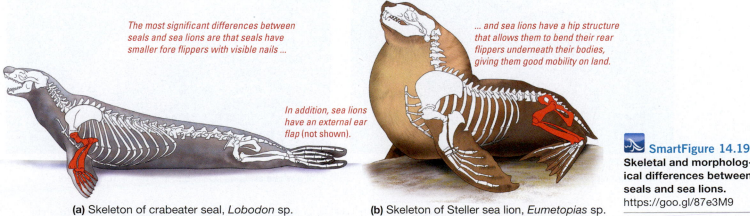

The most significant differences between seals and sea lions are that seals have smaller fore flippers with visible nails ...

... and sea lions have a hip structure that allows them to bend their rear flippers underneath their bodies, giving them good mobility on land.

In addition, sea lions have an external ear flap (not shown).

(a) Skeleton of crabeater seal, *Lobodon* sp.

(b) Skeleton of Steller sea lion, *Eumetopias* sp.

SmartFigure 14.19
Skeletal and morphological differences between seals and sea lions.
https://goo.gl/87e3M9

Order Sirenia

Animals of order **Sirenia** (*siren* = a mythical mermaidlike creature with an enticing voice) include the *manatees* and *dugongs*, collectively known as "sea cows."[10] Manatees are concentrated in coastal areas of the tropical Atlantic Ocean, while dugongs populate the tropical regions of the Indian and Western Pacific Oceans.

Both manatees and dugongs have a paddlelike tail and rounded front flippers (**Figure 14.20**). Their bodies are covered with sparse hairs, which are concentrated around the mouth. They are large animals that can reach lengths of up to 4.3 meters (14 feet) and weigh more than 1360 kilograms (3000 pounds). The land-dwelling ancestors of sirenians were elephantlike; in fact, the front flippers of manatees have prominent nails that bear a striking resemblance to the nails on elephant feet.

Sirenians eat only shallow-water coastal grasses and are thus the only vegetarian marine mammals. They spend most of their lives in coastal waters that are heavily used by humans for commerce, recreation, development, and waste disposal, and so a major concern for sirenian survival is habitat destruction. A scientific study of biologically sensitive seagrass habitat—which is vital for sirenians—indicates that it is being destroyed so rapidly that it is one of the most threatened ecosystems on Earth. The rate of seagrass loss, in fact, is similar to that of other endangered ecosystems, such as mangroves, coral reefs, and rainforests.

In addition, there have been many accidents with motorboats that run over these slow-moving animals. In 2002, for example, the Florida Fish and Wildlife Commission reported the deaths of 305 Florida manatees, of which 31% were attributed to boat collisions. Scientists have recently developed a forward-looking sonar system called a "manatee finder" to help boaters locate and avoid the animals. The continuing decline in populations of manatees and dugongs, however, has led to their being classified as endangered species.

Order Cetacea

The order **Cetacea** (*cetus* = whale) includes the whales, dolphins, and porpoises (**Figure 14.21**). The cetacean body is more or less cigar shaped and insulated with a thick layer of blubber. Cetacean forelimbs are modified into flippers that move only at the "shoulder" joint. The hind limbs are vestigial (rudimentary), not attached to the rest of the skeleton, and are usually not visible externally. All cetaceans share the following characteristics:

- An elongated (telescoped) skull
- Blowholes on top of the skull

Figure 14.20 Marine mammals of order Sirenia. Photos of representative marine mammals of order Sirenia, including **(a)** West Indian manatees and **(b)** Indian Ocean dugong.

(a) West Indian manatees, which have a rounded tail fin and nails on their front flippers.

(b) Indian Ocean dugong, which has a fluked tail fin similar to that of a whale and has no visible nails on its front flippers.

[10]Sea cows also include the cold-water Steller's sea cow (*Hydrodamalis gigas*), which was driven to extinction in 1768 by early whalers only 27 years after its discovery.

- Very few hairs
- A horizontal tail fin called a *fluke* (*flok* = to be flat) that is used for propulsion by vertical movements

These characteristics make cetaceans' bodies very streamlined, allowing them to be excellent swimmers.

MODIFICATIONS TO INCREASE SWIMMING SPEED Cetaceans' muscles are not a great deal more powerful than those of other mammals, so their ability to swim fast must result from modifications that reduce frictional drag. The muscles of a small dolphin, for example, would need to be five times stronger than they are to swim at 40 kilometers (25 miles) per hour in turbulent flow.

In addition to a streamlined body, cetaceans improve the flow of water around their bodies with a specialized skin structure. Their skin consists of a soft outer layer that is 80% water and has narrow canals filled with spongy material, and a stiffer inner layer composed mostly of tough connective tissue. The soft layer decreases the pressure differences at the skin–water interface by compressing when pressure is high and expanding when pressure is low, reducing turbulence and drag.

MODIFICATIONS TO ALLOW DEEP DIVING Humans can free-dive to a maximum depth of 130 meters (428 feet) and hold their breath in rare instances for up to six minutes. In contrast, sperm whales (*Physeter macrocephalus*) dive deeper than 2800 meters (9200 feet), and northern bottlenose whales (*Hyperoodon ampullatus*) can stay submerged for up to two hours. These remarkable feats require unique adaptations, such as special structures that allow them to use oxygen efficiently, muscular adaptations, and an ability to resist nitrogen narcosis, all of which are discussed below.

Oxygen Usage Figure 14.22 shows the internal structures that allow cetaceans to remain submerged for extended periods. Inhaled air finds its way to tiny terminal chambers, the *alveoli* (*alveus* = small hollow). Alveoli are lined with a thin alveolar membrane that is in contact with a dense bed of capillaries. The exchange of gases between the inhaled air and the blood (oxygen in, carbon dioxide out) occurs across the alveolar membrane. Some cetaceans have an exceptionally large concentration of capillaries surrounding the alveoli (Figure 14.22b), which have muscles that move air against the membrane by repeatedly contracting and expanding.

Interdisciplinary

Relationship

Cetaceans take from one to three breaths per minute while resting, compared with about 15 in humans. Because they hold the inhaled breath much longer, and because of the large capillary mass in contact with the alveolar membrane and the circulation of the air by muscular action, cetaceans can extract almost 90% of the oxygen in each breath, whereas terrestrial mammals extract only 4 to 20%.

To use oxygen efficiently during long dives, cetaceans store it and limit its use. The storage of so much oxygen is possible because prolonged divers have such a large blood volume per unit of body mass.

Some cetaceans have twice as many red blood cells per unit of blood volume and up to nine times as much myoglobin in their muscle tissue as terrestrial animals. As a result, large supplies of oxygen can be stored chemically in **hemoglobin** (*hemo* = blood, *globus* = sphere) within red blood cells and in myoglobin within muscles.

Muscular Adaptations Cetaceans' muscles are well adapted for deep dives. One adaptation is that their muscle tissue is relatively insensitive to high levels of carbon dioxide, which builds up in the body through respiration, especially during deep

Figure 14.23 Jawbone of a killer whale. The lower jawbone of a killer whale (*Orcinus orca*), showing large teeth that end in points. Thus, killer whales are members of the dolphin family.

toothed whales, sounds are picked up by the thin, flaring jawbone and passed to the inner ear via the connecting, fat-filled body.[13] The signals are then sent to the brain, where the sounds are interpreted (Figure 14.24).

There is growing concern in the scientific community that noise pollution in the ocean is affecting cetaceans. This increased noise comes from the greater number of ships plying the world's oceans as well as the larger size, higher speeds, and enhanced propulsion power of individual ships. For example, scientific studies show that the world's commercial marine trade fleet has contributed to a doubling of low-frequency ship noise every decade for the past 60 years in some of the most intensely used parts of the ocean. The impact of this increased underwater noise on cetacean hearing, behavior, and communication is unknown.

Interdisciplinary
Relationship

How Intelligent Are Toothed Whales? The question of cetacean intelligence is a topic of much debate. Although there may not be a definitive answer, the following facts about toothed whales imply a certain level of intelligence:

- They can communicate with each other by using sound.
- They have large brains relative to their body size.[14]

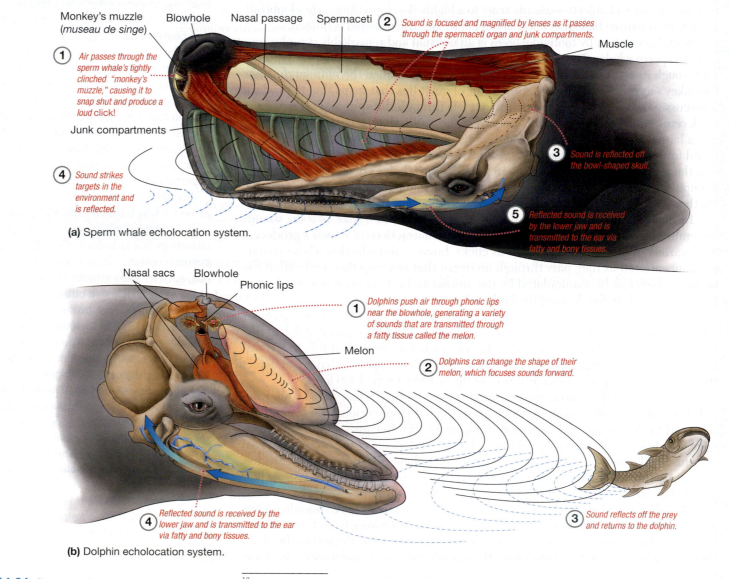

(a) Sperm whale echolocation system.

Monkey's muzzle (*museau de singe*) Blowhole Nasal passage Spermaceti

① Air passes through the sperm whale's tightly clinched "monkey's muzzle," causing it to snap shut and produce a loud click!

② Sound is focused and magnified by lenses as it passes through the spermaceti organ and junk compartments.

Muscle

Junk compartments

③ Sound is reflected off the bowl-shaped skull.

④ Sound strikes targets in the environment and is reflected.

⑤ Reflected sound is received by the lower jaw and is transmitted to the ear via fatty and bony tissues.

(b) Dolphin echolocation system.

Nasal sacs Blowhole Phonic lips

① Dolphins push air through phonic lips near the blowhole, generating a variety of sounds that are transmitted through a fatty tissue called the melon.

Melon

② Dolphins can change the shape of their melon, which focuses sounds forward.

③ Sound reflects off the prey and returns to the dolphin.

④ Reflected sound is received by the lower jaw and is transmitted to the ear via fatty and bony tissues.

SmartFigure 14.24 Cutaway views showing the echolocation system of a sperm whale and a dolphin.
https://goo.gl/0GhVHb

[13]To simulate this, try pushing the end of a vibrating tuning fork into your chin. The sound is transmitted through your jaw directly to your ear.

[14]In fact, sperm whales have the largest brain of any animal on the planet—it has been reported to weigh up to 9 kilograms (20 pounds), which is over six times the weight of a typical human brain.

- Their brains are highly convoluted—a characteristic shared by many organisms that are considered to have highly developed intelligence (such as humans and other primates).
- Some wild dolphins have been reported to assist drowning humans in the ocean.
- Some dolphins have been trained to respond to hand signals and do tricks on command (such as retrieve objects).

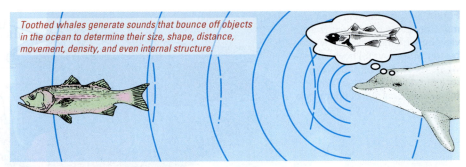

Toothed whales generate sounds that bounce off objects in the ocean to determine their size, shape, distance, movement, density, and even internal structure.

Figure 14.25 **Echolocation.**

Although toothed whales have remarkable abilities, this does not necessarily imply high intelligence. Pigeons, for instance, which are not known for being highly intelligent, have also been trained to retrieve objects by using hand signals. Even crows have demonstrated the problem-solving ability to craft tools to pick locks. Perhaps many of us would like to think that whales and dolphins are more intelligent than they really are because humans feel an attachment to these charismatic, seemingly ever-smiling, air-breathing creatures. It is interesting to note that even experts in the field of animal intelligence disagree on how to assess *human* intelligence accurately, let alone that of a marine mammal. In fact, experts on dolphin behavior suggest that public enthusiasm for dolphin smarts has outpaced the evidence for dolphin intelligence.

If the large brain that toothed whales possess is not an indication of intelligence, then why is their brain so large? Leading whale researchers don't exactly know, but it might be because toothed whales need a large brain to process the wealth of information they receive from the sound echoes they transmit. Because intelligence is difficult to measure, perhaps it is best to say that animals of suborder Odontoceti are tremendously well adapted to the marine environment.

SUBORDER MYSTICETI Suborder **Mysticeti** (*mystic* = moustache, *cetus* = whale)—also known as the *baleen whales*—includes the world's largest whales (the blue whale, finback whale, and humpback whale) and the gray whale (a bottom feeder).

Baleen whales are generally much larger than toothed whales because of differences in food sources. Baleen whales eat lower on the food web (including zooplankton, such as krill and small nektonic organisms), which are relatively abundant in the marine environment. How are the largest whales in the world able to survive on eating such small prey, especially when these smaller organisms are widely dispersed in the marine environment?

Use of Baleen To concentrate small prey items and separate them from seawater, baleen whales have parallel rows of **baleen** (*balaena* = whale) plates in their mouths instead of teeth (**Figure 14.26a**). These baleen plates hang from the whale's upper jaw and, when the whale opens its mouth, the baleen resembles a moustache (except that it is on the *inside* of their mouths), which is why these whales are sometimes called *moustached whales* (**Figures 14.26c** and **14.27**). Baleen is made of flexible keratin—the same as human nails and hair—and can be up to 4.3 meters (14 feet) long (**Figure 14.26d**).[15]

To feed, the largest baleen whales fill their mouths with a body-weight of water and prey (**Figure 14.26b**), allowing their pleated lower jaw to balloon in size (Figure 14.27). The whales force the water out between the fibrous plates of baleen, trapping small fish, krill, and other plankton inside their mouths. For example, humpback whales feed at or near the surface, sometimes working together in large groups to produce a circular curtain of bubbles to concentrate prey and then surfacing as a group within the *bubblenet* in a coordinated vertical lunge (**Figure 14.28**). The gray whale, on the other hand, has short baleen slats and feeds by straining benthic organisms such as amphipods and shellfish from bottom sediment.

STUDENTS SOMETIMES ASK . . .

In a battle between a killer whale and a great white shark, which one would win?

Although many people who are fascinated with large and powerful wild animals have often wondered which of the two would win such a fight, there was little evidence to settle the dispute until recently. A remarkable video was taken in waters off northern California in 1997, documenting a battle between a 6-meter (20-foot) juvenile killer whale (*Orcinus orca*) and a 3.6-meter (12-foot) adult great white shark (*Carcharodon carcharias*). The video clearly shows the killer whale biting and completely severing the shark's head! If this is representative of the way these two animals interact in the wild, then the killer whale is the top carnivore in the ocean. It is believed that the killer whale's superior maneuverability and use of echolocation helped it defeat the shark.

[15]Baleen (also called *whalebone*) was used for such items as buggy whips, umbrella ribs, and corset stays before synthetic materials—mostly plastics—were substituted.

In the mid-1800s, they were traced to their birthing and breeding lagoons and were hunted to the brink of extinction. A common strategy was to harpoon a calf, which then lured its vigilant mother to the whalers, too. During this time, gray whales were known as "devilfish" because the adults would often capsize small whaling boats when they came to the aid of their young. By the late 1800s, the number of gray whales had diminished to the point that they were difficult to find during their annual migration. Fortunately, these low numbers also made it difficult for whalers to hunt them successfully.

In 1938, the International Whaling Treaty banned the taking of gray whales, which were thought to be nearly extinct. This protection has allowed them to steadily increase in number to this day and to become the first marine creature to be removed from endangered status. Their repopulation is truly one of the most impressive success stories of how protecting animals can ensure their continued survival. What is perhaps most surprising is how friendly they are now toward people in boats (Figure 14.31) in the same lagoons where they were hunted to near extinction over 150 years ago. Devilfish, indeed!

Whaling and the International Whaling Commission

From the mid-1800s to the mid-1900s, many stocks of large whales in the world were overhunted and their numbers severely depleted (see Figure 14.30). A detailed scientific study published in 2015 reveals that since 1900, nearly 3 million whales were killed by the whaling industry; the number could be much higher due to illegal whaling and unreported catch. The most commonly killed whales during this time were hundreds of thousands of fin, sperm, and blue whales, but thousands of right, sei, humpback, and minke whales were also taken.

In 1946, the **International Whaling Commission (IWC)** was established in an effort to manage the subsistence and commercial hunting of large whales. In 1986, the 72 nations that are members of the IWC passed a ban on commercial whaling in order to allow whales to recover from overhunting and to give researchers time to develop methods for assessing whale populations. Today, the ban on whaling is still in effect, although some countries such as Japan, Norway, and Iceland have killed more than 35,000 whales since the 1986 whaling moratorium. Despite concerns about the practice of killing whales, some nations have proposed to end the ban on whaling and establish annual whale kill quotas. The ban on killing whales can be reversed only by a three-quarters majority vote of IWC member nations.

According to the IWC, there are currently three ways to engage in legal hunting of whales:

Figure 14.31 Gray whale friendly behavior. Gray whales (*Eschrichtius robustus*) exhibit friendly behavior by approaching a boat and initiating contact in Scammon's Lagoon, Baja California, Mexico.

1. *Whaling by objection.* Countries can continue to hunt whales legally under an objection to the IWC ban on commercial whaling. (*Notable countries doing this:* Norway, Iceland.)

2. *Scientific whaling.* Countries may decide to kill whales for scientific research. Under the IWC convention, meat from whales hunted for this research may be sold commercially. (*Notable country doing this:* Japan.[19])

3. *Aboriginal subsistence whaling.* Subsistence whaling by native cultures is allowed by the IWC in special cases where there is a long-standing cultural tradition of whaling and where whale meat satisfies the nutritional needs of the community. (*Notable countries doing this*: Greenland, Russia, the United States, and the Caribbean nation of Saint Vincent and the Grenadines.)

RECAP

Gray whales undertake the longest migration of any mammal, traveling from high-latitude Arctic summer feeding areas to low-latitude winter birthing and breeding lagoons in Mexico.

CONCEPT CHECK 14.5 | Demonstrate an understanding of why gray whales migrate.

1 Discuss reasons why gray whales leave their cold-water feeding grounds during the winter season.

2 Why did the International Whaling Commission (IWC) invoke a ban on commercial whaling?

3 What three ways exist to legally hunt whales? Which countries are doing each?

[19]Japan is permitted to kill nearly 1000 humpback, fin, and minke whales for scientific purposes each year, but international pressure in recent years has reduced the take to less than 500 whales. In 2014, the International Court of Justice found that Japan's whaling program was not for scientific purposes and ordered Japan to halt its scientific whaling program.

ESSENTIAL CONCEPTS REVIEW

14.1 How are marine organisms able to stay above the ocean floor?

▶ *Pelagic animals that comprise the majority of the ocean's biomass remain mostly within the upper surface waters of the ocean*, where their primary food source exists. Those animals that are not planktonic (floating forms such as microscopic zooplankton) depend on *buoyancy* or their *ability to swim* to help them remain in food-rich surface waters.

▶ *The rigid gas containers in some cephalopods and the expandable swim bladders in some fish help increase buoyancy.* Other organisms maintain their positions near the surface with *gas-filled floats* (such as those of the Portuguese man-of-war) and *soft bodies that lack high-density hard parts* (such as the jellies).

▶ *Nekton—squid, fish, and marine mammals—are strong swimmers* that depend on their swimming ability to avoid predators and obtain food. Squid swim by trapping water in their body cavities and forcing it out through a siphon. Most fish swim by creating a wave of body curvature that passes from the front of the fish to the back and provides a forward thrust.

▶ *The caudal (rear) fin provides the most thrust, while the paired pelvic and pectoral (chest) fins are used for maneuvering. The dorsal (back) and anal fins serve primarily as stabilizers.* A rounded caudal fin is flexible and can be used for maneuvering at slow speeds. The lunate fin is rigid and is of little use in maneuvering but produces thrust efficiently for fast swimmers such as tuna.

Dorsal blood vessels
Gas is added
Gas gland
In fish with no pneumatic duct, all gas must be added or removed through the blood, which requires more time.
Esophagus
Stomach Swim bladder Gas is removed

(b) Adaptations that only allow slow buoyancy changes.

Study Resources
MasteringOceanography Study Area Quizzes, Web Video

Critical Thinking Question
Explain how a swim bladder works.

Active Learning Exercise
Pair up with another student in class. For each pair, have each person decide on which swim bladder adaptation they would like to describe: (1) adaptations that allow for rapid buoyancy changes or (2) adaptations that allow for slow buoyancy changes. Be sure to include how the buoyancy changes are achieved and determine which method is more advantageous. Share your analysis with each other and then report out to the class.

14.2 What adaptations do pelagic organisms possess for seeking prey?

▶ *Fish can be categorized as lungers* (such as groupers) *or cruisers* (such as tuna). Lungers sit motionless and lunge at passing prey. They have mostly white muscle tissue, which fatigues more quickly than red muscle tissue. Cruisers swim constantly in search of prey and possess mostly red, myoglobin-rich muscle tissue.

▶ *Fish swim slowly when cruising, fast when hunting for prey, and fastest when trying to escape from predators.* Although *most fish are cold-blooded*, the fast-swimming tuna *Thunnus* is homeothermic, meaning that it maintains its body temperature well above water temperature.

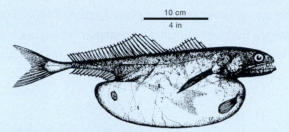

10 cm
4 in

(b) Ingestion capability of *Chiasmodon niger*, with a curled-up fish in its stomach that is longer than it is.

▶ *Deep-water nekton have special adaptations*—such as *good sensory devices* and *bioluminescence*—that allow them to survive in this still and completely dark environment. *Bioluminescence*—the ability to organically produce light—*has many uses in the deep ocean.*

Study Resources
MasteringOceanography Study Area Quizzes, Web Video

Critical Thinking Question

What are the major structural and physiological differences between fast-swimming cruisers and lungers that patiently lie in wait for their prey?

Active Learning Exercise

Working with another student in class, list the advantages that marine fish have by being warm-blooded as compared to cold-blooded fish. Then compare your list with the list from another group of students in class.

14.3 What adaptations do pelagic organisms possess to avoid being prey?

(b) Mutualism occurs when both participants benefit, such as this clown anemonefish and sea anemone.

▶ *Many marine organisms such as fish, squid, and crustaceans exhibit schooling*, probably because it increases their chances of avoiding predation compared to swimming alone and serves to preserve the species. Some organisms live closely together in *symbiotic relationships*.

Study Resources
MasteringOceanography Study Area Quizzes

Critical Thinking Question

Evaluate the advantages and disadvantages for a brightly colored yellow fish schooling within a large group of gray-colored fish.

Active Learning Exercise

Working with another student in class, list the advantages of schooling. Then compare your list with the list from another group of students in class.

14.4 What characteristics do marine mammals possess?

▶ Good fossil evidence shows that *marine mammals evolved from land-dwelling animals* about 50 million years ago. *Marine mammals are warm-blooded, breathe air, have hair or fur, bear live young, and the females have mammary glands.* Marine mammals belong to orders *Carnivora, Sirenia,* and *Cetacea.*

▶ *Marine mammals in order Carnivora have prominent canine teeth and include sea otters, polar bears,* and the *pinnipeds (walruses, seals, sea lions,* and *fur seals). Marine mammals of order Sirenia,* which include *manatees* and *dugongs* ("sea cows"), have toenails (manatees only) and sparse hairs covering their bodies, and they are vegetarians.

▶ The mammals best adapted to life in the open ocean are those of the *order Cetacea,* which includes *whales, dolphins,* and *porpoises.* Cetaceans have *highly streamlined bodies* so that they are fast swimmers. Other adaptations—such as being able to absorb 90% of the oxygen they inhale, storing large quantities of oxygen, reducing the use of oxygen by noncritical organs, and having collapsible ribs and lungs)—*allow them to dive deeply* and minimize the effects of nitrogen narcosis and decompression sickness, although research suggests that they are not immune to it.

Continued on next page...

...Continued from previous page.

▶ *Cetaceans are divided into suborder Odontoceti (the toothed whales) and suborder Mysticeti (the baleen whales). Odontocetes use echolocation* to find their way through the ocean and locate prey. They emit clicking sounds and can determine the size, shape, internal structure, and distance of the objects from the nature of the returning signals and the time elapsed.

▶ *Mysticetes,* which include the largest whales in the world, *separate their small prey from seawater using their baleen plates as a strainer.* Baleen whales include the *gray whale,* the *rorqual whales,* and the *right whales.*

Study Resources
MasteringOceanography Study Area Quizzes, MasteringOceanography Web Diving Deeper 14.1, MasteringOceanography Web Animations, Web Videos

Critical Thinking Question

List the modifications that are thought to give some cetaceans the ability to (a) increase their swimming speed, (b) dive to great depths without suffering the bends, and (c) stay submerged for long periods.

Active Learning Exercise

Pair up with another student in class. For each pair, have each person decide on which table they would like to construct: (1) a table on the physical characteristics that differentiate true seals from eared seals (sea lions and fur seals) or (2) a table on the physical characteristics that differentiate dolphins from porpoises. Be sure to include the names of representative organisms. Share your table with each other and then compare with other students' tables that were created in class.

(c) A rack of baleen from a bottom-feeding gray whale (*Eschrichtius robustus*).

14.5 An example of migration: Why do gray whales migrate?

▶ *Gray whales migrate from their cold-water summer feeding grounds in the Arctic to warm, low-latitude lagoons in Mexico during winter for breeding and birthing purposes.* This behavior may have evolved to allow their young to be born into warm water during the last ice age, when lower sea level eliminated today's highly productive Arctic feeding areas.

▶ *The International Whaling Commission (IWC) was established in 1946 to manage the subsistence and commercial hunting of large whales,* which had experienced severe population decreases. *Commercial whaling was banned* in 1986, although *whaling can still be done under objection to the ban, for scientific research, or by native cultures.*

Study Resources
MasteringOceanography Study Area Quizzes

Critical Thinking Question

Using a calendar and tying your answer to productivity considerations, explain the seasonal cycle of migration for gray whales.

Active Learning Exercise

Working with another student in class, discuss how you would establish a marine protected area for gray whales, considering that they migrate such long distances through international waters. Consult Chapter 11 for inclusion of the Law of the Sea.

MasteringOceanography™

www.masteringoceanography.com

Looking for additional review and test prep materials? With individualized coaching on the toughest topics of the course, MasteringOceanography offers a wide variety of ways for you to move beyond memorization and deeply grasp the underlying processes of how the oceans work. Visit the Study Area in **www.masteringoceanography.com** to find practice quizzes, study tools, and multimedia that will improve your understanding of this chapter's content. Sign in today to enjoy the following features: Self Study Quizzes, SmartFigures, SmartTables, Oceanography Videos, Squidtoons, Geoscience Animation Library, RSS Feeds, Digital Study Modules, and an optional Pearson eText.

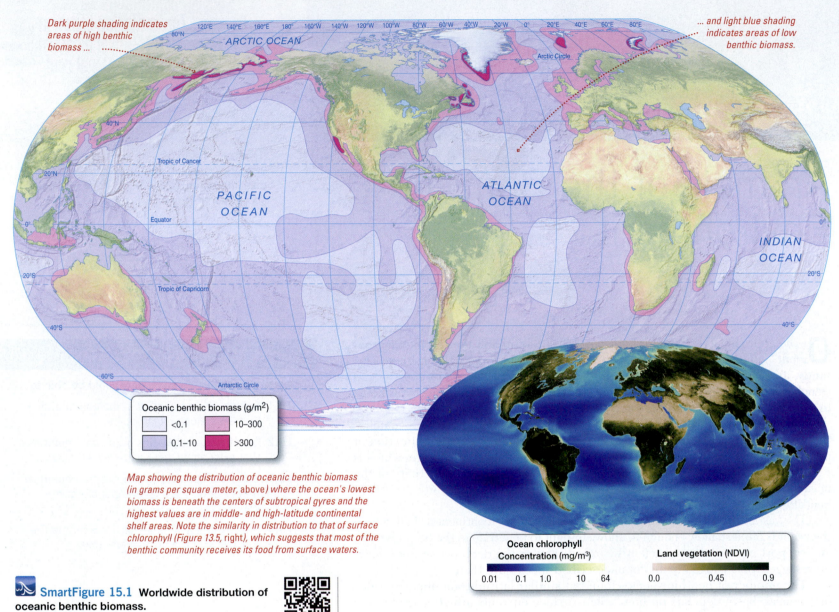

Dark purple shading indicates areas of high benthic biomass ...

... and light blue shading indicates areas of low benthic biomass.

Oceanic benthic biomass (g/m²)

	<0.1		10–300
	0.1–10		>300

Map showing the distribution of oceanic benthic biomass (in grams per square meter, above) where the ocean's lowest biomass is beneath the centers of subtropical gyres and the highest values are in middle- and high-latitude continental shelf areas. Note the similarity in distribution to that of surface chlorophyll (Figure 13.5, right), which suggests that most of the benthic community receives its food from surface waters.

Ocean chlorophyll Concentration (mg/m³)					Land vegetation (NDVI)		
0.01	0.1	1.0	10	64	0.0	0.45	0.9

SmartFigure 15.1 **Worldwide distribution of oceanic benthic biomass.**
https://goo.gl/9x2XXm

to the bottom (for example, marine algae) or move over it (for example, crabs). **Table 15.1** lists some of the special adaptations these organisms have to withstand the rigors of life on rocky shores.

Diversity of species that inhabit rocky shores varies widely. Overall, rocky intertidal ecosystems have a moderate diversity of species compared to other benthic environments. The greatest animal diversity is in the lower (tropical) latitudes, while the diversity of algae is greater in the middle latitudes, probably because of better availability of nutrients.[2]

Intertidal Zonation

Most shorelines exhibit *intertidal zonation*, which describes the natural organization of ecosystems relative to sea level that are caused by varying environmental conditions. For example, a typical rocky shore (**Figure 15.2a**) can be divided into a **spray zone**, which is above the spring high tide line and is covered by water only

[2]As discussed in Chapter 13, this increased nutrient supply is a result of the lack of a permanent thermocline in the middle latitudes.

SMARTTABLE 15.1	ADVERSE CONDITIONS OF ROCKY INTERTIDAL ZONES AND ORGANISM ADAPTATIONS	
Adverse conditions of rocky intertidal zones	**Organism adaptations**	**Examples of organisms**
Drying out during low tide	• Ability to seek shelter or withdraw into shells • Thick exterior or exoskeleton to prevent water loss • External surfaces covered with rock or shell fragments to prevent water loss • Physiologically adapted to periodic drying out	Sea slugs, snails, crabs, kelp
Strong wave activity	• In algae: Strong holdfasts to prevent being washed away • In animals: Seeking shelter or employing strong attachment threads, biological adhesives, a muscular foot, multiple legs, or hundreds of tube feet to allow them to attach firmly to the bottom • In both: Hard structures adapted to withstand wave energy; clustering closely together	Kelp, snails, sea stars, mussels, sea urchins
Predators occupy area during low tide/high tide	• Firm attachment of body parts, including a hard shell • Stinging cells • Camouflage • Inking response • Ability to break off body parts and regrow them later (regenerative capability)	Mussels, sea anemones, sea slugs, octopuses, sea stars
Difficulty finding mates for attached species	• Release of large numbers of eggs/sperm into the water column during reproduction • Long organs to reach others for sexual reproduction	Abalones, sea urchins, barnacles
Rapid changes in temperature, salinity, pH, and oxygen content	• Ability to withdraw into shells to minimize exposure to rapid changes in environmental conditions • Ability to exist in varied temperature, salinity, pH, and low-oxygen environments for extended periods	Snails, limpets, mussels, barnacles
Lack of space or attachment sites	• Overtake another organism's space • Attach to other organisms • Planktonic larval forms that inhabit new areas, which limits parental and offspring competition for the same space	Bryozoans, coral, barnacles, limpets

SmartTable 15.1 Adverse conditions of rocky intertidal zones and organism adaptations.
http://goo.gl/bGYy23

during storms, and an **intertidal zone**, which lies between the high and low tidal extremes. Along most shores, the intertidal zone can be clearly separated into the following subzones (Figure 15.2a):

- The **high tide zone**, which is relatively dry and is covered only by the highest high tides
- The **middle tide zone**, which is alternately covered by all high tides and exposed during all low tides
- The **low tide zone**, which is usually wet but is exposed during the lowest low tides

Organisms living within these intertidal zones have different environmental conditions that they must be adapted to. For example, physical stress (such as drying out) is much more important in higher tide zones. Wave energy and predation by other marine organisms are larger stresses in lower tide zones. And competition between intertidal organisms for attachment space is the biggest stress in middle tide zones. As a result, intertidal organisms have evolved specific adaptations to cope with the environmental conditions they face. Not surprisingly, then, the subzones of the intertidal zone can be delineated based on characteristic populations of benthic organisms found within each zone.

Although the zonation of marine organisms along rocky shores produces some of the most finely delineated biozones in the marine environment, the intertidal zone can have remarkably different characteristics from place to place. For example, some of the physical characteristics that change in a matter of just a few vertical centimeters are the amount of wave energy, exposure to the atmosphere, and

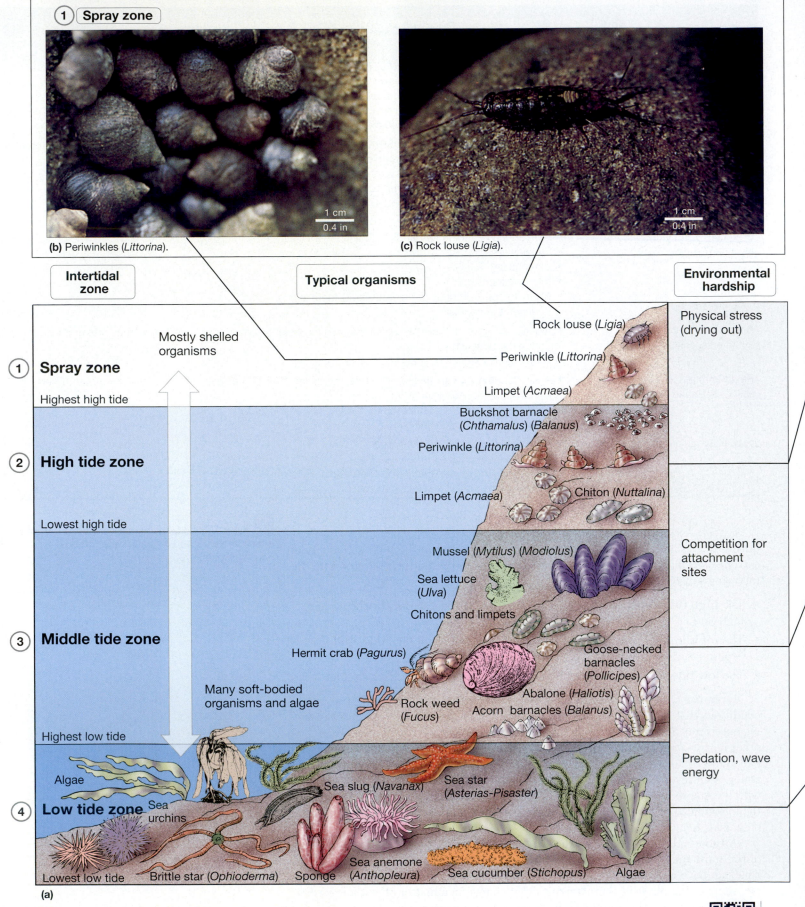

① **Spray zone**

(b) Periwinkles (*Littorina*).

(c) Rock louse (*Ligia*).

Intertidal zone	Typical organisms	Environmental hardship

Mostly shelled organisms

Rock louse (*Ligia*)

Periwinkle (*Littorina*)

① **Spray zone**

Highest high tide

Limpet (*Acmaea*)

Physical stress (drying out)

Buckshot barnacle (*Chthamalus*) (*Balanus*)

Periwinkle (*Littorina*)

② **High tide zone**

Limpet (*Acmaea*) Chiton (*Nuttalina*)

Lowest high tide

Mussel (*Mytilus*) (*Modiolus*)

Competition for attachment sites

Sea lettuce (*Ulva*)

Chitons and limpets

③ **Middle tide zone**

Hermit crab (*Pagurus*)

Goose-necked barnacles (*Pollicipes*)

Many soft-bodied organisms and algae

Rock weed (*Fucus*) Abalone (*Haliotis*)

Acorn barnacles (*Balanus*)

Highest low tide

Algae

Sea slug (*Navanax*) Sea star (*Asterias-Pisaster*)

Predation, wave energy

④ **Low tide zone** Sea urchins

Lowest low tide Brittle star (*Ophioderma*) Sponge Sea anemone (*Anthopleura*) Sea cucumber (*Stichopus*) Algae

(a)

SmartFigure 15.2 **Rocky shore intertidal zonation and common organisms. (a)** Diagrammatic view of rocky shore tide zones, including representative organisms (not to scale), general patterns of organisms, and environmental conditions. **(b–i)** Photos of some common rocky shore intertidal organisms. https://goo.gl/BYuluH

2 High tide zone

(d) Limpets (*Acmaea*).

(e) Buckshot barnacles (*Chthamalus*).

(f) Chiton (*Nuttalina*).

3 Middle tide zone

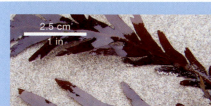

(g) Rock weed (*Fucus*).

(h) Acorn barnacles (*Balanus*).

(i) Blue mussel bed (*Mytilus*), including gooseneck barnacles (*Pollicipes*) and acorn barnacles (*Balanus*).

4 Low tide zone

(j) Sea anemone (*Anthopleura*).

(k) Sea slug (*Navanax*).

(l) Sea star (*Asterias-Pisaster*).

changes in temperature and salinity. Overall, the intertidal zone is a difficult place to live, but the organisms that inhabit the intertidal zone have specific adaptations to overcome the challenges of living there. Let's examine some of these organisms and their adaptations.

The Spray (Supratidal) Zone: Organisms and Their Adaptations

The spray zone, which is also known as the **supratidal zone** (*supra* = above, *tidal* = the tides), is above even the highest high tides and so it is continually exposed above sea level. As a result, drying out is a significant challenge to organisms that live within the spray zone. Many animals, such as the periwinkle snail (genus *Littorina*; **Figure 15.2c**), have shells, and only a few species of marine algae are found in this zone.

One common organism that is found in the spray zone well above the high tide line is called rock louse or sea roach (isopods of the genus *Ligia*) (**Figure 15.2b**), which lives on exposed rocks or is found among the cobbles and boulders that typically cover the floors of sea caves. These scavengers reach lengths of 3 centimeters (1.2 inches) and scurry about at night, feeding on organic debris. During the day, they mostly hide in crevices.

A distant relative of the periwinkle snail, the limpet is also found in the spray zone (genus *Acmaea*; **Figure 15.2e**). Both limpets and periwinkle snails feed on marine algae. The limpet has a flattened conical shell and a muscular foot with which it clings tightly to rocks.

The High Tide Zone: Organisms and Their Adaptations

Like animals within the spray zone, most animals that inhabit the high tide zone have a protective covering to prevent them from drying out. Periwinkles, for example, have a protective shell and can move between the spray zone and the high tide zone. Buckshot barnacles (**Figure 15.2d**) are crustaceans that have a protective shell, too, but they cannot live above the high-tide shoreline because they filter-feed from seawater, and their larval form is planktonic.

The most conspicuous algae in the high to middle tide zone are rock weeds, members of the genus *Fucus* that live in colder latitudes (**Figure 15.2g**) and *Pelvetia* that live in warmer latitudes. Both have thick cell walls to reduce water loss during periods of low tide.

Observations of newly created rocky shorelines or other recently disturbed regions of the coast indicate that rock weeds are among the first organisms to colonize a rocky shore. Later, **sessile** (*sessilis* = sitting on) animal forms—those attached to the bottom, such as barnacles and mussels—begin to establish themselves, competing for attachment sites with the rock weeds.

The Middle Tide Zone: Organisms and Their Adaptations

Seawater constantly bathes the middle tide zone, so more types of marine algae and soft-bodied animals can live there. The total biomass is much greater than in the high tide zone, so there is much greater competition for rock space among sessile forms.

Shelled organisms inhabiting the middle tide zone include mussels (genera[3] *Mytilus* and *Modiolus*) (**Figure 15.2f**); gooseneck barnacles (*Pollicipes*) (Figure 15.2f), which attach themselves to rocks with a long, muscular stalk; and acorn barnacles (*Balanus*) (**Figure 15.2h**). Mussels attach to bare rock, algae, or barnacles during their planktonic stage and remain in place by means of strong *byssus* threads.

[3]The term "genera" is plural of genus, as in the genus and species (scientific name) of an organism.

Figure 15.3 **Middle tide zone mussel bed and sea star.**

Mussel bed

(a) The distinctive black band covering the rock surface is a bed of mussels (*Mytilus*), which is a common feature of middle tide zones along rocky shores such as this one in Glacier Bay, Alaska. Note how the diffuse band can also be seen on the beach to the left.

(b) An ochre sea star (*Pisaster*), feeds on a mussel by first enveloping a mussel with its tube feet and prying apart the mussel's shell. The star can then digest the mussel's soft tissues with its own everted stomach.

Mussels are often grouped together into a distinctive *mussel bed* that appears as a pronounced band or layer (**Figure 15.3a**) and can often be one of the most recognizable features of middle tidal zones along rocky coasts. The mussel bed thickens toward the bottom until it reaches an abrupt bottom limit, where physical conditions restrict mussel growth. Often protruding from the mussel bed are numerous gooseneck barnacles; crowded in with the mussels are less-conspicuous species, such as acorn barnacles, other crustaceans, marine worms, rock-boring clams, sea stars, and algae.

Carnivorous snails and sea stars (such as genera *Pisaster* and *Asterias*) feed upon mussels in the mussel beds. To pry open the mussel shell, sea stars pull on either side with hundreds of tubelike feet. The mussel eventually becomes fatigued and can no longer hold its shell halves closed. When the shell opens ever so slightly, the sea star turns its stomach inside out, slips it through the crack in the mussel shell, and digests the edible tissue inside (**Figure 15.3b**).

Where the rock surface flattens out within the middle tidal zone, tide pools trap water as the tide goes out. These pools support microecosystems containing a wide variety of organisms. The most conspicuous member of this community is often the sea anemone (see **Figure 15.2i**), which is a relative of jellies.

Shaped like a sack, an anemone has a flat foot disk that provides a suction attachment to the rock surface. Directed upward, the open end of the sack is the mouth, which leads directly to the gut cavity and is surrounded by rows of tentacles (**Figure 15.4**). The tentacles are covered with stinging needlelike cells called **nematocysts** (*nemato* = thread, *cystis* = bladder) (Figure 15.4, *inset*), which inject the victim with a potent neurotoxin. When an organism brushes against a sea anemone's tentacles, the nematocysts are automatically released (except for certain organisms such as clownfish that live in a symbiotic relationship with a sea anemone).

Hermit crabs (*Pagurus*) inhabit tide pools, too. They have a well-armored pair of claws and upper body but a soft, vulnerable abdomen, which they protect by inhabiting an abandoned snail shell (**Figure 15.5a**). They can often be seen scurrying around the tide pool area or fighting with other hermit crabs for new shells. Their abdomen has even evolved a curl to the right to make it fit properly into snail shells. Once in the snail shell, the crab can further protect itself by closing off the shell's opening with its large claws.

STUDENTS SOMETIMES ASK . . .

What is the most venomous marine organism?

Many types of marine organisms are armed with potent venom. For example, certain species of jellies, stingrays, stonefish, snails, and octopuses all have venom strong enough to kill humans. Experts agree that the world's most venomous marine organism is the Australian box jelly (*Chironex fleckeri*), which has tentacles covered with poisonous darts called nematocysts that can inject poison into a person and kill them in minutes.

Coming in a close second is the cone snail, a small tropical snail with over 700 species and elaborately decorated shells that lives in coral debris among coral reefs. Cone snails immobilize their prey by using a modified radular tooth with barbs that is loaded with highly venomous compounds and is launched out of its mouth like a harpoon. The venom causes muscle paralysis and respiratory failure in its victim, which can lead to death. Because cone snails produce a mixture of toxins, there is currently no antivenom available. Researchers are analyzing the cone snail's complex venom in the hope of producing medicines to treat diabetes and a wide variety of human neurological diseases such as Parkinson's, Alzheimer's, and alcoholism.

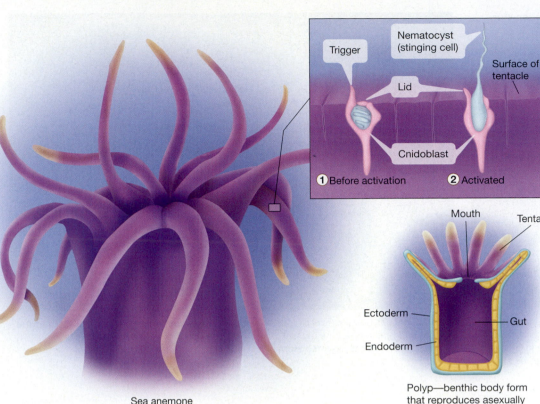

Sea anemone
(polyp body form)

Polyp—benthic body form
that reproduces asexually

Figure 15.4 Sea anemone structure and operation of its sting-ing cells. Sea anemone morphology (*left*), with detail of its stinging nematocysts (*inset*), which are used to sting and incapacitate prey. An internal view of the basic sea anemone body form is shown at right.

RECAP

Rocky shores are divided into the spray zone and high, middle, and low tide (intertidal) zones. Many shelled organisms inhabit the upper zones, while more soft-bodied organisms and algae inhabit the lower zones.

In tide pools near the lower limit of the middle tide zone, sea urchins may be found feeding on algae (**Figure 15.5b**). A sea urchin has a five-toothed mouth centered on the bottom side of its hard spherical shell, consisting of fused calcium carbonate plates perforated to allow tube feet and water to pass through. Resembling a pincushion, the shell of a sea urchin has numerous spines used for protection and to scrape out protective holes in rocks.

The Low Tide Zone: Organisms and Their Adaptations

The low tide zone is almost always submerged, and an abundance of algae is typically present. A diverse community of animals live here, too, but they are hidden by the great variety of marine algae and surf grass (*Phyllospadix*) (**Figure 15.6**). Various types of encrusting red algae (such as *Lithophyllum* and *Lithothamnium*), which are also seen in middle-zone tide pools, become very abundant in lower tide pools. In temperate latitudes, moderate-sized red and brown algae provide a drooping canopy beneath which much of the animal life can hide during low tide.

Scampering from crevice to crevice and in and out of tide pools across the full range of intertidal zones are various species of shore crabs (**Figure 15.7**). These scavengers help keep the shore clean. Shore crabs spend most of the day hiding in cracks or beneath overhangs. At night, they eat algae as rapidly as they can tear them from the rock surface with their large front claws, called *chelae* (*khele* = claw). Their hard exoskeleton prevents them from drying out too quickly, so they can spend long periods of time out of water.

Figure 15.5 Middle tide zone hermit crab and sea urchins.

(a) Hermit crab (*Pagurus*) that has taken up residence in a *Maxwellia gemma* shell.

(b) Purple sea urchins (*Echinus*) that have burrowed into the bottom of a rocky tide pool in the middle tide zone.

Figure 15.6 Surf grass. Green surf grass (*Phyllospadix*) and various species of brown algae are exposed during an extremely low tide in this California tide pool. When the tide rises, the anchored surf grass floats and provides a protective hiding place for many low tide zone organisms.

(a) Sally lightfoot crab (*Grapsus grapsus*), which is a brightly colored and speedy crab found in tropical regions.

(b) Shore crab (*Pachygrapsus crassipes*). This female is carrying eggs (orange mass under her abdomen).

Figure 15.7 Shore crabs.

CONCEPT CHECK 15.1	Specify characteristics of the communities that exist along rocky shores.

1 What are some adverse conditions of rocky intertidal zones? What are some organisms' adaptations for those adverse conditions? Which conditions seem to be most important in controlling the distribution of life?

2 One of the most noticeable features of the middle tide zone along rocky coasts is a mussel bed. Describe general characteristics of mussels and include a discussion of other organisms that are associated with mussels.

3 In which intertidal zone of a rocky shore would you typically find each of the following organisms: sea anemones, sea lettuce, rock louse, abalones, brittle stars, and buckshot barnacles?

15.2 What Communities Exist along Sediment-Covered Shores?

Even though most sediment-covered shores have intertidal zones similar to rocky shores, life on and in sediment-covered shores requires very different adaptations than on rocky shores. Sediment-covered shores, for example, are composed of unconsolidated materials that often change shape and so require specific adaptations for organisms. In addition, there is much less species diversity in sediment-covered shores, but the organisms are usually found in great numbers. In the low tide zone of some beaches and on mud flats, for example, as many as 5000 to 8000 burrowing clams have been counted in only 1 square meter (10.8 square feet).

Nearly all large organisms that inhabit sediment-covered shores are called **infauna** (*in* = inside, *fauna* = animal) because they can burrow into the sediment. Sediment-covered shores also contain large numbers of microbial communities, particularly in quiet environments such as salt marshes and mud flats that tend to accumulate organic matter.

STUDENTS SOMETIMES ASK . . .

I've been at a tide pool and seen sea anemones. When I put my finger on one, it tends to gently grab my finger. Why does it do that?

The sea anemone is trying to kill you with poison and devour you whole (seriously!). Disguised as a harmless flower, the sea anemone is actually a vicious predator that will attack any unsuspecting animal (even a human) that its stinging tentacles entrap. Fortunately, the skin on your hands is thick enough to resist the stinging nematocyst and its neurotoxin. A couple of unsuspecting people, however, were interested in finding out if the sea anemone grabbed other things with its tentacles, so they put their *tongues* into a sea anemone. After a short time, their throats swelled almost completely closed, and they had to be rushed to a hospital. They lived, but the moral of this story is: *NEVER* put your tongue into a sea anemone!

Physical Environment of the Sediment

Sediment-covered shores include *coarse boulder beaches*, *sand beaches*, *salt marshes*, and *mud flats*, which represent progressively lower-energy environments and are consequently composed of progressively finer sediment. The energy level that a shore experiences is related to the strength of waves and longshore currents. Along shores that experience low energy levels, particle size becomes smaller, the sediment slope decreases, and overall sediment stability increases. Thus, the sediment in a fine-grained mud flat is more stable than that of a high-energy sandy beach.

Along high-energy sandy beaches, a large quantity of water from breaking waves rapidly sinks into the sand and brings a continual supply of nutrients and oxygen-rich water for the animals that live there. This supply of oxygen also enhances bacterial decomposition of dead tissue. The sediment in salt marshes and mud flats, on the other hand, is not nearly as rich in oxygen, however, so decomposition occurs more slowly.

Intertidal Zonation

The intertidal zone of the sediment-covered shore consists of supratidal, high tide, middle tide, and low tide zones, as shown in **Figure 15.8**. These zones are best developed on steeply sloping, coarse-sand beaches and are less distinct on the more gentle sloping, fine-sand beaches. On mud flats, the tiny, clay-sized particles form a deposit with essentially no slope, so zonation is not possible in this protected, low-energy environment.

The species of animals differ from zone to zone. As in intertidal rocky shores, however, the maximum *number* of species and the greatest *biomass* in intertidal sediment-covered shores are found near the low tide shoreline, and both diversity and biomass decrease toward the high tide shoreline.

Sandy Beaches: Organisms and Their Adaptations

Most animals at the beach burrow into the sand because there is no stable, fixed surface (as on rocky shores) to which they can attach. As a result, organisms are much less obvious at beaches than in other environments. By burrowing only a few centimeters beneath the surface, they encounter a much more stable environment where they are not affected by fluctuations of temperature and salinity or the threat of drying out.

BIVALVE MOLLUSKS A **bivalve** (*bi* = two, *valva* = a valve) is an animal that has two hinged shells, such as a clam or a mussel. A **mollusk** is a member of the phylum Mollusca (*molluscus* = soft), characterized by a soft body and either an internal or external hard calcium carbonate shell.

Figure 15.8 Intertidal zonation and common organisms of sediment-covered shores.

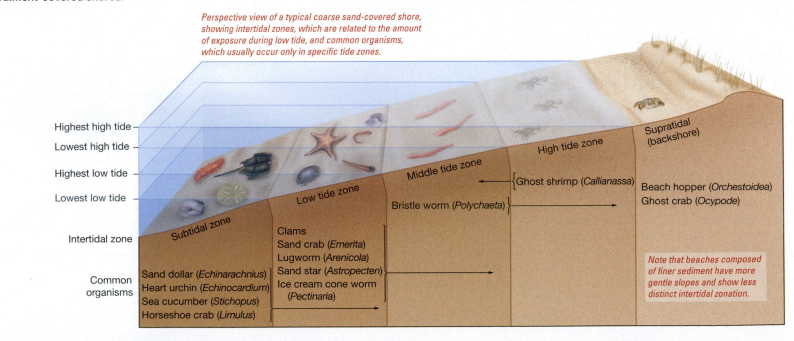

Perspective view of a typical coarse sand-covered shore, showing intertidal zones, which are related to the amount of exposure during low tide, and common organisms, which usually occur only in specific tide zones.

Highest high tide
Lowest high tide
Highest low tide
Lowest low tide

Intertidal zone

Common organisms

Supratidal (backshore)

High tide zone

Middle tide zone

Low tide zone

Subtidal zone

Ghost shrimp (*Callianassa*)

Bristle worm (*Polychaeta*)

Beach hopper (*Orchestoidea*)
Ghost crab (*Ocypode*)

Clams
Sand crab (*Emerita*)
Lugworm (*Arenicola*)
Sand star (*Astropecten*)
Ice cream cone worm (*Pectinaria*)

Sand dollar (*Echinarachnius*)
Heart urchin (*Echinocardium*)
Sea cucumber (*Stichopus*)
Horseshoe crab (*Limulus*)

Note that beaches composed of finer sediment have more gentle slopes and show less distinct intertidal zonation.

Bivalve mollusks are well adapted to living within sediment. A single foot digs into the sediment to pull the creature down into the sand. The method by which clams bury themselves is shown in **Figure 15.9**. How deeply a bivalve can bury itself depends on the length of its siphons, which must reach above the sediment surface to pull in water for food (plankton) and oxygen. Indigestible matter is forced back out the siphon periodically by quick muscular contractions.

The greatest biomass of clams is burrowed into the low-tide region of sandy beaches and decreases where the sediment becomes muddier.

ANNELID WORMS A variety of **annelids** (*annelus* = ring)—worms—are well adapted to life in the sediment. The lugworm (*Arenicola*), for example, constructs a vertical, U-shaped burrow, the walls of which are strengthened with mucus. The worm feeds by extending its proboscis (snout) up into the shaft of the burrow to loosen sand with quick, pulsing movements. A cone-shaped depression forms at the surface over the head end of the burrow as sand continually slides into the burrow and is ingested by the worm. As the sand passes through the worm's digestive tract, the sand's *biofilm* (coating of organic matter) is digested, and the processed sand is deposited back at the surface.

CRUSTACEANS **Crustaceans** (*crusta* = shell)—such as crabs, lobsters, shrimps, and barnacles—include predominately aquatic animals that are characterized by a segmented body, a hard exoskeleton, and paired, jointed limbs. On most sandy beaches, numerous crustaceans called *beach hoppers* feed on kelp cast up by storm waves or high tides. A common genus is *Orchestoidea*, which is only 2 to 3 centimeters (0.8 to 1.2 inches) long but can jump more than 2 meters (6.6 feet) high. Laterally flattened, beach hoppers usually spend the day buried in the sand or hidden in kelp. They become particularly active at night, when large groups may hop at the same time and form clouds above the piles of kelp on which they feed.

Sand crabs (*Emerita*) (**Figure 15.10**) are a type of crustacean common to many sandy beaches. Ranging in length from 2.5 to 8 centimeters (1 to 3 inches), they move up and down the beach near the shoreline. They bury their bodies in the sand and leave their long, curved, V-shaped antennae pointing up the beach slope. These little crabs filter food particles from the water and can be located by looking in the lower intertidal zone for a V-shaped pattern in the swash as it runs down the beach face.

ECHINODERMS **Echinoderms** (*echino* = spiny, *derma* = skin) that live in beach deposits include the sand star (*Astropecten*) and heart urchin (*Echinocardium*). Sand stars prey on invertebrates that burrow into the low-tide region of sandy beaches. The sand star is well designed for moving through sediment: It has a smooth back and five tapered legs with spines.

More flattened and elongated than the sea urchins of the rocky shore, heart urchins live buried in the sand near the low tide line (**Figure 15.11**). They gather sand grains into their mouths, where the biofilm of organic matter that coats sand grains is scraped off and ingested.

MEIOFAUNA **Meiofauna** (*meio* = lesser, *fauna* = animal) are small marine organisms that live in the spaces between sediment particles. These organisms, generally only 0.1 to 2 millimeters (0.004 to 0.08 inch) long, feed primarily on bacteria removed from the surface of sediment particles. Meiofauna include polychaetes, mollusks, arthropods, and nematodes (**Figure 15.12**) and live in sediment from the intertidal zone to deep-ocean trenches.

Mud Flats: Organisms and Their Adaptations

Eelgrass (*Zostera*) and turtle grass (*Thalassia*) are widely distributed in the low tide zone of mud flats and the adjacent shallow coastal regions. Numerous openings at the surface of mud flats attest to a large population of bivalve mollusks and other invertebrates.

Fiddler crabs (*Uca*) dig burrows within mud flats to live in; these burrows may extend 1 meter (3 feet) or more below the surface. Relatives of the shore

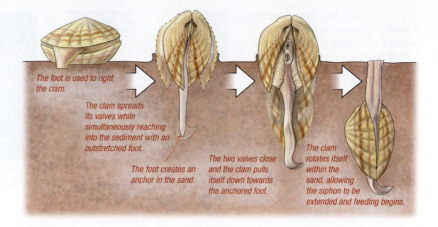

The foot is used to right the clam.

The clam spreads its valves while simultaneously reaching into the sediment with an outstretched foot.

The foot creates an anchor in the sand.

The two valves close and the clam pulls itself down towards the anchored foot.

The clam rotates itself within the sand, allowing the siphon to be extended and feeding begins.

Figure 15.9 How a clam burrows.

1 cm
0.4 in

Figure 15.10 Sand crab. A sand crab (Emerita) emerges from the sand. Sand crabs can often be found just beneath the surface of sandy beaches within the lower intertidal zone.

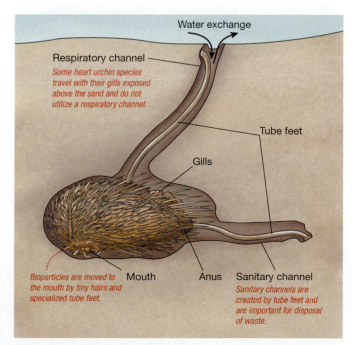

Water exchange

Respiratory channel

Some heart urchin species travel with their gills exposed above the sand and do not utilize a respiratory channel.

Tube feet

Gills

Bioparticles are moved to the mouth by tiny hairs and specialized tube feet. Mouth Anus Sanitary channel

Sanitary channels are created by tube feet and are important for disposal of waste.

Figure 15.11 Heart urchin. Feeding and respiratory structures of a heart urchin (*Echinocardium*), which feeds on the biofilm of organic matter that covers sand grains.

Figure 15.12 Scanning electron micrographs of meiofauna. Various examples of meiofauna, which are small marine organisms that live in the spaces between sediment particles.

100 microns

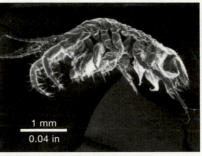

1 mm
0.04 in

0.3 mm
0.01 in

(a) Head of a nematode; the pit on its left side and numerous projections are sensory structures.

(b) Amphipod, which is a crustacean related to shrimp and krill.

(c) Polychaete worm, shown with its proboscis (mouth) extended (*left*).

Figure 15.13 Fiddler crab. A male fiddler crab (*Uca*), which uses its large claw for protection and for attracting mates.

RECAP

Sediment-covered shores—including sandy beaches and mud flats—have an intertidal zonation similar to that of rocky shores but also contain many organisms that live within the sediment (infauna).

crabs, they usually measure no more than 2 centimeters (0.8 inch) in width. A male fiddler crab has one small claw and one oversized claw, which is up to 4 centimeters (1.6 inches) long (**Figure 15.13**). Fiddler crabs get their name because they wave around their large claws as if they were playing imaginary fiddles. The females have two normal-sized claws. The large claw of the male is used to court females and to fight competing males.

CONCEPT CHECK 15.2 | Specify characteristics of the communities that exist along sediment-covered shores.

1 Describe how sandy and muddy shores differ in terms of energy level, particle size, sediment stability, and oxygen content.

2 How does the diversity of species on sediment-covered shores compare with that of the rocky shore? Suggest at least one reason why this occurs.

3 In which intertidal zone of a steeply sloping, coarse-sand beach would you typically find each of the following organisms: clams, beach hoppers, ghost shrimp, sand crabs, and heart urchins?

15.3 What Communities Exist on the Shallow Offshore Ocean Floor?

The shallow offshore ocean floor extends from the spring low-tide shoreline to the seaward edge of the continental shelf. It is mainly sediment covered, but rocky exposures may occur locally near shore. Rocky exposures feature many types of marine algae, which have adaptations (such as gas-filled floats) for reaching from the shallow sea floor to near the sunlit surface waters.

The sediment-covered shelf has moderate to low species diversity. Surprisingly, the diversity of benthic organisms is *lowest* beneath upwelling regions. This is because upwelling waters that are rich in nutrients produce high pelagic production, so large amounts of dead organic matter are produced. When this matter rains down on the bottom and decomposes, it consumes oxygen, so the oxygen supply can be locally depleted, thereby limiting benthic populations. However, kelp beds associated with rocky bottoms are a specialized shallow-water community with higher diversity.

Rocky Bottoms (Subtidal): Organisms and Their Adaptations

A rocky bottom within the shallow inner **subtidal zone** (*sub* = under, *tidal* = the tides) is usually covered with various types of marine macro algae.

KELP AND KELP FORESTS Along the North American Pacific coast, the giant brown bladder **kelp** (*Macrocystis*) uses a rootlike anchor called a *holdfast* (**Figure 15.14a**) to

Figure 15.14 **Kelp and kelp forests.**

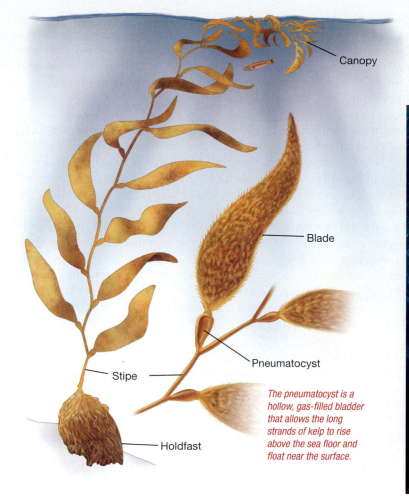

Canopy

Blade

Pneumatocyst

Stipe

The pneumatocyst is a hollow, gas-filled bladder that allows the long strands of kelp to rise above the sea floor and float near the surface.

Holdfast

(a) Structure and features of the giant brown bladder kelp (*Macrocystis*), a common species in kelp forests.

(b) Underwater photo of a kelp forest, which provides food, shelter, living space, spawning grounds, and nursery areas for many marine organisms.

Kelp with air bladders (red shading) includes larger species such as Macrocystis *and bull kelp* (Nereocystis).

Shrub kelp (orange shading) includes smaller species of kelp such as Sargassum *and rock weed* (Fucus, Pelvetia).

Web Video

Underwater Kelp Forests
https://goo.gl/Bkva2R

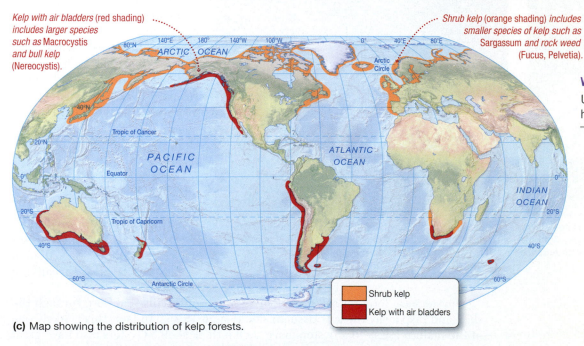

| | Shrub kelp |
| | Kelp with air bladders |

(c) Map showing the distribution of kelp forests.

attach to rocky bottoms as deep as 30 meters (100 feet). The holdfast is so strong that only large storm waves can break the algae free. The *stipes* and *blades* of the algae are supported by gas-filled floats called *pneumatocysts* (*pneumato* = breath, *cystis* = bladder), which allow the algae to grow upward and extend for another 30 meters (100 feet) along the surface to allow for good exposure to sunlight. Under ideal conditions, *Macrocystis* can grow up to 0.6 meter (2 feet) per day, making it the fastest-growing algae in the world.

The giant brown bladder kelp and bull kelp (*Nereocystis*), another fast-growing kelp, often form beds called **kelp forests** along the Pacific coast (**Figure 15.14b**). Smaller species of kelp that are generally less than 0.6 meter (2 feet) tall are known as shrub kelp. Examples of brown algae shrub kelp include *Sargassum* and rock weed (*Fucus, Pelvetia*). Even smaller tufts of red and brown algae are found on the bottom and also live on the kelp blades.

Kelp forests are highly productive ecosystems that provide shelter for a wide variety of organisms living within or directly upon the kelp as epifauna. These organisms are an important food source for many of the animals living in and near the kelp forest, including mollusks, sea stars, fishes, octopuses, lobsters, and marine mammals. Surprisingly, very few animals feed directly on the living kelp. Among those that do are sea urchins and a large sea slug called the sea hare (*Aplysia*). The distribution of kelp forests is shown in **Figure 15.14c**.

LOBSTERS Large crustaceans—including lobsters and crabs—are common along rocky bottoms. The spiny lobsters are named for their spiny covering and have two very large, spiny antennae (**Figure 15.15a**). These antennae serve as feelers and are equipped with noise-making devices near their base that are used for protection. The genus *Panulirus*, which reaches lengths to 50 centimeters (20 inches), is considered a delicacy and lives in water deeper than 20 meters (65 feet) along the European coast. The Caribbean lobster (*Panulirus argus*) sometimes exhibits a remarkable behavior of migrating single file across the sea floor in lines that are several kilometers long.

Panulirus interruptus is the spiny lobster of the American West Coast. All spiny lobsters are taken for food, but none are as highly regarded as the so-called true lobsters (genus *Homarus*), which include the American (Maine) lobster, *Homarus americanus* (**Figure 15.15b**). Although they are scavengers like their spiny relatives, the true lobsters also feed on live animals, including mollusks, crustaceans, and other lobsters.

(a) Spiny lobster (*Panulirus interruptus*), which lacks large claws but has long spiny antennae. It is found along rocky bottoms in the Caribbean and along the West Coast of North America.

(b) American, or Maine, lobster (*Homarus americanus*), which has large claws that are used for feeding and for defense. It is found along the East Coast of North America from Labrador, Canada, to North Carolina in the United States.

Figure 15.15 Spiny and American lobsters.

OYSTERS Oysters are thick-shelled organisms that grow best where there is a steady flow of clean water to provide plankton and oxygen.

Oysters are food for sea stars, fishes, crabs, and snails that bore through the shell and rasp away the soft tissue inside (**Figure 15.16**). In fact, this is likely one of the main reasons that oysters have such a thick shell.[4] Oysters also have great

(a) Visual wavelength image of the Sun from the Helioseismic and Magnetic Imager (HMI), showing sunspots.

(b) Matching extreme ultraviolet wavelength image of the Sun from the Atmospheric Imaging Assembly (AIA), showing the ejection of solar particles.

Figure 16.5 Sunspots. Matching images of the Sun, showing sunspots and accompanying ejection of particles taken on March 5, 2012.

the Sun's luminosity has increased by a small amount (0.04%), the observed changes were not large enough to account for the warming recorded during the same period. Even proxy data of solar brightness over the past 1000 years do not show a correlation with changes in climate.

Several proposals for climate change are based on solar variability related to **sunspots**, which are cooler dark areas that occur periodically on the Sun's surface (**Figure 16.5a**). **Faculae** (*facula* = bright torch), which are bright spots on the Sun, are also more abundant when sunspots are most active. Sunspots and faculae are associated with huge magnetic storms that extend from the Sun's interior to its surface and cause the Sun's ejection of particles (**Figure 16.5b**). These particles can disrupt satellite communications and also produce the *aurora* (*Aurora* = Roman goddess of dawn), which is a phenomenon caused by charged solar particles that interact with Earth's magnetic field and produce lights in the sky. In the Northern Hemisphere, these lights are known as the *aurora borealis*, or *northern lights*, and they have a matching component in the Southern Hemisphere, called the *aurora australis*, or *southern lights*.

When solar activity increases, as it does every 11 years or so, both sunspots and faculae become more numerous. But during the peak of a cycle, the faculae brighten the Sun more than sunspots dim it. The number of sunspots and faculae affects the amount of *total solar irradiance*, which is a measure of the amount of sunlight striking Earth (**Figure 16.6**, *blue curve*). For example, a *solar maximum* occurred in 2014, but had a below-average number of sunspots and was ranked among the weakest on record. During the stronger solar maximum that occurred in 2001, sunspots produced powerful solar flares that caused large geomagnetic disturbances and disrupted some space-based technology on Earth.

Figure 16.6 Earth temperature and solar activity since 1880. Graph showing average Earth temperature (*red curve*) and total solar irradiance (*blue curve*), which is a measure of the amount of sunlight striking Earth. The red and blue curves show smoothed data on an 11-year average. Note the 11-year cycle of total solar irradiance (*green curve*), which is affected by sunspots. Total solar irradiance is in watts per square meter.

STUDENTS SOMETIMES ASK . . .

In lay terms, what is a proxy?

A proxy is a substitution or an approximation for something else. A good example of a proxy is an avatar, which is a cyber-representation of you that competes in video games. There are even proxies at school. For example, if you send someone to attend class in your absence, that person is acting as a proxy for you. And it works both ways—a substitute teacher is a proxy who conducts your class when your regular teacher can't be there.

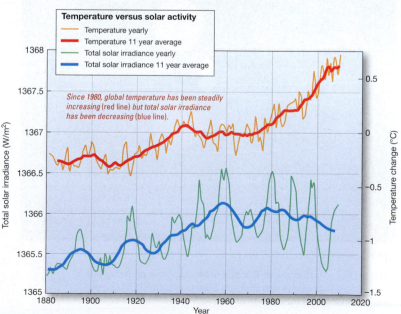

Since 1980, global temperature has been steadily increasing (red line) but total solar irradiance has been decreasing (blue line).

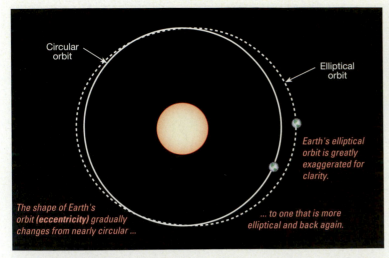

The shape of Earth's orbit (eccentricity) gradually changes from nearly circular ...

Circular orbit

Elliptical orbit

Earth's elliptical orbit is greatly exaggerated for clarity.

... to one that is more elliptical and back again.

(a) Eccentricity cycle: 100,000 years.

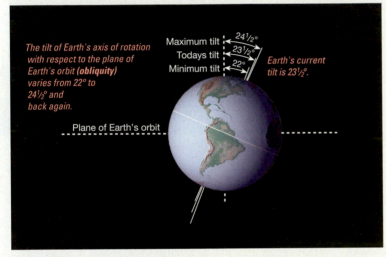

The tilt of Earth's axis of rotation with respect to the plane of Earth's orbit (obliquity) varies from 22° to 24½° and back again.

Maximum tilt — 24½°
Today's tilt — 23½°
Minimum tilt — 22°

Earth's current tilt is 23½°.

Plane of Earth's orbit

(b) Obliquity cycle: 41,000 years.

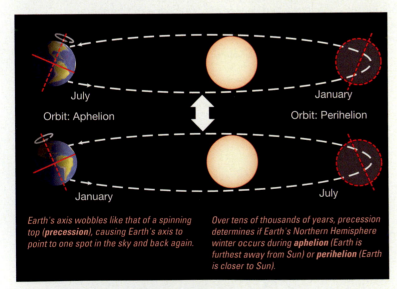

July
Orbit: Aphelion

January
Orbit: Perihelion

January

July

Earth's axis wobbles like that of a spinning top (precession), causing Earth's axis to point to one spot in the sky and back again.

Over tens of thousands of years, precession determines if Earth's Northern Hemisphere winter occurs during aphelion (Earth is furthest away from Sun) or perihelion (Earth is closer to Sun).

(c) Precession cycle: 23,000 years.

Figure 16.7 **Variations in Earth's orbit.** Diagram showing the three factors that cause variations in Earth's orbit and occur over tens to hundreds of thousands of years.

The graph in Figure 16.6 also shows the lack of correlation between solar activity (*blue curve*) and average Earth temperate since 1880 (*red curve*). Since 1980, in fact, average solar activity has been decreasing. In addition, many studies have shown that there is no significant correlation between solar activity and climate on such short timescales.

VARIATIONS IN EARTH'S ORBIT Another natural mechanism of climate change involves changes in Earth's orbit. Changes in (1) the shape of the orbit (*eccentricity*), (2) the variations in the angle that Earth's axis makes with the plane of its orbit (*obliquity*), and (3) the wobbling of Earth's rotational axis (*precession*) cause fluctuations in the seasonal and latitudinal distribution of solar radiation reaching Earth (**Figure 16.7**). These variations have cycles of about 100,000 years, 41,000 years, and 23,000 years, respectively; when they coincide with one another, they tend to amplify each other and cause climate variations on Earth. This is especially true for the Northern Hemisphere, which contains more landmasses and thus has a greater influence on the development of continental ice ages. For example, the following combined effects of Earth's orbit create conditions that are likely to initiate an ice age: (1) a slightly more elliptical orbit that takes Earth further from the Sun during a Northern Hemisphere winter (eccentricity), (2) a maximum tilt (obliquity) that causes the Northern Hemisphere to be tilted further away from the Sun, and (3) the wobbling of Earth's axis causing the Northern Hemisphere summer to coincide with perihelion,[4] which causes warmer summers but cooler winters (precession). When all three of these factors act in unison, it diminishes the solar radiation received in Earth's Northern Hemisphere and tends to produce major glaciations on northern landmasses that persist for tens of thousands of years or longer. This idea, first developed by Serbian astrophysicist Milutin Milankovitch, is called the *Milankovitch cycle*. It is now well established that these variations have contributed to the alternating glacial and interglacial episodes that characterize the most recent ice age, which occurred during the past few million years.

Currently, the combined factors of the three cycles of orbital changes are in a state of *cooling* for Earth's climate. Although it is well established that Milankovitch cycles drive long-term climate change associated with the most recent ice age on Earth, these changes take many thousands of years to manifest themselves. In contrast, the dramatic and rapid climate change that is occurring on our planet cannot be explained by these long-term variations in Earth's orbit.

VOLCANIC ERUPTIONS Explosive volcanic eruptions emit huge quantities of gases and fine-grained debris into the atmosphere (**Figure 16.8**). The largest eruptions are sufficiently powerful to inject material high into the atmosphere, where it spreads around the globe and remains aloft for many months or even years. As was seen with historic eruptions such as Mount Tambora in Indonesia (1815), Krakatoa in Indonesia (1883), El Chichón in Mexico (1982), and Mount Pinatubo in the Philippines (1991), volcanic material ejected into the atmosphere filters out a portion of the incoming solar radiation, which in turn cools the planet. For example, the year after the 1815 eruption of Mount Tambora became widely known as the *Year without Summer* because of its effect

[4]For a discussion of Earth's orbital cycle and perihelion (Earth's closest approach to the Sun), see Section 9.2 in Chapter 9.

on North American and European weather. However, the gases emitted during a volcanic eruption react with other components of the climate system and the volcanic dust eventually settles out. Thus, the cooling effect of a single eruption, no matter how large, is relatively small and short lived.

Which emits more heat-trapping carbon dioxide: Earth's volcanoes or human activities? Analysis of worldwide human and volcanic emissions over the past several decades indicate that human activities release into the atmosphere at least 130 times more heat-trapping carbon dioxide than volcanoes do (**Figure 16.9**). As discussed later in this chapter, carbon dioxide emissions make the greatest human contribution to warming Earth's atmosphere.

If volcanism is to have a pronounced impact over an extended period, many great eruptions closely spaced in time would need to occur. If this happened, the upper atmosphere would be loaded with enough gases to alter the composition of the atmosphere and enough volcanic dust to seriously diminish the amount of solar radiation reaching the surface. Because no such period of explosive volcanism is known to have occurred in the past few hundred years, it is unlikely to be responsible for the recent changes in climate. In the distant past, however, large and long-lasting volcanic eruptions may have been influential in contributing to Earth's climate shifts. For example, the Deccan Traps in India were created by extensive volcanism that started about 66 million years ago and may have caused global climatic changes that contributed to the demise of the dinosaurs.

MOVEMENT OF EARTH'S TECTONIC PLATES As described in Chapter 2, Earth's tectonic plates have moved great distances. During the geologic past, plate movements have accounted for many dramatic climate changes as landmasses shifted in relation to one another and moved to different latitudinal positions. As landmasses have moved, they have changed ocean circulation, altering the transport of heat and moisture and consequently the climate. For example, the opening of the Drake Passage between South America and Antarctica about 41 million years ago caused a fundamental reorganization of ocean currents in the Southern Hemisphere, leading to the isolation of Antarctica, which caused it to become much cooler and develop a permanent ice cap. However, the rate of plate movement is very slow—only a few centimeters per year—and so appreciable changes in the positions of continents occur only over great spans of geologic time. Thus, climate changes triggered by shifting plates are extremely gradual and happen on a scale of millions of years.

CAN NATURAL CLIMATE CHANGE FACTORS EXPLAIN RECENTLY OBSERVED CLIMATE CHANGES? It is clear that natural factors have changed Earth's climate in the past and that they will undoubtedly change it in the future. For example, natural climate change has been definitively linked to global climate shifts such as the Pleistocene Ice Age, the Medieval Warm Period, and the Little Ice Age.

During the past million years, the biggest temperature swings on Earth have been the ice ages. When ice ages have ended, it has taken about 5000 years for the planet to warm between 4°C and 7°C (7°F and 13°F). For comparison, in the 20th century alone, Earth's average temperature climbed about 0.7°C (1.3°F), which is roughly eight times faster than previous warming. In fact, proxy studies show that humanity is altering Earth's climate 5000 times faster than the pace of the most rapid natural warming episode in our planet's past.

Over the past century, scientists from all over the world have been collecting data on natural factors that influence climate—such as changes in the Sun's brightness, variations in Earth's orbit, major volcanic eruptions, and other factors. These observations have failed to show any long-term changes that could fully account for the recent rapid warming of Earth's temperature. Scientists agree that the observed warming in recent decades is occurring more quickly and with greater magnitude than can be explained by any natural factors. In essence, the release of human-caused emissions is the only viable explanation for the documented and

Figure 16.8 Volcanic eruptions spew volcanic debris and gases into the atmosphere. This 1991 eruption of Mount Pinatubo in the Philippines shows that volcanoes have the ability to inject into the atmosphere large quantities of volcanic dust and gases, which can circle the globe and block incoming solar radiation, thereby cooling the planet.

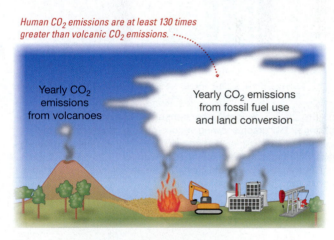

Human CO₂ emissions are at least 130 times greater than volcanic CO₂ emissions.

Yearly CO₂ emissions from volcanoes

Yearly CO₂ emissions from fossil fuel use and land conversion

Figure 16.9 A comparison of human and volcanic carbon dioxide emissions. Over the past several decades, human carbon dioxide emissions have been at least 130 times greater than volcanic emissions. Note that human-caused carbon dioxide emissions make the greatest human contribution to warming Earth's atmosphere.

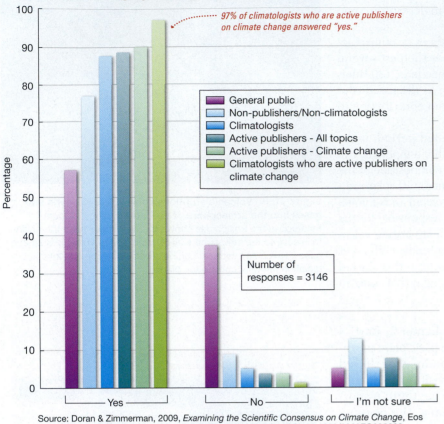

Do you think human activity is a significant contributing factor in changing mean global temperatures?

97% of climatologists who are active publishers on climate change answered "yes."

- General public
- Non-publishers/Non-climatologists
- Climatologists
- Active publishers - All topics
- Active publishers - Climate change
- Climatologists who are active publishers on climate change

Number of responses = 3146

Yes No I'm not sure

Source: Doran & Zimmerman, 2009, *Examining the Scientific Consensus on Climate Change*, Eos Transactions American Geophysical Union Vol. 90 Issue 3, 22; DOI: 10.1029/2009EO030002.

Figure 16.10 Scientific consensus on human-caused climate change. Bar graph showing various groups' responses to the question "Do you think human activity is a significant contributing factor in changing mean global temperatures?" Note that climatologists who are active publishers on climate change had the highest percentage of affirmative response.

observable recent climate changes—including the increased average temperature of Earth's surface.

IS THERE SCIENTIFIC CONSENSUS ABOUT HUMAN-CAUSED CLIMATE CHANGE? Today, there is a clear scientific consensus that human-induced emissions are responsible for the observed warming on Earth. In fact, the warming trends observed since 1950 cannot be explained without accounting for human-caused emissions.

Human-caused climate change has also received broad support by major scientific organizations, including the U.S. National Academy of Sciences (NAS), the U.S. National Research Council (NRC), the U.S. National Science Foundation (NSF), the American Meteorological Society (AMS), the National Center for Atmospheric Research (NCAR), the American Association for the Advancement of Science (AAAS), and the U.S. federal government agencies NOAA and NASA. In fact, more than 200 worldwide scientific organizations support the position that humans are causing climate change.

Scientific consensus that human activities are causing climate change was clearly established more than 20 years ago and has not changed since then. However, a recent poll (**Figure 16.10**) shows that even though the vast majority of scientists—especially climatologists—accept human-caused climate change, social acceptance of the idea by the U.S. general public lags far behind. Figure 16.10 also shows that the most highly trained experts in climate science (those who are active publishers on climate change) have the highest level of confidence that human activities are causing the recent changes in Earth's climate.

Although it makes interesting news stories to suggest that there is disagreement among scientists about the reality of human-caused climate change, there is overwhelming scientific consensus that human emissions are causing changes in Earth's climate. The existing scientific disagreements are, in fact, about specific future impacts and consequences of these climate changes.

The IPCC: Documenting Human-Caused Climate Change

In 1988, the United Nations Environment Programme and the World Meteorological Organization sponsored the **Intergovernmental Panel on Climate Change (IPCC)**, a worldwide group of atmospheric and climate scientists that began studying the human effects on climate change and global warming. The IPCC utilizes peer-reviewed literature to analyze all aspects of climate change—including science, impacts, adaptation, and mitigation—to provide independent scientific advice about climate change. Since 1990, the group has published a series of assessment reports (**Figure 16.11**) that are highly regarded by both scientists and policymakers and have sparked international movement on climate change.

THE IPCC ASSESSMENT REPORTS The IPCC's first assessment report, released in 1990, became the basis for the United Nations Framework Convention on Climate Change, an international treaty in which signatories agreed to the idea of reducing concentrations of greenhouse gases in the atmosphere. The IPCC's second assessment report, published in 1995, states that *"the balance of evidence suggests a discernible human influence on global climate"* and that global warming *"is unlikely to be entirely due to natural causes."*

In 2001, a third IPCC assessment report was published under the guidance of 426 scientists and was unanimously accepted by more than 160 delegates from 100 countries. The report states: *"There is new and stronger evidence that most of the warming observed over the past 50 years is attributable to human activities."* The report notes that recent regional climate changes already have affected many physical and biological systems on Earth and that projected climate change—as well as changes in climate extremes—could have major consequences. The report also revised the estimate of the world's expected temperature increase for the period between 1990 and 2100. Previously, the amount of predicted warming had been 1.0°C to 3.5°C (1.8°F to 6.3°F); the new report revised it upward, to 1.4°C to 5.8°C (2.5°F to 10.4°F), based on new climate models.

In 2005, an international consortium of science academies, including the U.S. National Academy of Sciences, issued this statement: *"The scientific understanding of climate change is now sufficiently clear to justify nations taking prompt action.... As the United Nations Framework Convention on Climate Change (UN-FCCC) recognizes, a lack of full scientific certainty about some aspects of climate change is not a reason for delaying an immediate response that will, at a reasonable cost, prevent dangerous [human-induced] interference with the climate system."*

In 2007, a fourth IPCC assessment report was published and included the work of more than 600 authors from 40 countries. It was reviewed by more than 600 individuals, and accepted and approved by representatives from 113 countries. The fourth IPCC report confirmed that human-caused climate change is already altering Earth. In fact, the report states that *climate change models can mimic present-day conditions only if human emissions are taken into account.* Some of the documented changes specifically mentioned in the report include the warming of oceans and land, temperature extremes, melting of snow and ice, changing wind patterns, changing water patterns, and a variety of changes to a large assortment of organisms. The report also states that the temperature increases observed since the mid-20th century are very likely due to human-caused emissions, with the probability of human influence upgraded to greater than 90% certainty. This IPCC report clearly documents the fact that by adding emissions to the atmosphere, humans are altering global climate and are producing significant impacts on physical and biological systems worldwide.

In 2014, a fifth IPCC assessment report was published; it included the work of 831 experts from scientific fields such as meteorology, physics, oceanography, engineering, and ecology. The report summarized 9200 recently published, peer-reviewed scientific studies on climate change and determined that warming of the atmosphere and ocean system is "unequivocal." The report states that many of the associated impacts such as sea level rise and increased global temperatures have occurred since 1950 at rates unprecedented in the historical record. The report also states that there is a clear human influence on climate and that it is extremely likely that human activities have been the dominant cause of observed warming since 1950, with the level of confidence raised to 95–100% certainty. In addition, the report points out that the longer humans wait to reduce our emissions, the more expensive it will become to counteract the resulting climate changes.

The IPCC assessment reports provide strong documentation of the planet's human-induced climate changes, such as global warming. In recognition of that fact, the IPCC was named a co-recipient of the 2007 Nobel Peace Prize (**Figure 16.12**), along with former U.S. Vice President Al Gore, Jr., for his work on the documentary film *An Inconvenient Truth.* When it bestowed the award, the Nobel Committee noted, *"Through the scientific reports it has issued over the past two decades, the IPCC has created an ever-broader informed consensus about the connection between human activities and global warming."*

In summary, the IPCC's five reports spanning over two decades have documented the broad scientific consensus of leading climate experts from around the globe, affirming the reality of human-caused climate change—now no longer as a

STUDENTS SOMETIMES ASK . . .

I've heard news reports that there's been a recent pause in the rise of Earth's average global temperature, which contradicts global warming. Are the reports true?

No; in actuality, scientific studies are now reporting that there was never a slowdown. The presumed pause (or false pause) in the rise of Earth's average global surface temperature in the past few decades—called *the global warming hiatus*—has been shown to be a result of measurement bias. The apparent hiatus resulted from a shift during the past couple of decades to greater use of buoys for measuring sea surface temperatures. According to experts at NOAA, buoys tend to give cooler reading than measurements taken from ships. Uncorrected, those discrepancies led to an apparent pause in the long-observed rise of Earth's average surface temperature and the belief that there was an apparent slowdown of global warming since 1998. Some of the missing heat can be accounted for by the fact that the surface Pacific Ocean has recently cooled but the upper Indian and Southern Oceans have warmed, causing a redistribution of heat in the ocean. In addition, increased heat storage has now been detected deeper within the ocean beneath surface waters.

Despite the apparent hiatus, global warming as a result of human activities is still happening and the rate of warming is increasing. In fact, climate change researchers state that global warming since the year 2000 is the strongest it's been since the latter half of the 20th century. The upward trend of global temperatures resumed in 2014.

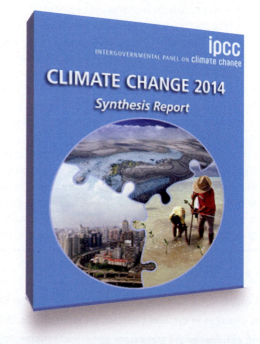

Figure 16.11 The fifth IPCC assessment report. The Intergovernmental Panel on Climate Change (IPCC), an international consortium of climate scientists, has published five assessment reports since 1990, confirming that human-induced emissions are altering Earth's climate.

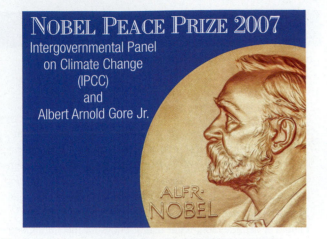

Figure 16.12 The IPCC and Al Gore Jr. share the 2007 Nobel Peace Prize. The Norwegian Nobel Committee awarded the 2007 Nobel Peace Prize to the Intergovernmental Panel on Climate Change (IPCC) and Albert Arnold (Al) Gore Jr. for "their efforts to build up and disseminate greater knowledge about man-made climate change, and to lay the foundations for the measures that are needed to counteract such change."

STUDENTS SOMETIMES ASK . . .

What's the big deal about a few degrees of warming? I normally experience a 5–10 degree temperature change during the course of a day, so why is the predicted increase in global temperature such a big deal?

For one thing, humans are highly adaptable organisms that can withstand a wide temperature range. Many plants and animals can be greatly affected by temperatures that are even a few degrees above or below their normal temperature range (this is especially true of marine organisms). Also, remember that the projected increase in temperature is on top of whatever normal (or elevated) temperatures exist. For example, let's suppose that there is a dangerous heat wave predicted. The small increase of a few degrees would make that heat wave all the more deadly. And that's exactly what is predicted for a warmer world: more extreme and prolonged weather events such as heat waves, hurricanes, tornadoes, droughts, and floods.

One other thing to consider is the geologic record. During the last ice age, the globally averaged temperature difference between a thick ice sheet on land and a warm interglacial period was only 4°C to 6°C (7.2°F to 10.8°F). Globally, just a few degrees of higher average temperatures can have a huge impact on ice sheets, sea levels, and many other aspects of climate.

RECAP

Earth's climate has changed in the past due to natural causes such as changes in the Sun's output, variations in Earth's orbit, volcanic activity, or the movement of tectonic plates. Multiple lines of evidence show that the current climate changes are due primarily to human activities that release heat-trapping emissions into the atmosphere.

prediction but as an observational reality. With new data contained in each subsequent assessment report, the message becomes increasingly certain: Humans are altering Earth's climate and actions need to be taken to reduce human-caused emissions. In addition, the IPCC reports state that global warming poses substantial risks to societies and ecosystems on all continents.

OTHER SCIENTIFIC REPORTS CONFIRM IPCC FINDINGS A host of subsequent reports have confirmed the findings of the IPCC. In 2009, for example, the U.S. Global Change Research Program issued a 190-page interagency report entitled *Global Climate Change Impacts in the United States*. The report states that *global warming is unequivocal and primarily human-induced.* The report also notes that *global average temperature has risen by about 1.5°F [0.8°C] since 1900. By 2100, it is projected to rise another 2°F to 11.5°F [1.1°C to 6.4°C]. Increases at the lower end of this range are more likely if global heat-trapping gas emissions are cut substantially. If emissions continue to rise at or near current rates, temperature increases are more likely to be near the upper end of the range.* The report warns that climate change will have numerous impacts on water resources, ecosystems, agriculture, coastal areas, human health, and other sectors.

In 2011, the U.S. National Research Council issued the final report of a five-volume series about climate change, titled *America's Climate Choices*. The reports, which were requested by the U.S. Congress, reaffirm the preponderance of scientific evidence and points to human activities—especially the release of carbon dioxide and other greenhouse gases into the atmosphere—as the most likely cause of the vast majority of global warming that has occurred over the past several decades. The reports also reiterate the pressing need for substantial action to limit the magnitude of climate change and to prepare to adapt to its impacts. The reports indicate that actions taken now *can reduce the risk of major disruptions to human and natural systems; inaction could serve to increase these risks, especially if the rate or magnitude of climate change is particularly large.*

In 2014, the U.S. National Climate Assessment Report was published by a team of more than 300 experts guided by a 60-member Federal Advisory Committee. The report was extensively reviewed by the public and scientific experts, including federal agencies and a panel of the U.S. National Academy of Sciences. The report affirms that humans are the primary cause of Earth's recent climate changes and summarizes the observable impacts of climate change on all regions of the United States. The report states that *Climate change is already affecting the American people in far-reaching ways. Certain types of extreme weather events with links to climate change have become more frequent and/or more intense, including prolonged periods of heat, heavy downpours, and, in some regions, floods and droughts.* The report also states that many climate change impacts that are already affecting Americans are expected to become increasingly disruptive across the nation throughout this century and beyond.

CONCEPT CHECK 16.2 | Examine the evidence that shows how Earth's recent climate change is caused by human activities, not a natural cycle.

1 What are proxy data? List several examples. Why are such data necessary for paleoclimatology studies?

2 List several examples of natural climate change. Do natural climate change mechanisms account for the recent climate changes that Earth is experiencing? Explain.

3 Is there scientific consensus about human-caused climate change? Explain.

4 What is the IPCC? What role does it have in documenting human-caused climate change?

16.3 What Causes the Atmosphere's Greenhouse Effect?

Numerous scientific studies indicate that human-caused emissions are responsible for the recent and dramatic climate changes experienced on Earth, including the increase in average worldwide temperature, which is called **global warming**. Although the **greenhouse effect** is a natural process that influences the temperature of Earth's surface and atmosphere, it is now being altered by human emissions, a phenomenon that is often referred to as the *anthropogenic greenhouse load* or the *enhanced greenhouse effect*.

The greenhouse effect gets its name because it keeps Earth's surface and lower atmosphere warm in a way similar to a greenhouse that keeps plants warm enough to grow, regardless of outside conditions (**Figure 16.13**). Energy radiated by the Sun covers the full electromagnetic spectrum, but most of the energy that reaches Earth's surface is short wavelengths, in and near the visible portion of the spectrum. In a greenhouse, shortwave sunlight passes through the glass or plastic covering, where it strikes the plants, the floor, and other objects inside and is re-radiated as longer-wavelength infrared radiation (heat). Some of this heat energy escapes from the greenhouse and some is trapped for a while by the glass or plastic covering, which keeps the greenhouse nice and snug—much like what happens in Earth's atmosphere.[5]

Interdisciplinary Relationship

Earth's Heat Budget and Changes in Wavelength

Figure 16.14 shows the various components of Earth's **heat budget**, which describes all the ways in which heat is added to and subtracted from Earth. Although Earth's atmosphere blocks some forms of solar radiation, it is transparent to most wavelengths of visible light, which is able to pass through the atmosphere like sunlight coming through greenhouse glass. However, only about 47% of the solar radiation that is directed toward Earth reaches Earth's surface and is absorbed by the oceans and continents. Of the 53% of solar radiation that isn't absorbed by land or water, about 23% is absorbed by molecules in the atmosphere, dust, and clouds, and about 30% is reflected back into space by atmospheric backscatter, clouds, and reflective regions of Earth's surface.

Figure 16.15 shows that most of the energy coming to Earth from the Sun is within the visible part of the spectrum and peaks at a wavelength of 0.48 micron (0.00002 inch).[6] When this radiation is absorbed by water and rocks at Earth's surface, these materials absorb some of the energy and warm up and then emit radiation away from Earth's surface toward space as longer-wavelength infrared (heat) radiation, with a peak at a wavelength of 10 microns (0.0004 inch). Atmospheric gases such as water vapor, carbon dioxide, and other gases absorb radiation at these longer

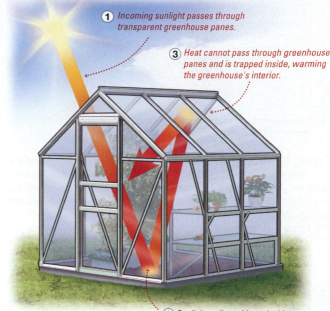

Atmospheric gases such as water, carbon dioxide, and methane act just like the glass of a greenhouse, allowing sunlight to pass through but trapping outgoing heat.

① *Incoming sunlight passes through transparent greenhouse panes.*

③ *Heat cannot pass through greenhouse panes and is trapped inside, warming the greenhouse's interior.*

② *Sunlight strikes objects inside greenhouse, loses energy, and is converted to heat.*

Figure 16.13 How a greenhouse works.

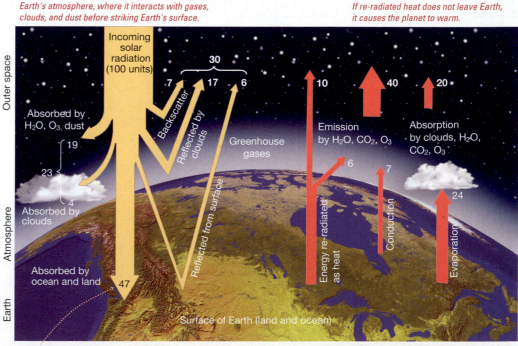

Shorter-wavelength visible light passes through Earth's atmosphere, where it interacts with gases, clouds, and dust before striking Earth's surface.

If re-radiated heat does not leave Earth, it causes the planet to warm.

Incoming solar radiation (100 units)

Outer space

Absorbed by H_2O, O_3, dust

Backscatter

Reflected by clouds

Greenhouse gases

Emission by H_2O, CO_2, O_3

Absorption by clouds, H_2O, CO_2, O_3

Atmosphere

Absorbed by clouds

Reflected from surface

Energy re-radiated as heat

Conduction

Evaporation

Earth

Absorbed by ocean and land

Surface of Earth (land and ocean)

30 7 17 6 10 40 20 19 23 4 6 7 24 47

Of all solar radiation able to penetrate Earth's atmosphere, only 47% is absorbed by the ocean and land.

Longer-wavelength infrared radiation either radiates back into space or is trapped in Earth's atmosphere.

Figure 16.14 Earth's heat budget. In this example, 100 units of solar radiation from the Sun (mostly shorter-wavelength visible light) are reflected, scattered, and absorbed by various components of the Earth–atmosphere system. The absorbed energy is re-radiated back into space from Earth as longer-wavelength infrared radiation (heat). If this infrared radiation does not leave Earth, global warming will occur.

[5]Recent studies have indicated that an additional factor in keeping a greenhouse warm is that the greenhouse covering prevents mixing of warm air inside with cooler air outside. Although this is different from how the atmosphere works, the term *greenhouse effect* is still commonly used to describe the atmosphere's warming process.

[6]A micron (μ), which is actually a micrometer, is one-millionth of a meter.

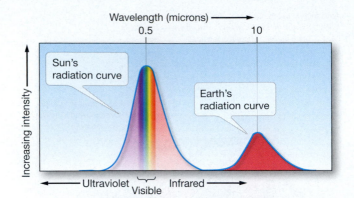

Wavelength (microns) →
0.5 10

Sun's radiation curve

Earth's radiation curve

Increasing intensity →

← Ultraviolet Visible Infrared →

Figure 16.15 Energy radiated by the Sun and Earth. The intensity of energy radiated by the Sun peaks at a wavelength of 0.48 microns (0.00002 inch), which is in the visible part of the spectrum. Some of this energy is absorbed or reflected while some reradiates from Earth in the infrared (heat) range at a wavelength of 10 microns (0.0004 inch).

Web Animation
Global Warming
http://goo.gl/Uz8rmm

Web Animation
Atmospheric Energy Balance
http://goo.gl/imdsCN

wavelengths, effectively intercepting the heat radiation that attempts to leave the planet, thus heating the atmosphere. This trapping of heat radiation and heating of the atmosphere is known as the *greenhouse effect.*

In summary, most of the solar radiation that is not reflected back to space passes through the atmosphere and is absorbed at Earth's surface. Earth's surface, in turn, re-emits longer-wavelength infrared radiation (heat). A portion of this energy is absorbed by certain heat-trapping gases in the atmosphere, thus producing the greenhouse effect. Thus, *the change of wavelengths from visible to infrared at Earth's surface is the key to understanding how the greenhouse effect works.*

Interdisciplinary
Relationship

Which Gases Contribute to the Greenhouse Effect?

Earth's greenhouse effect is caused by an array of atmospheric gases, many of which have both natural and human-caused sources. Take, for example, water vapor, which contributes more to the greenhouse effect than any other gas. In fact, water vapor is the single most important absorber of heat—its contribution to the greenhouse effect is between 36% and 66% of the greenhouse effect, and together with clouds, it comprises about 75% of the greenhouse effect.

Interdisciplinary
Relationship

Water vapor enters the atmosphere mostly through evaporation and other natural processes. Although atmospheric water vapor concentrations fluctuate regionally, studies suggest that human activity does not significantly affect water vapor concentrations except at local scales, such as near irrigated fields. And even then, the water vapor does not stay in the atmosphere very long. In essence, human activities do not directly affect the amount of water vapor in the atmosphere on a global scale. However, humans may be indirectly affecting the amount of water vapor in the atmosphere: Satellite measurements of atmospheric water vapor show a 4% increase in global specific humidity at sea level since 1970, which researchers have tied to human-caused climate change.

Still, atmospheric water vapor plays an important role in warming. For example, a recent study suggests that an increase in stratospheric water vapor due to natural processes between 1980 and 2000 may have amplified the rapid warming of that period by as much as 30%. Conversely, a 10% decrease in the amount of water vapor in the stratosphere since 2000 may have slowed global warming by causing global temperatures to level off despite the continued rise in human emissions.

Table 16.1 shows the concentration of **greenhouse gases**—so called because of their heat-trapping capacity—that have been increasing as a result of human activities. Remarkably, these gases exist in very small amounts in the atmosphere, yet they have a profound effect on heating. And unlike water vapor that does not stay in the atmosphere very long, many of these gases stay in the atmosphere for long periods and continue to trap heat. Some of these greenhouse gases are released by both human and natural sources. Others, however, have no natural source and thus are clearly a product of human activities.

CARBON DIOXIDE Of all the human-caused gases, carbon dioxide makes the greatest relative contribution to increasing the greenhouse effect (Table 16.1). Carbon dioxide enters the atmosphere as a result of combustion of carbon compounds with oxygen. It is a colorless and odorless gas that is the same one we exhale from our lungs. The conversion of **fossil fuels** (oil and natural gas) into energy by cars, factories, and power plants accounts for the majority of the annual human contribution to carbon dioxide emissions, with industrialized nations contributing the most. As a result of human activities, atmospheric concentration of carbon dioxide has increased more than 40% over the past 250 years (**Figure 16.16**). The direct measurement of carbon dioxide concentration in the atmosphere was initiated by Charles David Keeling in 1958 and is continued today by his son, Ralph Keeling. The iconic curve showing the steady increase in atmospheric carbon dioxide now bears their name (**Diving Deeper 16.1**).

Interdisciplinary
Relationship

SMARTTABLE 16.1 HUMAN-CAUSED GREENHOUSE GASES AND THEIR CONTRIBUTION TO INCREASING THE GREENHOUSE EFFECT

Atmospheric gas	Human-caused sources of gas	Pre-industrial (circa 1750) concentration (ppbv[a])	Present concentration (ppbv[a])	Current rate of increase or decrease (% per year)	Relative contribution to increasing the greenhouse effect (%)	Infrared radiation absorption per molecule (number of times greater than CO_2)
Carbon dioxide (CO_2)	Combustion of fossil fuels	280,000	401,000	+0.5	60	1
Methane (CH_4)	Leakage, domestic cattle, rice agriculture	700	1825	+1.0	15	25
Nitrous oxide (N_2O)	Combustion of fossil fuels, industrial processes	270	315	+0.2	5	200
Tropospheric ozone (O_3)	Byproduct of combustion	0	10–80	+0.5	8	2000
Chlorofluorocarbon (CFC-11)	Refrigerants, industrial uses	0	0.26	−1.0	4	12,000
Chlorofluorocarbon (CFC-12)	Refrigerants, industrial uses	0	0.54	0.0	8	15,000
Total					100	

[a]ppbv = parts per billion by volume (not by weight).

SmartTable 16.1 Human-caused greenhouse gases and their contribution to increasing the greenhouse effects.
https://goo.gl/m78XyV

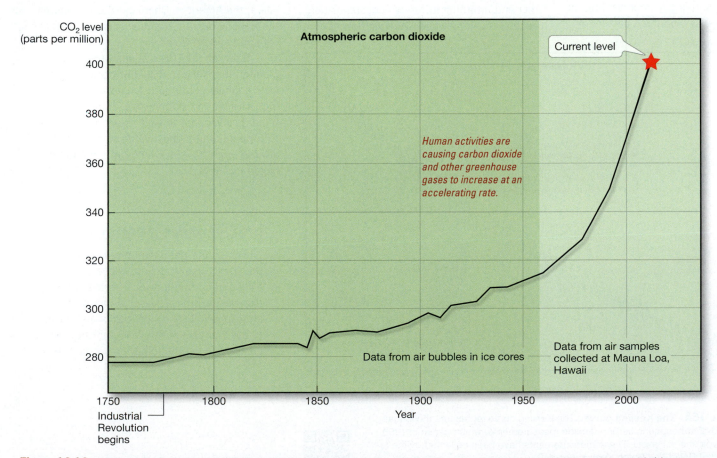

Figure 16.16 Amount of carbon dioxide in the atmosphere since 1750. Graph shows the dramatic increase of average worldwide atmospheric carbon dioxide since the Industrial Revolution began in the late 1700s. Values for 1958 to the present are from laboratory measurements of carbon dioxide in air samples collected at Mauna Loa Observatory in Hawaii; values prior to 1958 are from measurements of carbon dioxide in air bubbles preserved in polar ice cores.

THE ICONIC KEELING CURVE OF ATMOSPHERIC CARBON DIOXIDE AND THE FATHER–SON TEAM WHO CREATED IT

In the first part of the 20th century it was suspected that the concentration of atmospheric carbon dioxide (CO_2) might be increasing because of fossil fuel combustion. However, there were relatively few measurements of this important greenhouse gas and the measurements varied widely. In an effort to determine the atmospheric concentration of carbon dioxide, Charles David (Dave) Keeling in 1958 began measuring atmospheric carbon dioxide atop Mauna Loa volcano in Hawaii, where researchers can access very clean, high altitude air. Those measurements continue to this day, representing the longest continuously monitored dataset of

atmospheric carbon dioxide in the world. The graphical representation of atmospheric CO_2 is one of the most iconic graphs in support of human-caused global warming and is named in his honor (**Figure 16.A**).

Under the direction of the U.S. Weather Bureau (now part of NOAA) and Scripps Institution of Oceanography, Keeling installed a gas analyzer at Hawaii's Mauna Loa station in March 1958; on the first day of operation it recorded an atmospheric CO_2 concentration of 313 ppm. In April 1958, to Keeling's surprise, the CO_2 concentration at Mauna Loa had risen by 1 part per million (ppm) and then even higher in May, after which it began to decline, reaching a minimum in October. In the following months, the concentration increased again and repeated the same seasonal pattern, which is now a distinct component of the Keeling curve and is explained by the natural cycle of plants in the Northern Hemisphere withdrawing CO_2 from the air for plant growth via photosynthesis during summer and returning it through decomposition each succeeding winter. Keeling described it

as a seasonal "breathing cycle" on a global scale.

Keeling's measurements showed the first significant evidence of rapidly increasing CO_2 levels in the atmosphere. Many scientists credit Keeling's graph with first bringing the world's attention to the increase of atmospheric CO_2. In addition, researchers have documented that the steady increase in atmospheric CO_2 is a result of human emissions.

Keeling continued to direct the measurement of atmospheric CO_2 in Mauna Loa from his lab at Scripps until his death in 2005. Supervision of the measuring project was taken over by his son, Ralph Keeling, a Professor of Geochemistry at Scripps. The long-term measurement of atmospheric CO_2 initiated by David Keeling and continued today by his son Ralph Keeling is an important piece of climate data that clearly shows how humans are contributing to climate change. To commemorate the work done by the Keelings and to honor the importance of the curve, the American Chemical Society named the Keeling curve a National Historic Chemical Landmark in 2015.

Today, about 100 stations worldwide monitor atmospheric CO_2, but none of them have a longer continuous record than Keeling's Mauna Loa station. Most importantly, the data show that today's value of more than 400 ppm is the highest value of measured atmospheric CO_2 since long-term monitoring began in 1958.

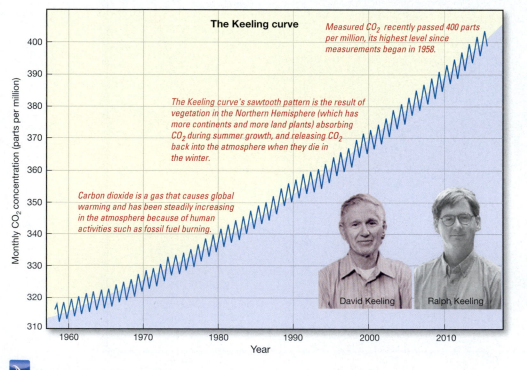

The Keeling curve

Measured CO_2 recently passed 400 parts per million, its highest level since measurements began in 1958.

The Keeling curve's sawtooth pattern is the result of vegetation in the Northern Hemisphere (which has more continents and more land plants) absorbing CO_2 during summer growth, and releasing CO_2 back into the atmosphere when they die in the winter.

Carbon dioxide is a gas that causes global warming and has been steadily increasing in the atmosphere because of human activities such as fossil fuel burning.

David Keeling Ralph Keeling

GIVE IT SOME THOUGHT

1. Why does the Keeling curve, which shows the steady rise in the atmospheric concentration of carbon dioxide over the course of decades, exhibit a seasonal pattern where atmospheric CO_2 concentration increases in the first part of the year, and then decreases during the following six months?

SmartFigure 16A **The Keeling curve.** The Keeling curve represents the longest continuous record of atmospheric carbon dioxide measurements, which began in 1958 atop Mauna Loa volcano in Hawaii. These measurements have been collected by the father and son team of David and Ralph Keeling (*insets*).
https://goo.gl/ffsCXD

What concerns scientists is that over the past 250 years—and especially in the past 50 years—human activities have been responsible for raising the concentration of greenhouse gases in the atmosphere at an ever-increasing rate. As of 2016, the average yearly concentration of atmospheric carbon dioxide is 401 parts per million and is increasing by about 2 parts per million each year; this rate of increase is double what occurred only 50 years ago. In terms of sheer numbers, humans are now pumping into the atmosphere more than 33 billion metric tons (73 trillion pounds) of carbon dioxide each year.[7] And the last time that carbon dioxide levels were as high as they are today was about 15 million years ago during the Miocene warm period, when temperatures were at least 3°C to 5°C (5°F to 9°F) warmer than today.

Interdisciplinary

Relationship

METHANE Methane is the second-most-abundant human-caused greenhouse gas (Table 16.1). It is produced by leakage from decomposing trash in landfills, by methane-belching domestic cattle, and by agriculture (particularly the cultivation of rice). Even though methane has a lower concentration in the atmosphere than carbon dioxide, it has a much greater ability to produce warming on a per-molecule basis. Since the Industrial Revolution began in about 1750, when the concentration of methane in the atmosphere was about 700 parts per billion by volume (ppbv), methane has increased in the atmosphere by more than two and a half times, to a value of 1825 ppbv. Even though methane is present in very tiny amounts in the atmosphere, scientists are concerned about this dramatic increase because, pound for pound, the comparative impact of methane on climate change is 25 times greater than that of carbon dioxide over a 100-year timescale.

OTHER GREENHOUSE GASES The other trace gases shown in Table 16.1—nitrous oxide, tropospheric ozone, and chlorofluorocarbons—are present in far lower concentrations than carbon dioxide and methane. Yet they are still very important because they absorb many times more infrared radiation per molecule than carbon dioxide or methane (Table 16.1, *last column*), thus making them very potent contributors to warming. Still, these gases have a smaller overall contribution to increasing the greenhouse effect because their concentrations are so low. Nevertheless, all these gases must be taken into account when considering the total amount of greenhouse warming.

GREENHOUSE GASES: PAST AND FUTURE In 2005, researchers recovered a nearly 3.2-kilometer (2-mile) continuous ice core from Antarctica that contains a record of past atmospheric concentrations of carbon dioxide and methane—two important greenhouse gases—that get trapped as ice accumulates. Analysis of the core, which extends back in time 800,000 years ago (**Figure 16.17**), shows that the average level of carbon dioxide (*red curve*) naturally varied from about 180 parts per million to about 280 parts per million. During the same time, methane (Figure 16.17, *green curve*) varied from about 350 parts per billion to about 750 parts per billion, in sync with carbon dioxide levels. In addition, the chemical make-up of the ice provides a proxy of the past average temperature on Earth (Figure 16.17, *black curve*), which shows a strong correlation with atmospheric methane and carbon dioxide concentrations: When carbon dioxide and methane are low, Earth experiences cooler temperatures (glacial periods), and when carbon dioxide and methane are high, Earth experiences warmer temperatures (interglacial periods). The graph shows that our planet has passed through a cycle of glaciation and deglaciation every 100,000 years or so and that carbon dioxide and methane have varied in step with it. This is all part of Earth's natural climate cycle.

Interdisciplinary

Relationship

[7]On average, each person on Earth emits more than 4.9 metric tons (10,800 pounds) of carbon dioxide per year. In addition, this number is several times greater for those living in industrial nations (such as the United States) than for those living in developing countries (such as China).

STUDENTS SOMETIMES ASK . . .

Is carbon dioxide causing the hole in the ozone layer?

Here's the short answer: definitely not! The ozone layer occurs within the atmosphere's stratosphere and is composed of ozone molecules (O_3) that absorb most of the Sun's ultraviolet radiation. Without it, unhealthy levels of ultraviolet radiation would reach Earth's surface, making the planet largely uninhabitable. The main ozone hole (actually, a seasonal thinning of the ozone layer) occurs above the South Pole, with a smaller one above the North Pole. Both are caused by chemical reactions with natural and human-generated compounds, particularly the now-banned chlorofluorocarbon chemicals CFC-11 and CFC-12, but not carbon dioxide. CFCs are also strong greenhouse gases (see Table 16.1), so CFCs are thus double threats to the environment: Their buildup in the atmosphere leads to the destruction of the ozone layer and, at the same time, contributes to global warming. Notice also from Table 16.1 that tropospheric (lower atmosphere) ozone, which is a byproduct of combustion, is a potent greenhouse gas. With the ban of CFCs firmly in place, scientists predict that Earth's protective ozone layer will achieve its normal thickness by the middle of the 21st century. However, new research reports that human-generated nitrous oxide, which also destroys ozone, is now the single greatest ozone-depleting substance.

Most importantly, the graph also shows the unusually high levels of carbon dioxide and methane in today's atmosphere (*colored stars*). In fact, the current atmospheric concentrations of carbon dioxide (401 parts per million) and methane (1825 parts per billion) are at their highest levels by far in the past 800,000 years—and perhaps for the first time in millions of years, according to geologic evidence.

Analysis of proxy data of Earth's geologic past shows that Earth has experienced times of much higher average surface temperatures than it is experiencing today. One of the largest-known events of this type is called the *Paleocene–Eocene Thermal Maximum*, or PETM, which occurred about 56 million years ago, when Earth's climate was much warmer than it is today. Chemical indicators point to a huge release of carbon dioxide and methane that caused Earth to warm by at least 5°C (9°F) over a few thousand years. The PETM is marked by massive marine ecosystem changes that were caused by an increase in ocean temperatures, which caused the depletion of dissolved oxygen in surface waters because warmer water can't hold as much oxygen. In addition, higher amounts of carbon dioxide from the atmosphere dissolved in the oceans, making them more acidic (which is also currently happening; this is discussed later in this chapter). As a result, these ancient warmer periods such as the PETM are typically marked by the extinction of many marine organisms.

Interdisciplinary

Relationship

What will the level of carbon dioxide—with its known ability to trap heat in Earth's atmosphere—be like in the future? Based on sophisticated computer models, the IPCC has made predictions to the year 2100 based on various scenarios (**Figure 16.18**). The predictions are based on greenhouse gas emission scenarios, which are affected by factors such as population growth, economic development, technological changes, and cultural and social interactions. For example, in scenario A2 (Figure 16.18, *red curve*), population increases at the current growth rate

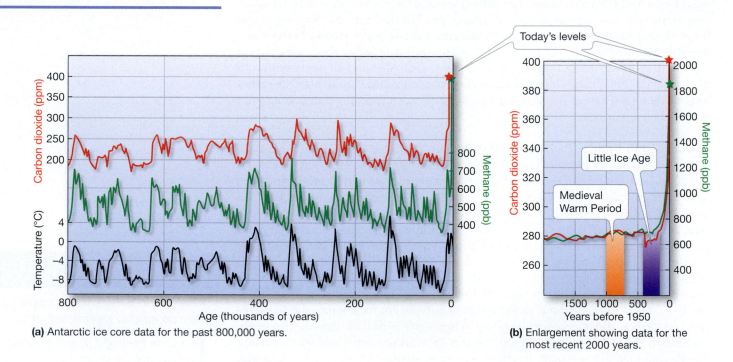

(a) Antarctic ice core data for the past 800,000 years.

(b) Enlargement showing data for the most recent 2000 years.

RECAP

The greenhouse effect is caused by gases that allow sunlight to pass through the atmosphere but trap heat energy before it is radiated back to space. Carbon dioxide is foremost in an array of gases from human activity that increase the atmosphere's ability to trap heat.

SmartFigure 16.17 Ice core data of atmospheric composition and global temperature. **(a)** Antarctic ice core data for the past 800,000 years, showing atmospheric carbon dioxide (*red curve*), methane (*green curve*), and average global temperature (*black curve*). Today's carbon dioxide and methane levels are shown by the red and green stars. Atmospheric composition is from analysis of air trapped in ice bubbles; temperature reconstruction is derived from the chemical make-up of the ice. **(b)** Enlargement showing data for the most recent 2000 years. Medieval Warm Period and Little Ice Age are also shown. https://goo.gl/eyqrTr

accompanied by an economic theme of self-reliance and preservation of local identities. In this "business-as-usual" scenario, carbon dioxide reaches its highest levels and global surface temperature is expected to increase by 4°C (7.2°F). Not since the end of the last ice age 10,000 years ago has the average temperature changed as dramatically as the change that this scenario predicts in less than 100 years.

In scenario A1B, (Figure 16.18, *blue curve*), there is very rapid economic growth, global population peaks in midcentury and declines thereafter, and there is rapid introduction of new and more efficient technologies (such as fuel cells, solar panels, and wind energy) that replace much of today's fossil fuel combustion. This "middle-of-the-road" scenario results in a moderate amount of atmospheric carbon dioxide that equates to a warming of about 2.5°C (4.5°F).

In scenario B1 (Figure 16.18, *green curve*), there is the same global population as in the A1B scenario but with rapid changes in economic structures toward a service and information economy, accompanied by reduction in material intensity and the widespread adoption of clean and resource-efficient technologies. In this scenario, countries come together to use both technology and general environmental controls to decrease emissions. This "best case" scenario results in the lowest amounts of atmospheric carbon dioxide and a warming of only about 1.0°C (1.8°F).

Even if greenhouse gas concentrations stabilized today with no additional increase, the planet would continue to warm by about 0.6°C (1.1°F) over the next century because it takes years for Earth's climate system to fully react to increases in greenhouse gases (including the stabilization of various feedback loops). This future warming is often referred to as Earth's *commitment to warming*. Most importantly, it is clear that the choices made now will determine the amount of atmospheric carbon dioxide and associated warming for the remainder of the 21st century and beyond.

Other considerations: Aerosols

Aerosols are suspended particles in the atmosphere that can affect the atmosphere's reflectivity and its ability to trap heat, thus contributing to climate change. One of the most important human-caused aerosols is *black carbon*, which is composed of tiny airborne particles of carbon and is simply known as *soot*. Black carbon enters the atmosphere through the incomplete combustion of organic matter such as wood in a cook-stove, coal in a power plant, diesel fuel in cars and trucks, or trees charred by a wildfire. Once it gets into the atmosphere, black carbon can remain afloat for weeks and increase the atmosphere's ability to absorb heat. The contribution to warming of 1 gram (0.035 ounce) of black carbon has been estimated to be anything from 100 to 2000 times higher than that of an equal amount of carbon dioxide. In addition, black carbon both aids in the formation of clouds and sometimes inhibits the formation of clouds, which is why climate models have difficulty in predicting the effect of black carbon on climate. Black carbon that settles onto land, however, has a predictable effect: It lowers most Earth surface material's **albedo** (*albus* = white), especially if it falls on snow or ice, where the dark color of soot increases warming and enhances further melting.

Much of the black carbon produced by humans enters the oceans and ends up on the bottom of the sea, but scientific studies have shown that it can hover in seawater for thousands of years. What concerns scientists is the constant and

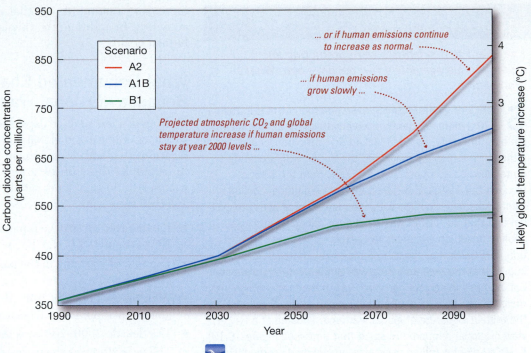

SmartFigure 16.18 **Scenarios for future atmospheric carbon dioxide levels and corresponding global temperature increase.** Graph showing the predicted atmospheric carbon dioxide levels and likely global temperature increase for various scenarios. Higher carbon dioxide and temperature values are associated with the A2 scenario (*red curve*), which assumes that human emissions will increase. Intermediate carbon dioxide and temperature values are associated with the A1B scenario (*blue curve*), which assumes greenhouse gas emissions will grow slowly. The lowest carbon dioxide and temperature values are associated with the B1 scenario (*green curve*), which assumes that greenhouse gases will stay at year 2000 levels.
https://goo.gl/mGGvn6

RECAP

Trapped air bubbles in ancient ice reveal that today's levels of heat-trapping carbon dioxide and methane are greater than at any time within the past 800,000 years. Scenarios show that human emissions will cause continued warming on Earth, but the choices we make now will determine the amount of future warming.

I've heard that scientists were predicting an oncoming ice age only a few decades ago. What's changed?

Skeptics of global warming have pointed out that in the 1970s, climate scientists were warning of an imminent ice age. In reality, the idea lacked scientific consensus, although the media popularized the notion with sometimes quite alarmist accounts. It is true that from the 1950s through the 1970s, there was an observed global cooling trend. This fact, coupled with the idea that ice ages are cyclical and that our planet was poised for millennia of global cooling based on variations in Earth's orbit, led to speculation that Earth may return to an ice age. However, it is now known that the recent cooling was probably due to a substantial increase in aerosol particles, which made the atmosphere more reflective to incoming sunlight and at the time masked the telltale signature of global warming. Looking at a longer view of global temperature since that time (see Figure 16.19), Earth's recent warming trend can be clearly discerned.

increasing amount of soot that is released into the environment through human activities. In addition, the inhalation of soot has a substantial negative impact on human health.

What Documented Changes Are Occurring Because of Global Warming?

Melting glaciers and ice caps, shorter winters, shifts in species distribution, and a steady rise in average global and sea surface temperatures are just some of the indications that additional human-induced greenhouse warming is occurring. Take, for example, these observations about Earth's temperature, which are based on weather stations on land, satellite data, and, for earlier measurements, proxy data and data from ships:

- Earth's average surface temperature has risen 0.6°C (1.1°F) over the past 30 years and 0.8°C (1.4°F) since 1865 (**Figure 16.19**).
- The rate of warming in the past 50 years is double the rate observed over the past 100 years.
- With the exception of 1998, the ten warmest years in the instrumental record (dating to 1880) have all occurred since 2000.
- The decade 2000–2010 was the warmest decade on record.
- The year 2014 was the warmest year on record.
- Over the past century, the planet has experienced the largest increase in surface temperature in at least 1300 years.
- Sea surface temperatures have increased worldwide (see the next section).
- There have been increasing occurrences of severe heat waves, such as the 2014 heat wave in India that killed more than 2300 people, the 2012 heat wave in the United States (with one of the hottest Julys on record), and the 2010 heat wave in eastern Europe and Russia (the region's greatest heat wave in the past 500 years). Such extreme events have been scientifically linked to human-caused climate warming, and models project that severe heat waves will become more likely in the future.

Figure 16.19 Thermometer recordings of surface air temperature show increase since 1865. The record of global average surface air temperature from thermometer readings indicates a global warming of at least 0.8°C (1.4°F) since 1865. The peaks and troughs indicate the natural year-to-year variability of climate.

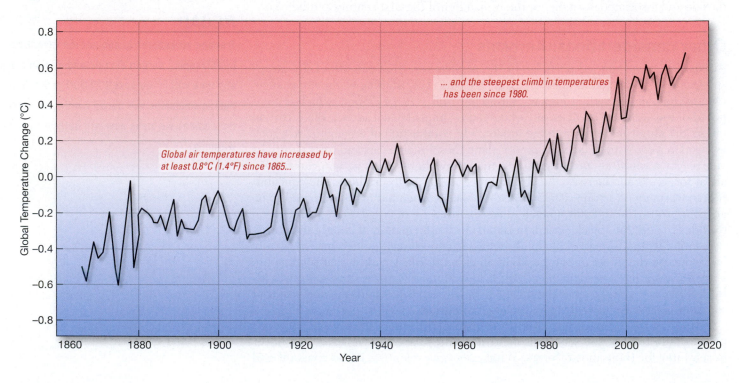

As global temperatures continue to increase, researchers use sophisticated climate models to forecast what changes will occur on Earth. Because of the complexity of the climate system and its feedback loops, not all models agree about what or how severe those changes will be. However, there are points on which the models do agree: a strong warming of high northern latitudes, a moderate warming of middle latitudes, and relatively little warming in low latitudes. Other predicted changes—many of which are already happening—include the following:

- Earlier summertime seasons, with higher summer temperatures, including longer and more intense heat waves.
- More extreme precipitation events, such as severe droughts in certain areas and increased chances of flooding in other areas.
- The worldwide retreat of ice fields and mountain glaciers, which is already being observed.
- Water contamination issues that lead to larger outbreaks of waterborne infectious diseases such as malaria, yellow fever, and dengue fever.
- Shifts in the distribution of plant and animal communities that affect entire ecosystems and may drive certain species to extinction.

Yet not all predicted changes have negative consequences. For instance, increased warming is expected to provide a longer growing season for some crops, increased atmospheric carbon dioxide should help promote biological **productivity** in plants, and ice-free Arctic waters tend to have higher oceanic productivity. However, most research suggests that the negative impacts of a changing climate far outweigh the positive impacts. In addition, a number of uncertainties remain in understanding regional effects of climate change, and various components of the climate system may respond to these changes in unanticipated and surprising ways.

STUDENTS SOMETIMES ASK . . .

Why was this winter so cold when there's supposed to be global warming?

One of the documented changes of a warmer world is increased variation in both temperature and precipitation extremes. This means that while the climate is warming, not only will there be a wider range of temperatures—including both warmer and colder temperature extremes—but also wetter or drier conditions. In essence, global warming increases the chances of extreme events such as heat waves, record cold snaps, storms, droughts, intense rains, unusual amounts of snow, and lack of snow altogether. It's no wonder that many scientists prefer the term global weirding.

Also, remember that climate is the long-term average of weather, so although it might be colder during one season, the climate can still be warming. What matters is not what happens on any given day or season but what the trend is over a period of years. On this, the data are clear: Earth is experiencing long-term global warming.

| CONCEPT CHECK 16.3 | Demonstrate an understanding of how the atmosphere's greenhouse effect works. |

1 Describe the fundamental difference between solar radiation absorbed at Earth's surface and the radiation that is primarily responsible for heating Earth's atmosphere.

2 Discuss the greenhouse gases in terms of their relative concentrations in the atmosphere and their relative contributions to global warming.

3 Why has the carbon dioxide level of the atmosphere been rising for more than 150 years?

4 Explain what a commitment to warming means. What commitment to warming will Earth experience in the future?

5 Describe how atmospheric temperatures and other factors are likely to change as carbon dioxide levels continue to increase.

16.4 What Changes Are Occurring in the Oceans as a Result of Global Warming?

The oceans are a key component in the global climate system that is currently experiencing dramatic changes. Let's examine some of the observed and predicted effects of global warming in the oceans.

Increasing Ocean Temperatures

Studies have revealed that the oceans have absorbed the majority of the increased heat in the atmosphere. Indeed, millions of ocean temperature

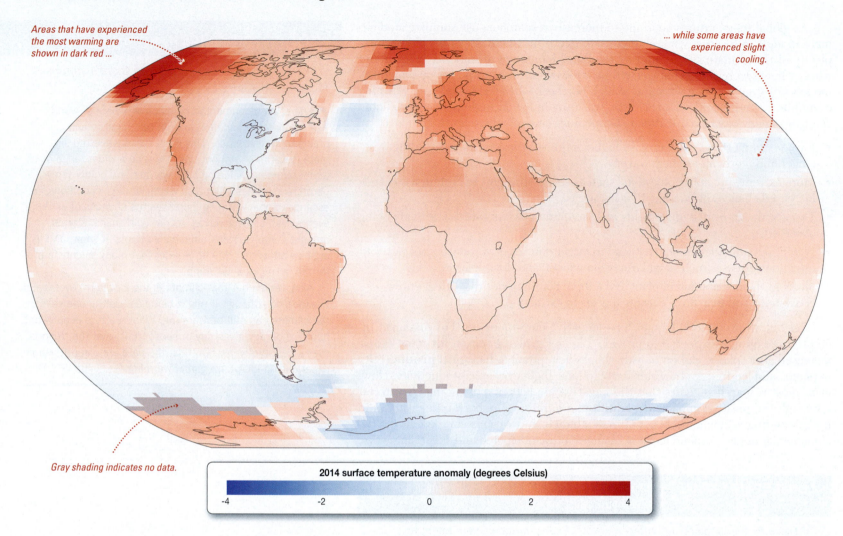

Areas that have experienced the most warming are shown in dark red ...

... while some areas have experienced slight cooling.

Gray shading indicates no data.

2014 surface temperature anomaly (degrees Celsius)

-4 -2 0 2 4

Figure 16.20 **Global temperature change.** Map showing the change in land and ocean surface temperature in 2014 compared to the 1950–1980 baseline period. Most of the globe is anomalously warm (*red areas*) as a result of greenhouse warming, with the greatest temperature increases in the Arctic Ocean, Alaska, Siberia, and West Antarctica. Gray areas indicate no data.

Web Video
Global Temperatures 1880–2010
https://goo.gl/rmPn8E

observations at various depths reveal that there has been an overall increase in surface temperature (**Figure 16.20**). These measurements indicate that global sea surface temperatures have risen by about 0.6°C (1.1°F) because of global warming, mainly since about 1970. However, this warming has not been uniform throughout the ocean. The greatest temperature increases have been experienced in the Arctic Ocean, near the Antarctic Peninsula, and in tropical waters. Even deep waters are showing signs of warming: In some places, warming has been documented to a depth of 0.8 kilometer (0.5 mile) or more. And deeper waters are warming faster than expected. To determine the extent of warming experienced in the oceans, scientists have initiated a program to monitor changes in ocean temperature using the ocean's ability to transmit sound (MasteringOceanography **Web Diving Deeper 16.1**).

The impacts of a warmer ocean are far-reaching and will persist for several centuries. For example, increased seawater temperatures are likely to affect temperature-sensitive organisms such as corals. Coral reefs are already in decline, and as discussed in Chapter 15, warmer sea surface temperatures have been implicated in widespread coral bleaching events. Warmer waters could also shift or disrupt the spawning cycles of corals. In addition, studies of coral distribution have found that some corals are already migrating with the spread of warmer waters to areas where they haven't been seen before.

Other studies have documented how warmer ocean waters are changing the physical characteristics of the ocean. For example, research using the Argo

Interdisciplinary

Relationship

program of free-drifting floats (see Section 7.1 in Chapter 7) has shown that the world's hydrologic cycle is accelerating as warming speeds up evaporation of ocean surface waters, thereby affecting ocean salinity. As mentioned earlier in this chapter, the melting of sea ice forms a positive-feedback loop that accelerates ocean warming in high latitudes because ice-free water absorbs much more solar radiation than reflective sea ice. In addition, increased seawater temperatures are likely to affect the ocean's deep-water circulation pattern, enhance warm-water El Niño events (and reduce cooler water La Niña events), and aid in the development of hurricanes.

HAVE INCREASED OCEAN TEMPERATURES CAUSED INCREASED HURRICANE ACTIVITY? Many scientists have suggested that warmer oceans would most certainly cause an increase in the general level of storminess because additional heat accelerates evaporation, which fuels hurricanes. In addition, the ultimate intensity of a hurricane largely depends on the temperature of deep seawater (which has also increased) that is churned upward as the storm passes overhead. The interactions between global warming and hurricane activity, however, are complex, with rising ocean temperatures, changing energy distributions, and altered atmospheric dynamics all having some effect. For example, as atmospheric temperature increases, so does atmospheric stability, a shift that limits convective transport and thus reduces the formation of tropical hurricanes.

In recent years, the frequency and severity of hurricanes—especially those in the Atlantic Ocean (see Section 6.5 in Chapter 6)—have led some to speculate that global warming has already enhanced the formation of hurricanes. For example, the recent landfall of several large, destructive Atlantic hurricanes such as Hurricane Katrina in 2005 and Hurricane Sandy in 2012 (**Figure 16.21**) has led to a general impression that hurricanes have increased, although there have been conflicting reports on the topic in the scientific literature. In fact, some research articles have attributed increases in hurricane intensity, numbers, wind speeds, and rainfall to warmer sea surface temperatures, while others have claimed that changes in data-gathering methods and instrumentation are responsible for the trends. Other studies suggest that the apparent increase in hurricanes falls within the statistical limits of normal.

Although scientists are not able to say definitively if the *number* of tropical storms has increased worldwide, the scientific consensus is that global warming has likely led to more *intense* hurricanes. In the most comprehensive study of recent hurricane activity to date, researchers demonstrate that there have been significant increases in tropical storm intensity and duration around the world since 1970 and that these trends are strongly related to rising sea surface temperatures. Another study of historical Atlantic hurricanes over the past 1500 years suggests that times of peak hurricane activity are related to increased sea surface temperatures and the reinforcing effects of La Niña–like climate conditions. Other research explicitly shows that the most energetic storm levels—those with Category 4 and 5 designations—have already increased significantly, particularly in the North Atlantic and northern Indian Oceans. Still other studies have detected a pronounced poleward migration in the average latitude at which tropical cyclones have achieved their lifetime-maximum intensity over the past 30 years as a result of human-caused climate change. In addition, sophisticated climate models suggest that the number of Category-4 and Category-5 storms in the western tropical Atlantic could double by the end of the century, despite a drop in the overall number of storms.

Figure 16.21 Flooding from Hurricane Sandy in New Jersey in 2012. The storm surge from Hurricane Sandy reached a height of 2.8 meters (9 feet), which caused severe flooding along the U.S. East Coast. Damage from Hurricane Sandy totaled more than $75 billion, making it the second costliest hurricane in U.S. history after Hurricane Katrina.

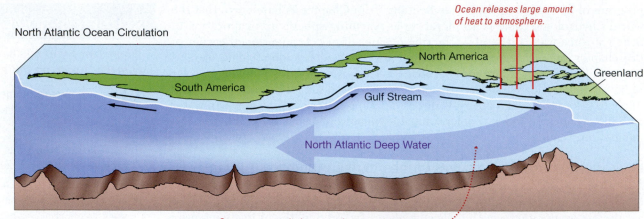

North Atlantic Ocean Circulation

Ocean releases large amount of heat to atmosphere.

North America

Greenland

South America

Gulf Stream

North Atlantic Deep Water

Ocean water cools, becomes denser and sinks to join a powerful, deep southward current.

Figure 16.22 North Atlantic Ocean circulation. Perspective view of circulation in the North Atlantic Ocean, showing that the Gulf Stream carries a tremendous amount of heat northward that warms the North Atlantic region. As this water cools, it generates a huge volume of cold, salty, dense water called North Atlantic Deep Water that sinks into the deep-ocean basin and flows southward. Disruption of this circulation pattern could have severe effects on global climate.

Web Animation
North Atlantic Deep-Water Circulation
https://goo.gl/bXwxdY

Changes in Deep-Water Circulation

Evidence from deep-sea sediments and computer models indicates that changes in the global deep-water circulation pattern can dramatically and abruptly affect climate. Circulation in the North Atlantic Ocean, which provides an important source of deep water (**Figure 16.22**), is particularly sensitive to these changes. What drives deep-water circulation is the sinking of cold, high-salinity, dense surface waters at high latitudes, particularly in the North Atlantic. If surface waters stopped sinking because they were too warm and/or too diluted by melting ice (and thus low in density), then the oceans would absorb and redistribute heat from solar radiation much less efficiently. This would likely cause even warmer surface water temperatures and much higher land temperatures than are experienced now.

Many scientific studies suggest that the buildup of greenhouse gases in the atmosphere will change ocean circulation. One way in which this could happen is that warmer air temperatures will increase the rate at which glaciers in Greenland melt, forming a pool of fresh, low-density surface water in the North Atlantic Ocean. This freshwater could inhibit the downwelling that generates North Atlantic Deep Water, reorganizing global circulation patterns and causing a corresponding change in climate. Many climate experts warn that a large outflow of freshwater from Greenland could take the North Atlantic system of currents to a tipping point, causing rapid reorganization of deep-water currents and related changes in climate. Indeed, evidence suggests that an outburst from a North American ice-dammed lake flooded the North Atlantic with freshwater about 8000 years ago, causing rapid global climate change. This event may be the type of scenario that the North Atlantic could experience again because of increased precipitation and melting of ice.

Melting of Polar Ice

Computer models have predicted that global warming will affect Earth's polar regions in a very dramatic way. One fundamental difference between the two polar regions is that in the Northern Hemisphere, the polar region is dominated by the Arctic Ocean and its cover of drifting sea ice (an ocean surrounded by land), whereas the South Pole is dominated by the continent of Antarctica and its thick ice cap, including shelf ice that extends into the ocean (land surrounded by an ocean).

The Arctic is one of the locations where the effects of global warming are being most keenly felt (see Figure 16.20) and likely will experience quite dramatic changes in the future; this phenomenon is called *Arctic amplification*. Since 1978, satellite analysis of the extent of Arctic Ocean sea ice indicates that it is dramatically shrinking and thinning at an accelerating rate (**Figure 16.23**). In the past decade alone, there has been a loss of over 2 million square kilometers (800,000 square miles) of Arctic sea ice. In fact, measurements of the ice cap in 2012 revealed that it had shrunk to

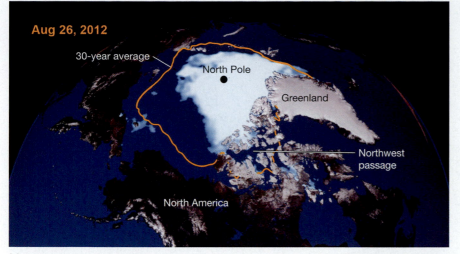

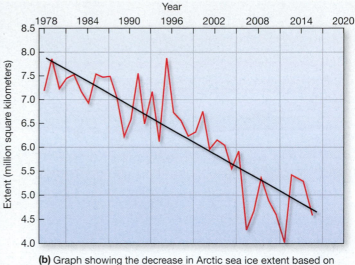

(a) Perspective view of the Arctic based on satellite data, showing the extent of Arctic sea ice in August 2012 compared with the 30-year average extent of sea ice (*yellow line*). Note the opening of a new ice-free Northwest Passage.

(b) Graph showing the decrease in Arctic sea ice extent based on satellite data.

Figure 16.23 **Arctic sea ice decline.** The substantial decrease in extent and thickness of Arctic Ocean sea ice is due to human-induced warming of the Arctic.

Web Animation

Arctic Sea Ice Decline
https://goo.gl/ynv8dU

Web Video

Loss of Arctic Sea Ice
https://goo.gl/ziV50U

its smallest size since researchers began collecting satellite measurements; summer Arctic sea ice is now about half the size it was only 30 years ago. In addition, thick multiyear ice in the Arctic has been disappearing and is being replaced by thinner first-year ice that is less likely to survive the summer melt season. As a result, Arctic ice is now unusually thin and spread out, causing wide patches of ice-free ocean during the summer—even at the North Pole. A recent study of proxy data showed that the current decline of Arctic sea ice is unprecedented for at least the past 1450 years.

Climate models are in general agreement that one of the strongest signals of greenhouse warming will be a loss of Arctic sea ice. Indeed, during the past 15 years, the decline in Arctic sea ice has occurred much faster than models had predicted. In fact, the Arctic has been warming twice as quickly as the Northern Hemisphere average. Cycles of natural variability are known to play a role in Arctic sea ice extent, but the sharp decline observed in the past two decades cannot be explained by natural variability alone. The accelerated Arctic sea ice melting appears to be linked to shifts in Northern Hemisphere atmospheric circulation patterns that have caused the region to experience unusually rapid warming. As a result, ocean temperatures in the Arctic Ocean have also increased, causing sea ice to melt from below. Disappearance of sea ice is likely to enhance future warming in the region because lower amounts of sea ice will reflect less of the Sun's radiation back into space, creating a positive-feedback loop and exacerbating the problem as heat is absorbed by the newly uncovered ocean. Researchers fear that the Arctic may be on the verge of a fundamental transition, or "tipping point," that will lead to the Arctic having only seasonal ice cover. Some models, for example, suggest that the Arctic could experience the complete disappearance of summer sea ice as early as 2030.

The decrease of sea ice in the Arctic has already had profound effects on Arctic ecosystems. Polar bears (*Ursus maritimus*; **Figure 16.24**), for example, are excellent swimmers but do not hunt in the water. Instead, they require a platform of floating sea ice to capture their prey, which are mainly ringed and bearded seals. As the Arctic Ocean becomes more ice-free and the ice habitat shrinks, polar bears will have more difficulty finding adequate food and making dens. As a result, polar bear breeding and survival rates may decline below the point needed to maintain the population. Their habitat destruction has been so severe that polar bears were listed as a threatened species in 2008, according to the U.S. Endangered Species Act. Studies reveal that polar bears are likely to lose nearly half of their summer sea ice habitat by the middle of the 21st century, which would in turn reduce the world's polar bear population—currently estimated at 25,000—by two-thirds.

Interdisciplinary

Relationship

STUDENTS SOMETIMES ASK . . .

I've heard that Antarctica is actually gaining ice. Is that true?

Technically, yes it is, but it may not last long. And even though Antarctic sea ice reached a new record maximum in September 2014, global sea ice is still decreasing worldwide. That's because the large decrease in Arctic sea ice far exceeds the small gain in Antarctic sea ice. Overall, polar ice is still shrinking as a result of higher temperatures in the atmosphere and oceans. But in the future if West Antarctica catastrophically loses some of its large floating ice shelves (which are thinning and losing volume), both polar regions will be in a state of rapid ice loss.

Figure 16.24 Habitat destruction threatens polar bears. Polar bears (*Ursus maritimus*) rely on floating sea ice as a feeding platform and for building dens. As a result of habitat destruction caused by shrinking Arctic sea ice, polar bears were designated a threatened species in 2008.

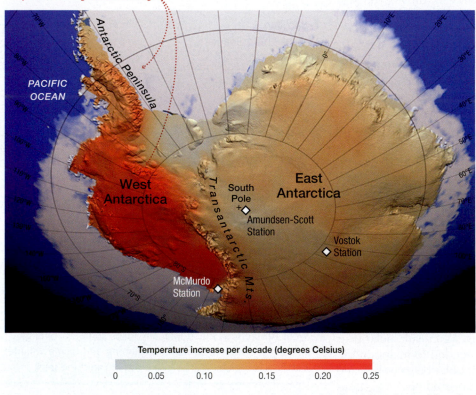

West Antarctica and the Antarctic Peninsula have experienced the greatest warming.

Temperature increase per decade (degrees Celsius)

0 0.05 0.10 0.15 0.20 0.25

Figure 16.25 Antarctic warming trends. This satellite image shows the amount of warming that Antarctica has experienced since 1957. The data are from satellites, which are calibrated with weather station measurements.

In addition, human inhabitants of the Arctic who depend on subsistence living are being affected by the lack of sea ice. This is because some of the inhabitants' food sources are more difficult to obtain now that marine species live further from shore to be near the ice edge. Arctic residents have also claimed that their weather is changing, and a recent study supports their observations. The study determined that years with the least amount of Arctic sea ice have had significantly stronger Arctic storms. One new opportunity that now exists, however, involves the creation of a new "Northwest Passage" shipping lane linking the North Pacific and North Atlantic Oceans through the largely ice-free portions of the Arctic Ocean (Figure 16.23a).

While all these changes are taking place in the Arctic, different but equally dramatic ones are taking place in Antarctica, particularly in the western part of Antarctica that includes the Antarctic Peninsula. As discussed in Chapter 6, Antarctica produces many icebergs from glaciers on land. The rate at which Antarctica is producing icebergs—especially large icebergs the size of small U.S. states—has increased in recent years. For example, the Antarctic Peninsula's Larsen Ice Shelf has decreased by more than 40% over the past decade, including a huge release of 3250 square kilometers (1250 square miles) of ice during two months in 2002 (see Figure 6.27d). In 2006, Antarctica lost nearly 200 billion metric tons (441 trillion pounds) of ice; during 10 days in 2008, the Wilkins Ice Shelf lost over 400 square kilometers (160 square miles) of ice. The rate of thinning of the Pine Island Glacier—the largest stream of fast-moving ice on the West Antarctic Ice Sheet—quadrupled from 1995 to 2006. And, in the past 30 years, there have been 10 major Antarctic ice shelf collapses—including the disappearance of the ice shelves knows as Jones, Larsen A, Muller, and Wordie—after some 400 years of relative stability. Scientists attribute this catastrophic retreat to warming in Antarctica; West Antarctica and the Antarctic Peninsula have experienced some of the greatest warming worldwide (see Figure 16.20). In fact, Antarctica has warmed at a rate of about 0.12°C (0.22°F) per decade since 1957, for a total average temperature rise of 0.5°C (1.0°F) (Figure 16.25). Some areas of West Antarctica are warming several times faster. For example, at Byrd Station in West Antarctica, analysis of temperature records that have been kept since 1958 reveal that it has warmed by 2.4°C (4.0°F), making it one of the fastest-warming places on Earth.

Recent Increase in Ocean Acidity

The human-induced increase in the amount of carbon dioxide in the atmosphere has some severe implications for ocean chemistry and for marine life. Recent studies show that a little less than half of the carbon dioxide released by the burning of fossil fuels stays in the atmosphere and about one-third currently ends up in the oceans, readily dissolving into seawater at the ocean surface. This "sink" reduces global warming—but at the expense of acidifying the sea.

When large amounts of carbon dioxide enter the ocean, it overwhelms the ocean's natural ability to buffer itself.[8] This absorbed carbon dioxide in seawater forms carbonic acid, which is a relatively weak acid (people drink it all the time in carbonated beverages). But when carbonic acid forms in seawater, it lowers the ocean's pH (that is, increases its acidity in a process called **ocean acidification**) and changes the balance of carbonate and bicarbonate ions. In fact, the oceans have already absorbed enough carbon dioxide for surface waters to have experienced a pH decrease of 0.1 pH unit since pre-industrial times. Although 0.1 pH unit sounds like a small amount, pH is a logarithmic scale (like the Richter scale that records earthquake intensity), so a difference of 0.1 pH unit represents about a 30% increase in hydrogen ion concentration; every decrease of 1 pH unit is equivalent to a *tenfold* increase in hydrogen ion concentration. Another study has confirmed a 0.04 pH unit decrease in the North Pacific Ocean during just the past two decades.

Moreover, this shift toward increased acidity and the ensuing changes in ocean chemistry makes it more difficult for certain marine creatures to build and maintain hard parts out of easily dissolved calcium carbonate.[9] The decline in pH thus threatens a diverse assortment of calcifying organisms—creatures that grow calcium carbonate skeletons or shells—such as coccolithophores, foraminifers, pteropods, calcareous algae, sea urchins, mollusks, and corals (**Figure 16.26**). These

 SmartFigure 16.26 Examples of marine organisms that are affected by increased ocean acidity. Various organisms make their skeletons or shells out of easily-dissolved calcium carbonate. As ocean acidity increases, these and many other types of organisms will have a more difficult time building and maintaining calcified hard parts.
https://goo.gl/ywiOoA

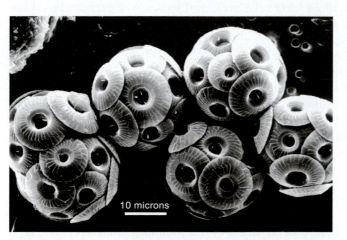

(a) Coccolithophores, which are a type of phytoplankton.

(b) Pteropod, which is a type of zooplankton that is a small swimming snail with a shell.

(c) Sea urchins, which crawl across the sea floor.

(d) Corals, which grow in tropical regions.

[8]For a discussion of the ocean's buffering system and the pH scale, see Section 5.5 in Chapter 5.
[9]Chemically, more acidic waters have more hydrogen ions that bind with carbonates. That leaves fewer carbonates available for carbonate-forming marine organism to make their shells.

Normal shell at start of experiment.

Shell after 6-week exposure to acidic seawater.

1 mm
0.04 in

Figure 16.27 Pteropods dissolved by acidic waters. Six-week time-lapse sequence of shell dissolution of the Antarctic pteropod *Limacima helicina* in water simulating surface ocean acidity projected for the Southern Ocean by 2100 if carbon dioxide emissions continue to increase.

Web Video

Ocean Acidification—The Other Carbon Dioxide Problem
https://goo.gl/OOAagg

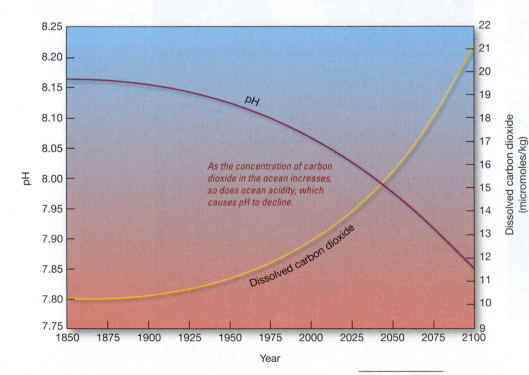

As the concentration of carbon dioxide in the ocean increases, so does ocean acidity, which causes pH to decline.

Figure 16.28 Historical and projected dissolved carbon dioxide and ocean pH.

organisms provide essential food and habitat to many other species, so their demise could affect entire ocean ecosystems. For example, research shows that the projected future level of ocean acidity prevents Antarctic krill eggs from hatching, which would affect entire Antarctic food webs. Other studies have shown that in the past 20 years, ocean acidification has already caused a 15% decrease in the growth rate of corals in Australia's Great Barrier Reef. And, laboratory experiments that expose organisms to conditions that mimic the ocean's future acidity clearly show the negative impact on the shells of calcite-secreting organisms (**Figure 16.27**).

Interdisciplinary

Relationship

Ocean acidification affects more than just organisms that make hard parts out of calcium carbonate. Studies show that increasing ocean acidity can interfere with basic bodily functions for all marine animals, shelled or not. For example, research has shown that the larvae of Atlantic cod (*Gadus morhua*) exposed to acidic waters experience severe tissue damage in many internal organs, thus increasing their mortality rate. Other studies of reef-dwelling clownfish and damselfish report that they are exposed to CO_2-infused waters, the fish exhibit learning problems and odd behavioral issues, including a tendency to seek out predators' odors. Still other studies of marine organisms exposed to acidic waters show a decrease in reproductive success. By disrupting processes as fundamental as growth, behavior, and reproduction, ocean acidification threatens marine animals' health and even the survival of species.

Interdisciplinary

Relationship

The graph in **Figure 16.28** shows the projected increase of carbon dioxide in the ocean and the resulting decrease in ocean pH (rise in acidity). The graph shows that if the current trend of human-induced carbon dioxide emissions continues, by 2100, the ocean will experience a pH decrease of at least 0.3 pH unit; some studies indicate that pH could decrease by as much as 0.6 pH unit. Even at the lower value of 0.3 pH unit, this reduction of ocean pH represents an increase in hydrogen ion concentration of 100% above what it was in pre-industrial times. An additional concern is that deep currents will eventually transmit this increase in acidity to the deep-ocean floor, where it will likely affect deep-sea marine organisms, whose stable environment leaves them ill-equipped to adapt to change.

Interdisciplinary

Relationship

Geologic evidence shows that comparable ocean pH changes wiped out many species of sea life, especially bottom-dwelling organisms. Scientific studies show that today's alteration of ocean chemistry is unprecedented on Earth: An analysis of the past 300 million years of Earth history fails to find a time when ocean chemistry is changing as fast as it is today.

Several processes influence the amount of carbon dioxide that is absorbed by the ocean and how much remains in the atmosphere. The ocean and atmosphere are two reservoirs for carbon dioxide storage, but carbon dioxide is not equally divided between the two. This is because carbon dioxide gas is readily absorbed by the ocean and, as a result, the ocean contains a much larger amount of carbon dioxide than the atmosphere.[10] The amount of carbon dioxide from the atmosphere that is dissolved in the ocean varies with the chemistry of seawater but is also controlled by positive-feedback loops. For example, one positive-feedback loop is that as the ocean approaches saturation of carbon dioxide,

[10]Of the three places where carbon dioxide resides—atmosphere, ocean, and land biosphere—approximately 93% is found in the ocean. The atmosphere, in contrast, contains the smallest amount.

it will absorb less of the gas, which means that more will remain in the atmosphere. Another positive-feedback loop is that as the ocean warms, it will further reduce the amount of carbon dioxide that goes into the oceans because warmer water can't hold as much dissolved gas. Yet another positive-feedback loop involves the rate at which deep waters mix with surface waters: The more rapid the mixing, the more it facilitates the uptake of carbon dioxide from the atmosphere. If deep-water circulation slows as predicted, this will also slow the uptake of carbon dioxide from the atmosphere. In addition, another positive-feedback loop is that ocean acidification inhibits the ability of marine organisms to make calcium carbonate hard parts, so less carbon dioxide in the form of calcium carbonate can be stored by organisms and effectively removed from the environment. All of these positive-feedback loops cause more of the carbon dioxide being released now to stay in the atmosphere, where it will likely create additional human-caused warming.

Rising Sea Level

Analysis of worldwide tide records indicates that there has been a rise in global sea level of between 10 and 25 centimeters (4 and 10 inches) over the past 100 years. At certain tide-recording stations where data go back well into the 19th century, there has been an increase in relative sea level of as much as 40 centimeters (16 inches) over the past 150 years (**Figure 16.29**). More recently, satellite altimeter data since 1993 indicate a global increase in sea level of about 3 millimeters (0.1 inch) per year (**Figure 16.30**). Studies have shown that the current rate of sea level rise is occurring faster than at any other time over the past 4000 years, and the rate is expected to increase with additional warming.

Two main factors contribute to the global rise in sea level: (1) thermal expansion of ocean water as it warms and (2) an increase in the amount of water in the ocean from the melting of ice on land. Note that the melting of floating sea ice (such as in the Arctic Ocean) or floating ice shelves (such as those that fringe Antarctica) does not contribute to sea level rise because that ice/water is already in the ocean. More specifically, in order of their overall contribution to the observed global rise in sea level (**Figure 16.31**), the main contributors are:

1. The melting of the Antarctic and Greenland Ice Sheets
2. The thermal expansion of ocean surface waters
3. The melting of land glaciers and small ice caps
4. The thermal expansion of deep-ocean waters

An additional factor that affects global sea level is the storage of freshwater on land in reservoirs. A recent study indicates that the amount of this storage has varied but has generally increased since 1900; the study suggests that without reservoirs, sea level would have risen even more.

Although the current rate of sea level rise might seem inconsequential, even a small amount of sea level rise can severely affect regions that have a gently sloping shoreline, such as the U.S. Atlantic and Gulf coasts. Perhaps surprisingly, sea level rise does not occur equally everywhere. For example, a recent study identified a zone along the U.S. East Coast from Cape Cod to Cape Hatteras that has experienced rates of sea level rise three to four times higher than the global average since 1950. Hazards associated with sea level rise include drowning of beaches, accelerated coastal erosion, small-scale and permanent inland flooding, alteration of coastal ecosystems, loss of protective coastal wetlands, and increased damage from destructive storms. In addition, if global warming increases the intensity of hurricanes (as discussed above), flooding and damage to coastal regions will be even greater. For example, rising sea levels exacerbated Hurricane Sandy's storm surge along the U.S. East Coast in 2012, establishing a direct link between global warming and storm damage.

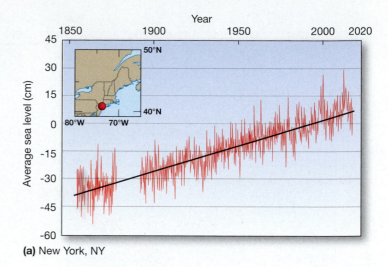

(a) New York, NY

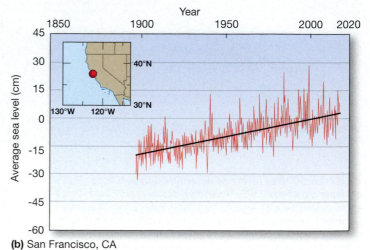

(b) San Francisco, CA

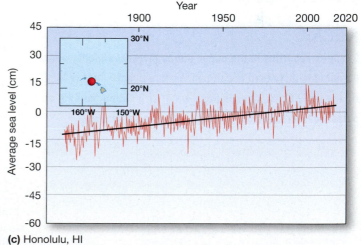

(c) Honolulu, HI

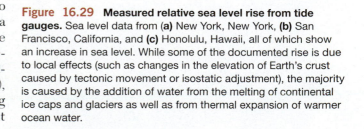

Figure 16.29 Measured relative sea level rise from tide gauges. Sea level data from **(a)** New York, New York, **(b)** San Francisco, California, and **(c)** Honolulu, Hawaii, all of which show an increase in sea level. While some of the documented rise is due to local effects (such as changes in the elevation of Earth's crust caused by tectonic movement or isostatic adjustment), the majority is caused by the addition of water from the melting of continental ice caps and glaciers as well as from thermal expansion of warmer ocean water.

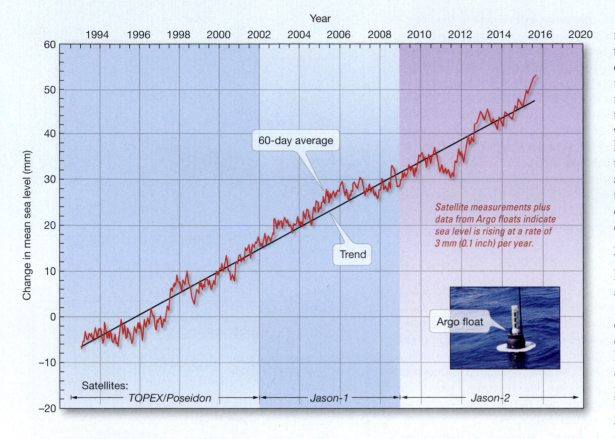

Figure 16.30 **Sea level rise determined by satellites.** Radar altimeter data from *TOPEX/Poseidon*, *Jason-1*, and *Jason-2* satellites combined with Argo drifting floats (*inset photo*; see also Chapter 7) reveal that sea level has risen 3 millimeters (0.1 inch) per year since 1993. Scientists attribute about half of this increase to melting ice and the other half to thermal expansion as the ocean absorbs excess heat from the atmosphere.

RECAP

Changes occurring in the ocean due to increased global warming include increased ocean temperatures, more intense hurricane activity, changes in deep-water circulation, melting of polar ice, increased ocean acidity, and rising sea level.

The current rate of global sea level rise, though small, is likely to increase as the Greenland and Antarctic Ice Sheets experience additional warming and make a greater contribution to sea level rise. In fact, detailed studies of Antarctic ice thickness reveal that the rate at which the ice is thinning near the coast has doubled since the 1990s. Neither the Greenland nor Antarctic Ice Sheets are likely to disappear before 2100, but there is the danger that global warming could trigger massive and catastrophic discharges of ice from these ice sheets. A recent study determined that if the West Antarctic Ice Sheet were to collapse, it would raise global sea level by about 3.2 meters (10.5 feet).

According to models, the rate of sea level rise will increase with increased global warming. Studies of sea level rise that include both thermal expansion and ice-sheet contributions project a rise in sea level by 2100 of between 0.6 and 1.6 meters (2.0 and 5.2 feet), which is particularly worrisome for low-lying coastal regions. In addition, increased amounts of human settlement and development in coastal regions compound the problem. Looking farther ahead, projections suggest that sea level will likely rise by several meters in the next two centuries, which will affect *all* coastal cities worldwide.

Other Predicted and Observed Changes

Several other ocean changes are predicted to result from global warming, and some of these changes are being observed in the ocean right now.

DISSOLVED OXYGEN IN SEAWATER One predicted change is lower dissolved oxygen in seawater. As discussed in Chapter 12, Dissolved oxygen in seawater is vital for most marine animals, which extract dissolved oxygen directly from seawater. As the ocean warms, its ability to hold and carry dissolved oxygen is diminished. At the same time, higher water temperatures cause chemical reactions to happen more quickly, resulting in higher metabolic rates for marine organisms, which means they need higher levels of dissolved oxygen. In addition, the warming of surface waters will limit the critical overturning process that brings oxygen to deep waters. At current rates of carbon dioxide emissions, studies predict severe oxygen-depleted zones both at the surface and into deeper waters for thousands of years. Reduced oxygen levels will likely have dramatic consequences for marine ecosystems and coastal regions that already experience oxygen-depleted "dead" zones (see Section 13.2 in Chapter 13). Reduced dissolved oxygen in seawater has already been documented in both coastal waters and open-ocean environments.

Interdisciplinary

Relationship

WIND SPEEDS AND WAVE HEIGHTS Wind speeds and wave heights have been affected by climate change, too. A recent analysis of 23 years of satellite altimeter measurements suggests that windiness has been increasing on a global scale, with severe winds increasing the most. Although the study found that wave heights have not increased globally, the heights of larger waves had a slight increase in higher latitudes.

MIGRATION OF MANGROVES Mangrove forests in Florida are on the move as well. A study that compared satellite images from 1984 to 2011 concluded that Florida's mangroves have migrated north along the U.S. East Coast by about 12 square kilometers (4.6 square miles). The mangroves' gains come mainly at the expense of salt marshes, which thrive in areas normally too cold for mangroves. The researchers found that mangroves have expanded into places where it had been historically too cold in the winter for mangroves to survive, but winter lows are now warm enough to support the growth of mangroves.

OCEANIC PRODUCTIVITY Still another effect of ocean warming is the alteration of oceanic productivity, which has implications for the distribution of essentially all marine organisms. As ocean surface waters warm, ocean stratification will increase and a stronger **thermocline** (*thermo* = heat, *cline* = slope) will develop.[11] This stratification would make it harder for nutrients to reach the surface and also limit nutrient recycling with deeper waters. As a result, productivity is expected to decrease because of diminished **upwelling**, and the water that is brought to the surface by upwelling would be more nutrient depleted. Recall from Chapter 13 that **phytoplankton** (*phyto* = plant, *planktos* = wandering)—which include marine algae such as *diatoms* and *coccolithophores*—comprise the base of most marine food webs and thus support other larger organisms in the oceans, including commercial fish species. In fact, studies already show a decline of global phytoplankton biomass related to ocean warming. Researchers estimate that, by 2100, productivity could decrease by as much as 20% relative to pre-industrial times.

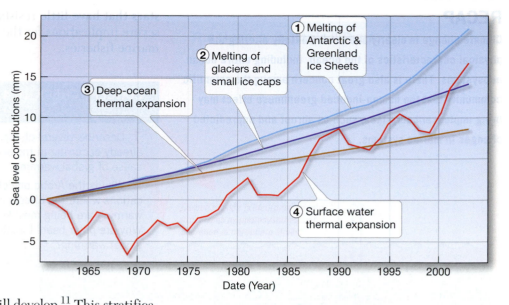

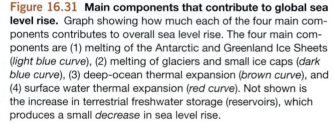

Figure 16.31 Main components that contribute to global sea level rise. Graph showing how much each of the four main components contributes to overall sea level rise. The four main components are (1) melting of the Antarctic and Greenland Ice Sheets (*light blue curve*), (2) melting of glaciers and small ice caps (*dark blue curve*), (3) deep-ocean thermal expansion (*brown curve*), and (4) surface water thermal expansion (*red curve*). Not shown is the increase in terrestrial freshwater storage (reservoirs), which produces a small *decrease* in sea level rise.

EFFECT ON MARINE ORGANISMS Warmer waters can also affect marine organisms directly. Many types of phytoplankton and other organisms are very sensitive to changes in water temperature. For example, a large study of phytoplankton in the North Atlantic shows that ocean warming has increased the abundance of phytoplankton in regions formerly associated with cooler waters, and decreased them in warmer waters. Another example is a recent study in waters off California that shows a decrease in cool-water organisms caused by a deep, penetrative warming not observed in the past 1400 years. In these organisms' place, there has been a boost in the population of 25 fish groups that prefer warmer waters as surface waters have shifted from cold to warm temperatures over recent decades.

In response to rising ocean temperatures, marine organisms have also begun to migrate into deeper waters and toward the poles. As verification of this trend, a recent study of fish species in the North Sea suggests that many commercially important fish, such as cod, whiting, and anglerfish, have shifted northward as much as 800 kilometers (500 miles). The report notes that if these climate trends continue, some species of fish may withdraw completely from the North Sea by 2050. Another report suggests that high-latitude regions are expected to benefit from anticipated changes in ocean fisheries, but the warm-water tropics are likely to suffer declines in fish stocks. In other cases, warm-water species are moving into frigid waters that have been previously inaccessible for millions of years. For example, king crabs (*Neolithodes yaldwyni*) have invaded Antarctic waters, where they are feeding on fragile creatures such as sea cucumbers, sea lilies, and brittle

[11]For more details on the thermocline (the layer of rapidly changing temperature), see Section 5.7 in Chapter 5.

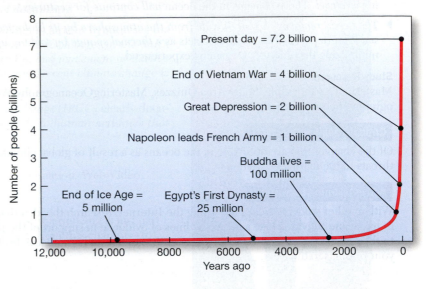

"In the end, we will conserve only what we love. We love only what we understand. We will understand only what we are taught."

—Baba Dioum, Senegalese conservationist (1968)

At the end of our journey through this book together, it seems fitting to examine people's perceptions of the ocean. Many people describe the ocean as "powerful," "awe inspiring," "moving," "serene," "abundant," and "majestic." Others call it "vast," "infinite," or "boundless." These are all appropriate descriptions because even though humans have been exploring and studying the ocean for centuries, the ocean still holds many secrets. The oceans are constantly surprising researchers with unusual features, newly discovered species, and geologic wonders that exist within its watery world. In fact, less than 5% of the ocean has been seen by humans, and even less has been scientifically explored.

Despite the ocean's impressive size, it is beginning to feel the effects of human activities. For instance, every ocean contains large areas of floating plastic, and even remote beaches are littered with trash. Organisms living in the ocean also feel the effect of humankind's use of the ocean. Whaling in the 19th and 20th centuries, for example, pushed many great whale populations to near extinction. Restrictions on whaling and the development of substitutes for whale products helped whales survive this threat, but now they face destruction of their feeding and breeding grounds. Overfishing has degraded entire marine ecosystems, and human-caused climate change is altering marine chemistry on a global scale. The oceans are approaching irreversible, potentially catastrophic change, suggesting that the ocean is not quite as "vast," "infinite," or "boundless" as most people believe.

Humans have become one of the most important agents of change on the planet. This is largely due to our rapidly expanding human population, which is increasing at an exponential rate (**Figure Aft.1**). Today, more than 7.2 billion people occupy the planet, and births now exceed deaths by about two-and-a-half times. At the present rate of population increase—which sounds small enough, at about 1.1%, but produces truly staggering numbers—there are more than 2 new people on the planet each second, or about 9000 more people each hour. Each year there are about 77 million more of us, a total equal to the combined populations of the six largest cities in the world (Mumbai, Shanghai, Karachi, Delhi, Istanbul, and São Paulo). A recent study showed that there is an 80% probability that world population will increase to between 9.6 and 12.3 billion people in 2100. Some scientists, in fact, have proposed calling today's time period the *Anthropocene* (*anthro* = human, *cene* = new), or a "sixth mass extinction," in recognition of human activities that are degrading vital Earth systems and the evidence of those impacts that will be preserved in the geologic record. Clearly, our burgeoning human population is the greatest environmental threat of all.

Human-induced changes in the marine environment are broad and far-reaching. Examples include pollution, shoreline development, overfishing, introduction of non-native species, biodiversity decline, ecosystem degradation, and perhaps most serious of all, climate change. A 2008 study of the cumulative impact of 17 of the most urgent land- and ocean-based threats to worldwide marine ecosystems shows that no areas of the ocean are untouched by human activities. The study revealed that fully one-third of the oceans are strongly affected by multiple factors and that the most heavily impacted ecosystems are continental shelves, rocky reefs, coral reefs, seagrass beds, and deep-sea seamounts. More recently, a 2012 study reported that one-fifth of all invertebrate species worldwide are at risk of extinction, including almost one-third of reef-building corals.

Several commissions have suggested ways to protect the marine environment. In 2003, for example, the Pew Oceans Commission recommended significant changes aimed at guiding the way in which the federal government should manage American's marine environment. In 2004, the U.S. Commission on the Oceans submitted 212 recommendations for a coordinated and comprehensive national ocean policy. In 2008, the U.S. Joint Ocean Commission issued a report card assessing the nation's collective progress in ocean policy during 2007 (its overall grade: a "C," with some categories receiving a "D"). A 2012 study of the health and benefits of the global ocean produced an index of 10 diverse public goals for a healthy coupled human–ocean system, including an integrative assessment of factors such as food provision, carbon storage, tourism value, and biodiversity. The study gave the world ocean a score of 60 out of 100, with developed countries generally performing better than developing countries. (The United States achieved a score of 63.) These reports urge an integrated research and education ecosystem-based management approach to support ocean environmental health and resource sustainability. Another recommendation is the establishment of marine protected areas.

Figure Aft.1 Human population growth. Graph showing world population, which has experienced rapid growth in recent decades. It took 4 million years for humanity to reach the 2 billion mark but only 50 years to double that total. The current world population recently surpassed 7.2 billion people and is increasing at a rate of more than 1% per year.

What Are Marine Protected Areas?

Concerns about the health and longevity of the oceans have led to the establishment of *marine protected areas (MPAs)*, which cover a wide range of marine areas with some level of restriction to protect living, nonliving, cultural, and/or historic resources (**Figure Aft.2**). MPAs can be established for a multitude of reasons: to protect a certain species, to benefit fisheries management, or to protect full ecosystems, rare habitat, or nursing grounds for fish. MPAs are also established to protect historical sites such as shipwrecks and important cultural sites such as aboriginal fishing grounds. MPAs can be very large (for example, Australia's 2000-kilometer [1200-mile] Great Barrier Reef) or very small (for example, Italy's 13,500-hectare [33,360-acre] coastal Marina Protetta Capo Rizzuto). Because the term *MPA* has been used widely around the globe, its meaning in any one region may be quite different than in another. MPAs offer various levels of protection that include areas designated as marine sanctuaries, marine reserves, special protected areas, marine parks, no-take refuges, and areas of special conservation, all of which have specific types of restrictions associated with them, as defined by the laws of the governing body.

Studies have shown that MPAs can help increase biodiversity, protect habitats, and aid in the recovery of fish populations when placed and managed effectively. While MPAs may not solve all the issues that threaten various parts of the ocean, they have been shown to help restore marine ecosystems on a local scale and act as buffers to larger-scale impacts.

In 1972, the U.S. Congress began establishing a type of MPA called a *national marine sanctuary* to protect vital pockets of the ocean from further degradation. Today, 14 national marine sanctuaries cover more than 47,000 square kilometers (18,000 square miles) in areas such as the Florida Keys, Stellwagen Bank off Massachusetts, Monterey Bay off central California (largest national marine sanctuary in the United States), the Channel Islands off Southern California, the Flower Garden Banks in the Gulf of Mexico, and the Hawaiian Islands (**Figure Aft.3**). Many activities that degrade the marine environment, however, are still allowed in marine sanctuaries, such as fishing, recreational boating, and even mining of some resources.

Many scientists who recognize the importance of preserving vital marine habitats are calling for governments to more fully protect marine areas by establishing large no-take refuges called *marine reserves*, in which fishing and other activities are prohibited. Such action would allow the recovery of heavily overfished stocks and protect sea floor communities decimated by trawling the sea floor with nets. In 2006, the United States established the 360,000-square-kilometer (139,000-square-mile) Northwest Hawaiian Islands Marine National Monument, creating the world's largest marine reserve at the time. The reserve, which in 2006 was renamed the Papahānaumokuākea Marine National Monument, encompasses 362,073 square kilometers (139,797 square miles) of the Pacific Ocean—an area larger than all U.S. National Parks combined.

Other countries are beginning to realize the economic benefit of fully protecting ocean resources, too. A healthy coral reef, for example, can be worth more as a tourist draw than it might be worth as a source of seafood. In 2008, for example, the Republic of Kiribati, an ocean nation in the central Pacific approximately midway between

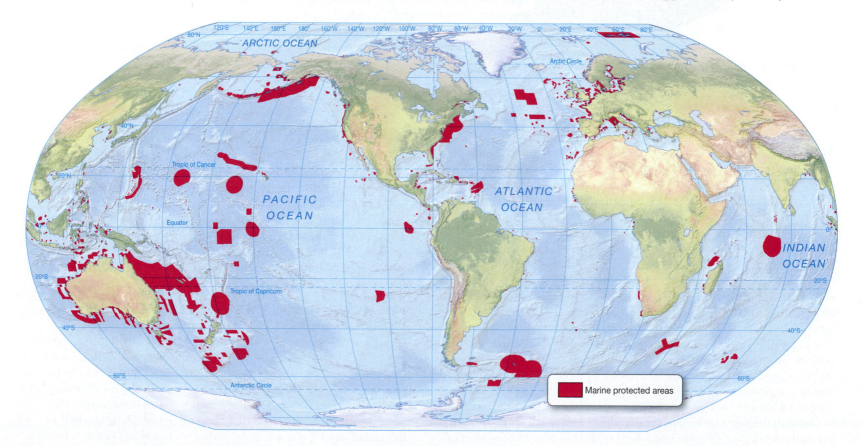

Figure Aft.2 Marine protected areas. Worldwide, more than 4000 marine protected areas (MPAs) have been established, offering various levels of protection to living, nonliving, cultural, and/or historic resources. MPAs cover about 2.8% of the world ocean.

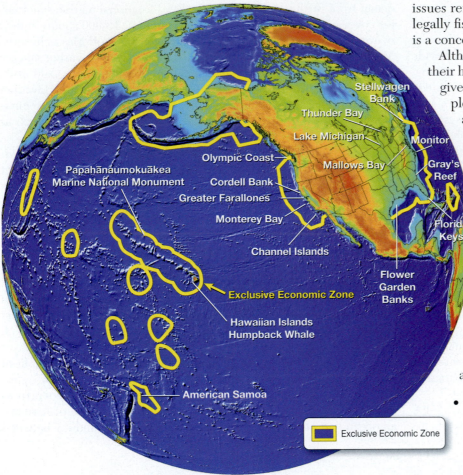

Figure Aft.3 **U.S. national marine sanctuaries.** The United States has established 14 national marine sanctuaries, which protect natural and/or cultural resources. Also shown is the boundary of the 200-nautical-mile (370-kilometer) exclusive economic zone (*yellow lines*).

Australia and Hawaii, established the Phoenix Islands Protected Area, which is about the same size as the state of California, making it the largest MPA in the Pacific Ocean.

Today, more than 4000 marine reserves exist worldwide, but they cover only about 2.8% of the world ocean. (In 2010, the United Nations Convention on Biological Diversity set an ambitious target of 10% ocean protection by 2020.) Although the commercial and recreational fishing industries have historically opposed creating marine reserves, research indicates that marine reserves "spill over" and provide benefits to surrounding regions by acting as nurseries for fish larvae that are later spread by ocean currents, thereby boosting certain populations of commercially important fish outside their borders. For example, studies show that within two decades of the prohibition of fishing at Cabo Pulmo, a small Mexican marine reserve off the southern end of the Baja California Peninsula, the total mass of marine organisms quintupled, demonstrating the tremendous potential of habitat protection. Other research has used DNA analysis to show that marine reserves supply about half of all young fish that live outside marine reserve boundaries. Marine reserves are also important in protecting large fish that help restore the natural predator–prey relationships of healthy, self-sustaining ecosystems. And yet,

issues remain. For example, the temptation exists for poachers to illegally fish within these "no-take" reserves, and proper enforcement is a concern.

Although the oceans are feeling the effects of human activities, their huge size and physical properties make them very resilient and give them a tremendous ability to withstand change. For example, studies reveal that when human impact is reduced, ocean areas often return to a nearly pristine state because natural ocean processes tend to disperse and eventually remove many types of pollutants. However, the ocean's natural resistance is beginning to be seriously compromised. Pollution, habitat loss, and overfishing are dangerous threats on their own, but when these factors converge, they can destroy marine ecosystems.

What Can I Do?

Because the ocean is vast and capable of absorbing many substances, it has been used as a dumping ground for many of society's wastes. Even today, humans are adding pollutants to the ocean at staggering rates. What can each of us do to help? Some ways to help the environment in general and the ocean in particular include the following:

- **MINIMIZE YOUR IMPACT ON THE ENVIRONMENT.** Reduce the amount of waste you generate by making wise consumer choices. Avoid products with excessive packaging and support companies that have good environmental records. Use nontoxic or less hazardous products around the house. Conserve resources. Reuse and recycle items and also help close the recycling loop by buying goods made of recycled materials. Do simple things that make a positive impact on the environment (**Diving Deeper Aft.1**).

- **BECOME POLITICALLY AWARE.** Many ocean-related issues come before the public and require a majority of voters to approve a proposal before it is enacted into law. A common political saying is that the "majority rules," and this situation is true for local as well as national and international issues.

 Within our lifetimes, for instance, we may very well decide whether to spend large amounts of money to add finely ground iron to the ocean to reduce the amount of carbon dioxide in the atmosphere. Many political issues in the future will involve the ocean.

- **EDUCATE YOURSELF ABOUT HOW THE OCEAN WORKS.** A recent poll indicates that more than 90% of the American public consider themselves scientifically illiterate. With our society becoming more scientifically advanced, people need to understand how science operates. Science is not meant to be comprehended by an elite few. Rather, science is for everyone. By studying oceanography, you have begun to understand how the ocean works. It is our hope that you will be a lifelong student of the ocean. And if you are considering participating in the scientific study of the oceans, look for more details on working in marine science in Appendix V, "Careers in Oceanography."

DIVING DEEPER AFT.1 FOCUS ON THE ENVIRONMENT

TEN SIMPLE THINGS YOU CAN DO TO HELP PREVENT MARINE POLLUTION

"Nobody made a greater mistake than he who did nothing because he could only do a little."
—*Edmund Burke (circa 1790)*

There are many simple things you can do every day to help prevent marine pollution, such as the following:

1. **Snip plastic six-pack rings.** Plastic six-pack rings entangle many marine organisms, so snip each circle with scissors before you toss them into the garbage—or, better yet, recycle them along with the cans or bottles to which they are attached. If you find any at a beach, pick them up, snip them, and recycle them, too.

2. **Use lawn fertilizers, yard chemicals, and laundry detergent sparingly.** Lawn fertilizers contain nitrates and phosphates, which cause harmful algal blooms when runoff from land enters the ocean through storm drains. Detergents also contain phosphates, so read detergent labels to find one that is phosphate free and use a smaller amount than recommended. Use yard chemicals such as pesticides and weed killers only when absolutely necessary.

3. **Clean up after your pet.** Dog and cat feces have high bacteria levels. When they wash into a stream or storm drain and eventually to the ocean, they also provide nitrates and phosphates, which create harmful algal blooms.

4. **Make sure your car doesn't leak oil.** Oil that leaks from automobiles is responsible for a large percentage of the oil that gets into the ocean. Annually, the amount of oil that enters the ocean as runoff from road sources (nonpoint source pollution) through storm drains is greater than a major oil spill. If you change your car's oil yourself, be sure to recycle the used oil at an appropriate recycling center.

5. **Drive less and carpool more.** Reducing the amount of gasoline you use reduces the amount of oil that must be transported across the ocean, which minimizes the potential for oil spills.

6. **Take your own bags to the grocery store.** Paper bags are biodegradable, but plastic bags are not, and plastics are an increasing problem in the ocean, especially for animals like sea turtles that eat plastic bags when they mistake them for jellyfish. That's why plastic bags—as well as Styrofoam containers—have been banned in certain countries and coastal communities.

7. **Don't release balloons.** Balloons that are released far from the ocean can still wind up there, where they deflate, quickly lose their color, and resemble drifting jellyfish that marine animals can ingest.

8. **Don't litter.** Any material that is carelessly discarded on land can become nonpoint source pollution when it washes down a storm drain, into a stream, and eventually into the ocean.

9. **Pick up trash at the beach or volunteer for an organized beach cleanup.** It is important to remove trash that washes up at the beach so it can't endanger marine organisms. Trash arrives on beaches from nonpoint source pollution, ships, recreational boaters, beachgoers, and other sources. It is truly surprising (and also somewhat horrifying) to discover what ends up on beaches as the result of human activities.

10. **Inform and educate others.** Many people are unaware that their actions have a negative influence on the environment—especially the marine environment.

Figure Aft.4 Sunset at the ocean in the Gulf of California, Mexico.

How many inches are there in a mile? How many cups in a gallon? How many pounds in a ton? In our daily lives, we often need to convert between units. Worldwide, the metric system of measurement is the most widely used system. Besides the United States, only *two other countries in the world*—Liberia and Myanmar (formerly Burma)—still use English units as their primary system of measurement. The metric system has many advantages over the English system. It is simple, logical, and makes conversion between units easy. For those of us in the United States, it is only a matter of time before the change to the metric system occurs.

Benjamin Franklin was one of the first to propose that the United States adopt the sensible metric system. Interestingly, the United States was officially declared a metric nation by the Secretary of the Treasury in 1893, but Americans have never embraced the idea.

On December 23, 1975, U.S. President Gerald R. Ford signed the Metric Conversion Act of 1975. It defined the metric system as the International System of Units (officially called the *Système International d'Unités* [SI]), as interpreted in the United States by the secretary of commerce. The Trade Act of 1988 and other legislation declared the metric system the preferred system of weights and measures for U.S. trade and commerce, called for the federal government to adopt metric specifications, and mandated the Commerce Department to oversee the program.

Although the metric system has not become the system of choice in most Americans' daily use, and there is great resistance to using it, the United States must change over to remain competitive in world markets. Many of the sciences are leading the way in this changeover. In fact, oceanographers all over the world have been using the metric system for years.

The English System

The English system actually consists of two related systems—the U.S. Customary System[1] (used in the United States and its dependencies) and the British Imperial System (used throughout Great Britain). Ironically, Great Britain has now largely converted to the metric system. The basic unit of length in the English system is the *yard*; the basic unit of weight (not mass) is the *pound*.[2]

In the English system, the units of length were initially based arbitrarily on dimensions of the body. The yard as a measure of length, for example, can be traced to the early Saxon kings. They wore a sash around their waists that could be used as a convenient measuring device. Thus, the word *yard* comes from the Saxon word *gird* (like a girdle), in reference to the circumference of a person's waist. The circumference of a person's waist varies from person to person, however, and even varies from time to time on the *same* person, so it is of limited use. It had to be standardized to be useful, so King Henry I of England decreed that the yard should be the distance from the tip of his nose to the end of his thumb!

Romans initially defined the mile (*mille passuum* = 1000 paces) as an even 5000 feet. The early Tudor rulers in England, however, defined a *furlong* ("furrow-long") as 220 yards, based on the length of agricultural fields. To facilitate the conversion between miles and furlongs, Queen Elizabeth I declared in the 16th century that the traditional Roman mile of 5000 feet would be forever replaced by one of 5280 feet, making the mile exactly 8 furlongs. Today, a furlong is an archaic unit of measurement (although still used in horse races), but a mile is still 5280 feet.

The Metric System

The metric system of weights and measures was devised in France and adopted there in 1799. It is based on a unit of length called the *meter* and a unit of mass called the *kilogram*. Originally defined as one ten-millionth the distance from the North Pole to the equator, the meter is now defined as the distance light travels through a vacuum in $1/_{299,792,458}$ second. The kilogram was originally related to the volume of 1 cubic meter of water. It is now defined as the mass of the International Prototype Kilogram, a precision-fabricated cylinder of platinum–iridium alloy about the size of a plum that is kept at Sèvres (near Paris), France, but work is currently under way to base the kilogram on a fundamental property of nature. Other metric units can be defined in terms of the meter and the kilogram.

Fractions and multiples of the metric units are related to each other by powers of 10, allowing conversion from one unit to a multiple of it simply by shifting the decimal point. The lengthy arithmetic operations required with English units can be avoided. Even the names of the metric units indicate how many are in a larger unit. For instance, how many cents are there in a dollar? It is the same as the number of *centi*grams in a gram. The prefixes listed in **Table A1.1** have been accepted for designating multiples and fractions of the meter, the gram (which equals $1/_{1000}$ of a kilogram), and other units.

Despite many fears that changing to the metric system will be an economic burden on businesses in the United States, the conversion has already begun. Examples include 35-millimeter film, 2-liter soft drink bottles, 750-milliliter wine bottles, computers with gigabytes of memory, and 10-kilometer runs. American cars are now manufactured using metric measurements in order to compete with other countries. Most people in the United States will experience the increasing use of the metric system in their daily lives. Some ways to ease yourself into this unfamiliar (yet sensible) system of units include noticing distances on road signs in kilometers, using a meterstick instead of a yardstick, and charting your weight in kilograms.

[1] The names of the units and the relationships between them are generally the same in the U.S. Customary System and the British Imperial System, but the sizes of the units differ, sometimes considerably.

[2] "Mass" is commonly equated with "weight," but technically the "mass" of an object refers to the amount of matter in it, whereas its "weight" is caused by the gravitational attraction between the object and Earth. Furthermore, the basic unit of mass in the English system is the slug, although it is rarely used.

TABLE A1.1 SOME COMMON PREFIXES FOR BASIC METRIC UNITS

Factor	Name	Prefix name
$10^{12} = 1,000,000,000,000$	trillion	tera
$10^9 = 1,000,000,000$	billion	giga
$10^6 = 1,000,000$	million	mega
$10^3 = 1000$	thousand	kilo
$10^2 = 100$	hundred	hecto
$10^1 = 10$	ten	deka
$10^{-1} = 0.1$	tenth	deci
$10^{-2} = 0.01$	hundredth	centi
$10^{-3} = 0.001$	thousandth	milli
$10^{-6} = 0.000001$	millionth	micro
$10^{-9} = 0.000000001$	billionth	nano
$10^{-12} = 0.000000000001$	trillionth	pico

Temperature Scales

The Celsius (centigrade) temperature scale was invented in 1742 by a Swedish astronomer named Anders Celsius. The scale was designed so that the freezing point of pure water—the temperature at which it changes state from a liquid to a solid—is set as 0 degrees. Similarly, the boiling point of pure water—the temperature at which it changes state from a liquid to a gas—is set at 100 degrees. Doing so made the liquid range of water—0 degrees to 100 degrees—an even 100 units, giving the scale its name: centigrade (*centi* = 100, *grade* = graduations).

The Fahrenheit temperature scale, on the other hand, sets water's freezing and boiling points at 32 degrees and 212 degrees, respectively, for a spread of 180 units. This scale was devised by a German-born physicist named Gabriel Daniel Fahrenheit, who invented the mercury thermometer in 1714. He initially designed the scale so that normal body temperature was set at 100°F. However, because of inaccuracies in early thermometers or how the scale was initially devised, normal human body temperature is now known to be 98.6°F(37°C).

Conversion Tables

The following tables show conversions between various units:

Length	
1 micrometer (micron)	0.001 millimeter 0.0000394 inch
1 millimeter (mm)	1000 micrometers (microns) 0.1 centimeter 0.001 meter 0.0394 inch
1 centimeter (cm)	10 millimeters 0.01 meter 0.394 inch
1 meter (m)	100 centimeters 39.4 inches 3.28 feet 1.09 yards 0.547 fathom
1 kilometer (km)	1000 meters 1093 yards 3280 feet 0.62 statute mile 0.54 nautical mile
1 inch (in)	25.4 millimeters 2.54 centimeters
1 foot (ft)	12 inches 30.5 centimeters 0.305 meter
1 yard (yd)	3 feet 0.91 meter
1 fathom (fm)	6 feet 2 yards 1.83 meters
1 statute mile (mi)	5280 feet 1760 yards 1609 meters 1.609 kilometers 0.87 nautical mile
1 nautical mile (nm)	1 minute of latitude 6076 feet 2025 yards 1852 meters 1.15 statute miles
1 league (lea)	5280 yards 15,840 feet 4805 meters 3 statute miles 2.61 nautical miles

Area

1 square centimeter (cm^2)	0.155 square inch 100 square millimeters
1 square meter (m^2)	10,000 square centimeters 10.8 square feet
1 square kilometer (km^2)	100 hectares 247.1 acres 0.386 square mile 0.292 square nautical mile
1 square inch (in^2)	6.45 square centimeters
1 square foot (ft^2)	144 square inches 929 square centimeters

Volume

1 cubic centimeter (cc;cm^3)	1 milliliter 0.061 cubic inch
1 liter (l)	1000 cubic centimeters 61 cubic inches 1.06 quarts 0.264 gallon
1 cubic meter (m^3)	1,000,000 cubic centimeters 1000 liters 264.2 gallons 35.3 cubic feet
1 cubic kilometer (km^3)	0.24 cubic mile 0.157 cubic nautical mile
1 cubic inch (in^3)	16.4 cubic centimeters
1 cubic foot (ft^3)	1728 cubic inches 28.32 liters 7.48 gallons

Mass

1 gram (g)	0.035 ounce
1 kilogram (kg)	2.2 pounds[†] 1000 grams
1 metric ton (mt)	2205 pounds 1000 kilograms 1.1 U.S. short tons
1 pound[†] (lb)	16 ounces 454 grams 0.454 kilogram
1 U.S. short ton (ton; t)	2000 pounds 907.2 kilograms 0.91 metric ton

Pressure

1 kilogram per square centimeter (at sea level)	1 atmosphere (atm) 760 millimeters of mercury 101,300 Pascal 1013 millibars 14.7 pounds per square inch 29.9 inches of mercury 33.9 feet of freshwater 33 feet of seawater

Speed

1 centimeter per second (cm/s)	0.0328 foot per second
1 meter per second (m/s)	2.24 statute miles per hour 1.94 knots 3.28 feet per second 3.60 kilometers per hour
1 kilometer per hour (kph)	27.8 centimeters per second 0.62 mile per hour 0.909 foot per second 0.55 knot
1 statute mile per hour (mph)	1.61 kilometers per hour 0.87 knot
1 knot (kt)	1 nautical mile per hour 51.5 centimeters per second 1.15 miles per hour 1.85 kilometers per hour

Oceanographic Data

Velocity of sound in 34.85‰ seawater	4945 feet per second 1507 meters per second 824 fathoms per second
Seawater with 35 grams of dissolved substances per kilogram of seawater	3.5 percent (%) 35 parts per thousand (‰) 35,000 parts per million (ppm) 35,000,000 parts per billion (ppb)

[†]The pound is a weight unit, not a mass unit, but it is often used as such.

Temperature

Exact formula	Approximation (easy way)
$°C = \dfrac{(°F - 32)}{1.8}$	$°C = \dfrac{(°F - 30)}{2}$
$°F = (1.8 \times °C) + 32$	$°F = (2 \times °C) + 30$

Some useful equivalent temperatures

100°C = 212°F	boiling point of pure water
40°C = 104°F	heat wave conditions
37°C = 98.6°F	normal body temperature
30°C = 86°F	very warm—almost hot
20°C = 68°F	room temperature
10°C = 50°F	a warm winter day
3°C = 37°F	average temperature of deep water
0°C = 32°F	freezing point of pure water

In degrees centigrade

Thirty is hot, twenty is pleasing;
Ten is not, and zero is freezing.

APPENDIX III Latitude and Longitude on Earth

Suppose that you find a great fishing spot or a shipwreck on the sea floor. How would you remember where it is—far from any sight of land—so that you might return? How do sailors navigate a ship when they are at sea? Even using sophisticated equipment such as the Global Positioning System (GPS), which uses satellites to determine location, how is this information reported?

To solve problems like these, a navigational grid—a series of intersecting lines across a globe—is used. Once starting places are fixed for a grid, a location can be identified based on its position within the grid. Some cities, for example, use the regular numbering of streets as a grid. Even if you have never been to the intersection of 5th Avenue and 42nd Street in New York City, you would be able to locate it on a map, or know which way to travel if you were at 10th Avenue and 42nd Street. Although many grid (or coordinate) systems are used today, the one most universally accepted uses latitude and longitude.

A series of north–south and east–west lines that together comprise a grid system can be used to locate points on Earth's surface whether at sea or on land. The north–south lines of the grid are called meridians and extend from pole to pole (**Figure A3.1**). The meridian lines converge at the poles and are spaced farthest apart at the equator. The east–west lines of the grid are called parallels because they are parallel to one another. The longest parallel is the equator (so called because it

divides the globe into two equal hemispheres), and the parallels at the poles are a single point (Figure A3.1).

Latitude and Longitude

Latitude is the angular distance (in degrees of arc) measured north or south of the starting line (the equator) from the center of Earth. All points that lie along the same parallel are an identical distance from the equator, so they all have the same latitude. The latitude of the equator is 0 degrees, and the North and South Poles lie at 90 degrees north and 90 degrees south, respectively. **Figure A3.2** shows that the latitude of New Orleans is an angular distance of 30 degrees north from the equator.

Web Animation
Latitude and Longitude on Earth
https://goo.gl/MMxrN7

Web Video
Latitude and Longitude on Earth
https://goo.gl/6fBMFO

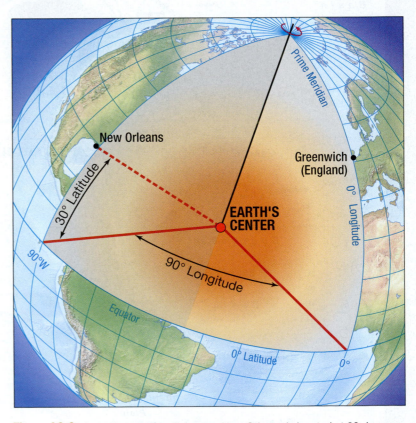

Figure A3.1 Earth's grid system. Earth's grid system is composed of parallels (which extend east–west and are measured north or south of the equator) and meridians (which extend north–south and are measured west or east of the Prime Meridian). Parallels are lines of latitude, while meridians are lines of longitude. Note that while all meridians are the same length, the parallels are not (the longest parallel is the equator).

Figure A3.2 Location of New Orleans. New Orleans is located at 30 degrees north latitude and 90 degrees west longitude. Note the angles of each relative to the center of Earth.

Longitude is the angular distance (in degrees of arc) measured east or west of the starting line from the center of Earth. All meridians are identical (none is longer or shorter than the others, and they all pass through both poles), so the starting line of longitude has been chosen arbitrarily. Before it was agreed upon by all nations, different countries used different starting points for 0 degrees longitude. Examples include the Canary Islands, the Azores, Rome, Copenhagen, Jerusalem, Pisa, Paris, and Philadelphia. At the 1884 International Meridian Conference, it was agreed that the meridian that passes through the Royal Observatory at Greenwich, England, would be used as the zero or starting point of longitude. This zero degree line of longitude is also called the Prime Meridian, and the longitude for any place on Earth is measured east or west from this line. Longitude can vary from 0 degrees along the Prime Meridian to 180 degrees (either east or west), which is halfway around the globe and is known as the International Date Line. As shown in Figure A3.2, New Orleans is 90 degrees west of the Prime Meridian.

A particular latitude and longitude defines a unique location on Earth, provided that the direction from the starting point is given, too. For instance, 42 degrees north and 120 degrees west defines a unique spot on Earth because it is different from 42 degrees south and 120 degrees west and from 42 degrees north and 120 degrees east.

A degree of latitude or longitude can be divided into smaller units. One degree (°) of arc (angular distance) is equal to 60 minutes (′) of arc. One minute of arc is equal to 60 seconds (″) of arc. When using a small-scale map or a globe, it may be difficult to estimate latitude and longitude to the nearest degree or two. When using a large-scale map, however, it is often possible to determine the latitude and longitude to the nearest fraction of a minute.

Determination of Latitude and Longitude

Today, latitude and longitude can be determined very precisely by using satellites that remain in orbit around Earth in fixed positions. How did navigators determine their position before this technology was available?

Latitude was determined from the positions of particular stars. Initially, navigators in the Northern Hemisphere measured the angle between the horizon and the North Star (Polaris), which is directly above the North Pole. The latitude north of the equator is the angle between the two sightings (**Figure A3.3**). In the Southern Hemisphere, the angle between the horizon and the Southern Cross was used because the Southern Cross is directly overhead at the South Pole. Later, the angle of the Sun above the horizon, corrected for the date, was also used.

There was no method of determining longitude until one based on time was developed at the end of the 18th century. As Earth turns on its rotational axis, it moves through 360 degrees of arc every 24 hours (one complete rotation on its axis) (**Figure A3.4a**). Earth, therefore, rotates through 15 degrees of longitude per hour (360 degrees ÷ 24 hours = 15 degrees per hour). As a result, a navigator needed only to know the time at the prime or Greenwich meridian (**Figure A3.4b**) at exactly the same time the Sun was at its highest point (local noon) at the ship's location. In this way, a navigator aboard a ship could calculate the ship's longitude each day at noon. That's why the development of

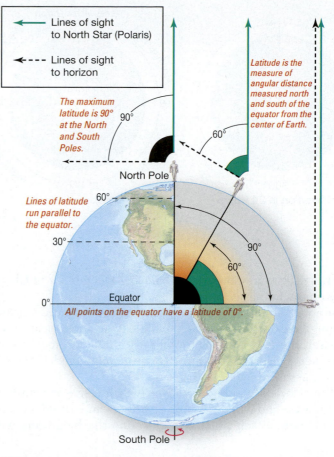

Figure A3.3 Determining latitude based on the North Star. Latitude can be determined by noting the angular difference between the horizon and the North Star, which is directly over the North Pole of Earth.

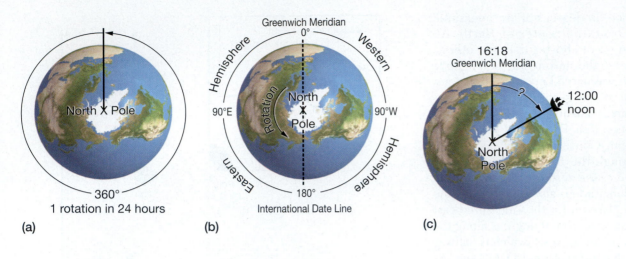

(a) 360°
1 rotation in 24 hours

(b)

(c)

Figure A3.4 **Determining longitude based on time.** View of Earth as seen from above the North Pole. **(a)** Earth rotates 360 degrees of arc every 24 hours. **(b)** The Greenwich Meridian is set as 0 degree longitude, which divides the globe into Eastern and Western Hemispheres. The International Date Line is 180 degrees from the Greenwich Meridian. **(c)** An example of how a ship at sea can determine its longitude using time.

John Harrison's chronometer (see Diving Deeper 1.1 in Chapter 1) was so crucial for navigation.

Suppose a ship sets sail west across the Atlantic Ocean from Europe, checking its longitude each day at noon local time. One day when the Sun is at the noon position (noon local time), the chronometer reads 16:18 hours (0:00 is midnight and 12:00 is noon). What is the ship's longitude (**Figure A3.4c**)?

If the clock is keeping good time (which Harrison's chronometer did), then we know that the ship is 4 hours and 18 minutes behind (west of) Greenwich time. To determine the longitude, we must convert time into longitude. Each hour represents 15 degrees of longitude based on the rotation of Earth. Thus, the four hours represents 60 degrees of longitude (4 hours × 15 degrees of longitude = 60 degrees). One degree is divided into 60 minutes of arc, so Earth rotates through $1/4$ degree (15') of arc per minute of time. Thus, 18 minutes of time multiplied by 15 minutes of arc per minute of time equals 270 minutes of arc. To convert 270 minutes of arc into degrees, it must be divided by 60 minutes per degree, which gives 4.5 degrees of longitude. Therefore, the answer is 60 degrees plus 4.5 degrees, or 64.5 degrees west longitude.

APPENDIX V Careers in Oceanography

Many people think a career in oceanography consists of swimming with marine animals at a marine life park or snorkeling in crystal-clear tropical waters, studying coral reefs. In reality, these kinds of jobs are extremely rare, and there is intense competition for them. Most oceanographers use science to answer questions about the ocean, such as the following:

- What is the role of the ocean in limiting human-caused climate change?
- What kinds of pharmaceuticals can be found naturally in marine organisms?
- How does sea floor spreading relate to the movement of tectonic plates?
- What economic deposits are there on the sea floor?
- Can rogue waves be predicted?
- What is the role of longshore transport in the distribution of sand on the beach?
- How does a particular pollutant affect organisms in the marine environment?

Oceanography is considered an *interdisciplinary* science (see Section 1.3 in Chapter 1) because it utilizes all the disciplines of science as they apply to the oceans. Some of the most exciting work and best employment opportunities combine multiple disciplines.

Job Duties of Oceanographers

There has been and will continue to be enormous expansion in the number of ocean-related jobs. Job opportunities for oceanographers exist with scientific research institutions (universities), various government agencies, and private companies that are engaged in searching for economic sea floor deposits, investigating areas for sea farming, and evaluating natural energy production from waves, currents, and tides. The duties of oceanographers vary from job to job but generally can be described within the subfields of oceanography as follows:

- *Geological oceanographers and geophysicists* explore the ocean floor and map submarine geologic structures. Studies of the physical and chemical properties of rocks and sediments give us valuable information about Earth's history. The results of their work help us understand the processes that created the ocean basins and the interactions between the ocean and the sea floor.
- *Chemical oceanographers and marine geochemists* investigate the chemical composition of seawater and its interaction with the atmosphere and the sea floor. Their work may include analysis of seawater components, desalination of seawater, and study of the effects of pollutants. They also examine chemical processes operating within the marine environment and work with biological oceanographers to study living systems. Their study of trace chemicals in seawater helps us understand how ocean currents move seawater around the globe and how the ocean affects climate.
- *Physical oceanographers* investigate ocean properties such as temperature, density, wave motions, tides, and currents. They study ocean–atmosphere relationships that influence weather and climate, the transmission of light and sound through water, and the ocean's interactions with its boundaries at the sea floor and the coast.
- *Biological oceanographers, marine biologists, and fisheries scientists* study marine plants and animals. They want to understand how marine organisms develop, relate to one another, adapt to their environment, and interact with their environment. They develop ecologically sound methods of harvesting seafood and study biologic responses to pollution. New fields associated with biological oceanography include marine biotechnology (the use of natural marine resources in the development of new industrial and biomedical products) and molecular biology (the study of the structure and function of bioinformational molecules—such as DNA, RNA, and proteins—and the regulation of cellular processes at the molecular level). Because marine biology is the best-known field of oceanography (and because the larger marine animals have such wide appeal), it is currently the most competitive, too.
- *Marine and ocean engineers* apply scientific and technical knowledge to practical uses. Their work ranges from designing sensitive instruments for measuring ocean processes to building marine structures that can withstand ocean currents, waves, tides, and severe storms. Subfields include acoustics, robotics, naval architecture, and electrical, mechanical, civil, and chemical engineering. These engineers often use highly specialized computer techniques.
- *Marine archaeologists* systematically recover and study material evidence, such as shipwrecks, graves, buildings, tools, and pottery remaining from past human life and culture that is now covered by the sea. Marine archaeologists use state-of-the-art technology to locate underwater archaeological sites.
- *Marine policy* and *communication experts* combine their knowledge of oceanography and social sciences, media, law, or business to develop guidelines and policies for the wise use of the ocean and coastal resources and communicate those items effectively to specialized groups and the general public.

Other job opportunities for oceanographers include work as science journalists specializing in marine science, teachers at various grade levels, and aquarium and museum curators.

Preparation for a Career in Oceanography

Individuals working in oceanography and marine-related fields need a strong background (typically an undergraduate degree) in at least one area of basic science (for example, geology, physics, chemistry, or biology) or engineering. In almost all cases, mathematics is required as well. Marine archaeology requires a background in archaeology or anthropology; marine policy studies require a background in at least one of the social sciences (such as law, economics, or political science).

The ability to speak and write clearly and critical thinking skills are prerequisites for any career. Fluency in computers and the ability to debug computer systems is a necessity. Because many job opportunities in oceanography require trips on research vessels (**Figure A5.1**), any shipboard experience is also desirable. Mechanical ability (such as the ability to fix equipment with limited materials on board a vessel without having to return to port) is a plus. Depending on the type of work that is required, other traits that may be desirable include the ability to speak one or more foreign languages, certification as a scuba diver, the ability to work for long periods of time in cramped conditions, physical stamina, physical strength, and a high tolerance to motion sickness.

Oceanography is a relatively new science with much room left for discovery, so most people enter the field with an advanced degree (master's or doctorate). Work as a marine technician, however, usually requires only a bachelor's degree or applicable experience. Although it does take a large commitment to achieve an advanced degree, don't feel overwhelmed. Many others like you have faced a similar journey. In the end, what makes all the hard work worthwhile is the incredible satisfaction of discovering new things about the ocean as well as all the interesting people you'll meet along the way.

Sources of Career Information

For sources of career information, consult the catalog of any college or university that offers a curriculum in oceanography or marine science. Online pamphlets that feature oceanography career information include:

- *Marine Science Careers: A Sea Grant Guide to Ocean Opportunities* by NOAA's Sea Grant College; available at **http://marinecareers.net**

- *Preparing for a Career in Oceanography* by the Scripps Institution of Oceanography at the University of California, San Diego; available at **www.sio.ucsd.edu/img/pdfs/careers.pdf**

- *The SeaWorld/Bush Gardens Guide to Zoological Park Careers* by SeaWorld's Education Department; available at **www.seaworld.org/infobooks/zoocareers/home.html**

- *Careers in Science and Engineering: A Student Planning Guide to Graduate School and Beyond* by the National Academies Press; available at **www.nap.edu/openbook.php?record_id=5129**

Other online sources of oceanography career information include the following:

- The Oceanography Society (TOS), which has a listing of oceanography career profiles at **http://tos.org/career-profiles**

- The U.S. Navy's Website, which has information on careers in oceanography at **www.oc.nps.edu/careers.html**

- The Website of Peter Brueggeman of the Scripps Institution of Oceanography, which offers a comprehensive list of information about careers in oceanography, marine science, and marine biology, including many links, at **http://ocean.peterbrueggeman.com/career.html**

- The Palomar College Oceanography Program Careers in Oceanography Website, which includes this appendix, at **www2.palomar.edu/pages/oceanography/careers-in-oceanography/**

- The SUNY Stony Brook Biological Sciences Department, which offers a list of worldwide marine laboratories and institutions at **http://life.bio.sunysb.edu/marinebio/mblabs.html**

- Woods Hole Oceanographic Institution's Website on women in oceanography, which is devoted to the achievements of women oceanographers and includes unique perspectives of women scientists, at **www.womenoceanographers.org**

Figure A5.1 **Oceanographers at work on a research vessel.** Oceanographers deploy a sonar device on a cable using the A-frame on Scripps Institution of Oceanography's research vessel R/V *Melville*.

DIVING DEEPER A5.1 OCEANS AND PEOPLE
REPORT FROM A STUDENT/OCEANOGRAPHER

One of the true pleasures of teaching is that some of your students become so interested in the subject matter you teach that they pursue their own career in it. One of Al Trujillo's former students, Joseph Fell-McDonald (**Figure A5**), works as a ranger and interpretive specialist on Maui for the State of Hawaii's Department of Land and Natural Resources, Division of Forestry and Wildlife, Natural Area Reserve System. Joe writes,

Aloha:

My career in oceanography-related work has followed a long and sinuous path. After high school I attended a trade school for mechanical engineering, where I learned how to fix cars and build houses. Even though I worked in industry for many years, I had a yearning for a work environment that would be satisfying, enjoyable, and well-paid. I decided fixing and building things was not for me. So I looked into colleges and attended nearby Palomar Community College in California, where I took many courses in geography, geology, and biology. I also enrolled in an oceanography course and realized how important the oceans are to life in general. It was during this time that it hit me: I longed for a career in ocean science!

Because I was formally trained in mechanical engineering and construction, I knew I could get a job just about anywhere. I decided to give the island of Maui a try just after finishing my courses in oceanography. There was no formal work in oceanography here so I applied for a job as a vessel mechanic at a local eco-tour company. The company offered volunteer opportunities, including photo identification and public interpretation of the seasonal humpback whales that migrate to Maui each year. I became fascinated with my surroundings and volunteered at many places, realizing my best attribute was an understanding of Earth's terrain and oceanic interactions.

Figure A5A Former oceanography student Joseph Fell-McDonald (*at far right on vessel* and *inset*) at work in the Hawaiian Islands Humpback Whale National Marine Sanctuary, Maui, Hawaii.

Soon after I was recruited to help restore a high erosion area at the Hawaiian Islands Humpback Whale National Marine Sanctuary (sister sanctuary to Monterey Bay). I was asked to assist with the National Marine Fisheries Service in helping stranded marine life, particularly green sea turtles (*Chelonia mydas*), and habitat restoration and monitoring of the critically endangered hawksbill turtle (*Eretmochelys imbricata*). I also became involved with rescuing stranded Hawaiian monk seals (*Monachus schauinslandi*) and entangled humpback whales (*Megaptera novaeangliae*).

I currently work as a ranger for the State of Hawaii's Department of Land and Natural Resources, Division of Forestry and Wildlife, Natural Area Reserve System. I work on Maui at the Ahihi-Kinau Natural Area Reserve where we follow recommendations by residents and local native Hawaiian kapuna (elders) to bring conservation back to Ahihi-Kinau. Much of my work is enforcement of protection laws, although I do a fair amount of public education, restoration, cultural inventory, biologic

surveying, and other protection of natural resources. The reserve is approximately 2000 land acres and about 800 marine protected acres out to a water depth of 100 fathoms. I patrol the area via four-wheel drive vehicle, jet ski, ATV, and even on horseback.

The most common question I get asked is, "How does someone get a job like yours?" Truthfully, I wake up with a smile on my face every day and enjoy the location, pay, and benefits. Sometimes it can be a tough and demanding job but I actually enjoy it. Due to my compassion and deep understanding of a broad range of sciences, I was able to apply my education, experience, and curiosity into the career of my dreams.

—*Joseph Fell-McDonald*
Ahihi-Kinau Ranger, State of Hawaii Department of Land and Natural Resources, Division of Forestry and Wildlife 54 High Street, Room 101, Wailuku, Maui, HI 96793
Joseph.C.Fell-McDonald@hawaii.gov

GLOSSARY

A

Abiotic Pertaining to nonliving factors, not life.

Abiotic environment The nonliving components of an ecosystem including chemical and physical factors (such as food, water, air, shelter, and temperature) that are essential for life.

Abyssal clay Deep-ocean (oceanic) deposits containing less than 30% biogenous sediment. Often oxidized and red in color, thus commonly termed red clay.

Abyssal hill A volcanic peak rising less than 1 kilometer (0.6 mile) above the ocean floor.

Abyssal hill province A deep-ocean region, particularly in the Pacific Ocean, where oceanic sedimentation rates are so low that abyssal plains do not form and the ocean floor is covered with abyssal hills.

Abyssal plain A flat depositional surface extending seaward from the continental rise or oceanic trenches.

Abyssal storm Stormlike occurrences of rapid current movement affecting the deep-ocean floor. They are believed to be caused by warm- and cold-core eddies of surface currents.

Abyssal zone The benthic environment between 4000 and 6000 meters (13,000 and 20,000 feet).

Abyssopelagic The open-ocean (oceanic) environment below 4000 meters (13,000 feet) in depth.

Acid A substance that releases hydrogen ions (H^+) in solution.

Acoustic Thermometry of Ocean Climate (ATOC) The measurement of oceanwide changes in water properties such as temperature by transmitting and receiving low-frequency sound signals.

Active margin A continental margin marked by a high degree of tectonic activity, such as those typical of the Pacific Rim. Types of active margins include convergent active margins (marked by plate convergence) and transform active margins (marked by transform faulting).

Adiabatic Pertaining to a change in the temperature of a mass resulting from compression or expansion. It requires no addition of heat to or loss of heat from the substance.

Aerosol A type of suspended particle in the atmosphere that can affect the atmosphere's reflectivity and its ability to trap heat.

Age of Discovery The 30-year period from 1492 to 1522 when Europeans explored the continents of North and South America and the globe was circumnavigated for the first time.

Agulhas Current A warm surface current that carries Indian Ocean water around the southern tip of Africa and into the Atlantic Ocean.

Air mass A large volume of air that has a definite region of origin and distinctive characteristics.

Alaskan Current A cool surface current that carries water in a counterclockwise fashion in the Gulf of Alaska.

Albedo The fraction of incident electromagnetic radiation reflected by a surface.

Algae Primarily aquatic, eukaryotic, photosynthetic organisms that have no root, stem, or leaf systems. Can be microscopic or macroscopic.

Alkaline Having a pH greater than 7, indicating a substance that releases an excess of hydroxide ions (OH^-) into solution. Also called basic.

Alveoli A tiny, thin-walled, capillary-rich sac in the lungs where the exchange of oxygen and carbon dioxide takes place.

Amino acid One of more than 20 naturally occurring compounds that contain NH_2 and $COOH$ groups. They combine to form proteins.

Amnesic shellfish poisoning Partial or total loss of memory resulting from poisoning caused by eating shellfish with high levels of domoic acid, a toxin produced by a diatom.

Amphidromic point A nodal, or "no-tide," point in the ocean or sea around which the crest of the tide wave rotates during one tidal period.

Amphipoda A crustacean order containing laterally compressed members such as the "beach hoppers."

Anadromous Pertaining to a species of fish that spawns in freshwater and then migrates into the ocean to grow to maturity.

Anaerobic Requiring or occurring in the absence of free oxygen (O_2).

Anaerobic respiration Respiration carried on in the absence of free oxygen (O_2). Some bacteria and protozoans carry on respiration this way.

Anchovy (anchoveta) A small silvery fish (*Engraulis ringens*) that swims through the water with its mouth open to catch its food.

Andesite A gray, fine-grained volcanic rock composed chiefly of plagioclase feldspar.

Anhydrite A colorless, white, gray, blue, or light purple evaporite mineral (anhydrous calcium sulfate, $CaSO_4$) that usually occurs as layers associated with gypsum deposits.

Animalia A kingdom of many-celled animals.

Anion An atom that has gained one or more electrons and has an electrical negative charge.

Annelida A phylum of elongated segmented worms.

Anomalistic month The time required for the Moon to go from perigee to perigee, 27.5 days.

Anoxic Without oxygen.

Antarctic Bottom Water A water mass that forms in the Weddell Sea, sinks to the ocean floor, and spreads across the bottom of all oceans.

Antarctic Circle The latitude 66.5 degrees south.

Antarctic Circumpolar Current The eastward-flowing current that encircles Antarctica and extends from the surface to the deep-ocean floor, it is the largest volume current in the world. Also called the West Wind Drift.

Antarctic Convergence The zone of convergence along the northern boundary of the Antarctic Circumpolar Current where the southward-flowing boundary currents of the subtropical gyres converge on the cold Antarctic waters.

Antarctic Divergence The zone of divergence separating the westward-flowing East Wind Drift and the eastward-flowing Antarctic Circumpolar Current.

Antarctic Intermediate Water Antarctic zone surface water that sinks at the Antarctic convergence and flows north at a depth of about 900 meters (2950 feet) beneath the warmer upper-water mass of the South Atlantic Subtropical Gyre.

Antarctic Ocean The ocean that surrounds the continent of Antarctica and is located south of about 50 degrees south latitude; also called the Southern Ocean.

Anthophyta Seed-bearing plants.

Anticyclone An atmospheric system characterized by the rapid, outward circulation of air masses about a high-pressure center that is associated with sinking air. Anticyclones circulate clockwise in the Northern Hemisphere and counterclockwise in the Southern Hemisphere and are usually accompanied by dry, clear, fair weather.

Anticyclonic flow The flow of air around a region of high pressure clockwise in the Northern Hemisphere.

Antilles Current A warm current that flows north seaward of the Lesser Antilles from the North Equatorial Current of the Atlantic Ocean to join the Florida Current.

Antinode A zone of maximum vertical particle movement in standing waves where crest and trough formation alternate.

Aphelion The point in the orbit of a planet or comet where it is farthest from the Sun.

Aphotic zone A zone without light. The ocean is generally in this state below 1000 meters (3280 feet).

Apogee The point in the orbit of the Moon or an artificial satellite that is farthest from Earth.

Aragonite A form of $CaCO_3$ that is less common and less stable than calcite. Pteropod shells are usually composed of aragonite.

Archaea One of the three major domains of life. The domain consists of simple microscopic bacterialike creatures (including methane producers and sulfur oxidizers that inhabit deep-sea vents and seeps) and other microscopic life-forms that prefer environments of extreme conditions of temperature and/or pressure.

Archaebacteria A kingdom of organisms that do not have nuclear material confined within a sheath but spread throughout the cell. Includes archaea.

Archipelago A large group of islands.

Arctic Circle The latitude 66.5 degrees north.

Arctic Convergence A zone of converging currents similar to the Antarctic Convergence but located in the Arctic.

Arctic Ocean The ocean located in the Northern Hemisphere polar region; the smallest ocean in the world.

Argo A global array of free-drifting profiling floats that move vertically and measure the temperature, salinity, and other water characteristics of the upper 2000 meters (6600 feet) of the ocean.

Aspect ratio The index of propulsive efficiency obtained by dividing the square of fin height by fin area.

Asthenosphere A plastic layer in the upper mantle 80 to 200 kilometers (50 to 124 miles) deep that may allow lateral movement of lithospheric plates and isostatic adjustments.

Atlantic Ocean The ocean located between South America, North America, Europe, and Africa; the second largest ocean in the world.

Atlantic meridional overturning circulation (AMOC) The part of conveyer belt circulation in the North Atlantic that includes both surface and deep currents.

Atlantic-type margin See *passive margin*.

Atmospheric wave A type of wave that occurs in the atmosphere. See *wave*.

Atoll A ring-shaped coral reef growing upward from a submerged volcanic peak. It may have low-lying islands composed of coral debris.

Atom A unit of matter, the smallest unit of an element, having all the characteristics of that element and consisting of a dense, central, positively charged nucleus surrounded by a system of electrons.

Atomic mass The mass of an atom, usually expressed in atomic mass units.

Atomic number The number of protons in an atomic nucleus.

Autotroph Algae, plants, and bacteria that can synthesize organic compounds from inorganic nutrients.

Autumnal equinox The passage of the Sun across the equator as it moves from the Northern Hemisphere into the Southern Hemisphere, approximately September 23. During this time, all places in the world experience equal lengths of night and day. Also called fall equinox.

B

Backshore The inner portion of the shore, lying landward of the mean spring tide high water line. Acted upon by the ocean only during exceptionally high tides and storms.

Backwash The flow of water down the beach face toward the ocean from a previously broken wave.

Bacteria One of the three major domains of life. The domain includes unicellular, prokaryotic microorganisms that vary in terms of morphology, oxygen and nutritional requirements, and motility.

Bacterioplankton Bacteria that live as plankton.

Bacteriovore An organism that feeds on bacteria.

Baleen A fibrous substance made of keratin found in parallel rows of long plates that hang down from the upper jaw of baleen whales.

Bar-built estuary A shallow estuary (lagoon) separated from the open ocean by a bar deposit such as a barrier island. The water in these estuaries usually exhibits vertical mixing.

Barrier flat An area that lies between the salt marsh and dunes of a barrier island and that is usually covered with grasses or forests if protected from overwash for a sufficient length of time.

Barrier island A long, narrow, wave-built island separated from the mainland by a lagoon.

Barrier reef A coral reef separated from the nearby landmass by open water.

Barycenter The center of mass of a system.

Basalt A dark-colored volcanic rock characteristic of the ocean crust. Contains minerals with relatively high iron and magnesium content.

Base A substance that releases hydroxide ions (OH^-) in solution. See *alkaline*.

Bathyal zone The benthic environment between the depths of 200 and 4000 meters (660 and 13,000 feet). It includes mainly the continental slope and the oceanic ridges and rises.

Bathymetry The measurement of ocean depth.

Bathypelagic zone The pelagic environment between the depths of 1000 and 4000 meters (3300 and 13,000 feet).

Bathyscaphe A specially designed deep-diving submersible.

Bathysphere A specially designed deep-diving submersible that resembles a sphere.

Bay barrier A marine deposit attached to the mainland at both ends and extending entirely across the mouth of a bay, separating the bay from the open water. Also known as a bay-mouth bar.

Beach Sediment seaward of the coastline through the surf zone that is in transport along the shore and within the surf zone.

Beach compartment A series of rivers, beaches, and submarine canyons involved in the movement of sediment to the coast, along the coast, and down one or more submarine canyons.

Beach face The wet, sloping surface that extends from the berm to the shoreline. Also known as the low tide terrace.

Beach replenishment The addition of beach sediment to replace lost or missing material. Also called beach nourishment.

Beach starvation The interruption of sediment supply and resulting narrowing of beaches.

Beaufort Wind Scale A standardized wind scale that describes the appearance of the sea surface from dead calm conditions to hurricane-force winds.

Benguela Current The cold eastern boundary current of the South Atlantic Subtropical Gyre.

Benthic Pertaining to the ocean bottom.

Benthos The forms of marine life that live on the ocean bottom.

Berm The dry, gently sloping region on the backshore of a beach at the foot of the coastal cliffs or dunes.

Berm crest The area of a beach that separates the berm from the beach face. The berm crest is often the highest portion of a berm.

Bicarbonate ion (HCO_3^-) An ion that contains the radical group (HCO_3^-).

Bioaccumulation The accumulation of a substance, such as a toxic chemical, in various tissues of a living organism.

Bioerosion Erosion of reef or other solid bottom material by the activities of organisms.

Biofilm Coating of organic matter such as that found on sand grains.

Biogenous sediment Sediment containing material produced by plants or animals, such as coral reefs, shell fragments, and housings of diatoms, radiolarians, foraminifers, and coccolithophores; components can be either macroscopic or microscopic.

Biogeochemical cycle The natural cycling of compounds among the living and nonliving components of an ecosystem.

Biological pump The movement of carbon dioxide that enters the ocean from the atmosphere through the water column to the sediment on the ocean floor by biological processes—photosynthesis, secretion of shells, feeding, and dying.

Bioluminescence Light organically produced by a chemical reaction. Found in bacteria, phytoplankton, and various fishes (especially deep-sea fish).

Biomagnification Concentration of impurities as animals are eaten and the impurity is passed through food chains.

Biomass The total mass of a defined organism or group of organisms in a particular community or in the ocean as a whole.

Biomass pyramid A representation of trophic levels that illustrates the progressive decrease in total biomass at successive higher levels of the pyramid.

Bioremediation The technique of using microbes to assist in cleaning toxic spills.

Biotic community The living organisms that inhabit an ecosystem.

Biozone A region of the environment that has distinctive biological characteristics.

Bivalve A mollusk, such as an oyster or a clam, that has a shell consisting of two hinged valves.

Black smoker A hydrothermal vent on the ocean floor that emits a black cloud of hot water filled with dissolved metal particles.

Body wave A longitudinal or transverse wave that transmits energy through a body of matter.

Boiling point The temperature at which a substance changes state from a liquid to a gas at a given pressure.

Bore A steep-fronted tide crest that moves up a river in association with an incoming high tide.

Boundary current The northward- or southward-flowing currents that form the western and eastern boundaries of each subtropical gyre.

Brackish Low-salinity water caused by the mixing of freshwater and saltwater.

Brazil Current The warm western boundary current of the South Atlantic Subtropical Gyre.

Breaker zone A region where waves break at the seaward margin of the surf zone.

Breakwater An artificial structure constructed roughly parallel to shore and designed to protect a coastal region from the force of ocean waves.

Brittle Descriptive term for a substance that is likely to fracture when force is applied to it.

Bryozoa A phylum of colonial animals that often share one coelomic cavity. Encrusting and branching forms secrete a protective housing (zooecium) of calcium carbonate or chitinous material.

Buffering The process by which a substance minimizes a change in the acidity of a solution when an acid or base is added to the solution.

Buoyancy The ability or tendency to float or rise in a liquid.

Bycatch Marine organisms that are caught incidentally by fishers seeking commercial species.

C

Calcareous Containing calcium carbonate.

Calcite A mineral with the chemical formula $CaCO_3$.

Calcite compensation depth (CCD) The depth at which the amount of calcite ($CaCO_3$) produced by the organisms in the overlying water column is equal to the amount of calcite the water column can dissolve. No calcite deposition occurs below this depth, which, in most parts of the ocean, is at a depth of 4500 meters (15,000 feet).

Calcium carbonate ($CaCO_3$) A chalklike substance secreted by many organisms in the form of coverings or skeletal structures.

California Current The cold eastern boundary current of the North Pacific Subtropical Gyre.

Calorie A unit of heat, defined as the amount of heat required to raise the temperature of 1 gram of water 1°C.

Calving The process by which a glacier breaks at an edge, so that a portion of the ice separates and falls from the glacier.

Canary Current The cold eastern boundary current of the North Atlantic Subtropical Gyre.

Capillarity The action by which a fluid, such as water, is drawn up in small tubes as a result of surface tension.

Capillary wave An ocean wave whose wavelength is less than 1.74 centimeters (0.7 inch). The dominant restoring force for such waves is surface tension.

Carapace (1) A chitinous or calcareous shield that covers the cephalothorax of some crustaceans. (2) The dorsal portion of a turtle shell.

Carbohydrate An organic compound containing the elements carbon, hydrogen, and oxygen with the general formula $(CH_2O)_n$.

Carbonate ion ($CO_3{}^{2-}$) An ion that contains the radical group $CO_3{}^{2-}$.

Caribbean Current The warm current that carries equatorial water across the Caribbean Sea into the Gulf of Mexico.

Carnivora The order of marine mammals that includes the sea otter, polar bear, and pinnipeds.

Carnivore An animal that depends on other animals solely or chiefly for its food supply.

Carnivorous feeding The process by which an organism feeds solely or chiefly on other animals as food items.

Carotin An orange-yellow pigment found in plants.

Cation An atom that has lost one or more electrons and has an electrical positive charge.

Caulerpa taxifolia A tropical seaweed; a cold-water clone was introduced into the aquarium industry and found its way into coastal waters of the Mediterranean and Southern California. It continues to spread in the Mediterranean, but has been eradicated in Southern California.

Centigrade temperature scale A temperature scale based on the freezing point (0°C = 32°F) and boiling point (100°C = 212°F) of pure water. Also known as the Celsius scale after its founder.

Centripetal force A center-seeking force that tends to make rotating bodies move toward the center of rotation.

Cephalopoda (cephalopod) A class of the phylum Mollusca whose members have a well-developed pair of eyes and a ring of tentacles surrounding the mouth. The shell is absent or internal on most members. The class includes the squid, octopus, and nautilus.

Cetacea (cetacean) An order of marine mammals that includes the whales, dolphins, and porpoises.

Chalk A soft, compact form of calcite, generally gray-white or yellow-white in color and derived chiefly from microscopic fossils.

Chemical energy A form of potential energy stored in the chemical bonds of compounds.

Chemosynthesis A process by which bacteria or archaea synthesize organic molecules from inorganic nutrients using chemical energy released from the bonds of a chemical compound (such as hydrogen sulfide) by oxidation.

Chloride ion (Cl^-) A chlorine atom that has become negatively charged by gaining one electron.

Chlorinity The amount of chloride ion and ions of other halogens in ocean water expressed in parts per thousand (‰) by weight.

Chlorophyll A group of green pigments that make it possible for plants to carry on photosynthesis.

Chlorophyta Green algae. Characterized by the presence of chlorophyll and other pigments.

Chloroplast Photosynthetic structures containing chlorophyll found in plants, algae, and other photosynthetic organisms.

Chondrite A stony meteorite composed primarily of silicate rock material and containing chondrules (spheroidal granules). They are the most commonly found meteorites.

Chronometer An exceptionally precise timepiece.

Chrysophyta An important phylum of planktonic algae that includes the diatoms. The presence of chlorophyll is masked by the pigment carotene that gives the plants a golden color.

Ciguatera A type of seafood poisoning caused by ingestion of certain tropical reef fish (most notably barracuda, red snapper, and grouper) that have high levels of naturally occurring dinoflagellate toxins.

Cilium A short, hairlike structure common on single-celled or simple multicelled animals. Beating in unison, cilia may create water currents that carry food toward the mouth of an animal or may be used for locomotion.

Circadian rhythm Behavioral and physiological rhythms of organisms related to the 24-hour day. Sleeping and waking patterns are an example.

Circular orbital motion The motion of water particles caused by a wave as the wave is transmitted through water.

Circumpolar Current An eastward-flowing current that extends from the surface to the ocean floor and encircles Antarctica.

Clay (1) A particle size between silt and colloid. (2) Any of various hydrous aluminum silicate minerals that are plastic, expansive, and have ion exchange capabilities.

Climate The meteorological conditions, including temperature, precipitation, and wind that characteristically prevail in a particular region; the long-term average of weather.

Climate system Refers to all factors that influence the climate of a region, including both natural and human-induced climate changes and their interactions.

Cnidaria A phylum that contains some 10,000 species of predominantly marine animals with a sacklike body and stinging cells on tentacles that surround the single opening to the gut cavity. There are two basic body forms. The medusa is a pelagic form represented by the jellyfish. The polyp is a predominantly benthic form found in sea anemones and corals. Previously named Coelenterata.

Cnidoblast A stinging cell of phylum Cnidaria that contains the stinging mechanism (nematocyst) used in defense and capturing prey.

Coast A strip of land that extends inland from the coastline as far as marine influence is evidenced in the landforms.

Coastal geostrophic current See *geostrophic current*.

Coastal plain estuary An estuary formed by rising sea level flooding a coastal river valley.

Coastal upwelling The movement of deeper nutrient-rich water into the surface water mass as a result of windblown surface water moving offshore.

Coastal waters The relatively shallow-water areas that adjoin continents or islands.

Coastal wetland A biologically productive region bordering estuaries and other protected coastal areas; typically as salt marshes in latitudes greater than 30° and as mangrove swamps in lower latitudes.

Coastline The landward limit of the effect of the highest storm waves on the shore.

Coccolith Tiny calcareous discs averaging about 3 microns (0.00012 inch) in diameter that form the cell wall of coccolithophores.

Coccolithophore A microscopic planktonic form of algae encased by a covering composed of calcareous discs (coccoliths).

Coelenterata A phylum of radially symmetrical animals that includes two basic body forms, the medusa and the polyp. Includes jellyfish (medusoid) and sea anemones (polypoid). Preferred name is now Cnidaria.

Cohesion The intermolecular attraction by which the elements of a body are held together.

Cold core ring A circular eddy of a surface current that contains cold water in its center and rotates counterclockwise in the Northern Hemisphere.

Cold front A weather front in which a cold air mass moves into and under a warm air mass. It creates a narrow band of intense precipitation.

Cold-blooded See *poikilothermic* and *ectothermic*.

Colonial animal An animal that lives in groups of attached or separate individuals. Groups of individuals may serve special functions.

Columbia River estuary An estuary at the border between the states of Washington and Oregon that has been most adversely affected by the construction of hydroelectric dams.

Comb jelly Common name for members of the phylum Ctenophora. See also *ctenophore*.

Commensalism A symbiotic relationship in which one party benefits and the other is unaffected.

Compensation depth for photosynthesis The depth at which net photosynthesis becomes zero; below this depth, photosynthetic organisms can no longer survive. This depth is greater in the open ocean (up to 100 meters or 330 feet) than near the shore due to increased turbidity that limits light penetration in coastal regions.

Compound A substance containing two or more elements combined in fixed proportions.

Condensation The conversion of water from the vapor to the liquid state. When it occurs, the energy required to vaporize the water is released into the atmosphere. This is about 585 calories per gram of water at 20°C.

Condensation point The point at which condensation occurs.

Conduction The transmission of heat by the passage of energy from particle to particle.

Conjunction The apparent closeness of two heavenly bodies. During the new moon phase, Earth and the Moon are in conjunction on the same side of the Sun.

Constant proportions, principle of A principle which states that the major constituents of ocean-water salinity are found in the same relative proportions throughout the ocean, independent of salinity.

Constructive interference A form of wave interference in which two waves come together in phase, for example, crest to crest, to produce a greater displacement from the still-water line than that produced by either of the waves alone.

Consumer An animal within an ecosystem that consumes the organic mass produced by the producers.

Continent About one-third of Earth's surface that rises above the deep-ocean floor to be exposed above sea level. Continents are composed primarily of granite, an igneous rock of lower density than basaltic oceanic crust.

Continental accretion Growth or increase in size of a continent by gradual external addition of crustal material.

Continental arc An arc-shaped row of active volcanoes produced by subduction that occurs along convergent active continental margins.

Continental borderland A highly irregular portion of the continental margin that is submerged beneath the ocean and is characterized by depths greater than those characteristic of the continental shelf.

Continental drift A term applied to early theories supporting the idea that the continents move across Earth's surface through time.

Continental effect Describes areas that are less affected by the sea and therefore having a greater range of temperature differences (both daily and yearly).

Continental margin The submerged area next to a continent comprising the continental shelf, continental slope, and continental rise.

Continental rise A gently sloping depositional surface at the base of the continental slope.

Continental shelf A gently sloping depositional surface extending from the low water line to the depth of a marked increase in slope around the margin of a continent or island.

Continental slope A relatively steeply sloping surface lying seaward of the continental shelf.

Convection Heat transfer in a gas or liquid by the circulation of currents from one region to another.

Convection cell A circular-moving loop of matter involved in convective movement.

Convergence The act of coming together from different directions. There are polar, tropical, and subtropical regions of the oceans where water masses with different characteristics come together. Along these lines of convergence, the denser masses sink beneath the others.

Convergent plate boundary A lithospheric plate boundary where adjacent plates converge, producing ocean trench–island arc systems, ocean trench–continental volcanic arcs, or folded mountain ranges.

Conveyer-belt circulation An integrated deep-water and surface current circulation pattern that resembles a large conveyer belt.

Copepoda An order of microscopic to nearly microscopic crustaceans that are important members of zooplankton in temperate and subpolar waters.

Coral A group of benthic anthozoans that exist as individuals or in colonies and secrete $CaCO_3$ external skeletons. Under the proper conditions, corals may produce reefs composed of their external skeletons and the $CaCO_3$ material secreted by varieties of algae associated with the reefs.

Coral bleaching The loss of color in coral reef organisms that causes them to turn white. Coral bleaching is caused by the expulsion of the coral's symbiotic zooxanthellae algae in response to high water temperatures or other adverse conditions.

Coral reef A calcareous organic reef composed significantly of solid coral and coral sand. Algae may be responsible for more than half of the $CaCO_3$ reef material. Found in waters where the minimum average monthly temperature is 18°C or higher.

Core (1) The deep, central layer of Earth, composed primarily of iron and nickel. It is subdivided into a liquid outer core 2270 kilometers (1410 miles) thick and a solid inner core with a radius of 1216 kilometers (756 miles). (2) A cylinder of sediment and/or rock material usually obtained by drilling.

Coriolis effect An apparent force resulting from Earth's rotation that causes particles in motion to be deflected to the right in the Northern Hemisphere and to the left in the Southern Hemisphere.

Cosmogenous sediment Sediment derived from outer space.

Cotidal line A line connecting points where high tide occurs simultaneously.

Counterillumination Camouflaging by using bioluminescence to match the color and intensity of dim filtered sunlight from above and obliterate a telltale shadow.

Countershading Protective coloration in an animal or insect, characterized by darker coloring of areas exposed to light and lighter coloring of areas that are normally shaded.

Covalent bond A chemical bond formed by the sharing of one or more electrons, especially pairs of electrons, between atoms.

Crepuscular A term that describes animals that are active primarily during twilight (dawn and dusk).

Crest (wave) The portion of an ocean wave that is displaced above the still-water level.

Cruiser Fish (such as the bluefin tuna) that constantly cruise pelagic waters in search of food.

Crust (1) The uppermost outer layer of Earth's structure that is composed of basaltic oceanic crust and granitic continental crust. The average thickness of the crust ranges from 8 kilometers (5 miles) beneath the ocean to 35 kilometers (22 miles) beneath the continents. (2) A hard covering or surface layer of hydrogenous sediment.

Crustacea (crustacean) A class of subphylum Arthropoda that includes barnacles, copepods, lobsters, crabs, and shrimp.

Crystalline rock Igneous or metamorphic rocks. These rocks are made up of crystalline particles with orderly molecular structures.

Ctenophore A member of the phylum of gelatinous organisms that are more or less spheroidal with biradial symmetry. These exclusively marine animals have eight rows of ciliated combs for locomotion, and most have two tentacles for capturing prey.

Current A physical movement of water; if in the ocean, called an ocean current, which can be further subdivided into surface currents and deep currents.

Cyclone An atmospheric system characterized by the rapid, inward circulation of air masses about a low-pressure center that is associated with rising air. Cyclones circulate counterclockwise in the Northern Hemisphere and clockwise in the Southern Hemisphere and are usually accompanied by stormy, often destructive, weather.

Cyclonic flow The flow of air around a region of low pressure counterclockwise in the Northern Hemisphere.

D

Davidson Current A northward-flowing surface current along the Washington–Oregon coast that is driven by geostrophic effects on a large freshwater runoff.

DDT An insecticide (dichlorodiphenyltrichloroethane) that caused damage to marine bird populations in the 1950s and 1960s. Its use is now banned throughout most of the world.

Dead zone A region of hypoxic conditions that kills off most marine organisms that cannot escape. It is usually the result of eutrophication caused by runoff from land-based fertilizer applications.

Decay distance The distance over which waves change from a choppy "sea" to uniform swell.

Declination The angular distance of the Sun or Moon above or below the plane of Earth's equator.

Decomposer Primarily bacteria that break down nonliving organic material, extract some of the products of decomposition for their own needs, and make available the compounds needed for primary production.

Decompression sickness A serious condition that occurs in divers when they ascend too rapidly, causing nitrogen bubbles to form in the blood and tissue, resulting in great pain and sometimes death. Also known as the bends.

Deep biosphere The microbe-rich region beneath the sea floor.

Deep boundary current A relatively strong deep current flowing across the continental rise along the western margin of ocean basins.

Deep current A density-driven circulation that is initiated at the ocean surface by temperature and salinity conditions that produce a high-density water mass, which sinks and spreads slowly beneath surface waters.

Deep-ocean Assessment and Reporting of Tsunamis (DART) A system that utilizes sea floor sensors capable of picking up the small yet distinctive pressure pulse from a tsunami at the surface.

Deep-ocean basin Areas of the ocean floor that have deep water, are far from land, and are underlain by basaltic crust.

Deep scattering layer (DSL) A layer of marine organisms in the open ocean that scatter signals from an echo sounder. It migrates daily from depths of slightly over 100 meters (330 feet) at night to more than 800 meters (2600 feet) during the day.

Deep Sea Drilling Project (DSDP) A sea floor drilling program initiated in 1968 that was designed to obtain cores from the deep sea; it was the predecessor to the Ocean Drilling Program (ODP) and the Integrated Ocean Drilling Program (IODP).

Deep-sea fan A large fan-shaped deposit commonly found on the continental rise seaward of such sediment-laden rivers as the Amazon, Indus, or Ganges-Brahmaputra. Also known as a submarine fan.

Deep-sea fish Any of a large group of fishes that lives within the aphotic zone and has special adaptations for finding food and avoiding predators in darkness.

Deep-sea system The bathyal, abyssal, and hadal benthic environments.

Deep water The water beneath the permanent thermocline (and resulting pycnocline) that has a uniformly low temperature.

Deep-water wave An ocean wave traveling in water that has a depth greater than one-half the average wavelength. Its speed is independent of water depth.

Delta A low-lying deposit at the mouth of a river, usually having a triangular shape as viewed from above.

Density The mass per unit volume of a substance. Usually expressed as grams per cubic centimeter (g/cm^3) For ocean water with a salinity of 35‰ at 0°C, the density is 1.028 g/cm^3.

Density stratification A layering based on density, where the highest density material occupies the lowest space.

Deposit feeding The process by which an organism feeds on food items that occur as deposits, including detritus and various detritus-coated sediment.

Depositional shore A shoreline dominated by processes that form deposits (such as sand bars and barrier islands) along the shore.

Desalination The removal of salt ions from ocean water to produce pure water.

Destructive interference A form of wave interference in which two waves come together out of phase, for example, crest to trough, and produce a wave with less displacement than the larger of the two waves would have produced alone.

Detritus (1) Any loose material produced directly from rock disintegration. (2) Material resulting from the disintegration of dead organic remains.

Diatom A member of the class Bacillariophyceae of algae that possesses a wall of overlapping silica valves.

Diatomaceous earth A deposit composed primarily of the tests of diatoms mixed with clay. Also called diatomite.

Diffraction A change in the direction or intensity of a wave after passing an obstacle that cannot be interpreted as refraction or reflection.

Diffusion A process by which materials move through fluids by random molecular movement from areas of high concentration to areas in which they are in lower concentrations, thus becoming evenly distributed.

Dinoflagellate A single-celled microscopic planktonic organism that may possess chlorophyll and belong to the phylum Pyrrophyta (autotrophic) or may ingest food and belong to the class Mastigophora of the phylum Protozoa (heterotrophic).

Dipolar Having two poles. The water molecule possesses a polarity of electrical charge with one pole being more positive and the other more negative in electrical charge.

Discontinuity An abrupt change in a property such as temperature or salinity at a line or surface.

Disphotic zone The dimly lit zone, corresponding approximately to the mesopelagic, in which there is not enough light to support photosynthetic organisms; sometimes called the twilight zone.

Disruptive coloration A marking or color pattern that confuses prey.

Dissolved oxygen Oxygen that is dissolved in ocean water.

Distillation A method of purifying liquids by heating them to their boiling point and condensing the vapor.

Distributary A small stream flowing away from a main stream. Such streams are characteristic of deltas.

Disturbing force The energy that causes waves to form.

Diurnal inequality The difference in the heights of two successive high tides or two successive low tides during a lunar (tidal) day.

Diurnal tidal pattern A tidal pattern exhibiting one high tide and one low tide during a tidal day; a daily tide.

Divergence A horizontal flow of water from a central region, as occurs in upwelling.

Divergent plate boundary A lithospheric plate boundary where adjacent plates diverge, producing an oceanic ridge or rise (spreading center).

Doldrums A global belt of light, variable winds near the equator, resulting from the vertical flow of low-density air masses upward within this equatorial belt. Associated with much precipitation.

Dolphin (1) A brilliantly colored fish of the genus *Coryphaena*. (2) The name applied to the small, beaked members of the cetacean family Delphinidae.

Dorsal Pertaining to the back or upper surface of most animals.

Downwelling In the open or coastal ocean, where Ekman transport causes surface waters to converge or impinge on the coast, surface water that moves down beneath the surface.

Drift bottle Equipment used to study ocean surface current movement by drifting with currents.

Driftnet A fishing net made of monofilament fishing line that catches organisms by entanglement.

Drifts Thick sediment deposits on the continental rise produced where a deep boundary current slows and loses sediment when it changes direction to follow the base of the continental slope.

Drowned beach An ancient beach now beneath the coastal ocean because of rising sea level or subsidence of the coast.

Drowned river valley The lower part of a river valley that has been submerged by rising sea level or subsidence of the coast.

Dune (coastal) A coastal deposit of sand lying landward of the beach and deriving its sand from onshore winds that transport beach sand inland.

Dynamic topography A surface configuration resulting from the geopotential difference between a given surface and a reference surface of no motion. A contour map of this surface is useful in estimating the nature of geostrophic currents.

E

Earthquake A sudden release of energy inside Earth that creates seismic waves and is usually caused by either faulting (movement along a fracture in Earth's crust) or volcanic activity.

East Australian Current The warm western boundary surface current of the South Pacific Subtropical Gyre.

East Pacific Rise A fast-spreading divergent plate boundary extending southward from the Gulf of California through the eastern South Pacific Ocean.

East Wind Drift The coastal surface current driven in a westerly direction by the polar easterly winds blowing off Antarctica.

Eastern boundary current Equatorward-flowing cold, sluggish movements of surface water on the eastern side of each subtropical gyre.

Eastern Pacific Garbage Patch An area in the eastern part of the North Pacific Gyre that serves as an accumulation area for floating plastic and other trash; it is about twice the size of Texas.

Ebb current The flow of water seaward during a decrease in the height of the tide.

Ebb tide An outgoing, falling tide.

Echinodermata (echinoderm) A phylum of animals that have bilateral symmetry in larval forms and usually a five-sided radial symmetry as adults. Benthic and possessing rigid or articulating exoskeletons of calcium carbonate with spines, this phylum includes sea stars, brittle stars, sea urchins, sand dollars, sea cucumbers, and sea lilies.

Echo sounder A device that transmits sound from a ship's hull to the ocean floor where it is reflected back to receivers. The speed of sound in the water is known, so the depth can be determined from the travel time of the sound signal.

Echolocation A sensory system in odontocete cetaceans in which usually high-pitched sounds are emitted and their echoes interpreted to determine the direction and distance of objects.

Ecliptic The plane of the center of the Earth–Moon system as it orbits around the Sun.

Ecosystem All the organisms in a biotic community and the abiotic environmental factors with which they interact.

Ectothermic Of or relating to an organism that regulates its body temperature largely by exchanging heat with its surroundings; cold-blooded.

Eddy A current of any fluid forming on the side of a main current. It usually moves in a circular path and develops where currents encounter obstacles or flow past one another.

Ekman spiral A theoretical consideration of the effect of a steady wind blowing over an ocean of unlimited depth and breadth and of uniform viscosity. The result is a surface flow at 45 degrees to the right of the wind in the Northern Hemisphere. Water at increasing depth below the surface will drift in directions increasingly more slowly and to the right, until at about 100 meters (330 feet) depth it may move in a direction opposite to that of the wind.

Ekman transport The net transport of surface water set in motion by wind. Due to the Ekman spiral phenomenon, it is theoretically in a direction 90° to the right and 90° to the left of the wind direction in the Northern Hemisphere and Southern Hemisphere, respectively.

El Niño A southerly flowing warm surface current that generally develops off the coast of Ecuador around Christmastime. Occasionally it will move farther south into Peruvian coastal waters and cause the widespread death of plankton, fish, and other organisms such as marine mammals that depend on fish for food.

El Niño–Southern Oscillation (ENSO) The correlation of El Niño events with an oscillatory pattern of pressure change in a persistent high-pressure cell in the southeastern Pacific Ocean and a persistent low-pressure cell over the East Indies.

Electrical conductivity The ability or power to conduct or transmit electricity.

Electrolysis A separation process by which salt ions are removed from saltwater through water-impermeable membranes toward oppositely charged electrodes.

Electromagnetic energy Energy that travels as waves or particles with the speed of light. Different kinds possess different properties based on wavelength. The longest wavelengths belong to radio waves, up to 100 kilometers (60 miles) in length. At the other end of the spectrum are cosmic rays with greater penetrating power and wavelengths of less than 0.000001 micron.

Electromagnetic spectrum The spectrum of radiant energy emitted from stars, ranging from cosmic rays with wavelengths of less than one millionth of a micron, to very long waves with wavelengths in excess of 100 kilometers (60 miles).

Electron A subatomic particle that orbits the nucleus of an atom and has a negative electric charge.

Electron cloud The diffuse area surrounding the nucleus of an atom where electrons are found.

Electrostatic force The force between charged particles; also called electrostatic attraction or electrostatic repulsion, depending on if the particles have different or the same charge.

Element One of a number of substances, each of which is composed entirely of like particles—atoms—that cannot be broken into smaller particles by chemical means.

Emerging shoreline A shoreline resulting from the emergence of the ocean floor relative to the ocean surface. It is usually rather straight and characterized by marine features usually found at some depth.

ENSO See *El Niño–Southern Oscillation (ENSO)*.

ENSO index An index showing the relative strength of El Niño and La Niña conditions.

Entropy The degree of randomness or disorder in a system.

Environment The sum of all physical, chemical, and biological factors to which an organism or community is subjected.

Environmental bioassay An environmental assessment technique that determines the concentration of a pollutant that causes 50% mortality among a specific group of test organisms.

Epicenter The point on Earth's surface that is directly above the focus of an earthquake.

Epifauna Animals that live on the ocean bottom, either attached or moving freely over it.

Epipelagic zone A subdivision of the oceanic province that extends from the surface to a depth of 200 meters (660 feet).

Equator The imaginary great circle around Earth's surface, equidistant from the poles and perpendicular to Earth's axis of rotation. It divides Earth into the Northern Hemisphere and the Southern Hemisphere.

Equatorial Pertaining to the equatorial region.

Equatorial countercurrent Eastward-flowing currents found between the North and South Equatorial Currents in all oceans but particularly well developed in the Pacific Ocean.

Equatorial current Westward-flowing surface currents that travel along the equator in all ocean basins, caused by the trade winds. They are called North or South Equatorial Currents, depending on their position north or south of the equator.

Equatorial low A band of low atmospheric pressure that encircles the globe along the equator.

Equatorial upwelling The movement of deeper nutrient-rich water into the surface water mass as a result of divergence of currents along the equator.

Erosion The group of natural processes, including weathering, dissolution, abrasion, corrosion, and transportation, by which material is worn away from Earth's surface.

Erosional shore A shoreline dominated by processes that form erosional features (such as cliffs and sea stacks) along the shore.

Estuarine circulation pattern A flow pattern in an estuary characterized by a net surface flow of low-salinity water toward the ocean and an opposite net subsurface flow of seawater toward the head of the estuary.

Estuary A partially enclosed coastal body of water in which salty ocean water is significantly diluted by freshwater from land runoff. Examples of estuaries include river mouths, bays, inlets, gulfs, and sounds.

Eubacteria A kingdom of organisms that do not have nuclear material confined within a sheath but spread throughout the cell. Includes bacteria and blue-green algae; previously known as Kingdom Monera.

Eukarya One of the three major domains of life. The domain includes single-celled or multicellular organisms whose cells usually contain a distinct membrane-bound nucleus.

Euphotic zone A layer that extends from the surface of the ocean to a depth where enough light exists to support photosynthesis, rarely deeper than 100 meters (330 feet).

Euryhaline A descriptive term for organisms with a high tolerance for a wide range of salinity conditions.

Eurythermal A descriptive term for organisms with a high tolerance for a wide range of temperature conditions.

Eustatic sea level change A worldwide raising or lowering of sea level.

Eutrophic A region of high productivity.

Eutrophication The enrichment of waters by a previously scarce nutrient; if caused by humans, called cultural eutrophication.

Evaporation The process of changing from the liquid to the vapor state at a temperature below the boiling point of a substance.

Evaporite A sedimentary deposit that is left behind when water evaporates; also known as evaporite minerals, which include gypsum, calcite, and halite.

Evolution The change of groups of organisms with the passage of time, mainly as a result of natural selection, so that descendants differ morphologically and physiologically from their ancestors.

Exclusive economic zone (EEZ) A coastal zone that generally extends 200 nautical miles (370 kilometers) from shore and establishes coastal nation jurisdiction including mineral resources, fish stocks, and pollution. If the continental shelf extends beyond the 200-mile EEZ, the EEZ is extended to 350 nautical miles (648 kilometers) from shore.

Extrusive rock Igneous rock that flows out onto Earth's surface before cooling and solidifying (lava).

Eye (1) An organ of vision or of light sensitivity. (2) The circular low-pressure area of relative calm at the center of a hurricane.

F

Fact Something having real, demonstrable existence. A scientific fact is an occurrence that has been repeatedly confirmed.

Faculae A bright spot on the Sun that is associated with magnetic storms.

Fahrenheit temperature scale (°F) A temperature scale whereby the freezing point of water is 32° and the boiling point of water is 212°.

Falcate Curved and tapering to a point; sickle-shaped.

Falkland Current A northward-flowing cold surface current that occurs off the southeastern coast of South America.

Fall bloom A middle-latitude bloom of phytoplankton that occurs during the fall and is limited by the availability of sunlight.

Fall equinox See *autumnal equinox*.

Fan A gently sloping, fan-shaped feature normally located near the lower end of a canyon. Also known as a submarine fan.

Fat An organic compound formed from alcohol, glycerol, and one or more fatty acids; a lipid, it is a solid at atmospheric temperatures.

Fathom (fm) A unit of depth in the ocean, commonly used in countries using the English system of units. It is equal to 1.83 meters (6 feet).

Fault A fracture or fracture zone in Earth's crust along which displacement has occurred.

Fault block A crustal block bounded on at least two sides by faults. Usually elongate; if it is down-dropped, it produces a graben; if uplifted, it is a horst.

Fauna The animal life of any particular area or of any particular time.

Fecal pellet The excrement of planktonic crustaceans that assist in speeding up the descent rate of sedimentary particles by combining them into larger packages.

Feedback loop A self-repeating processes in which an initial change is modified. A positive-feedback loop amplifies an initial change, and a negative-feedback loop counteracts an initial change.

Ferrel cell The large atmospheric circulation cell that occurs between 30 and 60 degrees latitude in each hemisphere.

Ferromagnesium Describes minerals rich in iron and magnesium.

Fetch (1) Pertaining to the area of the open ocean over which the wind blows with constant speed and direction, thereby creating a wave system. (2) The distance across the fetch (wave-generating area) measured in a direction parallel to the direction of the wind.

Filter feeding The process by which an organism obtains its food by filtering seawater to collect floating organisms to ingest. Also known as suspension feeding.

Fishery Fish caught from the ocean by commercial fishers.

Fishery management The organized effort directed at regulating fishing activity with the goal of maintaining a long-term fishery.

Fissure A long, narrow opening; a crack or cleft.

Fjord A long, narrow, deep, U-shaped inlet that usually represents the seaward end of a glacial valley that has become partially submerged after the melting of the glacier.

Flagellum A whiplike living structure used by some cells for locomotion.

Floe A piece of floating ice other than fast ice or icebergs. May range in maximum horizontal dimension from about 20 centimeters (8 inches) to more than 1 kilometer (0.6 mile).

Flood current A tidal current associated with increasing height of the tide, generally moving toward the shore.

Flood tide An incoming, rising tide.

Flora The plant life of any particular area or of any particular time.

Florida Current A warm surface current flowing north along the coast of Florida that merges into the Gulf Stream.

Flotsam Any floating debris, particularly from wrecked ships.

Folded mountain range A mountain range formed as a result of the convergence of lithospheric plates. Folded mountain ranges are characterized by masses of folded sedimentary rocks that formed from sediments deposited in the ocean basin that was destroyed by the convergence.

Food chain The passage of energy materials from producers through a sequence of a herbivore and a number of carnivores.

Food web A group of interrelated food chains.

Foraminifer An order of planktonic and benthic protozoans that possess protective coverings, usually composed of calcium carbonate.

Forced wave A wave that is generated and maintained by a continuous force such as the gravitational attraction of the Moon.

Foreshore The portion of the shore lying between the normal high and low water marks; the intertidal zone.

Fossil Any remains, print, or trace of an organism that has been preserved in Earth's crust.

Fossil fuel A natural fuel such as petroleum, gas, or coal that formed in the geological past from the remains of living organisms.

Fracture zone An extensive linear zone of unusually irregular topography of the ocean floor, characterized by large seamounts, steep-sided or asymmetrical ridges, troughs, or long, steep slopes. Usually represents ancient, inactive transform fault zones.

Free wave A wave created by a sudden rather than a continuous impulse that continues to exist after the generating force is gone.

Freeze separation The desalination of seawater by multiple episodes of freezing, rinsing, and thawing.

Freezing The process by which a liquid is converted to a solid at its freezing point.

Freezing point The temperature at which a liquid becomes a solid under any given set of conditions. The freezing point of water is 0°C at one atmosphere pressure.

Frequency See *wave frequency*.

Fringing reef A reef that is directly attached to the shore of an island or continent. It may extend more than 1 kilometer (0.6 mile) from shore. The outer margin is submerged and often consists of algal limestone, coral rock, and living coral.

Fucoxanthin The reddish-brown pigment that gives brown algae its characteristic color.

Full moon The phase of the Moon that occurs when the Sun and Moon are in opposition; that is, they are on opposite sides of Earth. During this time, the lit side of the Moon faces Earth.

Fully developed sea The maximum average size of waves that can be developed for a given wind speed when it has blown in the same direction for a minimum duration over a minimum fetch.

Fungi Any of numerous eukaryotic organisms of the kingdom Fungi, which lack chlorophyll and vascular tissue and range in form from a single cell to a body mass of branched filamentous structures that often produce specialized fruiting bodies. The kingdom includes the yeasts, molds, lichens, and mushrooms.

Fur seal Any of several eared seals of the genera *Callorhinus* or *Arctocephalus*, having thick, soft underfur.

G

Galápagos Rift A divergent plate boundary extending eastward from the Galápagos Islands toward South America. The first deep-sea hydrothermal vent biocommunity was discovered here in 1977.

Galaxy One of the billions of large systems of stars that make up the universe.

Gas hydrate A latticelike compound composed of water and natural gas (usually methane) formed in high-pressure and low-temperature environments such as those found in deep-ocean sediments. Also known as clathrates because of their cagelike chemical structure.

Gaseous state A state of matter in which molecules move by translation and only interact through chance collisions.

Gastropoda A class of mollusks, most of which possess an asymmetrical, spiral one-piece shell and a well-developed flattened foot. A well-developed head will usually have two eyes and one or two pairs of tentacles. Includes snails, limpets, abalone, cowries, sea hares, and sea slugs.

Geostrophic current A surface current that is the result of a near balance between gravitational force and the Coriolis effect.

Ghost fishing Any lost or discarded fishing gear that continues to catch marine organisms after it has been abandoned.

Gill A thin-walled projection from some part of the external body or the digestive tract used for respiration in a water environment.

Gill net See *driftnet*.

Glacial deposit A sedimentary deposit formed by a glacier and characterized by poor sorting.

Glacier A large mass of ice formed on land by the recrystallization of old, compacted snow. It flows from an area of accumulation to an area of wasting where ice is removed from the glacier by melting.

Glauconite A group of green hydrogenous minerals consisting of hydrous silicates of potassium and iron.

Global engineering Deliberately manipulating one or more of Earth's systems for the benefit of humankind. Also called geoengineering.

Global Positioning System (GPS) A system of satellites that transmit microwave signals to Earth, allowing people at the surface to accurately determine their locations.

Global warming The warming of Earth's global surface temperature because of heat-trapping components in the atmosphere; the consensus of scientific opinion is that recent warming is caused by human emissions into the atmosphere.

***Globigerina* ooze** An ooze that contains a large percentage of the calcareous tests of the foraminifer *Globigerina*.

Goiter An enlargement of the human thyroid gland, visible as a swelling at the front of the neck, often associated with iodine deficiency.

Gondwanaland A hypothetical protocontinent of the Southern Hemisphere named for the Gondwana region of India. It included the present continental masses Africa, Antarctica, Australia, India, and South America.

Graded bedding Stratification in which each layer displays a decrease in grain size from bottom to top.

Gradient The rate of increase or decrease of one quantity or characteristic relative to a unit change in another. For example, the slope of the ocean floor is a change in elevation (a vertical linear measurement) per unit of horizontal distance covered. Commonly measured in meters per kilometer.

Grain size The average size of the grains of material in a sample. Also known as fragment or particle size.

Granite A light-colored igneous rock characteristic of the continental crust that is rich in nonferromagnesian minerals such as feldspar and quartz.

Gravitational force The force of attraction that exists between any two bodies in the universe that is proportional to the product of their masses and inversely proportional to the square of the distance between the centers of their masses.

Gravity wave A wave for which the dominant restoring force is gravity. Such waves have a wavelength of more than 1.74 centimeters (0.7 inch), and their speed of propagation is controlled mainly by gravity.

Gray whale A slow-moving, coastal-dwelling baleen whale (*Eschrichtius robustus*) of northern Pacific waters, having grayish-black coloring with white blotches. It undertakes the longest yearly migration of any mammal.

Greenhouse effect The heating of Earth's atmosphere that results from the absorption of infrared radiation from Earth's surface by components of the atmosphere, such as water vapor, methane, and carbon dioxide.

Greenhouse gas Any gas in the atmosphere that causes additional greenhouse effect.

Groin A low artificial structure built perpendicular to the shore and designed to interfere with longshore transportation of sediment so that it traps sand and widens the beach on its upstream side.

Groin field A series of closely spaced groins.

Gross ecological efficiency The amount of energy passed on from a trophic level to the one above it divided by the amount it received from the one below it.

Gross primary production The total carbon fixed into organic molecules through photosynthesis or chemosynthesis by a discrete autotrophic community.

Grunion A small fish (*Leuresthes tenuis*) of coastal waters of California and Mexico that spawns at night along beaches during the high tides of spring and summer. Grunion time their reproductive activities to coincide with tidal phenomena and are the only fish that comes completely out of water to spawn.

Gulf Stream The high-intensity western boundary surface current of the North Atlantic Subtropical Gyre that flows northward off the East Coast of the United States.

Guyot See *tablemount*.

Gypsum A colorless, white, or yellowish evaporite mineral, $CaSO_4 \cdot 2H_2O$.

Gyre A large, horizontal, circular-moving loop of water. Used mainly in reference to the circular motion of water in each of the major ocean basins centered in subtropical high-pressure regions.

H

Habitat A place where a particular plant or animal lives. Generally refers to a smaller area than environment.

Hadal Pertaining to the deepest ocean environment, specifically that of ocean trenches deeper than 6 kilometers (3.7 miles).

Hadal zone Pertaining to the deepest ocean benthic environment, specifically that of ocean trenches deeper than 6 kilometers (3.7 miles).

Hadley cell The large atmospheric circulation cell that occurs between the equator and 30° latitude in each hemisphere.

Half-life The time required for half the atoms of a sample of a radioactive isotope to decay to an atom of another element.

Halite A colorless or white evaporite mineral, NaCl, which occurs as cubic crystals and is used as table salt.

Halocline A layer of water in which a high rate of change in salinity in the vertical dimension is present.

Hard stabilization Any form of artificial structure built to protect a coast or to prevent the movement of sand along a beach. Examples include groins, breakwaters, and seawalls.

Harmful algal bloom (HAB) See *red tide*.

Headland A steep-faced irregularity of the coast that extends out into the ocean.

Heat Energy moving from a high temperature system to a lower temperature system. The heat gained by the one system may be used to raise its temperature or to do work.

Heat budget (global) The equilibrium that exists on the average between the amounts of heat absorbed by Earth and that returned to space.

Heat capacity The amount of heat energy required to raise the temperature of a substance by 1°C. See also *specific heat*.

Heat energy Energy of molecular motion. The conversion of forms of energy such as radiant or mechanical energy to heat energy within a system increases the heat energy within the system and the temperature of the system.

Heat flow (flux) The quantity of heat flow to Earth's surface per unit of time.

Hemoglobin A red pigment found in red blood corpuscles that carries oxygen from the lungs to tissue and carbon dioxide from tissue to lungs.

Herbivore An animal that relies chiefly or solely on plants for its food.

Hermatypic coral Reef-building corals that have symbiotic algae in their ectodermal tissue. They cannot produce a reef structure below the euphotic zone.

Heterotroph Animals and bacteria that depend on the organic compounds produced by other organisms as food. Organisms not capable of producing their own food by photosynthesis.

High slack water The period of time associated with the peak of high tide when there is no visible flow of water into or out of bays and rivers.

High tide zone The portion of the intertidal zone that lies between the lowest high tides and highest high tides that occur in an area. It is, on average, exposed to desiccation for longer periods each day than it is covered by water.

High water (HW) The highest level reached by the rising tide before it begins to recede.

Higher high water (HHW) The higher of two high waters occurring during a tidal day where tides exhibit a mixed tidal pattern.

Higher low water (HLW) The higher of two low waters occurring during a tidal day where tides exhibit a mixed tidal pattern.

Highly stratified estuary A relatively deep estuary in which a significant volume of marine water enters as a subsurface flow. A large volume of freshwater stream input produces a widespread low surface–salinity condition that produces a well-developed halocline throughout most of the estuary.

Holding capacity The amount of a pollutant water can hold without being made unusable.

Holoplankton Organisms that spend their entire life as members of the plankton.

Homeothermic Of or relating to an animal that maintains a precisely controlled internal body temperature using its own internal heating and cooling mechanisms; warm-blooded.

Horse latitudes The global belts between about 30 and 35 degrees north and south latitude, where winds are light and variable as a result of the vertical movement of air masses downward at these latitudes, causing a hot, dry climate. These belts are associated with the major continental and maritime deserts of the world.

Hotspot The relatively stationary surface expression of a persistent column of molten mantle material rising to the surface.

Hurricane A tropical cyclone in which winds reach speeds in excess of 120 kilometers (74 miles) per hour. Generally applied to such storms in the North Atlantic Ocean, eastern North Pacific Ocean, Caribbean Sea, and Gulf of Mexico. Such storms in the western Pacific Ocean are called typhoons and those in the Indian Ocean are known as cyclones.

Hydration The condition of being surrounded by water molecules, as when sodium chloride dissolves in water.

Hydrocarbon Any organic compound consisting only of hydrogen and carbon. Crude oil is a mixture of hydrocarbons.

Hydrocarbon seep biocommunity A deep bottom-dwelling community of organisms associated with a hydrocarbon seep from the ocean floor. The community depends on methane and sulfur-oxidizing bacteria as producers. The bacteria may live free in the water, on the bottom, or symbiotically in the tissues of some of the animals.

Hydrogen bond An intermolecular bond that forms within water because of the dipolar nature of water molecules.

Hydrogenous sediment Sediment that forms from precipitation from ocean water or ion exchange between existing sediment and ocean water. Examples are manganese nodules, metal sulfides, and evaporites.

Hydrologic cycle The cycle of water exchange among the atmosphere, land, and ocean through the processes

of evaporation, precipitation, runoff, and subsurface percolation. Also called the water cycle.

Hydrothermal spring Vents of hot water found primarily along the spreading axes of oceanic ridges and rises.

Hydrothermal vent Ocean water that percolates down through fractures in recently formed ocean floor is heated by underlying magma and surfaces again through these vents. They are usually located near the axis of spreading along the mid-ocean ridge.

Hydrothermal vent biocommunity A deep bottom-dwelling community of organisms associated with a hydrothermal vent. The hot water vent is usually associated with the axis of a spreading center, and the community is dependent on sulfur-oxidizing bacteria that may live free in the water, on the bottom, or symbiotically in the tissue of some of the animals of the community.

Hydrozoa A class of cnidarians that characteristically exhibit alternation of generations, with a sessile polypoid colony giving rise to a pelagic medusoid form by asexual budding.

Hypersaline Waters that are highly or excessively saline.

Hypersaline lagoon A shallow lagoon that may reach high salinity levels due to little tidal flushing, high evaporation rates, and low freshwater input.

Hypersaline seep A seep of high-salinity water that trickles out of the sea floor.

Hypersaline seep biocommunity A bottom-dwelling community of organisms associated with a hypersaline seep that depends on methane and sulfur-oxidizing bacteria as producers. The bacteria may live free in the water, on the bottom, or symbiotically in the tissue of some of the animals within the biocommunity.

Hypertonic Pertaining to the property of an aqueous solution having a higher osmotic pressure (salinity) than another aqueous solution from which it is separated by a semipermeable membrane that will allow osmosis to occur. The hypertonic fluid will gain water molecules through the membrane from the other fluid.

Hypothesis A tentative, testable statement about the general nature of the phenomenon observed.

Hypotonic Pertaining to the property of an aqueous solution having a lower osmotic pressure (salinity) than another aqueous solution, from which it is separated by a semipermeable membrane that will allow osmosis to occur. The hypotonic fluid will lose water molecules through the membrane to the other fluid.

Hypsographic curve A curve that displays the relative elevations of the land surface and depths of the ocean.

I

Ice Age The most recent glacial period, which occurred during the Pleistocene Epoch of geologic time.

Iceberg A massive piece of glacier ice that has broken from the front of the glacier (calved) into a body of water. It floats with its tip at least 5 meters (16 feet) above the water's surface and at least four-fifths of its mass submerged.

Ice floe See *floe*.

Ice rafting (1) The movement of trapped sediment within or on top of ice by flotation. (2) The forcing of some ice flows on top of others as a result of expanding

floes running out of open water on which to float as they grow.

Ice sheet An extensive, relatively flat accumulation of ice.

Igneous rock One of the three main classes into which all rocks are divided (igneous, metamorphic, and sedimentary). Igneous rock forms from the solidification of molten or partly molten material (magma).

In situ In place; *in situ* density of a sample of water is its density at its original depth.

Incidental catch See *bycatch*.

Indian Ocean The ocean located between Africa, India, and Australia; it exists mostly in the Southern Hemisphere and is the third largest ocean in the world.

Indian Ocean Subtropical Gyre The large, counter-clockwise-flowing subtropical gyre that exists in the Indian Ocean.

Inertia Newton's first law of motion. It states that a body at rest will stay at rest and a body in motion will remain in uniform motion in a straight line unless acted on by some external force.

Infauna Animals that live buried in the soft substrate (sand or mud).

Infrared radiation Electromagnetic radiation lying between the wavelengths of 0.8 micron (0.000003 inch) and about 1000 microns (0.04 inch). It is bounded on the shorter wavelength side by the visible spectrum and on the longer wavelength side by microwave radiation.

Inner sublittoral zone The zone on the inner continental shelf, above the intersection with the euphotic zone, where attached plants grow.

Insolation The rate at which solar radiation is received per unit of surface area at any point at or above Earth's surface.

Integrated Ocean Drilling Program (IODP) A drilling program that replaced the Ocean Drilling Program in 2003 with a new drill ship that has riser technology, enabling cores to be collected from deep within Earth's interior.

Interface A surface separating two substances of different properties, such as density, salinity, or temperature. In oceanography, it usually refers to a separation of two layers of water with different densities caused by significant differences in temperature and/or salinity.

Interface wave An orbital wave that moves along an interface between fluids of different density. An example is ocean surface waves moving along the interface between the atmosphere and the ocean.

Interference pattern (wave) The overlapping of different wave groups, either in phase (constructive interference, which results in larger waves), out of phase (destructive interference, which results in smaller waves), or some combination of the two (mixed interference).

Intergovernmental Panel on Climate Change (IPCC) A worldwide group of atmospheric and climate scientists that compile research and make policy recommendations on human effects of climate change and global warming.

Intermolecular bond A relatively weak bond that forms between molecules of a given substance. The hydrogen bond and the van der Waals bonds are intermolecular bonds.

Internal wave A wave that develops below the surface of a fluid, the density of which changes with increased depth. This change may be gradual or occur abruptly at an interface.

International Whaling Commission (IWC) An international body that was established in 1946 to manage the subsistence and commercial hunting of large whales. In 1986, the IWC passed a ban on commercial whaling in order to allow whales to recover from overhunting and to give researchers time to develop methods for assessing whale populations.

Intertidal zone The ocean floor within the foreshore region that is covered by the highest normal tides and exposed by the lowest normal tides, including the water environment of tide pools within this region.

Intertropical Convergence Zone (ITCZ) The global zone where northeast trade winds and southeast trade winds converge. It occurs close to the equator but averages about 5 degrees north latitude in the Pacific and Atlantic Oceans and 7 degrees south latitude in the Indian Ocean.

Intraplate feature Any feature that occurs within a tectonic plate and not along a plate boundary.

Intrusive rock An igneous rock such as granite that cools slowly beneath Earth's surface.

Invertebrate An animal without a backbone.

Ion An atom that becomes electrically charged by gaining or losing one or more electrons. The loss of electrons produces a positively charged cation, and the gain of electrons produces a negatively charged anion.

Ionic bond A chemical bond formed as a result of the electrical attraction.

Irminger Current A warm surface current that branches off from the Gulf Stream and moves up along the west coast of Iceland.

Iron hypothesis A hypothesis that states that an effective way of increasing productivity in the ocean is to fertilize the ocean by adding the only nutrient that appears to be lacking—iron. Adding iron to the ocean also increases the amount of carbon dioxide removed from the atmosphere.

Iron meteorite A meteorite consisting essentially of iron, but it may also contain up to 30% nickel.

Island arc A linear arrangement of islands, many of which are volcanic, usually curved so that the concave side faces a sea separating the islands from a continent. The convex side faces the open ocean and is bounded by a deep-ocean trench.

Island mass effect An effect that occurs as surface current flows past an island and causes surface water to be carried away from the island on the downcurrent side. This water is replaced in part by upwelling of water on the downcurrent side of the island.

Isohaline Of the same salinity.

Isopoda An order of dorsoventrally flattened crustaceans that are mostly scavengers or parasites on other crustaceans or fish.

Isopycnal Of the same density.

Isostasy A condition of equilibrium, comparable to buoyancy, by which Earth's brittle crust floats on the plastic mantle.

Isostatic adjustment The adjustment of crustal material due to isostasy.

Isostatic rebound The upward movement of crustal material due to isostasy.

Isotherm A line connecting points of equal temperature.

Isothermal Of the same temperature.

Isotonic Pertaining to the property of having equal osmotic pressure. If two such fluids were separated by a semipermeable membrane that will allow osmosis to occur, there would be no net transfer of water molecules across the membrane.

Isotope One of several atoms of an element that has a different number of neutrons, and therefore a different atomic mass, than the other atoms, or isotopes, of the element.

J

Jellyfish (1) Free-swimming, umbrella-shaped medusoid members of the cnidarian class Scyphozoa. (2) Also frequently applied to the medusoid forms of other cnidarians.

Jet stream An easterly moving air mass at an elevation of about 10 kilometers (6 miles). Moving at speeds that can exceed 300 kilometers (185 miles) per hour, the jet stream follows a wavy path in the middle latitudes and influences how far polar air masses may extend into the lower latitudes.

Jetty A structure built from the shore into a body of water to protect a harbor or a navigable passage from being closed off by the deposition of longshore drift material; plural is jetties.

Juan de Fuca Ridge A divergent plate boundary off the Oregon–Washington coast.

K

K–T event An extinction event marked by the disappearance of the dinosaurs that occurred 65 million years ago at the boundary between the Cretaceous (K) and Tertiary (T) Periods of geologic time.

Kelp Large varieties of Phaeophyta (brown algae).

Kelp forest An extensive bed of various species of macroscopic brown algae that provides a habitat for many other types of marine organisms.

Key A low, flat island composed of sand or coral debris that accumulates on a reef flat.

Kinetic energy Energy of motion, which increases as the mass or speed of the object in motion increases.

Knot (kt) A unit of speed equal to 1 nautical mile per hour, approximately 1.15 statute (land) miles per hour.

Krill A common name frequently applied to members of crustacean order Euphausiacea (euphausiids).

Kuroshio Current The warm western boundary surface current of the North Pacific Subtropical Gyre.

Kyoto Protocol An agreement among 60 nations to voluntarily limit greenhouse gas emissions, signed in 1997 in Kyoto, Japan.

L

La Niña An event where the surface temperature in the waters of the eastern South Pacific falls below average values. It often follows an El Niño event.

Labrador Current A cold surface current flowing south along the coast of Labrador in the northwest Atlantic Ocean.

Lagoon A shallow stretch of seawater partly or completely separated from the open ocean by an elongate narrow strip of land such as a reef or barrier island.

Laguna Madre A hypersaline lagoon located landward of Padre Island along the south Texas coast.

Laminar flow The manner in which a fluid flows in parallel layers or sheets such that the direction of flow at any point does not change with time. Also known as nonturbulent flow.

Land breeze The seaward flow of air from the land caused by differential cooling of Earth's surface.

Langmuir circulation A cellular circulation set up by winds that blow consistently in one direction with speeds in excess of 12 kilometers (7.5 miles) per hour. Helical spirals running parallel to the wind direction are alternately clockwise and counterclockwise.

Larva An embryo that has a different form before it assumes the characteristics of the adult of the species.

Latent heat The quantity of heat gained or lost per unit of mass as a substance undergoes a change of state (such as liquid to solid) at a given temperature and pressure.

Latent heat of condensation The heat energy that must be removed from 1 gram of a substance to convert it from a vapor at a given temperature below its boiling point. For water, it is 585 calories at 20°C.

Latent heat of evaporation The heat energy that must be added to 1 gram of a liquid substance to convert it to a vapor at a given temperature below its boiling point. For water, it is 585 calories at 20°C.

Latent heat of freezing The heat energy that must be removed from 1 gram of a substance at its melting point to convert it to a solid. For water, it is 80 calories.

Latent heat of melting The heat energy that must be added to 1 gram of a substance at its melting point to convert it to a liquid. For water, it is 80 calories.

Latent heat of vaporization The heat energy that must be added to 1 gram of a substance at its boiling point to convert it to a vapor. For water, it is 540 calories.

Lateral line system A sensory system running down both sides of fishes to sense subsonic pressure waves transmitted through ocean water.

Latitude Location on Earth's surface based on angular distance north or south of the equator. Equator = 0 degrees; North Pole = 90 degrees north; South Pole = 90 degrees south.

Laurasia An ancient landmass of the Northern Hemisphere. The name is derived from Laurentia, pertaining to the Canadian Shield of North America, and Eurasia, of which it was composed.

Lava Fluid magma coming from an opening in Earth's surface, or the same material after it solidifies.

Law of gravitation See *gravitational force*.

Law of the sea See *United Nations Conference on the Law of the Sea*.

Leeuwin Current A warm surface current that flows south out of the East Indies along the western coast of Australia.

Leeward The direction toward which the wind is blowing or waves are moving.

Levee (1) Natural low ridges on either side of river channels that result from deposition during flooding. (2) Artificial ridges built by humans to control water flow.

Light Electromagnetic radiation that has a wavelength in the range from about 4000 (violet) to about 7700 (red) angstroms and may be perceived by the normal unaided human eye; also called visible light.

Light-year The distance traveled by light during one year at a speed of 300,000 kilometers (186,000 miles) per second. It equals 9.8 trillion kilometers (6.2 trillion miles).

Limestone A class of sedimentary rocks composed of at least 80% carbonates of calcium or magnesium. Limestones may be either biogenous or hydrogenous.

Limpet A mollusk of the class Gastropoda that possesses a low conical shell that exhibits no spiraling in the adult form.

LIMPET 500 The world's first commercial wave power plant that can generate up to 500 kilowatts of power. It is located on Islay, a small island off the west coast of Scotland, and began generating electricity in November 2000.

Linnaeus, Carolus Latinized name of Swedish botanist Carl von Linné, the father of taxonomic classification and binomial nomenclature.

Liquid state A state of matter in which a substance has a fixed volume but no fixed shape.

Lithify A process by which sediment becomes hardened into sedimentary rock.

Lithogenous sediment Sediment composed of mineral grains derived from the weathering of rock material and transported to the ocean by various mechanisms of transport, including running water, gravity, the movement of ice, and wind.

Lithosphere The outer layer of Earth's structure, including the crust and the upper mantle to a depth of about 200 kilometers (124 miles). Lithospheric plates are the major components involved in plate tectonic movement.

Lithothamnion ridge A feature common to the windward edge of a reef structure, characterized by the presence of the red algae *Lithothamnion*.

Littoral zone The benthic zone between the highest and lowest spring tide shorelines; also known as the intertidal zone.

Lobster A large marine crustacean considered a delicacy. *Homarus americanus* (American or Maine lobster) possesses two large chelae (pincers) and is found offshore from Labrador to North Carolina. Various species of *Panulirus* (spiny lobsters or rock lobsters) have no chelae but possess long, spiny antennae effective in warding off predators. *P. argus* is found off the coast of Florida and in the West Indies, whereas *P. interruptus* is common along the coast of Southern California.

Loihi The name of the volcanically active submerged seamount that is located southeast of the island of Hawaii.

Longitude Location on Earth's surface based on angular distance east or west of the Prime (Greenwich) Meridian (0 degrees longitude). 180 degrees longitude is the International Date Line.

Longitudinal wave A wave phenomenon when particle vibration is parallel to the direction of energy propagation; also known as a push-pull wave.

Longshore bar A deposit of sediment that forms parallel to the coast within or just beyond the surf zone.

Longshore current A current located in the surf zone and running parallel to the shore as a result of waves breaking at an angle to the shore.

Longshore drift The load of sediment transported along the beach from the breaker zone to the top of the swash line in association with the longshore current. Also called longshore transport or littoral drift.

Longshore trough A low area of the beach that separates the beach face from the longshore bar.

Lophophore A horseshoe-shaped feeding structure bearing ciliated tentacles characteristic of the phyla Bryozoa, Brachiopoda, and Phoronidea.

Low slack water The period of time associated with the peak of low tide when there is no visible flow of water into or out of bays and rivers.

Low tide terrace See *beach face*.

Low tide zone The portion of the intertidal zone that lies between the lowest low tide shoreline and the highest low tide shoreline.

Low water (LW) The lowest level reached by the water surface at low tide before the rise toward high tide begins.

Lower high water (LHW) The lower of two high waters occurring during a tidal day where tides exhibit a mixed tidal pattern.

Lower low water (LLW) The lower of two low waters occurring during a tidal day where tides exhibit a mixed tidal pattern.

Lunar day The time interval between two successive transits of the Moon over a meridian, approximately 24 hours and 50 minutes of solar time. Also called a tidal day.

Lunar month The time it takes for one orbit of the Moon around Earth; $29\frac{1}{2}$ days. Also known as the lunar cycle or synodic month.

Lunar tide The part of the tide caused solely by the tide-producing force of the Moon.

Lunger Fish (such as grouper) that sit motionless on the ocean floor waiting for prey to appear. A lunger uses quick bursts of speed over short distances to capture prey.

Lysocline The level in the ocean at which calcium carbonate begins to dissolve, typically at a depth of about 4000 meters (13,100 feet). Below the lysocline, calcium carbonate dissolves at an increasing rate with increasing depth until the calcite compensation depth (CCD) is reached.

M

Macroplankton Plankton larger than 2 centimeters (0.8 inch) in their smallest dimension.

Magma Fluid rock material from which igneous rock is derived through solidification.

Magnetic anomaly A distortion of the regular pattern of Earth's magnetic field resulting from the various magnetic properties of local concentrations of ferromagnetic minerals in Earth's crust.

Magnetic dip The dip of magnetite particles in rock units of Earth's crust relative to sea level. It is approximately equivalent to latitude. Also called magnetic inclination.

Magnetic field A condition found in the region around a magnet or an electric current, characterized by the existence of a detectable magnetic force at every point in the region and by the existence of magnetic poles.

Magnetic inclination See *magnetic dip*.

Magnetite The mineral form of black iron oxide, Fe_3O_4, that often occurs with magnesium, zinc, and manganese and is an important ore of iron.

Magnetometer A device used for measuring the magnetic field of Earth.

Manganese nodule A concretionary lump containing oxides of manganese, iron, copper, cobalt, and nickel found scattered over the ocean floor.

Mangrove swamp A marshlike environment dominated by mangrove trees. They are restricted to latitudes below 30 degrees.

Mantle (1) The zone between the core and crust of Earth; rich in ferromagnesian minerals. (2) In pelecypods, the portion of the body that secretes shell material.

Mantle plume A rising column of molten magma from Earth's mantle.

Marginal sea A semienclosed body of water adjacent to a continent and floored by submerged continental crust.

Mariculture The application of the principles of agriculture to the production of marine organisms.

Marine effect Describes locations that experience the moderating influences of the ocean, usually along coastlines or islands.

Marine Mammals Protection Act An act by the U.S. Congress in 1972 that specifies rules to protect marine mammals in U.S. waters.

Marine Protected Area (MPA) An area of the ocean in which there is some level of restriction to protect living, nonliving, cultural, and/or historic resources.

Marine reserve A type of MPA in which all species and their habitat are fully protected.

Marine sanctuary A type of MPA in which biologic or cultural resources are protected. In some marine sanctuaries, fishing, recreational boating, and mining are allowed.

Marine terrace A gently seaward-sloping horizontal platform cut by waves (wave-cut bench) that has been uplifted above sea level.

MARPOL (Marine Pollution) An international treaty that banned the disposal of all plastics and regulated the dumping of most other garbage at sea.

Marsh An area of soft, wet, flat land that is periodically flooded by saltwater and common in portions of lagoons.

Maximum sustainable yield (MSY) The maximum fishery biomass that can be removed yearly and still be sustained by the fishery ecosystem.

Mean high water (MHW) The average height of all the high waters occurring over a 19-year period.

Mean higher high water (MHHW) The average height of the daily higher of the high waters occurring over a 19-year period where tides exhibit a mixed tidal pattern.

Mean low water (MLW) The average height of all the low waters occurring over a 19-year period.

Mean lower low water (MLLW) The average height of the daily lower of the low waters occurring over a 19-year period where tides exhibit a mixed tidal pattern.

Mean sea level (MSL) The mean surface water level determined by averaging all stages of the tide over a 19-year period, usually determined from hourly height observations along an open coast.

Mean tidal range The difference between mean high water and mean low water.

Meander A sinuous curve, bend, or turn in the course of a current.

Mechanical energy Energy manifested as work being done; the movement of a mass over some distance.

Mediterranean circulation Circulation characteristic of bodies of water with restricted circulation with the ocean that results from an excess of evaporation as compared to precipitation and runoff similar to the Mediterranean Sea. Surface flow is into the restricted body of water with a subsurface counterflow as exists between the Mediterranean Sea and the Atlantic Ocean.

Medusa A free-swimming, bell-shaped cnidarian body form with a mouth at the end of a central projection and tentacles around the periphery.

Meiofauna Small species of animals that live in the spaces among particles in a marine sediment.

Melon A fatty organ located forward of the blowhole on certain odontocete cetaceans that is used to focus echolocation sounds.

Melting point The temperature at which a solid substance changes to the liquid state.

Mercury A silvery white poisonous metallic element, liquid at room temperature and used in thermometers, barometers, vapor lamps, and batteries and in the preparation of chemical pesticides.

Meridian of longitude Great circles running through the North and South Poles.

Meroplankton Planktonic larval forms of organisms that are members of the benthos or nekton as adults.

Mesopelagic zone That portion of the oceanic province 200 to 1000 meters (660 to 3300 feet) deep. Corresponds approximately with the disphotic (twilight) zone.

Mesosaurus An extinct, presumably aquatic, reptile that lived about 250 million years ago. The distribution of its fossil remains helps support plate tectonic theory.

Mesosphere The middle region of Earth below the asthenosphere and above the core.

Metal sulfide A compound containing one or more metals and sulfur.

Metamorphic rock Rock that has undergone recrystallization while in the solid state in response to changes of temperature, pressure, and chemical environment.

Meteor A bright trail or streak that appears in the sky when a meteoroid is heated to incandescence by friction with Earth's atmosphere. Also called a falling or shooting star.

Meteorite A stony or metallic mass of matter that has fallen to Earth's surface from outer space.

Methane hydrate A white compact icy solid made of water and methane. The most common type of gas hydrate.

Microplankton Plankton not easily seen by the unaided eye but easily recovered from the ocean with the aid of a silk-mesh plankton net.

Microplastic Microscopic pieces of plastic from the use of human health care products and other applications.

Mid-Atlantic Ridge A slow-spreading divergent plate boundary running north–south and bisecting the Atlantic Ocean.

Mid-ocean ridge A linear, volcanic mountain range that extends through all the major oceans, rising 1 to 3 kilometers (0.6 to 2 miles) above the deep-ocean basins. Averaging 1500 kilometers (930 miles) in width, rift valleys are common along the central axis. Source of new oceanic crustal material.

Middle latitudes The region of Earth between approximately 30 and 60 degrees north and south latitudes; also called the midlatitudes.

Middle tide zone The portion of the intertidal zone that lies between the highest low tide shoreline and the lowest high tide shoreline.

Migration Long journeys undertaken by many marine species for the purpose of successful feeding and reproduction.

Minamata Bay, Japan The site of the occurrence of human poisoning in the 1950s by mercury contained in marine organisms that were consumed by victims.

Minamata disease A degenerative neurological disorder caused by poisoning with a mercury compound found in seafood obtained from waters contaminated with mercury-containing industrial waste.

Mineral An inorganic substance occurring naturally on Earth and having distinctive physical properties and a chemical composition that can be expressed by a chemical formula. The term is also sometimes applied to organic substances such as coal and petroleum.

Mixed interference A pattern of wave interference in which there is a combination of constructive and destructive interference.

Mixed surface layer The surface layer of the ocean water mixed by wave and tide motions to produce relatively isothermal and isohaline conditions.

Mixed tidal pattern A tidal pattern exhibiting two high tides and two low tides per tidal day with a marked diurnal inequality. Coastal locations that experience such a tidal pattern may also show alternating periods of diurnal and semidiurnal patterns. Also called mixed semidiurnal.

Mixotroph An organism that depends on a combination of autotrophic and heterotrophic behavior to meet its energy requirements. Many coral reef species exhibit such behavior.

Mohorovičíc discontinuity (Moho) A sharp compositional discontinuity between the crust and mantle of Earth. It may be as shallow as 5 kilometers (3 miles) below the ocean floor or as deep as 60 kilometers (37 miles) beneath some continental mountain ranges.

Molecular motion Molecules move in three ways: vibration, rotation, and translation.

Molecule A group of two or more atoms bound together by ionic or covalent bonds.

Mollusca (mollusk) A phylum of soft, unsegmented animals usually protected by a calcareous shell and having a muscular foot for locomotion. Includes snails, clams, chitons, and octopi.

Mononodal Pertaining to a standing wave with only one nodal point or nodal line.

Monsoon A name for seasonal winds derived from the Arabic word for season, *mausim*. The term was originally applied to winds over the Arabian Sea that blow from the southwest during summer and from the northeast during winter.

Moraine A deposit of unsorted material deposited at the margins of glaciers. Many such deposits have become important economically as fishing banks after being submerged by the rising level of the ocean.

Mud Sediment consisting primarily of silt and clay-sized particles smaller than 0.06 millimeters (0.002 inch).

Mutualism A symbiotic relationship in which both participants benefit.

Myoglobin A red, oxygen-storing pigment found in muscle tissue.

Mysticeti The baleen whales.

N

Nadir The point on the celestial sphere directly opposite the zenith and directly beneath the observer.

Nannoplankton Plankton less than 50 microns (0.002 inch) in length that cannot be captured in a plankton net and must be removed from the water by centrifuge or special microfilters.

Nansen bottle A device used by oceanographers to obtain samples of ocean water from beneath the surface.

National Flood Insurance Program (NFIP) A program financed by the U.S. government that was intended to prevent costly federal aid after a natural disaster but instead encourages building in risk-prone areas.

Natural selection The process in nature by which only the organisms best adapted to their environment tend to survive and transmit their genetic characters in increasing numbers to succeeding generations while those less adapted tend to be eliminated.

Nauplius A microscopic, free-swimming larval stage of crustaceans such as copepods, ostracods, and decapods. Typically has three pairs of appendages.

Neap tide Tides of minimal range occurring about every two weeks when the Moon is in either first- or third-quarter moon phase.

Nearshore The zone of a beach that extends from the low tide shoreline seaward to where breakers begin forming.

Nebula A diffuse mass of interstellar dust and/or gas.

Nebular hypothesis A model that describes the formation of the solar system by contraction of a nebula.

Negative-feedback loop See *feedback loop*.

Nektobenthos Those members of the benthos that can actively swim and spend much time off the bottom.

Nekton Pelagic animals such as adult squids, fish, and mammals that are active swimmers to the extent that they can determine their position in the ocean by swimming.

Nematath A linear chain of islands and/or seamounts that are progressively older in one direction. It is created by the passage of a lithospheric plate over a hotspot.

Nematocyst The stinging mechanism found within the cnidoblast of members of the phylum Cnidaria.

Neritic province That portion of the pelagic environment from the shoreline to where the depth reaches 200 meters (660 feet).

Neritic sediment Sediment composed primarily of lithogenous particles and deposited relatively rapidly on the continental shelf, continental slope, and continental rise; also called neritic deposits.

Net primary production The primary production of producers after they have removed what is needed for their metabolism.

Neutral A state in which there is no excess of either the hydrogen or the hydroxide ion.

Neutron An electrically neutral subatomic particle found in the nucleus of atoms that has a mass approximately equivalent to that of a proton.

New moon The phase of the Moon that occurs when the Sun and the Moon are in conjunction; that is, they are both on the same side of Earth. During this time, the dark, unlit side of the Moon faces Earth.

New production Primary production supported by nutrients supplied from outside the immediate ecosystem by upwelling or other physical transport.

Newton's law of universal gravitation An equation that quantifies gravitational force between two bodies; it states that the gravitational force is directly proportional to the product of the masses of the two bodies and is inversely proportional to the square of the distance between the two masses.

Niche The ecological role of an organism and its position in the ecosystem.

Niigata, Japan The site of mercury poisoning of humans in the 1960s by ingestion of contaminated seafood.

Nitrogen narcosis A sickness that affects divers. It results from too much nitrogen gas being dissolved in the blood and reducing the flow of oxygen to tissues. The threat of this problem increases with increasing pressure (depth).

Node The point on a standing wave where vertical motion is lacking or minimal. If this condition extends across the surface of an oscillating body of water, the line of no vertical motion is a nodal line.

Non-native species Species that are introduced into waters in which they are alien and often cause severe problems by displacing native species. Also called exotic, alien, or invasive species.

Non-point source pollution Any type of pollution entering the ocean from multiple sources rather than from a single discrete source, point, or location. Examples include urban runoff, trash, pet waste, lawn fertilizer, and other types of pollution generated by a multitude of sources. Also called poison runoff.

North Atlantic Current The northernmost surface current of the North Atlantic Subtropical Gyre.

North Atlantic Deep Water A deep-water mass that forms primarily at the surface of the Norwegian Sea and moves south along the floor of the North Atlantic Ocean.

North Atlantic Subtropical Gyre The large, clockwise-flowing subtropical gyre that exists in the North Atlantic Ocean.

North-East Pacific Time-series Undersea Networked Experiments (NEPTUNE) A cutting-edge sea floor observatory system designed to monitor tectonic activity along the Juan de Fuca tectonic plate in the northeast Pacific Ocean.

North Pacific Current The northernmost surface current of the North Pacific Subtropical Gyre.

North Pacific Subtropical Gyre The large, clockwise-flowing subtropical gyre that exists in the North Pacific Ocean.

Northern boundary current The northern boundary surface current of Northern Hemisphere subtropical gyres.

Northeast Monsoon A northeast wind that blows off the Asian mainland onto the Indian Ocean during the winter season.

Norwegian Current A warm surface current that branches off from the Gulf Stream and flows into the Norwegian Sea between Iceland and the British Isles.

Nucleus The positively charged central region of an atom, composed of protons and neutrons and containing almost all of the mass of the atom.

Nudibranch A sea slug. A member of the mollusk class Gastropoda that has no protective covering as an adult. Respiration is carried on by gills or other projections on the dorsal surface.

Nurdle A small, rounded piece of pre-production plastic.

Nutrient Any of a number of organic or inorganic compounds used by primary producers. Nitrogen and phosphorus compounds are important examples.

O

Observation An occurrence that can be measured with one's senses.

Ocean The entire body of saltwater that covers 70.8% of Earth's surface.

Ocean acidification The process by which the ocean's pH is lowered, which increases its acidity.

Ocean acoustical tomography A method by which changes in water temperature may be determined by changes in the speed of transmission of sound. It has the potential to help map ocean circulation patterns over large ocean areas.

Ocean beach The beach on the open-ocean side of a barrier island.

Ocean Drilling Program (ODP) A program that replaced the Deep Sea Drilling Project in 1983, focusing on drilling the continental margins using the drill ship *JOIDES Resolution*.

Ocean thermal energy conversion (OTEC) A technique that involves generating energy by using the difference in temperature between surface waters and deep waters in low latitude regions.

Ocean wave A type of wave that occurs in the ocean. See *wave*.

Oceanic Common Water Deep water found in Pacific and Indian Oceans as a result of mixing of Antarctic Bottom Water and North Atlantic Deep Water.

Oceanic crust A mass of rock with a basaltic composition that is about 5 kilometers (3 miles) thick.

Oceanic province The division of the pelagic environment where the water depth is greater than 200 meters (660 feet).

Oceanic ridge A portion of the global mid-ocean ridge system that is characterized by slow spreading and steep slopes.

Oceanic rise A portion of the global mid-ocean ridge system that is characterized by fast spreading and gentle slopes.

Oceanic sediment The inorganic abyssal clays and the organic oozes that accumulate slowly on the deep-ocean floor.

Oceanography The scientific study of the floor of the ocean, the water itself, physical processes such as waves, currents, and tides, and the organisms contained within the ocean. Also known as oceanology.

Ocelli A light-sensitive organ around the base of many medusoid bells.

Odontoceti The toothed whales.

Offset Separation that occurs due to movement along a fault.

Offshore The comparatively flat submerged zone of variable width extending from the breaker line to the edge of the continental shelf.

Oligotrophic Exhibiting low levels of biological production, such as the centers of subtropical gyres.

Omnivore An animal that feeds on both plants and animals.

Oolite A deposit formed of small spheres from 0.25 to 2 millimeters (0.01 to 0.08 inch) in diameter. Each oolite is composed of concentric layers of calcite.

Ooze A pelagic sediment containing at least 30% skeletal remains of pelagic organisms, the balance being clay minerals. Oozes are further defined by the chemical composition of the organic remains (siliceous or calcareous) and by their characteristic organisms (e.g., diatomaceous ooze, foraminifer ooze).

Opal An amorphous form of silica ($SiO_2 \cdot nH_2O$) that usually contains from 3 to 9% water. It forms the shells of radiolarians and diatoms.

Ophiolite An assemblage of mafic and ultramafic igneous rocks from Earth's oceanic crust and the underlying upper mantle that has been uplifted and exposed above sea level and often emplaced onto continental crustal rocks by plate tectonic processes. Ophiolites are generally composed of green-colored metamorphic rocks such as peridotite and serpentinite.

Opposition The separation of two heavenly bodies by 180 degrees relative to Earth. The Sun and Moon are in opposition during the full moon phase.

Orbital wave A wave phenomenon in which energy is moved along the interface between fluids of different densities. The wave form is propagated by the movement of fluid particles in orbital paths.

Orthogonal line A line constructed perpendicular to a wave front and spaced so that the energy between lines is equal at all times. Orthogonals are used to help determine how energy is distributed along the shoreline by breaking waves.

Osmosis The process by which water molecules move through a semipermeable membrane from higher water molecule concentration (lower salinity) to lower water molecule concentration (higher salinity).

Osmotic pressure A measure of the tendency for osmosis to occur. It is the pressure that must be applied to the more concentrated solution to prevent the passage of water molecules into it from the less concentrated solution.

Osmotic regulation Physical and biological processes used by organisms to counteract the osmotic effects of differences in osmotic pressures of their body fluids and the water in which they live.

Ostracoda An order of crustaceans that are minute and compressed within a bivalve shell.

Otocyst Gravity-sensitive organs around the bell of a medusa.

Outer sublittoral zone The continental shelf below the intersection with the euphotic zone where no plants grow attached to the bottom.

Outgassing The process by which gases are removed from within Earth's interior.

Overfishing A situation that occurs when adult fish in a population are harvested faster than their natural rate of reproduction.

Oxygen compensation depth The depth in the ocean at which marine plants receive just enough solar radiation to meet their basic metabolic needs. It marks the base of the euphotic zone.

Oxygen minimum layer (OML) A zone of low dissolved oxygen concentration that occurs at a depth of about 700 to 1000 meters (2300 to 3280 feet).

P

Pacific Decadal Oscillation (PDO) A natural and cyclic pattern of ocean–atmosphere variability in the Pacific Ocean that lasts 20 to 30 years and influences sea surface temperatures.

Pacific Ocean The ocean located between Australia, Asia, North America, and South America; the largest ocean in the world.

Pacific Ring of Fire An extensive zone of volcanic and seismic activity that coincides roughly with the borders of the Pacific Ocean.

Pacific Tsunami Warning Center (PTWC) A tsunami warning center that was established in Hawaii after the devastating 1946 tsunami; it coordinates information from 25 Pacific Rim countries.

Pacific Warm Pool A large region of warm surface water on the western side of the Pacific Ocean.

Pacific-type margin See *active margin*.

Paleoclimatology The study of ancient climates on Earth.

Paleoceanography The study of how the ocean, atmosphere, and land have interacted to produce changes in ocean chemistry, circulation, biology, and climate.

Paleogeography The study of the historical changes of shapes and positions of the continents and oceans.

Paleomagnetism The study of Earth's ancient magnetic field.

Pancake ice Circular pieces of newly formed sea ice from 0.3 to 3 meters (1 to 10 feet) in diameter that form in the early fall in polar regions.

Pangaea An ancient supercontinent of the geologic past that contained all Earth's continents.

Panthalassa A large, ancient ocean that surrounded Pangaea.

Paralytic shellfish poisoning (PSP) Paralysis resulting from poisoning caused by eating shellfish contaminated with the toxic dinoflagellate *Gonyaulax*.

Parasitism A symbiotic relationship between two organisms in which one benefits at the expense of the other.

Parts per thousand (‰) A unit of measurement used in reporting salinity of water equal to the number of grams of dissolved substances in 1000 grams of water. For example, 1‰ is equivalent to 0.1%, or 1000 ppm.

Passive margin A continental margin that lacks a plate boundary and is marked by a low degree of tectonic activity, such as those typical of the Atlantic Ocean.

PCBs A group of industrial chemicals (polychlorinated biphenyls) used in a variety of products; responsible for several episodes of ecological damage in coastal waters.

Peat deposit Partially carbonized organic matter found in bogs and marshes that can be used as fertilizer and fuel.

Pelagic sediment Sediment composed primarily of fine lithogenous and biogenous particles that is deposited slowly on the deep ocean floor; also called pelagic deposits.

Pelagic environment The open-ocean environment, which is divided into the neritic province (water depth 0 to 200 meters or 656 feet) and the oceanic province (water depth greater than 200 meters or 656 feet).

Pelecypoda A class of mollusks characterized by two more or less symmetrical lateral valves with a dorsal hinge. These organisms filter feed by circulating water through their bodies and over their gills through posterior siphons. Many possess a hatchet-shaped foot used for locomotion and burrowing. Includes clams, oysters, mussels, and scallops.

Perigee The point on the orbit of an Earth satellite (Moon) that is nearest Earth.

Perihelion That point on the orbit of a planet or comet around the Sun that is closest to the Sun.

Permeability Capacity of a porous rock or sediment for transmitting fluid.

Peru Current The cold eastern boundary surface current of the South Pacific Subtropical Gyre.

Petroleum A naturally occurring liquid hydrocarbon.

Pfiesteria piscicida A species of toxic dinoflagellate that has been known to cause fish kills.

pH scale A measure of the acidity or alkalinity of a solution, numerically equal to 7 for neutral solutions, increasing with increasing alkalinity and decreasing with increasing acidity. The pH scale commonly in use ranges from 0 to 14.

Phaeophyta Brown algae characterized by the carotenoid pigment fucoxanthin. Contains the largest members of the marine algal community.

Phosphate Any of a number of phosphorus-bearing compounds.

Phosphorite A sedimentary rock composed primarily of phosphate minerals.

Photic zone The upper ocean in which the presence of solar radiation is detectable. It includes the euphotic and disphotic zones.

Photophore One of several types of light-producing organs found primarily on fishes and squids inhabiting the mesopelagic and upper bathypelagic zones.

Photosynthesis The process by which plants and algae produce carbohydrates from carbon dioxide and water in the presence of chlorophyll, using light energy and releasing oxygen.

Phycoerythrin A red pigment characteristic of the Rhodophyta (red algae).

Phytoplankton Algal plankton. One of the most important communities of primary producers in the ocean.

Picoplankton Small plankton within the size range of 0.2 to 2.0 microns (0.000008 to 0.00008 inch) in size. Composed primarily of bacteria.

Pillow basalt A basalt exhibiting pillow structure. See *pillow lava*.

Pillow lava A general term for those lavas displaying discontinuous pillow-shaped masses (pillow structure) caused by the rapid cooling of lava as a result of underwater eruption of lava or lava flowing into water.

Ping A sharp, high-pitched sound made by the transmitting device of many sonar systems.

Pinniped A group of marine mammals that have prominent flippers; includes the sea lions/fur seals, seals, and walruses.

Plane of the ecliptic The surface connecting all points in Earth's orbit.

Plankter Informal term for plankton.

Plankton Passively drifting or weakly swimming organisms that are not independent of currents. Includes mostly microscopic algae, protozoa, and larval forms of animals.

Plankton bloom A very high concentration of phytoplankton, resulting from a rapid rate of reproduction as conditions become optimal during the spring in high latitude areas. Less obvious causes produce blooms that may be destructive in other areas.

Plankton net A plankton-extracting device that is cone shaped and typically of a silk material. It is towed through the water or lifted vertically to extract plankton down to a size of 50 microns (0.0002 inch).

Planktonic Having the characteristics of plankton.

Plantae A kingdom of many-celled plants.

Plastic (1) Capable of being shaped or formed. (2) Composed of plastic or plastics.

Plate tectonics Global dynamics having to do with the movement of a small number of semirigid sections of Earth's crust, with seismic activity and volcanism occurring primarily at the margins of these sections. This movement has resulted in changes in the geographic positions of continents and the shape and size of ocean basins.

Plume A rising column of molten mantle material that is associated with a hotspot when it penetrates Earth's crust.

Plunging breaker Impressive curling breakers that form on moderately sloping beaches.

Pneumatic duct An opening into the swim bladder of some fishes that allows rapid release of air into the esophagus.

Poikilothermic An organism whose body temperature varies with and is largely controlled by its environment; cold-blooded.

Polar Pertaining to the polar regions.

Polar cell The large atmospheric circulation cell that occurs between 60 and 90 degrees latitude in each hemisphere.

Polar easterly wind belt A global wind belt that moves away from the polar regions toward the polar

front at about 60 degrees north or south latitude in each hemisphere. These winds move from a northeasterly direction in the Northern Hemisphere and from a southeasterly direction in the Southern Hemisphere.

Polar front The boundary between the global wind belts' prevailing westerlies and polar easterlies that is centered at about 60 degrees latitude in each hemisphere and is characterized by rising air and much precipitation.

Polar high The region of high atmospheric pressure that occurs at the poles in both hemispheres.

Polar wandering path A path that shows the change in position of a pole through time. Sometimes called polar wandering curve.

Polarity Intrinsic polar separation, alignment, or orientation, especially of a physical property (such as magnetic or electrical polarity).

Pollution (marine) The introduction of substances that result in harm to the living resources of the ocean or humans who use these resources.

Polychaeta A class of annelid worms that includes most of the marine segmented worms.

Polyp A single individual of a colony or a solitary attached cnidarian.

Population A group of individuals of one species living in an area.

Porifera A phylum of sponges. Supporting structure composed of $CaCO_3$ or SiO_2 spicules or fibrous spongin. Water currents created by flagella-waving choanocytes enter tiny pores, pass through canals, and exit through a larger osculum.

Porosity The ratio of the volume of all the empty spaces in a material to the volume of the whole.

Positive-feedback loop See *feedback loop*.

Potential energy The energy of a particle or system of particles derived from position or condition rather than motion.

Precession Describes the change in the attitude of the Moon's orbit around Earth as it slowly changes its direction. The cycle is completed every 18.6 years and is accompanied by a clockwise rotation of the plane of the Moon's orbit that is completed in the same time interval.

Precipitate To cause a solid substance to be separated from a solution, usually due to a change in physical or chemical conditions.

Precipitation In a meteorological sense, the discharge of water in the form of rain, snow, hail, or sleet from the atmosphere onto Earth's surface.

Precision depth recorder (PDR) An early type of sonar device that was developed in the 1950s.

Pressure ridge A ridge produced on floating sea ice by buckling or crushing under lateral pressure by wind or movement of ice.

Prevailing westerly wind belt A global wind belt that moves from a subtropical high-pressure belt at about 30 degrees north or south latitude toward the polar front at about 60 degrees north or south latitude. These winds move from a southwesterly direction in the Northern Hemisphere and from a northwesterly direction in the Southern Hemisphere.

Primary productivity The rate at which energy is stored by organisms through the formation of organic matter (carbon-based compounds) using energy

derived from solar radiation (photosynthesis) or chemical reactions (chemosynthesis). Also known simply as productivity.

Prime meridian The meridian of longitude 0 degrees used as a reference for measuring longitude that passes through the Royal Observatory at Greenwich, England. Also known as the Greenwich Meridian.

Producer The autotrophic component of an ecosystem that produces the food that supports the biocommunity.

Productivity See *primary productivity*.

Progressive wave A wave in which the waveform progressively moves.

Propagation The transmission of energy through a medium.

Protein A very complex organic compound made up of large numbers of amino acids. Proteins make up a large percentage of the dry weight of all living organisms.

Protista A kingdom of organisms that includes any of the unicellular eukaryotic organisms and their descendant multicellular organisms. Includes single-celled and multicelled marine algae as well as single-celled animals called protozoa.

Proto-Earth The young, early developing Earth.

Proton A positively charged subatomic particle found in the nucleus of atoms that has a mass approximately equivalent to that of a neutron.

Protoplanet Any planet that is in its early stages of development.

Protoplasm The self-perpetuating living material making up all organisms, mostly consisting of the elements carbon, hydrogen, and oxygen combined into various chemical forms.

Protozoa A phylum of one-celled animals with nuclear material confined within a nuclear sheath.

Proxigean A tidal condition of extremely large tidal range that occurs when spring tides coincide with perigee. Also called "closest of the close moon" tides.

Proxy A substitute value or measurement that takes the place of something else.

Pseudopodia An extension of protoplasm in a broad, flat, or long needlelike projection used for locomotion or feeding. Typical of amoeboid forms such as foraminifers and radiolarians.

Pteropoda An order of pelagic gastropods in which the foot is modified for swimming and the shell may be present or absent.

Purse seine net A style of large fishing net that resembles a purse. It is set around a grouping of organisms (such as tuna) and the bottom is drawn tight to capture the organisms.

Pycnocline A layer of water in which a high rate of change in density in the vertical dimension is present.

Pyrrophyta A phylum of dinoflagellates that possess flagella for locomotion.

Q

Quadrature The state of the Moon during the first- and third-quarter moon phases when the Sun and the Moon are at right angles relative to Earth.

Quarter moon First- and third-quarter moon phases, which occur when the Moon is in quadrature about

one week after the new moon and full moon phases, respectively. The third-quarter moon phase is also known as the last quarter moon phase.

Quartz A very hard mineral composed of silica, SiO_2.

R

Radiata A grouping of phyla with primary radial symmetry—phyla Cnidarian and Ctenophora.

Radioactivity The spontaneous breakdown of the nucleus of an atom resulting in the emission of radiant energy in the form of particles or waves.

Radiolaria An order of planktonic and benthic protozoans that possess protective coverings usually made of silica.

Radiometric age dating A technique that involves the use of radioactive half-lives to determine the age of a rock.

Ray A cartilaginous fish in which the body is dorsoventrally flattened, eyes and spiracles are on the upper surface, and gill slits are on the bottom. The tail is reduced to a whiplike appendage. Includes electric rays, manta rays, and stingrays.

Recreational beach The area of a beach above shoreline, including the berm, berm crest, and the exposed part of the beach face.

Red clay See *abyssal clay*.

Red muscle fiber Fine muscle fibers rich in myoglobin that are abundant in cruiser-type fishes.

Red tide A reddish-brown discoloration of surface water, usually in coastal areas, caused by high concentrations of microscopic organisms, usually dinoflagellates. It normally results from increased availability of certain nutrients. Toxins produced by the dinoflagellates may kill fish directly; decaying plant and animal remains or large populations of animals that migrate to the area of abundant plants may also deplete the surface waters of oxygen and cause asphyxiation of many animals.

Reef A strip or ridge of rocks, sand, coral, or human-made objects that rises to or near the surface of the ocean and creates a navigational hazard.

Reef flat A platform of coral fragments and sand on the lagoon side of a reef that is relatively exposed at low tide.

Reef front The upper seaward face of a reef from the reef edge (seaward margin of reef flat) to the depth at which living coral and coralline algae become rare, 16 to 30 meters (50 to 100 feet).

Reflection (wave) The process in which a wave has part of its energy returned seaward by a reflecting surface.

Refraction (wave) The process by which the part of a wave in shallow water is slowed down, causing the wave to bend and align itself nearly parallel to the shore.

Regenerated production The portion of gross primary production that is supported by nutrients recycled within an ecosystem.

Relict beach A beach deposit laid down and submerged by a rise in sea level. It is still identifiable on the continental shelf, indicating that no deposition is presently taking place at that location on the shelf.

Relict sediment A sediment deposited under a set of environmental conditions that still remains unchanged

although the environment has changed, and it remains unburied by later sediment. An example is a beach deposited near the edge of the continental shelf when sea level was lower.

Relocation The strategy of moving a structure that is threatened by being claimed by the sea.

Residence time The average length of time a particle of any substance spends in the ocean. It is calculated by dividing the total amount of the substance in the ocean by the rate of its introduction into the ocean or the rate at which it leaves the ocean.

Respiration The process by which organisms use organic materials (food) as a source of energy. As the energy is released, oxygen is used and carbon dioxide and water are produced.

Restoring force A force such as surface tension or gravity that tends to restore the ocean surface displaced by a wave to that of a still water level.

Resultant force The difference between the provided gravitational force of various bodies and the required centripetal force on Earth. The horizontal component of the resultant force is the tide-producing force.

Reverse osmosis A method of desalinating ocean water that involves forcing water molecules through a water-permeable membrane under pressure.

Reversing current The tide current as it occurs at the margins of landmasses. The water flows in and out for approximately equal periods of time separated by slack water where the water is still at high and low tidal extremes.

Rhodophyta A phylum of algae composed primarily of small encrusting, branching, or filamentous plants that receive their characteristic red color from the presence of the pigment phycoerythrin. With a worldwide distribution, they are found at greater depths than other algae.

Ring A circular-moving surface flow of water that can be either a warm-core ring, which contains warm water in its center, or a cold-core ring, which contains cold water in its center.

Rift valley A deep fracture or break, about 25 to 50 kilometers (15 to 30 miles) wide, extending along the crest of a mid-ocean ridge.

Rifting The movement of two plates in opposite directions such as along a divergent boundary.

Right whale A surface-feeding baleen whale of the family Balaenidae that were a favorite target of early whalers.

Rip current A strong narrow surface or near-surface current of short duration and high speed flowing seaward through the breaker zone at nearly right angles to the shore. It represents the return to the ocean of water that has been piled up on the shore by incoming waves.

Rip-rap Large blocky material used to armor coastal structures.

Rogue wave An unusually large wave that occurs unexpectedly amid other waves of smaller size. Also known as a superwave, monster wave, sleeper wave, or freak wave.

Rorqual whale A large baleen whale with prominent ventral groves (rorqual folds) of the family Balaenopteridae: the minke, Baird's, Bryde's, sei, fin, blue, and humpback whales.

Rotary current A tidal current that is observed in the open ocean and makes one complete rotation during a tidal period.

Rotary drilling Drilling involving the use of a long, hollow pipe with a drill bit on its end that is rotated to crush the rock around the outside and retain a cylinder of rock (a core sample) on the inside of the pipe.

S

Saffir–Simpson scale A scale of hurricane intensity that divides tropical cyclones into categories based on wind speed and damage.

Salinity A measure of the quantity of dissolved solids in ocean water. Formally, it is the total amount of dissolved solids in ocean water in parts per thousand (‰) by weight after all carbonate has been converted to oxide, the bromide and iodide to chloride, and all the organic matter oxidized. It is normally computed from conductivity, refractive index, or chlorinity.

Salinometer An instrument that is used to determine the salinity of seawater by measuring its electrical conductivity.

Salpa A genus of pelagic tunicates that are cylindrical, transparent, and found in all oceans.

Salt A substance that yields ions other than hydrogen or hydroxyl. Salts are produced from acids by replacing the hydrogen with a metal.

Salt marsh A relatively flat area of the shore where fine sediment is deposited and salt-tolerant grasses grow. One of the most biologically productive regions of Earth.

Salt wedge estuary A very deep river mouth with a very large volume of freshwater flow beneath which a wedge of saltwater from the ocean invades. The Mississippi River is an example.

San Andreas Fault A transform fault that cuts across the state of California from the northern end of the Gulf of California to Point Arena north of San Francisco.

Sand Particle size of 0.0625 to 2 millimeters (0.002 to 0.08 inch). It pertains to particles that lie between silt and granules on the Wentworth scale of grain size.

Sargasso Sea A region of convergence in the North Atlantic lying south and east of Bermuda where the water is a very clear, deep blue color, and contains large quantities of floating *Sargassum*.

Sargassum A brown alga characterized by a bushy form, substantial holdfast when attached, and a yellow-brown, green-yellow, or orange color. The two dominant species of macroscopic algae in the Sargasso Sea are *S. fluitans* and *S. natans*.

Scarp A linear steep slope on the ocean floor separating gently sloping or flat surfaces.

Scavenger An animal that feeds on dead organisms.

Schooling A well-defined large groups of fish, squid, and crustaceans that apparently aid them in survival.

Scientific method The principles and empirical processes of discovery and demonstration considered characteristic of or necessary for scientific investigation, generally involving the observation of phenomena, the formulation of a hypothesis concerning the phenomena, experimentation to demonstrate the truth or falseness of the hypothesis, and a conclusion that validates or modifies the hypothesis.

Scuba An acronym for *s*elf-*c*ontained *u*nderwater *b*reathing *a*pparatus, a portable device containing compressed air that is used for breathing underwater.

Scyphozoa A class of cnidarians that includes the true jellyfish in which the medusoid body form predominates and the polyp is reduced or absent.

Sea (1) A subdivision of an ocean, generally enclosed by land and usually composed of saltwater. Two types of seas are identifiable and defined. They are the Mediterranean seas, where a number of seas are grouped together collectively as one sea, and adjacent seas, which are connected individually to the ocean. (2) A portion of the ocean where waves are being generated by wind.

Sea anemone A member of the class Anthozoa whose bright color, tentacles, and general appearance make it resemble flowers.

Sea arch An opening through a headland caused by wave erosion. Usually develops as sea caves are extended from one or both sides of the headland.

Sea breeze The landward flow of air from the sea caused by differential heating of Earth's surface.

Sea cave A cavity at the base of a sea cliff formed by wave erosion.

Sea cow See *Sirenia*.

Sea cucumber A common name given to members of the echinoderm class Holothuroidea.

Sea floor spreading A process producing the lithosphere when convective upwelling of magma along the oceanic ridges moves the ocean floor away from the ridge axes at rates between 2 to 12 centimeters (0.8 to 5 inches) per year.

Sea ice A form of ice originating from the freezing of ocean water.

Sea lion Any of several eared seals with a relatively long neck and large front flippers, especially the California sea lion *Zalophus californianus* of the northern Pacific. Along with the fur seals, these marine mammals are known as eared seals.

Sea otter A seagoing otter that has recovered from near extinction along the North Pacific coasts. It feeds primarily on abalone, sea urchins, and crustaceans.

Sea snake A reptile belonging to the family Hydrophiidae with venom similar to that of cobras. They are found primarily in the coastal waters of the Indian Ocean and the western Pacific Ocean.

Sea stack An isolated, pillarlike rocky island that is detached from a headland by wave erosion.

Sea turtle Any of the reptilian order Testudinata found widely in warm water.

Sea urchin An echinoderm belonging to the class Echinoidea possessing a fused test (external covering) and well-developed spines.

Seabeam The first multibeam echo sounder.

Seaknoll See *abyssal hill*.

Seal (1) Any of several earless seals with a relatively short neck and small front flippers. Also known as true seals. (2) A general term that describes any of the various aquatic, carnivorous marine mammals of the families Phocidae and Otariidae (true seals and eared seals), found chiefly in the Northern Hemisphere and having a sleek, torpedo-shaped body and limbs that are modified into paddlelike flippers.

Seamount An individual volcanic peak extending over 1000 meters (3300 feet) above the surrounding ocean floor.

Seasonal thermocline A thermocline that develops due to surface heating of the oceans in middle to high latitudes. The base of the seasonal thermocline is usually above 200 meters (656 feet).

Seawall A wall built parallel to the shore to protect coastal property from the waves.

SeaWiFS An instrument aboard the SeaStar satellite launched in 1997 that measures the color of the ocean with a radiometer and provides global coverage of ocean chlorophyll levels as well as land productivities every two days.

Secchi disk A light-colored disk-shaped device that is lowered into water in order to measure the water's ability to transmit light.

Sediment Particles of organic or inorganic origin that accumulate in loose form.

Sediment maturity A condition in which the roundness and degree of sorting increase and clay content decreases within a sedimentary deposit.

Sedimentary rock Rock resulting from the consolidation of loose sediment, or rock resulting from chemical precipitation, such as sandstone and limestone.

Seep An area where water of various temperatures and/or salinities trickles out of the sea floor, such as in warm seeps, cold seeps, and hypersaline seeps. Also includes locations where hydrocarbons seep from the sea floor, which create hydrocarbon seeps.

Seiche A standing wave of an enclosed or semienclosed body of water that may have a period ranging from a few minutes to a few hours, depending on the dimensions of the basin. The wave motion continues after the initiating force has ceased.

Seismic Pertaining to an earthquake or Earth vibration, including those that are artificially induced.

Seismic moment magnitude (M_w) A scale used for measuring earthquake intensity based on energy released in creating very long-period seismic waves.

Seismic reflection profile A profile view of the structure beneath the sea floor produced by the energy generated from explosions or air guns.

Seismic sea wave See *tsunami*.

Seismic surveying The use of sound-generating techniques to identify features on or beneath the ocean floor.

Semidiurnal tidal pattern A tidal pattern exhibiting two high tides and two low tides per tidal day with small inequalities between successive highs and successive lows; a semidaily tide.

Sequester To isolate, hide away, or remove a particular item from the local environment.

Sessile Permanently attached to the substrate and not free to move about.

Sewage sludge The semisolid material precipitated by sewage treatment. See also *treatment*.

Shallow-water wave A wave on the surface having a wavelength of at least 20 times water depth. The bottom affects the orbit of water particles and speed is determined by water depth.

Shelf break The depth at which the gentle slope of the continental shelf steepens appreciably. It marks the boundary between the continental shelf and continental rise.

Shelf ice Thick shelves of glacial ice that push out into Antarctic seas from Antarctica. Large tabular icebergs calve at the edge of these vast shelves.

Shoal To become shallow.

Shore The area seaward of the coast, which extends from the highest level of wave action during storms to the low water line.

Shoreline The line marking the intersection of water surface with the shore. Migrates up and down as the tide rises and falls.

Side-scan sonar A method of mapping the topography of the ocean floor along a strip up to 60 kilometers (37 miles) wide using computers and sonar signals that are directed away from both sides of the survey ship.

Silica Silicon dioxide (SiO_2).

Silicate A mineral whose crystal structure contains SiO_4 tetrahedra.

Siliceous A condition of containing abundant silica (SiO_2).

Sill A submarine ridge partially separating bodies of water such as fjords and seas from one another or from the open ocean.

Silt A particle size of 0.008 to 0.0625 millimeter (0.0003 to 0.002 inch). It is intermediate in size between sand and clay.

Siphonophora An order of hydrozoan cnidarians that form pelagic colonies containing both polyps and medusae. An example is *Physalia*.

Sirenia An order of large, vegetarian marine mammals that includes dugongs and manatees, which are also known as sea cows.

Slack water A situation that occurs when a reversing tidal current changes direction at high tide (high slack water) or low tide (low slack water). During slack water, the speed of the current is zero.

Slick A smooth patch on an otherwise rippled surface caused by a monomolecular film of organic material that reduces surface tension.

Slightly stratified estuary An estuary of moderate depth in which marine water invades beneath the freshwater runoff. The two water masses mix so the bottom water is slightly saltier than the surface water at most places in the estuary.

Sodium ion A sodium atom that has become positively charged by losing one electron.

SOFAR channel An acronym for the *so*und *f*ixing *a*nd *r*anging channel, which is a low-velocity sound travel zone that coincides with the permanent thermocline in low and middle latitudes.

Solar day The 24-hour period during which Earth completes one rotation on its axis.

Solar distillation A process by which ocean water can be desalinated by evaporation and the condensation of the vapor on the cover of a container. The condensate then runs into a separate container and is collected as freshwater. Also called solar humidification.

Solar humidification See *solar distillation*.

Solar system The Sun and the celestial bodies, asteroids, planets, and comets that orbit around it.

Solar tide The partial tide caused by the tide-producing forces of the Sun.

Solid state A state of matter in which the substance has a fixed volume and shape. A crystalline state of matter.

Solstice The time during which the Sun is directly over one of the tropics. In the Northern Hemisphere, the summer solstice occurs on June 21 or 22 as the Sun is directly above the Tropic of Cancer, and the winter solstice occurs on December 21 or 22 when the Sun is directly above the Tropic of Capricorn.

Solute A substance dissolved in a solution. Salts are the solute in saltwater.

Solution A state in which a solute is homogeneously mixed with a liquid solvent. Water is the solvent for the solution that is ocean water.

Solvent A liquid that has one or more solutes dissolved in it.

Somali Current A surface current that flows north along the Somali coast of Africa during the southwest monsoon season.

Sonar An acronym for *s*ound *n*avigation *and r*anging, a method that uses sound to determine the distance of objects in the ocean.

Sorting A texture of sediments, where a well-sorted sediment is characterized by having great uniformity of grain sizes.

Sounding A measured depth of water beneath a ship.

South Atlantic Subtropical Gyre The large, counterclockwise-flowing subtropical gyre that exists in the South Atlantic Ocean.

South Pacific Subtropical Gyre The large, counterclockwise-flowing subtropical gyre that exists in the South Pacific Ocean.

Southern boundary current The southern boundary surface current of Southern Hemisphere subtropical gyres.

Southern Ocean See *Antarctic Ocean*.

Southern Oscillation The periodic change in the pressure differential between the Southeastern Pacific high pressure and the Western Pacific equatorial low pressure that occurs in concert with El Niño–Southern Oscillation events.

Southwest Monsoon A southwest wind that develops during the summer season. It blows off the Indian Ocean onto the Asian mainland.

Southwest Monsoon Current During the southwest monsoon season, an eastward-flowing surface current that replaces the west-flowing North Equatorial Current in the Indian Ocean.

Space dust Micrometeoroid space debris.

Spawn To reproduce, including releasing eggs and sperm directly into seawater (broadcast spawning) or by internal fertilization.

Species A fundamental category of taxonomic classification, ranking below a genus or subgenus and consisting of related organisms capable of interbreeding.

Species diversity The number or variety of species found in a subdivision of the marine environment.

Specific gravity The ratio of density of a given substance to that of pure water at 4°C and at 1 atmosphere pressure.

Specific heat The amount of heat energy required to raise the temperature of 1 gram of a substance by 1°C. Also known as specific heat capacity.

Spermaceti organ A large fatty organ located within the head region of sperm whales (*Physeter macrocephalus*) that is used to focus echolocation sounds.

Spermatophyta See *Anthophyta*.

Spherule A cosmogenous microscopic globular mass composed of silicate rock material (tektites) or of iron and nickel.

Spicule A minute, needlelike calcareous or siliceous projection found in sponges, radiolarians, chitons, and echinoderms that acts to support the tissue or provide a protective covering.

Spilling breaker A type of breaking wave that forms on a gently sloping beach, which gradually extracts the energy from the wave to produce a turbulent mass of air and water that runs down the front slope of the wave.

Spit A small point, low tongue, or narrow embankment of land commonly consisting of sand deposited by longshore currents and having one end attached to the mainland and the other terminating in open water.

Splash wave A long-wavelength wave created by a massive object or series of objects falling into water; a type of tsunami.

Sponge See *Porifera*.

Spray zone The shore zone lying between the high tide shoreline and the coastline. It is covered by water only during storms.

Spreading center A divergent plate boundary.

Spreading rate The rate of divergence of plates at a spreading center.

Spring bloom A middle-latitude bloom of phytoplankton that occurs during the spring and is limited by the availability of nutrients.

Spring equinox See *vernal equinox*.

Spring tide A tide of maximum range occurring about every two weeks when the Moon is in either new or full moon phase.

Stack An isolated mass of rock projecting from the ocean off the end of a headland from which it has been separated by wave erosion.

Standard seawater Ampules of ocean water for which the chlorinity has been determined by the Institute of Oceanographic Services in Wormley, England. The ampules are sent to laboratories all over the world so that equipment and reagents used to determine the salinity of ocean water samples can be calibrated by adjustment until they give the same chlorinity as is shown on the ampule label.

Standing stock (crop) The mass of fishery organisms present in an ecosystem at a given time.

Standing wave A wave, the form of which oscillates vertically without progressive movement. The region of maximum vertical motion is an antinode. On either side are nodes, where there is no vertical motion but maximum horizontal motion.

Stenohaline Pertaining to organisms that can withstand only a small range of salinity change.

Stenothermal Pertaining to organisms that can withstand only a small range of temperature change.

Stick chart A device made of sticks or pieces of bamboo that was used by early navigators at sea.

Still water level The horizontal surface halfway between crest and trough of a wave. If there were no waves, the water surface would exist at this level. Also known as zero energy level.

Storm An atmospheric disturbance characterized by strong winds accompanied by precipitation and often by thunder and lightning.

Storm surge A rise above normal water level resulting from wind stress and reduced atmospheric pressure during storms. Consequences can be more severe if it occurs in association with high tide.

Strait of Gibraltar The narrow opening between Europe and Africa through which the waters of the Atlantic Ocean and Mediterranean Sea mix.

Stranded beach deposit An ancient beach deposit found above present sea level because of a lowering of sea level.

Streamlining The shaping of an object so it produces the minimum of turbulence while moving through a fluid medium. The teardrop shape displays a high degree of streamlining.

Stromatolite A calcium carbonate sedimentary structure in which algal assemblages trap sediment and bind it into forms that are often dome shaped. They are known to form only in shallow-water environments.

Subduction The process by which one lithospheric plate descends beneath another as they converge.

Subduction zone A long, narrow region beneath Earth's surface in which subduction takes place.

Subduction zone seep biocommunity Animals that live in association with seeps of pore water squeezed out of deeper sediments. They depend on sulfur-oxidizing bacteria that act as producers for the ecosystem.

Submarine canyon A steep, V-shaped canyon cut into the continental shelf or slope.

Submarine fan See *deep-sea fan*.

Submerged dune topography Ancient coastal dune deposits found submerged beneath the present shoreline because of a rise in sea level or submergence of the coast.

Submerging shoreline A shoreline formed by the relative submergence of a landmass in which the shoreline is on landforms developed under subaerial processes. It is characterized by bays and promontories and is more irregular than a shoreline of emergence.

Subneritic province The benthic environment extending from the shoreline across the continental shelf to the shelf break. It underlies the neritic province of the pelagic environment.

Suboceanic province The benthic environments seaward of the continental shelf.

Subpolar Pertaining to the oceanic region that is covered by sea ice in winter. The ice melts away in summer.

Subpolar gyre A small, circular-moving loop of water that is centered at about 60 degrees latitude in both hemispheres. Subpolar gyres rotate in a counterclockwise orientation in the Northern Hemisphere and clockwise in the Southern Hemisphere.

Subpolar low A global belt of low atmospheric pressure located at about 60 degrees north or south latitude that is associated with upward vertical flow of low-density air and abundant precipitation.

Subside To sink to a low or lower level.

Substrate The base on which an organism lives and grows.

Subsurface current A current usually flowing below the pycnocline, generally at slower speed and in a different direction from the surface current.

Subtidal zone That portion of the benthic environment extending from low tide to a depth of 200 meters (660 feet); considered by some to be the surface of the continental shelf.

Subtropical Pertaining to the oceanic region poleward of the tropics (about 30 degrees latitude).

Subtropical Convergence The zone of convergence that occurs within all subtropical gyres as a result of Ekman transport driving water toward the interior of the gyres.

Subtropical gyre A large, circular-moving loop of water that is centered at about 30 degrees latitude and is initiated by the trade winds and the prevailing westerlies. A total of five subtropical gyres exist, with rotation clockwise in the Northern Hemisphere and counterclockwise in the Southern Hemisphere.

Subtropical high A region of high atmospheric pressure located at about 30 degrees latitude.

Sulfur A yellow mineral composed of the element sulfur. It is commonly found in association with hydrocarbons and salt deposits.

Sulfur-oxidizing bacteria Bacteria that support many deep-sea hydrothermal vents and cold-water seep biocommunities by using energy released by oxidation to synthesize organic matter chemosynthetically.

Summer solstice In the Northern Hemisphere, it is the instant when the Sun moves north to the Tropic of Cancer before changing direction and moving southward toward the equator approximately June 21.

Summertime beach A beach that is characteristic during summer months. It typically has a wide sandy berm and a steep beach face.

Sunspot A dark spot on the Sun that is associated with magnetic storms.

Superwave See *rogue wave*.

Supralittoral zone The splash or spray zone above the spring high tide shoreline; also called the supratidal zone.

Surf beat An irregular wave pattern caused by mixed interference that results in a varied sequence of larger and smaller waves.

Surf zone The nearshore zone of breaking waves.

Surface tension The tendency for the surface of a liquid to contract owing to intermolecular bond attraction.

Surfing The sport of riding on the crest or along the tunnel of a wave, especially while standing or lying on a surfboard.

Surging breaker A compressed breaking wave that builds up over a short distance and surges forward as it breaks. It is characteristic of abrupt beach slopes.

Suspension feeding See *filter feeding*.

Suspension settling The process by which fine-grained material that is being suspended in the water column slowly accumulates on the sea floor.

Sverdrup (Sv) A unit of flow rate equal to 1 million cubic meters per second. Named after Norwegian oceanographer Harald Sverdrup.

Swash A thin layer of water that washes up over exposed beach as waves break at the shore.

Swell A free ocean wave by which energy put into ocean waves by wind in the sea is transported with little energy loss across great stretches of ocean to the margins of continents where the energy is released in the surf zone.

Swim bladder A gas-containing, flexible, cigar-shaped organ that aids many fishes in attaining neutral buoyancy.

Symbiosis A relationship between two species in which one or both benefit or neither or one is harmed. Examples are commensalism, mutualism, and parasitism.

Syzygy Either of two points in the orbit of the Moon (full or new moon phase) when the Moon lies in a straight line with the Sun and Earth.

T

Tablemount A conical volcanic feature on the ocean floor resembling a seamount except that it has had its top truncated to a relatively flat surface.

Taxonomy The classification of organisms in an ordered system that indicates natural relationships.

Tectonic estuary An estuary, the origin of which is related to tectonic deformation of the coastal region.

Tectonics Deformation of Earth's surface by forces generated by heat flow from Earth's interior.

Tektite See *spherule*.

Temperate Pertaining to the oceanic region where pronounced seasonal change occurs (about 40 to 60 degrees latitude). Also known as the middle latitudes.

Temperature A direct measure of the average kinetic energy of the molecules of a substance.

Temperature of maximum density The temperature at which a substance reaches its highest density. For water, it is 4°C.

Temperature–salinity (T–S) diagram A diagram with axes representing water temperature and salinity, whereby the density of the water can be determined.

Terrane A distinct geologic fragment of crustal material broken off from one tectonic plate and accreted or sutured onto another plate. Different than terrain, which is the lay of the land.

Terrigenous sediment Sediment produced from or of the Earth. Also called lithogenous sediment.

Territorial sea A strip of ocean, 12 nautical miles wide, adjacent to land over which the coastal nation has control over the passage of ships.

Test The supporting skeleton or shell (usually microscopic) of many invertebrates.

Tethys Sea An ancient body of water that separated Laurasia to the north and Gondwanaland to the south. Its location was approximately that of the present Alpine–Himalayan mountain system.

Texture The general physical appearance of an object.

Theory A well-substantiated explanation of some aspect of the natural world that can incorporate facts, laws (descriptive generalizations about the behavior of an aspect of the natural world), logical inferences, and tested hypotheses.

Thermal contraction The reduction in size as a result of lowering of temperature.

Thermocline A layer of water beneath the mixed layer in which a rapid change in temperature can be measured in the vertical dimension.

Thermohaline circulation The vertical movement of ocean water driven by density differences resulting from the combined effects of variations in temperature and salinity; produces deep currents.

Thermonuclear fusion A high temperature process in which hydrogen atoms are converted to helium atoms, thereby releasing large amounts of energy.

Thermostatic effects Refers to the natural thermostat that regulates temperature on Earth and is largely controlled by the properties of water.

Tidal bore A steep-fronted wave that moves up some rivers when the tide rises in the coastal ocean.

Tidal bulge The theoretical mound of water found on both sides of Earth caused by the relative positions of the Moon (lunar tidal bulges) and the Sun (solar tidal bulges).

Tidal day See *lunar day*.

Tidal period The time that elapses between successive high tides. In most parts of the world, it is 12 hours and 25 minutes.

Tidal range The difference between high tide and low tide water levels over any designated time interval, usually one lunar day.

Tide The periodic rise and fall of the surface of the ocean and connected bodies of water resulting from the gravitational attraction of the Moon and Sun acting unequally on different parts of Earth.

Tide-generating force The magnitude of the centripetal force required to keep all particles of Earth having identical mass moving in identical circular paths required by the movements of the Earth–Moon system is identical. This required force is provided by the gravitational attraction between the particles and the Moon. This gravitational force is identical to the required centripetal force only at the center of Earth. For ocean tides, the horizontal component of the small force that results from the difference between the required and provided forces is the tide-generating force on that individual particle. These forces are such that they tend to push the ocean into bulges toward the tide-generating body on one side of Earth and away from the tide-generating body on the opposite side of Earth.

Tide wave A long-period gravity wave generated by tide-generating forces and manifested as rising and falling tides.

Tissue An aggregate of cells and their products developed by organisms for the performance of a particular function.

Tombolo A sand or gravel bar that connects an island with another island or the mainland.

Topography The configuration of a surface. In oceanography it refers to the ocean bottom or the surface of a mass of water with given characteristics.

Toxic compound A poisonous substance capable of causing injury or death, especially by chemical means.

Trade winds A global wind belt that moves from a subtropical high-pressure belt at about 30 degrees north or south latitude toward the equatorial region.

These winds move from a northeasterly direction in the Northern Hemisphere and from a southeasterly direction in the Southern Hemisphere.

Transform fault A fault with side-to-side motion that offsets segments of a mid-ocean ridge. Oceanic transform faults occur wholly on the ocean floor, while continental transform faults occur on land.

Transform faulting The process by which a transform fault moves with side-to-side motion.

Transform plate boundary The boundary between two lithospheric plates formed by a transform fault.

Transitional wave A wave moving from deep water to shallow water that has a wavelength more than twice the water depth but less than 20 times the water depth. Particle orbits are beginning to be influenced by the bottom.

Transverse wave A wave in which particle motion is at right angles to energy propagation; also known as a side-to-side wave.

Treatment In reference to sewage at a sewage treatment plant, there is primary treatment where solids are allowed to settle and separate from the liquid, and secondary treatment where sewage is exposed to bacteria-killing chlorine or other means of disinfection.

Trench A long, narrow, and deep depression on the ocean floor with relatively steep sides that is caused by plate convergence; often referred to as an ocean trench.

Trophic level A nourishment level in a food chain. Plant producers constitute the lowest level, followed by herbivores and a series of carnivores at the higher levels.

Tropic of Cancer The latitude 23.5 degrees north, which is the furthest location north that receives vertical rays of the Sun.

Tropic of Capricorn The latitude 23.5 degrees south, which is the furthest location south that receives vertical rays of the Sun.

Tropical Pertaining to, characteristic of, occurring in, or inhabiting the tropics.

Tropical Atmosphere and Ocean (TAO) A scientific program that monitors the equatorial Pacific and studies how El Niño events develop.

Tropical cyclone See *hurricane*.

Tropical Ocean–Global Atmosphere (TOGA) A scientific program that was initiated in 1985 to monitor the equatorial Pacific and study how El Niño events develop; predecessor to the TAO program.

Tropical tide A tide that occurs twice monthly when the Moon is at its maximum declination near the Tropic of Cancer or the Tropic of Capricorn.

Tropics The region of Earth's surface lying between the Tropic of Cancer and the Tropic of Capricorn. Also known as the Torrid Zone.

Troposphere The lowermost portion of the atmosphere, which extends from Earth's surface to 12 kilometer (7 miles). It is where all weather is produced.

Trough (wave) The part of an ocean wave that is displaced below the still-water level.

Tsunami A seismic sea wave. A long-period gravity wave generated by a submarine earthquake or volcanic event. Not noticeable on the open ocean but builds up to great heights in shallow water.

Turbidite deposit A sediment or rock formed from sediment deposited by turbidity currents characterized by both horizontally and vertically graded bedding.

Turbidity A state of reduced clarity in a fluid caused by the presence of suspended matter.

Turbidity current A gravity current resulting from a density increase brought about by increased water turbidity. Possibly initiated by some sudden force such as an earthquake, the turbid mass continues under the force of gravity down a submarine slope.

Turbulent flow Flow in which the flow lines are confused heterogeneously due to random velocity fluctuations.

Typhoon See *hurricane*.

U

Ultraplankton Plankton for which the greatest dimension is less than 5 microns (0.0002 inch). Because of their small size, they are very difficult to separate from water.

Ultrasonic Sound frequencies above those that can be heard by humans (above 20,000 cycles per second).

Ultraviolet (UV) radiation Electromagnetic radiation shorter than visible radiation and longer than X-rays. The approximate wavelength range is 0.001 to 0.4 micron.

United Nations Conference on the Law of the Sea A series of meetings to establish legal rights in the sea, particularly in regards to sea floor mining.

Upper water An area of the ocean near the surface that includes the mixed layer and the permanent thermocline. It is approximately the top 1000 meters (3300 feet) of the ocean.

Upwelling The process by which deep, cold, nutrient-laden water is brought to the surface, usually by diverging equatorial currents or coastal currents that pull surface waters away from a coast.

V

Valence The combining capacity of an element measured by the number of hydrogen atoms with which it will combine.

van der Waals force A weak attractive force between molecules resulting from the interaction of one molecule and the electrons of another.

Vapor The gaseous state of a substance that is liquid or solid under ordinary conditions.

Vent An opening on the ocean floor that emits hot water and dissolved minerals; vent type is based on water temperature and includes hydrothermal (hot water) vents and warm-water vents.

Ventral Pertaining to the lower or under surface.

Vernal equinox The passage of the Sun across the equator as it moves from the Southern Hemisphere into the Northern Hemisphere, approximately March 21. During this time, all places in the world experience equal lengths of night and day. Also known as the spring equinox.

Vertebrata The subphylum of chordates that includes those animals with a well-developed brain and a skeleton of bone or cartilage; includes fish, amphibians, reptiles, birds, and land animals.

Vertically mixed estuary Very shallow estuaries such as lagoons in which freshwater and marine water are totally mixed from top to bottom so that the salinity at the surface and the bottom is the same at most places within the estuary.

Virioplankton Viruses that live as plankton.

Viscosity A property of a substance to offer resistance to flow caused by internal friction.

Volcanic arc An arc-shaped row of active volcanoes directly above a subduction zone. Can occur as a row of islands (island arc) or mountains on land (continental arc).

W

Walker Circulation Cell The pattern of atmospheric circulation that involves the rising of warm air over the East Indies low-pressure cell and its descent over the high-pressure cell in the southeastern Pacific Ocean off the coast of Chile. It is the weakening of this circulation that accompanies an El Niño event, which has led to the development of the term *El Niño–Southern Oscillation event*.

Walrus A large marine mammal (*Odobenus rosmarus*) of Arctic regions belonging to the order Pinnipedia and having two long tusks, tough wrinkled skin, and four flippers.

Waning crescent The Moon when it is between third-quarter and new moon phases.

Waning gibbous The Moon when it is between full moon and third-quarter phases.

Warm-blooded See *homeothermic*.

Warm front A weather front in which a warm air mass moves into and over a cold air mass producing a broad band of gentle precipitation.

Water mass A body of water identifiable from its temperature, salinity, or chemical content.

Wave A disturbance that moves over the surface or through a medium with a speed determined by the properties of the medium.

Wave base The depth at which circular orbital motion becomes negligible. It exists at a depth of one-half wavelength, measured vertically from still-water level.

Wave-cut bench A gently seaward-sloping horizontal platform produced by wave erosion and extending from the base of the wave-cut cliff out under the offshore region. See also *marine terrace*.

Wave-cut cliff A cliff produced by landward cutting by wave erosion.

Wave dispersion The separation of waves as they leave the sea area by wave size. Larger waves travel faster than smaller waves and thus leave the sea area first, to be followed by progressively smaller waves.

Wave frequency (f) The number of waves that pass a fixed point in a unit of time (usually one second). A wave's frequency is the inverse of its period.

Wave height (H) The vertical distance between a crest and the adjoining trough.

Wave period (T) The elapsed time between the passage of two successive wave crests (or troughs) past a fixed point. A wave's period is the inverse of its frequency.

Wave speed (S) The rate at which a wave travels. It can be calculated by dividing a wave's wavelength (*L*) by its period (*T*).

Wave steepness Ratio of wave height (*H*) to wavelength (*L*). If a 1:7 ratio is ever exceeded by the wave, then the wave breaks.

Wave train A series of waves from the same direction. Informally known as a wave set.

Wavelength (L) The horizontal distance between two corresponding points on successive waves, such as from crest to crest.

Waxing crescent The Moon when it is between new moon and first-quarter phases.

Waxing gibbous The Moon when it is between first-quarter and full moon phases.

Weather The state of the atmosphere at a given time and place, with respect to variables such as temperature, moisture, wind velocity, and barometric pressure.

Weathering A process by which rocks are broken down by chemical and mechanical means.

Wentworth scale of grain size A logarithmic scale for size classification of sediment particles.

West Australian Current A cold surface current that forms the eastern boundary of the Indian Ocean Subtropical Gyre. It is separated from the coast by the warm Leeuwin Current except during El Niño–Southern Oscillation events when the Leeuwin Current weakens.

West Wind Drift See *Antarctic Circumpolar Current*.

Western boundary current Poleward-flowing warm surface current on the western side of each subtropical gyre.

Western boundary undercurrent (WBUC) A bottom current that flows along the base of the continental slope eroding sediment from it and redepositing the sediment on the continental rise. It is confined to the western boundary of deep-ocean basins.

Western intensification Pertaining to the intensification of warm western boundary currents of each subtropical gyre that are faster, narrower, and deeper than their corresponding eastern boundary currents.

Wetland See *coastal wetland*.

Whirlpool A rapidly rotating current of water; a vortex.

White muscle fiber Thick muscle fibers with relatively low concentrations of myoglobin that make up a large percentage of the muscle fiber in lunger-type fishes.

White smoker A hydrothermal vent feature similar to a black smoker but emitting water of a lower temperature that is white in color.

Wilson cycle A model that uses plate tectonic processes to show the distinctive life cycle of ocean basins during their formation, growth, and destruction.

Wind The movement of air, usually as a result of pressure differences.

Wind-driven circulation A movement of ocean water that is driven by winds. This includes most horizontal movements in the surface waters of the world's oceans.

Windward The direction from which the wind is blowing.

Winter solstice The instant the southward-moving Sun reaches the Tropic of Cancer before changing direction and moving north back toward the equator, approximately December 21.

Wintertime beach A beach that is characteristic during winter months. It typically has a narrow rocky berm and a flat beach face.

Z

Zebra mussel A non-native species released into U.S. and Canadian waters of the Great Lakes region.

Zenith The point on the celestial sphere directly over the observer.

Zooplankton Animal plankton.

Zooxanthellae A form of algae that lives as a symbiont in the tissue of corals and other coral reef animals and provides varying amounts of their required food supply.

CREDITS AND ACKNOWLEDGMENTS

All uncredited figures: International Mapping/Pearson Education, Inc.

Chapter 1
Opener NASA. **Figure 1A** Alan P. Trujillo. **Figure 1B** Alan P. Trujillo. **Figure 1C** Science and Society/Superstock. **Figure 1D (left inset)** HIP/Art Resource, New York. **Figure 1D (right inset)** David Parry/PA Photos/Landov. **Figure 1.1** NASA. **Figure 1.6** Office of Naval Research, US Navy. **Figure 1.7** Mark Thiessen/AFP/Getty Images/Newscom. **Figure 1.9** AP Images. **Figure 1.10** World History Archive/Alamy. **Figure 1.13** US Naval Historical Center. **Figure 1.16** Masa Ushioda/AGE Fotostock. **Figure 1.18** NASA, ESA & Mohammad Heydari.Malayeri (Observatoire de Paris, France). **Figure 1.20** Stocktrek Images, Inc./Alamy. **Figure 1.26b** Reuters. **Figure 1.30** After Zimmer, C., 2001, How old is it? *National Geographic* 200:3, 92. **Figure 1.31** Based on Tarbuck, E. J. and Lutgens, F. K., *The Earth: An Introduction to Physical Geology*, 7th ed. (Fig. 8.15), Prentice Hall, 2002. Data from the Geological Society of America, the U.S. Geological Survey, and the International Commission on Stratigraphy.Pearson Education, Inc. **Quote, page 3** Loren Shriver, NASA astronaut (2008). **Quote, page 30** Dobzhansky, Theodosius. "Nothing in Biology Makes Sense Except in the Light of Evolution", *American Biology Teacher* vol. 35 (March 1973).

Chapter 2
Opener Alan P. Trujillo. **Figure 2A** Based on Tarbuck, E. J. and Lutgens, F. K., *Earth Science*, 8th ed. (Fig. 7.17), Prentice Hall, 1997. **Figure 2.1** Bpk Berlin/Art Resource. **Figure 2.2** Based on Dietz, R. S. and Holden, J. C., 1970, Reconstruction of Pangaea: Breakup and dispersion of continents, Permian to present, *Journal of Geophysical Research* 75:26, 4939–4956. **Figure 2.3** Based on Continental Drift by Don and Maureen Tarling, 1971 by G. Bell & Sons, Ltd. **Figure 2.4** Based on Tarbuck, E. J. and Lutgens, F. K., *Earth Science*, 8th ed. (Fig. 7.6), Prentice Hall, 1997. **Figure 2.5** Based on Tarbuck, E. J. and Lutgens, F. K., *Earth Science*, 8th ed. (Fig. 7.4), Prentice Hall, 1997. **Figure 2.6** Based on Tarbuck, E. J. and Lutgens, F. K., *Earth Science*, 8th ed. (Fig. 7.4), Prentice Hall, 1997. **Figure 2.7** Reprinted by permission from Tarbuck, E. J. and Lutgens, F. K., *The Earth: An Introduction to Physical Geology*, 3d ed. (Fig. 18.8 and Fig 18.9), Merrill Publishing Company, 1990. **Figure 2.11** Based on Tarbuck, E. J. and Lutgens, F. K., *The Earth: An Introduction to Physical Geology*, 3d ed. (Fig. 19.16), Merrill Publishing Company, 1990. After The Bedrock Geology of the World, by R. L. Larson et al., Copyright © 1985 by W. H. Freeman. **Figure 2.13** NASA/ Goddard Space Flight Center. **Figure 2.15** Based on Tarbuck, E. J. and Lutgens, F. K., *Earth Science*, 8th ed. (Fig. 7.10), Prentice Hall, 1997. **Figure 2.16 (Inset)** Alan P. Trujillo. **Figure 2.18** Based on Tarbuck, E. J. and Lutgens, F. K., *Earth Science*, 8th ed. (Fig. 7.11), Prentice Hall, 1997. **Figure 2.18c** Copyright © 2013 Cornelis Klein and Tony Philpotts. Reprinted with the permission of Cambridge University Press. **Figure 2.19.1** Anthony R. Philpotts. **Figure 2.19.2** Scripps Institution of Oceanography, UCSD. **Figure 2.19.3** US Geological Survey Library (USGS). **Figure 2.21.2** US Geological Survey Library (USGS). **Figure 2.22c** Frank Bienewald/ImageBroker/AGE Fotostock. **Figure 2.24** Peter Essick/Aurora Photos/Alamy. **Figure 2.29.1** Google Earth. **Figure 2.29.2** Google Earth. **Figure 2.29.3** Google Earth. **Figure 2.31** Plate tectonic reconstructions by Christopher R. Scotese, PALEOMAP Project, University of Texas at Arlington. **Figure 2.32** Based on Dietz, R. S. and Holden, J. C., 1970, The breakup of Pangaea, *Scientific American* 223: 4, 30–41. **Figure 2.33** Adapted from Wilson, J. T., *American Philosophical Society Proceedings* 112, 309–320, 1968; Jacobs, J. A., Russell, R. D., and Wilson, J. T., *Physics and Geology*, McGraw. Hill, New York, 1971. **Quote, page 39** Wegener, Alfred. The Origins of Continents and Oceans. Friedrich Vieweg & Sohn, Braunschweig. 1929.

Chapter 3
Opener Planetary Visions, Ltd./Science Source. **Figure 3A** After Tarbuck, E. J. and Lutgens, F. K., *Earth Science*, 6th ed. (Fig. 10.2), Macmillan Publishing Company, 1991. **Figure 3B** Woods Hole Oceanographic Institution. **Figure 3B (Inset)** Woods Hole Oceanographic Institute. **Figure 3.2** Erika Mackay/National Institute of Water and Atmospheric Research. **Figure 3.1** Courtesy Peter A. Rona, Hudson Laboratories of Columbia University. **Figure 3.2** Erika Mackay/National Institute of Water and Atmospheric Research. **Figure 3.3a** Courtesy Daniel J. Fornari, Lamont. Doherty Geological Observatory, Columbia University. Reprinted with permission of the American Geophysical Union. **Figure 3.3b** American Geophysical Union. **Figure 3.4** After Gross, M. G., *Oceanography*, 6th ed. (Fig. 16.10), Prentice Hall, 1993. **Figure 3.5** Scripps Institution of Oceanography, UCSD. **Figure 3.6** Scripps Institution of Oceanography, UCSD. **Figure 3.7** Courtesy of the Deep Sea Drilling Project, Scripps Institution of Oceanography, UCSD; with thanks to Jerry Bode. **Figure 3.8** After Tarbuck, E. J. and Lutgens, F. K., *The Earth: An Introduction to Physical Geology*, 5th ed. (Fig. 19.3), Prentice Hall, 1996. **Figure 3.12a** After Tarbuck, E. J. and Lutgens, F. K., *The Earth: An Introduction to Physical Geology*, 5th ed. (Fig. 19.6), Prentice Hall, 1996. **Quote, page 81** Maury, Matthew Fontaine. 1855. *The Physical Geography of the Sea*. New York: Harper & Brothers. **Figure 3.12b** Martin Strmiska/Alamy. **Figure 3.12c** Alan P. Trujillo. **Figure 3.14B** James V. Gardner/University of New Hampshire. **Figure 3.20** ALCOA Aluminum Company. **Figure 3.21b** Scripps Institution of Oceanography, UCSD. **Figure 3.21c** Jeffrey Grover. Cuesta College. **Figure 3.22a (Inset)** Verena Tunnicliffe. **Figure 3.22b** Fred N. Spiess. **Figure 3.25** Atsushi Taketazu/AP Images.

Chapter 4
Opener Alfred Wegener Institute for Polar and Marine Research. **Figure 4A** World Minerals Inc. **Figure 4.1** NOAA. **Figure 4.2a** Courtesy of the Ocean Drilling Program, Texas A&M University. **Figure 4.2** Ocean Drilling Program. . **Figure 4.3** Alan P. Trujillo. **Figure 4.4** Alan P. Trujillo. **Figure 4.5** Integrated Ocean Drilling Program. **Figure 4.6a** NASA. **Figure 4.6b** Marine Corps University. **Figure 4.6c** Alan P. Trujillo. **Figure 4.6d** Alan P. Trujillo. **Figure 4.7** Mack Walker. **Figure 4.8** NASA. **Figure 4.9a** CSIRO Plant Industry. **Figure 4.9b** Scripps Institution of Oceanography. **Figure 4.9c** World Minerals Inc. **Figure 4.10a** CSIRO Plant Industry. **Figure 4.10b** CSIRO Plant Industry. **Figure 4.10c** Scripps Institution of Oceanography. **Figure 4.10d** Scripps Institution of Oceanography. **Figure 4.11** David Hughes/Robert Harding World Imagery. **Figure 4.11 (inset)** Steve Gschmeissner/Science Source. **Figure 4.12b** Jarrod Boord/Shutterstock. **Figure 4.12c** Francois Gohier/Science Source. **Figure 4.16** After Biscaye, P. E., et al., 1976; Berger, W. H., et al., 1976; and Kolla V. and Biscaye, P. E., 1976. **Figure 4.17a** Charles D. Winters/Science Source. **Figure 4.17b** Charles D Winters/Getty Images. **Figure 4.17c** Institute of Oceanographic Sciences/NERC/Science Source. **Figure 4.18** Alan P. Trujillo. **Figure 4.19** Integrated Ocean Drilling Program. **Figure 4.21** After Sverdrup, H. U., et al., 1942. **Figure 4.22** Modified after Sverdrup, H. U., et al., 1942. **Figure 4.23** Woods Hole Oceanographic Institute. **Figure 4.24** Divins, D. L., NGDC Total Sediment Thickness of the World's Oceans & Marginal Seas, Retrieved date 4/10/2006, http://www.ngdc.noaa.gov/mgg/ http://www.ngdc.noaa.gov/mgg/sedthick/sedthick.html. **Figure 4.25** Scott Gibson/Corbis. **Figure 4.26a** Alan P. Trujillo. **Figure 4.26b** Alan P. Trujillo. **Figure 4.28** Alan P. Trujillo. **Figure 4.29** Sunoco, Inc. **Figure 4.30** After Cronan, D. S., 1977, Deep sea nodules: Distribution and geochemistry, in Glasby, G. P., ed., *Marine Manganese Deposits*, Elsevier Scientific Publishing Co. **Quote, page 105** Wolf, Berger: Oceans: Reflections on a Century of Exploration (c) 2009 by the Regents of the University of California. Published by the University of California Press. **Quote, page 115** Darwin, Charles. 1859. *Origin of the Species*. London: John Murray. **Quote, page 137** Caglioti, Luciano. 1985. *The Two Faces of Chemistry*. MIT Press.

Chapter 5
Figure 5.1 After Tarbuck, E. J. and Lutgens, F. K., *The Earth: An Introduction to Physical Geology*, 5th ed. (Fig. 2.4), Prentice Hall, 1996. **Figure 5A** Bruce Coleman Inc./Alamy. **Figure 5.4** Rafael Ben.Ari/Chameleons Eye Witness/Newscom. **Figure 5.11** After data from Jet Propulsion Laboratory, NASA. **Figure 5.13** Electron Microscopy Laboratory, Agricultural Research Service, USDA. **Figure 5.14** Martin F. Chillmaid/Science Source. **Figure 5.17** Howard Shooter/Getty Images. **Figure 5.18** Rafael Ben.Ari/Chameleons Eye Witness/Newscom. **Figure 5.24** After Sverdrup, H.U., Martin W. Johnson and Richard H. Fleming. The oceans, their physics, chemistry, and general biology. New York, Prentice.Hall, inc., 1942. ASIN: B00103AKBK. **Figure 5.25**

Norman Kuring, Goddard Space Flight Center/NASA. **Figure 5.26** After Pickard, G. L., *Descriptive Physical Oceanography*, Pergamon Press Ltd., 1963. **Poem, page 171** Samuel Taylor Coleridge, about ships getting stuck in the horse latitudes, Rime of the Ancient Mariner (1798).

Chapter 6

Opener Justin Hoffman/Pearson Education, Inc. **Figure 6A** Morgan P. Sanger/The Columbus Foundation, British Virgin Islands. **Figure 6.7** Chris James/Alamy. **Figure 6.20a** NASA. **Figure 6.21b** Michelle McLoughlin/Reuters. **Figure 6.22** Communications and Education Division. **Figure 6.23** US Air Force photo/Master Sgt. Mark C. Olsen. **Figure 6.24a** Communications and Education Division. **Figure 6.24b** Vincent Laforet/Pool/EPA/Newscom. **Figure 6.26a** Alan Trujillo. **Figure 6.26b** Alan Trujillo. **Figure 6.26c** Sue Flood/Nature Picture Library. **Figure 6.27a** William Bacon/Science Source. **Figure 6.27c** Josh Landis/National Science Foundation. **Figure 6.27d** NASA Earth Observatory. **Figure 6.28** NASA Earth Observatory. **Figure 6.29** Chris James/Alamy. **Quote, page 207** Anonymous, but often attributed to Mark Twain; said in reference to San Francisco's cool summer weather caused by coastal upwelling.

Chapter 7

Opener NASA Earth Observatory. **Figure 7A** Courtesy of Eos Transactions AGU 73: 34, 361 (1992). Copyright by the American Geophysical Union. **Figure 7A.1** Rachel Youdelman/Pearson Education, Inc. **Figure 7A.2** Reuters. **Figure 7A.3** Jack Sullivan/Alamy. **Figure 7A.4** Alan P. Trujillo. **Figure 7B.1** Benjamin Franklin, 1706.1790,Library of Congress Prints and Photographs Division [LC. USZ62.41888]. **Figure 7B.2** Communications and Education Division. **Figure 7.1** Alan P. Trujillo. **Figure 7.1a** Scripps Institution of Oceanography, UCSD. **Figure 7.1b** Dann S Blackwood/USGS, U.S. Geological Survey Library. **Figure 7.2** NASA. **Figure 7.3a** Based on data collected and made freely available by the International Argo Program and the national programs that contribute to it. (http://www.argo.ucsd.edu, http://argo.jcommops.org). The Argo Program is part of the Global Ocean Observing System. **Figure 7.03b** Alan P. Trujillo. **Figure 7.06** Siemens Press Pictures. **Figure 7.17** Communications and Education Division. **Figure 7.20a** NASA Earth Observatory. **Figure 7.20b** NASA Earth Observatory. **Figure 7.23 a–b** International Research Institute for Climate and Society. **Figure 7.24** Communications and Education Division. **Figure 7.26** International Research Institute for Climate and Society. **Figure 7.30** Siemens Press Pictures. **Quote, page 245** Fanny Crosby.

Chapter 8

Opener Rebecca Jackrel/Danita Delimont/Newscom. **Figure 8B** Mainichi Shimbun/Reuters. **Figure 8.1b** NASA. **Figure 8.4a** Based on the Tasa Collection: Shorelines. Published by Macmillan Publishing Co., New York. Copyright © 1986 by Tasa Graphic Arts, Inc. **Figure 8.6** Kyodo/AP Images. **Figure 8.6.2** Pelamis Wave Power. **Figure 8.11** NASA. **Figure 8.13** National Archives and Records Administration. **Figure 8.17a.b** Project Michelangelo. **Figure 8.19** Based on the Tasa Collection: Shorelines. Published by Macmillan Publishing Co., New York. Copyright © 1986 by Tasa Graphic Arts, Inc. **Figure 8.20a** EpicStockMedia/Shutterstock. **Figure 8.20b** Irabel8/Shutterstock. **Figure 8.20c** SweetParadise/Alamy. **Figure 8.21** Rich Reid/Getty Images. **Figure 8.22** Mark Rightmire/Newscom. **Figure 8.24** US Geological Survey Library (USGS). **Figure 8.25a.d** Joanee Davis/Newscom. **Figure 8.26** Corbis. **Figure 8.28a–b** US Geological Survey Library (USGS). **Figure 8.29** CNES, NOAA, EUMETSAT/NASA. **Figure 8.31** Alan P. Trujillo. **Figure 8.32** Wavegen. **Figure 8.33** Pelamis Wave Power. **Figure 8.30** Communications and Education Division. **Table 8.1** US Government Printing Office. **Quote, page 279** Sir Isaac Newton.

Chapter 9

Opener Laszlo Podor/Alamy. **Figure 9A** New Brunswick Department of Tourism. **Figure 9B** Jurgen Skarwan/Landov. **Figure 9.5** Peter McBride/Getty Images. **Figure 9.6.2** De Agostini Editore/C Sappa/AGE Fotostock. **Figure 9.11** Based on the Tasa Collection: Shorelines. Published by Macmillan Publishing Co., New York. Copyright © 1986 by Tasa Graphic Arts, Inc. **Figure 9.11 a–b** NASA. **Figure 9.12** NASA. **Figure 9.15** NASA. **Figure 9.20a** Laszlo Podor/Alamy. **Figure 9.22** Peter McBride/Getty Images. **Figure 9.23** Visual&Written SL/Alamy. **Figure 9.24** De Agostini Editore/C Sappa/AGE Fotostock. **Quote, page 305** Lord Byron. 1922. The Works of Lord Byron: Letters and Journals. John Murray.

Chapter 10

Opener Alan P. Trujillo. **Figure 10.A1** Alan P. Trujillo. **Figure 10.A2** Alan P. Trujillo. **Figure 10.2** Alan P. Trujillo. **Figure 10.3a.b** Alan P. Trujillo. **Figure 10.4** University of Washington Libraries. **Figure 10.6c** USDA/FSA. **Figure 10.7** Travel Picture/Alamy. **Figure 10.7.2** Ethan Daniels/Getty Images. **Figure 10.8** University of Washington Libraries. **Figure 10.9** University of Washington Libraries. **Figure 10.11a** USDA/FSA. **Figure 10.11b** Martin Beebee/Stock Photo/Alamy. **Figure 10.12c** USDA/FSA. **Figure 10.15b** NASA's Earth Observatory. **Figure 10.15a** NASA Earth Observatory. **Figure 10.15b** NASA. **Figure 10.21** University of Washington Libraries. **Figure 10.22** Pearson Education, Inc. **Figure 10.23** US Army Corps of Engineers. **Figure 10.24** Peter Titmuss/Alamy. **Figure 10.25** Kevin Steele/Getty Images. **Figure 10.26a** Fairchild Aerial Photography Collection. Whittier College Collection. **Figure 10.26b** Fairchild Aerial Photography Collection. Whittier College Collection. **Figure 10.28a** Alan P. Trujillo. **Figure 10.28b** Alan P. Trujillo. **Figure 10.29** Alan P. Trujillo. **Figure 10.33a** NASA. **Figure 10.33b** Glow Images. **Figure 10.33c** USDA/FSA. **Figure 10.33d** M Sat Ltd/Science Source. **Figure 10.40b** Alan P. Trujillo. **Figure 10.40c** Alan P. Trujillo. **Figure 10.41** Ethan Daniels/Getty Images. **Quote, page 347** Jane Lubchenco, marine ecologist (2002). **Quote, page 348** United Nations Department of Public Information.

Chapter 11

Opener Newscom. **Figure 11.1** Sam Chadwick/Shutterstock. **Figure 11A.2** Chuck Cook/AP Images. **Figure 11A2** Newscom. **Figure 11.2b** Natalie B. Fobes/National Geographic Stock. **Figure 11.2c** NOAA Central Library Photo Collection. **Figure 11.4** Tony Freeman/PhotoEdit, Inc. **Figure 11.4b** Bob Jordan/AP Images. **Figure 11.5** Alan P. Trujillo. **Figure 11.5** National Oceanic and Atmospheric Administration. **Figure 11.6** Pearson Education, Inc. **Figure 11.7** Charlie Riedel/AP Images. **Figure 11.10** John Trever. **Figure 11.11** Jack Smith/AP Images. **Figure 11.14** Age fotostock/SuperStock. **Figure 11.15** AP Images. **Figure 11.18** Alan P. Trujillo. **Figure 11.20** International Mapping/Pearson Education, Inc. **Figure 11.21** David W. Leindecker/Shutterstock. **Figure 11.22a** National Marine Fisheries Service. **Figure 11.22b** Photoshot. **Figure 11.22c** David Liittschwager/National Geographic/Getty Images. **Figure 11.22d** Photo by Susan Middleton © 2005. **Figure 11.23** Tony Freeman/PhotoEdit, Inc. **Figure 11.24** Alan P. Trujillo. **Figure 11.25** Brendan Bannon/Polaris/Newscom. **Figure 11.26** International Mapping/Pearson Education, Inc. **Figure 11.27** Alan P. Trujillo. **Figure 11.28** Tony Souter/Dorling Kindersley, Ltd. **Figure 11.29** AFP/Getty Images. **Quote, page 375** E.O. Wilson.

Chapter 12

Opener Jeff Rotman/Alamy. **Figure 12.A** Scripps Institution of Oceanography, UCSD. **Figure 12.2** The Art Archive/Stock Photo/Alamy. **Figure 12.8** Newscom. **Figure 12.13** CSIRO Plant Industry. **Figure 12.19** Cbimages/Alamy. **Figure 12.20** Aaltair/Shutterstock. **Figure 12.21a** Eugene Sim/Fotolia. **Figure 12.21b** Alan P. Trujillo. **Figure 12.23** Alan P. Trujillo. **Figure 12.27** Peter David/The Image Bank/Getty Images. **Figure 12.28** Woods Hole Oceanographic Institute. **Quote, page 405** Anonymous.

Chapter 13

Opener Robinson Ed/Getty Images. **Figure 13A.1** Monroe County Public Library. **Figure 13A.2** Monroe County Public Library. **Figure 13A.3** Scripps Institution of Oceanography. **Figure 13.2** Alan P. Trujillo. **Figure 13.2a** Scripps Institution of Oceanography, UCSD. **Figure 13.2b** Peter Parks/Image Quest Marine. **Figure 13.4** Patricia Deen. **Figure 13.5** GeoEye Inc. **Figure 13.7** Alan P. Trujillo. **Figure 13.8a** Alan P. Trujillo. **Figure 13.8b** Alan P. Trujillo. **Figure 13.8c** Harold V. Thurman. **Figure 13.8d** Alan P. Trujillo. **Figure 13.9** Ng Han Guan/AP Images. **Figure 13.10a** CSIRO Plant Industry. **Figure 13.10b** Scripps Institution of Oceanography, UCSD. **Figure 13.10c** CSIRO Plant Industry. **Figure 13.10d** CSIRO Plant Industry. **Figure 13.11** Photoshot Holdings/Alamy. **Figure 13.13** Everett Collection. **Figure 13.15** NASA. **Figure 13.17** Claire Ting/Science Source. **Figure 13.19b** Scripps Institution of Oceanography, UCSD. **Figure 13.23** Justin Hoffman/Pearson Education, Inc. **Figure 13.26** Walter E Harvey/Science Source/Getty Images. **Figure 13.34** National Marine Fisheries Service. **Figure 13.35** Stefan Jacobs/Alamy. **Figure 13.39** Trujillo, *Essentials of Oceanography*, 12th Ed., Pearson Education, Inc. **Figure 13.40** Trujillo, *Essentials of Oceanography*, 12th Ed., Pearson Education, Inc. **Quote, page 445** © Robert "Pete" Pederson. Used by permission.

Chapter 14

Opener Image Source/Alamy. **Figure 14.A** Michael Patrick O'Neill/Alamy. **Figure 14.2** Justin Hoffman. **Figure 14.3a** Kozo Takahashi, Kyushu University, Fukuoka, Japan. **Figure 14.3b** Kozo Takahashi, Kyushu University, Fukuoka, Japan. **Figure 14.3c** Kozo Takahashi, Kyushu University, Fukuoka, Japan. **Figure 14.3d** Kozo Takahashi, Kyushu University, Fukuoka, Japan. **Figure 14.4** Siim Sepp. **Figure 14.4c** Alan P. Trujillo. **Figure 14.5** Wilhelm Giesbrecht. **Figure 14.6.1** Peter Parks/Image Quest Marine. **Figure 14.6.2** Alan P. Trujillo. **Figure 14.7b** WaterFrame/Alamy. **Figure 14.9** Justin Hoffman. **Figure 14.10** Justin Hoffman. **Figure 14.11a** Tania Zbrodko/Shutterstock. **Figure 14.11b** lioneldivepix/Fotolia. **Figure 14.11c** Stephan Kerkhofs/Fotolia. **Figure 14.11d** NaturePL/SuperStock. **Figure 14.11e** Cbpix/Fotolia. **Figure 14.12a** Amar/Isabelle Guillen/Alamy. **Figure 14.12b** Mark Conlin/Alamy. **Figure 14.13h** Phil Hastings/Scripps Institution of Oceanography, UCSD. **Figure 14.15** Science Source. **Figure 14.16a** Jonathan Bird/Getty Images. **Figure 14.16b** Cbpix/Fotolia. **Figure 14.16c** Image Quest Marine. **Figure 14.18a** Photoshot. **Figure 14.18b** Elvele Images Ltd/Alamy. **Figure 14.18c** Photoshot/Alamy. **Figure 14.18d** M. Delpho/Arco Images/Alamy. **Figure 14.18e** Irina Mos/Shutterstock. **Figure 14.20a** Cornforth Images/Alamy. **Figure 14.20b** Helmut Corneli/Image Broker/Alamy. **Figure 14.23** Alan P. Trujillo. **Figure 14.24** Justin Hoffman. **Figure 14.26c** Alan P. Trujillo. **Figure 14.26d** Alan P. Trujillo. **Figure 14.27** NaturePL/SuperStock. **Figure 14.28a** Christian Valle/Robert Harding World Imagery. **Figure 14.28b** Justin Hofman/Pearson Education, Inc. **Figure 14.31** Justin Hoffman/Pearson Education, Inc. **Quote (Benchley 2000) page 454** National Geographic World. **Quote (Benchley 2005) page 454** Smithsonian Magazine.

Chapter 15

Opener Joel Rogers/Corbis. **Figure 15A** Robert R. Hessler. **Figure 15.2b** Harold V. Thurman. **Figure 15.2c** Harold V. Thurman. **Figure 15.2d** Alan P. Trujillo. **Figure 15.2e** Harold V. Thurman. **Figure 15.2f** Suzanne Long/Alamy. **Figure 15.2g** Alan P. Trujillo. **Figure 15.2h** Harold V. Thurman. **Figure 15.2i** Alan P. Trujillo. **Figure 15.2j** Alan P. Trujillo. **Figure 15.2k** Alan P. Trujillo. **Figure 15.2l** Alan P. Trujillo. **Figure 15.3a** Alan P. Trujillo. **Figure 15.3b** PhotoShot. **Figure 15.5a** James R. McCullagh. **Figure 15.5b** Harold V. Thurman. **Figure 15.6** Alan P. Trujillo. **Figure 15.7a** Alan P. Trujillo. **Figure 15.7b** Premaphotos/Alamy. **Figure 15.9** Justin Hoffman. **Figure 15.10** Science Source. **Figure 15.11** Justin Hoffman. **Figure 15.12a** Howard J. Spero. **Figure 15.12b** Howard J. Spero. **Figure 15.12c** Howard J. Spero. **Figure 15.13** Blickwinkel/Woike/Alamy. **Figure 15.14a** Justin Hoffman. **Figure 15.14b** Justin Hoffman/Pearson Education, Inc. **Figure 15.15a** Alan P. Trujillo. **Figure 15.15b** Andrew J. Martinez/Science Source. **Figure 15.16** Justin Hoffman. **Figure 15.17** Shin Okamoto/Shutterstock. **Figure 15.19a** Peter Leahy/123RF. **Figure 15.19b** PhotoShot. **Figure 15.19c** Sebastien Burel/Shutterstock. **Figure 15.19c** Sebastien Burel/Shutterstock. **Figure 15.21a** Kerry L. Werry/Shutterstock. **Figure 15.21b** Charles Stirling/Alamy. **Figure 15.22** Franco Banfi/Steve Bloom Images/Alamy. **Figure 15.23b** XL Caitlin Seaview Survey. **Figure 15.27a** J. Frederick Grassle. **Figure 15.27b** Woods Hole Oceanographic Institute. **Figure 15.27c** Scripps Institution of Oceanography. **Figure 15.29** National Oceanic and Atmospheric Administration (NOAA). **Figure 15.30c** Scripps Institution of Oceanography. **Figure 15.31b** Charles R. Fisher, Pennsylvania State University.

Figure 15.31c Charles R. Fisher, Pennsylvania State University. **Figure 15.32b** Japan Agency for Marine. Earth Science and Technology (JAMSTEC). **Quote, page 479** Thomas Dahlgren. **Quote, page 513** Source: Dr. Jane Lubchenco, marine ecologist and Under Secretary of Commerce for Oceans and Atmosphere and NOAA Administrator (2009), at a White House news conference announcing the release of the report Global Climate Change Impacts in the United States.

Chapter 16

Opener Corbis Premium RF/Alamy. **Figure 16.1** Source: Reprinted with the kind permission of Dennis Tasa; modified from Tarbuck, E. J. ed. and Lutgens, F. K., *Earth: An Introduction to Physical Geology*, 9th (Fig. 21.2), Pearson Prentice Hall, 2008. **Figure 16A** Pearson Education, Inc. **Figure 16A.1** Scripps Institution of Oceanography, UCSD. **Figure 16A.2** Scripps Institution of Oceanography, UCSD. **Figure 16.2** British Antarctic Survey/Science Source. **Figure 16.3** Alan P. Trujillo. **Figure 16.4** NASA Goddard Space Flight Center. **Figure 16.4 (Inset)** British Antarctic Survey/Science Source. **Figure 16.5.2** Woods Hole Oceanographic Institute. **Figure 16.5a–b** NASA. **Figure 16.7** Source: Reprinted with the kind permission of Dennis Tasa; from Tarbuck, E. J. and Lutgens, F. K., *Earth: An Introduction to Physical Geology*, 9th ed. (Fig. 18.33), Pearson Prentice Hall, 2008. **Figure 16.8** InterNetwork Media/Getty Images. **Figure 16.10** Based on Doran & Zimmerman, 2009, Examining the Scientific Consensus on Climate Change, *Eos Transactions American Geophysical Union* Vol. 90 Issue 3, 22; DOI: 10.1029/2009EO030002. **Figure 16.11** IPCC. **Figure 16.12** Akademie/Alamy. **Figure 16.21** Tim Larsen/State of New Jersey Office of the Governor. **Figure 16.23a** NASA. **Figure 16.23** Fritz Poelking/Alamy. **Figure 16.24** Kerstin Langenberger. **Figure 16.25** NASA. **Figure 16.26a** CSIRO Plant Industry. **Figure 16.26b** Sinclair Stammers/Science Source. **Figure 16.26c** Jeff Rotman/Alamy. **Figure 16.26d** Dobermaraner/Shutterstock. **Figure 16.27a** David Liittschwager/Getty Images. **Figure 16.29** NOAA. **Figure 16.30** NOAA. **Figure 16.30 inset** Alan P. Trujillo. **Figure 16.33** NOAA. **Figure 16.34** Woods Hole Oceanographic Institute. **Quote (There is new and stronger ...), page 521** The National Academies of **Sciences, Engineering, and Medicine. Quote (The scientific understanding ...), page 521** Nobel Media. **Quote (global warming is unequivocal ...), page 522** Global Climate Change Impacts in the United States. U.S. Global Change Research Program. 2009. **Quote (can reduce the risk ...), page 522** SOURCE: America's Climate Choices, U.S. National Research Council. **Quote (Climate change is already ...), page 522** US Global Change Research Program. **Quote (until there is an adequate ...), page 546** Climate.related **Geoengineering and Biodiversity, https://www.cbd.int/climate/geoengineering/. Quote (the current pledges ...), page 546** United Nations Climate Change Conference.

Afterword

Opener Ken Ilio/Getty Images. **Figure Aft.4** Alan P. Trujillo.

Appendix V

Figure A5.1 Alan P. Trujillo. **Figure A5A** Joseph Fell-McDonald. **Figure A5A (inset)** Joseph Fell-McDonald.

Note: page numbers followed by f or t refer to Figures or Tables